GLOSSARY OF WORD FORMS

The following word roots, prefixes, and suffixes are frequently used in the study of physiology and anatomy:

a-	without (aphasia, asexual)	hyster-	womb (hysterectomy, hysteria)
ab-	away from (abduct, abscess)	-iatr-	physician (iatrogenic, pediatric)
ad-	toward, at (adduct, adrenal)	idio-	personal, distinct (idiopathic, idiosyncracy)
an-	without (anaerobic, anoxia)	inter-	between (interstitial, interferon, intercostal)
ana-	up, back (anatomy, anaphase)	intra-	within (intramuscular, intrauterine)
ante-	before (anterior, antenatal)	-itis	disease, inflammation (hepatitis, bronchitis)
-arthr-	joint (arthritis, diarthrodial)		
-blast-	bud (blastula, osteoblast)	lig-	to tie (ligament, ligature)
		-lymph-	water (lymphatic, perilymph)
carcin-	cancer (carcinoma, carcinogenic)	-lys-	breaking down (lysosome, plasmolysis)
-cardi-	heart (cardiac, tachycardia)		
-cephal-	head (cephalic, encephalitis)	mal-	bad (malpractice, malaria)
cervic-	a neck (cervix, cervical)	mater-	womb (maternal, endometrium)
-chondr-	grain, cartilage (chondrocyte, mitochondrion)	med-	middle (media, mediastinum)
-cil-	hair (cilium, ciliary body)	meg-	large (megadose, acromegaly)
-clas-	break (osteoclast)	men-	month, cycle (menstrual, menopause)
coll-	glue (collagen, colloid)	mening-	membrane (meninges, meningioma)
-condyl-	knuckle (epicondyle, condyloma)	meta-	between (metaphase, metatarsal)
corp-	body (corpulent, corpus callosum)	-my-	muscle (endomysium, myosin)
-crine	pertaining to a gland (ectocrine, endocrine)	myel-	marrow (myelin, myeloid)
-cyst-	a bladder (cystoid, cholycystitis)		
-cyt-	cell (cytology, erythrocyte)	-nephr-	kidneys (nephritis, epinephrine)
de-	from, down (deliquesce, detoxify)	ob-	against, in the way of (obstruct, obnoxious)
-derm-	skin (epidermis, dermatology)	-oid	likeness (lymphoid, trapezoid)
dia-	through, between (diaphysis, diaphragm)	-oste-	bone (osteoblast, periosteum)
dis-	apart (disease, dislocate)	oxy-	sharp, swift (oxygen, oxytocin)
duct-	to convey (duct, oviduct)		
dys-	difficult (dyspnea, dysuria)	palp-	touch (palpate, palpitation)
		pan-	every, all (pancreas)
endo-	within (endocrine, endometrium)	para-	beside (parenchyma, parathyroid)
-entero-	gut (enterostomy, enterotoxin)	-path-	disease (pathology, neuropathy)
epi-	upon (epiglottis, epithelium)	-pep-	digest (pepsin, polypeptide)
-esth-	perception (anesthetic, esthesiometer)	peri-	around, through (pericardium, perineum, peristalsis)
ex-	out of, from (excision, extend)	-phag-	eat (esophagus, phagocytosis)
		-phil-	love (philosophy, hemophilia, eosinophil)
fasc-	band, strip (fascicle, fascia)	-pleura-	side (pleural, pleurisy)
-fer-	to carry (fertile, seminiferous)	post-	after (posterior, postaxial)
flex-	to bend (flexor)	pro-	before (protein, prophylaxis, prothrombin)
gam-	pertaining to marriage, sexual (gamete, syngamy)	sarc-	flesh (sarcoma, sarcolemma)
gangli-	ganglion (autonomic ganglion, gangliectomy)	scler-	hard (sclera, sclerosis)
-gastr-	belly (gastritis, enterogastrone)	seb-	hard fat (sebaceous, sebum)
-gen-	to bear or produce (genital, antigen)	ser-	watery (serum, serous)
-gli-	glue (glial, astroglia)	-som-	body (somatic, chromosome)
-glom-	ball (glomerulus, conglomerate)	spir-	breath (spirometer, respiration)
gyn-	female (gynecologist, androgynous)	stom-	mouth (anastomosis, stomatology)
		sub-	under, beneath (substernal, sublingual)
hemi-	half (hemiplegia)	super-	above, over (superior, superovulation)
hemo-	blood (hemoglobin, hemophilia)	supra-	upper, above (supraoptic, suprarenal)
hetero-	different (heterogeneous, heterosexual)	syn-	with, together (synthesis, synarthrodial)
hist-	tissue (histology, histamine)		
hydr-	water (hydrolysis, hydrophobic)	ten-	stretch (tension, tendon)
hyper-	excessive, above (hypercarbia, hyperventilation)	-tom-	cut (appendectomy, tomography)
hypo-	deficiency, below (hypoxia, hypodermic)	trans-	across, beyond (transfer, transformation)

Human Anatomy and Physiology

PRINCIPLES AND APPLICATIONS

ROY HARTENSTEIN
State University of New York, Syracuse

D. VAN NOSTRAND COMPANY
New York   Cincinnati   Toronto   London   Melbourne

D. Van Nostrand Company Regional Offices:
New York   Cincinnati   Millbrae

D. Van Nostrand Company International Offices:
London   Toronto   Melbourne

Copyright © 1976 by Litton Educational Publishing, Inc.

Library of Congress Catalog Card Number: 75-32652
ISBN: 0-442-23183-0

All rights reserved. No part of this work covered by the copyright
hereon may be reproduced or used in any form or by any means—graphic,
electronic, or mechanical, including photocopying, recording, taping,
or information storage and retrieval systems—without written permission
of the publisher. Manufactured in the United States of America.

Published by D. Van Nostrand Company
450 West 33rd Street, New York, N.Y. 10001

10 9 8 7 6 5 4 3 2 1

PREFACE

This text is intended to meet the needs of the beginning student who seeks a working knowledge of the human body—the student in liberal arts, nursing, physical education, or allied health. The material is so organized that no prerequisite science courses are necessary. The writing is as straightforward and nontechnical as the subject permits, designed to make the text easy and pleasant to read while maintaining adequate depth of coverage.

The text may be used in a one-semester or a two-semester course, depending on the requirements of curriculum, student, and instructor and the availability of laboratory facilities. The 35 chapters are organized into eight sections, which cover the traditional content of the human biology course and also describe (in Sections Seven and Eight) very recent developments in the study of aging and some other special topics in human physiology.

The major sections of the text are largely self-contained to allow the instructor the choice of beginning with the section or chapter that best suits the organization of the course. The order in which the topics are presented, however, reflects the author's philosophy and experience in teaching students with a wide range of educational backgrounds and aims.

In recent years, physiology and anatomy have become an increasingly dynamic and closely integrated area of study and research. The modern researcher and the modern student focus on the coordinated functioning of the human body as much as on the architecture and workings of more or less isolated organs and systems. The author has therefore found it valuable to focus early in the course on the integrative systems of the human body—the nervous and endocrine systems—before proceeding to the more traditional coverage of the subject matter. With this approach the students soon become familiar with the principles of the body's self-regulation. At the same time they are introduced to the broad features of the body systems, which are described in greater detail in the later chapters. The author's experience has been that this approach to the material has helped students to visualize the importance of each organ system to the overall well-being of the individual.

A number of special features are incorporated into the book to make it a particularly valuable educational tool. The illustrations are treated as an integral part of the text. Many of them are drawn in a flow-chart style, using visual design to explain or clarify complex biological processes. In the chapters that deal with structural anatomy special pains have been taken to provide unusually comprehensive illustrations, thus meeting the needs of students whose interest in the material will ultimately be practical as well as academic.

All technical terms are defined as they occur. Many such terms are further explained in a glossary that runs side-by-side with the text, supplying word derivations or supplementary definitions. This device helps the student acquire the technical vocabulary without feeling "swamped" with terminology.

The author has focused strongly on physiological explanations of the causes, effects, and treatments of injury and disease, from sprains to heart disease and cancer. This procedure serves to make the book more interesting and also provides the student with information that is often of personal as well as professional value.

The text lends itself to different sequences in which the topics are taught. Several alternative sequences of reading assignments are suggested in an Instructor's Manual that has been prepared to accompany this volume. The Manual is available from the publisher and it also provides supplementary course materials and a listing of valuable teaching aids.

Roy Hartenstein
Syracuse, New York

CONTENTS

**PART 1
Introduction to
the Human
Body, 1**

1 Some Tools from Chemistry and Physics 3
The metric system, 4 / Basic chemical concepts, 7 / Basic physical concepts, 20 / Summary, 23 / Review questions, 24

2 Cells 25
Cells, 25 / Cell reproduction, 32 / Summary, 37 / Review questions, 38

3 Tissues 39
Epithelial tissues, 40 / Connective tissues, 44 / Muscle tissues, 52 / Abnormalities of cells and tissues, 54 / Summary, 57 / Review questions, 58

4 Organs and Body Regions 59
Organs, 59 / Anatomical orientations, 61 / Body regions, 63 / Summary, 71 / Review questions, 72

**PART 2
The Nervous
System, 73**

5 Introduction to the Nervous System 75
Basic elements of the nervous system, 76 / Larger building blocks of the nervous system, 88 / Development of the nervous system: an overview of its parts and functions, 97 / Summary: functional divisions of the nervous system, 111 / Summary, 111 / Review questions, 113

6 The General Sensory System 114
Principles of the general sensory system, 115 / Evaluation of injury to the general sensory system, 119 / Some abnormalities of the general sensory system, 119 / Summary, 120 / Review questions, 120

Contents

7 The Skeletal Motor System — 121
Components of the skeletal motor system, 121 / Reflexes and their regulation, 123 / Evaluation of the skeletal motor system, 128 / Summary, 130 / Review questions, 130

8 The Visceral Sensory and Motor Systems — 132
The visceral sensory system, 132 / Visceral motor system, 134 / Summary, 138 / Review questions, 139

9 The Special Sensory Systems — 140
Hearing, 140 / Vision, 144 / Smell, 151 / Taste, 153 / Summary, 154 / Review questions, 155

10 The Proprioceptor System — 156
General proprioception, 156 / Special proprioception—equilibrium, 157 / Evaluation of the proprioceptor system, 159 / Summary, 161 / Review questions, 162

11 The Cortical Systems — 163
The cerebral cortex, 164 / Evaluation of the cerebral cortex, 174 / Summary, 176 / Review questions, 177

**PART 3
The Endocrine System, 179**

12 General Endocrinology — 181
General endocrine concepts, 183 / Regulation of hormone output, 194 / The hypothalamus and pituitary gland, 198 / The thyroid gland, 202 / The parathyroid glands, 205 / The adrenal glands, 207 / The pancreas, 212 / Summary, 215 / Review questions, 216

13 The Reproductive System — 217
The male reproductive system, 218 / The female reproduc-

tive system, 223 / Birth control, 234 / Summary, 236 / Review questions, 237

**PART 4
The Structural and Support Systems, 239**

14 The Skin 241
The epidermis, 242 / The dermis, 245 / Sweat glands, 247 / The nails, 250 / Hair, 251 / Sebaceous glands, 252 / Abnormalities of skin, 253 / Summary, 256 / Review questions, 257

15 The Skeletal System 258
The skeleton, 259 / The axial skeleton, 261 / The appendicular skeleton, 275 / The structure, function, and maintenance of bone, 280 / Development of bone, 281 / Disorders and diseases of bone 287 / Summary, 289 / Review questions, 290

16 The Joints 291
Structural materials of joints, 291 / Classification of joints, 297 / Disorders and diseases of the joints, 305 / Summary, 307 / Review questions, 307

17 The Physiology of Muscle 308
Properties of muscle tissue, 309 / Skeletal muscle, 310 / Smooth muscle, 320 / Cardiac muscle, 324 / Responsiveness of muscle to physical training, 328 / Summary, 330 / Review questions, 330

18 The Functional Anatomy of Skeletal Muscle 332
Structure of skeletal muscle, 333 / Functions of skeletal muscles, 335 / Summary, 374 / Review questions, 374

**PART 5
Digestion, Metabolism, and Nutrition, 375**

19 Digestion and Elimination 377
Components of the digestive system, 378 / Functional anatomy of the digestive tract, 380 / Preparatory digestive activities in the mouth, 382 / The pharynx, or throat, 382 / Passage of food from the esophagus into the stomach, 384 / The stomach, 386 / The pancreas, 389 / The liver and the gallbladder, 390 / The small intestine, 393 / The large intestine, 396 / Regulating the output of the digestive juices, 397 / Elimination, 398 / Summary, 399 / Review questions, 400

Contents

20 Absorption of Nutrients and Elimination of Cell Wastes 401
Intestinal absorption, 402 / Mechanisms of cellular transport, 405 / Faulty intestinal absorption, 410 / Summary, 412 / Review questions, 412

21 Principles of Metabolism 413
Carbohydrates, 414 / Proteins, 418 / Fats, 422 / Acetate, 425 / Vitamins, 426 / Minerals, 426 / Summary, 428 / Review questions, 428

22 The Principal Organs of Metabolism 429
Principal organs of metabolism, 430 / Hormonal regulation of metabolism, 435 / The principal regulatory hormones of metabolism, 440 / Translation of hormonal messages into action, 445 / Summary, 445 / Review questions, 446

23 Caloric Requirements 447
Calories and how they are measured, 447 / Caloric needs of an individual, 452 / An approximate determination of daily caloric needs, 455 / Regulation of food intake, 456 / The essence of designs for losing weight, 458 / Summary, 462 / Review questions, 462

24 Nutritional Requirements 464
Carbohydrates, 465 / Proteins, 466 / Fats, 470 / Vitamins, 473 / Minerals, 478 / Standards for a nutritious diet, 480 / Summary, 487 / Review questions, 488

25 Circulatory System: Functional Anatomy and Electrocardiography 491
Body fluids, 492 / Functional anatomy of the heart, 493 / The cardiac cycle, 497 / Electrocardiography, 500 / Functional anatomy of blood and lymph vessels, 509 / Circulatory routes, 515 / Summary, 523 / Review questions, 524

26 Dynamic Characteristics of the Circulatory System 525
Kinds of pressure in the circulatory system, 526 / Resistance encountered by blood in its flow, 531 / Exchange of nutrients and wastes, 532 / Return of venous blood to the heart, 532 / Varying the cardiac output to meet the body's

**PART 6
Transport Systems:
Circulation, Respiration, and Excretion, 489**

viii
Contents

need for oxygen, 534 / Shifting the blood from inactive to active regions of the body, 538 / The pulse—an indicator of general conditions in the circulatory system, 540 / Regulation of the circulatory system, 541 / Reflex regulation of the circulatory system, 545 / Summary, 546 / Review questions, 547

27 Functional Disorders of the Circulatory System — 549
Shock, 550 / Fainting, 550 / Dehydration, 551 / Water intoxication and edema, 551 / Hypertension, 552 / Arteriosclerosis and atherosclerosis, 553 / Heart disease, 555 / Heart failure, 561 / Summary, 562 / Review questions, 563

28 The Respiratory System — 564
Components of the respiratory system, 565 / Mechanism of breathing, 566 / Functions of the upper respiratory tract, 569 / Functions of the lower respiratory tract, 570 / Exchange of oxygen and carbon dioxide, 571 / Breathing capacity, 573 / Lung diseases and breathing problems, 574 / Blood buffering, 575 / Summary, 576 / Review questions, 577

29 The Red Blood Cell System — 579
Production of red blood cells, 579 / Assessing the physiology of red blood cells, 580 / Summary, 586 / Review questions, 587

30 The Excretory System — 588
The renal excretory system, 590 / Formation of urine, 596 / Evaluation of kidney function, 602 / Role of the kidneys in regulating the composition and quantity of body fluids, 603 / Endocrine functions of the kidneys, 609 / Urinalysis and kidney problems, 610 / Summary, 613 / Review questions, 613

PART 7 The Mechanisms of Defense, 615

31 Body Defense Systems: Nonimmunological — 617
The body's first line of defense, 618 / The body's second line of defense, 619 / The body's third line of defense, 623 / Summary, 628 / Review questions, 629

32 Body Defense Systems: Immunological — 630
The immune process, 631 / Immunity, 642 / Antigens, 648 / Antibodies, 649 / Type I hypersensitivity, 650 / Type II hypersensitivity, 650 / Type III hypersensitivity, 653 /

Contents

Type IV hypersensitivity, 654 / Drugs and hypersensitivity, 654 / Cancer and immunity, 655 / Summary, 655 / Review questions, 656

33 Aging   661
Characteristics of aging, 661 / Theories of aging, 662 / Control of aging factors, 666 / Summary, 667 / Review questions, 667

34 Pain, Anesthesia, and Acupuncture   668
Theory of pain, 668 / Relief of pain by drugs, 669 / Relief of pain by acupuncture, 471 / Summary, 475 / Review questions, 475

35 Stress and Its Relief   478
Mental attitudes, emotions, and stress, 478 / Unconscious relief of stress, 478 / Conscious relief of stress, 681 / Summary, 684 / Review questions, 684

Suggested Readings   686

Glossary   694

Index   709

**PART 8
Special Topics
in Physiology,
659**

Introduction to the Human Body

1. Some Tools from Chemistry and Physics
2. Cells
3. Tissues
4. Organs and Body Regions

PART ONE

Some Tools from Chemistry and Physics 1

THE METRIC SYSTEM
    Length, mass, and time
        Length
        Mass
        Time
    Interconversion of terms in the metric system
BASIC CHEMICAL CONCEPTS
    Elements, molecules, and compounds
    Nutrients, metabolites, and metabolism
    Kinds of compounds found in the body
    Chemical measuring units
        Atomic weight
        Molecular weight, moles, and molarity
    Chemical bonds
        Covalent bonds
        Ionic bonds
        Hydrogen bonds
    Biological significance of covalent bonds
    Biological significance of ionic bonds
        Bioelectricity
        Buffering the pH of blood
    Biological significance of hydrogen bonds
    Important biochemical reactions
BASIC PHYSICAL CONCEPTS
    Force and pressure
    Diffusion
    Osmosis and osmotic pressure

Anatomy and physiology are basic sciences which are fundamental to an understanding of health as well as the abnormalities of the human body. Often *anatomy* is offered in college as a separate course of study, allowing the student to examine the various *structures* of the body without giving

ANATOMY. *Literally, dissection; the study of the structure of an organism* (G. ana. *up;* tome, *a cutting*).

PHYSIOLOGY. *The science which deals with the functions and processes of a living organism* (G. *physis, nature;* logia, *study of*).

much attention to their functions. *Physiology* also is often offered as a separate course, in which the various *functions* of the body are studied with only minimal reference to their underlying structure.

In this text, structures and functions are integrated with each other. Thus, anatomy, a subject which can otherwise be very static, becomes more dynamic. With respect to physiology, the integration with anatomy provides a type of structural foundation that not only lends itself to a better understanding of normal functions but to an understanding of some of the body's pathological, or abnormal, functions.

Understanding the functions of the human body requires some background knowledge in many areas of science. To understand the material in this book, you will need some basic knowledge of cell structure, basic anatomy, chemistry, and physics.

Cell structure and basic anatomy are discussed in Chapters 2, 3, and 4. In this chapter we will discuss several concepts, or "tools," which are taken from chemistry and physics, and which are frequently used in the study of physiology.

THE METRIC SYSTEM

The metric system of measurement is used in all branches of science. When a nurse conducts a laboratory analysis of a blood sample and finds that there are five million red cells per *cubic millimeter*, she knows immediately that there are five million red cells per *microliter*. Because an adult's body contains about 5 liters of blood, the nurse can quickly figure out that the patient has about 25 trillion red blood cells—the normal total number for an adult.

The *metric system* is used to measure the basic dimensions, or parameters, of scientific phenomena. It is used in all of the medical, biological, and physical sciences, and its use is basic to a study of all biological and physical activities which have to be measured.

The chief advantage of the metric system over the English system—the system widely used throughout the United States—is that it allows the user to make quick calculations and translations between size units. In particular, it takes advantage of the "base-ten" number system, and it allows direct interconversions between solid, liquid, and gaseous measurements. In the English system it is difficult to translate ounces into pounds, pounds into quarts, and so forth. But in the metric system, which the United States may soon adopt as a standard, these conversions are easy to make.

Length, Mass, and Time

Any chemical, physical, or biological quantity, such as volume, pressure, velocity, weight, density, or force, can be measured with a *number*, which expresses magnitude, and a *term*, which is derived from *length, mass*, or *time*.

Length. Length is a linear dimension. It is usually meaningful to most people only in terms of inches, feet, yards, and miles. A scientist, however, must be able to visualize linear (1), square (1^2), and cubic (1^3) dimensions in the metric system. A practical starting point for such visualization is to consider the approximate metric equivalent of an inch, 2.5 centimeters.

Aside from centimeters, there are only two other metric units which are very commonly used by biologists for measuring length, area, and volume: the millimeter and the micrometer. A *millimeter* is approximately the thickness of a dime. A *micrometer* is so thin that a thousand of them make up a millimeter.

Centimeters, millimeters, and micrometers are subdivisions of a meter (1.094 yards). More exactly, they are base-ten subdivisions; *centi*, meaning 1/100th; *milli*, meaning 1/1000th; and *micro*, meaning 1/1,000,000th.

If we take the meter as having a length of unity, that is, 1, all measurements which are multiples of ten, either higher or lower—such as 100 or 1000 meters or 1/100th or 1/1000th of a meter—can be expressed as some power of ten. Thus, a decameter is 10 meters, and since there is one zero in ten, we may write it as 10^1 meters. A decimeter, in contrast, is 1/10th of a meter, and since there is only one zero in the denominator of this fraction we may write it as 10^{-1} meters. The most commonly used subdivisions and multiples of the metric system are:

```
pico   — a trillionth = 10⁻¹²
nano   — a billionth = 10⁻⁹
micro  — a millionth = 10⁻⁶
milli  — a thousandth = 10⁻³
centi  — a hundredth = 10⁻²
deci   — a tenth = 10⁻¹
deca   — ten = 10¹
kilo   — one thousand = 10³
mega   — one million = 10⁶
```

- pico — a trillionth = 10^{-12}
- nano — a billionth = 10^{-9}
- micro — a millionth = 10^{-6}
- milli — a thousandth = 10^{-3}
- centi — a hundredth = 10^{-2}
- deci — a tenth = 10^{-1}
- deca — ten = 10^1
- kilo — one thousand = 10^3
- mega — one million = 10^6

In each of these cases it should be noted that the superscript denotes the multiple of tens in the magnitude of the metric unit; a positive superscript is used for multiples greater than 1; a negative superscript is used for submultiples less than 1.

Mass. Any object which occupies space is said to have *mass*. Outside of the earth's gravitational field, the mass of an object cannot be weighed. When the object enters the earth's gravitational field, the force exerted on the object by gravity gives the object *weight*. At sea level, the *mass* of an object is equal to the *weight* of the object. Most people are familiar with ounces and pounds, but the biologist must be able to think in terms of their metric equivalents. A simple starting point is the *gram*: thirty grams make up an ounce. To help you visualize an ounce, think of two tablespoons of sugar.

Only three other metric units are used very commonly by biologists to

The Metric System

describe weight: kilogram (1000 grams), milligram (1/1000th gram), and microgram (1/1,000,000th gram). There are 2.2 pounds per kilogram, so a 150-pound person weighs about 70 kilograms. A single fine crystal of table salt weighs about 100 micrograms; and a coarse crystal of table sugar weighs about 1 milligram.

Time. *Time* is a duration, a period, an interval. Its basic unit of measurement in both the English and metric systems is the *second*. Physiologists often measure time in *milliseconds* (one thousandth of a second) as well as in minutes, days, weeks, and years.

Interconversion of Terms in the Metric System

We have discussed the descriptive terms (length, mass, and time) and the magnitudes (base-ten system) used to make measurements in the metric system. We now want to discuss how measurements of *volume* (a derivative of length) and *weight* (a derivative of mass) can be interconverted in the metric system. It is difficult to make these interconversions in the English system, but is desirable and necessary to be able to make them in biology. Sometimes one scientist or doctor will discuss oxygen consumption in terms of *cubic centimeters*, while another researcher will discuss oxygen consumption in terms of *milliliters*. A person who is familiar with the metric system will know that both units measure the same quantity of oxygen. Similarly, someone may refer to a *kilogram* of water, while another person may refer to a *liter* of water. Anyone who is familiar with the metric system knows that one liter of water weighs one kilogram at sea level.

How are volume and weight related to one another in the metric system? *Weight* (*w*) is related to *volume* (*v*) by the concept of *density* (*d*). The relationship is expressed by the formula

$$d = w/v$$

This relationship holds true at sea level and, for all practical purposes, is correct in most parts of the world, although it is not true in Death Valley or on top of Mount Everest (because of variations due to changes in air pressure). The formula is correct under most atmospheric conditions, except when there is a violent change in the weather and the air pressure is either very high or very low.

From this formula you can see that weight is equal to volume when density equals one. The density of pure water is one. For all practical purposes, the density of tap water and of aqueous solutions containing dissolved substances with densities of approximately one are also said to be one. When the density of a liquid is one, a *gram* of the liquid occupies a *milliliter* of volume, or a *cubic centimeter* of space. A *cubic centimeter* of water is equivalent to one *milliliter* of water. These are important relationships which should be memorized: one gram of water equals one milliliter of water equals one cubic centimeter of water; one kilogram of water equals one liter of water and occupies 1000 cubic centimeters of space.

To put this knowledge to use, let us return to the example given at the beginning of this section. A nurse analyzes a blood sample and finds that there are five million red blood cells per cubic millimeter. How does she know that the individual from whom this blood sample was taken has approximately 25 trillion red blood cells in his body? She knows that one cubic millimeter is equivalent to one microliter (one millionth of one liter) and that the average adult has about five liters of blood. Using the relationships outlined above, she knows that there are five million red cells per microliter, five trillion red cells per liter, and a total of 25 trillion red cells altogether.

Several metric units are often used in this text. They are listed below, together with their equivalents in the English system:

1 millimeter	0.0394 inches
1 gram	0.0352 ounces avoirdupois
	0.0321 troy (apothecaries') ounces
1 liter	1.0567 U.S. liquid quarts
1 milliliter	0.0338 U.S. fluid ounces
1 kilocalorie	1 kcal. = 1 Calorie = 3.0859 foot-pounds = work done to lift one pound of material 3.0859 feet into the air, or work done to lift 3.0859 pounds one foot into the air.

BASIC CHEMICAL CONCEPTS

Some concepts in chemistry are important in the study of human anatomy and physiology. Among these are the concepts of elements, molecules, and compounds.

Elements, Molecules, and Compounds

All substances are composed of one or more *elements*. Approximately 100 different kinds of elements have been discovered by chemists so far. When an element exists by itself, uncombined with any other element, it is called an *atom*. Some common examples of atoms are sodium (Na), carbon (C), and iron (Fe). Sometimes, two or more atoms of the same element combine to form a *molecule*. Examples of molecules are nitrogen (N_2), hydrogen (H_2), and oxygen (O_2). If two or more dissimilar atoms combine, a *compound* is formed. Sodium chloride (NaCl), water (H_2O), and glucose ($C_6H_{12}O_6$) are examples of compounds. Atoms, molecules, and compounds may exist as gases, liquids, or solids, depending on conditions of temperature and pressure.

Basic Chemical Concepts

Nutrients, Metabolites, and Metabolism

Atoms, molecules, and compounds which are taken into the body are classified as *nutrients* if they are used to supply energy, to regulate physiological activities, or to produce *protoplasm*.

After a nutrient has been chemically reacted in the body it may be referred to as a *metabolite* (a changeable substance). A metabolite may be used to build more protoplasm or to supply energy for the metabolic processes which go on in all living organisms.

There are many kinds of metabolic processes which go on in the body but they all fall into two categories. Some processes build up the body. These are the *anabolic processes* and all of these processes, taken together, are called *anabolism* (building up). Other processes break down parts of the body so that energy can be released. These processes are called *catabolic*, and their sum total is called *catabolism* (breaking down). All of the metabolic reactions which occur in our bodies at all times are a series of anabolic and catabolic processes in which protoplasm is being constructed and broken down—over and over—and energy is being used to bring about those qualities of physiology which we recognize as the characteristics of life.

> **PROTOPLASM.** The substance of which living organisms are formed (G. *protos*, *first*; *plasma*, *thing formed*).
>
> **METABOLISM.** The sum of chemical changes in the body (G. *metabole*, *change*).

Kinds of Compounds Found in the Body

A nutrient or metabolite may be an inorganic or organic compound. *Inorganic compounds* are substances which do not carry both carbon and hydrogen atoms. Therefore, they cannot be oxidized, or burned, in the body to form carbon dioxide and water, releasing energy in the process. A list of elements important in biology is shown, along with their chemical symbols, in Table 1.1. These elements occur in the body in the form in-

TABLE 1.1. Major Elements in the Body

Element	Symbol	Usual Form in the Body
Hydrogen[a]	H	Hydrogen ion (H^+)
Carbon[a]	C	Carbon dioxide (CO_2), bicarbonate ion (HCO_3^-), carbonic acid (H_2CO_3)
Nitrogen[a]	N	Ammonium ion (NH_4^+)
Oxygen[a]	O	Oxygen molecule (O_2)
Sodium	Na	Sodium ion (Na^+)
Magnesium	Mg	Magnesium ion (Mg^{++})
Chlorine	Cl	Chloride ion (Cl^-)
Potassium	K	Potassium ion (K^+)
Calcium	Ca	Calcium ion (Ca^{++})
Iron	Fe	Ferrous ion (Fe^{++}) or ferric ion (Fe^{+++})
Iodine	I	Iodide ion (I^-) or iodine molecule (I_2)

[a] These four elements make up the bulk of the organic matter found in living tissues.

dicated in the table. Table 1.2 lists the major inorganic compounds found in the body. Each of these compounds usually combines with a small number of other elements to form the compounds listed in the right-hand column of Table 1.2.

Organic compounds contain carbon and hydrogen and they can be oxidized to form water and carbon dioxide. Organic compounds may also contain other elements, such as nitrogen, sulfur, chlorine, and iron. Organic compounds may form *polymers* or they may exist as single molecules.

Biological polymers are compounds which can be broken down—by the process of hydrolysis—into their unit molecules, which are called *monomers*. Hydrolysis is a chemical reaction in which the elements of water are used to break up a particular kind of chemical bond. All of the chemical reactions that take place in the digestion of foods are hydrolytic reactions.

The principal biological polymers are: (a) proteins; (b) certain carbohydrates, including starch and glycogen (animal starch); (c) nucleic acids (the molecules from which genes are formed, and through which proteins are made); and (d) neutral fats. The chemical structures of the first three of these polymers may be extremely complex, because they are built up from long chains of monomeric units. Each of the four major types of biological polymers is constructed from "building block" monomers. Amino acids are the building blocks of proteins; glucose is the monomer of starch and glycogen; organic bases, ribose (a simple sugar), and phosphoric acid are the monomeric constituents of nucleic acids; and three fatty acids, together with glycerol (or glycerine) form neutral fats. In addition to neutral fats, the human body contains considerable amounts of phospholipids. Phospholipids resemble neutral fats, but they contain phosphorus and nitrogen. Lecithin is an example of a phospholipid.

Nonpolymeric organic compounds are small molecules, such as glucose, fatty acid molecules, and amino acids.

POLYMERS. *Molecules that can be broken into one or more groups of repeating parts (G. polys, many; meros, part).*

MONOMER. *A molecule that can be built up by repetition to form a polymer (G. monos, one; meros, part).*

TABLE 1.2 Major Inorganic Compounds in the Body

Compound	Symbol	Combines with	Resulting Compounds
Hydroxyl	OH^-	Sodium, potassium, and other ions with a positive charge	$NaOH$, KOH, $Ca(OH)_2$
Carbonate	$CO_3^=$	Mostly with hydrogen ions; also with other positive ions	H_2CO_3, $CaCO_3$
Bicarbonate	HCO_3^-	Mostly with hydrogen, sodium, or potassium ions	H_2CO_3, $NaHCO_3$, or $KHCO_3$
Phosphate	$HPO_4^=$ $H_2PO_4^-$	Mostly with sodium or potassium ions	$NaHPO_4$ or NaH_2PO_4
Sulfate	$SO_4^=$	Mostly with sodium or potassium ions	Na_2SO_4 or K_2SO_4

Basic Chemical Concepts

Chemical Measuring Units

Physiologists borrow some units of measure from chemistry. The units you will see most often are: *atomic weight*, *molecular weight*, *mole*, and *molarity*.

Atomic Weight. Individual atoms, molecules, and compounds are too light to allow us to weigh them individually with ordinary scales or balances. However, it is possible to measure the weight of a large number of atoms, molecules, or compounds. In order to make the measurement meaningful, it is necessary to weigh equal numbers of the various kinds of atoms and to relate the weight of each kind to a single arbitrary standard. Carbon is the standard of atomic weights. It has an atomic weight of 12. Oxygen has an atomic weight of 16; hydrogen, 1; nitrogen, 14; phosphorus, 31; sulfur, 32; chlorine, 35; sodium, 23; and potassium, 39.

When the atomic weight of some element is measured in grams, we have a *gram-atomic* weight of the substance. Thus, 32 grams of sulfur is a gram-atomic weight of sulfur.

GRAM ATOMIC WEIGHT. *The weight of a definite number of atoms (6.023×10^{23}), in grams.*

MOLECULAR WEIGHT. *The summed-up weight of each atom which makes up a molecule.*

MOLE. *The molecular weight of a substance, in grams.*

Molecular Weight, Moles, and Molarity. Very few kinds of chemical substances—for example, copper, silver, and platinum—are capable of existing freely as pure atomic substances. By far the majority of chemicals are combined as molecules or molecular compounds. For these substances we may talk about a molecular weight. It turns out that when the *molecular weight* of some molecule or compound is measured in grams, we have a *mole* of the substance. Thus, 2 grams of hydrogen gas (H_2), 28 grams of nitrogen gas (N_2), 18 grams of water (H_2O), and 58 grams of sodium chloride (NaCl) give one mole each of hydrogen, nitrogen, water, and sodium chloride. In each of these cases we merely add up the atomic weights of the individual atoms in the molecule to get the molecular weight.

In physiology it is often necessary to specify the concentration of a chemical substance in solution. Sometimes this is done in terms of *grams percent*, and we say there are X grams of substance Y in 100 milliliters of a solvent, such as water. At other times, however, concentration is expressed in molar terms.

Molarity is a unit of concentration. When a *mole* of any substance is dissolved in a liquid so as to make up to a total volume of one liter, the solution is said to be 1 molar with respect to the substance. Thus, 58 grams of sodium chloride in a final volume of 1 liter of water is a 1 molar solution of sodium chloride.

Chemical Bonds

All the chemical elements of which the human body is composed are bonded to one another to form molecules and compounds. Before we can begin to study chemical bonds, however, we should first understand the structure of the atom.

Each atom has a central nucleus. This nucleus contains one or more particles which have a positive charge. These particles are called *protons*. Traveling around the nucleus, in *orbitals*, are negatively charged particles called *electrons*. There are an equal number of positively charged protons and negatively charged electrons, so that each atom, by itself, is neutral. This can be seen with respect to hydrogen, oxygen, and carbon (Figure 1.1).

Under certain conditions, some atoms interact with other atoms to form molecules and compounds. Some atoms combine with one another when they are subjected to heat, pressure, or electricity. Others are extremely reactive and may form bonds without the intervention of any external force.

Three types of bonds are especially important in biological systems—covalent, ionic, and hydrogen bonds.

Covalent Bonds. A *covalent bond* is formed between two atoms when each atom contributes to its interacting atom one or more electrons to form electron-pairs. That is, atom A sends one, two, or three electrons into one or more orbitals of atom B, and atom B sends a corresponding number of electrons into orbitals of atom A. Within the interacting orbitals the electrons form pairs. One electron in each pair spins in an opposite rotational sense to the electron with which it is paired. For the reader who is unfamiliar with this topic of electron-sharing, it is helpful to think of each electron pair as two hands in a handshake. One electron (hand) from each atom (person) becomes firmly engaged with an electron of the other atom in the bond-pair.

An example of covalent bonding is shown in a diagram of a molecule of water (H_2O) in Figure 1.2.

A reaction equation illustrating covalent bonding is given here:

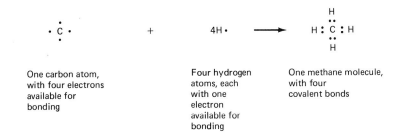

One carbon atom, with four electrons available for bonding

Four hydrogen atoms, each with one electron available for bonding

One methane molecule, with four covalent bonds

Ionic Bonds. An *ionic bond* is formed when an atom gives up one, two, or three electrons wholly to an interacting atom or group of atoms with which it forms a bond. The *electron-donor* acquires one, two, or three positive charges, and the *electron-acceptor* acquires a corresponding number of negative charges. Each member of the pair is now called an *ion*.

Table salt is made up of two ions—sodium (Na^+) and chloride (Cl^-). These ionic species came about in the first place because a sodium atom (Na) donated an electron to a chlorine atom (Cl). The sodium ion (Na^+) has one more proton in its nucleus than it has electrons in its orbits. The chloride ion (Cl^-) has one more electron than it has protons. When Na^+Cl^- is in the solid, dry form, it is said to be associated. When it is placed in water, it dissociates (breaks apart) into Na^+ and Cl^- ions, as shown here:

Hydrogen

Oxygen

Carbon

FIG. 1.1. *Some simple atoms: hydrogen (top), oxygen (center), and carbon*

COVALENT means equal strength. In a covalent bond each atom makes an equal contribution of electrons.

ION. *An electrically charged atom or group of atoms (G. go).*

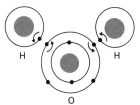

FIG. 1.2. *The water molecule. Two small hydrogen atoms are linked to a single oxygen atom by covalent bonds (arrows).*

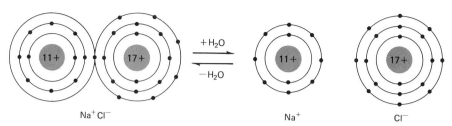

On the left side, the sodium and chlorine atoms are associated as a dry, solid form due to ionic bonding. On the right, they are dissociated. Moreover, they are in water and they are capable of conducting electricity. Hence, they are known as *electrolytes*.

ELECTROLYTE. *An ion in solution (G. elektron, electricity; lytos, soluble).*

A reaction equation which illustrates ionic bonding is shown here:

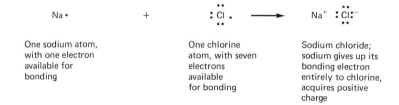

One sodium atom, with one electron available for bonding

One chlorine atom, with seven electrons available for bonding

Sodium chloride; sodium gives up its bonding electron entirely to chlorine, acquires positive charge

From a practical point of view, all of the sodium chloride and most of the other salts in the body are present as electrolytes. The notable exception is calcium phosphate, most of which is locked up (in an associated state) in the bones and teeth. Very little of it is present in a dissociated form in the blood.

Hydrogen Bonds. A *hydrogen bond* is a chemical bond which does not involve either a sharing of electrons or a complete surrender of electrons. In hydrogen bonding, a compound containing hydrogen atoms forms a weak bond with another compound, usually one containing oxygen or nitrogen atoms. Hydrogen bonds are best thought of as being magnetic in character. The hydrogen atom can be thought of as the north pole of a magnet and the oxygen or nitrogen atom as the south pole. A weak "magnetic" attraction is set up between the compound which contains hydrogen and the compound containing oxygen or nitrogen (Figure 1.3).

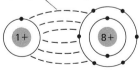

Hydrogen bond ("magnetic forces")

Hydrogen   Oxygen

FIG. 1.3. *The hydrogen atom, left, is joined to an atom of oxygen by the "magnetic force" of the hydrogen bond (broken lines).*

The formation of a hydrogen bond may be explained as follows. A hydrogen atom has a single proton in its nucleus and a single electron in its only orbital. As the electron of the hydrogen atom spins around the nucleus, it is unable to shield the proton at all times. The hydrogen atom, therefore, has a weak positive charge at some times. Oxygen and nitrogen atoms, in contrast, have many orbitals with electrons. These electrons shield the nucleus of oxygen and nitrogen atoms and give a negative character to the atoms. A hydrogen atom with a weak positive charge is thus attracted to an oxygen or nitrogen atom with a weak negative charge. As a result, a hydrogen bond is formed.

In the following figure note how water molecules (HOH) are held together by hydrogen bonds:

Biological Significance of Covalent Bonds

Covalent bonds are important to biological function in two ways. First, the making or breaking of a covalent bond is almost always catalyzed by an enzyme. Enzymes are proteins which catalyze metabolic reactions. They cause the chemical reaction to proceed more quickly, but they themselves remain unchanged when the reaction is over.

Covalent bonds are also important because they provide energy. This energy is present in the covalent bonds of carbohydrates, fats, and proteins. In the course of combusting (oxidizing) these compounds, a certain percentage of the total available energy is released as heat. The rest goes mainly into the production of so-called high energy compounds, the most important of which is ATP.

ATP is an abbreviation for adenosine triphosphate. When this compound is split into two of its component parts—ADP (adenosine diphosphate) and inorganic phosphate (H_3PO_4)—a certain amount of energy is released. This energy makes possible numerous kinds of physiological processes, such as the formation of proteins, the contraction of muscles, the propagation of nerve impulses, and the uptake of certain nutrients into cells. How this energy-release is coupled to the energy-consumption of the physiological process is not known. What is known is that ATP molecules are split, and *must* be split, if the physiological process is to occur.

CATALYST. *A substance which increases the velocity of a chemical reaction but is unchanged in the reaction* (G. kata, down; lysis, loosen).

ATP. *A substance which serves as an immediate source of energy for most physiological processes.*

Biological Significance of Ionic Bonds

Ionic bonds are essential in two major biological functions: in bioelectricity and in buffering the pH (acidity) of blood.

Bioelectricity. Since ions bear either positive or negative charges, they can be distributed across the outer membrane of a cell so that a *polarity* (biostatic electricity) is created. The polarity which is found across the membranes of nerve and muscle cells is especially important. When a nerve or muscle cell is not stimulated, the interior surface of the membrane is negatively charged with respect to the exterior surface.

What causes the membrane of a living cell to have an inner negative charge and an outer positive charge? Only a part of the answer is known.

To begin with, the inner and outer surfaces of the membrane are made up of nondiffusible proteins. These proteins, like all other kinds of proteins, contain two kinds of ionic groups as integral parts of the protein. These ionic groups—the negatively-charged carboxylic groups (COO^-) and the positively-charged amino groups (N^+H_3)—are able to attract small diffusible ions, such as chloride (Cl^-), sodium (Na^+), potassium (K^+), and hydrogen (H^+). Ionic bonds may be formed between the carboxylic groups (COO^-) and sodium (Na^+), potassium (K^+), or any other positively-charged diffusible ion. So too, ionic bonds may be formed between the positively-charged amino group (N^+H_3) and any negatively-charged diffusible ion, such as chloride (Cl^-).

This phenomenon is just one aspect of the partial explanation of the electric charge which exists across the membrane of a living cell. The second aspect concerns the distribution of the small diffusible ions across the cell membrane. For reasons which are not understood, the small, diffusible ions are attracted unequally to the inner and outer surfaces of the cell membrane: more potassium ions are attracted to the inner surface of the membrane, while more sodium ions are attracted to the outer surface. Other kinds of ions are also unequally distributed. The electrical charge across the cell membrane has a voltage of approximately 70 millivolts and it is thought that this voltage is caused by the unequal distribution of sodium and potassium ions across the membrane. How this unequal ion distribution gives rise to a measurable potential is not known.

Many kinds of electrically-charged cells are present in the body. Nerve cells and muscle cells are the most important of these, but many other cells, including gland, bone, and blood cells, are electrically charged.

When an electrically charged cell is stimulated, the polarity of its surface membrane changes. If the polarity increases and the inside of the cell membrane becomes more negative, the cell membrane is said to be *hyperpolarized*. If the polarity is decreased and the inside of the cell membrane becomes less negative, the cell membrane is said to be *depolarized*. Depolarization may continue until the interior surface of the cell acquires a transient (short-lived) positive charge with respect to the outer surface.

HYPERPOLARIZATION. *Making a cell membrane less electrically excitable.*

DEPOLARIZATION. *Making a cell membrane more electrically excitable.*

When the membrane of a nerve or muscle cell is hyperpolarized, the cell is said to *inhibited*. This happens whenever an inhibitory stimulus is applied to the cell membrane. This will be discussed in Chapter 5.

When the cell membrane is depolarized, the cell is said to be *excited*. There are two kinds of excitatory states in living cells. In one kind, the excitation arises in one region of a cell and dies out within the same region. This state is called *local potential*, and it is known to occur in most kinds of electrically-charged cells, including muscle cells and nerve cells.

LOCAL POTENTIAL. *A movement of bioelectricity over a very short distance on a cell membrane, but not from one end of the cell to the other.*

The second class of excitatory activity is found only in muscle and nerve cells. In these cells, when the membrane is depolarized to a *critical*, or *threshold, level*, the excitation spreads from the point of stimulation to the most distant regions of the cell surface. This activity is diagrammed in Figure 1.4. In both nerve and muscle cells, the reversal of polarity occurs all along the surface of the membrane, from the point of stimulation to the end region or regions of the cells, as illustrated in Figure 1.5.

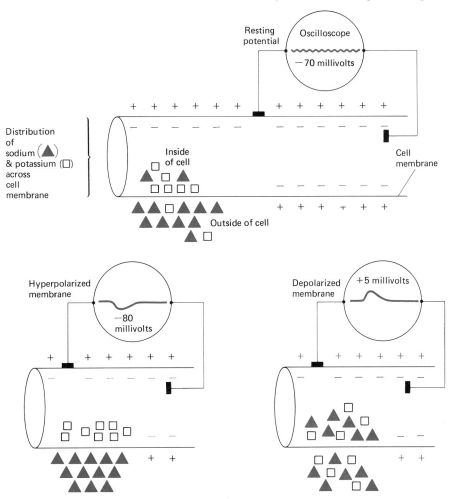

FIG. 1.4. Polarization of cell membranes: resting state (top), hyperpolarized (bottom left), and partly depolarized. If the positive and negative regions are connected, an oscilloscope will indicate the flow of current.

ACTION POTENTIAL. A movement of bioelectricity from one end of a cell membrane to the other end or ends; also called nerve impulse.

It is not electricity as such which travels in living cells. Rather, an ionic current, called an *action potential*, travels along the cell membrane. The action potential travels from about 0.5 to 130 yards per second, depending upon the characteristics of the cell membrane. (Electricity travels at a speed of 186,000 miles per second.) In most muscle cells and certain nerve fibers the action potential travels at the lower range of speed. In certain other nerve fibers it travels in the upper range.

Looking at Figure 1.5, the illustration of events in the *action potential*, you can see that a greater concentration of sodium ions (Na^+) is found outside the resting cell than inside. When an action potential is initiated, sodium ions rush in, and potassium (K^+) ions move out. This in-and-out movement of ions reverses itself within a few milliseconds after the stimulation occurred, restoring the resting conditions of the membrane. However, the in-and-out movement of ions continues from point to point along the surface of the excited membrane, traveling in a one-way direction from the point of stimulation to the ends of the cell membrane.

Finally, it should be mentioned that calcium (Ca^{++}), magnesium (Mg^{++}), and other ions, including the ionic components (COO^- and N^+H_3) of proteins, are important in maintaining membrane potentials. Ionic imbalances

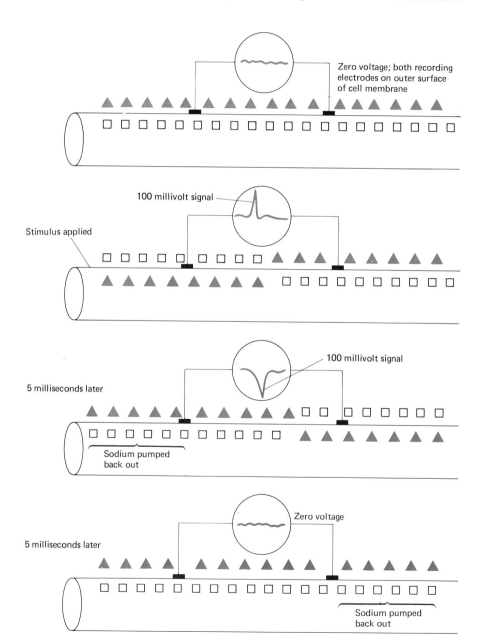

FIG. 1.5. Events of an action potential.

Buffering the pH of Blood. The human body cannot function properly when the blood contains too much or too little acid. An *acid* is any chemical substance which dissociates in water and gives off a hydrogen ion. When we speak of acid, then, we are really talking about the concentration of hydrogen ions (H^+).

Some ionic compounds are hydrogen ion acceptors, while others are hydrogen ion donors. A *hydrogen acceptor* takes up hydrogen ions and makes the blood less acidic. A *hydrogen donor* gives up a hydrogen ion and makes the blood more acidic. Taking up or giving off hydrogen ions to maintain a steady concentration of hydrogen ions is called *buffering*.

In the human body, buffering maintains a constant concentration of hydrogen ions in the blood. Hydrogen ion concentration is measured in pH units. *pH* means *p*ower of the *h*ydrogen ion (H^+). pH is measured in logarithmic units. A pH of 7.0 means there is a hydrogen ion concentration of 10^{-7} (0.0000001) moles per liter. At this pH there are as many hydroxyl ions (OH^-) in solution as there are H^+ ions. As the pH increases, fewer H^+ ions are in solution, and more OH^- ions are present. At a pH of 7.4, the blood contains fewer hydrogen than hydroxyl ions, so the blood is slightly alkaline, or basic.

A *buffer* is a weak acid or base, and a salt of the weak acid or base. An acid, as indicated above, is any chemical compound which, upon being ionized, gives off a hydrogen ion (H^+) as its positively charged ion. A *base* (alkali) is any molecule or ion which can add on a hydrogen ion. Common bases are the hydroxyl ion (OH^-), as in sodium hydroxide (Na^+OH^-), ammonia (NH_3), and the amino group (NH_2) in proteins.

A *salt* is a compound which is formed when an acid reacts with a base. In this reaction the hydrogen ion of the acid is exchanged for the positively-charged ion of the base. For example:

$$Na^+OH^- + H^+Cl^- \rightarrow Na^+Cl^- + H_2O$$

Here, sodium hydroxide reacts with hydrochloric acid to give sodium chloride, a salt.

Again, a buffer consists of either a weak acid *or* a weak base, and a salt of the weak acid or base. The word *weak* means the acid or base is not completely dissociated when it is placed in water. That is, only a certain proportion of the acid molecules ionize, and in doing so, release hydrogen ions. So too, only a certain proportion of the base molecules ionize, and in doing so, become capable of picking up, and becoming attached to, hydrogen ions. The remaining acid or base molecules remain associated and do not ionize. Carbonic acid and phosphoric acid are weak acids, and ammonium hydroxide is a weak base, inasmuch as they are only partially dissociated, as indicated in these reaction equations by reversible arrows:

$$H_2CO_3 \rightleftharpoons H^+ + HCO_3^-$$
$$H_3PO_4 \rightleftharpoons H_2PO_4^- + H^+ \rightleftharpoons HPO_4^= + 2H^+$$
$$NH_4OH \rightleftharpoons NH_4^+ + OH^-$$

ACID. *Any substance which can give up or release hydrogen ions.*

pH. *The concentration of hydrogen ions (acid) in solution.*

BUFFERING. *Keeping the hydrogen ion concentration constant.*

BASE. *Any substance capable of combining with hydrogen ions.*

Basic Chemical Concepts

Hydrochloric acid is a *strong acid* and sodium hydroxide is a *strong base*. Strong acids and bases dissociate completely in water, as indicated by the one-way arrows in these reaction equations:

$$H^+Cl^- \rightarrow H^+ + Cl^-$$
$$Na^+OH^- \rightarrow Na^+ + OH^-$$

The *principal weak organic acid* in the body occurs as a carboxyl group ($COO^-H^+ \rightleftharpoons COO^- + H^+$) on certain compounds. It is present on all amino acids (the building blocks of proteins) and on fatty acids. However, the fatty acid content of the body is very low, since most of the fatty acids are locked up with glycerol to form neutral fats or other fat derivatives. Hence, fatty acids are not important buffering agents in the body. In contrast to fatty acids, however, proteins have many free *carboxyl groups*. These groups may donate or accept hydrogen ions, as shown in this equation:

$$COO^-H^+ \rightleftharpoons COO^- + H^+$$

The free carboxyl groups of proteins are important buffering agents in the blood.

The *principal weak organic base* in the body occurs as an amino group ($NH_2 + H^+ \rightleftharpoons N^+H_3$) on certain compounds. It is present on many metabolites, but only proteins, because of their high concentration, are important sources of this buffering agent.

When weak acids react with bases and are converted into salts, they are transformed from hydrogen donors to hydrogen acceptors. For example:

$$COO^-H^+ + Na^+OH^- \rightarrow COO^-Na^+ + H_2O$$

When weak bases react with acids and are converted into salts, they are transformed from hydrogen acceptors into hydrogen donors. For example:

$$NH_2 + H^+Cl^- \rightleftharpoons N^+H_3Cl^-$$

The roles of the principal buffering systems in the body—the circulatory, respiratory, and excretory systems—are discussed in Chapters 26, 28, and 30 respectively.

Biological Significance of Hydrogen Bonds

Hydrogen bonding is related to biological processes in at least two ways. First, if a water-soluble substance is to go into solution (dissolve), it must be solvated by way of hydrogen bonding. That is, the substance must be *wetted* (surrounded by water molecules). This process, using sodium chloride as an example, is shown in Figure 1.6.

When a substance cannot form hydrogen bonds it must be *emulsified* before it can go into solution. Fats, for example, must be emulsified before they can be dissolved in body fluids.

The other way hydrogen bonding is important to biological function concerns the zippering up of long strands of certain polymeric molecules to form intertwined helixes, parallel strands, and other arrangements—

CARBOXYL GROUP. *An important portion of an organic molecule, which can give up a hydrogen ion (at high pH) or take up a hydrogen ion (at low pH).*

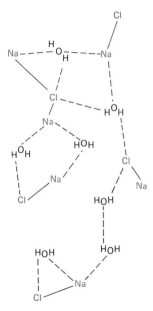

FIG. 1.6. *Solvation, or wetting. Hydrogen bonds link the sodium (Na^+) and chlorine (Cl^-) ions to atoms of hydrogen and oxygen.*

much like forming a yarn or a braid. The two helical strands of genetic material that make up chromosomes are hooked together this way. Hydrogen bonds also hold together the subunit polymers of certain proteins, such as the triple coiled collagen fibers present in connective tissues, various enzymes, and hemoglobin.

EMULSION. *Minute globules of one liquid dispersed throughout a second liquid (L. emulgere, to milk out).*

Important Biochemical Reactions

All of the known biochemical reactions can be classified into six groups. Two of these should be known to every biologist: hydrolysis and oxidation-reduction.

Hydrolytic reactions are involved in digestion. In hydrolytic reactions, a molecule of water is inserted between two building blocks of a dimeric, trimeric, or polymeric nutrient, and the bond which holds the two building blocks together is split, as may be seen here:

HYDROLYSIS. *A chemical reaction in which water is used to rupture a chemical bond (G. hydro, water; lysis, break apart).*

TRIMER. *A polymer made of three parts.*

$$CH_2NH_2\overset{O}{\overset{\|}{C}}-\overset{H}{\overset{|}{N}}CH_2COOH \xrightarrow[\text{enzyme}]{H_2O} 2\ CH_2NH_2COOH$$

Dipeptide — a compound → **Two amino acids**
made of two amino acids

A hydrolytic reaction also occurs every time ATP (adenosine triphosphate, the physiological storage unit of energy), is split into ADP (adenosine diphosphate) and phosphate. This reaction is shown here in simplified form:

$$ATP \rightarrow ADP + P$$

Oxidation-reduction reactions occur whenever one compound gives up one or more of its electrons to a second kind of compound. The donor compound thus becomes *oxidized*, and the recipient of the electrons becomes *reduced*, as shown here:

(1) $CH_3\overset{\overset{..}{\underset{..}{O}}}{C}COOH$ $\underset{\text{Oxidation}}{\overset{\text{Reduction}}{\rightleftharpoons}}$ $CH_3\overset{\overset{..}{\underset{..}{O}}}{\underset{H}{C}}COOH$

Pyruvic acid
(note 8 electrons)

Lactic acid
(note 10 electrons)

(2) $CH_3\overset{H}{\underset{H}{\overset{\cdot}{\underset{\cdot}{C}}}}OH$ $\xrightarrow[\text{(Loss of 2 electrons)}]{\text{Oxidation}}$ CH_3CHO $\xrightarrow[\text{(Gain of oxygen)}]{\text{Oxidation}}$ CH_3COOH

Alcohol   Acetaldehyde   Acetic acid

In the course of deriving energy from food, a nutrient, such as glucose, is ultimately oxidized to water and carbon dioxide. In the entire course of this process electrons are given up through various complex chemical reactions by (a) hydrogen to oxygen to form water, and (b) carbon to oxygen to form carbon dioxide. In this process, for one mole of glucose, about 32 moles of ATP are formed:

$$C_6H_{12}O_6 + 6O_2 \longrightarrow 6H_2O + 6CO_2 + 32\ ATP$$

(1 mole of glucose) (6 moles of oxygen) (6 moles of water) (6 moles of carbon dioxide) (6 moles of ATP)

One mole of glucose weighs 180 grams (about 6 ounces) and yields 672 Calories when completely combusted. Each of the 32 moles of ATP—which are formed at the expense of glucose in the present example—stores 10 Calories of energy. In the course of metabolism, this energy—320 Calories in all—serves as a direct, or immediate, source of energy for physiological processes. The remainder of the energy—352 Calories—is lost as heat, due to the inefficiency of the metabolic processes.

BASIC PHYSICAL CONCEPTS

Some concepts in physics are useful in understanding physiological concepts that will be discussed in this book. In particular, we need to understand the concept of *osmotic pressure*. To understand this concept we need to know about *force, pressure,* and the process of *diffusion*.

Force and Pressure

Air molecules surround the earth and are found about 15 miles up into the atmosphere. Because these molecules have mass, and because this mass is acted upon by gravity, air molecules have weight.

When an object having weight meets resistance, which happens, for example, when an object rests on any surface, the object is said to exert a force. This force may be measured by placing an inverted vacuum tube into a vessel containing mercury to make a simple barometer. At sea level, where there is one atmosphere of pressure, the mercury (Hg) will rise 760 millimeters (mm) in the tube. The mercury rises because the air molecules press down on the mercury in the vessel and force it to follow the path of least resistance—up the empty tube. The mercury rises 760 mm in the tube regardless of the tube's size. An object resting against your body likewise exerts a force, because it has mass which is acted upon by gravity, and therefore it has weight. (See Figure 1.7.)

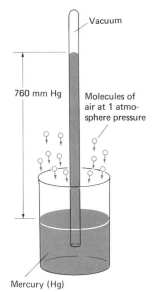

FIG. 1.7. *A simple barometer illustrates the concept of pressure.*

Pressure is defined (regardless of tube diameter) as the *force* (F) per unit of *area* (A). This relationship is expressed in the formula, $P = F/A$. Liquid at rest in a container exerts pressure on all sides of the container. This pressure is equal to the product of the *height* (h) and the *density* (d) of the liquid. The formula which gives this relationship is $P = hd$. Since mercury is 13.6 times as dense as water, an amount of mercury would exert 13.6 times as much pressure on the sides of the container as the same amount of water. The pressure generated in this way is called *static pressure*.

Figure 1.8 illustrates the importance of these concepts in understanding the factors which determine blood pressure. Here we are concerned with *dynamic* rather than with static pressure. As the container (the heart, in this case) contracts, high pressure is generated, forcing the blood into the arteries. The blood pressure throughout the system is determined by how much the different types of blood vessels resist the flow of blood. The highly elastic arteries of the general circulatory system are responsible for the normal blood pressure of 120/80; the contraction of the heart (systole) raises the pressure in the arteries by the equivalent of 40 mm of mercury. In the pulmonary circulatory system, which carries blood to the lungs, the re-

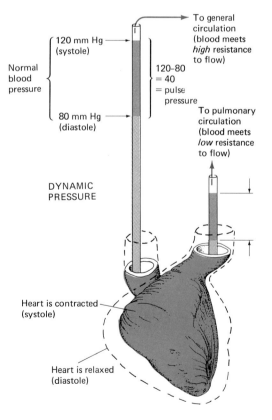

FIG. 1.8. *Schematic representation of blood pressure in the human body; see text for explanation.*

sistance, and therefore the blood pressure, is lower. In Figure 1.8, tubes of mercury are used to show this relationship graphically.

Diffusion

Diffusion is a movement of particles from a region of higher concentration to a region of lower concentration. Figure 1.9 shows the diffusion of dye particles in a beaker of water. When there is no difference in concentration, the rate of diffusion is zero.

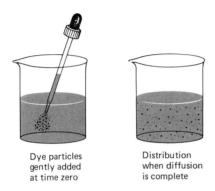

Dye particles gently added at time zero

Distribution when diffusion is complete

FIG. 1.9. *Diffusion.*

FICK'S LAW OF DIFFUSION states that the quantity of material diffusing per unit of time is proportional to (a) the area through which the particles are diffusing, and (b) the concentration gradient.

The process of diffusion is important to the biologist because it is the means by which oxygen and nutrients from the bloodstream get into living cells. It also allows carbon dioxide and other waste products to be carried away from the cells. It is also important because the body has special physiological mechanisms that prevent diffusion from upsetting the very uneven distribution of various electrolytes and metabolites which normally occurs at the surface of cell membranes. If these special mechanisms did not exist, diffusion would disrupt the normal—but unequal—distribution of electrolytes and metabolites across the cell membrane, and the cells would die.

Osmosis and Osmotic Pressure

OSMOSIS. *A form of pressure* (G. osmos, *a push*).

Osmosis (Figure 1.10) is a special type of diffusion. It involves the movement of water from a region of higher water concentration to a region of lower water concentration through a semipermeable membrane. A *semipermeable membrane* is one which allows some kinds of substances to pass through it easily while other kinds of substances are not allowed to pass through. The cell membrane of most living cells is a semipermeable membrane. This membrane allows water molecules to pass through it freely.

Certain other small molecules or ions (such as fats and chloride) also pass through this membrane easily. Most other molecules and ions, however, cannot pass through the cell membrane without the aid of special transport mechanisms. These mechanisms will be discussed later, in Chapter 20.

Osmosis occurs whenever there is a difference in the concentration of water molecules across a cell membrane. Ordinarily, the aqueous (water) solutions outside the cells contain less solute (dissolved) molecules and more water molecules than the aqueous solutions inside the cells. This means that osmosis generally brings water into the cells. It also means that an osmotic pressure is produced inside the cell. That is, more water tends to enter the cell than can be accommodated without a buildup in pressure. You can see an illustration of osmotic pressure in Figure 1.10.

A high osmotic pressure within living cells insures that there will be no excessive collection of water outside the cells in the intercellular spaces. The nervous, endocrine, and excretory systems provide the mechanisms which maintain a proper osmotic balance. When these mechanisms break down, *edema* occurs. Edema is a swelling of a part of the body. It occurs when too much fluid accumulates outside the body cells.

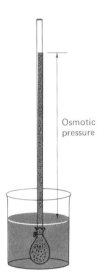

FIG. 1.10. *Osmosis. A concentrated glucose solution is confined within a semipermeable membrane, causing water to flow inward.*

SOLUTE. *The substance dissolved by solvent to form a solution (L. solutis, dissolved).*

SUMMARY

1. In all sciences, including anatomy and physiology, measurements are usually made in metric units. In the metric system, the formula $d = w/v$ relates the weight of an object to its volume. When $d = 1$, as is the case at sea level for pure water, 1 milliliter (or cubic centimeter) of the substance weighs 1 gram.
2. Protoplasm is composed of metabolites which are formed from inorganic and organic nutrients that can ultimately be broken down into compounds, molecules, and then atoms.
3. One special attribute of covalent bonds in living organisms is that they contain energy through which ATP—the direct source of energy for physiological operations—is formed.
4. Ionic bonds and ions are important in living organisms because they lend themselves to bioelectricity and buffers.
5. Hydrogen bonds are important in biology because they permit substances to be solvated (enter solution) and because they allow certain large molecules to unite with other large molecules.
6. Hydrolysis involves the insertion of a water molecule into a covalent bond, causing the bond to split.
7. Oxidation-reduction involves the removal (oxidation) of electrons from a donor-metabolite and the addition (reduction) of the electrons to an acceptor-metabolite.
8. Diffusion is the movement of a substance from a region of high to low concentration, while osmosis is the diffusion of water molecules across a membrane which is relatively impermeable to other types of molecules.

REVIEW QUESTIONS

1. How much does 1 milliliter of mercury weigh if its density is 13.6?
2. Vitamin B_{12} is present in human blood at a level of about 250 picograms per milliliter. How much B_{12} is present, in micrograms, in the body's five liters of blood?
3. Define: molecule, compound, nutrient, metabolite, protoplasm, organic compound, inorganic compound, ATP.
4. Describe how covalent, ionic, and hydrogen bonds differ from one another.
5. How are sodium and potassium involved in the flow of bioelectricity?
6. Explain what is meant by the following statement: a pH buffer consists of two kinds of compounds—a hydrogen donor and a hydrogen acceptor.
7. How much pressure—in grams-centimeter (or Calories)—has to be applied to the surface of a volume of water to force the water 1.7 meters upward into a vertical tube?
8. Write an equation for hydrolysis and an equation for oxidation–reduction. Explain what is involved in each of the two reactions.
9. How do diffusion and osmosis differ from each other?
10. Would it be proper to say that osmotic pressure is a negative pressure which draws water into a region of relatively low water concentration across a semipermeable membrane?

Cells 2

CELLS
    Cell organelles and their functions
        The plasma membrane
        The mitochondria
        The endoplasmic reticulum
        The cell nucleus
        Protein synthesis
        The Golgi apparatus
        The lysosomes
        The soluble fraction of the cell
CELL REPRODUCTION
    Reproduction of body cells: mitosis
    Production of sex cells: meiosis

Gross anatomy known also as *macroscopic anatomy*, deals with the structure of an organism as far as the structure can be studied without the use of a microscope. A subdivision of anatomy, called *microscopic anatomy*, includes *histology* which deals with the structure of tissues, and *cytology* which deals with the structure and biology of cells.

This chapter is concerned with cells. Chapter 3 deals with tissues, and Chapter 4 is concerned with organs and body regions. We begin with a description of a cell, its parts (organelles), and their functions. We will then discuss the reproduction of body cells (mitosis) and the reproduction of sex cells (meiosis).

HISTOLOGY. *Study of microscopic anatomy* (G. histos, *web;* logia, *study of*).

CYTOLOGY. *Study of cell biology* (G. kytos, *cell*).

CELLS

The body contains at least 200 distinct types of cells which differ from one another in shape, size, chemistry, structure, and function. Many of these cells are described in later chapters and some are described in this chapter. What follows now, however, is a description of cells in general, in terms of their components, or *organelles*.

ORGANELLES. *Organs within a cell* (G. organon, *a tool or instrument*).

Cell Organelles and Their Functions

CYTOPLASM. *Protoplasm that lies outside of the cell nucleus* (G. *kytos, cell;* plasma, *something formed*).

All living cells are bounded by an organelle called a plasma membrane. The plasma which this membrane surrounds is called *cytoplasm*.

The cytoplasm contains one or more of the following kinds of organelles: nucleus, nucleolus, mitochondria, endoplasmic reticulum, ribosomes, Golgi apparatus, centrioles, lysosomes, and the soluble fraction of the cell (Fig 2.1).

The Plasma Membrane. The plasma membrane, which has an overall thickness of about 10 nanometers, is a highly organized mosaic of fat, protein, and water. Some of the proteins and fats are structural; that is, they serve as a scaffolding for other cell structures. Other proteins and fats carry out specific roles, several of which are as follows:

ENZYME. *A catalytic substance* (G. en, *in;* zyme, *leaven*).

(1) Some of the proteins present in plasma membranes are *enzymes*. Enzymes are specialized organic molecules which catalyze specific biochemical reactions. In the small intestine, for example, several enzymes (lactase and several kinds of peptidases) are present in the plasma membrane to cleave, or split, molecules of lactose (a sugar) and peptides (small proteins) into smaller units. These smaller units can then be absorbed into the body. In the plasma membrane of skeletal muscle cells there is an enzyme called acetylcholinesterase. This enzyme destroys a special nerve-

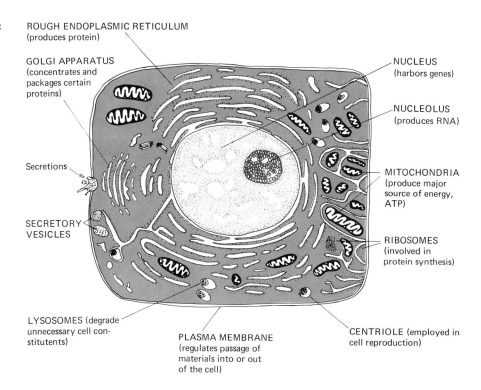

FIG. 2.1. Cell organelles and their functions.

derived molecule (acetylcholine), but it does not carry out its destructive action until acetylcholine has done its job of exciting the muscle cell, as described in Chapter 5.

(2) Other kinds of proteins in the plasma membrane—alone or complexed with fat (lipid) molecules as lipoproteins—are specialized for transporting materials into and out of the cell. They accomplish this within the confines of the plasma membrane; that is, they do not become detached from the membrane in carrying materials into or out of the cell.

(3) Still other kinds of proteins, complexed with lipid molecules as lipoproteins, or with carbohydrate molecules as glycoproteins, are in the plasma membrane. These proteins are capable of wrapping themselves around certain kinds of particles which come into contact with them, from either the inside or outside of the cell. In doing so, they become detached from the rest of the membrane and assume the form of a vesicle (a small sac). They then pass to the outside or inside of the cell. When the material passes into the cell, the process is called *endocytosis*. When it passes out of the cell the process is called *exocytosis*. The term *cytosis* is frequently used as a suffix in a word which describes certain features pertaining to cells, such as *phagocytosis, pinocytosis* or *emeiocytosis*. *Phago* means eating, *pino* means drinking, and *cmcio* means vomiting.

GLYCOPROTEINS.
Protein-carbohydrate compounds (G. glykys, sweet).

The general function of the plasma membrane is to allow or prohibit the passage of materials into or out of the cell. The plasma membrane also has other specific functions, but these are covered elsewhere in the text.

For most of the body's cells, the function of regulating the transport of substances into or out of the cell is controlled, in part, by what is already in the cell and what is needed or unwanted by the cell. For example, a sufficient quantity of amino acids in the cell may prevent the entry of other amino acids. Similarly, a lack of glucose may trigger the plasma membrane to transport glucose into the cell fluid (intracellular fluid) from the fluid outside the cell (extracellular fluid).

For many kinds of cells, the amount of material which enters or leaves is controlled by hormones or nerve cells. A *hormone* is a chemical substance which is produced in one region of the body and reacts with specific target cells in another region. Certain hormones stimulate the uptake of substances into cells. Others stimulate the release of intracellular substances. The hormone insulin, for example, stimulates the uptake of sugars and other nutrients into cells. Pancreozymin, in contrast, stimulates the release of digestive enzymes from exocrine cells in the pancreas.

With regard to nerves, the stimulation of certain target cells by a particular nerve cell may result in the release (secretion) of a cell product; the stimulation of the same target cells by a different kind of a nerve cell may inhibit the release of the same product. For example, the stimulation of stomach and intestinal glands by parasympathetic nerve fibers brings about the release of digestive enzymes and mucus. On the other hand, stimulation of these glands by sympathetic nerve fibers inhibits the release of these secretions.

We will see throughout this text that nerves and hormones regulate every physiological system in the body.

Cells

MITOCHONDRIA. *The energy-producing structures in the cell* (G. mitos, *thread*; chondros, *granule*).

The Mitochondria. Mitochondria are rod-shaped structures about as large as bacteria. Their major function is to generate energy from various metabolites, principally metabolites that are derived from fats, carbohydrates, and proteins. They do this by oxidizing the carbon of these metabolites to carbon dioxide and the hydrogen to water. More than 95% of the oxygen taken into the body by breathing is used here.

Mitochondria contain membrane-bounded enzymes which catalyze a process called *oxidative-phosphorylation*. In this process, ADP (adenosine diphosphate) is converted to ATP (adenosine triphosphate) while oxygen combines with hydrogen atoms to form water. Of chief concern here is that a molecule with two phosphoric acid units (ADP) gains an additional phosphate unit. Later, when that additional unit is cleaved off, energy will be released, and this energy will permit a certain physiological event to take place, such as the contraction of a muscle cell, the synthesis of a metabolite, or the secretion of a cellular substance.

ENDOPLASMIC RETICULUM. *Network of membranes within a cell* (G. endon, *within*; plastos, *formed material*; L. rete, *network*).

The Endoplasmic Reticulum. The *endoplasmic reticulum* is a membranous network found throughout most of the cytoplasm. When examined with an electron microscope, some portions of this network appear smooth. Other portions are studded with small particles, called *ribosomes*, and are referred to as the *rough endoplasmic reticulum*.

RIBOSOMES. *Organelles on which occur the terminal steps of protein production.*

The *smooth endoplasmic reticulum* is associated with the synthesis, or formation, of certain fats and carbohydrates. The rough endoplasmic reticulum is usually engaged in the production of protein in the ribosomes. Both the smooth and rough portions of the endoplasmic reticulum are also involved in conveying bioelectrical messages in skeletal and cardiac muscle cells.

NUCLEUS. *The region of a cell that contains chromosomes* (L. nucleus, *nut*).

The Cell Nucleus. The nucleus is the most conspicuous organelle in the cell. It is present in most cells; it is absent only in mature red blood cells, blood platelet cells, the outermost skin cells, and the surface cells of the eye lens.

CHROMOSOMES. *Cell structures that contain the genes; so-named because they pick up color from stains used in histological work* (G. chroma, *color*; somas, *body*).

Within the nucleus there are normally 46 chromosomes, one nucleolus, several forms of RNA (ribonucleic acid), and various kinds of enzymes. Collectively, these components form an integral part of the cellular machinery which determines how much and what kinds of proteins are to be produced by the cell. Some of the proteins are enzymes, others serve structural roles, and still others serve as nutritional reserves or as hormones. In being produced, these proteins express the genetic potential of a person's heredity, wherein an individual may acquire diabetes, be predisposed toward hardening of the arteries, inherit the potential to be a great athlete, or fall heir to the tendency toward obesity.

GENES. *The units through which hereditary characteristics are transmitted* (G. genesthai, *to be produced*).

Genes are segments of DNA (deoxyribonucleic acid). DNA is a polymer which is combined with proteins to form chromosomes. DNA itself is composed of a sugar called deoxyribose, four kinds of organic molecules called bases, and phosphoric acid. These materials are strung together to form a double-stranded helical molecule. Under special conditions, described in the following section, genes produce *messenger-RNA* molecules. RNA molecules are similar in many respects to DNA. However, they contain the sugar ribose instead of deoxyribose, and one of their organic bases is uracil instead of thymine. Also, they are single-stranded, not double-stranded, helices.

Protein Synthesis.   The process of producing any kind of protein is shown in Figure 2.2. The process begins within the nucleus, along a segment of chromosomal material (DNA). The chromosomal material contains several different kinds of genes which are needed for the production of a single kind of protein. At least three of these different kinds of genes are needed regardless of what kind of protein is to be made. These three kinds of genes are an operator gene, a structural gene, and a repressor gene.

Ordinarily, when the protein in question is not needed, its *operator gene* lies dormant; that is, it is inactive. The operator gene is said to be inactivated by an inhibitor—a protein whose production is dictated and initiated by the *repressor gene*.

HELIX. *A three-dimensional spiral or coil* (G. eilyein, *to roll*).

DEOXYRIBOSE. *A ribose sugar from which a single oxygen atom has been removed.*

ORGANIC BASES. *Nitrogen-bearing organic compounds.*

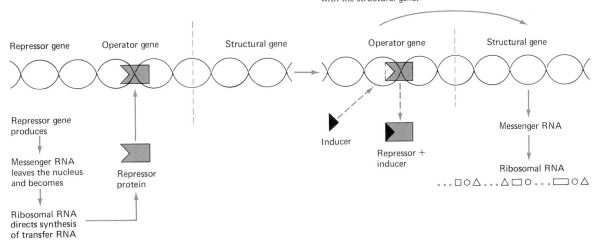

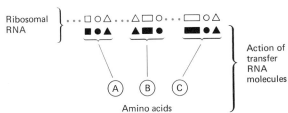

FIG. 2.2. *Protein synthesis.*

When the protein is needed, an inducer substance binds to the inhibitor and prevents the inhibitor from inactivating the operator gene. At this point, the *operator gene* allows a special enzyme, *RNA polymerase,* to work upon the *structural gene.* The structural gene serves as a *template.* (a pattern, much like a sewing pattern) upon which a *messenger-RNA* molecule is produced. The template is a series, or sequence, of organic bases, whose particular linear arrangement, or combination, constitutes genetic information. The messenger-RNA molecule then passes out of the nucleus through a pore, and becomes associated, like a bridgework, to ribosomal RNA. The ribosome may be attached to the endoplasmic reticulum, or it may be moving freely within the cytoplasm.

Two more steps must be taken to complete the manufacture of the protein. First, a third form of RNA, called *transfer-RNA,* comes into the picture. At one end of this polymer is an amino acid. At the other end is a *recognition triplet*—a trio of organic bases—which recognizes a corresponding trio of complementary bases on the messenger-RNA molecule. The recognition triplet then attaches itself, through hydrogen bonding, to the complementary trio on the messenger-RNA molecule.

RNA. *The material which carries instructions from the genes for protein production.*

In the final step of protein synthesis, numerous transfer-RNA molecules line up side by side on the messenger-RNA molecule. The amino acids carried by each transfer-RNA molecule then bond together. The bonds which are formed are called *peptide bonds,* and the amino acids are said to polymerize into a protein.

GOLGI APPARATUS. *A cell organelle which puts carbohydrates and proteins together to form glycoproteins.*

The Golgi Apparatus. The Golgi apparatus is named after Camillo Golgi, who first described it in 1898. When examined carefully with an electron microscope, it is seen as a stack of small swollen vesicles (sacs) which appear to be stemming from an underlying stack of flattened membranous plates.

What are the functions of the Golgi apparatus? One of them is to package intracellular digestive enzymes into membranous sacs. This function is accomplished immediately after the digestive enzymes are produced and released from ribosomes. In accomplishing this task the Golgi apparatus creates an organelle known as a *lysosome.* This organelle is described in the following section.

A second function of the Golgi apparatus is to produce specialized carbohydrates which are added to certain proteins as the proteins come off their assembly lines in the ribosomes. Known as *glycoproteins,* these carbohydrate-protein complexes are sent to the outside of the cell where they serve various purposes. Some of them adhere to the plasma membrane or become part of it; in both cases they serve as informational molecules. Other glycoproteins pass out of the cell into the space between cells (the *interstitial space*), where they contribute to the governing of tissue behavior in ways which scientists are only now beginning to study.

INTERSTITIAL SPACE. *The space between the cells in a tissue* (L. interstitium, a place between).

The informational molecules perform a variety of functions. Some of them tell the body's defense system that the cell is not diseased—as, for example, with cancer—and that the cell should not be disposed of by various kinds of scavenger cells, such as white blood cells. Other informational molecules perhaps tell adjoining cells in a tissue that the tissue has grown to its optimal size and that the cells should stop multiplying. Still another type of informational molecule is used by the plasma membrane of various cells to direct the processes of endocytosis or exocytosis wherein certain materials are brought into or sent out of the cell. In the case of sex cells—egg and sperm cells—it is believed that there are informational molecules that tell sex cells of opposite mating types that they are to unite biologically, as occurs in fertilization.

The third function of the Golgi apparatus is to enclose its glycoproteins within a membranous sac before discharging them out of the cell. This function is particularly widespread in the mucus cells lining the respiratory, digestive, excretory, and reproductive passageways of the body. The mucus secretions serve to lubricate the passageways, so that the surfaces of the passageways will not adhere to one another. In addition, the mucus secretions serve as a filtration gel which influences the kinds and amounts of metabolites that may enter the body cells from the passageway.

The Lysosomes. Lysosomes are membrane-bounded bodies which are similar in size to ribosomes (15 nanometers in diameter). Unlike ribosomes, however, lysosomes contain digestive enzymes. These enzymes, like all proteins, are produced in ribosomes, after which they are transferred to the Golgi apparatus for packaging into a membranous wrapper.

LYSOSOMES. *Small bodies which contain destructive enzymes* (G. lys, *destroy;* somas, *body*).

Lysosomes have at least four functions, one of which is not desirable:

(1) Intracellular digestion. During normal metabolism or cell division—when certain proteins, carbohydrates, nucleic acids, and other polymeric materials are not needed by a cell—the digestive enzymes of the lysosomes start disposing of the material. They do this by catalyzing the conversion of the polymers into their component building blocks.

(2) Function in injury. After a tissue has been injured, as in a wound or burn, it is necessary to dispose of the cellular materials. The lysosomal enzymes initiate this task by digesting the polymeric materials of the cell into metabolites. The metabolites then diffuse through extracellular fluids to healthy cells, where they may be used as a source of energy or in the building of protoplasm.

(3) Function in disease. When a cell becomes diseased it may undergo a process of "self-demolition" as a result of the digestive action of its own lysosomal enzymes.

(4) Function in aging. As the body ages, certain of its components begin to take on a foreign character. Tissues within joints, for example, or within a gland such as the thyroid gland, begin to "appear" different to certain kinds of surveyor cells in the body. These surveyor cells trigger an immunological response through which the "foreign" tissue becomes subject to attack, destruction, and disposal by the body's immunological system. This process is described more fully in Chapter 32, but for now we shall say only that the lysosomes are involved.

Through their digestive enzymes, lysosomes bring about a partial destruction of the "foreign" matter. Often, however, the "foreign" matter is not disposed of. Rather, it becomes more and more foreign in character and is attacked increasingly by the immunologic system. Throughout this destructive process pain is felt in the affected tissue—for example, in a rheumatic joint. Often, the pain can be relieved with aspirin, cortisone, or other drugs. It is believed that, among other things, these anti-inflammatory drugs prevent leakage of digestive enzymes from the lysosomes; in doing so, the drugs may reduce the amount of destruction by the lysosomal enzymes.

The Soluble Fraction of the Cell. The soluble fraction of the cell is the cytoplasm minus all the other organelles. It is obtained by mincing a piece of tissue and centrifuging the mince in an ultracentrifuge. With this centrifuge it is possible to obtain a centrifugal force which is at least 100,000 times the force of gravity. At this force, all of the particulate matter of the mince is spun to the bottom of the tube, leaving the soluble fraction of the cell as the overlying liquid, the *supernatant.*

> **ULTRACENTRIFUGE.** *A vacuum centrifuge which can achieve forces of about 300,000 times gravity.*

The soluble fraction of the cell contains numerous kinds of nutrients and metabolites, including inorganic ions and vitamins. It also contains many enzymes that are not present as components of the other organelles.

CELL REPRODUCTION

From the viewpoint of reproduction, the cells of the body may be classified as body (somatic) cells or sex (germ) cells.

Human *body cells* contain 23 pairs of chromosomes. Each cell is said to have a *diploid* number of chromosomes (two sets of genes), and the cell reproduces itself through *mitosis.*

Human egg and sperm cells begin their development with 23 pairs of chromosomes, but upon achieving maturity each germ cell contains 23 unpaired chromosomes. That is, each mature germ cell has a *haploid* number of chromosomes (one set of genes) which it has received through a process known as *meiosis.*

> **HAPLOID.** *Pertains to the single set of chromosomes of a sex cell, in contrast to the double (diploid) set in a body cell (G. haplous, simple or plain).*

Reproduction of Body Cells: Mitosis

Mitosis is the complex process of cell reproduction. Five stages of mitosis may be more or less arbitrarily distinguished under the microscope (Figure 2.3).

During the first stage, called *interphase*, the chromosomes are stretched out as long, thin strands. Each chromosome duplicates itself, but the double chromosomes are not yet visible through the microscope.

During the second stage, *prophase*, each chromosome becomes shorter and thicker and appears doubled. The visible replicate (duplicate) chromosomes are termed *chromatids* at this point.

In the third phase, *metaphase*, the paired chromatids are aligned on a single plane midway between two poles.

In the fourth phase, *anaphase*, one chromatid of each pair is drawn into a polar region. Each chromatid is now known as a daughter chromosome.

Finally, during the fifth phase, *telophase*, the cytoplasmic membrane constricts, giving rise to two daughter cells. These cells enter the interphase and a new cycle of mitosis may begin again.

The significance of mitosis is that each daughter cell receives the same genetic potential as its parent cell. Each daughter chromosome is reproduced as a duplicate copy of the chromosome from which it was derived in the original parent cell.

Most of the body's 200 different kinds of cells are replaced periodically through mitosis. There are two classes of cells through which such replacement is achieved: *mesenchymal cells* and *basal cells*. Mesenchymal cells are found in one of the body's four basic types of tissues—connective tissue. Basal cells are found in another basic type of tissue—epithelial tissue. Of the remaining basic types of tissue—nerve and muscle—it is perhaps correct to say that neither mesenchymal nor basal cells are present for their proliferation beyond the first few years of life, though some muscle cells may arise from mesenchymal cells under certain conditions. What is the significance of having mesenchymal cells and basal cells?

(1) The *mesenchymal* cells, known also as stem cells or blast cells, are embryonic in nature throughout most or all of our lifetime. They are highly plastic cells, capable of giving rise to any one of several different kinds of

MESENCHYME. *Primordial tissue which has the potential of giving rise to any one of several kinds of mature tissues* (G. mesos, *middle;* chyme, *infusion*).

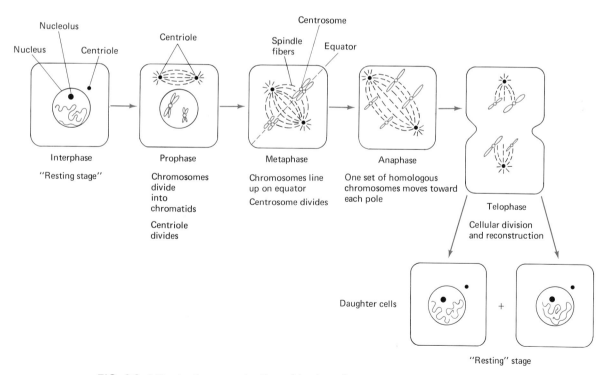

FIG. 2.3. *Mitosis, the reproduction of body cells.*

cells. Fat cells and all the various blood cells, for example, may all perhaps be derived from only one or two types of mesenchymal cells (see Figure 3.5 in Chapter 3). In addition to their importance as highly differentiable cells, mesenchymal cells are important as a basic material from which connective tissues (and to a lesser extent muscle tissue and possibly epithelial tissue) may be replaced in the course of wear and tear. Red blood cells, for example, are replaced every four months.

(2) *Basal cells* are more specialized than primitive mesenchymal cells. They do not differentiate into a wide variety of cell types. Instead, they give rise to additional cells of their own kind. They are important for renewing, and thus periodically replacing, the tissues to which they give rise. Basal cells are present in glands throughout the body, in the entire lining of the digestive tract, and in the outer layer (epidermis) of the skin. The individual cells lining the digestive tract are replaced every 5 to 7 days; the epidermal cells every 3 to 4 weeks.

Although most body cells can reproduce themselves, certain cells, such as nerve and possibly even fat cells, are unable to increase in number through mitosis. Early in life, at the age of a few days to a few years, we very likely find ourselves with all the nerve cells (and perhaps fat cells) we shall ever have. The specializations of these cells prevents them from undergoing mitosis, and neither mesenchymal cells nor basal cells can differentiate into such highly specialized cells.

Production of Sex Cells: Meiosis

Meiosis is a complex process through which haploid sex cells are produced from sex cells with a diploid number of chromosomes (Figure 2.4). Meiosis differs from mitosis in three fundamental ways: (1) There are two cycles of division rather than one, although the second interphase and prophase occur so briefly that they are seldom seen. (2) The total number of chromosomes is reduced from the diploid number, often written as $2n$, to the haploid number, or n. (3) The phenomenon of *crossing-over,* which allows different combinations of genes to be passed on from one cell generation to another, occurs during the first prophase.

The first division of meiosis, called the *reduction-division*, results in the formation of two daughter cells, each of which has a haploid number of chromosomes and a distinct combination of genes. The second division is called the *equation-division*. It is essentially a mitotic division. Each daughter cell divides again, and in doing so, one chromatid from each chromosome pair passes to one of the newly forming cells, while the other chromatid passes to the second newly forming cell.

From interphase I through prophase I (*reduction-division*) the following events happen. First, each chromosome becomes paired with a closely related, or *homologous*, chromosome. The paired homologous chromosomes form a single *bivalent* chromosome. One member of the homologous pair was originally contributed to the cell by the male parent, the other by the female parent. Next, each chromosome splits into so-called sister chroma-

BIVALENT. *Two paired homologous chromosomes, each split into two sister chromatids* (L. bi, *two;* valere, *to have power*).

tids, whereby the bivalent becomes a *tetrad,* with two pairs of bivalent chromosomes. At this time, segments of one chromatid from one chromosome usually *cross over* and become exchanged with segments from a chromatid of the other homologous chromosome.

Crossing-over (Figure 2.5) is an extremely important process because it allows an offspring to receive genetic traits which differ from those of either of its parents. It is the principal mechanism through which genes from the parent cell are passed on in different combinations to the daughter cells during the process of meiosis, as shown in Figure 2.4. Without crossing-over, it would be impossible for an offspring to receive genetic material which differs from that of its parents.

Following prophase I (and crossing-over), the cell enters metaphase I, wherein the tetrads line up in the equatorial plane.

During anaphase I, the tetrads split into dyads (half tetrads) and the two dyads from each tetrad move into opposite poles of the newly forming cells. Because of crossing-over, each dyad possesses a different combination of

CROSSING-OVER. *The process in heredity through which genes are shuffled to produce different combinations in the offspring.*

TETRAD. *A collection of four items. In heredity a tetrad is a bivalent chromosome that divides into four in meiosis (G. tetras).*

DYAD. *Literally, "two" (G. dyas).*

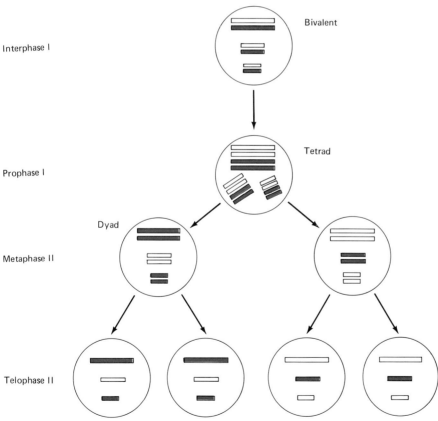

FIG. 2.4. *Meiosis, the production of sex cells, "shuffles" the genetic material in each generation. Genes are also recombined by crossing-over during prophase I (Figure 2.5).*

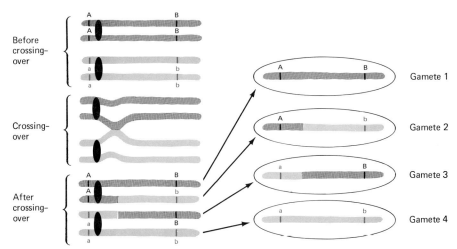

FIG. 2.5. The genes A, a, B, and b are recombined during meiosis by the process of crossing-over. Gametes 2 and 3 possess "new" chromosomes, which were not present in the parent organism.

genes—different, that is, from the combination that was present in the original chromosome (before the chromosome became split into sister chromatids). After anaphase I is complete telophase I takes place and two cells appear, each with a haploid number (n) of chromosomes.

In the equation-division of meiosis, the dyads in each of the two daughter cells become aligned on the equatorial plane during metaphase II. The chromatids that comprise the dyads separate during anaphase II, and following telophase II, four complete cells are seen.

The significance of meiosis lies in (1) a reshuffling of genetic material among chromosomes as a result of *crossing-over*, and (2) the production of sex cells that may combine with sex cells of another individual to form an offspring with a different set of genes from either parent.

In humans, there are 46 chromosomes in body cells and 23 chromosomes in sex cells. Two of the chromosomes in each body cell, and one of the chromosomes in each sex cell, are *sex chromosomes*. The remaining chromosomes are called *somatic chromosomes*.

SOMATIC. Relating to the body (G. somas).

The sex chromosomes are referred to as X and Y chromosomes. Ordinarily, females have two X chromosomes in each of their body cells and one X chromosome in each sex cell (egg cell, or ovum). Males have an X and a Y chromosome in each body cell, and either an X *or* a Y chromosome in each sex cell (sperm cell).

SEX CHARACTERS are those which distinguish a female from a male organism (L. sexus).

ZYGOTE. A fertilized egg, consisting of a fused male and female sex cell (G. zygotos, yoked).

Under abnormal conditions of meiosis, some sex cells may wind up with either two sex chromosomes or none at all. Depending upon the makeup of the sex cells which unite during fertilization in the fallopian tube (oviduct) of the female, the resulting *zygote* (fertilized egg) may either die, or give rise to (a) a normal individual; (b) an individual with Klinefelter's syndrome; or (c) an individual with Turner's syndrome. The results of these variations are shown in the table on page 37.

The most common ("normal") combinations are enclosed in heavy lines. A Klinefelter male usually has some secondary female sexual characteristics, such as sparse hair on the face and body, a high voice, and some

Cells

Male Sex Chromosomes in Sex Cell

	X	Y	XY	NONE
X	XX Normal female	XY Normal male	XXY Klinefelter male	X Turner female
XX	XXX Almost normal Female	XXY Klinefelter male	XXXY Klinefelter male	XX Normal female
NONE	X Turner female	Y Death	XY Normal male	none Death

(Female Sex Chromosomes in Sex Cell)

breast development. The testes are usually not well developed and sterility is common. Klinefelter boys often suffer severe emotional and social problems.

A Turner female is usually short, with a short, broad neck and low hairline. These individuals are sometimes deaf, are often mentally deficient, and usually have undeveloped female sex organs.

SUMMARY

1. Biological cells are contained in a plasma membrane whose primary function is to determine what shall pass into or out of the cell.
2. Mitochondria are cell organelles whose main function is to produce ATP, a compound that serves as an immediate and direct source of energy for most physiological processes.
3. The endoplasmic reticulum, a network of membranes within a cell, is involved in the synthesis of protoplasm in all cells and in the movement of bioelectricity in muscle cells of skeletal and heart muscle.
4. The cell nucleus contains the chromosomes of the cell, which in turn consist of genes through whose chemical actions proteins, including enzymes, are synthesized.
5. In protein synthesis, an operator gene directs a structural gene to produce a messenger RNA molecule. The messenger RNA molecule passes out of the nucleus to become ribosomal RNA on the endoplasmic reticulum. A transfer RNA molecule is attracted to the ribosomal RNA, bringing with it an amino acid. This amino acid becomes polymerized to an adjacent amino acid held by another transfer RNA molecule.
6. The Golgi apparatus adds a carbohydrate component to certain kinds of newly-synthesized protein. The resulting glycoprotein is usually packaged into a membranous sac by the Golgi apparatus.

7. Lysosomes are membranous sacs containing digestive enzymes that are used by the cell in cell division, repair of injury, and cell maintenance. Lysosomes can be destructive to the body in inflammatory processes.
8. Mitosis is a cell division in which identical sets of chromosomes are passed on to successive generations of the same type of cell. Meiosis is a division of sex cells through which, in the process of crossing-over, new combinations of genes arise in the new cells.
9. Mesenchymal cells are primitive cells that are capable of giving rise to any one of a large number of different kinds of specialized cells.
10. Basal cells are primitive cells that are capable of giving rise to epithelial cells in the body.

REVIEW QUESTIONS

1. Is the plasma membrane a cell organelle? What are some of its functions?
2. What is a gene? What is it made of?
3. Describe the roles of DNA, messenger RNA, microsomal RNA, and transfer RNA in protein synthesis.
4. Does the Golgi apparatus play a role in protein synthesis?
5. Of what importance are lysosomes to a cell? Are their actions always beneficial?
6. Where are the centrioles and centrosomes located in a cell? What are their functions?
7. Describe the five main stages of mitosis.
8. In what ways are mitosis and meiosis different from each other?
9. What happens in crossing-over? Why is crossing-over important from an evolutionary point of view?
10. Define: glycoprotein, mesenchyme, basal cell, haploid, diploid.

Tissues 3

EPITHELIAL TISSUES
    Types of epithelial tissues
    Epithelial cells
    Surface modifications of epithelial cells
    Functions of epithelia
    Occurrence of the various types of epithelia
CONNECTIVE TISSUES
    Structure of connective tissues
    Classification of connective tissues
    General types of connective tissues
        Loose connective tissue
        Dense connective tissue
    Special types of connective tissues
        Adipose tissue
        Blood tissue
        Myeloid tissue
        Lymphoid tissue
        Cartilage tissue
        Bone tissue
        Reticuloendothelial system
MUSCLE TISSUES
    Skeletal muscle
    Smooth muscle
    Cardiac muscle
ABNORMALITIES OF CELLS AND TISSUES
    Cancer
        Forms of cancer
        Diagnosis of cancer
    Inflammation

From a strictly anatomical viewpoint, the body is an assembly of <u>organs</u>, each performing one or more specialized or general functions. Organs, in turn, are made up of tissues.

Epithelial Tissues

A *tissue* is an assembly of cells and intercellular materials which share the task of carrying out a distinct physiological function. Although there are many different tissues in the body, each one can be classified into one of four groups, depending on the arrangement and appearance of its cells and other components. Before starting this chapter, look briefly at the pictures—the photomicrographs. They may have no meaning to you now, but as you go along ask yourself to identify the kinds of tissues in each photomicrograph.

This chapter is concerned with the identification and functions of three of the four classes of tissues—epithelial, connective, and muscle tissues. The fourth class, nerve tissue, is covered in Chapter 5. After our discussion of the three classes of tissue we will focus on two common abnormalities of tissues, cancer and inflammation.

EPITHELIAL TISSUES

If you are examining a histological slide or a photomicrograph, you cannot fail to distinguish epithelial tissue from tissues of other classes if you keep three points in mind. First, epithelial tissues are found as coverings on the surfaces of all organs of the body. They are also found lining all of the body's passageways or vessels, such as the respiratory, digestive, excretory, and reproductive tracts, and the blood and lymph vessels. Second, epithelial tissues constitute the *parenchyma* (main mass) of cells in every gland. Third, all epithelial cells, whether they are found as a surface lining or as the parenchyma of a gland, are close to one another; they are always similar to one another in appearance, and do not appear fibrous.

PARENCHYMA. *The essential or functional tissue of an organ, as distinguished from its connective tissue* (G. parenchein, *to pour in*).

Types of Epithelial Tissues

Epithelial tissues may be simple, pseudostratified, or stratified (Figure 3.1). A *simple epithelium* consists of a single layer of cells. A *pseudostratified epithelium* also consists of a single layer of cells, but the nuclei of the cells are located at various levels among the cells, not at the same level. As a result, it looks as if there are several layers of cells.

A *stratified epithelium* consists of two or more layers of epithelial cells. A special form of stratified epithelium is the *transitional epithelium*. This type of epithelium is found only in the excretory system of the body, where it lines the passageways which carry urine from the kidney through the ureter into the bladder, and from there out of the body by way of the urethra (Figure 3.1, bottom right-hand drawing). A transitional epithelium appears to have 5 or 6 layers of cells when the passageway is empty, and 2 to 3 layers when it is full and distended. The same number of cell layers, however, is always present. The "transitions" which appear merely reflect the differences in shape between the deflated and the expanded passageway.

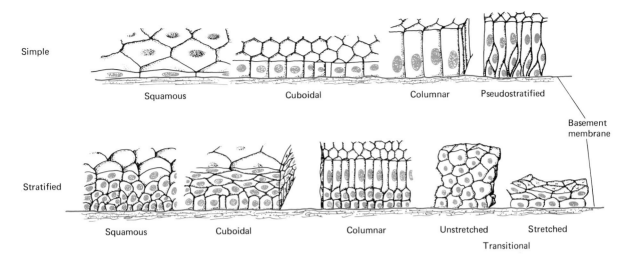

FIG. 3.1. *Epithelial tissue.*

Epithelial Cells

Epithelial tissues are made of one or more of the following kinds of cells: squamous, cuboidal, columnar, polyhedral, or myoepithelial; some of these are shown in Figure 3.2.

Squamous cells are flattened, somewhat resembling fried eggs. They are found lining blood and lymph vessels, and on the surface of the skin.

Cuboidal cells resemble cubes. However, they are somewhat larger at their base than at their top (pyramidal), and their nucleus is midway between the top and bottom of the cell. Cuboidal cells are commonly found lining certain small tubes (tubules), such as kidney tubules. They are also found in various glands, such as the ovaries, the liver, and the thyroid gland.

Columnar cells are distinctly taller than they are wide. Like cuboidal cells, they line various ducts, such as the intestinal tract, and they are common in various glands, such as glands in the uterus and intestinal glands.

Polyhedral cells are many-sided. They comprise the parenchyma (main mass) of the liver and the pituitary gland. Polyhedral cells also make up some portion of various other glands (such as the pancreas), the skin, and certain other body organs.

Myoepithelial cells are contractile epithelial cells. They lie between secretory cells in certain glands, such as sweat, tear, salivary, and mammary glands. They are sometimes classified as smooth muscle cells.

Surface Modifications of Epithelial Cells

The cells of simple epithelia have three kinds of surfaces—free, lateral, and basal. The *free surface* is exposed to the air, as in the outer surface of skin,

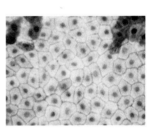
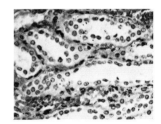
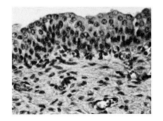

FIG. 3.2. *Epithelial tissues: top, squamous; center, cuboidal; bottom, transitional supported by connective tissue.*

VILLI. *Minute nipple-like projections from a surface (L. shaggy hair).*

CILIA. *Threadlike projections from the free surface of certain epithelial cells. Cilia beat rhythmically, causing fluids to be moved across the surface of the cell (L. cilium, eyelash).*

BASEMENT MEMBRANE. *A noncellular material on which an epithelium is seated.*

or it is bathed by a fluid that the cell secretes, as in the lining of the respiratory, digestive, excretory, and reproductive tracts. Sometimes, the free surface is modified with microvilli or cilia (Figure 3.3). *Microvilli* are tiny nipple-like projections which can be seen clearly only with an electron microscope. They serve to increase the surface area of the cell, and are prominent on cells which line the small intestine and certain kidney tubules. *Cilia* are fibrous, hairlike projections that drive fluids from the free surface toward an external opening of the body. They are found in the respiratory and reproductive tracts and in certain other regions of the body as well.

The *lateral surfaces* of epithelial cells are modified for attaching one cell to another. The modification may be in the form of fibrils, or it may be in the form of an intercellular cement (Figure 3.3).

The *basal surface* is modified in the sense that it secretes a *basement membrane*. This membrane, which is usually not easily seen with an ordinary microscope, is made up of sticky mucopolysaccharides—complex carbohydrate polymers containing nitrogen-bearing groups. The basement membrane serves to anchor the epithelial cell to the connective tissue. It is fastened to the connective tissue by microscopic fibers (fibrils) of collagen, which are produced by, and twisted among, the connective tissue. *Collagen* is a tough, triple-coiled protein of high tensile strength.

In stratified epithelia the cells beneath the outermost layer of cells have each surface attached to a neighboring cell. The lateral and apical surfaces of the lowest layer of cells are modified for attachment to epithelial cells, and the basal surface is modified for producing a basement membrane.

FIG. 3.3. *Surfaces of epithelial cells.*

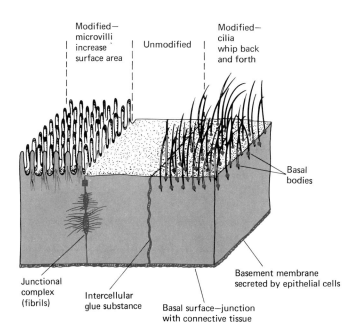

Functions of Epithelia

Most epithelial tissues have certain general functions. Several kinds also have specialized functions. The *general functions* of epithelia are absorption, excretion, and secretion. *Absorption* refers to the uptake of metabolites from extracellular fluids. *Excretion* is the elimination of products which are useless to the cell and body, and *secretion* is the release of metabolic products to the outer surface of a cell. The secreted material may be important to the cell or to the tissue to which it belongs. Alternatively, it may be carried into a blood or lymph vessel to be circulated to some other area of the body. Examples of such secretions are (a) hormones and (b) nutrients which are passed from liver or fat cells to other cells of the body.

The *special functions* of epithelia include protection and reproduction. *Protection* is rendered by epithelia in general, such as by the outer layer of skin and the lining of the digestive, respiratory, excretory, and reproductive tracts. *Reproduction* is a specialized function of the sex organs in which sperm and eggs are made.

HORMONE. *A chemical secretory product of one cell which stimulates another kind of cell (G. hormaein, to excite).*

Occurrence of the Various Types of Epithelia

It is worth repeating that epithelia are present in the body either as the parenchyma of glands or as the surface of an organ, a vessel, or a tract. *Simple epithelia* are present wherever there is little need of special protection for an organ. *Simple squamous epithelium* lines the tiny air sacs (alveoli) of the lungs, the blood vessels, the smallest ducts of many glands, the eardrum, the body cavities, and other structures where it is necessary to have an easy passageway for material (or sound waves in the case of the eardrum) into a tissue or organ. *Simple cuboidal epithelium* lines the kidney tubules, covers the ovary, forms the front surface of the eye's lens and the rear surface of the iris, and is present in many glands. In places where larger volumes of fluid must be secreted or absorbed, simple cuboidal epithelium is present instead of simple squamous epithelium. *Simple columnar epithelium* lines the stomach, intestines, gallbladder, uterus, oviducts (fallopian tubes), some of the smaller respiratory passageways (such as small bronchi and large bronchioles), and many glands. Like cuboidal epithelium, it is present where relatively large volumes of fluid must be secreted or absorbed.

A *pseudostratified epithelium* is found in some of the larger ducts of the respiratory, excretory, and male reproductive systems, and in certain other regions of the body, such as in the eustachian (auditory) tube and the olfactory (smell) region of the nose. The main activity of pseudostratified epithelium is to secrete a fluid and to move this fluid along the surface of the epithelium.

A *stratified epithelium* usually occurs wherever there is exposure to air, or possibility of harm, and wherever large quantities of secretions are needed. *Stratified squamous epithelium* lines the mouth, anus, urethra (the duct that carries urine to the outside), vagina, cornea, outer ear canal, and skin. *Stratified cuboidal epithelium* is present in the ovary's egg-producing follicles, the ducts of sweat glands and sebaceous glands, and in several other areas of the body. *Stratified columnar epithelium* usually occurs where a columnar or pseudostratified epithelium meets a stratified squamous epithelium. Examples include the transition region from voice box (larynx) to windpipe (trachea), the area where the urethra meets the outer skin, and the area where the conjunctiva (front surface of eyeball and inner free surface of eyelids) meets the outer skin of the eyelids.

Transitional stratified epithelium is found only in the lower urinary tract—from the ureters (which carry urine from the kidneys to the urinary bladder), through the urinary bladder, and into much of the urethra. The major function of the transitional epithelium is to accommodate wide variations in the size of the ducts and bladder, depending on how much fluid they contain.

The body also has three *specialized epithelia*, which do not fall into any of the four basic categories. *Neuroepithelia* have modified columnar cells that serve to receive smell, taste, and hearing sensations. The *syncytial trophoblast* is a multinucleated part of the placenta in which distinct cell membranes are absent. *Germinal epithelium* occurs in the sperm-producing tubules of the testes.

SYNCYTIAL. *Multinucleated protoplasmic mass without distinct cell boundaries* (G. syn, *with*; kytos, *hollow vessel*).

CONNECTIVE TISSUES

Connective tissues make up the framework of the organs and the body. When you look at a photomicrograph you will find connective tissue everywhere except where epithelial tissue, muscle cells, nerve cells, and satellite nerve cells are present. Keeping this in mind, it may be worth your while at this time to examine all of the photomicrographs in this chapter.

Structure of Connective Tissues

Connective tissues are made up of three basic elements—ground substance, fibers, and cells. *Ground substance* is an adhesive material made from carbohydrates and proteins. Together with connective tissue *fibers* it forms a *matrix* into which connective tissue *cells* may embed themselves. This is shown in Figure 3.4.

Three different kinds of fibers are present in connective tissues. Two of them, known as *reticular* and *collagenous fibers*, are made of *collagen*, a triple stranded, coiled protein which lends high tensile strength to a tissue. Tensile strength refers to the tissue's resistance to rupture and stress in the direction of fiber length. The third type of fiber, known as an *elastic fiber*, is a proteinaceous substance which has only about a tenth the tensile strength of collagen. However, it is considerably more elastic, that is, it is more capable of regaining its shape and size after being compressed, extended, or dis-

MATRIX. *Intercellular substance of a tissue* (L. a *breeding animal*).

RETICULAR. *Denotes a network* (L. reticulum, *a net*).

COLLAGEN. *Albumen-like protein. On boiling, collagen is converted to gelatin* (G. koila, *glue;* gen, *to form*).

torted in some other way. On the negative side, elastic fibers degenerate more rapidly than reticular and collagenous fibers in the course of aging.

Reticular and collagenous fibers are more abundant than elastic fibers in tissues which do not need to be stretched repeatedly. Reticular fibers are 0.2 to 1.0 micrometers in diameter. They are widely found throughout the body. Collagenous fibers are much thicker—1 to 30 micrometers—and although they too are extensively distributed in the body, they are most commonly found in such tissues as ligaments (which hold bone to bone), tendons (which hold muscle to bone, muscle to ligament, muscle to muscle, or muscle to other kinds of connective tissues), cartilage, and bone.

Elastic fibers are present wherever a tissue is subject to frequent or continuous distortions. For example, they are found in the large arteries, which expand considerably with each beat of the heart. They are also present in the voice box, in the outer ear, in the elastic ligaments of the spinal column, and in other structures of the body.

The third component of connective tissues—the cells—varies considerably in kind and quantity, depending on the function of the particular connective tissue. It is best to begin a description of connective tissue cells by

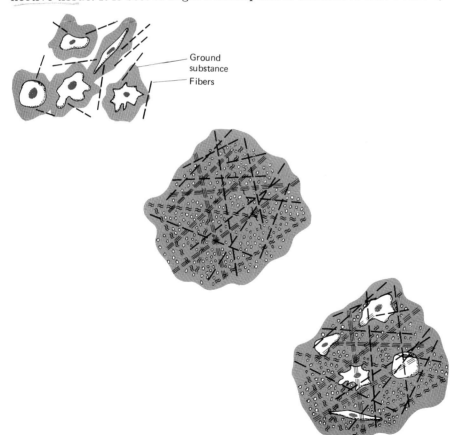

FIG. 3.4. *Structure of connective tissue. Fibroblasts, smooth muscle cells, and reticular cells (upper left) secrete several types of fibers and a sticky "ground substance." Together, these form a matrix (center) which serves to harbor connective tissue cells (lower right).*

FIBROBLAST. A connective tissue cell which produces fibers and gives rise to other kinds of cells (L. *fibra*, fiber; G. *blastos*, germ).

focusing on fibroblasts and their relationship to other kinds of connective tissue cells. Once this relationship is understood, we can proceed to a discussion of the various types of connective tissues and the kinds of cells in these tissues.

Fibroblasts are primitive, mesenchymal, stem cells which are present in all connective tissues. They are spindle-shaped or star-shaped (stellate), with an oval nucleus. They are not free to wander, but are fixed firmly in place. Their importance is twofold: (1) the formation (synthesis) of ground substance and the three types of connective tissue fibers that were mentioned above; and (2) the ability to differentiate into various other kinds of specialized tissue cells, such as white blood cells or muscle cells. See Figure 3.5.

FIG. 3.5. Formation of various kinds of body cells by differentiation of the primitive blast cell.

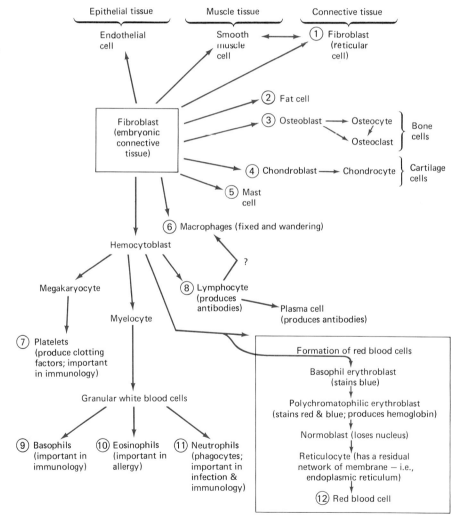

It is perhaps correct to say that every type of connective tissue has this cell type, though in some tissues it may have a different name. Fibroblasts are called *reticular cells* when they occur in the following three kinds of tissues: *lymphoid tissues*—thymus gland, spleen, tonsils, lymph nodes (see Figure 3.6); *myeloid tissue*—bone marrow—see Figure 3.7; the *reticuloendothelial system*—a collection of macrophage cells (phagocytic, or "eating," cells) throughout the body, especially as cells lining the openings in lymphoid tissue, bone marrow, and such organs as the liver and lungs. This system constitutes one of the body's major defense mechanisms against infection and foreign materials.

In cartilaginous tissue, fibroblasts are known as *chondroblasts* (cartilage stem cells) and they mature into *chondrocytes* (cartilage cells). In bone, they are called *osteoblasts* (bone stem cells) and they mature into *osteocytes* (bone cells).

It should be emphasized here that in each of these tissues, the fibroblasts (reticular cells, chondroblasts-chondrocytes, osteoblasts-osteocytes) produce and secrete two kinds of materials: ground substance (a cement substance) and fibers (collagen, elastic, reticular). In addition, fibroblasts or blast cells are capable of differentiating (developing and maturing) into other kinds of cells, as shown in Figure 3.5. In the bone marrow they differentiate into many kinds of cells, including all of the cells found in the blood. In the lymphoid tissues they differentiate into three kinds of white cells—lymphocytes, monocytes, and plasma cells—the first two of which are also found in the bloodstream.

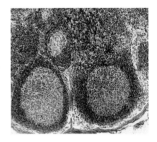

FIG. 3.6. *Tissue taken from a lymph gland. The circular lymph nodules are densely packed with lymphocytes and reticular cells.*

Classification of Connective Tissues

The various kinds of epithelial tissues, because of their regularity in appearance and ease of classification, are usually easy to identify, even by beginning students. Connective tissues, however, vary greatly from one kind of organ to another, often causing confusion and difficulty in identification. A helpful way around the problem is to realize at the outset that any connective tissue can be arbitrarily classified as a *general connective tissue* or a *special connective tissue*. There are basically two types of general connective tissues and seven types of special connective tissues.

General Types of Connective Tissue

Under this heading are included both loose connective tissues and dense connective tissues. The matrix of a *loose connective tissue* (Figure 3.8) is sparsely woven with fibers, while the matrix of a *dense connective tissue* is compactly woven. The two classes of tissues also differ from one another in their cell-to-fiber ratio. The loose connective tissues have a higher ratio of cells to fibers than the dense connective tissues.

Loose Connective Tissue. In general, loose connective tissue is found in every soft area of the body. It is found, for example, in the lower layer (der-

FIG. 3.7. *Myeloid tissue. As the developing bone grows (toward top of photograph), cartilaginous cells are gradually displaced by bone cells.*

48
Connective Tissues

mis) of the skin, in all of the glands of the body, and throughout the intestinal, respiratory, excretory, and reproductive tracts.

Ground substance, fibers, and fibroblasts are present in loose connective tissue, as are histiocytes and mast cells. *Histiocytes* are macrophage cells which literally eat materials that are foreign to the body. One of their functions is to eat dead cells, internal blood clots, bacteria, and perhaps newly arising cancer cells which might otherwise develop at the expense of normal cells. Another function of histiocytes is to eat certain kinds of substances, attach some unknown form of RNA on to them, and regurgitate these substances into the tissues as antigens. This process is described fully in the final chapter of this text. Mast cells are found in connective tissues near blood vessels, where they carry out the same functions as basophil white blood cells in the bloodstream. *Mast cells* and basophil cells produce and secrete heparin (which prevents blood coagulation) and histamine (which causes capillaries to become leaky at their seams). In addition, mast cells secrete serotonin, which causes the smooth muscle in the walls of blood vessels to contract, bringing about a constriction of the blood vessel. The actions of heparin, histamine, and serotonin are important in inflammation, as is described in Chapter 31.

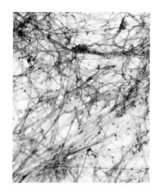

FIG. 3.8. *Loose connective tissue.*

In addition to the preceding kinds of cells, *melanocytes* are present in certain regions of the body. These color-carrying cells are located mainly in the skin and in the "white" of the eye. In the skin, where they are equally abundant in blacks, whites, and people of other races, they provide pigment to overlying epithelial cells in response to exposure to ultraviolet rays, perhaps as a means of protecting the skin from the harmful rays of the sun.

Dense Connective Tissue. Dense connective tissue may be found within or between strands or layers of loose connective tissue, as for example in the wall of the uterus, or within the *fascia*. Fascia is an organ that serves alone, or in combination with other structures, such as ligaments and tendons, to hold various parts of the body together. It holds skin to muscle, for example, and it helps certain tendons to attach muscle to bone.

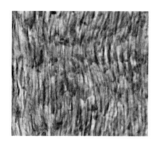

FIG. 3.9. *Ligament.*

Dense connective tissue may also stand more or less on its own, as for example in *ligaments* (Figure 3.9), which connects bone to bone, and *tendon* (Figure 3.10), which connects muscle to bone, ligament, fascia, or another muscle. Elastic fibers sometimes outnumber collagenous fibers, as for example in the vocal cords and walls of the large arteries. In most cases, however, collagenous fibers outnumber elastic, as in tendons, ligaments, fascia, and the coverings around bones.

Dense connective tissue is populated mainly with fibroblasts, and it has only occasional macrophages and even fewer cells of the other types mentioned so far.

Special Types of Connective Tissues

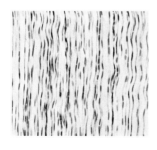

FIG. 3.10. *Tendon.*

There are seven special types of connective tissue: (1) adipose; (2) blood; (3) bone marrow tissue; (4) lymphoid tissue; (5) cartilage; (6) bone; and (7) the reticuloendothelial system.

Adipose Tissue. Adipose tissue (Figure 3.11) is composed chiefly of fat cells. It is located to a variable extent throughout most of the tissues of the body. It is especially abundant beneath the lower layer (dermis) of the skin. Elsewhere it is present in smaller amounts, as for example, around the kidneys and adrenal glands, in bone marrow, in the armpit (axilla), at the groin (the crease at the junction of thigh and trunk) and around muscle and nerve fibers.

Blood Tissue. Blood is not so much a connective tissue as it is a derivative of connective tissues. In embryonic life it is produced first in the yolk sac; then in the blood vessels; then in the liver; in the spleen, thymus, and lymph nodes; and finally in the bone marrow. At birth, bone marrow alone produces most of the blood, and as the person develops into an adult, fewer and fewer bones are called upon because of a decreasing need for proliferation of red blood cells.

Blood contains neither ground substance nor fibers. Also, with the exception of macrophages (monocytes) and lymphocytes, its cells are different from those of all other connective tissues, though they are all derived from one or more stem cells (Figure 3.5).

Blood consists of about 50 percent fluid, called *plasma*, and 50 percent cells. When the plasma is separated from the cells and allowed to clot, it exudes a liquid, called *serum*. Serum contains antibodies which are used in reactions against antigens. Plasma contains clotting factors.

The cells of the blood can be divided into three classes: red cells, white cells and platelet cells. The red blood cells are used in respiration (Chapter 28), the white blood cells are used in body defense mechanisms (Chapter 32), and the platelet cells (thrombocytes) are needed for proper blood clotting (Chapter 31).

White blood cells are also known as *leucocytes*, and there are five kinds normally found in the blood: *neutrophils, eosinophils, basophils, lymphocytes*, and *monocytes*. Each of these types is derived from a common blast cell, as seen in Figure 3.5. The neutrophils, eosinophils, and basophils are classified as *granulocytes*. They differ from lymphocytes and monocytes (*agranulocytes*) in having distinct granules in the cytoplasm outside the nucleus. These granules are visible when a smear of blood on a microscope slide is stained with Wright's stain. The granules of the neutrophils appear very weakly colored, at most a faint blue-gray. Those of the basophils show up intensely blue, and the granules of the eosinophils become a deep red-violet.

Neutrophils, known also as polymorphonuclear cells, comprise about 95% of the granulocytes, about 50% to 60% of the total leucocytes in the normal peripheral blood. *Peripheral blood* is the circulating blood, the blood found in arteries and veins. It does not include the blood in bone marrow and body organs. The chief function of neutrophils is *phagocytosis*. Like other kinds of phagocytes, they contain an enzyme called peroxidase, which is believed to initiate killing whenever a bacterium or other microbe is being consumed by the cell. In addition, the cell contains lysosomal digestive enzymes and a special digestive enzyme, called *lysozyme*, which selectively destroys vital components in the cell wall of bacteria.

PLASMA. *Blood serum, plus clotting factors (G. thing formed).*

THROMBOCYTES. *Platelet cells (G. thrombos, clot; kytos, cell).*

POLYMORPHONUCLEAR CELLS. *White blood cells whose nuclei have various shapes (G. polys, many; morphos, form; L. nucleus).*

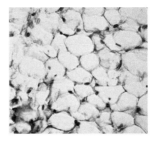

FIG. 3.11. *Adipose tissue.*

Connective Tissues

EOSINOPHIL. *White blood cells whose cytoplasmic granules pick up the acid stain eosin.*

BASOPHIL. *White blood cells whose cytoplasmic granules pick up a base stain.*

Eosinophils increase in number when the body is suffering an allergic response. They have been observed to phagocytize antigen-antibody complexes (Chapter 32).

Basophils are the least numerous of the five kinds of white cells in normal blood. In a *"differential white blood cell count"*—a procedure in which the percentage of each kind of white cell is determined—they usually comprise only about 1% of the total number. Their exact function in the body is not entirely known, but they are involved in inflammatory responses and immunological reactions, as discussed in Chapters 31 and 32.

Lymphocytes are the chief kind of agranulocyte in the blood. They comprise about 25% to 35% of the total white blood cells, about 80% of the agranulocytes. They are important in the immunological system (Chapter 32). There, as *B cells,* they differentiate into *plasma cells* and produce antibodies against various antigens, including certain infectious bacteria and viruses. In another form, known as *T cells,* they react directly with other kinds of antigenic material and thus initiate the destruction of the material.

MONOCYTE. *A white blood cell which, like lymphocytes, does not have large visible granules outside its nucleus.*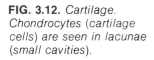

Monocytes resemble lymphocytes in appearance, but have a relatively larger amount of cytoplasm outside their nucleus. They are outnumbered by lymphocytes by about four to one. Often they pass into connective tissues, where they are indistinguishable from histiocytes.

Platelets, known also as thrombocytes, are tiny cells (1 to 3 micrometers) which lack a nucleus. When a blood vessel is cut, they aggregate in large quantities to start a clot (thrombus). At the same time, they release various chemicals called clotting factors to complete and solidify the clot (Chapter 31).

Myeloid Tissue. **Myeloid,** which means fatty, is a type of connective tissue that is present only in bone marrow. In the fetus and child it occurs as a

FIG. 3.12. *Cartilage. Chondrocytes (cartilage cells) are seen in lacunae (small cavities).*

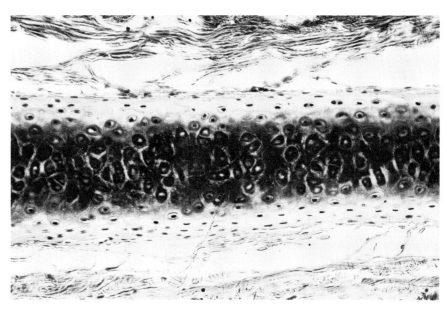

red marrow (pith) in all bones. Beginning around puberty, some of the red marrow is replaced by fat, giving rise to yellow marrow.

Red marrow consists of a framework of reticular cells, and a pulpy mass of cells bathed by the blood. Within the pulpy mass are all of the types of blood cells, in various stages of development.

Lymphoid Tissue. Lymphoid tissue is a type of connective tissue which is characterized chiefly by two kinds of cells: reticular cells and lymphocytes. The cells are densely packed into structures called lymph nodules. These structures are found in the tonsils (Figure 3.6), lymph nodes, spleen, thymus gland, and various other organs throughout the body. In the digestive tract, lymph nodules are known as *Peyer's patches*. The functions of lymphoid tissues, in addition to producing lymphocytes, are described in Chapters 31 and 32.

LYMPHOID. *Pertaining to the lymphatic system (L.* lympha, *clear water;* G. eidos, *resemblance).*

Cartilage Tissue. Only one type of cell is found in cartilage. Known in its mature form as a *chondrocyte*, it is derived from a chondroblast. Like fibroblasts, chondroblasts produce ground substance and fibers. These substances are secreted as the chondroblast matures, after which the chondrocyte is seen, either singly or paired with another chondrocyte, within a small cavity called a *lacuna* (Figure 3.12).

CHONDROCYTE. *A cartilage cell (G.* chondros, *cartilage;* kytos, *cell).*

Cartilage, a tough rigid substance, constitutes the gristle of the body. It is found in the windpipe, bronchi, voice box, the partition between the nostrils of the nose, and in the outer ear. It is also found in all of the joints as a covering on the articulating surfaces of the bones, and between the vertebrae as spinal discs.

Bone Tissue. Bone consists of ground substance, collagen fibers, osteoblasts (bone-forming cells), osteocytes (bone cells), and osteoclasts (bone-dissolving cells) (see Figure 3.13).

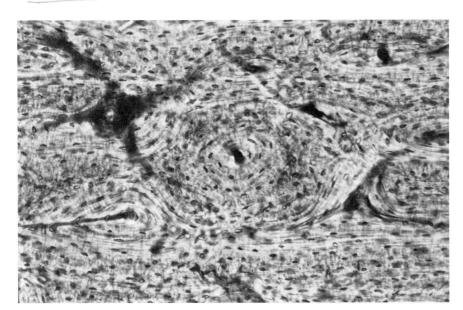

FIG. 3.13. *Bone.*

OSTEOBLAST. An embryonic bone cell (G. osteon, *bone;* kytos, *cell*).

Osteoblasts produce *osteoid,* an organic matrix which consists of ground substance and collagen. It is believed that osteoblasts use an enzyme, called *phosphatase,* to split phosphate (PO_4) from various organic metabolites in the cell's vicinity. This results in a localized increase in the numerical value of the following concentration-product:

calcium × phosphate

When this product exceeds the solubility of calcium phosphate in body fluids, some of the calcium phosphate precipitates, and a hydroxylapatite mineral is formed.

When an osteoblast is completely surrounded by mineral it is called an *osteocyte.* As such, it has access to nutrients from the blood, as described in Chapter 15, and it is able to take on or give off calcium to the blood, as needed.

OSTEOCYTE. A mature bone cell.

Many scientists believe that the giving up of calcium is due mainly to *osteoclasts,* cells which appear around bony material that is undergoing erosion. Osteoclasts are multinucleated cells. They contain the enzyme phosphatase, which may be used to erode phosphate from the osteoid matrix. They also contain *collagenase,* an enzyme which breaks down (*hydrolyzes*) the collagen fibers in the osteoid.

OSTEOCLAST. A cell which destroys and absorbs bony tissue (G. osteon, *bone;* klastos, *destroy*).

The Reticuloendothelial System. This system, which is also known as the *macrophage system,* is found throughout the entire body. It includes all tissues that are capable of giving rise to phagocytes or macrophages. Phagocytes and macrophages are cells which serve to cleanse the body of worn out tissue debris and foreign invaders, such as viruses and bacteria. The term "reticuloendothelial" indicates that "reticular" (network) phagocytes form an "endothelium" (covering) along large capillary spaces in most of the soft organs of the body.

RETICULOENDOTHELIAL. Pertaining to the network of endothelial cells, which line the body's blood passageways (L. reticulum, *network*)

The following is a listing of specific kinds of reticuloendothelial or macrophage cells which are either fixed or able to move (mobile) within connective tissues, or are circulating cells within the lymph fluids and blood: (1) *Kupffer cells* in the liver; (2) *littoral cells* in the spleen; (3) *mesangial cells* in the glomeruli of the kidney; (4) *nurse cells* in bone marrow; (5) *microglia* in the central nervous system; (6) *dust cells* in the alveoli of the lungs; (7) *histiocytes* in loose connective tissues; (8) *monocytes* in the bloodstream; and (9) *lymphocytes* in the lymph stream and bloodstream. Since all of the other white blood cells are phagocytic, we may also add them to this list.

For a long time the only function ascribed to macrophages was that of eating and digesting foreign particles, worn cells, and other debris. It is now known that they also play important roles in antigen-antibody reactions.

MUSCLE TISSUES

Muscle is a contractile tissue which imparts movement to the body and its tissues. There are three types of muscle tissue: skeletal, smooth, and cardiac.

FIG. 3.14. *Skeletal muscle.*

Skeletal Muscle

Skeletal muscle cells (Figure 3.14) are long, multinucleated fibers. They are joined together into the structures we call *muscles*, as described in Chapter 17. For now, two points are worth making: how skeletal muscles are controlled and how they are distributed throughout the body.

Skeletal muscle is controlled by the central nervous system. Some of this nervous input is delivered reflexly; that is, more or less involuntarily. The rest of the nervous input is delivered through volition; that is, through consciousness—through one's own free will. Neither smooth muscle nor cardiac muscle is ordinarily under voluntary control.

Skeletal muscle is located over the entire skeleton. It is present in the tongue; in the soft palate of the mouth; beneath the scalp; in the voice box; in the upper part of the gullet (esophagus); outside each eyeball as six extrinsic eye muscles; and in the eyelids.

Smooth Muscle

Smooth muscle cells (Figure 3.15) have a single nucleus and are spindle-shaped. They are about 5 micrometers in diameter and 75 micrometers long. Smooth muscle is also called involuntary or visceral muscle; *involuntary* because it is controlled autonomically, not voluntarily; and *visceral* because smooth muscle occurs in the soft tissues (viscera) of the body.

Smooth muscle sometimes occurs as sheets of fibers, as in the intestinal tract, blood vessels, and lymphatic vessels. Sometimes it forms bundles of fibers, as in the erector muscles that give rise to goose bumps. Sometimes, smooth muscle cells occur singly, as in the mammary glands, sweat glands, tear glands, and salivary glands. The cells which occur singly are present among epithelial tissues and they resemble the epithelial tissue. They are called *myoepithelial* cells.

Smooth muscle is found in all of the viscera—in the walls of the digestive, respiratory, excretory, and reproductive tracts; in the skin; in the blood and lymph vessels; in the gallbladder; and in the iris of the eye.

MYOEPITHELIAL CELLS. *Contractile cells found in glands (G. mys. muscle).*

MYOGENIC CELL. *A muscle cell which generates its own action potentials without the need of a neurotransmitter.*

Cardiac Muscle

Heart muscle consists of long, single-nucleus fibers which branch and meet with neighboring fibers. The individual fibers are striated (striped), as are skeletal muscle fibers, but unlike skeletal fibers, heart muscle can contract *myogenically* (without nervous input). This myogenic action can be regulated by the autonomic nervous system. It can also be influenced by free will through the voluntary centers of the central nervous system. Worry, tension, anxiety, and other undesirable attitudes have an adverse, or harmful, effect, whereas a productive attitude and happy spirit are beneficial.

In addition to cross striations (stripes), stained histological preparations of heart muscle display irregular cross bands. Known as *intercalated discs*,

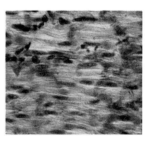

FIG. 3.15. *Smooth muscle from the digestive tract.*

INTERCALATE. *To insert; an intercalated disc occurs between the ends of two adjacent heart muscle cells.*

these bands are thick interdigitations, or interlockings, of plasma membranes from two adjoining cells (see Figure 3.16).

The presence of intercalated discs, and the location of the nuclei in the center of the fiber, help us to distinguish cardiac muscle from skeletal muscle. In skeletal muscle, the nuclei are located in the periphery of the cell.

ABNORMALITIES OF CELLS AND TISSUES

Throughout this text we will occasionally discuss certain abnormalities of cells, tissues, organs, or function. These discussions were selected on one of the following grounds: (1) the abnormality occurs so commonly that the college-educated student ought to know something about it, and/or (2) knowledge of the abnormality helps the student better understand normal physiological functions.

In this chapter a few introductory remarks are provided on cancer, a cell abnormality which leads to tissue, organ, and bodily misfunctions. The other abnormality discussed here is inflammation, a common abnormality that is characteristic of injured or diseased tissue.

Cancer

There are more than 100 billion cells in the skin and glands, more than 3 trillion cells in the lining of the intestinal tract, and more than 25 trillion cells in the blood. Each minute millions of these cells and cells from other tissues are in the process of cell division. Often, an error is made or the genetic structure of the cell is altered by some external agent—perhaps by a virus, chemical, bacterium, or some physical agent. In either case, the newly-formed cell is "foreign" to the body.

MALIGNANT. *Of a disease: severe, resistant to treatment, and often fatal (L.* maligno, *malicious).*

Most often the foreign cell is disposed of by the body's immunological system (see Chapter 32). In some people, the foreign cell may escape destruction and begin to reproduce. Sometimes, if this reproduction is not controlled, the cells are said to be *malignant*, or cancerous. The tissue made up of the cancerous cells is said to be a *neoplasia*.

Forms of Cancer. Cancer seldom occurs in nerve and muscle cells. It rarely occurs in nerve cells because these cells do not reproduce after the first few years of life, and there is little chance of an abnormal cell forming. Cancer does not occur commonly in muscle cells because these cells ordinarily proliferate only when cells are lost due to injury or disease.

Cancer arises commonly only in epithelial and connective tissues. Although there are more than 100 clinically distinct types of cancer, each type can be assigned to one of four major classes: leukemias, lymphomas, sarcomas, and carcinomas.

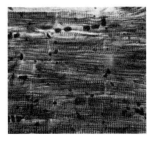

FIG. 3.16. *Cardiac muscle, showing extensive cell branching and intercalated discs. The discs appear as light streaks.*

Leukemias are cancers in which abnormally large numbers of white blood cells are produced by the bone marrow. They account for about 3%

of all cases of cancer. They are further described in Chapter 32. *Lymphomas* are cancers in which abnormally large numbers of lymphocytes are produced in the spleen and lymph nodes. They account for about 5% of all cancers.

Sarcomas are solid tumors which grow in connective tissues. They account for about 2% of diagnosed cancers. *Carcinomas* are solid tumors which grow in epithelial tissues, particularly in the breasts, ovaries, prostate gland, and skin. They account for about 85% of all diagnosed cases.

The remainder of malignancies, about 5%, are either not classifiable or are mixed cancers—combinations of the above types.

Diagnosis of Cancer. Cancer in epithelial cells can be diagnosed through a procedure known as *exfoliative cytology*. This term refers to the shedding

LYMPHOMA. *A tumor in one of the lymphoid tissues* (L. lympha, *clear water*; G. oma, *tumor*).

SARCOMA. *A malignant tumor in connective tissues* (G. sarkoma, *a fleshy growth*).

CARCINOMA. *A malignant tumor in epithelial tissue* (G. karkinoma, *cancer*).

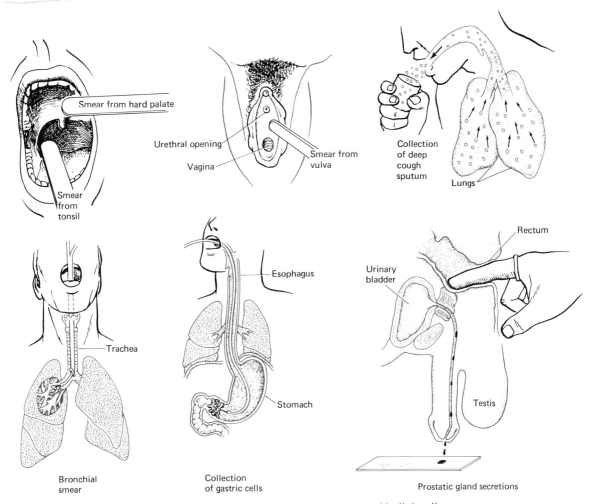

FIG. 3.17. *Methods of obtaining exfoliated epithelial cells.*

of cells—a common attribute of epithelial cells on surfaces of organs, ducts, tracts, and vessels. The cells may be obtained in various ways, examples of which are shown in Figure 3.17. They may be obtained from any surface, such as the mouth and the passageways of the digestive, respiratory, excretory, and reproductive systems. The sample which is obtained, known as a *biopsy*, is stained and examined under a microscope.

BIOPSY. *A sample of tissue taken from a living person to diagnose a disease (G. bios, life; opsy, vision).*

When studying the stained epithelial cells, the examiner asks certain questions to help determine if the cells are cancerous. If the questions are answered "yes," a malignancy may be suspected. There are many exceptions, however—some of which are indicated in parentheses below.

(1) Are many nuclei larger than ten micrometers? (Note that noncancerous cells may have larger nuclei when they are regenerating or inflamed).

(2) Are the nuclei abnormally large relative to the outer cytoplasmic compartment?

(3) Are the rims of the nuclei thickened relative to the remainder of the nucleus, and are they irregular?

(4) Are several nucleoli present in one nucleus?

(5) Is the nucleolus larger than five micrometers?

(6) Are several nuclei present?

(7) Do the nuclei have many lobes? (This may occur in noncancerous (benign) cells, such as transitional cells in the urinary tract, cells in body fluids, columnar cells of the bronchial tubes, and in histiocytes).

(8) Are there abnormal mitotic figures? (These are seen occasionally in benign histiocytes and cells in body fluids.)

(9) Do the nuclei display a variety of sizes and shapes?

(10) Are any cells eating other cells?

(11) Is there any clumping of cells?

Malignant cells are characterized by the features listed above, as well as other features, such as abnormal chromosome structure and abnormal numbers of chromosomes. Similar characteristics of cancer are present in cells from connective, nerve, and muscle tissues. Samples of these tissues may be obtained surgically. However, for any malignancy to be labeled as a true cancer, a histological examination of the cells would have to be made.

HYPERPLASIA. *An increase in the number of cells in a tissue (G. hyper, above; plasis, a molding).*

In general, there are three major characteristics of all cancers: hyperplasia, metastasis, and anaplasia. *Hyperplasia* is the result of an uncontrolled proliferation of cells. The cells fail to recognize their cues to halt their reproduction. As a result, a localized accumulation, or tumor, is produced.

METASTASIS. *Shifting of a disease from one part of the body to another (G. meta, a removing; stasis, a placing).*

Metastasis refers to the ability of cancer cells to detach themselves from their original site and pass through the bloodstream or lymph stream to another site, where they establish a new tumor. An example would be the metastasis of cancer cells from the prostate gland to the brain. *Anaplasia* refers to the failure of a tissue to reproduce itself faithfully. Tissues become mixed up or disoriented within an organ. For example, the epidermal cells of the skin, which happen to be epithelial cells, move into the lower, dermal layer, a layer of connective tissues.

ANAPLASIA. *A loss of the usual distinctive tissue characteristics (G. ana, again; plasis, a molding).*

The *Pap smear* (named for George Papanicolaou (1883–1962), who devised the test) is the most common example of exfoliative cytology. It is a smear of body secretions, especially from the cervix or vagina. The appearance of the cells in the smear may be used to detect cancer. The following scheme of classification is commonly used: I. Absence of atypical, or abnormal, cells. II. Atypical cells, but no evidence of cancer. III. Cytology (cell study) suggestive of cancer. IV. Cytology strongly suggestive of cancer. V. Cytology conclusive for cancer.

How Cancer Kills. It is generally believed that cancer kills through a process of *cachexia* — a generalized emaciation of the victim resulting from the cancer tissue receiving nourishment at the expense of other tissues. This is often the case, but more often cancer merely sets the stage for some other cause of death. Chief among these is infection, followed by other causes such as hemorrhaging, internal blood clotting, or the failure of one or more organs, such as the lungs or heart.

The **PAP TEST** for cancer is named for its inventor, the American scientist, George Papanicolaou (1883–1962).

HEMORRHAGE. An escape of blood from blood vessels (G. haima, blood; rrhagia, to burst).

Inflammation

An inflammation is a diseased condition produced in a tissue by an infection, injury, or irritant. A strained ligament, a drug, a bacterium, or an allergenic substance (such as pollen or bee venom), are examples of such causative agents. The chief characteristics of an inflamed tissue are swelling, redness, heat, pain, and a reduction or loss of function. One or more of these characteristics may be absent. Nevertheless, some discomfort is usually felt.

During an inflammation the reticuloendothelial system is actively engaged, through the process of phagocytosis, in cleaning up the tissue debris caused by the damage. In addition, it is active in ridding the body of infectious microbes or other physical or chemical agents which may have caused the inflammation.

An inflammatory response to a causative agent may be localized or widespread through the body. Pimples, warts, cold sores, and an arthritic finger are examples of local inflammations. An allergic attack or a fever following an inflammation of the tonsils or appendix are examples of widespread inflammatory responses.

An inflammation in a particular tissue is designated by the name of the tissue and the suffix *itis*. Thus, appendicitis is an inflammation of the appendix; arthritis (from the Greek *arthron*, "joint") is an inflammation of one or more joints.

SUMMARY

1. Epithelial tissues line the outer surface of an organ or the inner surface of a tubular structure. They also form the secretory cellular component of a gland.
2. Epithelial tissues are made of squamous, cuboidal, columnar, polyhedral, or myoepithelial cells. Squamous, cuboidal, or columnar cells

may occur as a single layer of tissue, called a simple epithelium, or as two or more layers, called a stratified epithelium. Polyhedral cells occur as large bunches of cells in glands. Myoepithelial cells occur singly as contractile cells in certain glands among secretory epithelial cells.
3. Connective tissue consists of fibers, a mucilaginous ground substance, and cells. The fibers and ground substance form a matrix in which connective tissue cells are embedded.
4. Loose connective tissue has a relatively high ratio of cells to fibers and is found in every soft portion of the body. Dense connective tissue has a low ratio of cells to fibers and is found as fascia, tendon, and ligament — as connective tissues which hold the body wall, skeleton, and viscera together.
5. Special types of connective tissues are: adipose (fat cells); blood (red and white blood cells, platelets, and plasma); lymphoid (lymph nodes, thymus, tonsils, spleen); myeloid (bone marrow); cartilage (chondroblasts, chondrocytes, lacunae); bone (osteoblasts, osteocytes, osteoclasts); and the reticuloendothelial system (macrophages).
6. Three types of muscle tissue (skeletal, smooth, cardiac) are found in the body. Each skeletal muscle cell is directly innervated and is under the control of the central nervous system. The smooth muscle cells are not all necessarily innervated and their contraction is activated either spontaneously or by the autonomic nervous system. Cardiac muscle cells are activated independently of nerves but are subject to nervous regulation.
7. Cancer is a disease that is characterized by an excessive proliferation of cells (hyperplasia), an abnormal tissue architecture (anaplasia), and the movement of cancerous cells from primary to secondary sites of growth (metastasis).
8. Inflammation is a tissue injury that is usually characterized by redness, swelling, an increase in temperature, pain, and a loss of function.

REVIEW QUESTIONS

1. Define "epithelium" by telling where epithelial tissues are found.
2. What is the difference between a simple and stratified epithelium? To which of these categories would you assign pseudostratified and transitional epithelia?
3. What are villi and cilia? Are they organelles? What is the major function of each?
4. What three kinds of components are commonly present in connective tissues?
5. What are the principal distinctions between collagenous and elastic fibers?
6. Define: myoepithelial cells, connective tissue matrix, chondroblast, osteoblast, osteoclast, reticuloendothelial system, carcinoma, sarcoma.
7. What are lymphoid tissues? Where are they located?
8. What are myeloid tissues? Where are they found?
9. How do cartilage, tendon, ligaments, and bone differ from each other histologically?
10. List five cytological characteristics of cancerous epithelial cells.
11. Name three principal characteristics of an inflammation.

Organs and Body Regions

ORGANS
ANATOMICAL ORIENTATIONS
BODY REGIONS
    The body
        Body cavities—their viscera and linings
        The body wall—its structure and function
    The head
    The neck
    The appendages

In the two preceding chapters we discussed cells and tissues. In this chapter we will first discuss a higher level of complexity in anatomy—organs and organ systems. We will then say something about the anatomical terms which are used in mapping out the body or its organs. It will then be possible to discuss the various regions of the body and their architectural arrangement. This is necessary to appreciate the following chapters in which various physiological systems—such as the nervous system and endocrine system—are discussed separately. It is necessary also for an appreciation of the chapters on the skeleton, joints, and functional anatomy of skeletal muscle.

ORGANS

An *organ* is an aggregate of tissues which exercises one or more general or specific functions, such as excretion or the storage of nutrients. The skeleton is an organ, as is the heart, the skin, and even the adipose tissue that occurs throughout the body.

Each organ of the body usually has one or more primary tissues and one or more secondary, or auxiliary, tissues. The primary tissues carry out the chief functions of the organ. The secondary tissues lend necessary support. For example, osteocytes within a collagenous matrix comprise the primary

OSTEOCYTES. *Mature bone cells (G. osteon, bone; kytos, cell).*

EPITHELIUM. *A protective, secreting membranous cellular tissue that covers an interior or exterior body surface (G. epi, upon; thele, nipple; thus, a soft tissue).*

STROMA. *The supporting framework of an organ (G. bed).*

tissue of bone, while the blood vessels, lymph vessels, and nerves are secondary tissues. Similarly, the epithelial lining and muscle of the stomach and intestines are the primary tissues of these organs, whereas their connective tissues and nerves serve auxiliary roles. The epithelial lining secretes digestive juices or absorbs the nutrients and the muscles propel the food through the intestinal tract toward the anus. The adrenal glands are organs that have two primary tissues. The medullary (inner) tissue is derived from nerve cells, while the cortical (outer) tissue is derived from epithelial cells. Within both the medulla and cortex are various kinds of secondary tissues.

A "typical" organ is usually composed of an external capsule, an internal framework, and an aggregate of primary tissue. The external capsule and inner framework (*stroma*) are composed of connective tissue, whereas the primary tissue (*parenchyma*) may be epithelial, as in glands; connective, as in bone; neural, as in the brain; or muscle, as in skeletal muscle.

An organ may be present in only one part of the body and may have a very limited function. The trachea (windpipe), for example, is located only in

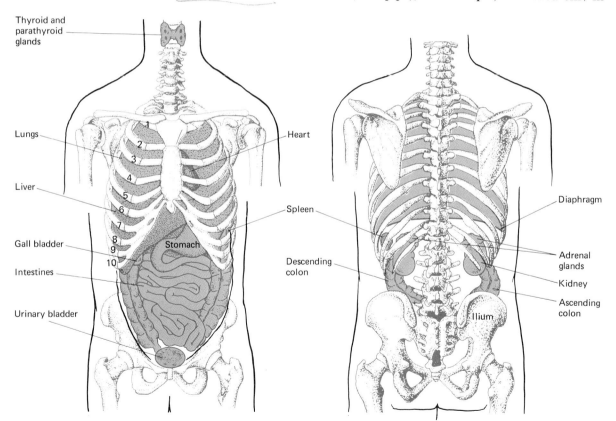

FIG. 4.1. *Location of the viscera, from the front (left) and the back. The first ten ribs are numbered in the left-hand drawing.*

Organs and Body Regions

the neck and upper chest where it serves to bring air into and out of the lungs.

Most often, an organ is part of what is known as an *organ system* or a *physiological system*. The components of these systems are usually widespread and the system itself performs more than one function. The trachea, for example, is part of the respiratory system, which serves not only for bringing oxygen into, and carbon dioxide out of, the body, but also serves to buffer the pH of the blood. The other components of the respiratory system include the lungs, the blood circulation through the lungs and heart (pulmonary circulation), the bones and muscles of the chest, and the nervous regulatory centers in the brain.

Although organs and organ systems are often discussed as though they operate independently of the remainder of the body, this is really never the case. The brain, for example, regulates most of the organ systems in the body. In turn, the brain itself is highly dependent upon many of the systems which it regulates. It depends upon the digestive system for nutrients, upon the respiratory and circulatory systems for oxygen, and upon the digestive, respiratory, circulatory, and excretory systems for getting rid of waste products.

Soft, internal organs of the body, in contrast to bone and skeletal muscle, are termed *viscera*. The lungs, heart, intestine, liver, and spleen—which lie in body cavities—are examples of viscera—as is the kidney, which is firmly attached to the body wall.

The locations of the major visceral organs are shown in Figure 4.1.

PARENCHYMA. *The essential or functional part of an organ (G. anything poured in).*

VISCERA. *Any one of the organs in one of the three great cavities (cranial, thoracic, and abdominal).*

ANATOMICAL ORIENTATIONS

In anatomy and physiology it is often difficult or impossible to describe a location, a direction, or a section of an organ or body with terms such as, "beneath," "on top of," and "inside of." On occasion it is absolutely necessary to use definitive locational, directional, and sectional terms. Failure to do so may lead to ambiguity and misunderstanding.

The terms described here are illustrated in Figure 4.2. They are used in reference to what is known as the *anatomical position*, in which the body is standing, eyes looking forward, arms held at the sides with palms forward, and the feet parallel to one another along the midline of the body.

(1) Anterior or ventral—front, or toward the front, of the body.
(2) Posterior or dorsal—back, or toward the back, of the body.
(3) Cranial or cephalad—head, or toward the head.
(4) Caudal—toward the feet.
(5) Superior—above a reference point; in a cranial direction.
(6) Inferior—below a reference point; in a caudal direction.
(7) Sagittal section—a vertical cut or plane that divides the body (or a longitudinally oriented organ) into right and left sections.

Anatomical Orientations

(8) Midsagittal section—a vertical cut or plane that divides the body into bilaterally equal right and left sections.

(9) Transverse or cross section—a horizontal cut that divides the body into superior and inferior sections. With respect to an organ, a cross section is made along the shortest plane of the organ, perpendicular to a longitudinal section.

(10) Coronal or frontal—a vertical cut that divides the body or an organ into anterior and posterior sections.

(11) Median or medial—middle plane, or toward the middle plane, of the body or organ.

(12) Lateral—away from the median of the body or an organ.

(13) Proximal—nearer to the center of the body (or some reference point) than some other distal point. For example, the arm is proximal to the forearm.

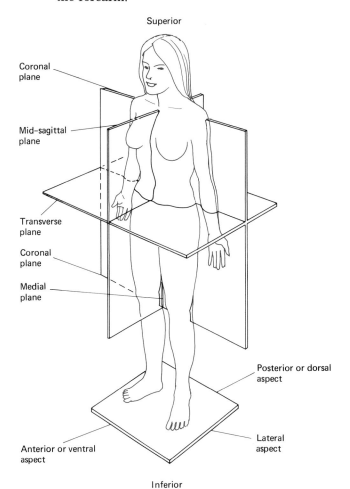

FIG. 4.2. Anatomical orientations and planes.

Organs and Body Regions

(14) Distal—further from the center of the body (or some reference point) than some other proximal point. For example, the leg is distal to the thigh.
(15) Superficial—close to the external body surface; usually visible or palpable (capable of being felt).
(16) Deep—extending beneath the external surface of the body; in contrast to superficial.

BODY REGIONS

Looked at anatomically, every person consists of a body, a head, a neck, and four jointed appendages. Before proceeding to describe these regions, the reader may profit from Figure 4.3, noting the differences between an

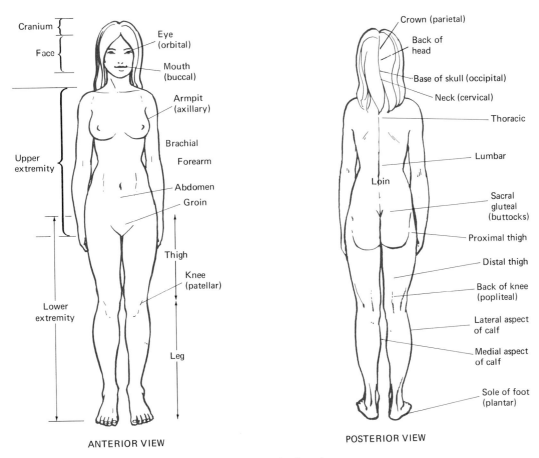

FIG. 4.3. *Anatomical regions.*

arm and forearm, a thigh and leg, a loin and groin, a cranium and face, the patellar and popliteal regions of the knee, and the three major sections of the body: thoracic, lumbar, and sacral.

The Body

The body consists of a thorax (chest) and abdomen. The thorax is separated from the abdomen internally by the *diaphragm* — a membranous structure comprised of muscle and connective tissue fibers. The diaphragm is used primarily in breathing, talking, and singing. During inspiration (inhalation), it contracts inferiorly (toward the abdomen), creating a vacuum in the thorax. This allows air to rush into the lungs. During expiration (exhalation) it relaxes, allowing air to leave the body.

When seen as a frontal section approximately midway between the an-

DIAPHRAGM. *Muscular tendinous partition between the thorax and abdomen. (G. diaphragma, partition).*

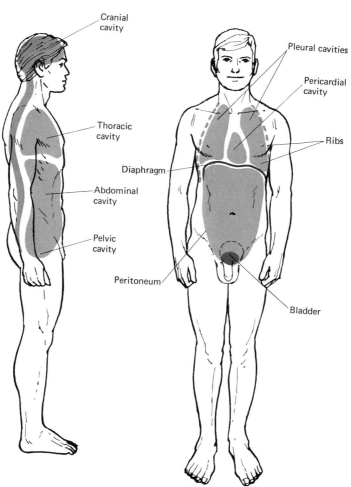

FIG. 4.4. *The major body cavities.*

terior and posterior surfaces, the body consists of two cavities (thoracic and abdominal, Figure 4.4) and a body wall.

Body Cavities—Their Viscera and Linings. The *thoracic cavity* is divided into two *pleural cavities* (each of which contains a lung), and a *pericardial cavity* (which contains the heart). The pericardial cavity and the heart are not the only structures which lie in the region between the pleural cavities. This region, known as the *mediastinum*, is bounded anteriorly by the sternum (breastbone) and posteriorly by the vertebral column (spine). In addition to the pericardial cavity and heart, the mediastinum contains the great vessels of the heart, part of the trachea (windpipe), the esophagus, the thoracic lymph duct, the thymus gland, and certain nerves. See Figure 4.5.

The *abdominal cavity* contains the stomach, gallbladder, intestines, liver, spleen, pancreas, and a so-called pelvic cavity. Actually, the pelvic cavity is the distal extension of the abdominal cavity. It houses the urinary bladder, colon, the rectum, and the reproductive organs.

Each of the body's true cavities (pleural, pericardial, and abdominal) is lined by a *serous membrane*. It is important to understand what this membrane is and how it is related to another important kind of membrane — the *mucous membrane*.

Serous membranes and mucous membranes, collectively, are referred

PLEURA. *The serous membrane which envelops a lung (G. rib).*

MEDIASTINUM. *The space in the chest between the two pleurae (L. medius, middle).*

SEROUS *pertains to serum or the serous membrane which lines the peritoneal, pleural, and pericardial cavities and covers their contents (L. serum, whey).*

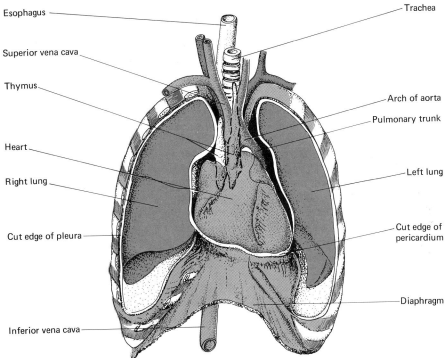

FIG. 4.5. *Viscera of the thoracic cavity.*

Body Regions

MUCOUS *pertains to mucus or the mucous membrane which lines the passageways from internal cavities to the exterior of the body, such as the digestive tract* (L. mucosus, *slimy*).

to as "moist membranes." Serous membranes produce a thin watery fluid, whereas mucous membranes produce a viscous, gummy fluid. (However, some mucous membranes, such as those of the tear glands, gastric glands of the stomach, and intestinal glands, produce a serous fluid, and certain mucous membranes, such as those of the urinary bladder and urinary passageway, produce very little, if any, secretory material.)

Serous membranes line the internal cavities of the body, while mucous membranes line the cavities or ducts of the hollow organs that communicate with the exterior of the body (respiratory, digestive, excretory, and reproductive tracts, and the ear canal and conjunctiva of the eyes).

LAMINA PROPRIA. *The connective tissue beneath a mucous membrane* (L. lamina, *layer or wall;* propria, *proper*).

Serous membranes consist of a simple squamous epithelium resting on a loose connective tissue called a *lamina propria;* mucous membranes consist of a cuboidal or columnar, simple, pseudostratified or stratified epthelium on a lamina propria beneath which is a layer or two of muscle (*muscularis mucosa*). The lamina propria in both membranes, in addition to being a support for the epithelium, contains lymphocytes and macrophages which help defend the body against foreign invaders, such as bacteria.

PARIETAL. *The wall of a cavity* (L. paries, *wall*).

Serous membranes are doubly-folded membranes, consisting of a *parietal portion*, which lines the external wall of a body cavity, and a *visceral portion* which reflects over the exposed surfaces of visceral organs. In contrast, mucous membranes are single-walled membranes, though they may be extensively pleated into folds, for example, stomach rugae and intestinal plicae (Chapter 19), or they may have an increased surface due to an outpocketing of cells (villi, for example) or an inpocketing of cells (as for example in the salivary glands).

VILLI. *Nipple-like projections from a surface, as from the mucosa of the small intestine* (L. *shaggy hair*).

Whereas lubrication and protection of the body against foreign invasion are functions of both serous and mucous membranes, each type of membrane has other functions. The serous membranes facilitate freedom of movement between the individual visceral organs which they cover,

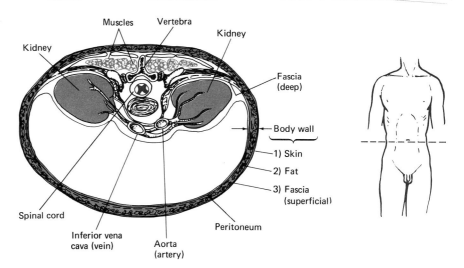

FIG. 4.6. *Cross section through the abdomen, showing structure of the body wall.*

whereas mucous membranes are constructed in a way as to vastly increase the surface area of a cavity, thus increasing the absorptive and/or secretory capacity of the organ they cover.

The Body Wall—Its Structure and Function. The body wall, from its external to its internal surface, consists of skin; subcutaneous fat; fascia; skeletal muscle; and either a peritoneal (abdominal) or pleural (thoracic) lining (Figure 4.6).

The *skin*, to which Chapter 14 is devoted exclusively, is composed of two layers—an external epidermis, and an internal dermis (Figure 4.7).

The *epidermis* is an epithelial tissue. It gives rise to the fingernails, which help in picking up objects; toenails, which may help in walking; hair, which may offer insulation on the scalp; sebaceous glands, which lubricate the hair; and sweat glands, which aid in heat regulation and excretion.

The *dermis* is a connective tissue. In many places it consists of loose connective tissue, and in other places of the dense type. Notable among its constituents are the following structures: (1) lymphatic vessels and lymphoid elements for phagocytic and immunological reactions against foreign invaders; (2) blood vessels; (3) the nails, hair, sebaceous glands and sweat glands, which are derived from the epidermis; and (4) nervous elements, which are discussed in the next chapter.

A typical square of skin from the body, about one-quarter of an inch on each side, contains about a foot each of blood and lymph vessels, five feet of nerve fibers, sixty sweat glands, one sebaceous gland, one hair, 200

EPIDERMIS. *The outer layer of skin* (G. epi, *above*; derma, *skin*).

DERMIS. *The layer of skin beneath the epidermis* (G. derma, *skin*).

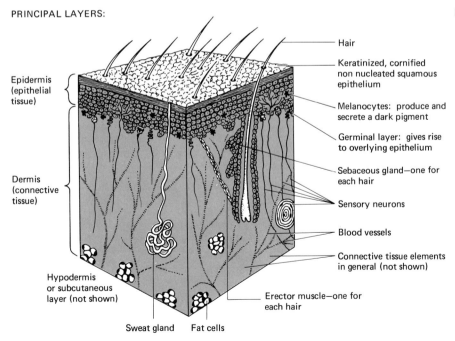

FIG. 4.7. *The skin.*

sensory fibers, and 125,000 melanocytes (pigment cells). About the same number of melanocytes are present in the skin, whatever the person's race. The apparent difference in skin color among races is due to the fact that various kinds of pigments are produced and also, perhaps most important, because the melanocytes of dark skin are more active in producing pigment than those of light skin. The pigment is secreted into the germinal, or basal, layer of the epidermis where it is picked up by newly-arising epidermal cells.

Subcutaneous fat is a connective tissue which lies partly within the dermis of the skin and partly within the third layer of the body wall—the fascia. It is virtually absent in the eyelids, back of the hands, ears, and penis. Where it occurs abundantly it is also called *panniculus adiposus*. Its chief functions include insulation; protection against physical injury; storage of metabolic fuel in the form of fat; and storage of food which is eaten in excess.

PANNICULUS ADIPOSUS. *The layer of fat beneath the skin (L.).*

Fascia, which literally means a band or ribbon, is a dense connective tissue. It occurs principally in two places. Beneath the skin it is present as a more or less continuous sheath known as the *superficial fascia*. This sheath extends itself into deeper regions of the body—as the *deep fascia*—which serves to enclose skeletal muscles and to separate groups of skeletal muscles from one another.

FASCIA. *A connective tissue which binds the skin to the body and wraps itself around muscles and other organs (L. band).*

The *superficial fascia* harbors a certain amount of fat—virtually none on the back of the hand and large quantities in the front wall of the abdomen. Where it contains considerable amounts of fat it is known variously as adipose, panniculus adiposus, or fascia. The primary functions of the superficial fascia are to connect the skin with the underlying tissues, and to serve as a repository for waste nutrients (fat) as well as a depot for reserve nutrients (fat).

The *deep fascia* generally contains little or no fat. Its primary functions are at least fourfold: to enclose and thus physically support muscle; to separate individual muscles, thus promoting independent movements by avoiding mutual interference; to attach muscle to bone or cartilage; and to carry blood and lymphatic vessels and nerves to the muscles.

Skeletal muscle is found next to bones in the body wall. In the thoracic region, the musculature surrounds the ribs, sternum (breastbone), and clavicles (collar bones). In the abdominal region, it encloses the pelvic (hip) bones. The spinal column too, throughout the length of the body, lies within a structure of skeletal muscle.

PERITONEUM. *A membrane which contains the abdominal viscera and lines the abdomen (G.* peri, *around;* teinein, *to stretch).*

The *peritoneum* (Figure 4.8) is a two-layered membranous sac. It lines the abdominal cavity and slips into this cavity to wrap around most of the viscera in it. It is a moist membrane of the serous type.

The outer *parietal layer* of the peritoneum is attached through connective tissue to fascia of the body wall, whereby it serves to anchor or support the viscera in the abdominal cavity. In addition, segments of the parietal layer are attached to one another at points where they pass into the cavity, and then back out, in the course of wrapping around various visceral organs and separating them from one another.

The inner, *visceral layer* of the peritoneum fits snugly against, and surrounds, much of the visceral organs. Between the parietal and visceral layers is a lubricant (serous), which allows the two layers to glide smoothly past one another as the visceral organs move within the abdominal cavity. Thus, whereas the peritoneum as a whole serves to support and separate various visceral organs, its arrangement—as a two-layered structure with serous fluid between—serves to allow a relatively friction-free movement within the viscera.

The visceral peritoneum is continuous with two important abdominal structures. One of these, known as *mesentery*, serves to anchor the small and large intestines to other abdominal viscera and to the parietal peritoneum of the body wall. The other structure is the *omentum*. This membrane is a single or double fold of peritoneum, like the mesentery. It serves to anchor the stomach to other abdominal organs (Figure 4.8).

The *pleura* is a two-layered membranous sac which lines the thoracic cavity and slips into this cavity where it wraps around each lung, separating

MESENTERY. *A fold of peritoneum that connects the intestine with the posterior abdominal wall* (G. mesos, *middle,* enteron, *intestine*).

OMENTUM. *A fold of peritoneum that connects the abdominal viscera with the stomach* (L. apron).

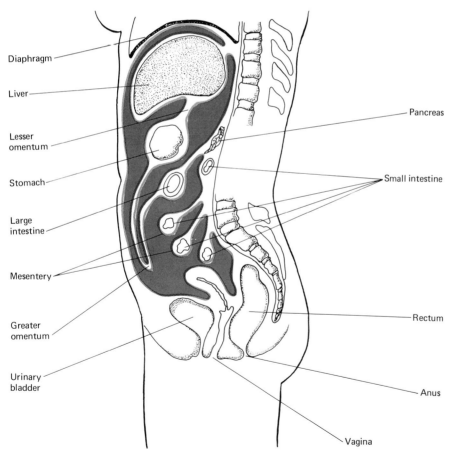

FIG. 4.8. *The abdominal cavity, showing the peritoneum.*

70
Body Regions

it from the other. In doing so, it leaves a cavity between the lungs—the pericardial cavity—which houses the heart within its pericardial sac.

Like the peritoneum, the pleura has an outer, *parietal,* and an inner, *visceral,* layer. The parietal layer is attached to the fascia of the body wall and to organs within the thoracic cavity, such as the trachea (windpipe) and esophagus (gullet). It serves to anchor and support the thoracic viscera. The *visceral layer* is closely wrapped around the lungs. Between the parietal and visceral layers is a pleural (serous) fluid which allows the lungs to expand and contract with a minimum of frictional resistance from the thoracic wall.

The Head

ORBITS. *The bony cavities that contain the eyes* (L. orbita, *track, course*).

The head contains four major cavities—cranial, orbital, nasal, and oral—within a cranial wall (cranium). The *cranial cavity* contains the brain. The *orbits* contain the eyes and associated structures. The *nasal cavity* consists of two nasal passages (fossae), which are separated from each other by a nasal *septum* (wall). The *oral cavity* is an opening which can be enlarged by the downward movement of the lower jaw (mandible) from the upper jaw (maxilla).

The cranial wall differs from the body wall in that, ordinarily, a consider-

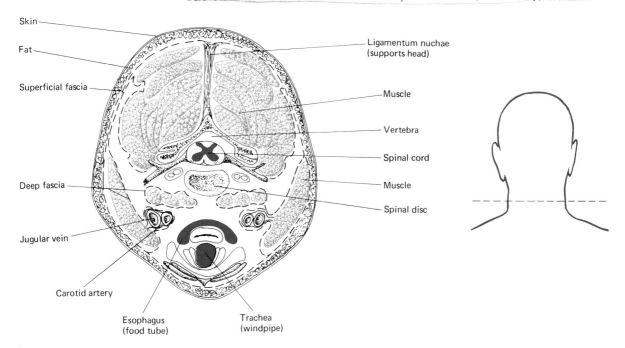

FIG. 4.9. *Cross section through the neck, showing the principal structures. The neck region does not have the serous membrane that lines the abdominal and thoracic cavities.*

able amount of hair grows from the skin of the scalp and (in males of certain ethnic groups) the face. In addition, there is no characteristically thick layer of subcutaneous fat, and there is no peritoneal or pleural membrane.

The Neck

The neck consists of a wall which is structured very much like the body wall, but is not lined by a pleural or peritoneal membrane. The cavity of the neck houses the larynx (voice box), trachea (windpipe), esophagus (gullet), thyroid gland, and parathyroid glands (Figure 4.9).

THYROID. *A shield-shaped gland or cartilage in the neck (G. thyreoeides, shield-shaped).*

The Appendages

The appendages—the arms and legs—display in cross section the basic structure of the body wall, with the difference that there are no cavities within, and thus there are no "moist membrane" linings.

SUMMARY

1. An organ has one or more primary tissues through which it carries out one or more primary functions, and one or more secondary tissues which provide support. The heart, for example, has a strong musculature and a muscular electrical system as its primary tissues. Its secondary tissues are an inner and outer lining, blood vessels, and nerves.
2. A "typical" organ generally has an external capsule, an internal connective tissue framework (stroma), and a primary tissue (parenchyma) which may be predominantly epithelial (gland), connective (bone), neural (brain), or muscle (skeletal muscle).
3. Organs are generally part of what is known as an organ system or a physiological system. The kidneys, for example, are organs which are part of the excretory system.
4. Soft internal organs are known as viscera.
5. Anatomical orientations are designated by words—such as dorsal and ventral—which must be understood and committed to memory by all biology, health science, and physical education majors.
6. The body contains a thoracic and abdominal cavity. The thoracic cavity is subdivided into two pleural cavities which are separated from one another by the mediastinum. The abdominal cavity has a pelvic subdivision at its lower extremity. The thoracic and abdominal cavities are lined by a serous membrane whereas all body cavities which eventually lead to the surface of the body are lined by a mucous membrane.
7. The body wall consists of skin, subcutaneous fat, fascia, skeletal

muscle, and either a peritoneal or pleural lining. Skin consists of an epithelial epidermis on a loose connective tissue dermis. Fascia is a dense connective tissue which holds the skin, muscles, skeleton, and viscera together. The peritoneum is a serous membrane that lines the abdomen and envelopes the abdominal viscera. The pleura are serous membranes that line the thoracic cavity and envelope the lungs.

8. The peritoneum consists of an outer parietal and an inner visceral layer. The visceral layer is continuous with the mesentery, which enters the body cavity to cover most of the abdominal organs, and with the omentum, which anchors the stomach to the other abdominal organs.

9. The pleura also consist of an outer parietal and an inner visceral membrane.

10. The head contains four kinds of cavities. The cranial cavity houses the brain, the orbits house the eyes, the nasal cavities consist of nasal passageways, and the oral cavity is the mouth and pharynx.

REVIEW QUESTIONS

1. Define "organ" and tell whether adipose, blood, fascia, and mesentery can be considered to be organs.
2. What are the primary and some secondary tissues of the following organs: skeleton, dermis, eyes, heart? Check other chapters of this book for some answers to this question.
3. Define: physiological system, popliteal, viscera, stroma, parenchyma, fascia, pleura, peritoneum, mesentery, omentum.
4. Define each of the anatomical orientations and directional terms listed in this chapter.
5. What principal organs are found in the mediastinum?
6. List the major parts of the body wall at the level of the thorax.
7. How is the visceral layer of the peritoneum related to mesentery?

The Nervous System

5 Introduction to the Nervous System
6 The General Sensory System
7 The Skeletal Motor System
8 The Visceral Sensory and Motor Systems
9 The Special Sensory Systems
10 The Proprioceptor System
11 The Cortical Systems

PART TWO

Introduction to the Nervous System 5

BASIC ELEMENTS OF THE NERVOUS SYSTEM
    Nerve cells
        Unipolar, bipolar, and multipolar nerve cells
    Propagation of electrical activity in a nerve cell
        The synapse: transmission of bioelectrical activity from a nerve cell to its target cells
        Inhibition at a synapse
    Nerve cell functions
        Afferent (sensory) nerve cells
        Efferent (motor) nerve cells
        Interneurons
        Endocrine nerve cells
    Satellite cells
LARGER BUILDING BLOCKS OF THE NERVOUS SYSTEM
    Ganglia
    Nerve nuclei, tracts, and peduncles
    Nerves
DEVELOPMENT OF THE NERVOUS SYSTEM: AN OVERVIEW OF ITS PARTS AND FUNCTIONS
    The Forebrain
        The internal capsule
        The basal ganglia
        The thalamus
        The reticular system
        The hypothalamus
    The midbrain
    The hindbrain
    The brain stem
    The spinal cord
    The meninges and cerebrospinal fluid
SUMMARY: FUNCTIONAL DIVISIONS OF THE NERVOUS SYSTEM

All of man's voluntary activities, such as seeing or walking, and many of the autonomic activities, such as digesting food and breathing, are controlled by the nervous system. The nervous system is very complicated, and its functioning is very complex; nevertheless, the basic principles of its organization and something about its functions can be summarized rather briefly. In this chapter we will begin our study of the nervous system with an overview of its anatomy, function, and development.

First, we will describe the cells of the nervous system and their functions. In doing so we will also discuss the concept of bioelectric signals, and how these signals originate and are transmitted. Next we will describe the larger functional units of the nervous system—ganglia, nuclei, tracts, and nerves. Using this information as background, we will then discuss the origin and development of the nervous system. Through this discussion you will acquire an orientation to the nervous system as a whole and its relation to the tissues it regulates.

Finally, the entire complex nervous system is subdivided into seven principal functioning units. The basic functions and operations of each of these units will be described in Chapters 6 through 11, along with some of their common problems and malfunctions.

Anatomically, the nervous system is divided into three main subdivisions: (1) the *central nervous system*, which is composed of the brain and spinal cord; (2) the *peripheral nervous system*, which is composed of all nervous tissues that lie outside the brain and spinal cord; and (3) the *autonomic nervous system*, composed of the nervous elements which regulate the viscera and all those functions of the body that are carried out involuntarily, except those of the skeletal muscle reflexes.

PERIPHERY. *The outer part or surface; the part of a body away from the center (G. peri, around; phero, to carry).*

BASIC ELEMENTS OF THE NERVOUS SYSTEM

The nervous system is an anatomical arrangement of nerve cells, synapses, and satellite cells.

Nerve cells, or *neurons*, are one of the two basic anatomic units of the nervous system. They number about 12 billion and have the following functions: (1) They respond to certain changes in their environment, called stimuli. Stimuli are changes in some form of energy which cause the nerve cell to respond bioelectrically. The energy may be derived from one or more sources: mechanical, electrical, chemical, thermal, or photo. (2) Nerve cells convert the energy of a stimulus into a bioelectrical signal which may develop into an *action potential (nerve inpulse)*. (3) Nerve cells propagate (send) this action potential along a fiber which extends from the cell body of the nerve cell to other cells, called *target cells*. Target cells are usually other nerve cells, gland cells, or muscle cells. (4) Nerve cells release a special chemical substance, called a *neurotransmitter*, from the tips of the nerve cell fiber. This substance is released when the action

ACTION POTENTIAL. *A pulse of bioelectricity which travels from one point on a cell to the extreme opposite end(s) of the cell.*

Introduction to the Nervous System

potential arrives at the tips of the fiber. Through this neurotransmitter substance the nerve cell stimulates its target cell(s).

Synapses are special junctions between a nerve cell and its target cell(s). The bioelectric signal is *transmitted* across the synapse from the nerve cell into the target cell(s) by the neurotransmitter.

Satellite cells are the other basic anatomic components of the nervous system. Numbering about 120 billion in all, satellite cells render a number of necessary services to the neurons. In general we can say that the satellite cells support nerve cells and modify their electrical activities. If satellite cells were not present, nerve cells would not function properly. The five types of satellite cells, and their specific functions, are described below.

CELLS WITH DISTINCT AXONS

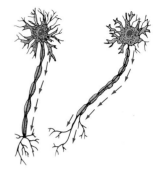

Nerve Cells

A nerve cell viewed in three dimensions appears as a distorted globular structure from which arise fibrous extensions. A nerve cell may have only one fiber or as many as several thousand. Usually, each fibrous extension is referred to either as an *axon* (main extension) or as a *dendrite* (branch). In some cases it is impossible to distinguish an axon from a dendrite (Figure 5.1).

Dendrites convert stimuli into bioelectrical activity, and the axon propagates this activity—in the form of action potentials—toward the target cells of the neuron. At its end, the axon divides into numerous branches, called *telodendria*. During a synapse the neurotransmitter is released through the tips of the telodendria.

The globular region of the neuron, known as the *cell body*, contains the nucleus and various other cell organelles. The chief functions of the cell body are to provide nourishment to the axon and dendrites, to help regenerate the axon and dendrites if they are cut or injured, and to synthesize neurotransmitter molecules or the enzymes which bring about the synthesis of neurotransmitter molecules.

Unipolar, Bipolar, and Multipolar Nerve Cells. A nerve cell which has a single fibrous extension is known as a *unipolar neuron* (Figure 5.2). In the body most of the unipolar neurons are sensory cells that are found near the spinal cord. (The cell bodies of these neurons are housed in *spinal ganglia*, or swellings, as shown in Figure 5.15.) The single fibrous extension of the unipolar neuron gives off two branches in the spinal ganglion. One

CELLS WITHOUT DISTINCT AXONS

Amacrine cells in retina of eye

From central nervous system

FIG. 5.1. *Interneurons, a class of multipolar neurons. Some of these cells have distinct axons, others do not.*

AXON. *The main process of a nerve cell (G. axon, axis).*

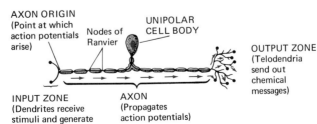

FIG. 5.2. *General sensory neurons or unipolar neurons.*

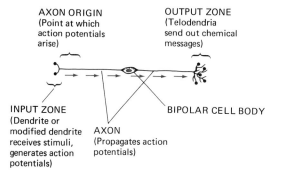

FIG. 5.3. *Special sensory neurons or bipolar neurons.*

DENDRITE. *One of many branching processes from a nerve cell. It is usually distinct from the axon (G. dendrites, pertaining to a tree).*

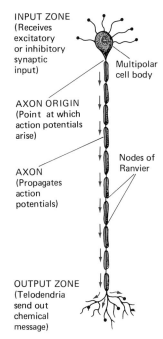

FIG. 5.4. *Multipolar motor neurons. These neurons innervate skeletal muscle, postganglionic cells of the autonomic nervous system, smooth muscle, heart muscle and gland cells.*

branch leaves the spinal ganglion and innervates a peripheral region of the body—a region outside of the brain and spinal cord. This branch receives sensory information about the inner and/or outer environment of the body. The other branch moves into the spinal cord, where its telodendria may form synapses with numerous target neurons.

A nerve cell with two fibrous extensions from its cell body is called a *bipolar neuron*. Bipolar nerve cells are found in various sensory organs, such as the eye, nose, ear, and tongue. One of the fibrous extensions receives environmental sensory information. The other extension gives off telodendria which innervate other neurons (Figure 5.3).

The largest number of nerve cells are the *multipolar* type. Most of the ten billion neurons of the brain and spinal cord fall into this category. All of the neurons which form synapses with skeletal muscle are multipolar neurons. Usually, one fibrous extension from this type of neuron is longer and smoother than the others, and branches distally into telodendria. The other fibrous processes, as dendrites, receive stimuli from other neurons (Figure 5.4).

Propagation of Electrical Activity in a Nerve Cell

A nerve cell that is not being stimulated has a negative charge on the interior of the cell membrane and a positive charge on the exterior surface. This membrane polarity is due to an unequal distribution of ions across the cell membrane. The concentration of sodium and chloride ions is much greater outside than inside the cell, while the concentration of potassium and negatively charged proteins is much greater inside than outside. The unequal distribution of these diffusible ions (sodium, potassium, and chloride) is maintained by the expenditure of metabolic energy in the cell.

This membrane polarity of a resting nerve cell is called the *resting potential*. Although the magnitude of the resting potential varies slightly from one kind of nerve cell to another, it is generally about −70 millivolts.

When a nerve cell is stimulated, the magnitude of the resting potential either increases or decreases, depending on the nature of the stimulus. If the resting potential increases toward a more negative value, the membrane

Introduction to the Nervous System

is said to be *hyperpolarized* and the stimulus is called an *inhibitory stimulus* (Figure 5.5).

If the resting potential decreases toward a positive value, the membrane is experiencing *depolarization*, and the stimulus is called an *excitatory stimulus* (Figure 5.5).

The voltage of the membrane potential can be measured with an oscilloscope. In what follows we will describe the events which lead to the propagation of bioelectricity from the dendrites of a neuron to the tip of its telodendria.

Let us consider a neuron whose dendrites are located in the skin. An excitatory stimulus, such as the application of heat, is applied to the dendrites. This stimulus can lead to one of two situations.

In the first situation the resulting bioelectric activity dies out locally, within the dendrites or within the initial length of the axon, which is called the *axon origin* (see Figure 5.2, 5.3, or 5.4). In the second situation the electrical activity travels all the way from the dendrites to the telodendria. Sometimes, the source of heat is hot enough to stimulate the neuron, but it is applied for only an extremely short period of time or over an infinitesimally small surface area of the dendrites. If this happens, the membrane of the dendrites will respond to the stimulus with slight depolarization. However, because of inadequate stimulation, the depolarization will spread only slightly away from the point of stimulation toward the axon. It will then die out. This type of depolarization is called a *receptor potential*

HYPERPOLARIZE. *To add negative charge to the interior surface of a membrane* (G. hyper, over; L. polus, charge).

DEPOLARIZE. *To take away some of the negative charges from the interior surface of a membrane* (L. de, away).

STIMULUS. *Anything that arouses into action* (L. a goad).

AXON ORIGIN. *The region on a nerve cell from which an action potential takes off.*

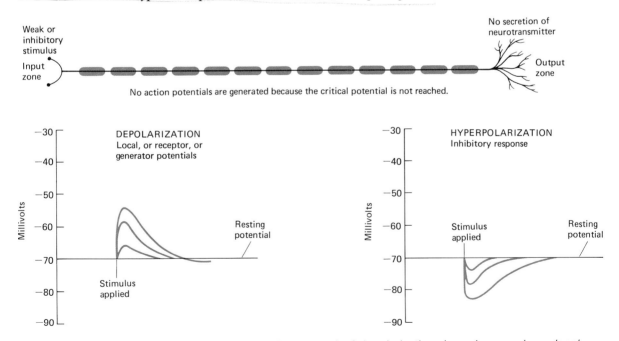

FIG. 5.5. *Depolarization and hyperpolarization. The strength of depolarization shown here produces local potentials and not action potentials.*

because it develops in a sensory nerve fiber. It also is called a *local potential* because it dies out locally. Finally, it is called a *generator potential* because it can give rise to excitatory electrical activity in the axon. Although this type of potential can be measured with an oscilloscope, it cannot be felt by us as a sensation. The excitation dies out near the point where it arose and is not transmitted to the regions of the brain in which conscious perceptions arise.

If, instead of considering a sensory nerve cell, we consider a target nerve cell, and further, we consider the stimulation of this target cell by an infinitesimally small quantity of neurotransmitter, the potential which arises is called a *synaptic potential*. Like the receptor potential, it is a local, or generator, potential.

SYNAPSE. *Junction between two neurons* (G. synapto, *to join*).

Now consider the second situation, where the stimulus is applied over a longer period of time (a sufficient period), or over a larger surface area of the dendrites (a sufficient area), or over a combination of a longer period of time and a larger surface area.

In this case also, generator potentials will arise. But now, a higher frequency of larger generator potentials will arise. In turn, they will induce a *critical structural change* in the membrane of the axon. As a result, action potentials arise in the axon (Figure 5.6). Unlike generator potentials, these action potentials do not die out locally. Instead, they pass from the axon

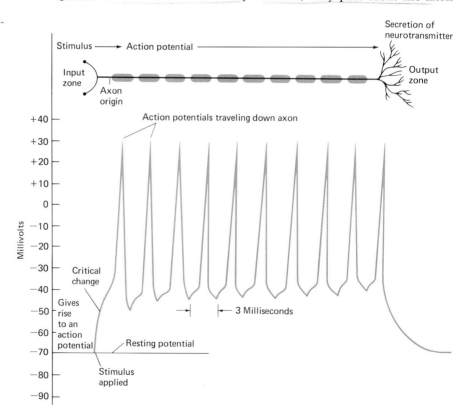

FIG. 5.6. *Axon depolarization leading to action potentials.*

origin to the terminal tips of the telodendria. As they travel they do not lose speed or strength.

When the critical structural change occurs in the membrane of the axon, the axon is said to have reached its *threshold*. At this point it *discharges*, or *fires*, action potentials.

THRESHOLD. *An entrance into something; the beginning point of some action or effect (Anglo-Saxon, therxold).*

The frequency at which an axon discharges action potentials is an inherent characteristic of the axon, but the total number of action potentials which it discharges is related to the strength of the stimulus, the duration of stimulation, and the surface area of the dendrites to which the stimulus is applied.

In the course of generating each action potential, the resting potential (−70 millivolts) of the membrane tends toward a magnitude of zero, and then reverses its polarity, becoming about +30 millivolts (see Figure 5.6). These are approximate values which vary from one type of nerve fiber to another. During an action potential, a change of about 100 millivolts occurs along the axonal membrane, fleetingly, as a nerve impulse passes from the axon origin to the tips of the telodendria.

The action potential is not an electric current which passes along the length of the axon. Instead, it is the result of a chemical chain reaction in which sodium ions rush into the axon while potassium ions rush out. This ionic current causes the membrane polarity to reverse from −70 millivolts to +30 millivolts. This reversal occurs successively at each point along the path of the axon over a period of about two or three milliseconds. Thereafter, the ionic flow is reversed at each of these successive points: the membrane pumps out sodium ions and pumps in potassium ions. The pumping action continues until the resting potential is achieved again. This also occurs over a period of milliseconds. When the resting potential is nearly or entirely restored, the nerve cell can be discharged again.

The action potential travels at a constant speed which varies from one axon to another. The range is from about 1 meter per second to 120 meters per second.

What about an inhibitory stimulus? What is its nature and purpose? The answers to these questions are provided best after considering the synapse.

The Synapse: Transmission of Bioelectrical Activity from a Nerve Cell to Its Target Cells. When an action potential arrives at the ends of an axon, there is a delay of about 5 milliseconds during which no change in the resting potential can be measured across the membrane of the target cells.

After this synaptic delay, a change in potential occurs in the target cells. This change is either a reversal in polarity, so that the inside of the cell membrane tends toward a positive charge (depolarization), or it is a tendency toward a larger negative charge on the interior surface of the membrane (hyperpolarization).

TELODENDRIA. *The distal, terminal branches of an axon (G. telos, distant; dendron, tree).*

What happens at a synapse? When an action potential arrives at the tip of a telodendrion, a certain number of small membrane-bounded vesicles (packets) open and release their neurotransmitter into a small gap (about 1.0 nanometers, or 0.000000001 meters, across) between the telodendrion and its target cell. The neurotransmitter reacts with the cell membrane of the target cell and, depending on the kind of neurotransmitter which

NEUROTRANSMITTER. *A chemical that transmits a stimulus from a nerve cell to one or more target cells.*

the neuron releases and the nature of the target cell, the membrane of the target cell becomes either hyperpolarized (inhibited) or depolarized (excited).

Consider the excitatory case first. Excitation occurs at many different kinds of synapses in the nervous system. The skeletal muscle synapse, also called the neuromuscular junction, is an example. The neurotransmitter at this kind of synapse is *acetylcholine*. When acetylcholine is released from a nerve terminal, it diffuses across a fluid gap and passes into the cell membrane of a muscle cell. In turn, the muscle cell membrane becomes more permeable to an inflow of sodium ions and an outflow of potassium ions (as is the case for an excited axon) and the membrane becomes slightly depolarized.

The magnitude of depolarization in the muscle fiber depends upon how much acetylcholine is liberated from the telodendria. This, in turn, depends on how many action potentials are propagated along the axon. If the neuron is stimulated only slightly, only a few action potentials are fired, and a small amount of acetylcholine is released. If the neuron is stimulated more intensely, or for a longer period of time, more action potentials arise, larger amounts of acetylcholine are released, and a greater amount of depolarization occurs at the muscle membrane.

Generator potentials also arise in the muscle cell membrane at the neuromuscular junction. When a critical potential (threshold) is reached, the muscle membrane fires action potentials along its surface. In this case the impulses travel from the point where they arise to the ends of the muscle fiber, causing the muscle to contract (Figure 5.7).

The membrane of the muscle cell cannot remain depolarized indefinitely or it could not be reset to generate another action potential to produce another contraction. Ordinarily, when acetylcholine arrives at the muscle membrane from the nerve endings, two things happen: (1) Acetylcholine reacts with the muscle membrane, making it more permeable to the inflow of sodium and the outflow of potassium. (2) An enzyme in the muscle membrane, called *acetylcholinesterase*, reacts with the acetylcholine and splits it into two parts, acetic acid and choline. The muscle membrane potential then returns to its original value and the muscle membrane is ready to fire again. In short, the effect of acetylcholinesterase is to reset the membrane for another action potential.

Are neurotransmitter molecules always destroyed after they have served to stimulate a target cell? Before this question can be answered it is necessary to define a neurotransmitter and to list the chemical substances which meet the requirements of the definition. A *neurotransmitter* is a substance which is produced in a neuron and released from nerve endings in response to the arrival of action potentials, is capable of increasing or decreasing the resting potential of a target cell, and is inactivated thereafter, thus terminating the synapse. Acetylcholine, adrenaline, noradrenaline, dopamine, and serotonin are known to be neurotransmitters. Glycine, glutamic acid, GABA (gamma-amniobutyric acid), and a few other substances are suspected to be neurotransmitters, and are being studied intensively to determine whether they can fully qualify.

Introduction to the Nervous System

There are two ways a neurotransmitter may be inactivated after it has served to stimulate a target cell. One way is to modify the chemical through the catalytic action of one or more enzymes in the course of metabolism. Acetylcholine is modified by acetylcholinesterase, as described above. Adrenaline, noradrenaline, and dopamine are modified by the action of either monoamine oxidase or catechol methyl transferase. Serotonin is modified by monoamine oxidase. Of these neurotransmitters, only acetylcholine is always inactivated only by the process of chemical modification.

The second way in which a neurotransmitter may be inactivated is by being taken back into the neuron from which it was released. In the central nervous system, for example, some of the noradrenaline which is used at a synapse is inactivated by enzymatic modification. Much of it, however, is inactivated by the process of *reuptake*. Presumably it is energetically cheaper to bring the molecules back into the presynaptic nerve fiber than it is to resynthesize the molecules. Inactivation by reuptake may be an important conservation feature, especially in the brain, where millions of synapses are occurring every millisecond.

Inhibition at a Synapse. Now we may ask what happens when a nerve fiber receives an inhibitory stimulus. Let us consider this question with

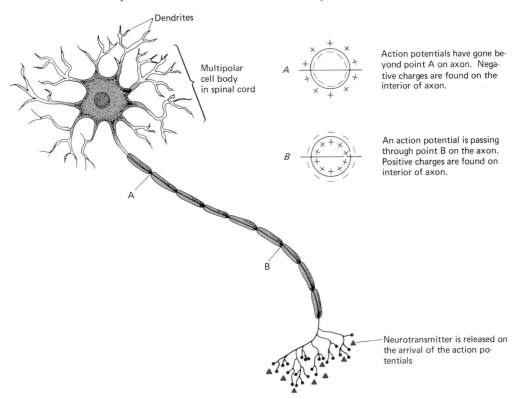

FIG. 5.7. *Propagation of bioelectricity.*

Basic Elements of the Nervous System

respect to an inhibitory stimulus at a synapse. In our discussion we will use as an example the parasympathetic nerves which innervate the heart muscle.

When action potentials arrive in the telodendria of these fibers, acetylcholine is released. Acetylcholine does not have the same effect here as it does at the neuromuscular junction. Whereas the cell membrane of the skeletal muscle cell depolarizes (becomes excited), the cell membrane of the heart muscle cell hyperpolarizes (becomes inhibited).

Why does the membrane of the heart muscle cell become inhibited by acetylcholine, while the skeletal muscle cell membrane becomes excited? Because there is a basic difference in the chemistry of these two membranes—a difference we know nothing about, except that it exists.

What happens to sodium and potassium ions at an inhibitory synapse? There is some evidence that the postsynaptic membrane (the heart muscle cell membrane in this example) becomes less permeable to the inflow of sodium than when it is at rest, and more permeable to the inflow of potassium. Hyperpolarization follows, the muscle becomes less contractile, and thereafter acetylcholinesterase destroys the acetylcholine and thus removes the inhibitory effect.

Why are there two kinds of reactions to a stimulus, excitatory and inhibitory? The answer is that many target cells, such as heart muscle fibers, smooth muscle fibers, and countless nerve cells in the central nervous system, are innervated by two or more nerve cells. One nerve cell makes an excitatory synaptic connection with the target cell. The other forms an inhibitory connection. In this way, the electrical activity of a target cell can be stepped up or down, as needed.

Nerve Cell Functions

AFFERENT. *Leading or bringing into* (L. affero).

Earlier we described nerve cells in terms of architecture: unipolar, bipolar, multipolar. Nerve cells may also be classified in terms of function: afferent (bring toward) neurons, efferent (take from) neurons, interneurons, and endocrine neurons.

Afferent (Sensory) Nerve Cells. *Afferent* neurons carry sensory information to the spinal cord and brain. Regardless of the nature of the sensation—taste, vision, pressure, or pain—a chain of at least three nerve cells is used to bring the sensory information from the receptor cell to the highest level of the central nervous system. A *first-order* afferent neuron picks up the sensory stimulus, generates action potentials, and stimulates a *second-order* neuron. The second-order neuron responds by generating action potentials and stimulating a *third-order* neuron and so on, finally into the cerebral cortex of the brain.

The sensory receptor end of a first-order afferent nerve cell may be a free nerve ending (an unmodified dendrite), a modified dendrite, or it may be encapsulated by non-nervous sensory cells.

First-order sensory neurons are classified as *receptors* according to the

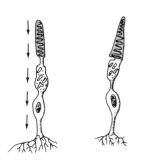

FIG. 5.8. *In the eye, rod cells (left) and cone cells (right) have modified dendrites for receiving visual information.*

type of energy they convert into action potentials. Visual receptors (Figure 5.8) convert light messages into nerve impulses; mechanical receptors convert pressure sensations (sound waves, osmotic pressure, touch pressure) into nerve impulses; thermal receptors convert infrared heat waves into nerve impulses; smell and taste receptors transform chemical energy into nervous activity.

Sensory neurons are also classified by position as exteroceptors or interoceptors. *Exteroceptors* are located in the periphery of the body, where they receive external information about temperature, light, sound, and pressure. *Interoceptors* are located internally where they monitor osmotic pressure, pH, chemical composition of body fluids, pain, and other internal conditions.

Sensory neurons are also classified on the basis of general functions, as proprioceptors and nociceptors. *Proprioceptors* signal information to the central nervous system about the positions and movements of the body. They are located in the muscles, joints, tendons, inner ear, and certain other tissues of the body. *Nociceptors* signal pain. These receptor cells, which are located throughout the body, have free nerve endings (dendrites, Figure 5.9) at their receiving end.

PROPRIOCEPTORS. *Receptors that allow us to take cognizance of our deeper body regions, such as tendons and joints (L. proprius, one's self; ceptus, perceive).*

First-order sensory neurons—together with all their circuitry to and within the brain—display four general properties:
(1) specificity for a particular stimulus (e.g., vision);
(2) response to a wide range in intensity of stimuli (e.g., dim to bright light);
(3) ability to discriminate between two different levels of stimulation within the range of response; and
(4) adaptability.

NOCICEPTOR. *A pain receptor (L. noceo, to injure; ceptus, perceive).*

Only the last property needs an explicit definition. *Adaptation* is said to occur in a nerve cell when, as stimulation continues, the axon discharges fewer impulses, and ultimately no impulses at all. A well-known example of adaptation is olfactory habituation, a situation we experience as we "get used to" a "new" odor.

All sensory units adapt to some extent. The speed and extent of adaptation depends on the nature of the sensory system. A sensory unit that responds to pain is usually a slow adapter. The afferents continue to discharge action potentials even after a prolonged stimulus. If a finger is placed on a hot stove, for example, the sensory system will continue to fire until the finger is removed, provided the sensory system is intact. If an individual does not have an intact thalamus in the brain, no pain will be felt, and the finger may be kept on the stove indefinitely.

Nerve cells which respond to changes in texture—such as sitting on a cushioned chair or going from air into water—are usually fast adapters. They cease to discharge after a while, thus allowing us to perceive other changes and sensations. So too, smell receptors adapt quickly, allowing the sensory cells to pick up new odors.

FIG. 5.9. *Free nerve endings in the skin serve as receptors of pressure and pain stimuli.*

Efferent (Motor) Nerve Cells. **Motor,** or *efferent,* neurons are multipolar cells whose function is to regulate muscle contractions and glandular secretions. Motor neurons may be classified as upper motor neurons, lower motor neurons, autonomic neurons, and cranial motor neurons. An *upper*

Basic Elements of the Nervous System

EFFERENT. *Conducting out of* (L. effero).

MOTOR. *Nerve cells which bring about glandular secretory activity or muscle activity* (L. a mover).

motor neuron has its cell body in the brain. Its axon terminates in the spinal cord. A *lower motor neuron* has its cell body in the spinal cord at a level where its axon will emerge in a spinal nerve. The telodendria of this axon terminate at fibers of skeletal muscle. Autonomic motor neurons are of two types: preganglionic and postganglionic. A *preganglionic neuron* has its cell body in the spinal cord and its telodendria in a ganglion outside the spinal cord. A *postganglionic neuron* has its cell body in the ganglion, where it synapses with preganglionic fibers; its telodendria are located on cardiac muscle, on smooth muscle, or in gland cells. A *cranial motor neuron* has its cell body in the brain. Its axon may innervate skeletal, smooth, or cardiac muscle fibers, another neuron, or a gland cell, depending on the particular cranial nerve referred to.

Interneurons. There are by far more *interneurons* (Figure 5.1) than all other kinds of neurons put together. They are found in all parts of the body, but occur mostly in the central nervous system. They may occur between two afferent neurons, between two efferent neurons, or between afferent and efferent neurons.

Interneurons serve several important functions. These include: amplifying a signal, changing an inhibitory message into an excitatory one, and sending signals into alternative paths. Abstract functions, such as thinking and memorizing, are made possible by the existence of pools of interneurons.

ENDOCRINE. *A gland which secretes a substance into blood; also, the secretory substance itself* (G. endon, *within;* krino, *to separate*).

Endocrine Nerve Cells. **Endocrine neurons** are nerve cells which secrete a chemical substance, called a hormone, from the tips of their axons. The *hormone* is sent into a bed of blood capillaries. From there it passes into the blood circulatory system. As it circulates with the blood, the hormone encounters many kinds of tissues and cells. It reacts, however, only with those cells (target cells) through which it exerts its special functions. (This is discussed more fully in Chapter 12.)

Satellite Cells

Nerve cells cannot reproduce themselves, as can epithelial cells, most connective tissue cells, and (to some extent) muscle cells. Nerve cells are high-premium cells and must therefore be given the greatest protection possible. As a step in this direction, nature has provided them with satellite cells.

Known also as *glial cells*, the satellite cells outnumber nerve cells by about ten to one. There are five known types: microglia, astrocytes, oligodendrocytes, Schwann cells, and ependymal cells.

The *microglia* are ovoid cells that have a number of fibrous extensions (Figure 5.10). They are phagocytic cells. When nervous tissue becomes inflamed or injured, these cells move to the site of injury and literally eat the waste particles. The particles are digested and converted into soluble waste materials, which then enter the body fluids.

In the event of a stroke, when a blood clot forms in the brain or the brain

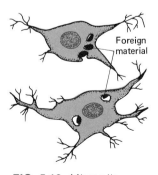

FIG. 5.10. *Microglia.*

suffers damage from insufficient oxygen, the microglia are especially important in recovery. They destroy the clot or damaged tissue and thus allow oxygen and nutrients to get into otherwise ill-nourished regions. Nerve cells that were injured but not destroyed may become functional again, permitting a return of certain abilities that were lost through the stroke.

The *astrocytes* occupy most of the space in the nervous system which is not taken up by nerve cells. They are star-shaped cells with numerous fibrous extensions. Many of these extensions broaden out at their terminal tip, where they butt up against a bit of protoplasm on a neuron or against a bit of protoplasm that makes up the wall of a blood capillary (Figure 5.11).

Astrocytes have five general functions. (1) In the brain, they allow only certain materials to pass from capillaries into the nerve tissue. Because of this, the chemical composition of the fluid around nerve cells in the brain is very different from the composition of blood. (2) Astrocytes also provide structural support to the nerve cells. (3) They also respond to an injury in the central nervous system by proliferating and forming scar tissue. (4) They pass nutrients into the spaces that lie between nerve cells. (5) They take up or give off certain ions (sodium and potassium, for example) and thus moderate the electrical activities in the central nervous system.

Oligodendrocytes and *Schwann cells* are believed to play similar roles. The only difference is that oligodendrocytes occur in the central nervous system while Schwann cells occur in the peripheral nervous system. During their development, these two types of cells come up against a nerve cell axon. At this contact point, the oligodendrocyte or Schwann cell begins to produce a cytoplasmic compartment which ultimately appears to be wrapped around the axon (Figure 5.12). This wrapping may be thin or thick, and it contains variable amounts of a fatty material called *myelin*.

Axons which are heavily myelinated have *nodes of Ranvier* (Figure 5.12) between their Schwann cells or oligodendrocytes. Messages travel much faster along such axons than along ones that are poorly myelinated or completely unmyelinated. The greater speed is due to the fact that the action potential jumps from node to node (*saltatory propagation*). At the nodes of Ranvier, resistance to ionic flow into and out of the axon is much lower than in the regions between the nodes.

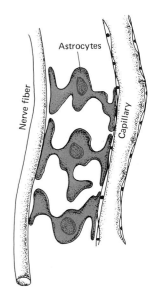

FIG. 5.11. Astrocytes filling space between nerve fiber (left) and blood capillary (right).

MYELIN. A mixture of lipids and protein material arranged in layers around an axon (G. myelos, marrow).

SALTATORY. Having the habit of dancing or leaping (L. saltatio, to move).

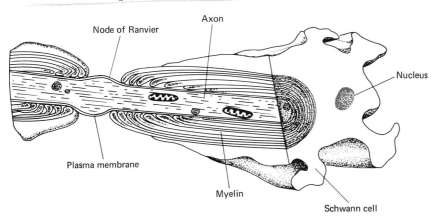

FIG. 5.12. Oligodendrocyte (central nervous system) and Schwann cell (peripheral nervous system) produce myelin around axon.

Schwann cells and oligodendrocytes have three other important functions. First, they help the cell body to maintain and nourish the axon. Second, when a nerve cell is injured they can assume part of a degenerating axon's function in propagating action potentials. Third, they serve as guides, or tracks, for the regenerating axon. In this connection it should be noted that although nerve cells cannot reproduce themselves, they can be repaired if the nerve cell body is not irreparably damaged. It is through the cell body that the fibrous extensions are regenerated following an injury, such as a cut.

EPENDYMA. *Cells which line the spinal canal and brain cavities (ventricles).*

Ependymal cells (Figure 5.13) are epithelial satellite cells. They line the ventricles (openings) in the brain and the central canal in the spinal cord. By secreting and absorbing various substances, they regulate the composition of the cerebrospinal fluid, which passes from the blood into the ventricles and central canal. Together with the astrocytes and the wall of blood capillaries the ependymal cells form a blood-brain barrier. That is, they keep certain substances out of the brain's nerve cells and allow certain other substances, such as nutrients, in.

LARGER BUILDING BLOCKS OF THE NERVOUS SYSTEM

In the preceding sections of this chapter we described the following anatomical components of the nervous system: *nerve cell* (neuron), *synapse* (the transmission of electrical activity from a nerve cell to its target cells), *satellite cell* (glial cell), *central nervous system* (spinal cord and brain), *peripheral nervous system* (all nervous system elements outside of the brain and spinal cord), and *autonomic nervous system* (nervous system elements which regulate involuntary physiological functions other than skeletal muscle reflex activities).

Four other terms are very important in visualizing the architecture and appreciating the function of the nervous system: nucleus, tract, peduncle, and nerve. In addition, a fuller description of ganglia (swellings) must be given.

Ganglia

When the cell body of a neuron is very large and the neuron is widely separated from any other neuron, the cell is usually called a *ganglion cell*. Such cells occur widely throughout the intestinal tract and in the retina of the eye.

The term *ganglion* is also used to describe a discrete collection of cell bodies in a sheath of connective tissue. Ganglia are usually found in the peripheral and autonomic nervous systems, but the term is applied also to certain aggregates located in the central nervous system as indicated below.

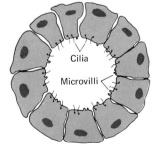

FIG. 5.13. *Ependymal cells. Note cilia and microvilli in lumen.*

Nerve Nuclei, Tracts, and Peduncles

When the term *nucleus* is used in conjunction with the nervous system, it refers to an aggregation of cell bodies in the central nervous system. As a rule, a nucleus performs a specific function. Two examples are: the paraventricular and supraoptic nuclei of the brain, which produce the hormones that are stored in and released from the posterior lobe of the pituitary gland (Figure 12.8); and the anterior nucleus of the spinal cord, which runs throughout the length of the spinal cord and gives off motor axons which innervate the skeletal muscles (Table 5.1).

As another rule, the cell bodies of a nucleus give rise to a tract of axons. A *tract* is a bundle of axons which run alongside each other over a certain distance in the central nervous system. Some tracts are classified as ascending tracts, others as descending tracts. *Ascending tracts* carry sensory information toward the brain; *descending tracts* carry motor information toward lower motor neurons.

TRACT. *A way, a path; a track.*

When a tract runs entirely perpendicular to the long axis of the brain it is called a *commissure*. Examples of commissures are the brain's corpus callosum and the anterior commissure, which is situated between the two hemispheres of the brain.

COMMISSURE. *A bundle of nerve fibers crossing over from one side of the body to the other side (L. joining together).*

When a tract begins on one side of the body, passes along the same side of the body for a while, and then crosses over to the other side, it is termed a *chiasma* (crosspiece). The optic chiasma is an example of this kind of tract.

All tracts are named according to their nucleus of origin, their direction of orientation, and the region where they end. Two examples are: the corticospinal tract, which runs from the cortex to the spinal cord, and the spinothalamic tract, which runs from the spinal cord to the thalamus.

Tables 5.2 and 5.3 show the arrangement of the ascending and descending tracts of our body, along with other pertinent information. Table 5.1 provides information on the major nuclei in the spinal cord. By comparing this Table with Tables 5.2 and 5.3, you should be able to understand how the spinal cord is structured. More than 100 nuclei are found in the brain, but it is far beyond the scope of this text to cover them all.

When a large mass of nerve fibers or nerve tracts runs from one large segment of the brain to another, such as from the cerebellum to the thalamus, the mass of fibers or tracts is called a *peduncle*. There are about eight peduncles in the brain.

PEDUNCLE. *A stalk of nerve fibers in the brain stem (L. pedunculus, foot).*

Nerves

A *nerve* consists of one or more fascicles (cables) of axons and lies outside the brain and spinal cord, in the peripheral nervous system. Each axon in the nerve is surrounded by a membrane called an *endoneurium*. A single fascicle is surrounded by a connective tissue membrane called a *perineurium*, and a group of fascicles is surrounded by an *epineurium*. The structural features of a nerve are seen (in cross section) in Figure 5.14. It should be noted that fat cells and blood vessels are present in the connective

TABLE 5.1. Nuclei in the Gray Matter of the Spinal Cord

Location within Cord	Name of Nucleus	Distribution Along Length of Cord	Function
	1. Substantia gelatinosa	Throughout length of spinal cord	Mediates pain and temperature
	2. Dorsal funicular	Throughout length of spinal cord	Mediates touch and proprioception, as well as pain and temperature
	3. Thoracic	From about the eighth cervical vertebra to the second lumbar vertebra	Gives rise to spinocerebellar tract
	4. Lateral	Thoracic and lumbar regions (to lesser extent in sacral region)	Receives sensory information from viscera and sends off preganglionic neurons of sympathetic nervous system
	5. Anterior	Throughout length of spinal cord	Cell bodies give off motor fibers to skeletal muscle

TABLE 5.2. Sensory Apparatus (Ascending Tracts) in the Spinal Cord

Location within Spinal Cord	Name of Tract	Location of Cell Bodies	Sensory Information Conveyed by Tract	Termination of Tract	Further Synapses (at Higher Levels)
	1. Fasciculus gracilis (begins at level below the mid-thorax)	Spinal ganglia	Proprioceptive	Nucleus gracilis (in medulla oblongata)	To thalamus on opposite side; then to cerebral cortex
	2. Fasciculus cuneatus (begins at level near the mid-thorax)	Spinal ganglia	Proprioceptive	Nucleus cuneatus (in medulla oblongata)	Same as (1)
	3. Spinocerebellar dorsal (found above level of second lumbar vertebra) ventral (found at all spinal levels)	Thoracic nucleus of spinal cord[a]	Proprioceptive	Cortex of cerebellum	Same as (1)
	4. Lateral spinothalamic	Substantia gelatinosa[a]	Pain and temperature	Thalamus; also in upper cord and medulla	To cerebral cortex
	5. Spinotectal	Substantia gelatinosa[a]	Pain, touch, and temperature	Tectum (roof) of midbrain, in superior colliculus of optic system	To cerebral cortex
	6. Ventral spinothalamic	Dorsal funicular[a]	Deep touch	Thalamus; also in upper cord and medulla	To cerebral cortex

[a] See Table 5.1.

Larger Building Blocks of the Nervous System

TABLE 5.3. Motor Apparatus (Descending Tracts) in the Spinal Cord

Location within Cord	Name of Tract	Location of Nucleus of Tract	Decussation (Point of Crossover)	Function
	Pyramidal tracts			
	1. Ventral corticospinal	Large areas of cerebral cortex, especially area four (the motor area)[a]	None	Involved in precise voluntary movements
	2. Lateral corticospinal	Same as (1)	In pyramids of medulla oblongata	Involved in precise voluntary movements
	Extrapyramidal tracts			
	3. Rubrospinal	Pons (nucleus ruber)	In roof of pons	Integrates information of cerebrum and cerebellum
	4. Lateral reticulospinal	Roof of medulla oblongata (nucleus giganticocellularis)	None	Inhibits extensor reflexes; facilitates flexor reflexes
	5. Medial reticulospinal	Roof of pons (nuclei pontis)	None	Opposite of (4)
	6. Lateral vestibulospinal	Upper medulla (vestibular nucleus, in floor of fourth ventricle)	None	Facilitates activities of alpha and gamma motor neurons in spinal reflexes
	7. Olivospinal	Medulla oblongata	None	Unknown
	8. Tectospinal	Midbrain (superior colliculus)	In roof of midbrain	Visual reflexes
	9. Medial vestibulospinal	Same as (6)	Some, in cord within neck	Same as (6)

[a] See Figure 5.22.

Introduction to the Nervous System

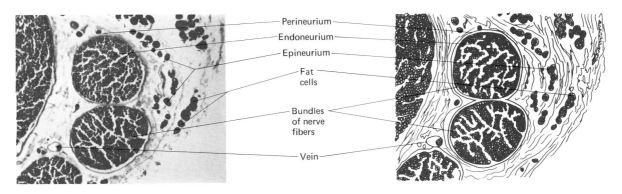

FIG. 5.14. *Cross section of a nerve.*

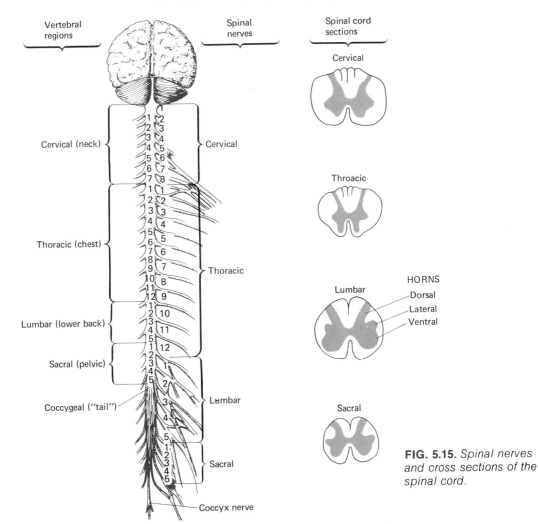

FIG. 5.15. *Spinal nerves and cross sections of the spinal cord.*

TABLE 5.4. Cranial Nerves

Position	Name	Function	Possible Result of Damage
Ventral view of brain stem	1st Olfactory	Sensory: odor	Loss of smell (anosmia)
	2nd Optic	Sensory: vision	Blindness
	3rd Oculomotor	Sensory–motor: to and from four of the six extrinsic muscles of the eyes	Side-eyed glance on side of lesion
		Motor: raises upper eyelid	Drooping of upper eyelid
		Autonomic motor: to smooth muscles that constrict pupil and focus lens	Pupils fail to constrict
	4th Trochlear	Sensory–motor: to and from one of the six extrinsic eye muscles	Difficulty in looking downward
	5th Trigeminal	Sensory: head and face	Anesthesia on side of injury
		Motor: to jaw and to ear drums (counteracts loud sounds)	Paralysis of jaws
	6th Abducens	Sensory–motor: to and from one of the six extrinsic eye muscles	Eye turns inward (squinting); double vision
	7th Facial	Sensory: taste from anterior two-thirds of tongue; facial glands	Loss of taste on side of injury
		Motor: facial expressions (closes eyelids)	Paralysis of face on side of injury
		Autonomic motor: secretions from tear glands, nose, sinuses, and mouth	Impaired secretory activity
	8th Vestibular acoustic	Sensory: body and joint movements	Deficiency in skeletal co-ordination
		Sensory: sound	Deafness

Introduction to the Nervous System

9th Glosso-pharyngeal	Sensory: taste from posterior third of tongue; general sensations from middle ear and throat; blood pressure from carotid sinus and carbon dioxide content from carotid body Motor: skeletal muscles used in swallowing Autonomic motor: salivary glands	Difficulty in swallowing and speaking Decreased salivation
10th Vagus	Sensory: digestive, circulatory, and respiratory organs; abdominal viscera; external ears Motor: to soft palate, larynx, and pharynx (in swallowing) Autonomic motor: to all thoracic and abdominal viscera	Anesthesia, loss of reflex control of heart, circulation, and respiration; poor digestion Difficulty in swallowing and speaking
11th Spinal accessory	Sensory: neck and shoulders Motor: partial supply of input to larynx and pharynx; shoulders and neck	Deficiency in skeletal coordination Difficulty in swallowing and speaking; lack of coordination
12th Hypo-glossal	Sensory: tongue muscles Motor: tongue and neck	Paralysis of tongue on side of injury

Larger Building Blocks of the Nervous System

tissue membranes, and there may be as many as a million nerve fibers in a single nerve. Clearly, a nerve is not the same thing as a neuron.

The human body contains 31 pairs of spinal nerves (Figure 5.15) and 12 pairs of cranial nerves (Table 5.4). There are also about 33 pairs of autonomic nerves (Figure 8.2). In the thoracic (chest) region, each pair of spinal nerves innervates a relatively specific segment of the body. Elsewhere in the spinal cord there is a considerable amount of branching and fusing of nerves. The collections of branching and fusing nerves are called *plexuses*

FIG. 5.16. *Spinal nerve plexuses.*

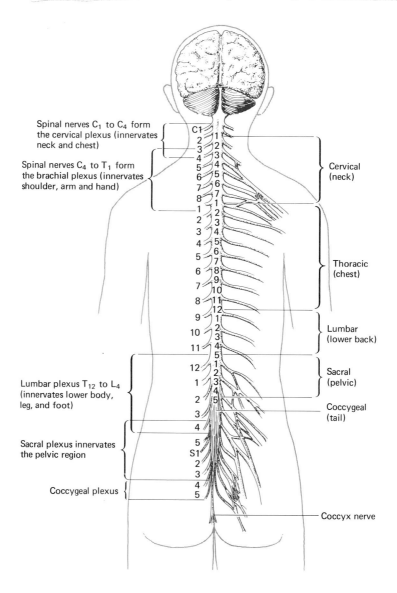

Introduction to the Nervous System

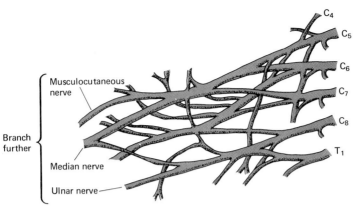

FIG. 5.17. Detailed structure of the brachial plexus, formed by an intermixing of fibers from spinal nerves C1 to T1.

(networks); see Figure 5.16. Each nerve in this kind of plexus can be identified and named (Figure 5.17). There is another kind of plexus which consists of such a complex profusion of nerve fibers that the individual nerve fibers cannot be identified. Examples of this type of plexus are Meissner's and Auerbach's plexuses, which are found in the wall of the digestive tract.

DEVELOPMENT OF THE NERVOUS SYSTEM: AN OVERVIEW OF ITS PARTS AND FUNCTIONS

One important way of providing an overview of the structure of the nervous system is to consider its development. In the first week of life a human embryo differentiates into three layers of tissue that extend throughout its length. In cross section these layers are seen as an inner, a middle, and an outer layer. Of importance here is the outer (ectodermal) layer, since a portion of this layer develops into a dorsal hollow tube with paired neural crests, which will differentiate into the central and peripheral nervous systems.

ENDODERM, MESODERM, and ECTODERM. Early distinguishable tissues which give rise to the organs of the body (G. endon, *within;* mesos, *middle;* ecto, *outer;* derm, *skin*).

By the third week of embryonic life, when organs have begun to form, certain ectodermal cells, which are located along the midpoint of the midline of the embryo's back, begin to form a thick flat layer of cells known as the *neural plate* (Figure 5.18). The cells within and around the neural plate undergo cell division, and as the number of cells increase and move out, a *neural tube* forms. Ultimately, this tube extends along the entire length of the embryo.

From along the two lateral edges of the neural tube a crop of cells bud out to form projections called *neural crests*. The neural tube with its attached pair of neural crests becomes pinched off from the skin that lies above it. The entire structure then becomes embedded within the embryonic connective tissue. Shortly after the neural tube and neural crests have formed, their primitive embryonic blast cells give rise to various kinds of nerve cells and satellite cells, as indicated in Figure 5.18.

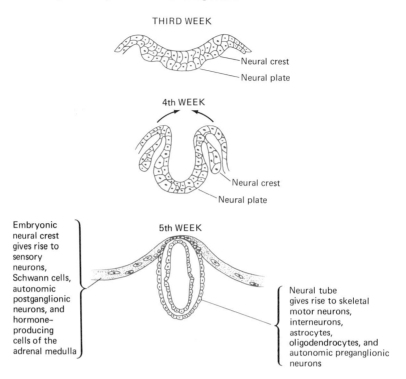

FIG. 5.18. Development of the nervous system.

As the neural tube grows further, its walls become thicker and acquire blood vessels. At this stage in development, certain cells escape from the blood capillaries and enter the nervous tissue to become microglia. The function of these cells, as mentioned previously, is to cleanse the nervous tissue of any extraneous material which enters it from the blood.

After a considerable amount of twisting, bending, bulging, and differentiation, four major landmarks of the brain are distinguishable: the forebrain, midbrain, hindbrain, and brain stem. You should examine Figure 5.19

FIG. 5.19. Arrangement of the nervous system.

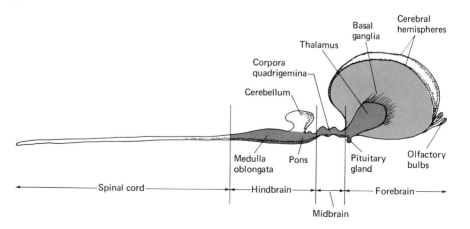

at this point and notice that two cerebral hemispheres project from the forebrain and the cerebellum projects from the hindbrain. When the cerebral hemispheres and cerebellum are removed from the brain, the remaining structure is called the brain stem.

The Forebrain

The forebrain, together with the midbrain, makes up the *cerebrum*. The most striking features of the cerebrum are the cerebral hemispheres (Figure 5.19). Each hemisphere consists of a thin outer layer (cortex) of gray matter (cell bodies) over a white body of myelinated nerve tissue which we might refer to as the "cerebral medulla."

The *cerebral cortex* is marked by folds (*gyri*) that are separated by crevices (*sulci*). The deepest crevices are termed *fissures*. Based on these folds and crevices, each hemisphere can be divided into lobes, known as the *frontal* (foremost) lobe, the *parietal* (side wall of brain cavities) lobe, the *temporal* (referring to temples) lobe, the *occipital* (back of head) lobe, and the *limbic* lobe (Figures 5.20 and 5.21).

Each of these lobes, except the limbic lobe, can be seen from the outside of the brain. The limbic lobe is situated medially. The word *limbic* means edge, or border. Thus, the limbic lobe borders the brainstem, and it includes four folds of brain tissue (Figure 5.21). These folds contain neurons which

CEREBRUM. (L. *brain*).

CEREBRAL CORTEX. The outer layer of the brain (L. cerebrum, *brain*; cortex, *bark*).

GYRUS. A convolution on the surface of the cortex (L. gyros).

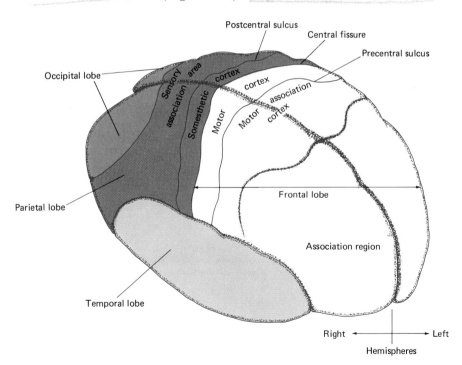

FIG. 5.20. *External lobes of the brain.*

100
Development of the Nervous System

receive olfactory (smell) stimuli and stimuli which are concerned with feelings, emotions, and behavior (Chapter 11).

The cerebral cortex is only about a quarter of an inch thick and it contains about ten billion cell bodies. It is a heavily pleated tissue, forming many convolutions and crevices.

Some regions of the cerebral cortex have been mapped out in terms of their function. This is shown in Figure 5.22. Most of the information presented in this figure has been discovered through three procedures: (1) direct electrical stimulation of exposed regions of the cerebral cortex in conscious individuals who are undergoing special neurological treatment; (2) surgical operations; and (3) postmortem examination (autopsy) of brains in individuals who were neurologically deficient in one way or another.

CORPUS CALLOSUM. *A large massive tract of nerve fibers which crosses the brain from side to side (L. corpus, body; callosus, hard).*

The cerebral hemispheres are connected to each other by a structure called the *corpus callosum*. In addition to connecting the cerebral hemispheres, the corpus callosum is a commissure, a tract of about 150 million myelinated fibers. Acting as a two-way bridge, the corpus callosum serves to integrate certain kinds of neural actions. This kind of integration would not be possible if the two hemispheres were severed from each other. This topic is discussed in Chapter 11.

Four major neural structures are present in the "cerebral medulla" of the forebrain: the internal capsule (which lies immediately below the cortex), the basal ganglia (which surround the thalamus), the thalamus (which lies beneath the internal capsule), and the hypothalamus (which lies beneath the thalamus). See Figures 5.23 and 5.25. We will now discuss these structures, as well as an important related structure, the reticular system, which is found throughout the brain stem.

The Internal Capsule. The *internal capsule* (Figure 5.23) consists of white matter which is made up of myelinated axons arising from the thalamus and the cerebral cortex. The axons relay every kind of sensory message, except smell, from the thalamus to the cerebral cortex. Messages also pass through the internal capsule from the cortex to the thalamus, to the nuclei of the cranial motor nerves, and to the brain stem (see also Figure 5.25).

FIG. 5.21. *Medial view of the left cerebral hemisphere.*

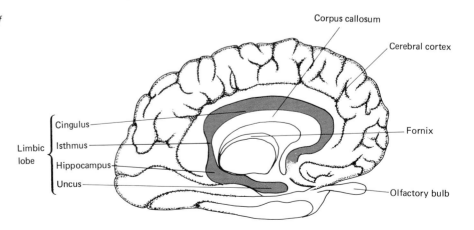

Introduction to the Nervous System

The Basal Ganglia. The paired *basal ganglia*—which are present in all vertebrate animals—constitute the highest motor center in most of the lower vertebrates. In higher vertebrates, including humans, this is a subordinate motor center under the skeletal motor cortex. The basal ganglia are nuclear masses which send axons into all of the nerve tracts that run

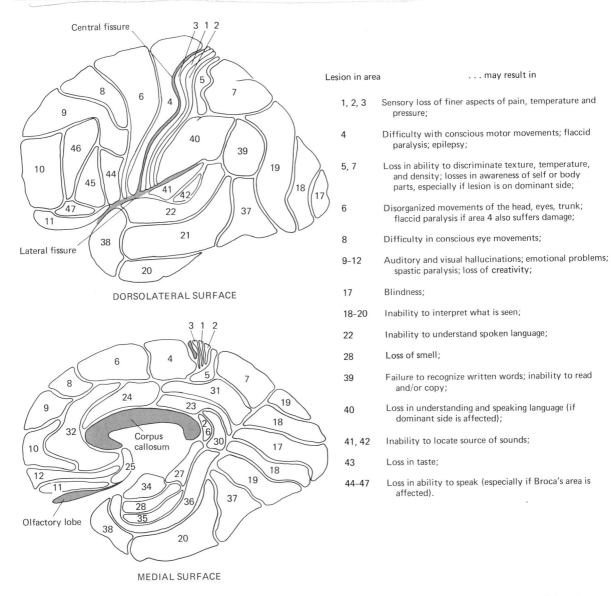

FIG. 5.22. *The functions of the different regions of the cerebral cortex can be determined by examining the effects of damage (lesions) in each of the various areas.*

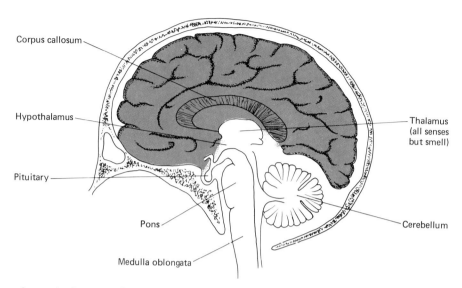

FIG. 5.23. Orientation of several major parts of the brain.

through the spinal cord except the corticospinal tract. The names and positions of the basal ganglia are shown in Figure 5.24.

Little is known about the exact role of the basal ganglia in man. Much of what we do know has come through medical observations about the loss of certain motor abilities in an individual, followed by confirmation of some degeneration in the basal ganglia during autopsy. In general, damage is seen in one or more basal ganglia nuclei in individuals who had displayed certain uncontrollable, purposeless movements, such as those in individuals suffering Parkinson's disease.

Some functions of the basal ganglia have been discovered from observations of behavior during electrical stimulation of separate nuclear masses in laboratory animals. Stimulation of certain basal ganglia (the caudate and

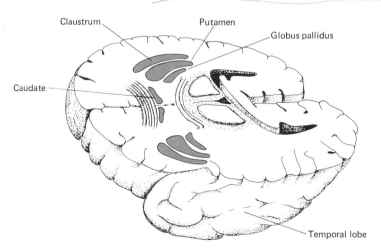

FIG. 5.24. The basal ganglia. Collectively, the caudate, globus pallidus and putamen nuclei constitute the corpus striatum.

putamen nuclei, for example) may cause an increase in locomotion and reflex actions. Stimulation of other ganglia (the globus pallidus, for example) may increase muscle tonicity and may produce tremors. So too, the removal of tissue from these nuclear masses modifies certain kinds of motor actions. The removal of the putamen, for example, may result in muscular hyperactivity (unrestrained and longer-lasting bodily movements). The removal of the globus pallidus may result in muscular hypoactivity (sluggish movements and locomotion). In general, the removal of any nucleus of the basal ganglia may somehow affect normal skeletal muscle movements.

The Thalamus. The *thalamus*, which is present in all vertebrates, is the highest sensory system in most of the lower vertebrate species, and is a sensory relay station to the sensory areas in the cerebral cortex of higher vertebrates. In man, it is a cluster of about 30 nuclear masses which lies deep within the cerebral "medulla," directly above the spinal cord (Figure 5.25). The nuclei may be classified into three groups: specific relay nuclei, specific association nuclei, and nonspecific nuclei.

A *specific relay nucleus* receives information from a specific sensory system in the body (visual, pain, pressure, etc.). The information which it

THALAMUS. *An ovoid mass of gray matter nestled in the brain as a sensory relay station to the cerebral cortex (G. bedroom).*

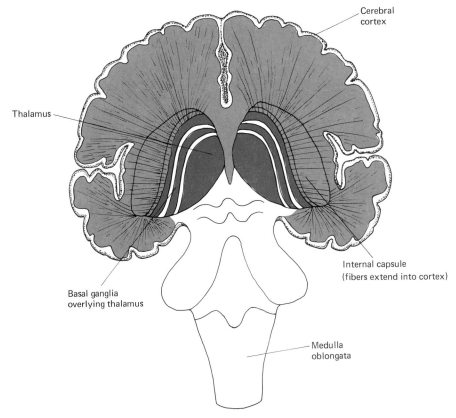

FIG. 5.25. *Frontal view of the brain.*

Development of the Nervous System

TOPOGRAPHY. A description of a place (G. graphe, describe; topos, place).

LEMNISCUS. A band of longitudinal fibers running up from the crossover point in the medulla and pons, carrying point-to-point information from the periphery of the body to the thalamus (G. lemniskos, band).

receives is *topographically preserved*. By this we mean that the information which is brought into the thalamus represents a spatial replica of the peripherally-derived stimuli. This point-to-point or area-to-area information is carried into the thalamus from the spinal cord by way of tracts called *lemnisci*. In turn, the information is carried—again topographically and with high accuracy—to the cerebral cortex.

A *specific association nucleus*, in contrast to the specific relay nuclei, does not receive sensory information from lemniscal tracts. Instead, a specific kind of sensory information (again, visual, pain, pressure, or the like) is received from other nuclei within the thalamus and from nuclei elsewhere in the brain. This sensory information is integrated and projected to sensory association areas in the cerebral cortex. Within these association areas the information is interpreted, and is used as the raw material for learning, memory, judgment, and so forth. It is also used to trigger the efferent paths of various reflexly- and consciously-directed motor actions.

The *nonspecific nuclei* of the thalamus receive sensory information from widespread regions of the brain, including other nuclei of the thalamus. This information is a mixture of the various sensations we perceive. It is used as the raw material for learning and memory, as well as to trigger the nerve pathways to muscles throughout the body. It is also incorporated into and modified by various kinds of nervous activities within the reticular system.

The Reticular System. The *reticular system* is a network (reticulum) of millions of neurons which are widely distributed throughout the brain stem and are interconnected with neurons from a wide variety of areas elsewhere in the brain and spinal cord (Figure 5.26). In contrast to the lemnisci, which serve to bring point-to-point sensory information into the thalamus, the reticular system serves in a nonspecific way to modulate and regulate ascending (sensory) and descending (motor) information. Specific pathways are not definable within the profusion of gray matter and fibers that make up the reticular system.

OPTIC CHIASMA. The crossover point in the brain where the optic nerves become the optic tracts (G. chiazein, crossing).

The Hypothalamus. Beneath the thalamus, at the junction of the cerebral hemispheres and brain stem, lies another important region of the brain— the *hypothalamus* (Figure 5.23). The boundary of the hypothalamus closest to the face is the *optic chiasma*, through which nerve fibers pass from the eyes to the brain (Chapter 9). The posterior boundary of the hypothalamus is marked by the mammillary bodies, from which nerve fibers pass into the thalamus carrying information about emotions.

The hypothalamus contains a number of nuclei, each of which has a relatively specific function. Four of these functions are described more fully in later chapters: (1) production and release of antidiuretic hormone, a hormone which restricts water output by the kidney; (2) production and release of oxytocin, a hormone which stimulates the muscle of the uterus during childbirth, and the muscle of the breasts during nursing; (3) regulation of hormonal output by the pituitary gland; and (4) regulation of food and water consumption.

Another function of the hypothalamus is the regulation of body tempera-

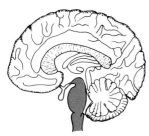

FIG. 5.26. *Sagittal view of the brain, showing location of reticular system.*

ture. As blood flows through the hypothalamus, the blood's temperature is monitored. If the temperature is higher than normal (98.6°F; 37°C), several mechanisms are brought into action to lower it. These mechanisms include: sweating, releasing less thyroxine and triiodothyronine (hormones which promote metabolic activity) from the thyroid gland, and regulating the openings of the arteries so that more blood flows through the skin and less blood flows elsewhere in the body. If the temperature of blood is lower than normal, the hypothalamus triggers other reactions: shivering, increasing the output of thyroxine and triiodothyronine, and shunting the blood from the skin into deeper regions of the body.

Another very important role of the hypothalamus is its function as an integration center for numerous emotional activities, as well as for activities that are performed autonomically (involuntarily). Through its connections with all regions of the brain and spinal cord, the hypothalamus serves to coordinate breathing, heartbeat, and circulation. It also acts as a center in which certain taste and smell sensations are brought to a level of consciousness, and a center through which emotions are partly felt and expressed.

The Midbrain

Behind the hypothalamus, toward the spinal cord, is a region of brain called the midbrain (Figure 5.19). In the roof (*tectum*) of the midbrain lies a structure known as the *corpora quadrigemina*. Two of these nuclear bodies, the *superior colliculi*, are larger than the other two, the *inferior colliculi*. The superior colliculi are centers through which visual reflexes are channeled. The inferior colliculi are auditory reflex centers. A sudden booming sound or a flash of light, for example, sends impulses through the midbrain, causing the individual to put his or her hands to the ears, to shudder, to blink, or turn the head.

The ventral, or basal, region of the midbrain is a relay station and an integration center for various kinds of sensory and motor activities. A portion of the reticular system is found here, as are some of the thalamic lemnisci which were discussed previously. In addition, a number of tracts and peduncles run along the inferior surface, carrying sensory and motor information to and from various regions of the central nervous system.

TECTUM. *A roof or covering (L.).*

CORPORA QUADRIGEMINA. *Four similar appearing nuclei (L. corpora, bodies; quad, four; gemini, turns).*

The Hindbrain

This anatomical landmark includes three important centers—the pons, medulla oblongata, and the cerebellum (Figures 5.19 and 5.23).

The *pons* connects the hindbrain to the midbrain. Together with certain centers in the medulla oblongata, it is important for the regulation of breathing, and it is somehow involved in regulating sleep and dreaming (Chapter 35).

PONS. *A bridge (L.).*

Development of the Nervous System

MEDULLA OBLONGATA. *An inch-long prolongation of the spinal cord into the medulla of the brain stem* (G. oblong, *prolong;* L. medulla, *middle*).

CEREBELLUM. *A coordination center for voluntary movements, posture, and equilibrium* (L. cerebellum, *small brain*).

The *medulla oblongata,* sometimes called just the medulla, lies between the pons and the spinal cord. It contains important reflex centers for regulating the force and rate of heartbeat, blood pressure, breathing, salivation, swallowing, and vomiting.

The *cerebellum* is an important center for the regulation of posture, body balance, and locomotory activities.

The Brain Stem

The brain stem is that part of the forebrain, midbrain, and hindbrain which remains after the cerebral hemispheres and cerebellum have been removed. It includes several structures mentioned so far: (1) the thalamus, which perceives pain and relays pain and other sensations to the cerebral cortex; (2) the basal ganglia, which initiate certain kinds of motor activities; (3) the corpus callosum, which sends messages from one cerebral hemisphere to the other; (4) the midbrain which functions in certain auditory (hearing) and visual (sight) reflexes; (5) the pons (breathing); (6) the medulla oblongata (breathing, heartbeat and force of heart contraction, blood pressure, and vomiting); (7) the lemniscal tracts, which carry sensory information topographically into the thalamus; and (8) the reticular formation, in which all kinds of sensory and motor information are integrated.

Three other vital structures are located in the brain stem: (1) the nuclei and tracts of fibers that shuttle information from various regions of the central and peripheral nervous system into the cerebellum; (2) all of the tracts that run between the brain and spinal cord; and (3) the nuclei of all of the *cranial nerves* except the olfactory nerve.

CRANIAL NERVES. *Nerves which originate in the brain.*

The brain stem is an extension of the spinal cord and a bridge to the higher nerve centers in the cerebrum and cerebellum. As such, it constitutes an important center for relaying and integrating messages from ultimately most, if not all, regions of the nervous system. Whereas a bullet wound to the cerebrum may or may not seriously harm an individual, a bullet into the brain stem—especially in the hindbrain—will most likely cause death.

The Spinal Cord

The development of the spinal cord proceeds together with the development of the brain, but involves much less bending, bulging, or twisting. The axons which develop within the spinal cord do not have a myelin sheath during early embryonic development. They pass up from the spinal cord to the brain or down from the brain to the spinal cord in the outer (marginal) layer of the embryonic spinal cord. Gradually, they acquire a myelin sheath from the oligodendrocytes.

OLIGODENDROCYTE. *A treelike cell which has few branching processes* (G. dendron, *tree;* kytos, *cell;* oligo, *few*).

Many axons do not receive their myelin coat until one to two years after birth. The slow development of the myelin coat is linked to the slow development of skeletal muscle movements in infancy.

When seen in cross section, the spinal cord looks very much like a gray butterfly upon a white background (Figure 5.27). The leading edges of the "wings" are called the *dorsal horns*. Here the first-order afferent nerve cells synapse with interneurons and second-order afferent cells. The trailing edges of the "wings" are called *ventral horns*. Most of the cell bodies of the motor neurons, as well as interneurons, are located here.

HORN. *A column of gray matter in the spinal cord.*

In the thoracic (chest) and lumbar (abdominal) regions, the butterfly has special lateral (side) extensions that lie between the leading and trailing tips of the "wings" (this can be seen in Figure 5.15). These extensions are called *lateral horns*. They harbor the cell bodies of the preganglionic neurons from the sympathetic division of the autonomic nerve system as well as numerous interneurons.

Nerve roots emerge from the dorsal and ventral horns. Called dorsal and ventral roots, they unite within the vertebral column before emerging from it. They are then seen as paired spinal nerves, one on each side of the body. The axons within the nerve roots are myelinated. Figure 5.28 shows how the spinal nerves branch out to innervate the periphery of the body. The same figure illustrates the anatomical pathways of the autonomic nervous system.

NERVE ROOT. *The beginning, or proximal, portion of a nerve.*

The two roots on the dorsal (back) side of the spinal cord contain the terminal portions of the unipolar sensory neurons, the cell bodies of which are located in paired *dorsal root* (or *spinal*) *ganglia*. The two roots on the ventral (belly) side of the spinal cord carry the axons of multipolar motor

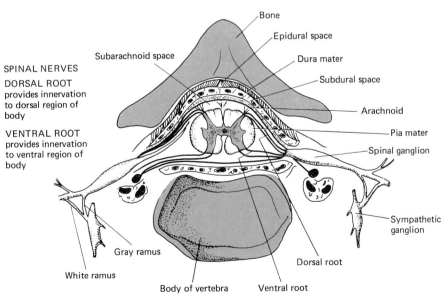

FIG. 5.27. *Spinal cord and its orientation to the spinal column.*

FIG. 5.28. Innervation of the body by spinal nerves.

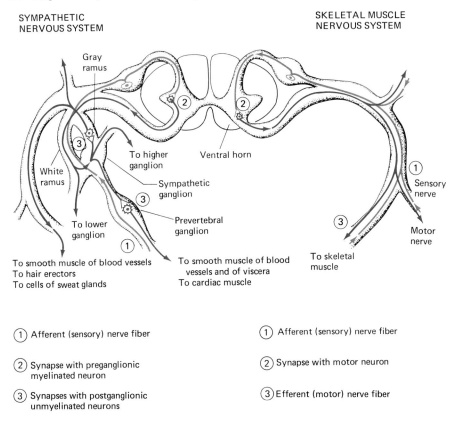

neurons which go to the skeletal muscle fibers and to the ganglia of the autonomic nervous system (Figure 5.28).

The Meninges and Cerebrospinal Fluid

The entire central nervous system is enclosed in a *meningeal sheath* which provides support, protection, and nourishment to the underlying nervous tissue (Figure 5.29).

The meningeal sheath consists of three layers of membranous material which lie next to three layers of fluid. The outermost membrane is called the *dura mater*. It lies next to the bone of the skull and spinal cord and is separated from the bony material by a fluid-filled *epidural space*.

In certain operations an anesthetic is injected into the epidural space at the base of the spinal column to produce *caudal anesthesia* (Figure 5.30). This type of anesthesia is sometimes used during childbirth to reduce pain. The anesthetic comes into contact with spinal nerves but does not pass into the brain. A small quantity of anesthetic is used to block the sensory nerves, but not the motor nerves, to the uterus. This means that the uterine con-

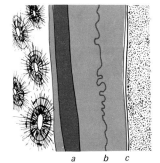

FIG. 5.29. Meningeal sheath separates bone tissue (left) from nerve tissue (right).
(a) dura mater;
(b) arachnoid; (c) pia mater.

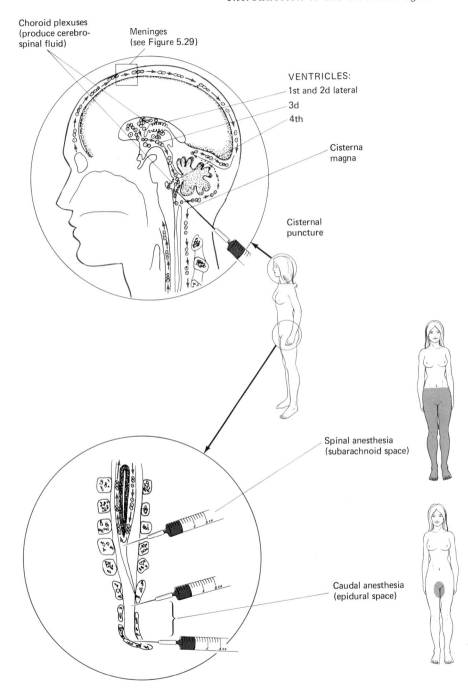

FIG. 5.30. *Flow of cerebrospinal fluid and two types of anesthetic procedures. Shaded area shows region of body affected by anesthetic.*

ARACHNOID. *The central of the three membranes which cover the brain and spinal cord* (G. anachnes, *spider;* eidos, *appearance*).

CHOROID PLEXUSES. *Tissues in which the cerebrospinal fluid is produced* (G. chorion, *skin;* L. plexus, *network*).

ARACHNOID VILLI. *Nipple-like projections into the brain from the arachnoid membranes* (L. villus, *hair*).

CISTERN. *A large subarachnoid space* (L. cisterna, *reservoir*).

tractions are not diminished but the woman in labor is unable to feel them. Caudal anesthesia is also used in other kinds of surgery, for men as well as for women.

The middle layer of the meningeal sheath is called the *arachnoid*. It is separated from the dura by a fluid-filled subdural space.

The third and innermost layer of the meninges—the *pia mater*—is separated from the arachnoid by the *subarachnoid space*. Rich with blood vessels, the pia mater closely covers every fold and crevice of the brain and the entire surface of the spinal cord.

In certain regions of the brain the pia mater contains membrane networks called *choroid plexuses* (Figure 5.30). These plexuses are folds of tissue, richly furnished with blood capillaries. From these blood capillaries the choroid plexuses produce a *cerebrospinal fluid*. This fluid is secreted into the cavities of the central nervous system. Unlike blood, from which it is derived, the cerebrospinal fluid contains no red blood cells and very little protein. In addition, there are major differences in the types of electrolytes of the two fluids. These differences are brought about by the ependymal cells and astrocytes whose actions determine what will pass from the blood into the central nervous system.

Five interconnected cavities are found in the central nervous system. Two of the cavities are a pair of *lateral ventricles* in the cerebral hemispheres. These cavities merge with the *third ventricle*, which lies in the brain stem of the forebrain. The third ventricle communicates by way of the *cerebral aqueduct* in the central core of the midbrain with the *fourth ventricle* in the pons. The cavity of the fourth ventricle merges with the *spinal canal* that runs centrally throughout the length of the spinal cord.

After the cerebrospinal fluid enters the cavities from the choroid plexuses, it passes into the subarachnoid space, as shown in Figure 5.30. From here the fluid flows to the tail end of the spinal cord, on to the front side of the cord, and then upward, to the upper surface of the brain. At this point the cerebrospinal fluid enters the *arachnoid villi* and is discharged into veins and then into the heart.

At the tail end of the spinal cord lies a cavity called the *lumbar cistern*. Figure 5.30 shows how this cavity, or the *great cistern* at the back of the neck, may be tapped to obtain a sample of cerebrospinal fluid for use in certain diagnostic procedures. The diagnostician will ask a number of specific questions when examining the cerebrospinal fluid: Are any blood cells present, indicating hemorrhage or injury to the brain? Are any bacteria or viruses present, indicating a disease such as meningitis? Is the concentration of solutes abnormal, indicating the effect of such things as syphilis or a metabolic disturbance?

The lumbar cistern can be injected with an anesthetic to produce *spinal anesthesia* for surgery in the lower extremities and in the pelvic and abdominal regions. Procaine, tetracaine, and other anesthetic agents may be used to block nervous impulses through the sensory, motor, and autonomic nerves. The individual is thus spared of pain, and the surgeon is aided by the lack of muscular movements.

SUMMARY: FUNCTIONAL DIVISIONS OF THE NERVOUS SYSTEM

We can sum up this introduction to the nervous system by considering the functional organization of the components of the nervous system. First, remember that the nervous system is used to monitor sensory information and to bring about appropriate motor responses. We can therefore make a major distinction between the *afferent* (inwardly directed, sensory) and the *efferent* (outwardly directed, motor) systems.

Second, nervous activities are associated either with the framework of the body, or with the soft organs within the body. These we may distinguish as the *skeletal* and the *visceral* systems, respectively.

Third, as humans we tend to think of some of our bodily functions as being *general* and widespread (for example, locomotion and feeding) and other functions as being *special* and limited in their distribution (hearing, vision).

Fourth, a system exists, called *proprioceptive*, through which we can monitor our positions and movements in space.

Finally, there are the highest, most complex actions, such as certain emotional reactions and speech, through which humans stand relatively distinct among other animals.

These distinct systems, and the functions of each, are summarized in the table below.

	System	General Functions
I	General sensory system	Sensing external mechanical and temperature stimuli
II	Skeletal motor system	Operation of skeletal muscles
III	Visceral sensory system	Sensing of internal stimuli
IV	Visceral motor system	Operation of smooth muscle, cardiac muscle, and glands
V	Special sensory system	Hearing, vision, taste, and smell
VI	Proprioceptor system	Equilibrium and coordination of skeletal movements
VII	Cortical systems	The highest regulators

The remaining chapters in this section, Chapters 6 through 11, describe the structure and functioning of the nervous system in terms of these functional divisions.

SUMMARY

1. Twelve billion units of the nervous system, called neurons, respond to changes in their environment, called stimuli. Upon reaching a threshold, the nerve cells discharge action potentials, the energy of

which is transmitted to target cells. Target cells may be other neurons, muscle cells, or gland cells.

2. There are three types of nerve cells: (a) unipolar neurons, which pick up general sensory information such as temperature, pressure, and pain from the periphery of the body; (b) bipolar neurons, which pick up special sensory information, such as vision and hearing; and (c) multipolar neurons, which are in themselves either endocrine neurons or motor neurons, or interneurons that integrate incoming sensory activity with outgoing motor activity.

3. Stimuli which excite nerve cells bring about a reversal in the sign of the charge which exists across the membrane of the nerve cell. In doing this they also bring about a movement of bioelectricity, in the form of action potentials, along the length of the axon. Stimuli which inhibit nerve cells cause the transmembrane potential to become even more negative on the interior surface than when the neuron is not being stimulated.

4. All nerve cells have dendrites, which serve as sensory receptors; an axon, which sends action potentials to the distal regions of the neuron; and telodendria, which release neurotransmitters into one or more target cells.

5. Synapses are functional junctions at which a target cell is stimulated by an excited nerve cell. The stimulus is in the form of a chemical messenger known as a neurotransmitter.

6. Following a synapse, the neurotransmitter is inactivated by a process of chemical destruction or by a process in which the neurotransmitter is taken back into the cells which formerly released it.

7. Afferent nerve cells carry sensory information to the central nervous system from the periphery of the body. Efferent nerve cells carry motor information from the central nervous system to gland cells or muscle cells. Interneural cells operate between afferent and efferent cells. Endocrine neurons secrete hormones into the bloodstream.

8. Approximately 120 billion glial cells are packed around the body's 12 billion nerve cells. The microglia are phagocytes. The astroglia are an important component of the blood-brain barrier. Oligodendrocytes and Schwann cells construct sheaths around axons, enabling the axons to propagate impulses very rapidly. Ependymal cells line the cavities in the brain and spinal cord and thus regulate the kinds and amounts of metabolites that will enter the brain.

9. The brain consists of a "medulla" which is overlaid by a cerebral cortex. The "medulla" of the forebrain contains (a) an internal capsule, which relays sensory and motor information between the thalamus and cortex and other regions of the brain; (b) a thalamus, which receives sensory information from the peripheral nervous system; (c) basal ganglia, which convey motor information from the cerebral cortex to the skeletal muscles; and (d) a hypothalamus, from which many autonomic and emotional behaviors are initiated, and in which various hormonal and visceral actions are monitored and regulated.

10. Sensory information is carried to the cerebral cortex by way of the

Introduction to the Nervous System

thalamus through point-to-point representation of the body's geography; and the reticular system, which integrates some of the sensory information with ongoing motor activities and other sensory information from other regions of the brain.

11. The midbrain is concerned with reflex actions that are triggered by visual and auditory stimuli, such as a ballet or rock musical performance.
12. The hindbrain contains the pons, which is important for regulating breathing and initiating dreams; and the medulla oblongata, which contains regulatory centers for various visceral functions, such as circulation and respiration.
13. In cross section the spinal cord appears as a gray butterfly on a white background. Sensory information comes into the leading edges of the wings while motor information leaves from the trailing and lateral edges. Nerve fibers ascend and descend in the white background. The ascending fibers deliver sensory messages to the brain. The descending fibers send motor messages out of the brain.
14. The brain and spinal cord are enclosed in a connective tissue capsule called the meninges. Cerebrospinal fluid is produced in this capsule to provide protection and nourishment to the nerve tissue.

REVIEW QUESTIONS

1. Define: unipolar neuron, dendrite, telodendria, stimulus, synapse, afferent, efferent, microglia, Schwann cell, central nervous system, peripheral nervous system, autonomic nervous system.
2. How is an action potential developed in a neuron?
3. How is neural information transmitted from a neuron to its target cells?
4. What is the difference between an inhibitory and excitatory stimulus? Between hyperpolarization and depolarization?
5. Name one major function of each of the following structures: basal ganglia, thalamus, hypothalamus, reticular system, internal capsule, midbrain, pons, medulla oblongata.
6. Describe the pathway of nerve impulses—at one level of the spinal cord—for effecting a muscle action.
7. Describe the pathway of nerve impulses at one level of the spinal cord for effecting an autonomic action.
8. How is cerebrospinal fluid produced? What kinds of information are obtained from an analysis of a sample of this fluid from a patient?
9. How does spinal anesthesia differ from caudal anesthesia?

The General Sensory System

PRINCIPLES OF THE GENERAL SENSORY SYSTEM
EVALUATION OF INJURY TO THE GENERAL SENSORY SYSTEM
SOME ABNORMALITIES OF THE GENERAL SENSORY SYSTEM

The general sensory system is mainly concerned with the perception and interpretation of stimuli that occur over most of the body's surface. In this chapter we will help you gain an understanding of this system by discussing it in terms of nine principles which pertain to it. Thereafter we will discuss how this system is evaluated in the event that it does not function properly.

In general, four kinds of sensory information are processed through the general sensory system: (1) *pain* – caused by mechanical, thermal, electrical, or chemical energy, or by some combination of these forms; (2) *temperature*; (3) *light touch* (also called simple, crude, or feather touch) – caused by stroking the skin without pressing it in; and (4) *deep touch* (caused by deforming the skin).

Many scientists believe that the foregoing sensory stimuli are received by the following nerve endings, which are illustrated in Figure 6.1: (1) pain by free nerve endings; (2) light touch by free nerve endings and various corpuscled fibers (Meissner's, Pacinian); and (3) deep touch by Meissner's corpuscles. No one can specify with certainty the kinds of nerve cells that sense temperature changes.

This belief – that the nature of the stimulus determines which one of a group of different nerve endings will serve as a receptor – with the further belief that the signal will pass true to its nature to the central nervous system, is called the *doctrine of specific nerve energies*. By and large, this doctrine is valid. Thus, the rods and cones in the retina of the eye are specifically geared to pick up photo information, and the Pacinian corpuscles are likewise specialized to receive and transmit mechanical stimuli.

In a more realistic sense, however, most kinds of sensory information we receive are combinations of various sources of energy. We experience rapidly-conducted sharp prickling pain, slowly-conducted burning pain, tickling (a form of pain or joy, depending on interpretation), itching (another form of pain), wet cold lips, dry warm lips, and so on. To account for a way in which these combinations of information are transmitted it is necessary to postulate the concept of an *adequate stimulus*. An adequate stimu-

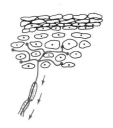

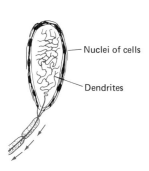

FIG. 6.1. *Receptor nerve endings in the skin and joints. Free dendrite endings (top) and Meissner's corpuscles (bottom).*

The General Sensory System

lus is a special combination of stimuli, received by the peripheral nervous system which produces a sensation that may vary from one individual to another, from one time to another, and one which is subject to the interpretation of the individual. The sensation is perceived in specialized regions of the cerebral cortex, called *sensory association areas*.

PRINCIPLES OF THE GENERAL SENSORY SYSTEM

In spite of man's ignorance on how simple, pure forms of stimuli (such as pressure) may be converted into complex perceptions (such as itching, tickling), nine principles have been deduced with regard to the operation of the general sensory system. These principles relate to the pathways taken by impulses in the course of either executing a reflex or registering a perception in the consciousness of the cerebral cortex.

(1) *A train of three or more sensory neurons*. The sensory pathway — from the periphery of the body to the sensory cortex in which the sensation is identified — consists of three afferent neurons. The receptor neuron (first-order neuron) has its cell body in a dorsal root ganglion (Figure 5.28) or in a ganglion in the root of the fifth cranial nerve. Figure 6.2 shows that the first-order neurons carrying sensations of pain, temperature, and light touch are observed to synapse at their level of entry in the spinal cord; whereas first-order neurons for deep touch synapse in the medulla oblongata, as do

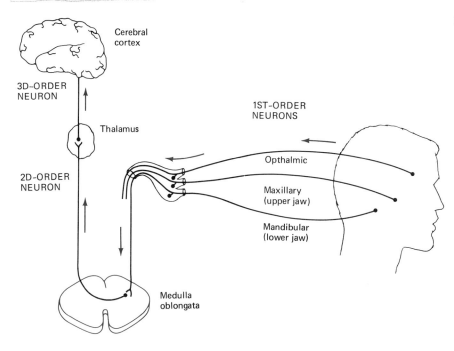

FIG. 6.2. *Train of three afferent neurons carries information from periphery of body to cerebral cortex.*

FIG. 6.3. *Specificity of ascending tracts in the spinal cord: lateral spinothalamic tracts; ventral spinothalamic tracts.*

CONTRALATERAL *relates to the opposite side, as when pain occurs on the side opposite to an injury* (L. contra, *opposite;* latus, *side*).

IPSILATERAL *refers to the same side, as when symptoms occur on the same side as the brain damage which causes them* (L. ipse, *same;* latus, *side*).

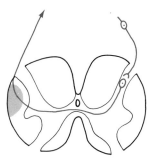

FIG. 6.4. *Contralateral dominance.*

Principles of the General Sensory System

the neurons from the fifth cranial nerve. In turn, most of the second-order neurons synapse in the thalamus, and the third-order neurons project their axons into the cerebral cortex, into a strip called the *somesthetic area*. Here, the sensation is identified, and through one or more sensory association areas within the cortex, as discussed later, the sensory information is perceived and brought to a higher level of consciousness.

(2) *Specificity of ascending tracts.* Ascending tracts are relatively specific for certain kinds of sensory information (Figure 6.3). Thus, whereas some sensations of pain and temperature are relayed through lateral spinothalamic tracts, light touch is relayed through ventral spinothalamic tracts. Other examples of this principle are given in Table 5.2.

(3) *Contralateral dominance.* Most of the sensory information which enters the body along one side of the body crosses over to the opposite side of the spinal cord (Figure 6.4). This occurs at the point where second-order neurons come into play. Most of these neurons pass from the side in which the sensation is received (ipsilateral) to the opposite side of the spinal cord (contralateral).

(4) *Ipsilateral representation.* Representation of some sensory information occurs on both sides of the thalamus and cerebral cortex. Not all of the sensory information is transmitted from first-order neurons to second-order neurons that pass to the opposite side of the spinal cord. Some sensory information is passed from first-order afferent neurons through side branches (collaterals) to second-order neurons which give rise to tracts on the same side of the spinal cord. In addition, fibers for certain kinds of information, such as deep touch and point discrimination, ascend into the brain without first crossing over.

(5) *Blending of sensations.* Some of the sensory information from first-order neurons may be passed through collateral branches into a second (or third) kind of tract which lies outside the tract into which the bulk of the pure sensation is directed. A mixing or blending of sensations may be initiated in this manner. It is possible that such mixing may explain why we are able to distinguish clearly such sensations as wetness, dryness, pleasure, tickling, and irritative tickling.

(6) *Sensory dissociation.* The principle of sensory dissociation concerns the level in the body to which first-order afferents are carried before they synapse: sensations of deep touch are carried all the way to the medulla oblongata by first-order neurons, whereas sensations of pain and temperature are transmitted into second-order afferents at the point where the first-order afferent enters the spinal cord (Figure 6.5).

(7) *Somatotopic registration.* As information passes into the central nervous system, a definite point-to-point relation is maintained between the peripheral and central nervous systems. For example, "heat information" from the leg is carried along the lateral edge of the appropriate tract whereas the same kind of information picked up by the arms is carried in the same tract, but by more centrally located fibers. Because of this spatial relation it is possible to map out the areas of the thalamus and the cerebral cortex that receive sensations from particular body regions. In this connection Figure 6.6 includes a map of the *somesthetic area* of the cerebral cortex. Note that the size of a cortical area is not proportional to the actual size of

the peripheral area which it represents, but rather to the number of fibers that innervate the peripheral area. This number, in turn, is related to the nature of the peripheral tissue. For example, the face and fingers have greater dexterity and variation in sensory as well as motor activities than the trunk or thigh. Accordingly, the face and fingers have proportionally greater representation in the brain.

Just as it is possible to map out the somesthetic area in terms of receptor areas in the periphery, so too it is possible to map out the skin receptor areas (*dermatomes*, Figure 6.7) in terms of spinal innervation. That is, if there is damage to some area of the spinal cord (as often happens in a car accident), even though this damage does not show on the outside of the body it is often possible to determine the specific site of damage within the spinal cord. This is possible because for each spinal sensory nerve, there is a corresponding region in the skin. Following an accident, for example, if there is a loss of sensation in dermatome number T8, it is very likely that the 8th thoracic nerve was damaged.

Sometimes, however, the damage sustained at one level of the spinal cord will not affect the corresponding dermatome. This safeguard lies in the fact that there are branches (collaterals) which pass from axons at one level of the spinal cord to make synapses with second-order afferents one or two levels higher or lower in the spinal cord.

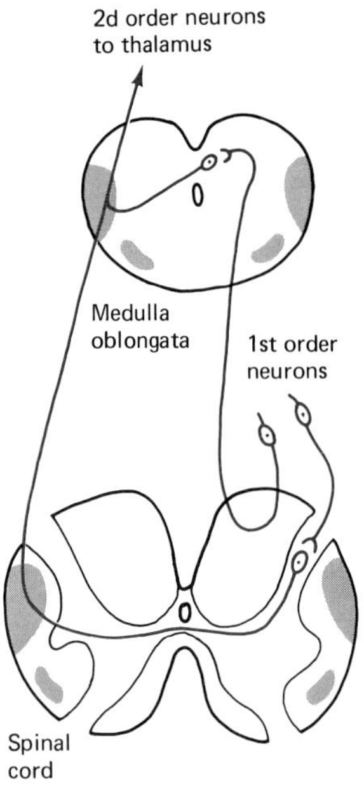

FIG. 6.5. *Sensory dissociation: some first-order neurons synapse in the medulla oblongata (top), rather than in the spinal cord.*

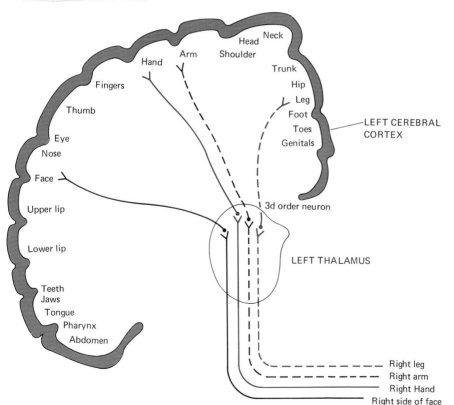

FIG. 6.6. *Somatotopic representation of body sensations.*

118
Principles of the General Sensory System

DERMATOME. *An area of the skin which receives sensory nerve fibers from a single spinal nerve (G. derma, skin; tomos, section).*

(8) *Reflex activities.* Reflex activities are involuntary physiological responses to stimuli applied at the periphery of the body. The mechanisms through which these reflex movements operate in conjunction with skeletal muscle are discussed in Chapter 7.

A reflex may occur at any level of the central nervous system. Let us consider what happens when a person suffers a painful blow to the solar plexus — the autonomic nerve plexus behind the stomach. Mechanical stimuli are sent from the skin to the brain, informing it of the location that was hit. Collateral branches along this route, from the first-order afferents, synapse at several levels in the spinal cord with interneurons and motor neurons. The abdominal wall there reflexly contracts. At the same time, the entire body reflexly curls in, the hands move protectively over the stomach, and

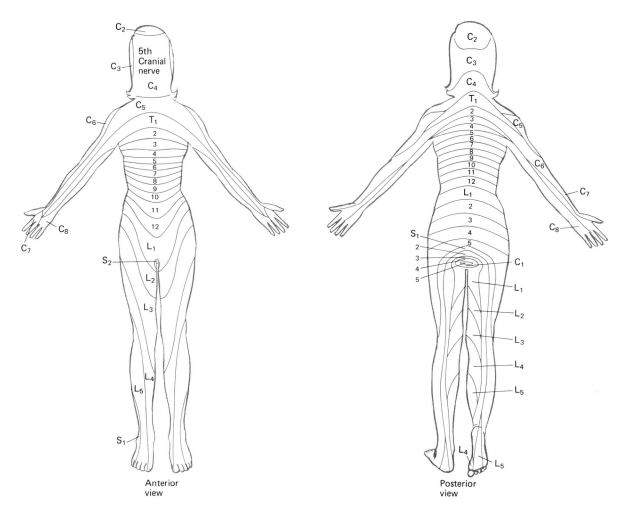

FIG. 6.7. *Skin receptor areas. Each region of the skin (dermatome) transmits sensory information to the central nervous system by way of a particular spinal nerve. See text for details.*

The General Sensory System

the face grimaces. The pain which is felt in the solar plexus is acknowledged in the brain, certain changes reflexly follow in the blood circulation, the brain may suffer a temporary loss of oxygen, and fainting may follow.

(9) *Conscious Activities.* Conscious awareness occurs after the sensory information has passed through the third-order afferents into the cerebral cortex. Possibly some conscious perception terminates in lower centers of the central nervous system. Pain, for example, is registered in the thalamus, where second-order afferents terminate. Most other sensations are believed to be perceived at a higher level of consciousness within one or more sensory association areas.

EVALUATION OF INJURY TO THE GENERAL SENSORY SYSTEM

When some region of the general sensory system becomes damaged—through accident, disease, or some other agent—it is often possible to determine the specificity, intensity, and extent of the resulting abnormality. The *specificity* of an injury may be tested by pinching (pain), touching the skin with a tube of hot or cold water (temperature), stroking the skin gently with tissue paper (light touch), or pressing deeply (crude touch).

The *intensity of sensory abnormality* is analyzed from the viewpoint of whether there is a total or partial loss or an increase in sensitivity with respect to one kind of sensation or another. Where pain is concerned, the appropriate descriptive terms are *anesthesia* (total loss of sensation), *hypoesthesia* (partial loss of sensation) and *hyperesthesia* (extreme sensitivity). Such terms as *thermoanesthesia* and the like, are used with regard to temperature. The *extent of abnormality* may be determined by testing various dermatomes and referring to appropriate charts.

ESTHESIA. *Capacity for perception or sensation (G. aisthesis).*

SOME ABNORMALITIES OF THE GENERAL SENSORY SYSTEM

Trigeminal Neuralgia (tic douloureux). This condition causes pain near the eyes (seldom to occasionally) and pain in the lower and upper jaw (commonly) when the trigeminal nerve is diseased. In severe cases, the pain may be relieved by injecting alcohol into one of the facial nerves or into the semilunar ganglion which houses the afferent cell bodies of the trigeminal nerve. In some cases, the sensory nerve is surgically severed.

Tabes Dorsalis. This condition is a form of syphilis in which the causative agent, *Treponema pallidum*, enters the dorsal horns of the spinal cord where it begins to destroy the sensory cells. Sensory capability is reduced or lost, where such destruction occurs. If the patient receives no treatment, the *Treponema* organism will continue its migration. Eventually it starts to destroy the brain. Walking may become abnormal, speech may be slurred, memory may be blurred, and all other higher faculties may be impaired.

Neuralgia. The suffix of this word, "algia," is derived from the Greek *algos*, pain. Neuralgia is any neural condition which is characterized by severe pain, usually of a throbbing or stabbing nature, such as trigeminal neuralgia or migraine headache. The abnormality may be caused by trauma, nerve damage, medicine, a metabolic disturbance, and the like.

Neuritis. Strictly speaking, "neuritis" should refer to an inflammation of one or more nerves. In most cases which are diagnosed as neuritis, however, there is no way of knowing whether the nerves actually are inflamed. Neuritis is characterized by neuralgia, hyperesthesia, paresthesia (abnormal sensations such as numbness, burning or pricking), anesthesia, paralysis, faulty reflexes associated with the affected nerve, and sometimes degeneration of the muscles involved. Neuritis may be caused by trauma, injury, drugs, and the like. Most often neuritis arises for unknown reasons, and commonly it may follow medication, including the administration of antitoxin to give immunity against tetanus toxins.

SUMMARY

1. The general sensory nervous system receives information from the skin of the body regarding temperature, pressure, and pain.
2. A train of three sensory neurons conveys the general sensory information into the cerebral cortex.
3. Most of the sensory information in one particular kind of ascending tract represents a single kind of sensation, such as pressure, although there is some mixing of different sensations, such as temperature and pressure.
4. Most of the sensory information which comes into the body on one side passes into the cerebral cortex on the opposite side.
5. The sensory information which comes into the body is represented in a point-to-point geography on the somesthetic strip of the cerebral cortex.

REVIEW QUESTIONS

1. Define the term "somesthetic." Describe how somesthetic stimuli are conveyed to the cerebral cortex.
2. Where, within a cross section of the spinal cord, are the ascending tracts located?
3. Where are the cell bodies of the second-order sensory neurons located?
4. Between what two regions of the brain are third-order sensory neurons located?
5. List nine principles which, collectively, describe or define the general sensory nervous system.
6. Define: paresthesia, neuralgia, tabes dorsalis, tic douloureux.

The Skeletal Motor System 7

COMPONENTS OF THE SKELETAL MOTOR SYSTEM
    Neuromuscular junctions
    Motor units
REFLEXES AND THEIR REGULATION
    Spinal reflexes
        Extensor reflexes
        Flexor reflexes
        Fine control over skeletal movements
    The pyramidal system
    The extrapyramidal system
    The final common path
EVALUATION OF THE SKELETAL MOTOR SYSTEM
    Abnormal reflexes
    Abnormal muscle behavior
    Abnormal skeletal movements

The skeletal motor system consists of all of the nerve cells that bring about the movements of skeletal muscles. To understand this system we must first say something about its basic components. After that we will describe the reflex, the functional unit of this system. In particular we will discuss spinal reflexes. Thereafter we will describe how spinal reflexes are modified by higher nerve centers through a so-called "final common path." In conclusion, we will consider how the performance of the skeletal motor system may be evaluated for diagnostic purposes.

COMPONENTS OF THE SKELETAL MOTOR SYSTEM

The skeletal motor system is made up of upper motor neurons, lower motor neurons, and certain motor neurons of the cranial nerves. The *cranial nerves* have their cell bodies in the brain. Their motor axons serve to move the eyes, eyelids, tongue, face, jaws, mouth, throat and voice box (see Table 5.4).

Components of the Skeletal Motor System

The *upper motor neurons* also have their cell bodies in the brain, but the axons of the upper motor neurons (unlike those of the cranial nerves) pass down from the brain to some level in the spinal cord. Here, upon discharging some of their neurotransmitter, the upper motor neurons serve either to excite or to inhibit lower motor neurons and thus to regulate the neural activities of lower motor neurons. *Lower motor neurons* have their cell bodies in the spinal cord. The axons of these neurons are carried out of the spinal cord by way of a nerve to outlying skeletal muscles, where they serve to activate the contraction of the muscles.

Neuromuscular Junctions

The only nerve axons which activate skeletal muscle contractions are the axons of the cranial nerves and the lower motor neurons. These axons branch out extensively upon entering a bundle of muscle fibers. The tip of each branch terminates on a *motor end plate* (Figure 7.1). A single motor end plate is intimately connected to the cell membrane of a single skeletal muscle cell. This connection, between the nerve fiber and the muscle fiber, is known as a *neuromuscular junction*. Skeletal muscle cells are activated to contract through synapses that occur at these neuromuscular junctions.

Motor Units

The entire collection of motor end plates from a single axon, together with all of the muscle fibers they innervate, is called a *motor unit*. A motor unit

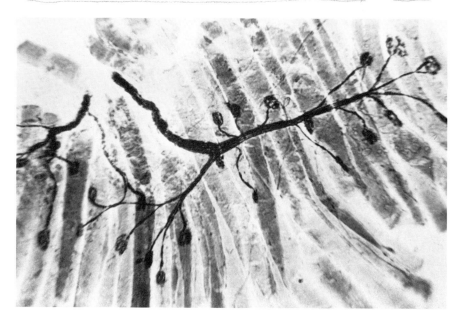

FIG. 7.1. *Motor end plates are visible at the ends of the nerve fibers, where they innervate skeletal muscle fibers.*

acts as an entity—a unit—in bringing about the contraction of a group of muscle fibers.

Motor units with only a few motor end plates innervate a correspondingly low number of muscle cells. Such motor units are used in movements that require delicate control, such as in eye or finger movements. Larger motor units—units which contain a large number of motor end plates—are used for gross movements that require relatively less precision, such as tilting the body or walking.

REFLEXES AND THEIR REGULATION

A reflex is essentially an involuntary movement of a muscle in response to a stimulus applied at the periphery of the body. A reflex may be executed through nerve pathways that lie entirely in the spinal cord, through nerve pathways that lie entirely in the brain, or through nerve pathways in both the brain and spinal cord.

Spinal Reflexes

A spinal reflex is executed entirely through the spine; the brain is not involved. The commonly tested spinal reflexes fall into two classes: extensor and flexor.

Extensor Reflexes. An *extensor reflex* is one in which a limb is extended anteriorly or posteriorly so that the angle in the joint is increased. See Figure 7.2.

The simplest spinal reflex is the *extensor reflex* (such as the *knee-jerk* or *stretch* reflex). The components of this reflex are: (1) a sensory receptor (*muscle spindle*); (2) an afferent neuron; (3) an alpha motor neuron which forms neuromuscular synapses with the skeletal muscle; and (4) the extensor muscle (quadriceps femoris).

The muscle spindle consists of two to ten modified skeletal muscle cells. These fibrous cells lie parallel to the unmodified skeletal muscle fibers. They lie deep within the belly of the muscle. Their modification is characterized by an accumulation of cell nuclei midway between the poles of the fiber in a region which lacks the usual crosswise stripes of skeletal muscle. In addition, they are innervated by afferent neurons and gamma motor neurons. Only the afferent neurons are directly concerned in the extensor reflex.

Upon stretching the muscle spindle—as happens, for example, when the kneecap (patella) is tapped while one knee is crossed over the other—afferent messages are transmitted across the spinal cord into the alpha motor neurons. A knee jerk follows (Figure 7.2).

An extensor reflex may be evoked from any extensor muscle of the body. All that needs to be done is to stretch the muscle suddenly, for example by striking either the muscle itself or its tendon.

In addition to the knee jerk, other commonly tested extensor reflexes include the Achilles reflex, in which tapping the back of the ankle causes the foot to extend; and the triceps reflex, in which tapping the back of the elbow causes the forearm to extend.

After a limb has been fully extended, the extensor muscle enters a state of relaxation. To prepare the muscle for another vigorous extension, a certain amount of muscle tone (a partial state of contraction) needs to be reestablished. This is done by activating the *gamma reflex loop* (Figure 7.3). The gamma reflex loop involves two additional anatomic components: a gamma motor neuron and the brain. As the muscle enters its state of relaxation, information is sent to the brain from the afferent fibers that are attached to the muscle spindle. In turn, information is sent from the brain to the gamma motor neurons and a certain amount of tension (tone) is reestablished.

Flexor Reflexes. **Flexion,** in contrast to extension, decreases the angle in a joint, the limbs of which are moving anteriorly or posteriorly (Figure 7.4).

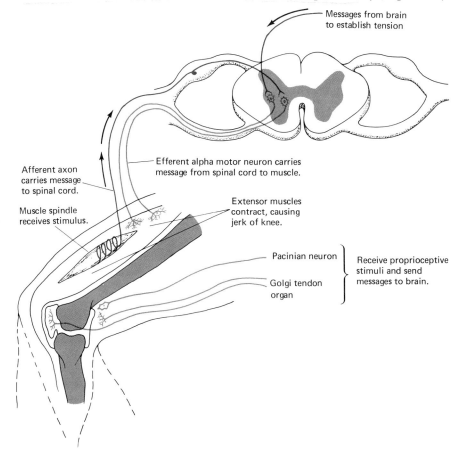

FIG. 7.2. *Extensor reflex. (The messages from the brain are described in Figure 7.3 and in the text.)*

The *flexor reflex* uses an arc made up of three consecutive neurons. Unlike the extensor reflex, which involves a two-neuron arc, the flexor reflex has an interneuron between the afferent and efferent neurons. The nature of the stimulus is also different. Pinching, pricking, sudden heat, or some noxious agent, rather than a sudden stretch of the muscle, initiates the reflex. The limb responds by withdrawing. Accordingly, the flexor reflex is also called the *withdrawal reflex*.

The peripheral stimulus in the flexor reflex is received by receptors in the skin and is carried by an afferent neuron to the spinal cord. In the spinal cord the impulse is transmitted to an interneuron, then to a motor neuron, and finally, to the flexor muscle (Figure 7.4).

Commonly tested flexor reflexes include flexion of the thigh and withdrawal of the foot, which is tested by pinching the foot; and flexion of the toes, which is produced by stroking the sole of the foot.

Fine Control Over Skeletal Movements. To produce smooth, coordinated moves in walking, running, and the like, the body relies upon a variety of fine control systems that are tuned in with flexor and extensor reflexes. We have already discussed one of these control systems, the gamma reflex loop, which re-establishes tension or tone in a muscle. Another control system—the *Golgi tendon reflex*—induces a state of relaxation and thus counteracts the gamma reflex loop. In actuality, these two control systems work together, cooperatively, to provide the appropriate muscle tone for a particular posture or movement.

The Golgi tendon reflex loop is made up of the following components: (1) an afferent neuron, with its receptor in the tendon of the joint (called a *Golgi tendon organ*); (2) an inhibitory interneuron in the spinal cord; and (3) an alpha motor neuron.

Its operation is as follows. When an extensor muscle in the leg begins to contract, the tendon to which it is attached is stretched. As a result, afferent

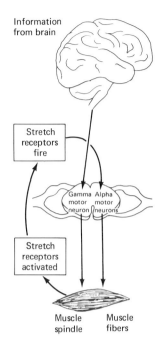

FIG. 7.3. The gamma reflex loop maintains a state of muscle tone.

FIG. 7.4. Flexor reflex circuit.

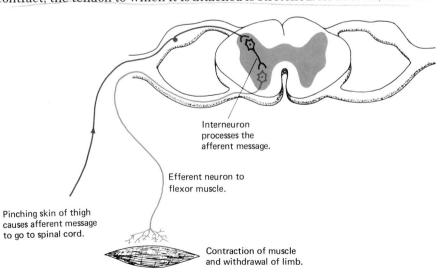

Reflexes and Their Regulation

impulses are generated in the Golgi tendon organ. These impulses pass into the spinal cord, where they activate the inhibitory interneuron. In turn, this interneuron sends inhibitory messages into the alpha motor neurons. As a result, fewer excitatory messages are sent along the alpha motor neurons to the muscle fibers. The muscle therefore contracts to a lesser extent than would otherwise be the case. It is believed that the chief function of the Golgi tendon reflex loop is to prevent overshortening of a muscle.

From our discussions on spinal reflexes it may be noted that these reflexes—whether triggered artificially, by testing; or naturally, as in walking—are executed primarily through the lower motor neurons. The same types of reflexes may be elicited—though in a more awkward form—from a decapitated animal.

Nevertheless, most of the body's natural skeletal movements, including spinal reflex activities, are regulated by upper motor neurons. These neurons are located entirely within the central nervous system. The tracts which they give rise to are classified either as pyramidal or extrapyramidal tracts (see Table 5.3).

The Pyramidal System

The *pyramidal system* consists of a pair of corticospinal tracts (Table 5.3) whose fibers stem primarily from large *Betz cells* which are shaped like

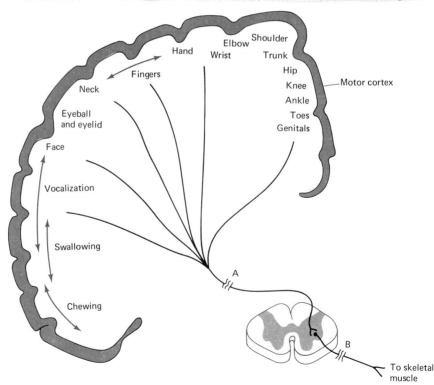

FIG. 7.5. *Somatotopic origin of motor action from the motor cortex. (A) and (B) are sites of possible damage; discussed in the text.*

pyramids and lie in the motor area of the cerebral cortex (Figure 5.20). To a much lesser extent, some of the fibers of the corticospinal tracts are derived from cell bodies which are widespread in other areas of the cerebral cortex.

From Figure 7.5 it may be noted that the cell bodies in the brain's motor area are organized somatotopically; that is, the regions of the body are each represented, in point-to-point fashion, in the motor area of the brain. Areas of the body that have considerable dexterity, such as the hands and face, are represented by more neurons than those body areas which do not perform fine and complex movements.

The pyramidal tracts leave the cerebral hemispheres and pass through the floor of the brain stem. As they journey through the medulla oblongata most of the million fibers from each pyramid cross over (form a chiasma) to the other side of the brain stem to form the paired *lateral corticospinal tracts*. About 10% of the fibers do not cross over, and continue on as the paired *ventral corticospinal tracts* (Table 5.3).

LATERAL CORTICOSPINAL TRACTS. *Either of two tracts which run along the edges of the spinal cord from the cerebral cortex.*

VENTRAL CORTICOSPINAL TRACTS. *Either of two tracts which run along the ventral side of the spinal cord from the cerebral cortex.*

The major function of the pyramidal system is to regulate skeletal movements that call for a great deal of dexterity. This system, which is highly developed only in mammals, allows the sort of finger and body movements one does not see in birds and lower vertebrates.

The Extrapyramidal System

The extrapyramidal portion of the upper motor system includes all the descending tracts except the corticospinal tracts. These extrapyramidal tracts may carry some fibers whose cell bodies lie in the motor area of the cerebral cortex, but most of their fibers are derived from nuclei located in the brain stem.

EXTRAPYRAMIDAL. *Refers to any descending nerve tract which lies outside the pyramids of the medulla oblongata (L. extra, outside).*

The exact role played by the extrapyramidal tracts in motor function is not known. Some authorities claim that fine control over muscular movements is their major role. Other authorities think that the extrapyramidal system provides a background motor pattern which sets up approximate positions of the body and appendages by adjusting large muscle groups, leaving the control of precision movements to the pyramidal system. Another view states that the extrapyramidal system regulates lower motor neurons which lead to the body, arms, and legs, whereas the pyramidal system controls lower motor neurons of the hands and feet. It has also been suggested that the extrapyramidal system is involved primarily with the maintenance of posture, and that coordination and voluntary movement are under the control of other parts of the nervous system.

The Final Common Path

Although the activities of skeletal muscle may be initiated from numerous, widespread nuclei, the lower motor neurons constitute the *final common path*. That is, all of the integrated information from the brain and spinal

Evaluation of the Skeletal Motor System

cord is funneled into the lower motor neurons, which innervate skeletal muscle.

The final common paths to the trunk and limbs originate in the ventral horns of the spinal cord. The final common paths to the eyelids, eyeballs, and tongue originate in the nuclei of their cranial nerves (see Table 5.4).

EVALUATION OF THE SKELETAL MOTOR SYSTEM

The skeletal motor system may be affected by a variety of disorders, including an injury, a stroke, an infection, a metabolic disturbance, or a drug reaction. The disorder usually shows up in one or more of three ways: abnormal reflexes, abnormal muscle behavior, and abnormal skeletal movements (Figure 7.5).

Abnormal Reflexes

Reflex activities may be entirely absent (*areflexia*), subdued (*hyporeflexia*), or exaggerated (*hyperflexia*). When they are subdued or absent, the cause may lie in the afferent or efferent path, insofar as spinal reflexes are concerned; or it may lie in the upper motor system, insofar as brain reflexes are concerned. Syphilis—which in its third stage destroys portions of the dorsal roots and brain stem—affects the afferent nervous system, and thus, indirectly, the motor system. The polio virus—which selectively destroys the cell bodies of the lower motor neurons and/or the medulla oblongata and other brain regions—affects the efferent paths directly.

Spinal reflex activities can be tested to tell whether the sensory and motor connections are intact. Other kinds of reflex activities can be tested to determine whether a disorder is present in the central nervous system. The *Babinsky reflex* is tested to determine whether the disorder lies in the pyramidal system. Using his finger or a stick, the tester draws a line along the lateral border of the sole of the subject's foot. In normal persons, the big toe bends downward. In babies—until the motor area of their cerebral cortex becomes well developed—and in adults suffering damage to the motor area, the toe is raised up instead.

Exaggerated reflexes may be due to a peripheral or central disorder. You may recall that the Golgi tendon reflex loop basically inhibits an extensor muscle from becoming overly contracted. Damage to this system—or damage to those upper motor neurons which carry inhibitory messages to the lower motor neurons (there are millions of these)—can bring about exaggerated reflexes.

Abnormal Muscle Behavior

TONE. *Muscle tension (G. tonos, tension).*

Normally, the skeletal muscle possesses *tone*—a condition in which many to most of the muscle cells are partially activated—even when there is no voluntary effort to contract them. Damage to the skeletal motor system may bring about partial to total loss in tone (*hypotonia* to *atonia*). In this situation the affected muscles are *flaccid*. If the tone is abnormally great (*hypertonia*), the affected muscles are *spastic*. In either case, paralysis sets in. *Paralysis* means that the ability to make certain voluntary movements is lost.

In *flaccid paralysis* there is no excitatory input from the afflicted motor neurons to their muscles. As a result, it is not possible to contract the corresponding muscles. This situation arises when damage occurs in the pyramidal system or in the lower motor neurons. Damage to the motor area of the cerebral cortex—from which the axons of the pyramidal system arise—may result in the loss of motor control over the corresponding regions of the body. Extensive damage to the motor cortex, brain stem, or upper spinal cord may result in total paralysis of the upper or lower extremities (*monoplegia*), the left or right side of the body (*hemiplegia*), both legs (*paraplegia*), or all four limbs (*quadraplegia*).

In *spastic paralysis* a number of *spasms* (involuntary contractions) are seen. The spasms may be persistent but weak (a *tonic* spasm); intermittent and short-lived (a *clonic spasm*); or persistent and strong (*tetanus*). Usually, these spasms are caused by damage arising in neurons that send inhibitory messages to the final common path. The damage may be in the Golgi tendon reflex loop or in the extrapyramidal system. In the latter case there are often exaggerated extensor activities in the legs and exaggerated flexor activities in the arms.

PARALYSIS. *A complete or partial loss of function (G. paralyein, to weaken at the side).*

FLACCID. *Relaxed, flabby, or without muscle tone (L. flaccidus).*

SPASTIC. *Characterized by spasmodic, convulsive movement (G. spastikos).*

Abnormal Skeletal Movements

Skeletal movements are ordinarily performed voluntarily. However, when damage arises in the skeletal motor system, various abnormal movements may occur (Figure 7.5). Among these abnormal movements are spasms, which were discussed above. Usually, when the extrapyramidal system is damaged, one or more of the following kinds of abnormalities may be observed. *Tremors* are alternate contractions of opposing sets of muscles, such as flexors and extensors, or abductors and adductors. *Choreiform motion* is rapid, jerky movements, such as jerking a shoulder upward, unintentional smiles, and facial twitches. *Athetosis* takes the form of successive, slow, writhing, involuntary movements—overextending or underextending the arm or fingers, for example, in reaching for an object. *Dystonia* is characterized by involuntary, grotesque postures.

A common extrapyramidal disorder is *Parkinson's disease,* a disease which usually appears only late in life. Spasticity and tremors are notable features, as are choreiform motions and athetosis. Usually there is a lesion (damaged area) in the basal ganglia and nearby centers. Similar lesions and abnormal skeletal movements characterize Sydenham's chorea, which occurs in children and adolescents, and Huntington's chorea, which occurs in people of middle age.

Abnormal movements are seen in the eyes and tongue when there is damage to their motor neurons. The tongue may deviate to one side or the

CHOREIFORM. *Resembling chorea, an involuntary, irregular muscular action of the extremities and face (G. choreia, dance).*

DIPLOPIA. *Double vision* (G. diploos, *double*; opsis, *vision*).

other, or remain outside the mouth. The eyes may fail to follow the movement of an object or may not move in unison, resulting in a squint (*strabismus*). Or the eyes may fail to align themselves simultaneously along a common axis, resulting in double vision (*diplopia*). Sometimes the jaws deviate from their natural positions, and the person may have difficulty in retaining saliva, in speaking, in blinking, and in achieving desired facial expressions.

SUMMARY

1. The skeletal motor system consists of cranial nerves, upper motor neurons, and lower motor neurons. The motor fibers of the cranial nerves send motor information to various regions of the body, including the eyes, nose, ears, mouth, throat, and viscera. The upper motor neurons send information down the spinal cord to the lower motor neurons. A lower motor neuron receives information from various kinds of sensory neurons and upper motor neurons. The lower motor neuron integrates this information and sends appropriate messages into a collection of skeletal muscle fibers.
2. A single lower motor neuron, together with all the muscle fibers it controls, is called a motor unit.
3. An extensor reflex brings about an increase in the angle of a joint around which a limb moves in an anterior-posterior direction; a flexor reflex effects a decrease in the angle of the joint.
4. The gamma afferent reflex loop helps re-establish tone in a muscle during the relaxation phase following a vigorous contraction.
5. The Golgi tendon reflex loop tends to induce a state of relaxation in a muscle which is undergoing a vigorous contraction.
6. The pyramidal system regulates skeletal muscle activity directly by way of nerve cells whose cell bodies lie in the motor strip of the cerebral cortex.
7. The extrapyramidal system modifies the actions of the pyramidal system.
8. Damage to the pyramidal or extrapyramidal system can bring about spastic paralysis, while damage to the lower motor neurons can induce flaccid paralysis.

REVIEW QUESTIONS

1. List the major anatomic components of the general skeletal motor system.
2. Describe the pathway of nerve impulses in a flexor reflex from the point of stimulation to the point of muscle action.
3. In what ways does an extensor reflex differ from a flexor reflex?

4. What is the role of the gamma afferent reflex loop?
6. What is meant by the "final common pathway" in the general skeletal motor system?
7. How does flaccid paralysis differ from spastic paralysis?
8. Define: pyramidal system, extrapyramidal system, Parkinson's disease.

8 The Visceral Sensory and Motor Systems

THE VISCERAL SENSORY SYSTEM
    Evaluation of the visceral sensory system
        Referred pain
        Phantom sensations
VISCERAL MOTOR SYSTEM (AUTONOMIC NERVOUS SYSTEM)
    Characteristics of the autonomic nervous system
        The sympathetic nervous system
        The parasympathetic nervous system
        Autonomic nerve centers
    Evaluation of the autonomic nervous system

The viscera of the body, such as the glands, blood vessels, heart, lungs, and intestines, are innervated by sensory and motor nerve fibers. We begin this chapter with a description of the visceral sensory system and a discussion of the evaluation of this system. We then discuss the visceral motor system, which is commonly called the *autonomic nervous system*.

THE VISCERAL SENSORY SYSTEM

The visceral sensory system is the afferent pathway of the autonomic nervous system. This pathway is used to carry various kinds of sensory information to the brain. Such information includes blood pressure, blood pH, rate and depth of breathing, fullness of the urinary bladder, and pain. With regard to pain, the visceral sensory system serves to monitor such things as heartburn, nausea, appendicitis, and the distension of the ureters by kidney stones.

As in the general sensory system, sensations from the viscera are integrated by interneurons in the spinal cord and are carried through tracts to higher regions in the central nervous system, including the thalamus and cerebral cortex. It is not possible to map these regions topographically, as

The Visceral Sensory and Motor Systems

is done for the general sensory system and skeletal motor system (described earlier). Nor is it possible to talk about mapping the visceral sensory system in terms of spinal nerves, as was done for dermatomes of skin. The autonomically-innervated tissues are simply too irregularly spread to permit mapping.

Evaluation of the Visceral Sensory System

No attempt is generally made by neurologists to evaluate the performance of the visceral sensory system. However, when an individual feels certain kinds of pain, the sensation can be used to locate the source of the pain and possibly to correct it (see Figure 8.1).

Referred Pain. In some cases, as Figure 8.1 suggests, pain is felt in an area other than where the damage has occurred. This kind of pain is called

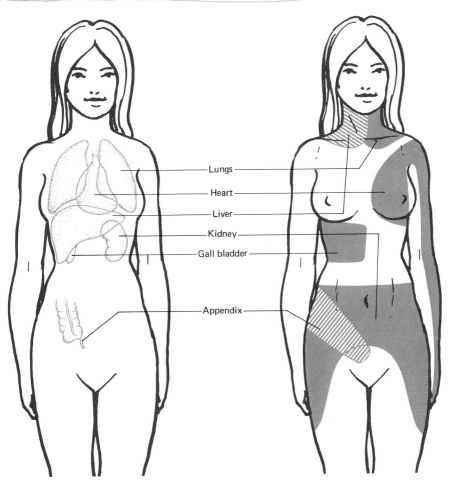

FIG. 8.1. *Disorders in the visceral organs (left) can cause referred pain — pain that is "sensed" in other body regions (right).*

referred pain. Heart pain, for example, is usually felt as though it is coming from the chest muscles. In angina pectoris, the pain commonly passes from the chest through the left arm to the pinky and ring finger.

The reason why pain from viscera may be "referred" to outer body regions may be explained as follows, using angina pectoris as an example. The afferent axons of the heart are carried in nerves which also contain the afferent and efferent axons that innervate the chest muscles and left arm. It is believed that the impulses that travel through the nerves from the painful viscera tend to lower the threshold of discharge in the neighboring sensory fibers—the sensory fibers from the chest wall and left arm. It is also believed that many interneurons are shared by the skeletal and visceral tissues in this case.

Phantom Sensations. Related to referred pain is the *phantom sensation*. Following an amputation, the amputee often "feels" the presence of the removed limb. A satisfactory explanation for this sensation is not available. It is thought, however, that certain sensory cells in the brain are activated. In this connection, we should note that the body regions with the greatest somatotopic representation in the brain give rise to longer lasting phantom sensations than regions with lesser representation (for example, fingers versus thigh).

VISCERAL MOTOR SYSTEM (AUTONOMIC NERVOUS SYSTEM)

The visceral motor system is usually referred to as the autonomic nervous system because most of its actions are not under conscious control. The activities of this nervous system occur whether we are awake or asleep.

VISCERA. *The soft internal organs in the abdominal and thoracic cavities* (L.).

Although the autonomic nervous system is concerned primarily with viscera, many of its functions are integrated with the activities of the skeletal motor system. The sight of a delicious meal, for example, may cause salivation (visceral system) and opening of the mouth (skeletal system). So too, emotional disturbances may be accompanied by an increased rate of breathing (requiring the contraction of skeletal muscle in the rib cage and diaphragm) as well as an increase in the rate and force of heartbeat (visceral system).

Characteristics of the Autonomic Nervous System

The autonomic nervous system has a number of features that distinguish it from the other subdivisions of the nervous system. First, it sends out two sets of fibers—one excitatory, the other inhibitory—to most of the organs it innervates.

Second, whereas in the skeletal motor system a single motor neuron extends from the spinal cord to the skeletal muscle, in the autonomic nervous system two motor neurons pass successively from the spinal cord to the target tissues (heart muscle, smooth muscle, glands). The first of these two neurons has its cell body in the spinal cord and sends its axon (its preganglionic fiber) into a ganglion outside the spinal cord. Here, telodendria of the *preganglionic fibers* form synapses with several target cells which, in turn, send their axons (*postganglionic fibers*) into the viscera. See Figure 8.2.

Another characteristic is that it is possible to subdivide the autonomic nervous system because of the way in which the ganglia are distributed. One of its divisions—the *sympathetic system*—has its ganglia located very near the spinal cord in the thoracic and lumbar regions of the body. See Figure 5.28. The second subdivision—the *parasympathetic system*—has its ganglia located near or within the viscera it innervates.

Whereas acetylcholine is the main neurotransmitter liberated from every preganglionic fiber of both the sympathetic and parasympathetic systems, a different situation holds true for the postganglionic fibers. In general, the major neurotransmitter from postganglionic fibers in the parasympathetic system is also acetylcholine. However, in the sympathetic system two kinds of postganglionic neurotransmitters are recognized and they help divide the system into two functional subdivisions. One subdivision is the adrenal gland. From a functional viewpoint this gland behaves like a sympathetic postganglionic nerve. Although it secretes small amounts of noradrenaline, its main secretion is adrenaline. (Adrenaline and noradrenaline elicit different kinds of responses from certain arteries of the body as shown in Table 8.1.) The other functional subdivision—the postganglionic nerves from the remainder of the sympathetic system—secrete mostly noradrenaline and only very small amounts of adrenaline.

Table 8.1 shows the effects of acetylcholine, noradrenaline, and adrenaline upon the viscera. Note that acetylcholine may be excitatory or inhibitory, as may noradrenaline.

In contrast to the skeletal motor system, where every skeletal muscle cell receives a nerve fiber, many of the target cells in the viscera are not directly innervated. Instead, they depend upon the diffusion of neurotransmitter from nearby postganglionic end branches.

Also in contrast to the skeletal motor system, the target cells of the postganglionic fibers (the heart muscle cells, smooth muscle cells, and gland cells) do not degenerate when their nerve supply is cut off. Skeletal muscle fibers temporarily atrophy (become weak and degenerate) when the axon which innervates them is cut or destroyed. When the cell body of the motor neuron is destroyed, they become permanently damaged, as, for example, in certain cases of polio.

ATROPHY. *Reduce in size* (G. atrophia, *want of food*).

The Sympathetic Nervous System. Generally speaking, the activities of the sympathetic system are related to excitation, physical exertion, or emotional disturbances—to the person on the go. This statement may be better appreciated if we consider the separate actions of sympathetic stimulation.

Several important things happen when large amounts of adrenaline are released from the adrenal gland and large amounts of noradrenaline are

SYMPATHETIC. *The thoraco-lumbar nerves of the autonomic nervous system* (G. sympathetikos).

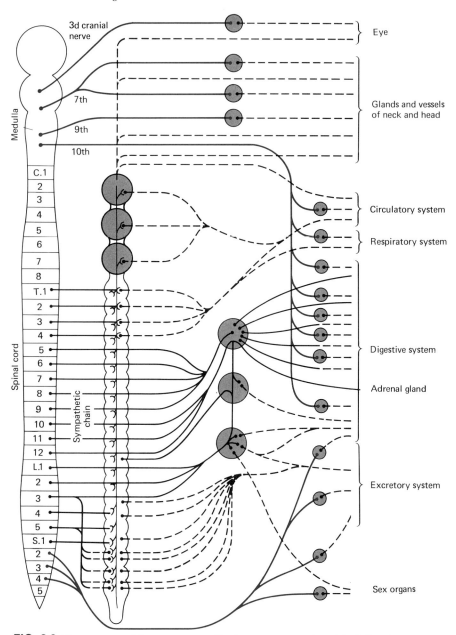

FIG. 8.2. *The nerves of the parasympathetic nervous system (dark lines) work cooperatively with those of the sympathetic nervous system. In both systems, preganglionic nerve fibers synapse within ganglia where they activate the postganglionic fibers which innervate the visceral systems. The small circles indicate parasympathetic ganglia; the larger circles at the center of the figure indicate the prevertebral ganglia (see also Figure 5.28).*

TABLE 8.1. Autonomic Nerve Functions

Target Tissue	Effect of Sympathetic System (Noradrenaline)	Effect of Parasympathetic System (Acetylcholine)
Heart	Increases heartbeat and force of contraction	Opposite effect
Blood vessels	Noradrenaline constricts arteries in skin and viscera. Adrenaline dilates arteries in heart and skeletal muscle	Dilates arteries, but the effect is small
Bronchial muscle	Enlarges the air passageway	Opposite effect
Stomach and intestine	Reduces motility and secretions	Opposite effect
Sweat glands	Increases secretory activity (acetylcholine is liberated)	Similar effect
Urinary bladder	Inhibits urination—relaxes bladder and closes outlet (internal sphincter)	Opposite effect
Pupil	Pupil opens up (radial muscles only are innervated)	Pupil closes (circular muscles only are innervated)
Lens	No effect	Lens thickens, due to contraction of ciliary muscles
Penis		Erection
Accessory sex glands	Ejaculation of semen	Increase in secretory activity

released from the rest of the sympathetic system. One is that the rate and force of the heartbeat increase. Also, the arteries in the heart and skeletal muscle dilate (widen), allowing more blood to flow through these areas. At the same time, the arteries in the skin and viscera constrict, thus allowing the blood to be shunted into the heart muscle (not the heart chambers) and skeletal muscles. The bronchioles (air tubes leading into the lungs) dilate, allowing more air to pass into and out of the lungs. In addition, the sugar concentration in the blood increases, providing a greater source of energy to the muscle cells of the heart and skeleton. Other actions of the sympathetic nervous system are listed in Table 8.1.

The Parasympathetic Nervous System. In contrast to the sympathetic system, the activities of the parasympathetic system are related to relaxation. When the parasympathetic system is active, the force and rate of heartbeat decrease, as does the level of sugar in the blood, and the rate and volume of breathing. Digestive activities, on the other hand, are promoted. The parasympathetic system suggests the picture of a relaxed person who has just eaten a heavy meal.

Strictly speaking, however, the two divisions of the autonomic nervous

PARASYMPATHETIC. *The cranial-sacral nerves of the autonomic nervous system (G. para, beside or near).*

CARDIOVASCULAR.
Relating to the circulatory system (G. kardia, *heart;* L. vasculum, *vessel*).

system work cooperatively, in an integrated fashion, usually in the direction of benefiting the economy of the body.

Autonomic Nerve Centers. Many of the activities of the autonomic nervous system are directed through centers in the brain. These centers are regions from which certain excitatory or inhibitory messages arise. Thus, a *cardiovascular center* in the brain's hypothalamus and medulla oblongata regulates the rate and force of heartbeat and the blood pressure. *Respiratory centers*—controlling the rate and depth of breathing—are situated in the medulla oblongata and pons. *Feeding* and *thirst centers* are present in the hypothalamus, as are *vomiting* and *salivation centers*. Also in the hypothalamus is a *thermoregulatory center*, which regulates the raising of goose bumps, shivering, sweating, and the shifting of blood to the skin or to the interior of the body in response to environmental changes.

From another point of view one might also speak of an *emotion center* located in the hypothalamus. This center is triggered by nerve impulses which arise in the limbic lobe following the integration of sensory information in various other regions of the brain. Regardless of the type of emotion an individual may express, the hypothalamus seems to serve as the last link —the final common path—to the expression itself.

Evaluation of the Autonomic Nervous System

Examining the autonomic nervous system is not a clear-cut procedure. This is because problems may exist elsewhere in the nervous system and merely seem to appear as though they are in the autonomic nervous system. High blood pressure and the overproduction of acid in the stomach might suggest a faulty control in the autonomic nervous system, but may in fact be due to some metabolic problem or to a problem elsewhere in the body, including the central nervous system.

In general, however, information may be obtained about the blood pressure and heartbeat, digestive problems, ability of the iris to respond to near and distant objects, and the ability to shed tears and salivate. A misfunction of the autonomic nervous system may be diagnosed this way.

SUMMARY

1. The visceral sensory nervous system sends afferent stimuli to the brain regarding the condition of the viscera. When the viscera are diseased, pain sensations arise, but these sensations are "referred" to certain surface areas of the body whose nerves harbor the visceral sensory fibers.
2. Phantom sensations are related to "referred pain." Thus an amputee who experiences a phantom sensation in the amputated limb probably has information carried to the brain in nerves which once carried nerve fibers to that limb.

3. The autonomic nervous system is the motor component of the visceral sensory system. Preganglionic fibers from the autonomic nervous system synapse with postganglionic nerve cells in ganglia that are located outside of the spinal cord.
4. The sympathetic nervous system generally puts an individual into a "go" state while the parasympathetic system tends to place a person into a "relaxed" state.
5. Adrenaline (a) increases the rate and force of heartbeat; (b) dilates the arteries of the heart and skeletal muscle; (c) constricts the arteries of the skin and viscera; (d) widens the bronchi; and (e) increases the availability of sugar to the heart and skeletal muscles.
6. Autonomic nerve centers are present in the brain stem through which emotions are sensed and expressed and through which breathing, circulation, eating, and other visceral activities are regulated.

REVIEW QUESTIONS

1. What might be some functions of the visceral sensory nervous system (other than to signal pain) with respect to the following: the circulatory system; the stomach; the urinary bladder.
2. What is referred pain? How is referred pain related, if at all, to phantom sensations?
3. To what region(s) of the body is pain referred in a typical case of angina pectoris?
4. What are the main distinctive characteristics of the autonomic nervous system?
5. What are the chief differences between the sympathetic and parasympathetic nervous systems?
6. List the five principal effects of adrenaline upon the visceral (autonomic) nervous system.

 The Special Sensory Systems

HEARING
    Conscious activities, reflexes, and hearing
    Evaluation of hearing ability
VISION
    Conscious activities, reflexes, and vision
    Evaluation of visual ability
        Non-nervous eye defects
SMELL
TASTE

In human beings, the special senses are those of hearing, vision, taste, and smell. Each of these senses involves the functioning of unique receptor organs and neural pathways. They are described in the rest of this chapter.

HEARING

The ear is made up of three components: the outer ear, middle ear, and inner ear (see Figure 9.1).

The *pinna*, or *outer ear* collects sound waves and directs them into the middle ear. The practical value of the pinna can be shown by a simple experiment. Blindfold yourself and twist your pinna so that it closes the opening to the middle ear. Now have someone produce a sound a foot or two away. Not only is the magnitude (loudness) of the sound reduced, but its location is practically indeterminable.

The *middle ear* transmits sound waves from the pinna to the inner ear. It consists essentially of three bones—the *auditory ossicles*—and a *tympanic membrane* (Figure 9.1).

The *inner ear* contains the nervous elements of the hearing system. These elements are housed in a fluid-filled coiled structure, called the *cochlea* (see Figure 9.2). Sound waves are directed from the *stapes*, the innermost bone of the middle ear, into an *oval window* at the base of the cochlea.

COCHLEA. *The sensory receptor organ in the inner ear (L. snail shell).*

STAPES. *The auditory ossicle which "steps" into the oval window of the inner ear (L. a step).*

The Special Sensory Systems

From here the sound waves travel to the apex of the cochlea, along an outermost passageway—the *vestibular duct*—and back to the base of the cochlea, along an innermost passageway—the *tympanic duct*. The vestibular and the tympanic duct are joined at the apex. In the course of their travel, the sound waves enter a third passageway—the *cochlear duct*—which is located between the vestibular and tympanic ducts.

As sound waves pass up the vestibular duct and then down the tympanic duct, they set up vibrations within the fluid in the ducts. The resulting mechanical energy easily deforms two very thin membranes—the vestibular and tectorial membranes. The movement of these membranes causes fine hairlike tips on auditory receptor cells, often called *hair cells*, to bend. As a result, receptor potentials are generated. These potentials, upon achieving

TYMPANIC. *Describes various anatomical components of hearing* (G. tympanon, *drum*).

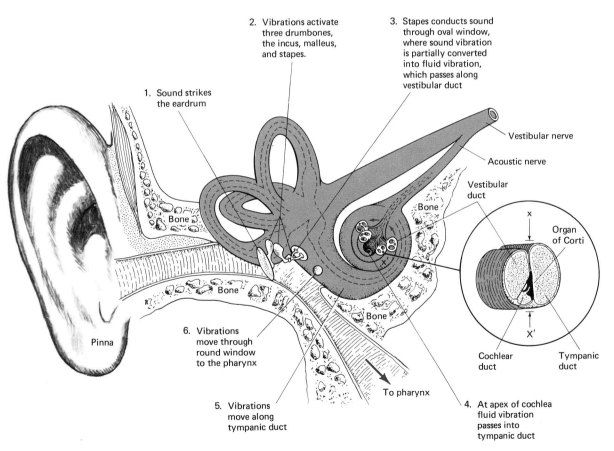

FIG. 9.1. *Anatomy of the ear.*

Hearing

EUSTACHIO. *An Italian anatomist (1520–1574) who first described the auditory tube.*

a critical value, set up action potentials within bipolar neurons whose axons (about 30,000 of them) are carried in the auditory nerve (the eighth cranial nerve) to the brain.

That amount of sound energy which is not transferred into the cochlear duct continues its travel along the tympanic duct to the base of the cochlea, where it enters the *eustachian tube* from the *round window*. From the eustachian tube, the residual sound energy is passed out of the nose or mouth, resulting in pressure equalization.

Conscious Activities, Reflexes, and Hearing

In the discussion of the general sensory system, mention was made of two responses to a sensory stimulus: a reflex response and a conscious response. This concept holds true for any kind of sensory stimulus; a stimulus may evoke an involuntary action—a reflex—or a voluntary action—a conscious activity. From the psychological point of view, the stimulus may enter some regions of the brain without reaching the individual's consciousness, but nothing is known about how this might take place.

In the conscious pathway for hearing, the afferent messages are carried to the medulla oblongata and then to the thalamus. From here they are relayed to the primary auditory cortex. Here, a conscious interpretation is made of the meaning that the sound conveys (see Figure 9.3).

In the reflex pathways for sound, the afferent messages are relayed from cochlear nuclei within the pons into regions elsewhere in the brain stem. Information, in turn, is sent efferently to the extrinsic eye muscles, the skeletal muscles of the body and face, and to the throat and voice box in the respiratory system. The appropriate motor responses are then made.

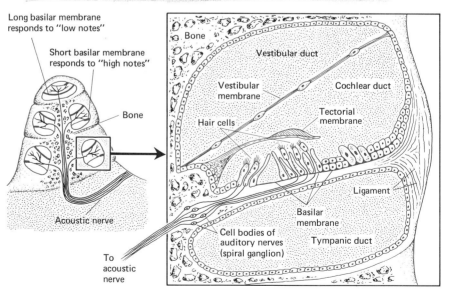

FIG. 9.2. *Cochlea of the ear. Detail shows the receptor cells of the hearing apparatus.*

Evaulation of Hearing Ability

Difficulty in hearing may be due to conduction deafness or nerve deafness. *Conduction deafness* may be due to a variety of causes. It may be caused by a blockage of the external ear by ear wax produced by cerumen glands. Blockage of any of the sound pathways by serum or pus due to infections (otitis) may also cause conduction deafness. Another cause is blockage of the middle ear passageways due to an abnormal growth of the adenoid tonsils, which ordinarily end below the eustachian tubes, along the upper extension of the pharynx. Scars or growths on the tympanic membrane, as well as hardening of the stapes (otosclerosis), may also cause conduction deafness.

Nerve deafness may be due to damage within the cochlea, the auditory nerves, or the hearing centers in the brain. It may be distinguished from conduction deafness by means of various tests. In the *Rinne test*, a vibrating tuning fork is placed on the mastoid bone of the skull, near the ear.

CONDUCTION DEAFNESS. *Loss of hearing due to some blockage within the passageway for sound waves.*

NERVE DEAFNESS. *Loss of hearing due to a defect in the neural region of the inner ear or the hearing centers in the brain.*

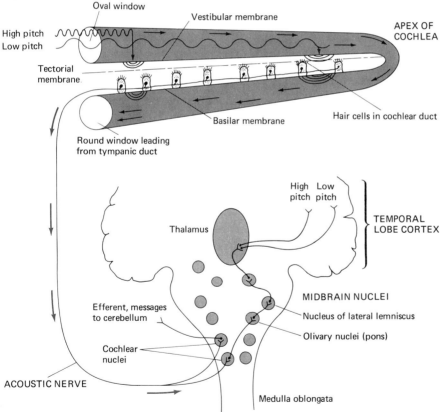

FIG. 9.3. *Sound pathways to and within the brain.*

Sound waves traveling through fluid press on vestibular and basilar membranes, causing hair cells to brush against tectorial membrane; impulses are generated in the acoustic nerve and are transmitted into the cerebral cortex.

When sound is no longer heard, the fork is held in the air close to the opening of the outer ear. If hearing is normal, the sound will be heard again, usually for about as long as it was heard before. In conduction deafness the sound is heard longer through bone than through air. In nerve deafness the sounds are diminished, but they are usually heard longer through air than through bone.

The *Weber test* is also used to distinguish conduction deafness from nerve deafness. In this test the vibrating fork is placed on the midline of the skull. In cases of nerve deafness, hearing (as would be expected) seems to be better in the normal ear. In conduction deafness, however, the individual has the illusion of normal hearing in the defective ear. The reason for this is that auditory messages received on one side of the body are sent mainly into the opposite side of the cerebral cortex. In conduction deafness the individual *thinks* he is hearing normally with the defective ear. In nerve deafness this illusion does not occur, because the individual does not "feel" the sound that is received through the deaf ear.

Conduction deafness can usually be cured. Wax may be removed, antibiotics may be used to treat infection, adenoid masses may be removed, and a hardened stapes bone may be replaced by a teflon substitute. Both nerve and conduction deafness may be helped in some cases by a hearing aid.

VISION

Rods (about 100 million of them) and *cones* (about 10 million) are the sensory receptors of sight. Rods are especially receptive to dim light and shadows; cones to bright light and color. Rods and cones are two of the many

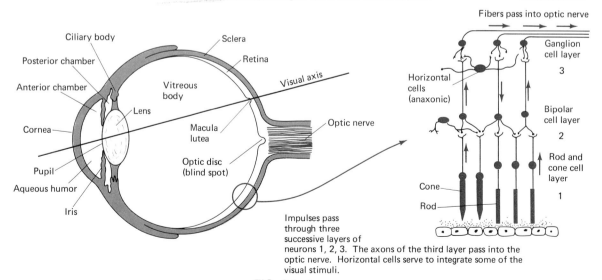

FIG. 9.4. *Structure of the eyeball and retina.*

kinds of cells in the retina (Figure 9.4), which is a component of the eyeball.

In the *retina*, a great deal of integration occurs with regard to color, shadow, texture, and other visual information. The retina is supported by the *sclera*, which constitutes the outermost wall of the eyeball. The sclera is visible as the "white" of the eye. In the front of the eye it merges with the *cornea*, a clear membranous structure though which light passes on its way to the retina.

SCLERA. *The "white" of the eye* (G. skleros, *hard*).

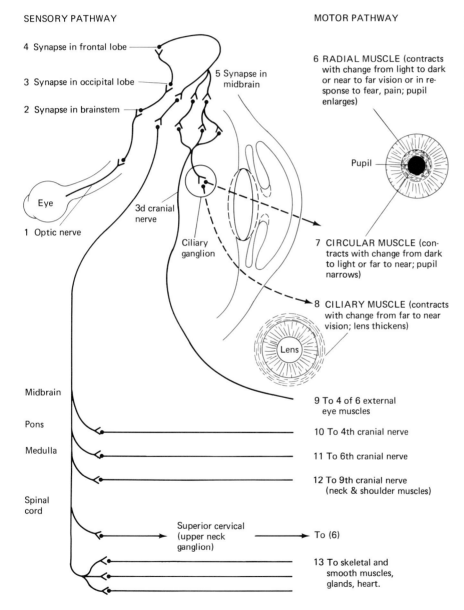

FIG. 9.5. *Adjusting the size of the pupil and accommodation of vision to near and far objects. Figure shows sensory pathway (1–4) and motor pathway (6–13).*

RADIAL MUSCLE. *A muscle arranged in spokelike fashion in the iris of the eye* (L. *radius,* spoke or ray).

CILIARY. *Describes certain structures of the visual apparatus, such as the ciliary body, the ciliary canal, and the ciliary ganglion* (L. *resembling an eyelid or eyelash*).

In very bright light, only a small fraction of light rays are allowed to continue toward the retina. In dim light, on the other hand, more of the incident light is permitted to pass into the eye. The structure that accounts for the regulation of light entering the eye is an adjustable diaphragm, the *iris* (see Figure 9.4). The contraction of a *radial muscle* in the iris increases the size of the pupil, whereas contraction of the circularly-arranged muscle fibers in the iris reduces the size of the pupil (see Figure 9.5).

Directly behind the iris lies the *lens*, a proteinaceous structure attached to the eyeball by fibers from the *ciliary body*. A muscle within the ciliary body, the *ciliary muscle*, regulates the thickness of the lens (Figures 9.4 and 9.5). When this muscle is relaxed—as happens when looking into the distance—the lens is relatively flat. In this case a thin ligament attached between the lens and ciliary body is relatively tensed up. When the ciliary muscle contracts—as happens autonomically when looking close-up—the lens becomes thicker, and the eye is said to *accommodate* for near, rather than far, vision.

The eye contains two kinds of fluid. The more voluminous fluid, called the *vitreous body* or *vitreous humor*, occupies a chamber behind the lens. The

FIG. 9.6. *Muscles of the eyeball and eyelids.*

EYELID MUSCLES	CRANIAL NERVE	ACTION
Levator palpebrae superioris	third	Raises upper eyelid
Orbicularis oculi	seventh	Closes both eyelids

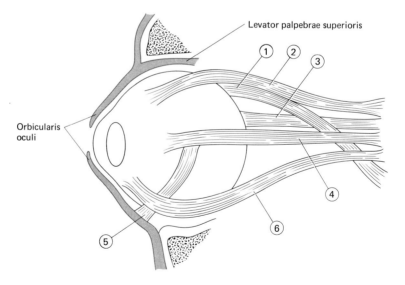

EXTERNAL EYE MUSCLES	CRANIAL NERVE	TURNS EYE:
1 Superior rectus	third	up and toward nose
2 Superior oblique	fourth	down and away from nose
3 Medial rectus	third	toward nose
4 Lateral rectus	sixth	away from nose
5 Inferior oblique	third	up and away from nose
6 Inferior rectus	third	down and toward nose

less abundant and less viscous fluid, called the *aqueous humor,* lies in front of the lens. The aqueous humor is a secretory product of the capillary networks in the ciliary body. Rich in glucose, the aqueous humor nourishes the cornea and the lens. (For this reason, diabetics, who fail to regulate their blood sugar in a normal way, often encounter visual problems.)

When aqueous humor is produced excessively, or the *canal of Schlemm* (the venous duct which drains the fluid) becomes blocked, the fluid pressure within the eyeball increases above normal. The resulting condition is called *glaucoma*, and it can be treated with certain drugs, such as physostigmine and pilocarpine.

SCHLEMM. *A German anatomist (1795–1858) who described the canal which lies in the junction of the sclera and cornea of the eye.*

The eyeballs are so important to man that they receive considerable anatomical protection. They are protected from physical harm by being embedded in bony sockets. They are protected from excessive light by the eyebrows. Protection from injury is afforded by the eyelids and eyelashes. The eyes are protected from wear and tear by the tear fluid from the lacrimal glands and by the *conjunctiva*, a transparent membrane that covers the cornea and the inner surface of the eyelids.

The movements of each eyeball are carried out by six muscles (Figure 9.6). Four of them (the *rectus muscles*) are located at positions corresponding to east, west, north, and south. The other two pass obliquely along the outer wall of the eyeball. Whereas the rectus muscles turn the eyeball up and down and side to side, the *oblique muscles* permit a smooth rotation of the eyeball.

RECTUS. *Straight (L.).*

OBLIQUE. *Slanting (L.).*

The six muscles are operated by cranial nerves. The fourth cranial nerves regulate the superior oblique muscles, which direct vision downwardly and outwardly. The sixth cranial nerves regulate the lateral rectus muscles for vision to the side. The third cranial nerves regulate the medial rectus (for looking inwardly), the superior rectus (for looking up and in), the inferior oblique (for looking up and out), and the inferior rectus (for looking down and inwardly).

The visual information received by the rods and cones is partially processed by interneurons within the retina (Figure 9.4). The visual information is passed from the rods and cones into second-order and third-order afferent neurons. The axons of the third-order neurons form the surface of the retina, and as such they can be seen with an opthalmoscope. All of their axons converge upon the *optic disc* and leave the eyeball as the *optic nerve* (Figure 9.7).

OPTIC DISC. *A circular area in the retina where the ganglion cells converge to form the proximal part of the optic nerve.*

The two optic nerves join each other in the brain and form the *optic chiasma*. This structure was mentioned in Chapter 5. Here many of the medially-located axons of each optic nerve cross over to the opposite side of the brain. Beyond this point, the optic nerve is known as the *optic tract*.

Conscious Activities, Reflexes, and Vision

As in the case for hearing, there are essentially two classes of pathways for visual information: conscious and reflex.

Two kinds of *conscious pathways* may be considered; (1) those involved

148
Vision

in registering the visual information in the cerebral cortex; and (2) those involved in moving the eyes, either voluntarily or involuntarily.

In the conscious pathway for vision, about 80% of the axons in the optic tract terminate in the thalamus. Here they synapse with fourth-order afferent neurons that pass further up into the cerebral cortex, into the *primary visual center* in the occipital lobe. In Figure 9.7 it may be seen that the visual information which is passed through this conscious pathway is represented topographically. That is, each point in one's visual field has a corresponding point in the primary visual center.

To provide for the movement of the eyes, a large number of axons from the optic tract first pass into various regions of the cerebral cortex, particularly the cortical areas 8 and 19 (Figure 5.22). From these areas, nerve fibers pass into the midbrain and form synapses with nerve fibers that pass into the nuclei of the third, fourth, and sixth cranial nerves. These nuclei, in turn, send out the motor fibers which operate the eyes.

In the event of damage to cortical area 19, there may be difficulty with automatic movements of the eyes, the kind of movements we routinely carry out in daily activities, and normally make involuntarily. It may or may

OPTIC TRACT. *Either of two cables of nerve fibers in the brain posterior to the optic chiasma, where the optic nerves cross over.*

FIG. 9.7. *Visual pathway in the brain; shaded area indicates visual region of occipital lobe. Lesions at the points indicated cause partial loss of the visual field.*

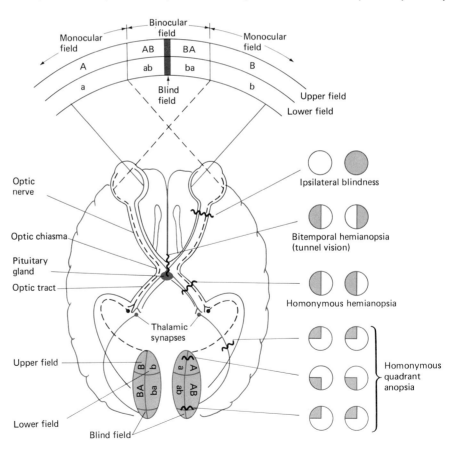

not be possible, in spite of the damage, to make voluntary movements, the kind which require a special effort or concentration. Damage to cortical area 8, on the other hand, usually impairs the ability to make voluntary eye movements.

Visual reflexes also occur through the cerebral cortex and midbrain (Figure 9.8). About 20% of the axons from the optic tract are led into the cerebral cortex and midbrain in order to provide for such reflex activities, which include: adjusting the size of pupil; adjusting the thickness of the lens (accommodation); and bringing about various movements of the body and face in response to various visual stimuli, such as a threatening gesture.

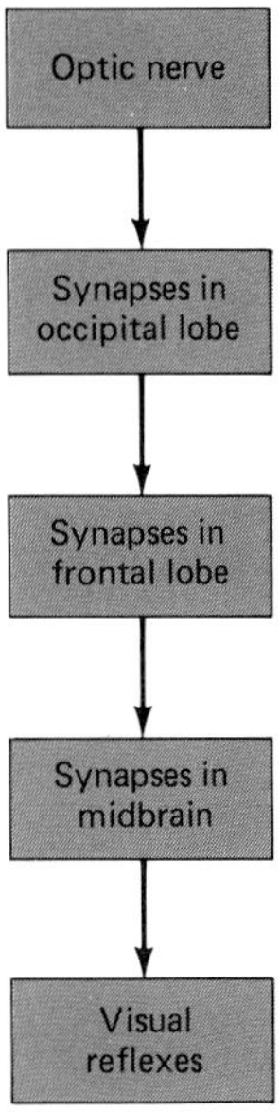

Adjustments of the pupil are made through pathways within the autonomic nervous system. To decrease the size of the pupil, the parasympathetic system activates the circular muscle of the iris; the contraction causes the pupil to become smaller. To increase the size of the pupil, the sympathetic system stimulates the radial muscle of the iris.

Accommodation is the process through which the lens of the eye is adjusted to see near or far. This process—like voluntary and involuntary movements of the eyes—employs nerve fibers from cortical areas 8 and 19. These fibers pass into the midbrain where—after the integration of a message by interneurons—motor messages are sent into the nucleus of the third cranial nerve, and then into the ciliary ganglion, as is the case for increasing the pupil size. When near objects are to be seen, certain postganglionic fibers from the ciliary ganglion release acetylcholine, causing the ciliary muscle to contract. As a result, the lens becomes thicker (see Figure 9.9).

The visual reflexes that involve movements of the body, limbs, face, and tongue are numerous. In general, motor messages are sent from various regions of the brain to bring about skeletal muscle movement.

FIG. 9.8. *Simplified outline of visual reflex pathways.*

Evaluation of Visual Ability

The visual system may be evaluated for proper functioning by a variety of tests. Sometimes, a visual problem is caused by a disturbance in the visual nervous system. Usually, however, there is another cause.

Proper eye movement may be tested by asking the person to look from side to side, upward, downward, and to execute additional movements, such as rotation. Failure to move one or both eyes in any particular direction usually signifies a lesion in the cerebral cortex or in the third, fourth, or sixth cranial nerve. See Table 5.4 for the functions of these nerves.

Each eye has four *visual fields:* upper and lower nasal (toward the nose) and upper and lower temporal (toward the outer edge of the head). Because the image we see is inverted as well as reversed by the lens, the upper nasal field is received by the lower temporal region of the retina. So too, the lower nasal field is received by the upper temporal region of the retina. These and other correspondences between retina and normal visual fields are shown in Figure 9.7.

Often, when the visual system is damaged by a stroke, tumor, accident, or metabolic defect, the site of the lesion (damage) may be determined

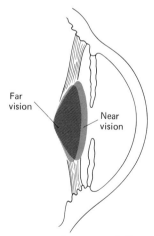

FIG. 9.9. *Accommodation to near and far vision.*

through an evaluation of the visual fields. A test of the visual fields may be started by asking the person to focus on an object straight ahead, with one eye covered. A wand, with a ball or light attached to its end, is gradually brought from the periphery into the center of one of the four visual fields. This is repeated for each eye. Failure to see the object in a given field permits the examiner to narrow down the probable site(s) of nerve damage. Figure 9.7 illustrates this principle.

A person's *visual acuity* is generally measured in terms of a fraction: the numerator is the distance in feet (usually 20) at which a line of type can be seen; the denominator designates the distance at which a normal person reads the smallest type capable of being read by the test subject. Thus, a rating of 20/100 indicates that the test subject can read a line at 20 feet which a person with normal vision can read at 100 feet.

Other eye actions — *accommodation, adjustment of the pupil,* and *convergence* — may be tested by bringing a finger slowly toward the nose from a distance of about two feet. If the ability to accommodate is normal, the ciliary muscle will contract and the pupil will become smaller. At the same time the eyes will turn inwardly, toward one another, showing normal convergence.

Reflex coordination of both pupils may be tested by flashing a beam of light into one eye and observing both eyes. When this is done both pupils should become smaller simultaneously.

Non-nervous Eye Defects. There are a variety of eye problems which do not involve the nervous system. The following are among the more common problems.

ASTIGMATISM. *Faulty vision due to an irregularity in the curvatures of the cornea or lens (G. a, not; stigma, a point).*

Astigmatism is a condition which everyone has to some degree, because the eyeball is not a perfect sphere and the lens and retina are never perfectly symmetrical. In severe cases, prescription lenses can correct this condition.

CATARACT. *Partial or complete opaqueness of the lens (G.).*

Cataracts are opaque regions on the lens due to degenerative changes in the proteins that make up the lens. This problem may be aggravated by antibodies which arise in response to "seeing" the abnormality of the lens tissue; the antibodies may react destructively with the lens tissue. Cataracts may sometimes be treated with drugs, such as cortisone, and sometimes by pricking the lens and thus allowing vitreous humor to enter the lens and reabsorb the opaque growths. In severe cases cataracts must be treated by removing both lenses and prescribing compensatory eyeglasses.

Cloudy cornea is due to injury or disease. Usually it calls for replacement of the cornea. *Conjunctivitis (pink eye)* is an inflammation of the conjunctiva, the membrane that covers the cornea and reflects back to form the inner surface of the eyelids. *Glaucoma* is excessive pressure (greater than about 25 mm Hg) in the eyeball. It arises because of an overproduction of aqueous humor by the ciliary body or an obstruction in the canal of Schlemm which drains the front of the eye.

Farsightedness (hyperopia) is a condition in which the image falls beyond the retina. People who are farsighted can see objects in the distance well but their near vision is poor (Figure 9.10).

Nearsightedness (myopia) is the opposite of farsightedness. It results when the eyeball is too long, so that the image is focused in front of the retina, or when the cornea and lens do not bend the light enough to make it fall on the retina. People who are nearsighted have poor far vision.

SMELL

Smelling (*olfaction*) is initiated when an airborne chemical becomes dissolved in the fluid that bathes a tiny cluster of bipolar hair cells located in the roof of the nose, along the median ridge. The hair cells are stimulated to set up generator potentials and action potentials. Approximately 100 million axons from these neurons carry the sensation of smell from the olfactory region of the nose to paired olfactory bulbs (first cranial nerve) which protrude from the forebrain. See Figure 9.11.

HAIR CELLS. *Epithelial cells with hairlike projections on their free surface for responding to sound waves in the ear, odors in the nose, or taste in the tongue.*

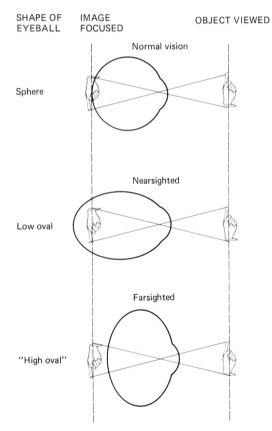

FIG. 9.10. *Nearsightedness and farsightedness.*

152
Smell

Following a certain amount of processing by interneurons, the partially integrated sensory information is transmitted through various regions of the brain, particularly the limbic lobe. As is the case for all other sensory information, the pathways for smell can be assigned into two classes: reflex and conscious. The pathways for *olfactory reflex actions* are extremely varied, complex, and little understood. Only the slightest degree of complexity is hinted by the nerve pathways indicated in Figure 9.12. A wide variety

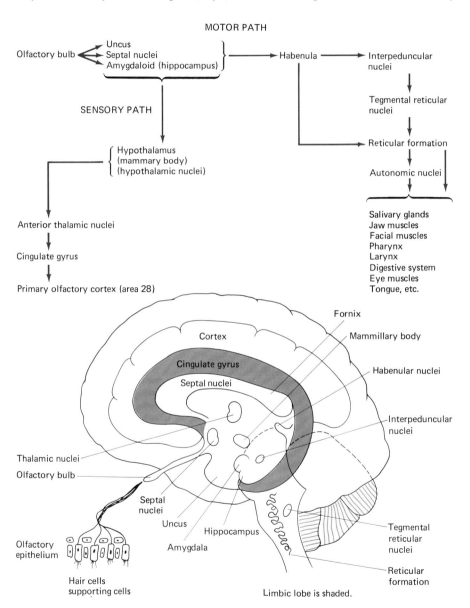

FIG. 9.11. *Olfactory apparatus in the brain.*

of reflex responses can be triggered by smell, including retching or vomiting, salivation, sparkling of the eyes, or an increased or decreased rate of breathing and heartbeat.

For the *conscious perception* and interpretation of odor, a large number of neurons in the olfactory bulb (first cranial nerve) convey messages about smell to the limbic lobe of the cerebral cortex. It may be noted in this connection that the sensations of smell, unlike those involved in the other sensory systems described so far, do not pass into the thalamus before reaching the highest center of consciousness—the limbic lobe of the cerebral cortex. In fact, the olfactory bulb is considered by some investigators to be an analogue of the thalamus. Just as the thalamus allows a crude awareness of pain, temperature, pressure, vision, hearing, and taste, so too, the olfactory bulb may crudely sense certain odors before the limbic lobe executes a highly discriminative judgment.

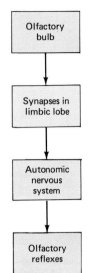

FIG. 9.12. *Simplified outline of olfactory reflex pathways.*

TASTE

Tasting starts when dissolved chemicals flow over taste receptors that are located: (a) sparsely and diffusely over the rear wall of mouth, and (b) abundantly and densely within the *taste buds* of the tongue. The chemical in-

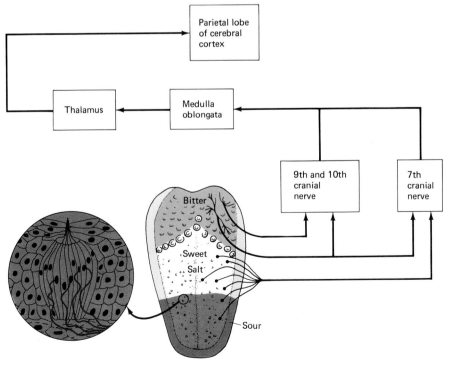

FIG. 9.13. *Pathway for taste.*

TASTE BUDS. *Barrel-shaped structures in the tongue consisting of stave-like support cells and hair cells which pick up taste stimuli.*

formation is converted into nerve impulses, which are carried in several cranial nerves, as shown in Figure 9.13. From these nerves, the taste sensations are transmitted into the medulla oblongata and pons. Following a certain amount of integration, the messages are directed along pathways which lead to reflex actions and/or conscious perception.

In the case of *reflex actions*, the efferent messages may be conveyed into the throat, voice box, and tongue for swallowing; into the face and jaws for chewing; into various regions of the body to bring about other reactions, as for example, shuddering when a very unpleasant taste is experienced. For the *conscious perception* of taste, second-order and third-order afferent neurons pass into the thalamus and into the parietal lobe of the cerebral cortex.

AGEUSIA. *Lack of a sense of taste* (G. a, *without*; geusis, *taste*).

PARAGEUSIA. *False sense of taste* (G. para, *beyond*; geusis, *taste*).

The sensations of taste may be tested to determine whether there is a complete absence of tasting ability (*ageusia*) in a given region of the tongue, a reduced tasting ability (*hypogeusia*), or a false interpretation of taste (*parageusia*). In the test, substances having a sweet (glucose), sour (lemon), bitter (quinine), or salty taste are applied to various regions of the tongue.

Loss of sensation over the front two-thirds of the tongue suggests damage to the seventh cranial nerve, whereas a loss in sensation over the posterior one-third of the tongue suggests damage to the ninth and tenth cranial nerves.

Differences in response to one of the four test substances may further suggest what portion of the system is at fault. The tip of the tongue is sensitive to all four tastes, the edge of the tongue is more sensitive to sour tastes, and the rear of the tongue is most sensitive to bitter tastes.

SUMMARY

1. The outer ear collects sound waves and directs them into the middle ear. Three bones in the middle ear transmit the waves to the inner ear, where the energy of sound is converted into action potentials that are sent to the brain.
2. Hearing difficulties can be classified as conduction deafness when the sound cannot be conducted properly into the inner ear, or as nerve deafness when the inner ear cannot convert the sound energy into action potentials or when the hearing centers in the brain are unable to respond to audio messages.
3. Light is received by rods and cones in the retina of the eyeball. After passing through at least one additional nerve cell, the visual information is passed by way of the optic nerve and optic tract to the visual center in the cerebral cortex, where interpretations are made.
4. The lens of the eye can be made thick or thin to see near or far. For near vision, it is thickened by parasympathetic stimulation which contracts the ciliary muscle.

5. The eyeballs are moved by external skeletal muscles which are regulated by cranial nerves 3, 4, and 6.
6. The size of the pupil is increased by contracting the radial muscle and decreased by contracting the circular muscle. The radial muscle is stimulated by sympathetic nerve fibers and the circular muscle by parasympathetic fibers.
7. Smell sensations pass into the limbic cortex of the brain. Unlike all other sensations we receive, smell sensations do not enter the thalamus before passing into the cortex.
8. Taste sensations are conveyed into the somesthetic cortex, where the tongue is represented, as well as into the limbic cortex.

REVIEW QUESTIONS

1. Describe the pathway by which sound waves travel from the outer ear to the eustachian tube.
2. At what place in the auditory apparatus is the energy of sound waves converted into bioelectric energy? What route do the action potentials travel when a sound is perceived?
3. Distinguish nerve deafness from conduction deafness.
4. Define: rod, cone, cornea, retina, sclera, ciliary body, cataract, astigmatism, nearsightedness, conjunctivitis, glaucoma, hyperopia.
5. Describe the pathway by which action potentials travel from the rods and cones in the retina to the visual cortex in the brain.
6. What happens in the visual system when a person shifts his gaze from a distant to a near object?
7. Heroin, morphine, and other narcotics cause a severe reduction in the size of the pupil. Explain this effect in terms of the muscles and nerves that are directly involved.
8. Is there anything unique about smell with respect to its pathway to the cerebral cortex?
9. Where do taste sensations originate? Where are they perceived?

10 The Proprioceptor System

GENERAL PROPRIOCEPTION
SPECIAL PROPRIOCEPTION—EQUILIBRIUM
EVALUATION OF THE PROPRIOCEPTOR SYSTEM
    Motion sickness and drunkenness
    Cerebellar problems

Proprioception refers to the perception of positional and locomotory information that originates as sensations within one's own skin, muscles, tendons, joints, and other internal tissues. The act of perceiving, however, is thought to be mediated not so much consciously, within association areas of the cerebral cortex, as within the cerebellum, which is attached to the brain stem in the hind brain.

Free nerve endings, Pacinian corpuscles, Golgi tendon organs, and muscle spindles are the major *general proprioceptors*—organs which inform the central nervous system about peripheral events. They are shown in Figure 10.1 along with their associated nerve pathways.

The vestibular system of the inner ear is a *special proprioceptor* through which the brain is informed about the directional movements of the head and the speed of these movements.

GENERAL PROPRIOCEPTION

CEREBELLUM. *A coordination center for voluntary movements, posture, and equilibrium* (L. cerebellum, *small brain*).

General proprioceptive information may be directed into a reflex action, and/or brought into a level of consciousness. Unlike all of the sensory systems described so far, the information is processed, in part, within the *cerebellum* (see Figure 10.1).

The *reflex execution* of proprioceptive messages involves the use of numerous afferent neurons, interneurons, and efferent neurons. Walking, for example—which involves extensor and flexor reflexes on both sides of the body—is in fact a coordinated, self-sustaining, or recurring, series of spinal reflexes that are modulated by the cerebellum.

The *conscious execution* of proprioceptive messages depends upon the

The Proprioceptor System

passage of the sensory information through the thalamus into the somesthetic (sensory) strip of the cerebral cortex and also into the cerebellum. Information about the position and movements of the head, trunk, and limbs is processed within these brain tissues and appropriate efferent messages are passed on to the skeletal muscles.

SPECIAL PROPRIOCEPTION—EQUILIBRIUM

It is usually taken for granted that we know whether we are looking up or down, whether we are walking or sitting, whether we are moving slow or fast. When we become seasick or drunk, however, we may become dizzy, fail to walk straight, and fail to know what position the body is in or how fast it is moving. In short, we may lose our sense of equilibrium.

Equilibrium is a state of balance which is regulated by the special proprioceptor system—the *vestibular system*. In particular, this system regulates the directional movements of the head and the direction of eye movements relative to head movements.

The proprioceptors of the vestibular system are located in the inner ear, in a structure that lies close to the cochlea (see Figure 10.2). They lie in five sensory organs: three ampullas; a utricle; and a saccule.

The *ampullas* are the expanded ends of three *semicircular canals* that lie at right angles to one another. They contain proprioceptors whose func-

AMPULLA. *Saccular dilation of a canal, as seen in the semicircular canals of the ears and in the milk-carrying ducts of the mammary glands (L. a two-handled bottle).*

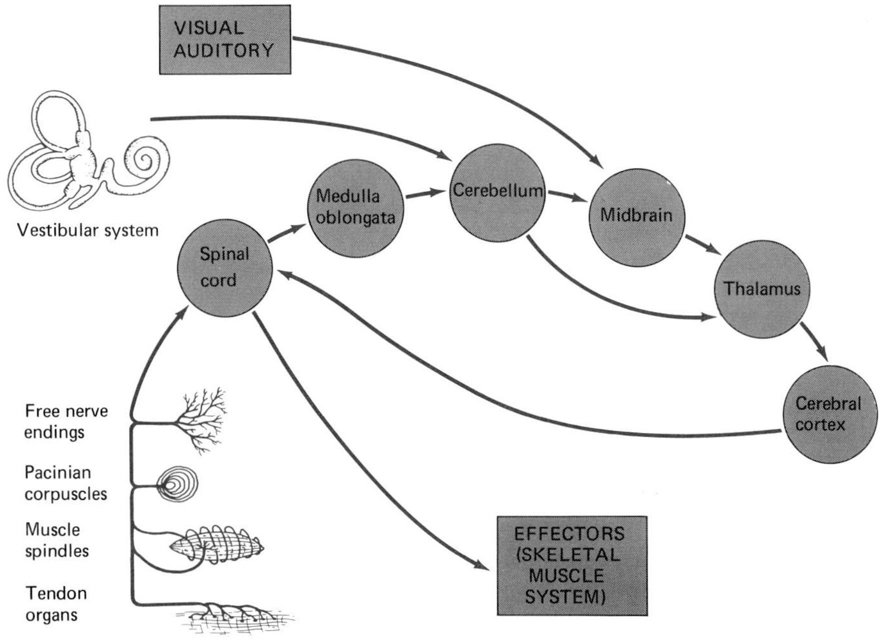

FIG. 10.1. *Pathways of the proprioceptor system.*

UTRICLE. *The larger of the two membranous sacs in the vestibule of the labyrinth of the inner ear (L.* saccus, *a sack).*

MACULA. *A small spot or what appears as a spot. In the ear, a macula is an oval, neuroepithelial sensory area in the medial wall of the saccule (L. spot or stain).*

tion is to signal the speed at which the head is moving in some particular direction. For example, as the head moves directly downward, fluid in the semicircular canal which runs along the long axis of the body moves special hairlike processes within the ampulla. These processes stem from hair cells which, in turn, are in contact with the dendrites of bipolar neurons. Action potentials are generated in the bipolar neurons and information regarding the speed of the downward motion is signalled to the brain through axons that enter the eighth cranial nerve.

The *utricle* and the *saccule,* perhaps together with the ampullas—and certainly by themselves—signal information to the brain about the position of the head; that is, whether it is upright, to the left or right side, or leaning back or forward. They also signal information about the direction in which the head is moving; that is, whether it is moving up or down, forward or backward, or if it is stationary.

The proprioceptors of the utricle and saccule are embedded in gelatinous masses, called *maculae,* which contain crystals of calcium carbonate. (The gelatinous masses, or *cristae,* in which the hair cells of the ampulla are

FIG. 10.2. *Vestibular system—the special organ for balance. Impulses generated in the hair cells are transmitted along the vestibular nerve to the cerebellum.*

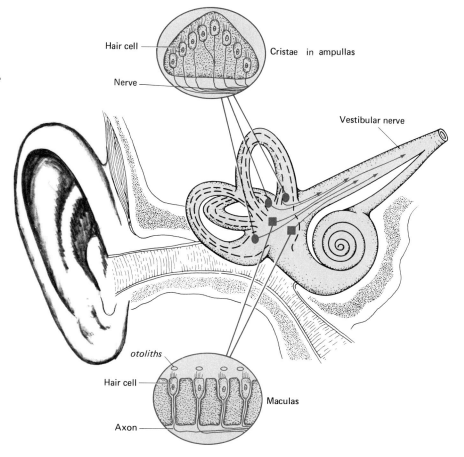

embedded do not contain such crystals.) As the head changes its position, the calcium carbonate crystals, called *otoliths*, shift accordingly and signal the appropriate information to various regions of the brain, especially to the cerebellum and cerebral cortex. Some of the pathways are illustrated in Figure 10.1.

Between the vestibular system and the cerebellum occurs the *vestibular ganglion*. Both reflex and conscious proprioceptive actions are directed from this ganglion. The reflex actions are triggered by messages which are passed into the nuclei of the third, fourth, and sixth cranial nerves, and elsewhere in the brain stem. In response, efferent messages are directed to the extrinsic muscles of the eye and the skeletal muscles of the body. In bringing about consciously-directed activities, information is sent from the vestibular ganglion into the cerebellum. Here, sensory messages from the muscles, tendons, joints, skin, eyes, and cochlea are integrated with the sensory messages from the vestibular system. Appropriate efferent messages are then sent to the cerebral cortex and throughout the rest of the brain and body.

The cerebellum is important primarily as a modulator of ongoing motor activities. It serves to integrate the motor messages from the cerebral cortex, basal ganglia, and brain stem with sensory messages from the peripheral nervous system.

OTOLITH. *A calcareous particle in the labyrinth of the ear* (G. ous, *ear;* lithos, *stone*).

EVALUATION OF THE PROPRIOCEPTOR SYSTEM

The effectiveness of the proprioceptor system may be evaluated by testing various responses, including the coordination of the arms, legs, and body; the ability to judge the texture (rough, smooth, soft, hard, etc.) of a mechanical stimulus; the ability to judge the location at which a stimulus is applied; and coordination of eye-to-head movements.

The coordination of the arms, legs, and body is evaluated by: (a) noting whether the person is aware of the position of a particular limb while lying down; (b) observing whether balance is possible—without moving the feet—when the eyes are closed (if balance is possible, is there excessive swaying?); (c) judging whether the person can walk properly when the eyes are closed. Failure to balance oneself or walk properly (*ataxia*), may likely be due to a lesion in the cerebellum or the general sensory system or the vestibular system of the inner ear.

Failure to judge texture (*agnosia*) or the point of application of a stimulus (*astereognosis*) also may be due to a lesion in one or more of the sites just mentioned.

The coordination of head and eye movements is evaluated by determining whether there is an abnormality in *nystagmus*, the rhythmic horizontal, rotary, and vertical oscillations of the eyeballs. In a normal individual, when the head turns along any plane to its extreme—for example, horizontally to the extreme left—the eyes are kept stationary. At the extreme end of the turn, however, the eyes should snap in the direction of movement. The

AGNOSIA. *A lack of sensory ability to recognize something* (G. a. *without;* gnosis, *knowledge*).

NYSTAGMUS. *Oscillatory movements of the eyeballs* (G. mystagmos, *drowsiness*).

Evaluation of the Proprioceptor System

stationary position (as fixing your vision upon a telephone pole while driving along) in actuality is a relative movement (of the eyes in relation to the body) which is controlled by the vestibular system. The snap of the eyes into the direction of movement (as into a position to look at the upcoming telephone pole) is a reflex action believed to be activated through the cerebral cortex. The snap is attributed to an attempt to locate a fixed point, in order to achieve a focal reference point.

Nystagmus may be tested by the *Barany caloric test*, in which cold (20°C) or hot (50°C) water is placed into the person's ear. If the individual is lying on his back, the cold water will cool the fluid in the semicircular canals. As a result, the fluid in the horizontal canal will move toward the floor (that is, away from the ampulla), creating a sensation of turning (see Figure 10.2). In turn, the eyes should move slowly toward the flooded ear and then rapidly to the opposite side. Hot water produces a similar response, but the eyes move in the opposite direction.

If the cold test is performed with the individual in an upright position, three things should happen: (1) rotatory or vertical nystagmus (2) a tendency of the body to fall toward the irrigated ear and (3) past pointing (see below). Failure to produce these responses suggests a nervous disorder. On the other hand, the presence of these three things in the absence of some provocation such as the cold water test also suggests a nervous disorder. The disorder may be characterized further by bouts of vomiting, nausea, dizziness, and ringing in the ear. The disorder may lie peripherally (in the inner ear) or centrally (in the cerebellum, brain stem, or cerebral cortex).

PAST POINTING. *Pointing beyond an intended object.*

Past pointing can be tested by rotating a person in a revolving chair ten times to the right with eyes closed. The person then holds his right arm horizontal and in front of himself with eyes open, while the examiner places an extended finger on the person's right index finger. Next the person raises his arm vertically and is told to touch the examiner's finger again. If the vestibular apparatus is normal, the index finger will be brought down several inches to the right of the extended finger because the person has the sensation of being rotated to the left.

Motion Sickness and Drunkenness

Motion sickness, which is usually characterized by nausea, dizziness, and vomiting (or retching), may be attributed to excessive stimulation of the utricle and saccule. Drunkenness, when characterized by similar symptoms, may possibly be due to a lowering of the threshold of nervous discharge in the vestibular system; that is, the neurons fire more readily.

In the case of motion sickness, the symptoms are relieved—or are not as likely to occur—when the individual is lying down. In this position the saccule, and particularly the utricle, are being stimulated minimally.

Individuals who are prone to motion sickness, whether in a car, boat, or plane, may reduce their tendency by supplying their vestibular system with appropriate visual cues—by focusing on the horizon at sea, for example, rather than allowing the eyes to drift repetitively and rhythmically in the direction of travel.

In the case of drunkenness, the symptoms of motion sickness happen more readily when the eyes are closed—regardless of whether the individual is standing, sitting, or lying down. In this case, it appears that the saccule and utricle are not so much concerned as are regions elsewhere in the vestibular system.

Drugs—such as dramamine (an antihistamine), scopolamine and atropine (both of which counteract the action of acetylcholine), and others—are useful in reducing the tendency toward motion sickness. They are also often effective against motion sickness after the symptoms have started.

Cerebellar Problems

Injury to the cerebellum usually brings about a reduction in its smoothing effect on various skeletal motor activities. The hand overshoots an object it seeks to grasp (past pointing), or falls short of the object, trembling in doing so. The spinal reflexes also usually suffer as a result of hypotonia, because the skeletal muscles receive less than normal nervous input. Jerky movements (*ataxia*) due to an inadequate coordination of the limbs, head, and body may make the person seem to be a staggering drunk.

Failure to repeatedly flip the palm of the hand up (supination) and down (pronation) also appears. This is due to inadequate regulation by the cerebellum of opposing muscle groups. In some cases, a flexor muscle, upon being restrained, will not move in its normal direction once the restraining force is removed. Instead, it will cease to move, due to excessive activity in the extensor muscles. Swaying, difficulty in standing upright, and faulty speech are other problems which may be seen when the cerebellum is damaged.

HYPOTONIA. *Underactive or subnormal muscular activity* (G. hypo, *below;* tonos, *strength*).

ATAXIA. *A loss of the power of muscular movement* (G. a, *without;* taxis, *order*).

PRONATE. *To bend forward or to place into a prone (lying face down) position* (L. pronus, *leaning forward*).

SUMMARY

1. General perception and regulation of body position and limb movements are carried out through the cerebellum in conjunction with other regions of the brain.
2. Special perception and regulation of balance, or equilibrium, is accomplished through the cerebellum and other regions of the brain by way of receptors in the labyrinth of the inner ear.
3. Ampullas within the labyrinth contain sensory nerve cells which signal information to the brain about the speed of our head movements along x, y, z coordinates.
4. Receptors within the utricle and saccule of the labyrinth signal information on the direction of the movements of the head.

5. Nystagmus refers to movements of the eyeballs which are normal or abnormal depending on the health of the labyrinth, cerebellum, and other regions of the brain concerned with balance.
6. Motion sickness reflects an unbalanced stimulation of the labyrinth receptors in which the utricle especially, and the saccule to some extent, are excessively stimulated.

REVIEW QUESTIONS

1. Distinguish general perception from special perception with respect to body position and limb movements.
2. What is the special function of the receptor cells in the ampullas of the inner ear?
3. What is the special function of the receptor cells in the utricle and saccule of the inner ear?
4. Define nystagmus. How does nystagmus relate to normal and abnormal special proprioception?
5. Describe motion sickness and tell how it may be relieved.

The Cortical Systems 11

THE CEREBRAL CORTEX
    The motor area and motor association areas
    Sensory areas and sensory association areas
        Vision
        Hearing
        Taste
        Smell
    Emotion areas in the brain
        Neurotransmitters in emotional feelings
        The neural circuitry of emotionality
    The centers for verbal communication
        Pure word deafness (acoustic aphasia)
        Pure word blindness (alexia)
        Pure speech aphasia
        Amnesic aphasia
EVALUATION OF THE CEREBRAL CORTEX
    Multiple sclerosis
    Epilepsy

The cerebral cortex can be pictured as a flat piece of tissue, about a quarter of an inch thick, a foot and a half in diameter, squeezed here and there to form folds and crevices. It is composed of ten billion cell bodies. Each one is somehow wired to the others. Each sends out several hundred to several thousand dendritic processes to neighboring cells, and axons to nearby neurons or to neurons within the spinal cord—or it receives dendrites or axons from neurons elsewhere in the brain.

Think of the infinite number of paths, or combinations of paths, a nervous signal might take if it were to move at random through this complex of fibrous material. And yet, incredibly, it is within this tissue that we perceive sensory information, initiate most motor messages, learn to read, write, and solve problems, memorize, ponder, dream, initiate emotional reactions, passively experience an emotional reaction, judge moral and ethical codes of conduct, and carry out all our creative activities.

CEREBRAL CORTEX. *The outer tissue of the brain (L. cerebrum, brain; cortex, bark).*

THE CEREBRAL CORTEX

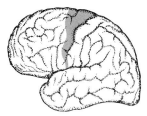

FIG. 11.1. *Motor area of cerebral cortex.*

It is possible to map a few of the functions of the cerebral cortex, as we described earlier and illustrated in Figure 5.22. Unfortunately, when a region of the cortex is destroyed—through disease (for example, syphilis or meningitis), stroke, or injury—the function of that area and the functions of other areas in the cortex may be lost, often permanently. Sometimes the function can be regained. Following a stroke, for example, there may be a partial or total loss of speech, hearing, vision, muscle movement, or the like, but in time those functions may be partially or totally restored.

How is this restoration brought about? In the first place, many of the neurons may have only been damaged, not destroyed. In this case it becomes possible for the neurons to recover and function properly again, largely through the effort of microglia in cleaning up the blood clot (thus allowing a greater circulation of oxygen and nutrients and disposal of wastes) and through the effort of other glia cells in adjusting the chemical environment around the damaged nerve cells. In the second place, if the damage has not been too severe, it is probable that new paths and circuits can become established. Through these new pathways, the "lost" functions are often "regained."

An overall view of cerebral cortex function—leaving aside for the moment the specific functional areas—can be gained through a discussion of the following functional units: (1) the motor area; (2) the motor association areas; (3) the sensory areas; (4) the sensory association areas; (5) the emotion areas; and (6) the station for verbal communication.

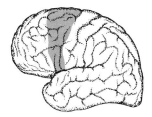

FIG. 11.2. *Premotor area.*

MOTOR. *That which causes movement* (L. movere, motion).

The Motor Area and Motor Association Areas

The motor area and its association areas lie mainly in the frontal lobe. The *motor area* is a strip of cortex which runs in front of the central fissure (see Figure 11.1). The cell bodies in this area send out axons which run uninterruptedly through the cerebral medulla and brain stem to the ventral horns of the spinal cord.

When the motor area is damaged, as often happens during a stroke, a certain amount of paralysis may follow. The particular muscle or muscles affected depends upon which portion of the motor area was damaged (Figure 7.5).

Slightly ahead of the motor area, toward the forehead, lies a motor association area, the *premotor area* (Figure 11.2). Here, information from various sensory areas is integrated, and impulses are sent out to the motor area to trigger an appropriate motor action.

Still further forward, within the frontal lobe, lies the *prefrontal area* (Figure 11.3). Intellectual and emotional activities, such as judgment, short-term memory and emotional drive, are believed to take place, at least in part, in this area. It is also believed that, whereas many routine motor actions are dictated from the premotor area, purposeful actions requiring

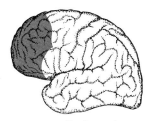

FIG. 11.3. *Prefrontal area.*

personal judgment are dictated from the prefrontal area and perhaps some other as yet unidentified coordinative centers.

PREFRONTAL. *The anterior part of the frontal lobe (L. prae, before; frons, forehead).*

Sensory Areas and Sensory Association Areas

There are five primary sensory areas into which various sensations make their initial appearance in the cerebral cortex (see Figure 11.4). The *somesthetic area* lies next to the motor area in the parietal lobe. This is the area into which general sensory information (including pressure, pain, and temperature) is brought to the highest level of identification. The *visual area* lies in the rear of the cerebrum, in the occipital lobe, where it wraps itself around the medial aspect. The *hearing center* lies deep in the temporal lobe, and the *smell area* is situated in the limbic lobe. The *taste area* is present in the tongue and jaw region of the somesthetic area.

SOMESTHETIC. *Pertains to general body feelings, such as temperature, as opposed to special feelings, such as sound (G. soma, body; aisthema, sensation).*

The locations of the five primary sensory areas are shown in Figure 11.4. Also shown in this figure are a number of corresponding sensory association areas in which the identified sensory information is integrated with other information, both current and memorized.

Whereas the primary sensory areas serve to receive and identify sensations, the association areas somehow allow for the integration of the adjacently-located primary sensations and other ongoing or past experiences. The sight of a meal, for example, may be integrated with the smell and taste of food, with a pleasant past experience, with the barking of a dog nearby. It is through the association areas that we learn, that we submit short-term memory to perhaps the prefrontal lobe and long-term memory to various regions of the cortex—including areas within the temporal lobe. The association areas enable us to carry out such functions as thinking, judging, and imagination.

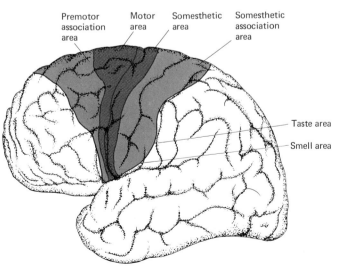

FIG. 11.4. *Motor areas and certain sensory areas of the cerebral cortex.*

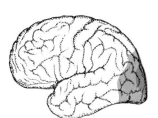

FIG. 11.5. *Visual area in occipital lobe. Light color shows secondary visual area.*

Vision. Visual information, which is carried first from the optic tract to the thalamus (see Figure 9.7), is then relayed to the primary visual center in the occipital lobe (Figure 11.5). At this point an individual may be able to see an object in reference to other points in space. In addition, the person may be aware of color, shadow, brightness, and shape. From the occipital lobe the information is led into an association area that permits the person to recognize the object and to discern its meaning.

The destruction of the primary visual area leads to blindness in the corresponding topographic visual fields. The destruction of the association area often results in failure to read—though the person may still be able to see—and often leads to the impairment of various motor functions that are tied in with vision, such as autonomic movements of the eyes in relation to movements of the head.

Hearing. The primary hearing center lies deep in the temporal lobe. Nerve impulses that are associated with high, intermediate, and low frequency sounds pass respectively from the apex, intermediate, and base regions of the cochlea into the thalamus. Here they are topographically presented and topographically relayed to the primary hearing center in the temporal lobe. See Figures 9.3 and 11.6.

Whereas the primary hearing center serves to receive sound, hearing association areas permit the discrimination and interpretation of the meaning of the sound. For example, disorganized sound (noise) may be distinguished from highly structured sound (speech, music).

Damage to the primary hearing center may result in deafness, whereas damage to the association area may result in failure to localize the source of a sound, as well as to interpret it. Furthermore, inasmuch as intimate neural connections exist between the hearing and speech areas, damage to either of these centers may affect the performance of the other center.

Taste. Taste sensations are first represented in the brain in the hypothalamus and thalamus, and then in various regions of the cerebral cortex, such as in the temporal lobe, the hippocampus, and the uncus (see Figure 5.21). The primary taste center lies in the somesthetic sensory area for the tongue and jaws (Figure 11.4). Association areas coexist with smell centers in the limbic lobe.

Smell. The sensation of smell is the only sensory function that passes from the periphery into the cerebral cortex without first entering the thalamus. Within the cortex, the sensations are carried first to the vicinity of the amygdaloid nucleus in the limbic lobe (see Figure 9.11). Olfactory association centers in the hippocampus (Figure 11.4) allow one to associate an odor with a visceral feeling or body activity.

Emotion Areas in the Brain

During each day of our lives we are subject to numerous emotional stimuli which evoke either an emotional feeling or an emotional feeling followed

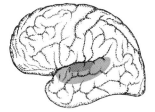

FIG. 11.6. *Hearing area in in the temporal lobe. Light color shows secondary hearing area.*

by an emotional action. The *emotional feeling* may be said to be the afferent limb of an *emotional experience;* the *emotional action* may be said to be the efferent limb.

Between the afferent and efferent limbs lies the domain of *emotional arousal,* in which the emotional feeling is appraised by a complex pool of interneurons and a particular emotional action is either initiated or inhibited. When any emotional action is initiated the action itself is called *emotional behavior.* Pleasure and anxiety are examples of emotional feelings. Fighting and singing are examples of emotional behavior.

Emotional feelings and behavior arise as a result of an *emotional process* in which the following events take place: a stimulus is perceived, bodily changes are sensed, visceral reactions are felt, the emotional stimulus is appraised in light of past experiences, an emotional attitude is formed, former activities in the cerebral cortex are temporarily disrupted, and a new affective state is created. When we say a new *affective state* is created we are saying one's feelings (emotions) are altered. Love, for example, is displaced by annoyance, or a feeling of depression is converted into a feeling of elation or a neutral mood. When we say former activities in the cerebral cortex are disrupted during an emotional process, we are spelling out the essence of emotionality. The word *emotion* means to take "out of motion," and this is exactly what we do in response to some motivating force, or emotional stimulus, and experience a newly-created mood. Paradoxically, emotion also means to put *into* motion, and this is exactly what happens when we laugh or cry, or express any other emotional behavior, following a strong change in affective state.

Now we may ask whether it is possible for us to explain emotional feelings and behavior solely in physiological terms. The answer is that it definitely is not. We can describe some of the neural pathways that are used, and we can say something about the neurotransmitters that are involved. However, little can be said about the cognitive factors which set up the emotional feelings and emotional behavior. *Cognitive factors* include perception, knowing, and affective memory. These are psychological factors which may trigger similar kinds of emotional actions in one individual or group of individuals, and quite another in a different individual or group.

Cognitive factors play a major role in guiding our emotional feelings and emotional actions. Neurotransmitters are ordinarily released in consequence of the cognitive factors. The particular kinds of neurotransmitters and the specific neural pathways that are activated in the brain determine the nature of our affective state and the behavior that will be initiated by our autonomic and skeletal motor systems.

Neurotransmitters in Emotional Feelings. The major known neurotransmitters in emotional feelings are dopamine, noradrenaline, and serotonin. We do not know the total function of any one of these substances, let alone the function of all of them put together. Nevertheless, enough observations have been made, systematically and by different scientists, to allow us to make several important generalizations.

The first generalization has to do with the roles of noradrenaline and

EMOTION. *A strong feeling, a mental feeling, a sentiment* (L. emotum, *to stir up*).

AFFECTIVE. *Related to emotional stimuli* (L. affectus, *to affect*).

COGNITIVE. *Pertaining to knowing, becoming, or perceiving; in contrast to "affective."*

The Cerebral Cortex

SEROTONIN. *A blood-pressure increasing substance produced in cells from the amino acid tryptophan.*

SCHIZOPHRENIA. *A mental disease or psychosis (G. schistos, split; phren, mind).*

dopamine in establishing the mood of euphoria (well-being, or happiness), and of serotonin in establishing the mood of depression. The following example will serve to illustrate this. In the early 1950's mental depression was commonly observed when individuals with high blood pressure were treated with a drug called reserpine. Then, it was soon learned that reserpine depletes nerve fibers of their supplies of noradrenaline and dopamine. It was therefore natural to suspect that high levels of noradrenaline or dopamine in the brain would lead to euphoria. That this is actually so has been shown by various kinds of experiments. Other experiments have shown that high levels of brain serotonin are associated with mental depression. Some scientists now suspect that the particular mood between depression and euphoria depends upon the ratio of noradrenaline and dopamine to serotonin in certain regions of the brain.

The second generalization about the relation between neurotransmitters and emotions is that abnormal metabolites are commonly found in individuals with certain forms of mental illness. We can illustrate this general statement by reference to a number of conditions, including Parkinson's disease, schizophrenia, and hallucinations. Parkinson's disease is a motor disorder in which there is, among other things, a deficiency of dopamine in certain basal ganglia of the brain. Schizophrenia is a mental disease which is characterized by thought disorders, feelings of persecution and/or omipotence, social withdrawal, and hallucinations. A hallucination refers to the perception of any type of sensation, such as a sound or vision, that is not really present.

Regarding the second generalization, it had been noted that when persons

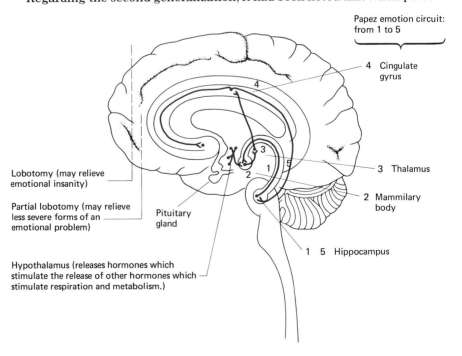

FIG. 11.7. *Limbic cortex and other emotional regions of the brain.*

The Cortical Systems

were treated with reserpine for high blood pressure, a Parkinson-like syndrome would often develop. It is now believed that this syndrome was caused by a depletion of dopamine. It is also known that when a Parkinson patient is treated with DOPA—which in turn is converted to dopamine in nerve cells of the brain—the patient often behaves like a person with schizophrenia.

It turns out, moreover, that schizophrenics often have hallucinogenic substances in their urine. One of these substances, called DMPEA, can be derived from dopamine, while another substance, called bufotenin, can be derived from serotonin. DMPEA resembles a well-known hallucinogen called mescaline, while serotonin resembles another well-known hallucinogen called LSD.

Finally, it has been observed that the behavior of a person who is under the influence of a hallucinogenic drug often resembles that of a person suffering from mental illness. Mental and emotional regression, diminished motor control, and compulsive laughing are commonly seen, and the individual often loses all sense of what is or is not socially acceptable behavior.

The Neural Circuitry of Emotionality. The limbic system is the highest neural region of emotional feelings and actions. The limbic cortex forms a border around the brain stem (Figure 11.7), and it also constitutes a border of cortical tissue beneath the sensory and motor regions of the cerebral cortex.

The limbic cortex consists of four folds of gray matter. These folds (which were shown in Figure 5.21) are the uncus, hippocampus, isthmus, and cingulus. The *uncus* hooks around the thalamus, the *hippocampus* lies beneath the thalamus, the *isthmus* connects the uncus to the cingulus, and the *cingulus* surrounds the corpus callosum.

In the region of the uncus and hippocampus there is an important nerve nucleus called the *amygdala* (Figure 11.8). The amygdala is concerned

DMPEA. *Dimethoxyphenylethylamine, an abnormal metabolite.*

LSD. *Lysergic acid diethylamide, a hallucinogenic drug.*

HIPPOCAMPUS. *A fold of gray matter in the floor of either lateral ventricle (G. seahorse).*

ISTHMUS. *A constricted part (G. isthmos, neck).*

AMYGDALA. *A nucleus of nerve cells in the floor of either lateral ventricle (G. amygdale, almond).*

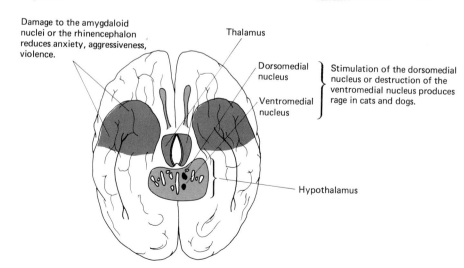

FIG. 11.8. *The amygdaloid nuclei are important in certain types of emotional disturbance. Other prominent areas of emotionality are shown in light tones.*

with emotional feelings and behavior which insure self-preservation. Axons from the amygdala discharge nervous activities which are involved in feeding, fighting, and self-protection. Because the amygdala is so close to the auditory and visual centers, damage to the nucleus generally interferes with auditory and visual discrimination.

Damage within the limbic system often leads to reduced vigor, a tendency to sulk, speak less, and be indifferent to pain, and in the extreme, to a state of stupor and coma. So too, the removal of portions of the frontal lobes (*lobotomy* or *leukotomy*) often results in reduced emotional expressivity (Figure 11.7). The same thing happens when the amygdaloid nucleus in the anterior portion of the prefrontal area is partly destroyed. In all three cases fewer reverberations of nerve impulses pass between the limbic and frontal lobes.

LOBOTOMY. *An operation in which one or more cerebral nerve tracts are sectioned (G. lobos, lobe; tome, cutting).*

LEUKOTOMY. *Cutting through the white matter which connects the gray matter of the frontal lobe to the thalamus (G. leukos, white; tome, cutting).*

At the anterior end of the cingulus of the limbic cortex, straddling both hemispheres, lies what is known as the *septal area*. Located here are septal nuclei, which are concerned with expressions and feelings concerned with sociability, grooming, sexual relations, and reproduction. Because of the nearness of the septal nuclei to the somesthetic and motor regions of the cortex, damage to the nuclei generally interferes with all sensory moods and motor memory.

A third important cluster of cell bodies concerned with emotionality is a group of nuclei called the *anterior thalamic nuclei*. These nuclei send millions of nerve axons from the thalamus into the cortex of the frontal lobe, especially to the anterior regions of the frontal lobe. Because of the close proximity of these nuclei to the limbic lobe and sensory cortex, damage here generally destroys sensory emotional (affective) memory.

How can we specify, in simple terms, the pathways which mediate emotionality? A partial answer to this question was given in Chapter 9, where we discussed emotional (reflex) behavior that is based on reactions to smell sensations. In more general terms, we may consider three components of the emotional process: sensory processing, affective (emotional) memory, and emotional behavior.

Emotional sensory processing requires us to be aware of how an event occurs in time and space. Any emotional event is perceived in context with the sequence of events that occur, the environment in which the event occurs, and the person's emotional mood (established by cognitive factors and mediated by neurotransmitters) at the time. The pathway which mediates sensory processing includes the limbic lobe, the midbrain (visual and auditory reflexes), the reticular formation, the hypothalamus, and the sensory cortex of the brain.

LIMBIC LOBE. *The region of the cerebral cortex which borders on various sensory, motor, and emotion regions of the brain (L. edge or border).*

Emotional memory may be stored in various lobes of the brain, particularly the temporal and frontal lobes. It is thought that certain feelings, such as withdrawal, repression, and unexplainable likes and dislikes, are stored in the limbic lobe. For example, a child who is severely punished during a visit to the seashore may feel a strong dislike of the beach for the rest of her life, even though she enjoys swimming. In this case it is likely that the limbic lobe retains the unpleasant association.

Emotional behavior, like emotional sensory processing and emotional memory, is mediated through the limbic cortex. In this case, however, the autonomic nervous system is activated by way of the hypothalamus, pons, and medulla oblongata, and the skeletal muscle system is activated through the motor cortex.

The Centers for Verbal Communication

One of man's earliest evolutionary challenges was to develop an effective system of communication. The assignment of spoken sounds and written symbols to objects, activities, and events was the vehicle through which languages arose. Certain regions of the cerebral cortex evolved as areas in which we could recognize written and spoken words and in which we could formulate the instructions to our motor system for writing and speaking.

Verbal communication refers to the ability of human beings to speak with words, to understand the spoken word, and to read and write. Our communication with one another is mainly verbal, although on occasion it is desirable or necessary to communicate nonverbally, as for example, when an individual suffers damage to the verbal machinery of the cerebral cortex as a result of disease, accident, or a stroke.

The *verbal machinery* is located mainly in *Broca's area* (cortex areas 44 and 45) and *Wernicke's area* (cortex areas 21 and 22), shown in Figures 11.9 and 5.22. These areas are functionally connected to one another by the *arcuate fascicle* and *angular gyrus*, as shown in Figure 11.10. In addition, Broca's and Wernicke's areas are connected through synapses to the visual and hearing centers, to the part of the motor cortex that operates the communication apparatus (tongue, voice box, pharynx, lungs, jaws, lips, and the hands, when the hands are used in writing or typing), and to various association areas. The verbal machinery is located in the left brain in most people. In about 15 percent of left-handed people the verbal machinery is in the right brain, and in about 2 percent of right-handed people it lies in both sides of the brain.

Figure 11.10 shows a simplified representation of how spoken words might be converted through the verbal machinery into "mental images" — how the spoken word "cat," for example, may be converted into a mental picture of a cat.

In Figure 11.11 we give a simplified representation of how a visualized word might be converted into a spoken word. You will find it very profitable to examine the figure carefully before proceeding with the text. In the same figure you should also take special notice of the *corpus callosum*. The proper functioning of the entire system for verbal communication depends upon the connections between the cerebral hemispheres by way of this important commissure.

Broca's area is located in the frontal lobe, just in front of the motor representation for face, tongue, lips, palate, and vocal cords. When this area is injured—usually as a result of a stroke or accident—language problems (*aphasia*) often arise. The person's speech is impaired, that is, words are uttered slowly, articulation is poor, and it is difficult to speak complete

ARCUATE FASCICULUS. *Nerve fibers which run in an arc-shaped path from the frontal lobe to the temporal lobe (L. arcus, arch; fasciculus, cable).*

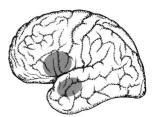

FIG. 11.9. *Approximate location of Brocas's area (above) and Wernicke's area (below).*

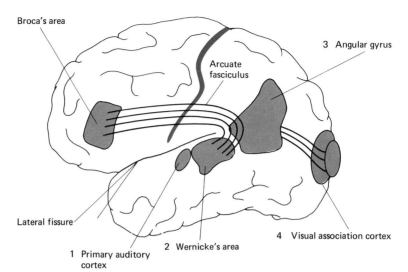

FIG. 11.10. *Circuit for understanding a verbal description. Words are heard at (1), interpreted in Wernicke's area (2), and sent to point (3). Visual association (image formation) occurs at point (4).*

sentences. A similar impairment occurs with regard to writing—words are written poorly and improperly, and sentences are difficult, if not impossible, to construct. Also, the body becomes paralyzed along the side opposite to the injured region in the brain. This is almost always the right side. However, even though the individual with damage to Broca's area can barely speak, the ability to understand the written and spoken word may be quite normal.

FIG. 11.11. *Mechanism of reading aloud.*

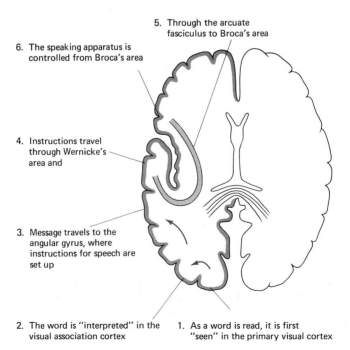

The Cortical Systems

Wernicke's area is located in the temporal lobe next to the hearing area. When this area alone is injured, the body usually suffers no paralysis, but speech and writing are impaired. In this case, the words are spoken rapidly and are written with well-formed letters. But the speech is nonsense, as is the writing: "I was in this one and here after that the yellow day they this way went." In addition, the person is unable to understand the spoken and written word. Still, he may be able to comprehend nonverbal language—gestures and sounds.

How are Broca's and Wernicke's areas related to each other and what happens when an injury arises in the arcuate fasciculus which connects the two areas? To answer these questions it is necessary to consider a third type of language problem, the repetition aphasia.

Repetition aphasia is a disorder in which one is unable to repeat certain words or sentences. It may be especially difficult for the person to repeat words like "if," "the," "yes," and "is." To say, "no ifs, ands, or buts" may be an impossible task. Speech and writing may be accomplished rapidly, which is not the case in Broca's disorder, and verbal language may be understood, which is not the case in Wernicke's disorder. Repetition aphasia usually follows an injury in the lower parietal lobe, an injury through which Broca's and Wernicke's areas have become substantially disconnected.

APHASIA. *Pertains to any language communication disorder* (G. a, *without;* phasis, *speech*).

To appreciate what has happened, it is important to realize that language is learned by either hearing or seeing words. When we hear or see a word, we send it into Wernicke's area, in which some association is made between the visual or auditory symbolism and other sensations we perceive from our environment. When we wish to speak, we send the verbal abstractions from Wernicke's area, through the arcuate fasciculus, to Broca's area. Here, instructions are sent out to the adjacent motor area to produce talking, singing, and the like.

Variations on Broca's, Wernicke's, and repetition aphasias are abundant. Four such examples will help further describe the verbal center for communication. These examples are: (1) pure word deafness (acoustic aphasia); (2) ability to copy the written word, but failure to understand its meaning (alexia); (3) ability to mimic someone's speech but failure to understand its meaning (pure speech aphasia); and (4) an inability to remember verbal information (amnesic aphasia).

Pure Word Deafness (Acoustic Aphasia). A person with pure word deafness *(acoustic aphasia)* is able to read and write. He is also able to hear, but is unable to understand the spoken word. On autopsy it may be seen that certain hearing pathways into Wernicke's area had been destroyed. The damage is seen along two routes: (1) in the left hemisphere along the hearing pathway to Wernicke's area; and (2) along the portion of the corpus callosum which carries hearing sensations from the right ear to Wernicke's area in the left hemisphere. Figure 11.12 is a representation of this situation, as might be seen by looking downward into the brain.

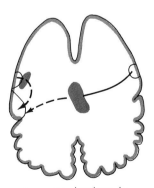

FIG. 11.12. *Lesions in the auditory circuit that result in pure word deafness. Nerve impulses do not reach Wernicke's area.*

Pure Word Blindness (Alexia). The ability to copy writing correctly without understanding its meaning *(alexia)* is another kind of aphasia caused by damage to the corpus callosum. In this case, the individual suffers damage along the pathway from the left visual cortex into Wer-

ALEXIA. *Word blindness* (G. *a*, *without;* lexis, *word*).

nicke's area, and along the portion of the corpus callosum which carries visual information from the right visual cortex to Wernicke's area. As a result, although words can be seen, the information they convey cannot be interpreted and understood. Figure 11.13 indicates the damaged pathway involved in this situation, as seen by looking downward into the brain.

Pure Speech Aphasia. Pure speech aphasia is the ability to repeat spoken words, but not to understand their meaning; it may be due to a destruction of nervous tissue that lies outside of the verbal center. Broca's area, Wernicke's area, and the arcuate fasciculus may be completely intact, and the individual may be able to "parrot" correctly. However, because of destruction in outlying association areas, the language input fails to activate the circuitry through which the complex of memory-recall-understanding is marshalled into action. The individual must now learn anew. If the damage is too extensive, this learning may not be possible, because it may not be possible to process new information. The individual is thus left permanently without the ability to initiate speech. Only mimicry may be possible.

AMNESIA. *Loss of memory* (G. amnestia, *forgetfulness*).

Amnesic Aphasia. Many cases of aphasia are complicated by a loss of memory (*amnesia*). In some amnesic aphasias damage is seen in the prefrontal area, and only recent memory is lost. In other amnesic aphasias, damage may occur extensively in the temporal lobe, and long-term memory, with or without recent memory, is lost. In still other cases an individual may lose the ability to store information for a certain period, as sometimes happens after a stroke or accident.

In time, these memory losses may slowly return. In many of these amnesic aphasias, an autopsy reveals damage in the left hippocampus of the limbic lobe. Inasmuch as the damage was never repaired during the life of the individual, how was it possible for him to regain memory-storage ability? One explanation is that the brain began to make use of the right hippocampus.

AGNOSIA. *Literally, not knowing* (G. a, *without;* gnosis, *knowing*).

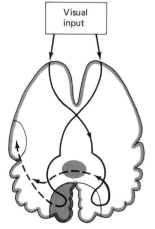

FIG. 11.13. *Certain lesions in the visual circuits result in pure word blindness.*

EVALUATION OF THE CEREBRAL CORTEX

Damage to the cerebral cortex may be caused by a stroke, accident, or disease. An individual who has suffered brain damage may be unable to recognize familiar objects (*agnosia*). He may be unable to perform previously learned tasks (*apraxia*), such as hitting a nail with a hammer. He may be unable to write (*agraphia*) even though the muscles used for writing are not affected. He may be unable to perceive peripheral touch stimuli (*anesthesia*) or he may experience difficulty in smelling (*anopsia*) or tasting (*ageusia*). In these disorders, the cause of the brain damage may be determined. However, there are numerous brain diseases arising from unknown causes. Examples of such disorders are multiple sclerosis and epilepsy.

Multiple Sclerosis

Multiple sclerosis is characterized by demyelination of nerve fibers and the occurrence of sclerotic patches in the brain and spinal cord. The patches represent a hyperplasia (excessive cell formation) of fibrous connective tissues around the nerve cells.

Multiple sclerosis rarely begins before age 20 or after 50. It occurs more often in women than men, and it sometimes appears to be inherited. The early symptoms may include short bouts of double vision, proprioceptive problems, dizziness, trembling, defective memory, and behavioral problems. These symptoms may be misdiagnosed and even overlooked for a period of several years. About ten years may elapse before the victim becomes severely incapacitated. In severe cases, the victim may suffer paralysis, disturbances of speech, proprioceptive disorders, and other malfunctions. The particular afflictions depend on where the lesions occur.

Epilepsy

This is a relatively common brain disorder, occurring in more than 0.5% of the population. Before age seven, epileptic seizures occur in about 5% of the population. The disorder has a strong hereditary basis and may also be due to such diverse causes as injury, metabolic disorders, an infection, or cancer.

The disease can be diagnosed by "seizure-discharge" patterns in electroencephalograms (EEG's), although there is absolutely no visible histological or cytological difference between a brain obtained on autopsy from an epileptic and nonepileptic individual. The "seizure-discharge" patterns are due to a sudden, excessive, and disorderly discharge of nerve cells in the motor region of the brain. Sometimes they occur in the temporal and limbic lobes as well.

There are various forms of epilepsy, the most common being petit mal, psychomotor seizure, and grand mal. *Petit mal*, which is usually not treated with drugs, occurs almost exclusively in children between the ages of 3 and 12. It usually disappears by age 9 or 10 and the individual may be spared any form of epilepsy for the rest of his or her life. During a petit mal seizure—which may occur many times an hour—the child blanks out for a few seconds, staring into space as if in a daydream. In addition, the child may display an uncontrollable facial gesture or a jerk or a meaningless movement of the head, eyes, and trunk. When petit mal persists in later childhood it may develop into psychomotor seizures, grand mal, or some other more serious form of epilepsy.

A *psychomotor seizure*, like a petit mal episode, is short-lived. For upwards to a few seconds there is a loss of consciousness along with some meaningless movement or act, such as smacking the lips or clapping the hands. These acts are accompanied by a brief lapse of memory; the individual cannot remember the seizure.

Grand mal, a more severe form of epilepsy, can be controlled in about

PETIT MAL. *A mild form of epilepsy (F.* petit, *small;* mal, *evil).*

GRAND MAL. *A major epileptic seizure.*

Summary

85% of the cases by Dilantin (diphenylhydantoin) or some other antiepileptic drug. The drugs cannot cure epilepsy, but they can reduce the frequency and severity of seizures and sometimes prevent them from occurring altogether.

Grand mal seizures are characterized by convulsions, unconsciousness, or both. Sometimes, in later stages, they lead to mental disturbances. The beginning of the seizure is recognized by the victim as an *aura*—a peculiar sensation which may include dizziness and paresthesia (burning, pricking, numbness, and similar sensations) in the upper central region of the abdomen or in the hand or leg ascending to the head. There may also be noises in the ear or flashes of light.

AURA. *A premonition that precedes a convulsion (G. breeze).*

SUMMARY

1. The motor area of the cerebral cortex has its neurons arranged in point-to-point correspondence to the body regions which they regulate.
2. Immediately anterior to the motor area are the premotor and prefrontal areas where some of the input for the motor area is set up on the basis of (a) emotional feelings, (b) judgment, and (c) other kinds of input arriving from other regions of the brain.
3. Immediately posterior to the motor area lies the somesthetic area. General sensory information about pain, pressure, and skin temperature is received here topographically, as is some of the special sensory information on taste and smell.
4. Four additional primary sensory areas are present in the cerebral cortex. Whereas the somesthetic area is a primary sensory area for general peripheral sensations, the four primary sensory areas referred to here are special areas for vision, hearing, smell, and taste. Vision is received in the occipital cortex, hearing in the temporal cortex, smell in the limbic cortex, and taste in the somesthetic cortex and limbic cortex.
5. Emotions are felt and expressed through various regions of the brain, especially through the limbic lobe and hypothalamus. Dopamine and noradrenaline are the key neurotransmitters which bring about feelings of emotional satisfaction. Other neurotransmitters or certain combinations of neurotransmitters are involved in other feelings. Special kinds of circuits are used to bring about certain kinds of emotional feelings. The amygdala is concerned with self-preservation and aggression, among other things; the septal area is related to social and sexual behavior; the anterior thalamic nuclei with emotional memory; and the foregoing and other regions of the brain with additional kinds of emotionality.
6. Verbal communication, the use and interpretation of written and spoken words, is accomplished through gray matter in Broca's and Wernicke's areas. These areas are connected with each other indirectly and with the visual and hearing centers as well as the motor cortex

The Cortical Systems

which operates the speech apparatus and the handwriting apparatus. Damage to these centers, as may happen through a stroke, brings about various kinds of aphasia, depending on what regions of the circuitry were damaged.

REVIEW QUESTIONS

1. Define the terms "somesthetic topography" and "general skeletal motor system topography." What is their importance in human physiology?
2. In what lobes of the brain are the primary sensory areas for each of the following: vision, hearing, taste, and smell?
3. Define: emotional feeling, emotional action, emotional behavior, affective state, mood, cognition.
4. Describe the limbic lobe. Based on its location, tell why this cortical region may be important in emotional state or in emotional behavior.
5. How are noradrenaline, dopamine, and serotonin implicated in mood?
6. What types of bodily and mental effects are likely to occur following a stroke in Broca's center only? in Wernicke's center only?
7. Why does the right side of the body commonly suffer paralysis when Broca's area is injured?
8. Where are lesions present when a person suffers pure word blindness? pure word deafness?

The Endocrine System

12 General Endocrinology
13 The Reproductive System

PART THREE

General Endocrinology 12

GENERAL ENDOCRINE CONCEPTS
    Two classes of endocrine cells
    Three classes of hormones
    Hormonal pathways
REGULATION OF HORMONE OUTPUT
    The negative feedback control system
    A second kind of control system
THE HYPOTHALAMUS AND PITUITARY GLAND
    Problems due to insufficient or excessive pituitary hormones
THE THYROID GLAND
    The effect of thyroxine
    Thyroid problems
    Another kind of thyroid hormone
THE PARATHYROID GLANDS
    The effect of parathyroid hormone
    Parathyroid problems
    Another regulator of calcium in the blood—a vitamin or hormone?
THE ADRENAL GLANDS
    The adrenal medulla
        The effect of adrenaline
        Misfunctions of the adrenal medulla
    The adrenal cortex
        Actions of glucocorticoids
        The effects of excessive glucocorticoids
        Actions of mineralcorticoids
        The effects of excessive mineralcorticoids
        The effects of a deficit in corticoids
        The effect of excessive adrenal sex hormones
THE PANCREAS
    The functions of insulin and glucagon
    Diabetes mellitus

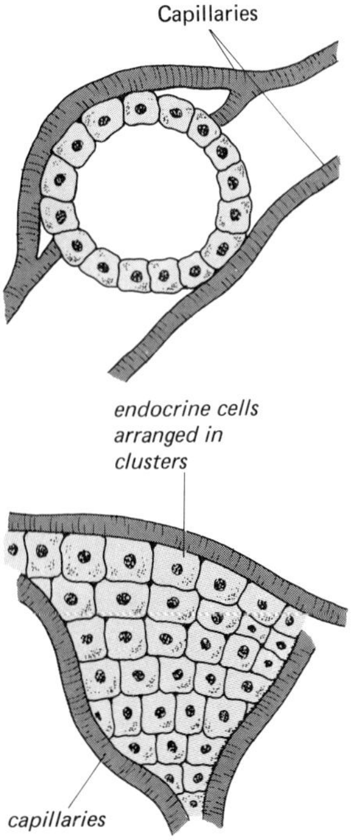

182
General Endocrinology

We have seen that many functions of the body are regulated by the nervous system. Others are regulated by *hormones*. Hormones are also called *endocrines,* a word which refers to their being cellular secretions that are passed directly from a cell into the bloodstream, as shown in Figure 12.1.

The opposite of endocrine is the word *exocrine*. Exocrine secretions are cellular secretions that are carried through ducts to the outside of the body. An exocrine gland is shown in Figure 12.2 for comparison. Sweat, sebum, tears, and even digestive juices, all of which lead eventually to the outside of the body, are exocrine secretions.

We begin this chapter by defining a hormone and describing two classes of endocrine cells and three classes of hormones. Each hormone travels along one, or two, or three kinds of pathway. Accordingly, the three basic pathways are described next, at which point the reader will have been introduced to every well-established endocrine system in the body. Thereafter, the text continues with a discussion on the basic mechanisms through which endocrine glands are regulated in terms of increasing or decreasing their hormonal output.

We will then discuss the basic functions and misfunctions of the various components of the endocrine system. In this chapter we shall limit these discussions to the following organs: (1) the hypothalamus of the brain; (2) the pituitary gland; (3) the thyroid gland; (4) the parathyroid glands; (5) the pancreas; and (6) the adrenal glands.

The following chapter discusses three additional endocrine organs—the ovaries, testes, and placenta. And in Chapter 19 we take up two others—the stomach and small intestine. Still later in the text, the reader will find information on the endocrine functions of the kidney (Chapter 30) and thymus gland (Chapter 32).

All of the major endocrine organs are shown in Figure 12.3. All of the known endocrines are listed in Table 12.1. The table indicates the sources of all the endocrines (that is, the glands or organs in which they are produced or secreted) and the primary function of each. The table also lists the major substances that are known to be involved in the regulation of the different hormone systems. These substances circulate in the bloodstream, and when their concentration in the blood reaches a particular level (which may be either a high or a low level) the endocrine tissue at the left-hand side of the table (such as the hypothalamus or the thyroid gland) is stimulated to stop secreting its hormone. You should also note that Table 12.1 describes the different hormones in terms of hormonal *systems*, or *pathways*, each of which involves a sequence of related endocrine glands and their products. In these pathways a vertical arrow leads from an endocrine tissue to the hormone it produces (shown in color beneath the tissue). Horizontal arrows lead to other tissues (some of which are also endocrine tissues) that are stimulated by the hormones; the colored arrows indicate the principal or primary function of the hormone system in the functioning of the body. This is discussed in greater detail on pages 193 to 198, and is illustrated in Figure 12.5.

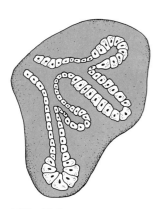

FIG. 12.1. *Basic structure of endocrine glands. In thyroid gland (top) cells are arranged in spherical follicles.*

FIG. 12.2. *One type of exocrine gland; secretions leave the gland by way of several ducts.*

GENERAL ENDOCRINE CONCEPTS

A *hormone* is a chemical substance produced by cells of one part of the body and passed from these cells into the bloodstream to circulate and eventually stimulate different kinds of cells elsewhere in the body. Numerous kinds of cells are encountered by each of the body's hormones as they circulate, but only specific kinds of cells are affected by any particular hormone. Inasmuch as the body's entire blood supply (about 5 liters in an average adult) makes a complete circulation in one minute when the body

HORMONE. *A cell stimulatory substance* (G. hormaein, *to excite*).

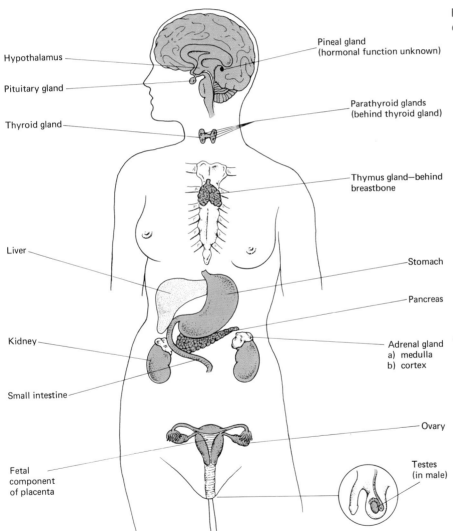

FIG. 12.3. *Endocrine organs of the body.*

General Endocrine Concepts

TABLE 12.1. Major Hormones (Color), Their Tissue Sources, and Their Primary Functions.

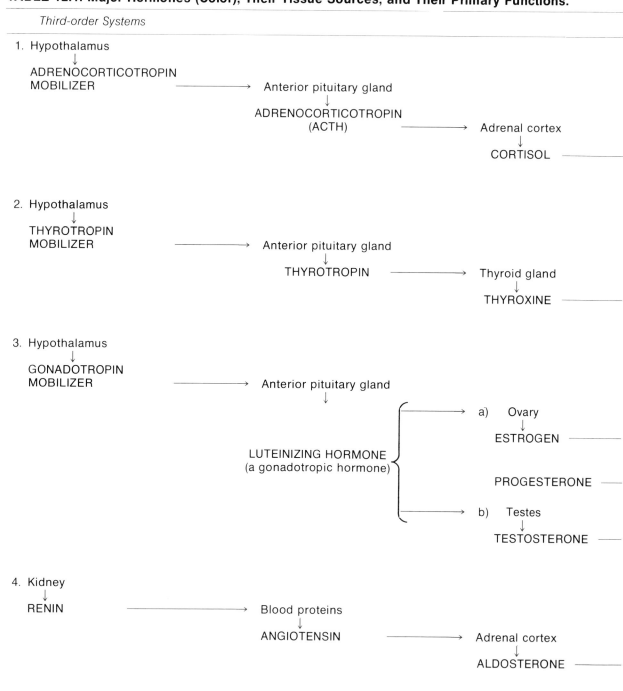

General Endocrinology

	Primary Function of System	*Substances Involved in Regulation*
→	Promotes production of sugar from proteins and fat	Excessive glucocorticoids (e.g., cortisol)
→	Promotes metabolism and respiration	Excessive thyroxine or T_3
→	Promotes development of female sex characteristics	Mechanism not known
→	Prepares uterus for childbirth	
→	Promotes development of male sex characteristics	Mechanism not known
→	Retains sodium in body and increases blood pressure	Excessive sodium

TABLE 12.1. (*Continued*)

Second-order Systems

1. Hypothalamus
   ↓
   GONADOTROPIN MOBILIZER ⟶ Anterior pituitary gland
   ↓
   FOLLICLE-STIMULATING HORMONE ⟶ a) Ovary
   (a gonadotropic hormone) ⟶ b) Testes

2. Hypothalamus
   ↓
   GROWTH-HORMONE MOBILIZER ⟶ Anterior pituitary gland
   ↓
   GROWTH HORMONE

3. Hypothalamus
   ↓
   PROLACTIN MOBILIZER ⟶ Anterior pituitary gland
   ↓
   PROLACTIN ⟶ Mammary glands

4. Kidney
   ↓
   RENIN ⟶ Blood proteins
   ↓
   ANGIOTENSIN ⟶ Blood vessels

5. Hypothalamus
   ↓
   MOBILIZER OF MELANOCYTE-STIMULATING HORMONE ⟶ Anterior pituitary gland
   ↓
   MELANOCYTE-STIMULATING HORMONE

General Endocrinology

Primary Function of System	Substances Involved in Regulation
a) production of eggs b) production of sperm	a) Excessive progesterone b) Mechanism not known
Promotes growth and periodic replacement of cells, extension of long bones	
Promotes lactation (production of milk)	
Increases blood pressure	Excessive sodium
Not known with certainty (see text)	

TABLE 12.1. (*Continued*)

First-order Systems

1. Hypothalamus
   ↓
   GROWTH HORMONE
   DEMOBILIZER ⟶ Anterior pituitary gland

2. Hypothalamus
   ↓
   PROLACTIN
   DEMOBILIZER ⟶ Anterior pituitary gland

3. Hypothalamus
   ↓
   OXYTOCIN ⟶ Uterus and mammary glands

4. Hypothalamus
   ↓
   ANTIDIURETIC HORMONE ⟶ Kidney tubules

5. Adrenal cortex
   ↓
   ALDOSTERONE ⟶ Kidney tubules

6. Adrenal cortex
   ↓
   TESTOSTERONE

7. Adrenal medulla
   ↓
   ADRENALINE ⟶ Circulatory, respiratory, and metabolic systems

8. Thyroid gland
   ↓
   CALCITONIN ⟶ Bone cells

9. Parathyroid glands
   ↓
   PARATHYROID HORMONE ⟶ Bone cells, kidney tubules, and intestinal lining

10. Stomach
    ↓
    GASTRIN ⟶ Stomach

General Endocrinology

	Primary Function of System	Substances Involved in Regulation
→	Inhibits release of growth hormone	
→	Inhibits release of prolactin	
→	Promotes action of smooth muscle in delivery and in breast-feeding.	?
→	Reduces volume of urine	Excessive body fluids
→	Reduces excretion of sodium; increases excretion of potassium	Excessive sodium
→	Increases libido	?
→	Shunts blood from viscera and skin to skeletal muscle and heart; dilates bronchi; increases metabolic rate.	Regulated by central nervous system
→	Reduces calcium in blood	Insufficient calcium
→	Increases calcium in blood	Excessive calcium
→	Increases release of digestive juice	Hypothetically, absence of food in lower region of stomach

General Endocrine Concepts

TABLE 12.1. (*Continued*)

First-order Systems

11. Small intestine
    ↓
    PANCREOZYMIN ⟶ { a) Pancreas
                     b) Gallbladder

12. Small intestine
    ↓
    SECRETIN ⟶ { a) Pancreas
                 b) Liver

13. Small intestine
    ↓
    ENTEROGASTRONE ⟶ Stomach

14. Hypothalamus
    ↓
    DEMOBILIZER FOR MELANOCYTE-STIMULATING HORMONE ⟶ Anterior pituitary gland

15. Pineal gland
    ↓
    MELATONIN

16. Thymus gland
    ↓
    THYMUS HORMONES

17. Kidney
    ↓
    ACTIVE VITAMIN D_3 ⟶ Small intestine

18. Kidney
    ↓
    ERYTHROPOIETIN ⟶ Bone marrow

	Primary Function of System	Substances Involved in Regulation
→	a) Release of digestive juice	Hypothetically, absence of food in duodenum
→	b) Contraction of gallbladder; release of bile	Hypothetically, absence of fat in duodenum
→	a) Release of an alkaline juice for neutralizing the stomach effluents as they enter the intestine.	
→	b) Production of bile	Hypothetically, absence of fat in duodenum
→	Regulates digestion in the stomach	
→	Inhibits release of melanocyte-stimulating hormone	
→	Unknown	
→	Stimulation of immunological system (see Chapter 32)	
→	Absorption of calcium	Hypothetically, excessive calcium.
→	Increases production of red blood cells	Excessive red blood cells

is at rest — and in as little as 15 or 20 seconds during exercise — the hormones reach their target sites fairly rapidly.

Two Classes of Endocrine Cells

Hormones are produced either in endocrine gland cells or in neuroendocrine cells. The *endocrine gland cells* are specialized epithelial cells which are widespread in the body. They occur in commonly known endocrine organs, such as the pancreas and thyroid gland, and in organs with specialized endocrine functions, such as the kidney and thymus gland.

Neuroendocrine cells are nerve cells which are known to be present only in the hypothalamus, a part of the brain. Hormones that are often called *neuroendocrines* are produced in the cell body of these cells and are thereafter transported through the axon of the cell to the tips of the nerve fiber, from which they are discharged into the bloodstream when needed.

HYPOTHALAMUS. *A region of the brain beneath the thalamus* (G. hypo, beneath).

Three Classes of Hormones

All of the known human hormones are derived either from cholesterol or from amino acids. The cholesterol derivatives, known as steroids, comprise one of the body's three classes of hormones. The other two classes of hormones are: polypeptides, which are small proteins made up of from about 10 to 100 amino acids; and amino acid derivatives, which are small hormones consisting of metabolically-modified amino acids.

Polypeptide hormones are small proteins made of amino acids that are linked to each other through peptide bonds. The peptide bonds can be split through the reaction of hydrolysis (see Figure 19.1 and page 19). The practical importance of this is that polypeptide hormones cannot be administered orally. To do so would subject them to the hydrolytic reaction of digestion. To be effective, polypeptide hormones must be administered intravenously (into a vein), intramuscularly (into a muscle), or subcutaneously (beneath the skin).

The following hormones are polypeptides: (1) all of the hormones from the hypothalamus and pituitary gland; (2) calcitonin from the thyroid gland; (3) parathyroid hormone from the parathyroid glands; (4) glucagon and insulin from the pancreas; and (5) chorionic gonadotropin, which is produced by the placenta immediately after pregnancy begins.

Modified amino acid hormones are produced from amino acids, and do not have a peptide bond. They include: thyroxine from the thyroid gland; adrenaline from the adrenal gland; and melatonin from the pineal gland. (Note that triiodothyronine, T3, is also produced by the thyroid gland. Chemically, it is very much like thyroxine, but T3 has three iodine atoms, whereas thyroxine has four. The actions of T3 are essentially identical to those of thyroxine so that when we speak of thyroxine in this book we are also speaking of T3. Also, the pineal gland and melatonin are not discussed

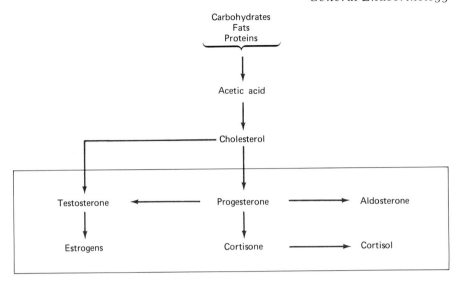

FIG. 12.4. *The steroid hormones (in box) are synthesized from various nutrients.*

further in this book because the function of this system in humans is not really known.)

None of these hormones can be destroyed by digestive enzymes while they are in the mouth, stomach, or intestine. Nevertheless, in order to get adrenaline quickly to its site of action, it is administered intravenously (when needed throughout the body), or as an inhalant (when needed for enlarging the passageway into the lungs). Thyroxine, on the other hand, is usually administered orally.

Steroid Hormones. There are more than 20 steroid hormones in the body, but only the six most important will be discussed in this text. They are: aldosterone, cortisol, cortisone, estrogen, progesterone, and testosterone. See Figure 12.4. These hormones can be taken orally, or, for quicker action, intravenously.

Hormonal Pathways

Many hormones, upon being secreted from the cell in which they were produced, travel through the blood and activate their target cells, causing a particular physiological effect. Other hormones stimulate their target cells to release a second hormone, which in turn passes into the blood to activate a second kind of target cell to bring about a physiological effect. Still other hormones react with target cells which release a second hormone. The second hormone reacts with a second kind of target cell to bring about the release of a third hormone. This hormone, in turn, acts upon a third and final kind of target cell in the body (see Figure 12.5). The relation between first-order, second-order, and third-order hormone pathways is shown in the following diagram:

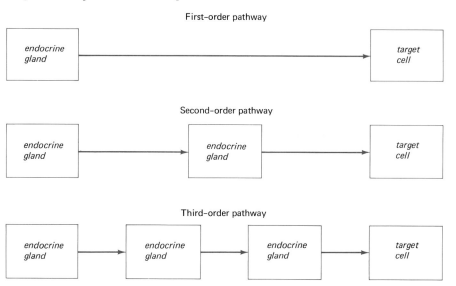

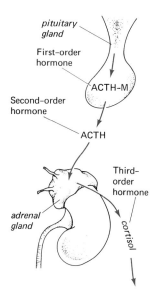

FIG. 12.5. *The principle of first-, second-, and third-order endocrine pathways (left), starting with ACTH-mobilizer. The intermediate endocrine glands (the adrenal gland, in the drawing above) can also be thought of as target cells.*

As the diagram indicates, a pathway is described as first-order, second-order, or third-order, depending upon the number of hormones released. Table 12.1 contains a list of these pathways and the sequence of hormones in each system. The reader is urged to examine all but the last column in this table at this point. (Further explanation is needed in order to appreciate the last column). In looking over the table, it should be noted that some hormones bear the suffix *tropin*. These hormones are chemicals which—in all cases—stimulate (turn on) a second-order or third-order gland to release a hormone.

REGULATION OF HORMONE OUTPUT

A given hormone is released from an endocrine tissue depending on the body's need for it. For example, when the level of sugar in the blood is very low, a specific amount of cortisone may be released to increase it. When the sugar level is too high, insulin is released to decrease it.

What determines how much of a given hormone should be released at any one time? Two regulatory mechanisms are known: negative feedback, and the use of releasing hormones and release-inhibiting hormones of the hypothalamus.

The Negative Feedback Control System

The basic principle of negative feedback control is that some endocrine tissue—call it A—which regulates the level of some substance in the blood—

call it B—is *inhibited* in its output of hormone. The inhibition may be due to either an excess or a deficiency of B, depending on what tissue A is and what substance B is. In either case, whether it is an excess or a deficiency, the point is that a critical level of B is what *feeds back* to the endocrine tissue to *negate* (inhibit) the hormonal output of tissue A. Hence the term *negative feedback*. Examples of this type of regulation are given in Table 12.2.

TROPIC. Acting upon (G. tropikos, *to turn*).

Why is the negative feedback system such an important control mechanism? To begin with, the body operates best under conditions where all of the substances in the blood are held at an optimum level. That is, the various blood-borne materials are not allowed to exceed or fall far short of some optimal "normal" level.

Second, in order to ensure this, all of the endocrine tissues A—such as the parathyroid, adrenal, and pancreatic glands—ordinarily put out one or more hormones which either increase or decrease the level of substances B in the blood.

Third, although there may be only one kind of endocrine tissue A which regulates—through negative feedback—the level of a substance B in the blood, in many cases there are two or more endocrine tissues operating through negative feedback to insure a steady-state level of blood-borne substance. An illustration of this is shown in Figure 12.6 with respect to the regulation of blood calcium.

Finally, two situations may be considered to illustrate the overall operation of the negative feedback principle. In one case, when a particular

TABLE 12.2. The Concept of Negative Feedback

Initial Condition	Primary Effect	Secondary Effect	Final Result of Feedback Mechanism
Too much calcium in blood	Reduces secretion of parathyroid hormone from parathyroid glands	Excretion of calcium by kidneys; less absorption of calcium from intestine; less release of calcium from bone into blood	Amount of calcium in blood is reduced to a "normal" level
Too much sodium in body	Inhibits release of renin from the kidney	Less release of aldosterone from the adrenal cortex	Sodium is excreted by the kidney and the level of sodium in the body is returned to a "normal" level
Too little sugar in blood	Inhibits release of insulin from the pancreas	Simultaneous release of glucagon, adrenalin, cortisol, and growth hormone from pancreas and other endocrine glands	Increase in blood sugar to "normal" level
Too much thyroxine in blood	Reduces secretion of thyrotropin-releasing hormone from the hypothalamus	Reduced stimulation of thyroid gland	Reduction of thyroxine in blood to "normal" level

substance B—such as calcium—is present at too *low* a level in the blood, it feeds back to several tissues A which produce the appropriate regulatory hormones. The tissues A in this case include the parathyroid glands in particular, and the thyroid gland, liver, and kidney to a lesser extent. The parathyroid glands are now induced to produce and secrete larger quantities of their hormone—the parathyroid hormone. This hormone operates predominantly on three kinds of cells: (a) bone cells, to bring calcium out of the bone tissue into the blood; (b) kidney tubular cells, to prevent calcium from passing out of the body; and (c) the intestinal lining (along with vitamin D) to bring calcium out of the intestinal canal into the bloodstream.

While parathyroid hormone is exercising its function, the hormone calcitonin, from the thyroid gland, may be inhibited—because of the low blood level of calcium—from entering the blood. The reduced level of calcitonin in the blood will result in less activity on the part of bone cells in incorporating blood-calcium into their organic matrix as a calcium phosphate mineral. And while the parathyroid glands and the thyroid gland are carrying out their regulatory role with respect to calcium, the liver and kidney may also be stimulated (by the low level of blood calcium) to produce increased quantities of active vitamin D (Figure 12.6). This will facilitate the uptake of calcium from the gut into the bloodstream. Thus, working together through negative feedback mechanisms, the tissues A ensure that the blood level of calcium will be optimal for the operation of the muscle cells and nerve cells in the body, as well as perhaps for all other kinds of cells in the body.

In the second case, if a particular substance B—such as glucose—is present at too *high* a level in the blood, it will feed back to several tissues A

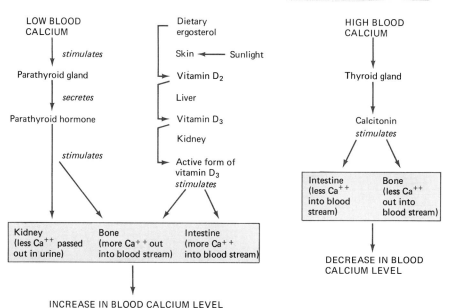

FIG. 12.6. *Mechanisms that regulate the level of calcium (Ca^{++}) in the blood.*

which put out the appropriate regulatory hormones. The tissues A in this case include the hypothalamus, adrenal cortex, adrenal medulla, and pancreas. These tissues, jointly, are stimulated via negative feedback to reduce the level of glucose in the blood.

In the two descriptions you have just read, two nutrients—calcium and glucose—were used as examples of some substance B which feeds back to regulate the hormonal output of tissue A. You are now urged to examine Table 12.2 again which will provide you with a better understanding of the concept of negative feedback. In this table you should note that both hormones and nutrients can play the role of substance B in regulating the hormonal output of tissue A.

You should now examine the last column of Table 12.1. For many of the endocrine systems, this column lists some key substance, which turns off the output of hormones from the tissue in the first column. Additional substances may also effect a negative feedback action; only the major substance is shown in the table.

A Second Kind of Control System

The use of releasing hormones and release-inhibiting hormones is another important mechanism for regulating hormonal output. This control system is confined entirely to the central nervous system, in particular to the hypothalamus.

In order to appreciate this concept it is necessary to realize that the hypothalamus produces nine different kinds of hormones, called neuroendocrines, which it sends into the anterior pituitary gland. Six of these neuroendocrines are releasing hormones. The other three are release-inhibiting hormones. In order to prevent our terminology from becoming too cumbersome, the releasing hormones will be called *mobilizers*, and the release-inhibiting hormones will be called *demobilizers*. Thus, for example, what is technically called growth hormone release-inhibiting hormone will be called growth hormone demobilizer.

What are the functions of the mobilizers and demobilizers? Let us consider this question with respect to growth hormone. Ordinarily, a certain amount of growth hormone mobilizer is secreted by the hypothalamus. This hormone is sent to the anterior pituitary gland by the bloodstream. In the anterior pituitary, it stimulates specific cells to release growth hormone. In consequence, growth hormone enters the bloodstream. As growth hormone travels through the body, it stimulates various cells to grow and divide.

There are times, however, when growth must be inhibited. This might be the case during a prolonged fast, when there may not be enough reserve nutrients in the body to provide for cell growth. It may also happen at the end of wound-healing, when growth hormone is not needed at extra high levels. It may also occur at the end of a pregnancy or when a person's long bones, at about age 21, cease to elongate. At such times a mechanism must be available for turning off the release of growth hormone. This mechanism makes use of the demobilizing hormone.

Growth hormone demobilizer, after it is secreted from neuroendocrine cells in the hypothalamus, passes into a stream of blood which is on its way to the anterior pituitary gland. Here, within the pituitary gland, the demobilizing hormone reacts with cells which specifically produce and secrete growth hormone. As a result of this reaction, less growth hormone is released into the bloodstream.

Thus, in this type of regulation—in contrast to negative feedback—the central nervous system triggers the secretion of either: a hormone which causes a gland to release some of its hormonal content; or a hormone which prevents the gland from releasing its hormone.

The reader may now wish to re-examine Table 12.1 and note which endocrine systems are regulated jointly by mobilizers and by demobilizers. In a sense these systems are analogous to the sympathetic-parasympathetic components of the autonomic nervous system, wherein the sympathetic-parasympathetic systems work cooperatively, yet in opposite directions, in regulating the operation of smooth muscle, cardiac muscle, and gland secretion.

THE HYPOTHALAMUS AND PITUITARY GLAND

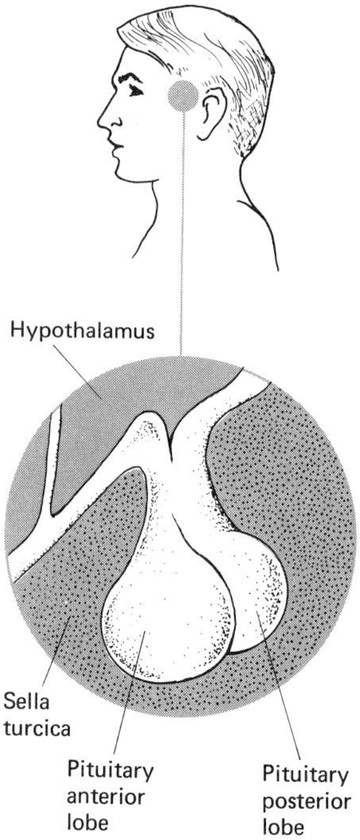

It was pointed out earlier that the hypothalamus monitors and regulates several physiological functions, including water intake, body temperature, and reproduction. Here we will see that these functions are mediated, at least in part, by hypothalamic endocrines.

The hypothalamus produces: six mobilizing hormones; three demobilizing hormones; and two hormones which are neither mobilizers nor demoblizers. All of these hormones are neuroendocrines. They are listed in Table 12.1, with a brief summary of their roles.

The known hormones that are produced in nerve cells within the hypothalamus are transported into the pituitary gland. Situated above the roof of the mouth, in the brain, encased in a protective bony structure, the sella turcica, the pituitary gland consists of two lobes, shown in Figure 12.7.

The *posterior pituitary gland*, or posterior lobe, is composed of gland cells (*pituicytes*) whose function is not known, and several hundred thousand nerve axons. The axons stem from cell bodies that are located in the supraoptic and paraventricular nuclei of the hypothalamus. The cell bodies produce *antidiuretic hormone* and *oxytocin*, which are then transported to the tips of the axons, where they are stored until needed (Figure 12.8). The functions of oxytocin and antidiuretic hormone are discussed below, after a brief description of the anterior pituitary gland.

The *anterior pituitary gland*, or anterior lobe, is composed of several distinct classes of cells. Each class is stimulated specifically by either a mobilizer or a combination of mobilizer and demobilizer.

FIG. 12.7. *Location and general structure of the pituitary gland.*

General Endocrinology

The mobilizers and demobilizers enter the pituitary gland by way of a *portal system*. This system consists of blood vessels which pick up blood from capillaries in the hypothalamus and send this blood into capillaries within the anterior pituitary gland. See Figure 12.9.

What is the function of this portal system?

The blood which enters the capillaries of the hypothalamus carries chemical "messages" in it from all regions of the body. These messages might say, "the body needs more blood sugar—send out adrenocorticotropin;" "the temperature of the body needs to be raised—send out thyrotropin;" or "an egg is ready to be released from the ovary—cut down on follicle stimulating hormone until it is seen whether pregnancy occurs."

In response to a message, the receptor neurons in the hypothalamus send out either a mobilizer or a demobilizer hormone. The hormone first enters a blood capillary in the hypothalamus. It is then swept into a portal vessel which conveys it to a second bed of capillaries in the anterior pituitary gland. Here it stimulates its target cells, causing those cells to release a hormone or hold back on releasing one.

With the preceding principles in mind, we will now list the nine hormones

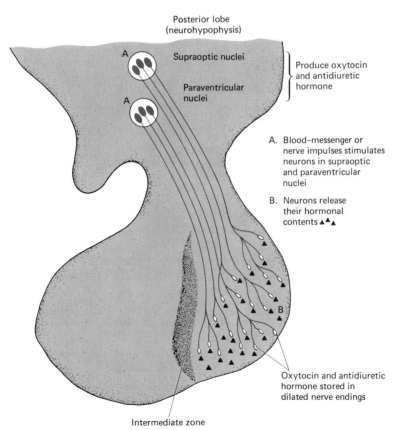

FIG. 12.8. *Mechanism of the posterior pituitary system.*

The Hypothalamus and Pituitary Gland

that are released from the pituitary gland, and summarize their actions. The first two—antidiuretic hormone and oxytocin—are released from the posterior lobe.

(1) *Antidiuretic hormone*, also known as *vasopressin*, is released when the body fluid is low in water, high in solutes. Its release is inhibited by a body fluid high in water concentration, and it acts on the kidney to prevent excessive urination.

Sometimes, as a result of an infection, a tumor, an inherited disorder, or some unknown (idiopathic) cause, the brain secretes inadequate levels of antidiuretic hormone. For similar reasons, the kidney sometimes fails to respond to antidiuretic hormone. In either case, a disease called *diabetes insipidus* may set in. Large volumes (4 to 10 quarts) of sugar-free water are urinated each day, volumes that must be compensated for by excessive drinking. If the posterior pituitary is at fault, this disease may be treated by injecting antidiuretic hormone (a polypeptide), into the skin or muscles, or by applying the hormone, as a powder, to the inner surface of the nose.

(2) *Oxytocin* is released in response to signals we know nothing about. It stimulates the uterus to contract during childbirth, and is often admin-

ANTIDIURETIC. *A substance that opposes or prevents the excretion of urine* (G. anti, *against*; diourein, *to pass urine*).

DIABETES INSIPIDUS. *A disease due to a deficiency of antidiuretic hormone* (G. diabetes, *to pass through*; L. insipidus, *tasteless*).

OXYTOCIN. *A hormone that facilitates childbirth* (G. oxys, *swift*; tokos, *birth*).

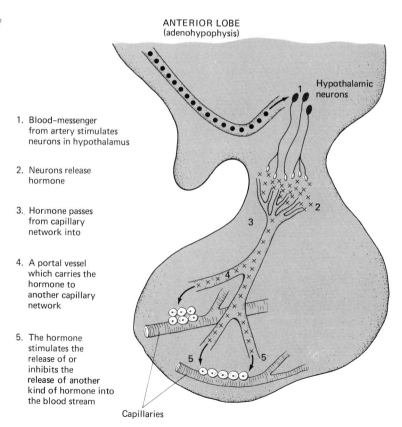

FIG. 12.9. Mechanism of the anterior pituitary system.

ANTERIOR LOBE (adenohypophysis)

Hypothalamic neurons

1. Blood-messenger from artery stimulates neurons in hypothalamus
2. Neurons release hormone
3. Hormone passes from capillary network into
4. A portal vessel which carries the hormone to another capillary network
5. The hormone stimulates the release of or inhibits the release of another kind of hormone into the blood stream

Capillaries

istered to women in labor to facilitate delivery. It also stimulates the release of milk from the mammary glands after the baby is born. Its role in the male is unknown.

The remaining pituitary hormones are produced in, and released from, the anterior pituitary gland.

(3) *Melanocyte-stimulating hormone* is a hormone whose function is not too clear. In some lower animals, such as certain lizards, it permits rapid changes in color, in response to environmental stimuli. In humans it is believed to exert a partial effect in the direction of increasing melanin production in the skin; melanin is the pigment which gives skin its color.

MELANOCYTE. *A melanin-producing cell (G. melanin, black; kytos, cell).*

(4) *Growth hormone*, mentioned earlier, is necessary for normal growth. Emotional disturbances can act through the hypothalamus to reduce its output. In such cases, as in other cases where growth hormone production is reduced, the effect is dwarfism. Dwarfism may also be due to other causes, such as thyroid deficiency, malnutrition, chronic illness, or a genetic deficiency—the lack of one X chromosome.

Excessive growth hormone—as occurs when the hormone-producing cells become tumorous—produces one of two effects, depending on whether the bones are fully mature: in growing children the effect is *gigantism;* in fully grown children and adults, acromegaly results. *Acromegaly* is a progressive enlargement of the skeleton and soft tissues without an increase in length. In particular, the head, hands, and feet become disproportionately large.

ACROMEGALY. *Enlargement of the viscera and the hands and feet due to excess growth hormone (G. akron, extremity; megas, large).*

The changes in physical structure that occur in gigantism or acromegaly are relatively unimportant. What is of concern are the secondary effects of the excessive hormone. Pressure of the tumorous pituitary gland on the nearby optic chiasma can produce visual problems, even blindness. Muscle weakness and joint problems appear in many cases. A large number of individuals who develop acromegaly also develop diabetes mellitus, probably because the excess amounts of growth hormone overtax the ability of the pancreas to produce insulin. Various disturbances in the metabolism of a wide variety of nutrients and minerals are not uncommon.

In some cases these difficulties can be resolved by removing part or all of the pituitary gland (subtotal or total hypophysectomy). In such cases it is necessary to compensate for other hormonal deficiencies that might arise. In particular, it is necessary to administer hormones that are ordinarily released from the thyroid gland, adrenal cortex, and gonads through the action of the tropic pituitary hormones.

HYPOPHYSIS. *The pituitary gland (G. undergrowth).*

HYPOPHYSECTOMY. *Removal of the pituitary gland (G. hypophysis, undergrowth; ektome, excision).*

(5) *Thyrotropin* stimulates the release of thyroxine from the thyroid gland. Thyroxine increases the metabolic rate and raises the body temperature. Like all other tropic hormones of the anterior pituitary gland, the release of thyrotropin is regulated through negative feedback. Adequate levels of thyroxine in the blood turn off the release of thyrotropin.

(6) *Adrenocorticotropin* (also called *ACTH*) stimulates the release of glucocorticoid hormones (for example, cortisol and cortisone) from the cortex of the adrenal gland. Its release is regulated by negative feedback—high levels of glucocorticoids in the blood turn it off.

CORTICOID. *A corticosteroid; any steroid which has properties characteristic of the hormones secreted by the adrenal cortex.*

The Thyroid Gland

(7) *Follicle-stimulating hormone,* or *FSH,* is also a tropic hormone. It stimulates the gonads of males to produce sperm and the gonads of females to produce eggs. (See Chapter 13).

(8) *Luteinizing hormone* stimulates the ovary and uterus to produce the female sex hormones estrogen and progesterone. In males it stimulates the testes to produce the male sex hormone testosterone. (See Chapter 13).

(9) *Prolactin* has two functions, both of which are carried out jointly with growth hormone and the female sex hormones. One action is to stimulate the enlargement of the mammary glands as the menstrual period progresses. Thereafter, should pregnancy occur, prolactin continues to stimulate the development of the mammary glands throughout childbirth and nursing. The second action of prolactin is to stimulate the production of milk just before childbirth and throughout the nursing period.

Prolactin is released in response to prolactin mobilizer through neural mechanisms that are not understood. It is released increasingly during the first half of each menstrual period and throughout pregnancy. The prolactin demobilizer is at work at other times.

LUTEINIZE. *To yellow, as an ovarian follicle as the follicular cells begin to produce increasing amounts of sex hormones* (L. luteus, *yellow*).

PROLACTIN. *A pituitary hormone which promotes lactation—the production of milk* (L. lactare, *to suckle*).

Problems Due to Insufficient or Excessive Pituitary Hormones

An overall general reduction in pituitary activity is called *hypopituitarism.* Hypopituitarism can be caused by a disease (for example, tuberculosis), an injury, a blood clot that blocks circulation in the hypothalamus, prolonged treatment with steroid drugs, or some other cause. Any of various problems may result.

There may be a reduced level of sugar in the blood, due to a lack of adrenocorticotropin, and the body temperature may drop, due to inadequate thyrotropin. Any other function regulated by the hypothalamus may also suffer. In some cases the problem may be remedied by administering the appropriate hormones.

Overactivity of the pituitary (*hyperpituitarism*) may also arise. Cushing's disease, for example, may arise due to a tumor in the pituitary gland or the adrenal cortex, as described below in the discussion of the adrenal gland.

THE THYROID GLAND

COLLOID. *A chemical solution of large molecular weight particles, as for example, a protein solution* (G. kolla, *glue;* eidos, *resembles*).

The thyroid gland is located in the neck, in front of the trachea, or windpipe (Figure 12.10). It differs histologically from all other tissues in the body because of its peculiar follicles. Each follicle is a sphere of cuboidal cells around a core of gelatinous substance called *colloid* (Figure 12.1).

The follicles, shown in the upper part of Figure 12.11, are the functional units of the thyroid gland. The follicle does three things no other tissue in

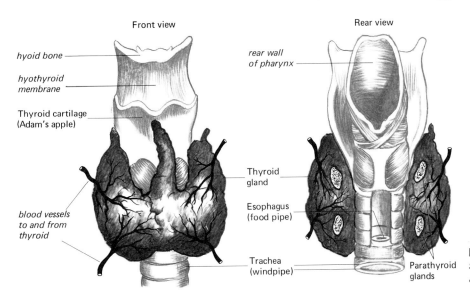

FIG. 12.10. Location and structure of the thyroid and parathyroid glands.

the body can do. First, the cuboidal cells of the follicle accumulate iodide from the bloodstream. The iodide is used in the production of thyroxine. When the cuboidal cells accumulate excessive quantitites of iodide, the condition may be treated in one of three ways. First, because iodide is a monovalent ion, it is possible to administer certain other monovalent ions which are known to compete with iodide for entrance into the cuboidal cells of the follicle. Thiocyanate (SCN^-) and perchlorate (ClO_4^-), for example, are effective competitors. These agents thus tend to reduce or prevent the development of a *goiter*, an enlarged thyroid gland. Also, inasmuch as this gland selectively accumulates iodide, it is possible to administer, by mouth, a large dose of a radioactive form of iodide, a dose that will destroy some of the follicular cells. Third, part of the thyroid gland may be removed surgically.

The second unique function of the thyroid follicle is to incorporate iodide into the amino acid tyrosine. The iodinated tyrosine molecules are then built into *thyroglobulin*, a protein that is stored in the liquid core of the follicle. When thyroxine is needed, the two iodinated building blocks used to produce it are released from thyroglobulin. Then, through several enzyme reactions, thyroxine molecules are assembled and secreted. All of this happens in a period of milliseconds after thyrotropin from the anterior pituitary gland stimulates the follicle cells of the thyroid gland.

When the thyroid gland is overactive (hyperthyroidal) in producing thyroxine, it is possible to reduce the activity as described above. Alternatively, it is possible to reduce its ability to incorporate iodide into tyrosine. Special drugs, such as thiourea and propylthiouracil, are used for this purpose.

The third function of the thyroid follicle is to release thyroxine. Thyroxine is released when the follicles are stimulated by thyrotropin from the pituitary gland.

GOITER. *Enlargement of the thyroid gland (L. guttur, throat).*

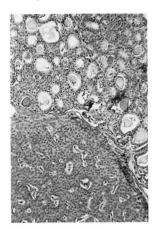

FIG. 12.11. *Tissue from the thyroid gland (top) and parathyroid gland (bottom).*

The Effect of Thyroxine

The major effect of thyroxine is to stimulate metabolism, and in doing so to help maintain the body temperature at 98.6°F (37°C). Although the mechanism is not entirely clear, much of the evidence favors the theory that this hormone does what is called the uncoupling of oxidation from phosphorylation. What does this mean?

In the course of oxidizing (combusting) carbohydrates, proteins, and fats, energy is released. A large percentage of the energy is invested into the production of ATP (adenosine triphosphate) through a process of phosphorylation. In this process, one phosphate ion is bonded to ADP (adenosine diphosphate) and it is within this bond that most of the energy of oxidation is stored.

Thyroxine tends to uncouple oxidation from phosphorylation. In the presence of thyroxine, less ATP is formed, and more of the energy of combustion goes into heat instead. This may explain why a hypothyroidal individual feels cold and a hyperthyroidal person feels very warm while other people in the same environment are comfortable.

Thyroid Problems

CRETINISM. *Stunting of physical and mental growth early in life due to a thyroid deficiency (French* cretin; *a kind of idiot found in the Alps).*

MYXEDEMA. *A thyroid deficiency disease; hypothyroidism* (G. myxa, *mucus;* oidema, *a swelling).*

Deficient quantities of thyroxine (*hypothyroidism*) may lead to the development of a *cretin* – a mentally and sexually defective dwarf. The hormone deficiency may be caused by insufficient iodide or iodine in the diet, a genetic defect, or a pituitary gland that fails to release an adequate amount of thyrotropin.

Hypothyroidism in adults may lead to myxedema and the formation of a goiter. *Myxedema* is characterized by swelling (edema), dryness of the scalp, loss of hair, low body temperature, a slow return of muscle to the "go" position after an extensor reflex, reduced heartbeat and cardiac output, and often an abnormal electrocardiograph.

Goiter is the enlargement of the thyroid gland. It is usually due to an iodide deficiency. As a result, the blood level of thyroxine is below normal. Usually, when this happens the pituitary is summoned to secrete thyrotropin. But in the hypothyroidal person the pituitary is being called continuously to secrete thyrotropin and this hormone is continuously stimulating the thyroid gland to produce thyroxine. The thyroid gland responds by producing more cells (*hyperplasia*) and larger cells (*hypertrophy*). Because of the iodide deficiency, however, these extra, larger cells cannot produce thyroxine.

Thyroid deficiencies in adults may cause trivial or serious consequences. Where the deficiency is minor the individual may scarcely be aware of the defect. The accompanying effects – difficulty in keeping the weight low despite a low caloric intake, fatiguing easily, a low metabolic rate, and feeling cool when others are comfortable – may go unnoticed. A blood analysis, however, will reveal the thyroid problem.

Hypothyroidism usually may be corrected by increasing the person's iodide intake, or by swallowing thyroxine tablets or thyroid powder.

Overactive thyroid glands (*hyperthyroidism*) produce excessive amounts of thyroxine. As a consequence, the rate of metabolism is increased above normal, weight is easily lost, the heart beats abnormally fast (tachychardia), reflex actions are increased, and the autonomic nervous system induces excess sweating. In about half the cases there is also a bulging of the eyes (*exophthalmos*). Exophthalmos is known to occur in cases of normal thyroid function and also in hypothyroidism. The condition is characterized by excessive fluid in the tissues of the eye socket and by swelling of the extrinsic eye muscles. There is evidence that a substance in the pituitary gland, called an exophthalmos-producing factor, is the cause of the condition. In serious cases, vision may be threatened. Although it usually is not possible to bring about a complete recovery, the disease can be treated in many cases. The use of adrenal steroids often is effective. In some cases the removal of the pituitary gland is necessary.

EXOPHTHALMOS. *Protrusion of the eyeballs from the orbits; considered to be due to a pituitary secretion (G. exo, outside; ophthalmos, eye).*

Another Kind of Thyroid Hormone

In addition to follicles, the thyroid gland contains epithelial cells outside and among the follicles. These cells produce another hormone—calcitonin. *Calcitonin* reacts with osteoblasts and osteocytes in bone. In doing so, it brings about the deposition of calcium phosphate, and/or it nullifies the action of another hormone—parathyroid hormone—which acts on bone to bring about the release of calcium phosphate into the blood.

CALCITONIN. *A thyroid hormone which tends to reduce the blood calcium level (G. tonos, tension).*

THE PARATHYROID GLANDS

Most people have four parathyroid glands, two on each side of the trachea (Figure 12.10). The parathyroid glands are embedded in the rear of the thyroid gland. The single function of the parathyroid glands is to regulate the levels of calcium and phosphorus in the body.

In the normal adult, calcium is held steady at a level of 9 to 11 milligrams per 100 milliliters of blood. Phosphorus is normally held at a level of 3 to 4 mg per 100 ml. These levels are necessary for the proper performance of nervous activities, muscle contraction, and membrane behavior in most of the cells of the body.

When calcium tends to drop below its normal level in the blood, the parathyroid gland becomes stimulated and releases *parathyroid hormone*, a polypeptide hormone which reacts with three kinds of tissue: the intestine, bone, and kidney.

The Effect of Parathyroid Hormone

In the intestine, parathyroid hormone cooperates with vitamin D to bring calcium from the intestinal canal into the bloodstream. In the kidney, it reacts with tubular cells to promote the excretion of excess phosphate ions and the retention of calcium ions. And in the bones of the body, it probably reacts with certain cells called osteoclasts to bring about the release of mineralized calcium. The overall effect of parathyroid hormone is to increase the availability of calcium to the tissues (Figure 12.6).

Parathyroid Problems

An overactive parathyroid gland (*hyperparathyroidism*) may arise as a result of excessive parathyroid growth. In *primary hyperparathyroidism* the excess growth may be due to a harmless or malignant tumor in the parathyroid gland. Or it may be due simply to excessive cell multiplication (hyperplasia). In *secondary hyperparathyroidism* the condition may result from a kidney problem, an excess of phosphate in the blood, a deficiency of calcium in the blood, or some other mineral imbalance.

An overactive parathyroid gland may be characterized by an excessive level of calcium in the blood and a deficiency in phosphorus. This imbalance often leads to: (1) kidney problems, such as the formation of kidney stones, excessive urination, and kidney failure; (2) bone pain; and (3) reduced appetite, vomiting, and thirst.

When the parathyroid gland is performing below normal (*hypoparathyroidism*), or when inadequate amounts of vitamin D and calcium are eaten, the levels of calcium in the blood fall too low. Inasmuch as the proper development of bone depends upon an adequate supply of phosphate and calcium, one result of hypoparathyroidism is *rickets*. This disease is characterized by soft bones, an enlarged liver and spleen, and tenderness of the body when touched. It occurs in children before skeletal maturation is complete. In adults the disease is usually called *osteomalacia*. Related to it is another disease, termed *osteoporosis*. In osteomalacia there is a normal quantity of bone but it is poorly mineralized. In osteoporosis there is a reduced amount of connective tissue (collagen), upon which calcium phosphate is deposited (mineralized). The causes of osteoporosis are numerous, including hyperthyroidism, acromegaly, excessive activity of the adrenal cortex (Cushing's disease), and rheumatoid arthritis.

OSTEOMALACIA. *A bone disorder characterized by failure to deposit calcium phosphate salts into the connective tissue matrix of bone* (G. osteon, *bone;* malakia, *soft*).

OSTEOPOROSIS. *Enlargement of the marrow and other soft spaces of bone at the expense of the hard mineralized regions* (G. osteon, *bone;* poros, *porous*).

RICKETS. *A childhood bone disease caused by a deficiency of vitamin D* (G. rachis, *rosary*).

Another Regulator of Calcium in the Blood—a Vitamin or Hormone?

In the last quarter of the nineteenth century it was observed that impoverished children who lived in dark alleys and poorly lit homes in Europe had a high incidence of rickets. Around the same time it was discovered that eating fish—particularly cod liver and other fish liver oils—could reduce the incidence of rickets. Certain investigators concluded that rickets was a dietary problem of the poor. With time, this conclusion gained support by investigations which led to the winning of a Nobel prize by Windaus for his research into the chemical structure of sterols and their connections with

General Endocrinology

Vitamin D. In fact, it became common to claim that, "since vitamin D cures rickets, rickets is a vitamin deficiency disease."

Curiously, in spite of the curative power of vitamin D, most authorities chose to ignore an important observation that was made in the last quarter of the nineteenth century: impoverished children living in regions of abundant sunshine (Manchuria, Java, Bombay), regions in which the children were skimpily clad and therefore exposed to many hours of sunlight, rarely, if ever, developed rickets. The suggestion made by several researchers—that a lack of sunlight, not a lack of some nutrient, was the cause of rickets—was simply dismissed from further consideration.

Was it a lack of sunshine? Or was it a lack of a nutrient? Recent investigations, beginning in the 1960's, have brought forth answers to these questions. It is now known that ample quantities of *ergosterol* are provided in an ordinary diet. This steroid, in the presence of sunlight, is converted within the skin into *calciferol* (vitamin D_2). Within the liver, calciferol is converted into vitamin D_3. Next, within the kidney, vitamin D_3 is converted into active vitamin D_3 (Figure 12.6). This substance is the only form of vitamin D that is effective in regulating blood calcium. It is secreted into the blood and is carried to all tissues of the body, many of which may be affected by it. In the intestine, for example, active vitamin D_3 induces the formation or activation of a protein which serves to bring calcium out of the intestinal canal into the bloodstream.

Inasmuch as the active form of vitamin D_3 is produced in one tissue (the kidney), and travels through the bloodstream to other tissues (for example, the intestine) where it exerts an action, it is correct, by definition, to call active vitamin D_3 a hormone.

ERGOSTEROL. *A fat occurring in yeast and other fungi.*

CALCIFEROL. *Vitamin D_2, resulting from irradiating ergosterol.*

THE ADRENAL GLANDS

The adrenal glands, known also as suprarenal glands, lie on the anterior medial surfaces of the kidneys. Each gland is made of two components: a medulla and a cortex.

The Adrenal Medulla

The adrenal medulla (Figure 12.12) is much like a postganglionic neuron in the autonomic nervous system. In fact, it is derived within the embryo from the same source of tissue—the neural crests.

When the adrenal medulla is stimulated by its preganglionic nerve fibers, large quantities of adrenaline (known also as epinephrine) are released into the bloodstream. At the same time, relatively insignificant amounts of noradrenaline are released. The actions of noradrenaline (also called norepinephrine) were discussed in Chapter 8.

MEDULLA. *Central part, as distinguished from the cortex (L.).*

GANGLION. *A group of nerve cell bodies (G.).*

The Adrenal Glands

The Effect of Adrenaline. The primary overall effect of adrenaline is to step up the body's ability to expend energy. This effect is mediated through the circulatory and respiratory systems and in the course of metabolism: In the circulatory system, adrenaline increases the force and the frequency of heart contraction. It also dilates (widens) the arteries that lie within the heart muscle and skeletal muscle, and it constricts the arteries that lie in the skin and viscera. By way of these actions adrenaline brings about an increase of blood flow through the heart and skeletal muscles. In consequence, the body is better able to cope with situations such as "fight, flight, or fright."

In the respiratory system, adrenaline hyperpolarizes the smooth muscle of the bronchi, causing them to relax, allowing for a larger passageway for incoming and outgoing air. The increased inflow of oxygen thus provides a greater capacity for the oxidation of metabolites and the derivation of energy.

In the course of metabolism, especially within the liver and skeletal muscle, adrenaline triggers a very complex sequence of enzyme reactions through which glucose (sugar) is released from glycogen (a large sugar-storage molecule). In the liver, the glucose is released into the bloodstream, and thus becomes available to most of the needy cells within the body. But not so in the skeletal muscle. Here, the glucose is released from glycogen and then it is converted into glucose-6-phosphate, as shown in Figure 21.1. As such, the glucose cannot leave the muscle cell, though it can be oxidized completely, as needed, to provide energy for skeletal movements.

Misfunctions of the Adrenal Medulla. When tumors develop in the adrenal medulla, or when the medulla becomes abnormally large due to excessive cellular division (hyperplasia), excessive amounts of adrenaline and noradrenaline are produced. A variety of symptoms follow, including pallor, sweating, headache, nervousness, an increase in blood protein, sugar in the urine, increased frequency in urination, and a number of serious

METABOLISM. *The chemical changes undergone by food in a living organism (G.* metabole, *change).*

GLYCOGEN. *A polymer made of glucose (G.* glykys, *sweet).*

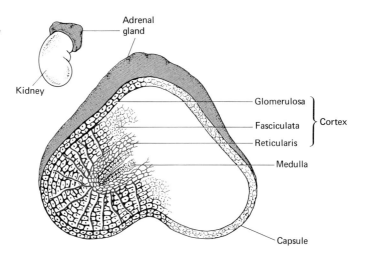

FIG. 12.12. *Microscopic anatomy of the adrenal gland.*

circulatory problems. The usual circulatory problems are high blood pressure, water on the lungs (pulmonary edema), heart failure, and stroke. The condition can be relieved by drugs, by removal of the medullary tissue, or by some combination of the two.

The Adrenal Cortex

Although the adrenal cortex lies over the surface of the adrenal medulla, it is derived from mesoderm, not ectoderm, it is not innervated, and its hormones bring about entirely different actions. Whereas the adrenal medulla is derived from the neural crests, the adrenal cortex comes from the embryonic tissue which also gives rise to the sex organs and kidneys.

A histological section of the adrenal cortex is shown in Figure 12.13. In this figure, three cortical regions can be noted. It turns out that three classes of hormones are produced within these regions. The *sex hormones* (testosterone, progesterone, and estrogen) are produced mainly in the innermost reticular layer. They are formed from cholesterol, which in turn, may be produced from sugars, fats, or proteins.

The *glucocorticoids* (cortisol, cortisone, among others) are produced from the sex hormones in the middle, fascicular layer. The *mineralcorticoids* (aldosterone, deoxycorticosterone, among others) are produced from glucocorticoids in the outermost, glomerular layer.

Figure 12.4 is a simplified version of the synthetic pathway for these hormones. Many intermediate compounds occur between the steps shown,

CORTEX. *The outer, or peripheral, portion of an organ (L. bark).*

GLUCOCORTICOIDS. *Cortisol, cortisone, and any similar compounds which affect glucose metabolism in like manner (G. glykys, sweet; cortico, cortex; eidos, resemblance).*

MINERALCORTICOID. *Aldosterone and any other similar natural or synthetic compound whose effect is to retain sodium in the body.*

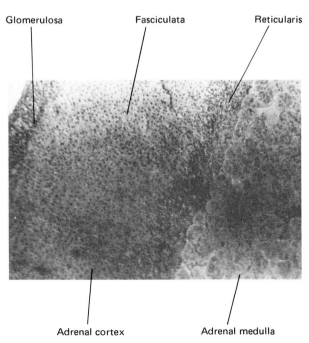

FIG. 12.13. *The three layers of the adrenal cortex.*

The Adrenal Glands

but our intention is to show only the basic interrelationships among the various hormones.

It should be pointed out that only negligible amounts of sex hormones are ordinarily produced in the adrenal cortex. By far the greatest fraction of these sex hormones is used to produce mineralcorticoids and glucocorticoids.

Actions of Glucocorticoids. The mechanisms through which glucocorticoids operate are not clearly known, but the resulting effects of these hormones are fairly well established:

(1) Glucocorticoids affect the metabolism of organic molecules whereby glucose is produced from proteins. One benefit of this action is that during fasting, when the level of glucose in the blood tends to decrease, glucocorticoids react with tissues to bring about the release of amino acids. These substances are sent to the liver where they may be converted, in part, into glucose.

(2) Glucocorticoids react in areas of inflammation to bring about relief. An inflammation is a reaction of a tissue to an injury. An inflammatory site is generally seen to be red and warm and/or swollen and painful. Diseases bearing the suffix *itis* are inflammations. Thus, tonsillitis and appendicitis are inflammatory diseases of the tonsils and appendix, respectively.

One of the mechanisms by which injured tissues break down is through the digestive activities of enzymes that are liberated from the damaged cells. Corticosteroids arrest the process, possibly by reacting with the membranes of the lysosomes in which the destructive enzymes are stored. They may "tighten" the membrane, and in doing so, prevent leakage of the enzymes into the inflamed area. In addition, they might counteract the painful action of prostaglandins, which are normally produced by various tissues, but abnormally produced in an inflammation. More about this process is described in Chapter 31.

(3) During stress, glucocorticoids are produced in larger amounts and somehow reduce the production of white blood cells and certain cells in connective tissues. The mechanism for this is not known, but it is thought that excessive amounts of proteins are converted into glucose, thus creating a deficiency of amino acids for producing the protein needed to construct white blood cells.

The Effects of Excessive Glucocorticoids. When excessive amounts of glucocorticoids are secreted over a prolonged period of time, *Cushing's syndrome* sets in. The excessive production may be due to a tumor in the pituitary gland (and therefore the release of excessive adrenocorticotropin) or to a tumor in the adrenal cortex. The syndrome is usually characterized by a large number of symptoms. Among these are muscle weakness due to a general destruction of proteins, and osteoporosis due to faulty formation of the protein matrix (collagen) in which mineralization takes place. Other symptoms include high blood sugar levels due to formation of excess glucose from proteins (gluconeogenesis), and diabetes mellitus due to depletion of insulin by the prolonged maintenance of high levels of glucose in the blood.

LYSOSOMES. *Membrane-bounded cell organelles which contain digestive enzymes* (G. lys, *rupture;* somas, *body*).

SYNDROME. *A group of symptoms* (G. syndromos, *a running together*).

General Endocrinology

Cushing's syndrome may be treated by destroying pituitary tissue or by removing part or all of the adrenal gland, depending on the cause of the problem. Certain drugs that are injurious to the adrenal cortex may be used in some cases. It is also necessary to administer hormones to correct any newly-arising deficiency.

Actions of Mineralcorticoids. The main function of *aldosterone*, the major mineralcorticoid, is the retention of sodium in the body. It does this by reacting with cells in the distal tubules of the kidney. Through its activity, sodium is brought into the bloodstream from fluid which is otherwise destined to be urine. In the course of this action small quantities of potassium are secreted into the urine.

ALDOSTERONE. *A hormone which retains sodium in the body.*

The retention of sodium increases the osmotic pressure of the fluid within cells of the kidney tubules. This increase tends to retain water in the body, but this tendency may be offset by the absence of antidiuretic hormone, as is described in Chapter 30.

The Effects of Excessive Mineralcorticoids. Excessive amounts of aldosterone may be produced when there is a cancerous or benign tumor in the adrenal cortex. Excessive amounts of aldosterone may also be produced in the absence of excessive adrenal cortical growth, as, for example, in liver diseases (hepatitis or cirrhosis), obesity, or heart failure. The pituitary gland does not seem to be involved, because its removal, or the administration of adrenocorticotropin, does not produce a noticeable increase in the output of aldosterone.

Excessive amounts of aldosterone in the blood bring about a variety of problems, including (1) excessive loss of potassium in the urine; excessive retention of sodium by the blood; (2) improper kidney function, diabetes insipidus, and kidney disease; (3) muscle paralysis due to excessive losses of potassium from the muscle fibers; (4) increased likelihood of diabetes mellitus; and (5) high blood pressure (hypertension).

The Effects of a Deficit in Corticoids. An overall deficiency in corticoids leads to *Addison's disease.* Addison's disease may be caused by destruction of the adrenal cortices by micro-organisms (as in tuberculosis) or by cancer, or it may be due to some unknown (idiopathic) cause. In this case it is called *primary hypoadrenocorticism.* Addison's disease also may arise as a result of damage to the pituitary gland or hypothalamus, whereby subnormal levels of adrenocorticotropin are released. In this case the deficiency is termed *secondary hypoadrenocorticism.*

ADDISON. *An English physician (1793–1860).*

Many of the metabolic symptoms of Addison's disease may be relieved through the use of appropriate hormones. Cortisone and hydrocortisone may be used for secondary hypoadrenocorticism. These same glucocorticoids and a mineralcorticoid (deoxycorticosterone) are used in the primary case. In men, testosterone can be administered, as needed in either case, to restore sex drive and strength.

Primary and secondary hypoadrenocorticism can usually be distinguished from each other. When the cause resides in the adrenal gland (primary hypoadrenocorticism) the administration of ACTH fails to bring about the production of 17-ketosteroids. Increased amounts of such steroids appear in the urine and plasma in cases of secondary hypoadrenocorticism. However, the distinction is not always possible, because as the adrenal cortex may fail to respond to ACTH in cases where the pituitary or hypothalamus had long been at fault.

Scarcely any tissue is spared of debilitating effects when the disease is prolonged and severe. Fatigue, weight loss, yellowing of nails, loss of hair, and emotional disturbances are only a part of a prominent outward display of the disease. Many other symptoms occur, including: (1) excessive lymphocytes and eosinophils may appear in the blood; (2) blood pressure may be low, along with an abnormal electrocardiograph; (3) blood sugar level is lower than normal between meals and the ability to respond to stress decreases; (4) kidney functions are improper; and (5) sexual drive (libido) decreases.

In addition, excessive amounts of sodium chloride are lost and potassium is retained in primary, but not secondary, hypoadrenocorticism. On the other hand, primary hypoadrenocorticism is associated with deficiencies of gonadotropins and thyroid stimulating hormone.

The Effect of Excessive Adrenal Sex Hormones. The sex hormones produced in the adrenal cortex of males and females are primarily male sex hormones (*androgens*). Ordinarily they may stimulate the sex drive (libido), but in the absence of gonadal sex hormones they are relatively ineffective in promoting sexual development and maintenance. However, in cases where excessive amounts are produced by the adrenal cortex, a precocious development of secondary sex characteristics arises in children. A boy, for example, tends to be a miniature Hercules. Girls may exhibit *virilism*. The breasts fail to develop, menstruation does not occur, and the girl develops the stature of a short adult. Adult women also display signs of virilism, such as hair on the face. In addition, they may suffer menstrual problems and infertility.

The excessive production of androgens by the adrenal cortex may be accompanied by a deficiency in glucocorticoids and/or mineralcorticoids. The overall situation may be remedied by hormonal treatment. Hydrocortisone is usually given. In addition to serving its function as a glucocorticoid, it turns off the need of ACTH. The adrenal cortex then reduces its output of androgens and increases its output of aldosterone.

PRECOCIOUS. *Developing earlier than usual* (L. praecox, *mature before its time*).

VIRILISM. *Development of masculine traits in a girl or woman* (L. virilis, *manly*).

THE PANCREAS

PANCREAS. *The abdominal gland which produces digestive enzymes, insulin, and glucagon* (G. pan, *all*; kreas, *flesh*).

The pancreas, like the adrenal gland, consists of two functionally distinct units. However, whereas the two units of the adrenal gland are also anatomically distinct, those of the pancreas are intermixed.

Figure 12.14 shows a histological section of the pancreas. The *islets of Langerhans* within the pancreas constitute the endocrine portion of the gland. The remainder of the gland is an exocrine structure that produces digestive enzymes, a topic which is discussed in Chapter 19.

The Functions of Insulin and Glucagon

The endocrine component of the pancreas produces insulin and glucagon. The *primary effect of insulin* is to increase the permeability (penetrability) of cell membranes to the inflow of glucose, fatty acids, and amino acids. Most important, it prevents the level of glucose from becoming excessive in the blood.

The *primary effect of glucagon* is to increase the level of blood glucose. Adrenaline, glucocorticoids, and growth hormone also increase the level of glucose in the blood. Together, all of these hormones normally act to ensure that the blood glucose will rarely, if ever, fall below a critical level. The specific roles of these hormones in regulating the blood sugar level are discussed in Chapter 22.

INSULIN. *An antidiabetic hormone, produced in small islands of endocrine tissue within the pancreas (L. insula, island).*

Diabetes Mellitus

Deficiencies of insulin lead to diabetes mellitus. This syndrome is marked by the following characteristics: an abnormally high level of glucose in the blood; the presence of glucose in the urine; and an "acetone breath," which smells something like nailpolish remover. This breath odor is caused by the production of acetone and acetoacetic acid (ketone bodies) in the course of faulty metabolism. Other symptoms include a deficient production of fat

KETONE. *A chemical substance with the characterizing group $C = O$.*

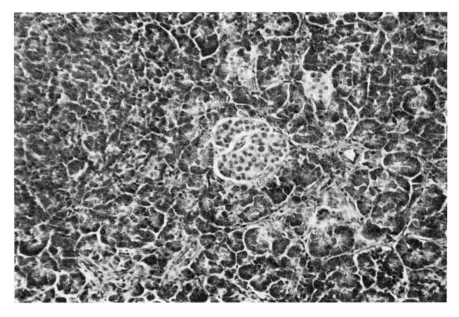

FIG. 12.14. *Tissue from the pancreas. The small cells of the islets of Langerhans (light area in center) are surrounded by digestive glands.*

CATARACT. *An opaque lens or region of the lens* (G. kataraktes).

and protein, resulting in weight loss, and excessive urination. In many cases additional problems arise, particularly in diabetes of many years standing, and especially in cases where regulation by an outside source of insulin is not entirely adequate. These problems include diseases of the retina, cataracts, atherosclerosis (deposition of fatty tissue in the inner wall of arteries), nervous problems, and kidney disorders.

It has been estimated that about 5% of the population is diabetic. However, the probability of acquiring diabetes at this time in our history may be as high as 25%. This situation arises partly as a result of having saved the lives of diabetics through insulin treatment since about 1930. Prior to the use of such treatment, diabetics usually died before they reached childbearing age. In addition, diabetes is predictable from genetic data regarding the frequency of homozygous and heterozygous diabetics in our population.

Three main types of diabetes mellitus are now recognized: juvenile diabetes, adult onset diabetes, and acquired diabetes. *Juvenile, or insulindependent diabetes,* appears primarily in children and young adults and is characterized eventually by an acetone breath if insulin is not used as a treatment. *Adult onset diabetes,* or *insulin nondependent diabetes*, appears in later life. It is associated with obesity, and usually is not characterized by an acetone breath if left untreated. *Acquired diabetes* appears in certain diseases, such as Cushing's syndrome (excess glucocorticoids), acromegaly (excess growth hormone), pheochromocytoma (tumors or excess growth of the adrenal gland with resulting excesses of adrenaline), and hyperthyroidism (excess thyroid hormone).

Diabetes is not always rapid in its onset, and in fact may be classified in stages in accord with the symptoms. These stages are not necessarily sequential and often one or more of them is never seen. One of the early stages appears when, during times of stress, as during an infection or trauma, the individual exhibits one or more symptoms of diabetes. A later stage is characterized by normal levels of glucose in the blood during fasting, but higher than normal levels two hours after a meal. The next stage is the most commonly recognized form. In this stage, the level of glucose in the blood is above normal at all times. In later stages *ketosis* comes into the picture. An acetone breath is a diagnostic feature of ketosis, as is the presence of ketone bodies in the blood and urine. In a still later stage, greater quantities of ketone bodies are found, there is an increase of acidity in the serum (blood pH goes down from a normal of 7.4 to 7.1), and rapid overbreathing sets in. The final stage is an extension of the preceding, and usually ends in coma.

COMA. *A state of deep unconsciousness* (G. koma, *deep sleep*).

Coma is a state of unconsciousness into which a diabetic or nondiabetic may lapse. For the individual without diabetes, the coma may be precipitated by other problems, such as a stroke, poisoning, or a serious disturbance in the balance of minerals, water, and metabolites in the body fluids. In the diabetic, coma may occur as a result of a shortage of insulin, an excess of insulin, or a deficiency of glucose in the blood.

HYPERGLYCEMIA. *Higher blood sugar than normal* (G. hyper, *above;* glykys, *sweet;* haima, *blood*).

In the case of insufficient insulin, the coma is associated with an excess of glucose in the blood (*hyperglycemia*) and urine (*glycosuria*); ketone bodies in the blood (*ketonemia*) and urine (*ketonuria*); and a blood pH that is less than normal (*acidosis*). Numerous other abnormalities may be detected through laboratory analyses, including an increase in blood protein, an increase in neutrophilic white blood cells, and a decrease in lymphocytic and eosinophilic white blood cells.

In the case of excess insulin, as happens on occasion during the administration of insulin intravenously, the level of blood sugar is reduced too rapidly. The diabetic then enters a state of *insulin shock*, the chief symptoms of which are sweating, tremor, dizziness, and double vision, followed by delirium, convulsions, and coma.

The symptoms of juvenile diabetes and acquired diabetes may be relieved with insulin. This polypeptide must be administered intradermally, subcutaneously, or intravenously throughout life for the true diabetic. Its mode of action can be summarized as follows: (1) it accelerates the entry of glucose into heart and skeletal muscle cells, fat cells, and the aqueous humor of the eye; (2) it promotes the conversion of glucose into glycogen within the cells of the liver, skeletal muscle, heart muscle, and certain other tissues; (3) it facilitates the entry of fatty acids into cells and their conversion into neutral fat—in the absence of insulin, or where deficient amounts are present, the fatty acids are broken down and converted into ketone bodies; (4) it promotes the conversion of glucose into fat; (5) it enhances the entry of amino acids into cells; and (6) it stimulates the formation of protein from glucose and nitrogenous compounds and from amino acids in liver and muscle cells.

GLYCOSURIA. *Sugar in the urine* (G. glykys, *sweet*; ouron, *urine*).

SUMMARY

1. Endocrines, or hormones, are cellular secretions that are liberated into the blood. In contrast, exocrine secretions, such as sweat and digestive juices, are secreted into ducts that lead to the outside of the body.
2. Endocrines are secreted from either epithelial cells or specialized nerve cells called neuroendocrine cells.
3. All endocrines can be assigned to one or more first-, second-, or third-order endocrine systems, depending on whether one, or two, or three hormones in succession are required to bring about a response to the stimulus which initiated the hormonal action.
4. Most endocrine secretory activities are regulated through negative feedback, while the release of three kinds of pituitary hormones is determined by whether a mobilizer or demobilizer is released from the hypothalamus. The principle of negative feedback is that low levels of a substance stimulate an endocrine tissue to secrete its hormone.
5. The hypothalamus regulates the release of the anterior pituitary hormones. The hypothalamus also produces antidiuretic hormone, which prevents the loss of excessive amounts of water, and oxytocin, which stimulates the smooth muscle of the uterus during childbirth and the smooth muscle of mammary glands during nursing.

6. The anterior pituitary gland secretes hormones that regulate the release of hormones from the thyroid gland, adrenal cortex, and gonads. Two other important anterior pituitary hormones are growth hormone, which regulates growth; and prolactin, which stimulates the development of the mammary glands.
7. The thyroid gland secretes thyroxine, which stimulates metabolism in the direction of helping the body to maintain a high body temperature (37°C). The thyroid gland also secretes calcitonin, which promotes the deposition of calcium salts into bone.
8. The parathyroid glands secrete parathyroid hormone. Together with vitamin D, parathyroid hormone tends to increase the blood calcium level.
9. The dominant endocrine secretion of the adrenal medulla is adrenaline.
10. The adrenal cortex produces steroid hormones that regulate carbon metabolism (glucocorticoids), sodium-potassium metabolism (mineral corticoids) and sex drive (sex hormones).
11. The pancreas secretes insulin, whose primary actions are to keep blood sugar levels low and to conserve body nutrients. The pancreas also secretes glucagon, whose actions are exactly opposite to those of insulin.

REVIEW QUESTIONS

1. Distinguish the terms "endocrine" and "exocrine."
2. Give an example of a first-order, a second-order, and a third-order endocrine system.
3. Define negative feedback. Give an example of the operation of negative feedback by a substance which is above an optimal level in the blood; by a substance which is below an optimal level in the blood.
4. Name a pituitary hormone whose release is regulated by a mobilizing–demobilizing system of hormones from the hypothalamus.
5. Where are antidiuretic hormone and oxytocin produced? What are their functions?
6. What are the actions of thyroxine. How is the release of thyroxine regulated?
7. Describe the actions of parathyroid hormone with regard to bones, the kidneys, and the intestines.
8. Name several endocrines whose secretion is not stimulated by a tropic hormone.
9. Where do glucocorticoids and mineralcorticoids come from? Give an example of each of these classes of hormones, and describe their functions in the body.
10. Name the two pancreatic hormones and describe their functions.

The Reproductive System 13

THE MALE REPRODUCTIVE SYSTEM
    Production of sperm and testosterone
    Transfer of sperm during intercourse
    Male reproductive problems
        Failure of the testes to descend
        Physiological sexual deficiencies
        Overactive tissues in the testes
        The prostate problem
THE FEMALE REPRODUCTIVE SYSTEM
    The external genitals
    The vagina and cervix
    The uterus
    The fallopian tubes
    The ovaries
        Production of mature egg cells
    Functions of the female sex hormones
    The menstrual cycle
    Pregnancy and childbirth
    The mammary glands
BIRTH CONTROL
    The intrauterine device
    The "pill"
    Vasectomy
    Tubal ligation

The very complex process of human reproduction includes five essential, and distinct, stages: (1) the *production of germ cells* (also called *sex cells* or *gametes*) by the male and female; (2) sexual intercourse, in which male sex cells—sperm—are transferred from the male gonads into the uterus of the female; (3) *fertilization*, in which a sperm fuses with an egg cell, the female gamete; (4) *pregnancy*; and (5) *childbirth*. Each of these stages is discussed in the following pages.

The first part of this chapter deals with the male reproductive system. The second part describes the female reproductive system. For both systems we begin by listing the anatomical components of the system and giving a brief description of each component. For the male, three additional topics are discussed: the production of the male sex hormone (testosterone) and sperm cells; the transfer of the sperm cells into the uterus of the female; and male reproductive problems.

For the female, there are four additional topics: the production of female sex hormones (estrogen and progesterone) and egg cells; the menstrual cycle; childbirth; and the development and functioning of the mammary glands.

This chapter concludes with a discussion of some of the most commonly used methods of birth control, for both men and women.

THE MALE REPRODUCTIVE SYSTEM

The male reproductive system consists of the following parts, all of which except the first two are shown in Figure 13.1: the hypothalamus, the anterior pituitary gland, the testes, the epididymes, the vas deferens, the seminal vesicles, the prostate gland, the urethra, Cowper's glands, and the penis.

COWPER. *An English surgeon (1660–1709).*

Nerve cells in the *hypothalamus* produce and secrete a mobilizing hor-

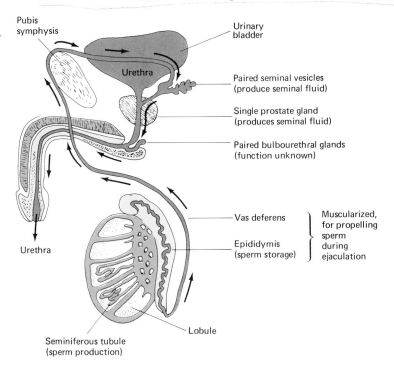

FIG. 13.1. *Male reproductive system (hypothalamus and anterior pituitary gland are not shown). Arrows indicate pathway of sperm.*

mone—the gonadotropin mobilizer. It is likely that this hormone is not released until puberty—when the brain has reached a certain critical point in its development. It is then released throughout life.

The *anterior pituitary gland* has certain cells that respond to the gonadotropin mobilizing hormone by secreting two tropic hormones—follicle stimulating hormone and luteinizing hormone. *Follicle stimulating hormone*, also called FSH, reacts with the seminiferous tubules of the testes to bring about the production of sperm. *Luteinizing hormone* reacts with the cells of Leydig within the testes to bring about the production of the male sex hormone, testosterone.

The *testes* are the male gonads. Sperm and testosterone are produced in these organs. Each *epididymis* constitutes the first portion of a sperm excretory duct. The epididymis consists of many yards of a convoluted duct in which the sperm are stored after being produced in the seminiferous tubules of the testes. At the lower end of each testis the duct of the epididymis turns upward and merges into the vas deferens.

TESTES. *Paired male reproductive glands (L.).*

EPIDIDYMYS. *The portion of the seminal duct in which sperm are stored (G. epi, upon; didymos, testis).*

The *vas deferens* are paired tubes with considerable amounts of smooth muscle in their wall. When they are stimulated by autonomic nerves, as during orgasm, the muscle contracts, as does the muscle of the epididymis, and the sperm are expelled from the body. The *seminal vesicles* are paired glands which secrete fructose and other nutrients for the sperm cells. They secrete these substances into a pair of *ejaculatory ducts*. These ducts are the terminal extensions of the vas deferens; they pass through the prostate gland and convey seminal fluid and sperm from the epididymis and seminal vesicles into the urethra.

A single *prostate gland* produces secretions that lubricate the ejaculatory ducts and provide additional nourishment to the sperm. In particular, they provide calcium, which increases the tendency of the seminal fluid to coagulate and thus to promote the retention of sperm in the female genital tract. (Calcium also tends to neutralize the acidity of the residual urine in the penis. The pH of urine normally varies from 4.5 to 8.0, averaging about 6.0).

PROSTATE. *The gland which "stands before" the beginning of the male urethra (G. prostates, standing before).*

The *urethra* is a passageway from the ejaculatory ducts, through the penis, to the outside of the body. The paired *Cowper's glands* are known also as *bulbourethral* glands because they are bulb-shaped and because they pass their secretions into the urethra. Nothing is known about their function. The *penis* makes it possible to transfer semen and sperm into the female. It also conveys urine to the outside of the body.

PENIS. *The male copulatory organ (L. tail).*

Production of Sperm and Testosterone

Each testis is a capsule that contains many *seminiferous tubules*. Outside and between these tubules, lies a connective tissue containing the epithelial *cells of Leydig*. These cells produce testosterone from cholesterol (See Figure 12.4).

SEMINIFEROUS TUBULES. *The tubules of the testes (L. semen, seed; fero, to carry).*

A microscopic examination of a seminiferous tubule before puberty reveals two types of cells: immature sperm and *Sertoli cells*. The Sertoli

SERTOLI. *An Italian histologist (1842–1910).*

cells, also known as sustentacular or supporting cells, are extremely irregularly-shaped columnar cells that extend from the basement membrane at the periphery of a tubule to the lumen (canal) of the tubule. Tucked within the numerous inpocketings of the Sertoli cells are the immature sperm, which the Sertoli cells nourish and support (see Figure 13.2).

At puberty, the gonadotropin mobilizing hormone of the hypothalamus reacts with certain cells in the anterior pituitary gland to bring about the release of follicle stimulating hormone and luteinizing hormone. Luteinizing hormone then triggers the cells of Leydig to produce testosterone. *Testosterone*, together with follicle stimulating hormone, interacts with the seminiferous tubules and causes the immature sperm to begin their maturation.

An immature sperm undergoes the two divisional processes of meiosis, producing four *spermatids*. The spermatids acquire a long whiplike structure, a flagellum. Now known as sperm, they become dislodged from deep crevices within the Sertoli cells, and pass into the lumen of the seminiferous tubule (Figure 13.2). From here they propel themselves with their flagella to the epididymis, where they are stored until ejaculation. If there is no ejaculation they are eventually destroyed and dissolved within the epididymis.

In addition to acting upon the seminiferous tubules in the maturation of sperm, testosterone also interacts with other tissues of the body to produce those qualities we think of as masculine. It reacts with the voice box to make it larger than that of the female and bring about a deep voice. It reacts with the skin of the face and chest, as well as other regions of the body, to bring about the growth of hair, and with the skeletal and muscles to shape the contour of the body.

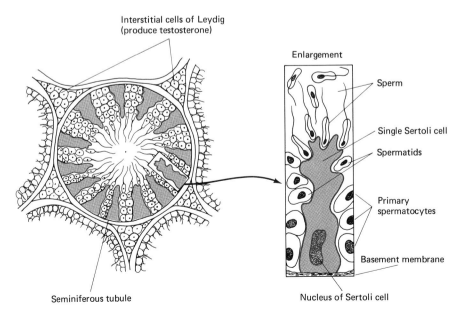

FIG. 13.2. *Cross-section of a seminiferous tubule. Primary spermatocytes are rarely seen because they are very short-lived.*

Transfer of Sperm During Intercourse

The *erection* of the penis is initiated autonomically by the parasympathetic system (Figure 13.3). A circulatory mechanism is involved. Numerous blood vessels run along and within three caverns that are situated along the length of the penis. Parasympathetic stimulation causes the veins in the penis to contract, and at the same time causes the arteries within the three caverns to relax. As a result, the caverns become engorged with blood permitting the erect penis to penetrate the vagina. The blood remains largely stagnant during the erectile process due to compression of veins by the enlarged caverns. The urethra, on the other hand, remains rigidly open throughout the sex act to allow the escape of seminal fluid during ejaculation. Following the sex act, the penis becomes flaccid. Parasympathetic stimulation is withdrawn, and the excess blood supply is retracted into the general blood circulation.

Ejaculation is under control of the sympathetic nervous system. During orgasm, the pudendal nerve fibers of this system usually reach a threshold, and then fire. The smooth muscle of the accessory glands and the epididymis contract, forcing the seminal fluid out of the penis. Approximately one spoonful of semen (3 milliliters, containing about 3 million sperm cells) is ejaculated.

In Figure 13.1 the reader may note a region of the penis known as the *glans penis*. Rich in sensory nerves, the glans is protected at birth by a double fold of skin, called the *prepuce*. Normally the prepuce can be retracted to expose the glans. This should be done to wash away a cheesy secretory product called smegma. The presence of smegma may be related to cancer, since cancer of the penis occurs more often in males with a prepuce than in males who have been circumcised.

GLANS PENIS. *The conical, distal end of the penis* (L. glans, *an acorn*).

PREPUCE. *The skin over the glans penis or clitoris* (L. praeputium, *foreskin*).

Male Reproductive Problems

A male may have normal sex chromosomes but still have problems with sexual development or maintenance of the reproductive system. In particu-

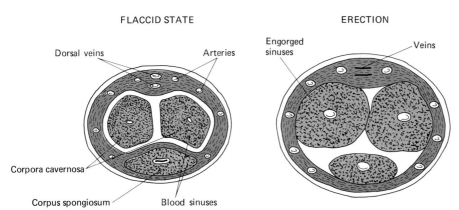

FIG. 13.3. *Cross-section of the penis, in the flaccid state (left) and erect. During erection the veins are closed, causing the corpora cavernosa, a pair of spongy sinuses, to become engorged with blood.*

The Male Reproductive System

lar, the testes may fail to descend from the pelvic cavity into the scrotum; there may be a deficiency of sperm, gonadotropins, or sex hormones (hypogonadism); there may be an excessive production of testosterone or estrogen; or the prostate gland may degenerate.

Failure of the Testes to Descend. The testes, epididymis, and part of the vas deferens are enclosed in a *scrotal sac*. During development, the testes may fail to descend through the inguinal canals of the pelvic bone into the scrotum. Instead, they remain in the body cavity, an abnormality termed *cryptorchism* (hidden testes). Where both testes are involved, failure to correct this situation leads ultimately to sterility. The testes decrease in size and the sperm-producing elements disappear.

If the problem is not caused by an anatomical obstruction, cryptorchism may be corrected with gonadotropins or testosterone. If there is an obstruction it must be corrected through surgery.

A constant temperature of about 33°C is required for proper development of sperm. The relative constancy of this temperature is promoted through the interplay of temperature receptors and a cremaster muscle within the scrotal wall. When the scrotum is exposed to cooler temperatures, as for example during swimming, the scrotum is brought closer to the body. Also, the scrotal wall becomes thicker because of the contraction of muscles in the scrotal wall, thus providing insulation against the cold. During a hot bath, or on a very hot summer day, the scrotum is extended away from the body and its wall thins out.

Physiological Sexual Deficiencies. A deficiency in follicle stimulating hormone, luteinizing hormone, testosterone, sperm, or a combination of these is termed *hypogonadism*.

The effect of hypogonadism varies according to whether it occurs before or after puberty. As a rule, its occurrence after puberty brings about sexual regression. When it occurs before puberty sexual development is inhibited or prevented—depending on the severity of the deficiency—and dwarfism may also occur.

When one or both of the gonadotropins (follicle stimulating hormone and luteinizing hormone) are absent or occur at very low levels, the problem usually resides in the pituitary gland or hypothalamus. Tumors and disease are among the possible causes.

A deficiency of follicle stimulating hormone alone results in a reduction of sperm production. A deficiency of luteinizing hormone alone leads to an inadequate development of the sex organs and sperm, and to a diminution of the male secondary sex characteristics.

Where the cells of Leydig fail to develop properly, or where they regress in their development because of a deficiency in luteinizing hormone, inadequate amounts of testosterone are produced. When the testosterone level is below normal, the seminiferous tubules produce a subnormal number of sperm, and the secondary sex characteristics either fail to develop or may regress if they are already present.

Overactive Tissues in the Testes. When tumors appear in the testes, abnormally large amounts of testosterone or estrogen may appear in the

---

SCROTUM. *The sac made of muscle and skin which encloses the testes (L.).*

CRYPTORCHISM. *The failure of the testes to descend into the scrotum (G. kryptos, concealed; orchis, testis).*

CREMASTER MUSCLES. *The muscles that suspend the testes (G. kremaster, suspend).*

HYPOGONADISM. *Underactive sex organs (G. hypo, under; gone, seeds).*

blood, depending on where the tumor is located. When the cells of Leydig are involved, excessive amounts of testosterone may be produced. In boys this may lead to premature or precocious puberty. When the Sertoli cells become tumorous, excessive amounts of estrogen may arise. You may remember from Figure 12.4 that estrogen can be produced from testosterone as well as from testosterone's precursors. Although the function of normal levels of estrogen in males is unknown, excessive amounts lead to the development of feminine characteristics.

The Prostate Problem. Of all the components in the male reproductive system, none is as prone to problems as the prostate gland. Often during the fourth and fifth decades of life the smooth muscle of this gland becomes increasingly replaced by fibrous scar tissue. This process occurs gradually, and frequently leads to the appearance of *amyloid bodies*. These glassy-appearing protein structures are usually deposited just beneath the inner walls of the capillaries and arterioles, where they develop from stagnating secretions and reduce the flow of blood. In addition, they contribute to the enlargement and hardening of the prostate gland. These degenerative changes can be felt by an examiner by inserting a finger into the rectum.

AMYLOID. *A carbohydrate-protein complex found in certain kinds of diseased tissue* (G. amylon, starch; oid, resembles).

As the prostate enlarges, it presses on the urethra, and prevents proper urination. The bladder consistently fails to empty itself in spite of its contractile actions. The bladder and kidney thus become easy prey to infection. To remedy the situation, and to offset the possibility of cancer, the prostate usually is partially or totally removed through surgery.

THE FEMALE REPRODUCTIVE SYSTEM

The female reproductive system consists of the following parts, whose relative position and form, except for the hypothalamus and pituitary gland,

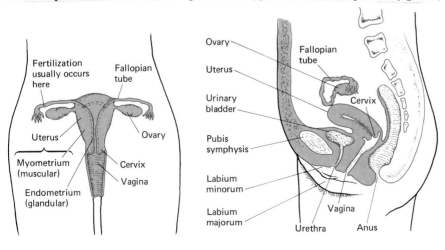

FIG. 13.4. *Female reproductive system (hypothalamus and anterior pituitary gland are not shown).*

The Female Reproductive System

are shown in Figure 13.4: hypothalamus, anterior pituitary gland, posterior pituitary gland, ovaries, fallopian tubes, uterus, vulva, vagina, and mammary glands.

Nerve cells in the *hypothalamus* secrete a mobilizing hormone—the gonadotropin mobilizer. Very likely this hormone is not released until the brain reaches some critical stage of development, beginning at puberty. It is released from then on throughout most of life, though it may turn off in very old age.

Certain cells in the *anterior pituitary gland* respond to the gonadotropin mobilizer by secreting two tropic hormones—follicle stimulating hormone and luteinizing hormone. Follicle stimulating hormone reacts with the ovaries to bring about the production of eggs. Luteinizing hormone reacts with the ovaries to bring about the production of the female sex hormones, estrogen and progesterone. (Estrogen is a generic term for a family of female steroid sex hormones: estradiol, estriol, and estrone.) In addition to secreting follicle stimulating hormone and luteinizing hormone, the anterior pituitary gland secretes prolactin. This hormone interacts with the epithelial tissues of the mammary glands to promote their development into milk-producing glands.

OXYTOCIN. *A hormone which facilitates child delivery (G. oxys, swift; tokos, childbirth).*

The *posterior pituitary gland* releases *oxytocin* from nerve fibers that stem from cell bodies in the hypothalamus. During childbirth, oxytocin stimulates the smooth muscle of the uterus to expel the baby. It also stimulates the smooth muscle of the mammary glands to contract and force milk out of the glands.

The *ovaries* produce and release eggs and the female sex hormones. The *fallopian tubes* are known also as oviducts. Every four weeks, beginning at puberty, one of these tubes conveys an egg from an ovary to the uterus. Fertilization usually occurs within this tube.

FETUS. *An offspring in a womb (L.).*

The *uterus* is a pear-shaped organ within which a fertilized egg (zygote) may develop through its fetal stages into a baby. In the course of this development, the fetus and mother both contribute tissues which form a *placenta*. A gonadotropin—chorionic gonadotropin—is secreted from the fetal portion of the placenta. This gonadotropin, like luteinizing hormone, stimulates the ovary to produce and secrete estrogen and progesterone.

The accessory sexual structures of the female include the external genital organs, known collectively as the *vulva;* a *vagina*, which facilitates the transfer of sperm from the male and also serves as the birth canal; and a pair of *mammary glands*, which produce and secrete milk for the baby.

The External Genitals

VULVA. *The female labia majora and the external sex structures which these labia enclose (L. a covering).*

The *vulva*, shown in Figure 13.5, includes the following components: labia majora, labia minora, clitoris, and vestibule.

The paired *labia majora* straddle the paired *labia minora* which, in turn, enclose the area known as the vestibule. Extending dorsally toward the anus, the labia majora merge into the perineum, an area that limits the vaginal opening.

The Reproductive System

During childbirth, it is often necessary to make an incision through the perineum. This operation, called an *episiotomy*, is used to expedite delivery, to prevent excessive trauma or damage to the brain of the newborn. Immediately after delivery the incision is stitched closed.

The *clitoris* is an organ which resembles the male penis. It contains two parallel cavernous bodies running through its length, but it lacks the third cavernous body and the urethra which are characteristic of the male penis. The clitoris can become slightly enlarged when it is erotically stimulated though it still is extremely small compared to the penis.

The *vestibule* is an area that lies between the labia minora. The vagina opens into it, as does the urethra and the openings of *Bartholin's glands*. These glands, which are counterparts to the male's Cowper's glands, lie on each side of the vagina, and provide a mucous secretion to lubricate the vulva.

EPISIOTOMY. *A surgical incision of the perineum to facilitate childbirth (G. episeion, vulva; tome, incision).*

BARTHOLIN. *A Danish anatomist (1616–1680).*

The Vagina and Cervix

The vagina is a canal which runs from the vulva to the uterus. It receives sperm and supports the uterus. Its entrance, known as the *introitus*, occupies the lower two thirds of the vestibule. The introitus in a virgin is covered by the *hymen*, a thin permeable membrane. This membrane usually allows the menstrual fluid to pass out of the body, but on rare occasion, where menstruation is painful and difficult, a girl must have it perforated. The hymen is sometimes ruptured during the initial period of intercourse (coition), but perhaps equally or more often it is broken earlier.

The vaginal wall, like the clitoris and labia, is innervated, and upon tactile

VAGINA. *The genital canal which extends from the uterus to the vulva (L. a sheath).*

HYMEN. *The thin membrane which partly or totally blocks the vaginal opening in a virgin (G.).*

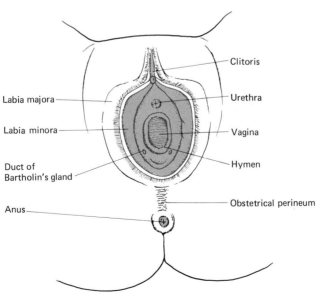

FIG. 13.5. *The vulva — the external genitals of the female.*

CERVIX. Any necklike structure (L.).

stimulation it gives rise to erotic sensations. The potential of evoking orgasms through these stimuli may be greater or lesser than those applied to the clitoris, depending on the individual.

The uppermost portion of the vagina is clasped around the cervix (neck) of the uterus. The function of the *cervix* at most times is to protect the uterus from invasion by external contaminants. Occasionally harmful bacteria get into the abdominal cavity by way of the vagina and fallopian tubes. The tubes may become inflamed (salpingitis) and/or the peritoneal lining may become inflamed (peritonitis). At childbirth, the cervix dilates (widens) and moves upward out of its confines within the vagina to facilitate the delivery of the newborn.

Small ulcers of the cervix (*cervical erosions*) often appear after childbirth. They sometimes give rise to discomfort and excessive vaginal secretions. Should they fail to clear spontaneously within several weeks, they may be removed by a physician. Cervical erosions may also arise from other causes.

The Uterus

UTERUS. The womb (L.).

The uterus lies in the pelvis of the abdomen, above the bladder. Ordinarily it is about three inches long, two inches wide, and one inch thick. Immediately after childbirth it is about six inches long, four inches wide, and three inches thick.

ENDOMETRIUM. The mucuous membrane lining the uterus (G. endon, *within;* metra, *uterus*).

The wall of the uterus is made up of three layers of tissue. The innermost (*endometrium*) is a mucous membrane which begins to increase in thickness each month, beginning shortly after menstruation has ceased and continuing until the menstrual fluid (blood and epithelial tissue) begins to flow (Figure 13.6). The outer covering of the uterus is a connective tissue which is called the *peritoneum* or *perimetrium*.

HYSTERECTOMY. Total or partial removal of the uterus (G. hystera, *womb;* ektome, *cut out*).

Sandwiched between the endometrium and perimetrium is the *myometrium*—a thick smooth muscle. This muscle grows as much as tenfold during pregnancy. Throughout this period it behaves more like an elastic than a contractile tissue. During labor, however, the myometrium suddenly becomes very sensitive to oxytocin and begins to contract vigorously.

During the early weeks of pregnancy the uterus is pear-shaped, but as it becomes distended by the growing fetus it becomes rounded. Its endometrial lining thickens rapidly and considerably during early pregnancy, as shown in Figure 13.8. The myometrium also undergoes considerable growth. Each muscle fiber increases about tenfold in length and threefold in width.

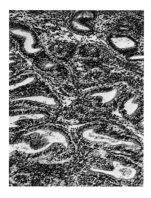

FIG. 13.6. *The endometrium.*

Whenever it is necessary to remove the uterus, as may happen in the event of cancer, the operation is called a *hysterectomy*. The removal of the body of the uterus, but not cervix, is called a *simple hysterectomy*. When the body and cervix are removed, the operation is called a *total hysterectomy*, and when the ovaries and fallopian tubes are removed also, a *radical hysterectomy*. The removal of the fallopian tubes only is a *salpingectomy*.

The Fallopian Tubes

The paired fallopian tubes, or oviducts, are about four inches long and one fourth inch in diameter. Each one has a free end that opens into the abdominal cavity. The opening, called the *infundibulum*, can draw an egg into the tube through the motile activities of its sticky fingerlike appendages. Once within the tube, the egg must journey through a tortuous, highly folded inner (epithelial) lining to the uterus. This journey is made possible by the muscular action of the walls of the tube, and by the action of cilia that line the lumen of the tube.

Fertilization usually takes place within a fallopian tube, usually within a day or so after ovulation (the release of an egg from the ovary). Around four days later the fertilized egg enters the uterus. Seven days later the fertilized egg becomes implanted within the endometrium.

Sometimes, the fertilized egg does not pass into the uterus but remains in the tube or enters the body cavity. This condition is called an *ectopic pregnancy* and it must be terminated, if not naturally then by an abortion.

FALLOPIUS. *An Italian anatomist (1523–1562).*

INFUNDIBULUM. *A funnel-shaped structure or passage (L.).*

ECTOPIC. *Out of place (G.).*

The Ovaries

Each ovary is about the size and shape of an almond, and is attached by ligaments to the uterus. An outer cortex and inner medulla are present, as shown in Figure 13.7. The cortex of each ovary contains several hundred thousand immature follicles at birth. These follicles consist of an immature egg cell covered by follicular cells. Only about 300 to 400 of these follicles mature during the reproductive life of the woman, which usually begins between 10 and 16 and continues into the fifth decade of life. The rest of the follicles ultimately degenerate. Beginning at puberty, the follicular cells, together with cells lying between the follicles, secrete estrogen. Midway through each menstrual cycle the follicular cells begin to secrete progesterone as well.

The medulla of each ovary is a fibrous and elastic tissue through which the blood vessels, lymph vessels, and nerve fibers enter the organ.

FOLLICLE. *A solid or hollow spherical mass of cells (L.* folliculus, *small bag).*

Production of Mature Egg Cells. The production of mature egg cells—called *oogenesis*—is initiated at puberty in response to follicle stimulating hormone from the anterior pituitary. Thereafter follicle stimulating hormone and luteinizing hormone, together with nutritional and environmental factors, regulate egg production and ovulation.

Every 28 days or so, at least one of the immature egg cells goes through two meiotic divisions to produce a mature egg cell. In metaphase of the first division, however, one of the daughter cells receives the bulk of the cytoplasm. The other daughter cell, called a *polar body*, is destined to degenerate. Prior to this fate, however, the rich daughter cell and impoverished polar body undergo the second divisional process of meiosis. The rich daughter cell divides unequally again, giving rise to an egg and another

The Female Reproductive System

polar body. Altogether, an immature egg cell divides twice to form a mature egg and three functionless polar bodies. The polar bodies degenerate and become enveloped within a mucous layer just outside of the egg, where they dissolve and enter the body fluids.

As a primary follicle matures, it changes sufficiently in character at certain points to allow the assignment of special names. In order of occurrence these include: Graafian follicle; corpus luteum; and corpus albicans.

The *Graafian follicle* is a fluid-filled sphere with walls composed of two layers of cells. The inner layer, the *theca interna*, gives rise to a mound of cells on top of which, surrounded by fluid, lies the rich daughter cell derived from the first divisional process of meiosis. The outer layer, the *theca externa*, is richly supplied with blood vessels and, together with the theca interna, produces increasing quantities of estrogen.

Midway through the menstrual cycle the Graafian follicle extrudes the

GRAAF. *A Dutch anatomist (1641–1673).*

THECA. *A sheath* (G. *theke*, a box).

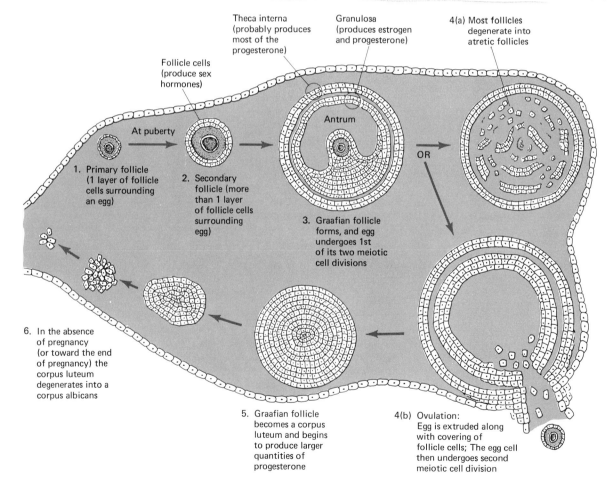

FIG. 13.7. *Production of eggs in the ovary.*

rich daughter cell, which now undergoes its second meiotic division. The follicle then becomes yellow in color and is known as a *corpus luteum*. The yellow color is due to estrogen and progesterone, which are produced in increasing amounts by the thecas interna and externa. The stages in this process are illustrated in Figure 13.6.

In the event of pregnancy, the corpus luteum increases in size for about three months, and then begins to wane. It is stimulated throughout pregnancy by luteinizing hormone from the anterior pituitary gland and by chorionic gonadotropic hormone from the placenta. Through mechanisms which are not known, these hormones prevent the corpus luteum from degenerating. At the same time they induce it to produce and secrete estrogen and progesterone.

At the end of pregnancy, or in the absence of pregnancy, the corpus luteum degenerates into a scar tissue known as the *corpus albicans*. It then disappears altogether.

CORPUS LUTEUM. *A yellow body which develops from a Graafian follicle* (L. luteus, *yellow;* corpus, *body*).

CORPUS ALBICANS. *A white body formed from a corpus luteum* (L. corpus, *body;* albico, *white*).

Functions of the Female Sex Hormones

Estrogen has three major functions. One is to produce the secondary sex characteristics, including the development of the mammary glands, the contour of the body, and the feminine voice and body movements. Estrogen's second function occurs when it achieves a critical level in the blood at the time of ovulation (Figure 13.8, on page 230). At this point, it works on the hypothalamus, together with rising levels of progesterone, to turn off signals that would otherwise bring about the release of follicle stimulating hormone. Alone, either estrogen or progesterone is effective in this regard, estrogen being required in much lower concentration than progesterone. Together, however, the two hormones work synergistically (more than just additively) and therefore much lower concentrations of each hormone are required than would be the case if either one were used alone. In preventing the release of follicle stimulating hormone, the process of developing another follicle is suspended until after the menstrual fluid flows again or a pregnancy is terminated.

Estrogen is also important in stimulating endometrial development in the uterus during the first half (the follicular phase) of the menstrual cycle.

Progesterone has two major functions, one of which was described above in conjunction with estrogen. The second function of progesterone is to promote gestation. In doing so it works together with estrogen on the uterus throughout pregnancy to maintain an ever-expanding chamber suitably responsive to the needs of the developing fetus. In addition, these hormones react with the tissues of the mammary glands to promote their development. If pregnancy does not occur, the corpus luteum begins to degenerate, progesterone and estrogen levels in the blood begin to drop, and menstruation begins.

Estrogen and progesterone are supplied mainly by the corpus luteum during the first three months of pregnancy. In the last six months they are supplied by both the placenta and the corpus luteum. The corpus luteum,

ESTROGEN. *Any substance which is capable of bringing a female dog, rabbit, or similar animal into estrus (heat)* (G. oistros, *mad desire*).

The Menstrual Cycle

The menstrual cycle recurs approximately every 28 days. It consists of four phases: (1) menstruation (bleeding); (2) proliferation (building up of the endometrium); (3) ovulation (release of an egg); and (4) progestation (release of increasing amounts of progesterone).

The *menstrual phase* begins on day one and continues until about day four. During this interval, the outer portion of the endometrium is sloughed off. This portion consists of numerous capillaries and glands that had developed since the preceding bleeding period.

The *proliferative phase*, also called the *follicular phase*, follows the menstrual phase and lasts about 10 days. During this period increasing

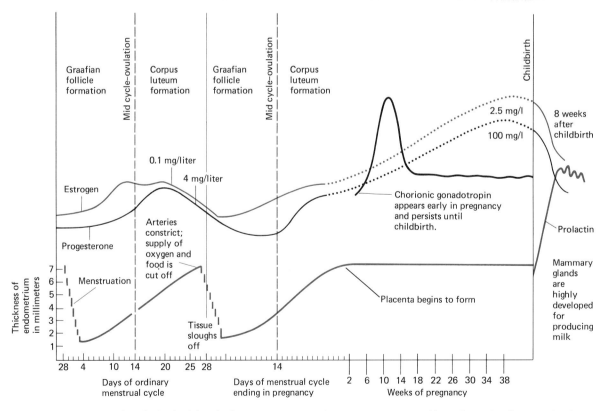

FIG. 13.8. *Hormone levels in the blood of nonpregnant and pregnant women. After about the first week of pregnancy the production of sex hormones begins to increase very rapidly, eventually reaching a concentration about 25 times greater than in the absence of pregnancy.*

amounts of estrogen and luteinizing hormone appear in the blood. At the same time, decreasing amounts of follicle stimulating hormone are seen. Throughout this period the endometrium increases at least twofold in thickness. In addition, a Graafian follicle develops towards its maturation point.

The *ovulatory phase* occurs midway through the cycle. Just prior to ovulation, the follicle stimulating hormone reaches a peak level, at which time it acts in conjunction with luteinizing hormone to cause the rupture of the follicle and the release of the egg. Also prior to ovulation there is a slight decrease in the output of estrogen and an increase in the output of progesterone. Now the uterus begins to proliferate even more, while the egg begins to journey through a fallopian tube. Follicle stimulating hormone continues to wane and the levels of luteinizing hormone and progesterone increase.

Progestation varies depending on whether pregnancy occurs. In the absence of pregnancy, there is a rapid rise of progesterone for about one week after ovulation, and then a decline. As indicated earlier, this hormone stimulates the development of the uterus; it also feeds back to the hypothalamus to prevent a rise in follicle stimulating hormone.

PROGESTATION. *The period beyond ovulation when pregnancy has, or can, set in (L. pro, for; gestatio, carrying).*

In the event of pregnancy, the progestational phase is extended for about nine months. Throughout this period the level of follicle stimulating hormone is low, luteinizing hormone levels are high, estrogen and progesterone levels increase continuously, and chorionic gonadotropin enters the picture.

Chorionic gonadotropin, a proteinaceous substance, is produced by certain cells in a portion of the placenta which is contributed by the embryo. The maternal portion of the placenta does not produce this hormone. It is a large molecule (molecular weight 38,000), but in spite of this, it passes out of the blood through the kidney and into the urine. Its presence in the urine is an early sign of pregnancy. The hormone can be detected by a chemical test as early as two weeks after pregnancy has started. During succeeding weeks, larger and larger amounts of this hormone are found in the urine. A peak is reached around the sixth week, after which the level declines over a period of another six weeks. Thereafter, low but relatively constant levels are present, until delivery. Throughout this period chorionic gonadotropin stimulates the ovaries to produce and secrete estrogen and progesterone.

CHORIONIC GONADOTROPIN. *A placental hormone which stimulates the ovary to produce and release sex hormones.*

Pregnancy and Childbirth

The development of a baby begins with the fertilization of an egg within a fallopian tube. The fertilized egg, called a zygote, undergoes a number of cellular divisions and is moved, by the musculature and cilia of the fallopian tube, to the uterus. Within the uterus, about three or four days after fertilization, the zygote may be composed of from 4 to 32 or more cells.

About seven more days pass, during which time the zygote develops into a fluid-filled sphere termed a *blastocyst*, which becomes implanted in the uterine wall. As the blastocyst enlarges, increased amounts of progesterone are released from the developing corpus luteum. In response to the negative

BLASTOCYST. *An inner embryonic cell mass, or blastula, and a thin outer trophoblast layer which, altogether, resembles a cyst (G. blastos, germ; kystis, bladder).*

TROPHOBLAST. *The outer cell layer around a fetus. It erodes the uterine mucosa and allows the embryo to receive its nourishment from the mother* (G. *trophe*, *nourishment*; *blastos*, *germ*).

feedback action of this hormone, the hypothalamus abstains from triggering further ovulation. Further preparations are made, instead, for the development of the embryo.

Shortly after implantation, the blastocyst appears as a sphere with a single layer of cells (a *trophoblast*) as its wall, and an inner cell mass from which the embryo will be formed. By the end of the second week the trophoblast is made of two layers of cells. The outer layer then begins to send out nipple-like structures, called *villi*. Some of these villi differentiate further to become the *chorion*, a membrane that surrounds the embryo and lies along the entire inner wall of the uterus. Other villi are invaded by blood vessels from the developing embryo and become known as chorionic villi (Figure 13.9).

The chorion constitutes the fetal contribution to the placenta. One of its main functions is to produce chorionic gonadotropin. As indicated above, this hormone stimulates the production of female sex hormones and is itself a sign of pregnancy when it is found in the urine.

The maternal contribution to the placenta is derived from the outer portion of the endometrium—the portion that normally is shed during a menstrual flow. Now known as a *decidua*, this portion of the endometrium becomes invaded by the chorionic villi. Spaces then appear around each villus and receive a supply of blood from the mother's circulatory system.

The capillaries of the fetal chorionic villi receive maternal blood from two arteries and pass their spent blood into one vein. The arteries and the vein lie within the placenta and are connected to the fetus by way of the *umbilical cord*.

Around the eighth week of pregnancy, painless contractions occur in preparation for labor. By the twelfth week, the uterus rises out of the pelvis and becomes a vertically-oriented abdominal organ.

Around the fourteenth week of pregnancy (Figure 13.10) the placenta is fully developed. It is disc-shaped and covers about one third of the surface

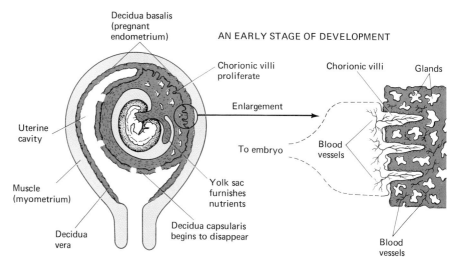

FIG. 13.9. *The uterus in an early stage of pregnancy. At right, an enlarged detail of the placental tissue.*

of the uterus. It is now about nine inches in diameter and about three fourths of an inch thick in its center, and thins out gradually toward the periphery.

Around the 40th week of pregnancy, the uterus enters *labor*. The first stage of this process is characterized by the onset of usually painful, recurring rhythmic contractions. As this stage proceeds, the cervix dilates. When the cervix has dilated sufficiently to allow the baby's head to pass into the vagina, the first stage of labor is said to be complete. The completion of the second stage of labor is marked by the passage of the baby into the outside world, and the completion of the third stage by the expulsion of the placenta. Thereafter, the uterus becomes smaller, returning to its original size about six weeks after delivery.

The Mammary Glands

The mammary glands are situated in the breasts over the pectoral muscles, where they are suspended by ligaments. The glandular tissue is divided into lobes. Each lobe contains numerous milk secreting cells (alveoli) and a ductwork which leads the milk into the nipple.

Until puberty, the breast tissue is confined to the nipple area. Thereafter, under the influence of estrogen, the ducts begin to proliferate, as they do during the first two weeks of each menstrual period from the onset (*menarche*) to the termination (*menopause*) of the reproductive years.

During the last two weeks of each menstrual period, increasingly greater amounts of progesterone stimulate the development of a correspondingly

ALVEOLUS. *A small saclike cell or cavity (L.).*

MENARCHE. *The beginning of a woman's menstrual life (G. men, month; arche, beginning).*

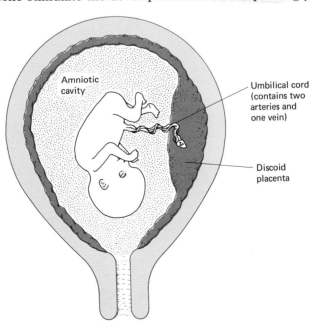

FIG. 13.10. *The uterus at fourteen weeks of pregnancy. At this stage the uterine cavity is filled.*

greater number of alveoli within the mammary glands. Both the ducts and the alveoli continue to proliferate in the event of pregnancy.

After about twelve weeks of pregnancy, a whitish opalescent fluid (*colostrum*) is produced. Colostrum is the first milk which is secreted at the termination of pregnancy. It is richer in sugar and protein than the milk that is produced about three days after birth.

Several days after delivery, the levels of estrogen and progesterone in the blood begin to diminish to levels that are characteristic of the first few days of the menstrual cycle. Their disappearance coincides with the appearance of prolactin from the pituitary gland.

Prolactin promotes the process of producing and—together with oxytocin—releasing milk. Prolactin is released from the anterior pituitary gland in response to the prolactin mobilizing hormone from the hypothalamus. When it is not required, the prolactin demobilizing hormone turns it off.

Prolactin mobilizing hormone is believed to be secreted in response to nerve impulses that are generated during nursing. The same triggering mechanism is believed to bring about the release of oxytocin. Oxytocin reacts with myoepithelial (smooth muscle) cells of the mammary glands to bring about their contraction and the release of milk.

To maintain a supply of milk it is necessary to stimulate the breasts periodically through suckling or by squeezing out the milk by hand. After each feeding the breast contents should be expelled by hand, in the event that the infant has not completely emptied it.

If it is necessary to suppress the production of milk immediately after birth, the mother is given estrogen, which at this time inhibits the action of prolactin and prevents the flow of milk. When the mother nurses the baby, prolactin normally begins to wane about eight weeks after birth, while the level of estrogen begins to increase. The mammary glands begin to diminish in size and productivity, ultimately achieving a condition similar to that before pregnancy.

COLOSTRUM. *The first milk secreted after pregnancy (L.).*

PROLACTIN. *A hormone which promotes the production of milk.*

BIRTH CONTROL

There are at least nine ways to prevent either conception or birth: (1) abstinence; (2) coitus interruptus, in which the penis is withdrawn prior to ejaculation; (3) the use of a rubber sheath (condom, prophylactic) over the penis; (4) the use of a diaphragm, a removable rubber cup which fits over the cervix; (5) the use of an intrauterine device (IUD); (6) the use of birth control pills; (7) vasectomy; (8) tubal ligation; and (9) the so-called morning-after pill. The first four of these procedures are achieved through psychological or physical means and need no further discussion. Nor will anything further be said about morning-after pills, which are drugs that destroy or abort a fertilized egg or make the environment of the uterus unfavorable for its survival or implantation.

CONDOM. *A sheath which is worn over the penis as a contraceptive device (perhaps named after Dr. Conton, an 18th-century English physician who is said to be its inventor).*

DIAPHRAGM. *In contraception, a dome-shaped plastic or rubber material.*

The Intrauterine Device

The IUD is usually a plastic or nylon device which may be of any shape (ring, serpentine, coiled). It is fitted snugly into the uterus by a physician. A string from the IUD extends to the vestibule, always available to assure the wearer that the device is still in place.

Although the exact mechanism of the IUD is unknown, it is beginning to appear that prostaglandins may be involved. These are chemical substances that are produced in tissues throughout the body. Much is known about these substances, but so complex are their actions and interactions, that little is understood. Clearly, however, one of their actions is to stimulate smooth muscle (such as the myometrium of the uterus). It is thought that the IUD irritates the uterine wall, causing more prostaglandins to be produced. This sets up muscular contractions, contractions which become exaggerated in the presence of the egg. In consequence the egg, whether fertilized or not, is expelled.

The "Pill"

Birth control pills, in contrast to the IUD, prevent ovulation. The pills contain a relatively large quantity of progesterone or progesterone-like substances, and a small quantity of estrogen. They are usually taken for 20 days during each menstrual cycle, beginning on day five. Shortly after day five, increasingly higher levels of progesterone are found in the blood. As mentioned earlier, one action of this hormone is to turn off the release of follicle stimulating hormone and another action is to prepare the uterus for implantation of a fertilized egg. Accordingly, in taking the pills the process of ovulation is arrested, and a thickened endometrium appears. On day 26, after the last of the pills is taken, there is a sudden, marked decrease in progesterone in the body. This drop continues for two or three days, after which a new menstrual cycle begins.

The choice of a 20-day regimen for taking the pill is based partly on the psychological needs of a woman. If the pill were to be taken indefinitely, there would be a corresponding arrest of the ovulatory process and a cessation of menstruation. By selecting a 20-day schedule the woman experiences a nonovulatory menstrual cycle which is, however, of normal length. The choice of a 20-day regimen is also based, in part, on the belief that a simulation of the normal menstrual period is better for the well-being of the body. Many hormonal events in our bodies are cyclic and synchronized. The cyclic release of gonadotropins and female sex hormones is a good example to support this statement. So too, the release of adrenocorticotropic hormone (ACTH) is a cyclic event. Maximum amounts are usually released between 4 and 8 in the morning. Correspondingly, maximum amounts of glucocorticoids are released between these hours, with attendant effects (notably an increase in blood sugar). Thus, rather than confound the complex metabolic, hormonal, and neural machinery by introducing an artificial time-variable, the regimen of taking birth control tablets is set at 20 days, resulting in a 28-day cycle.

Although the pill is relatively safe for most women, some experience side effects, such as nausea, abdominal cramps, or clotting in the legs (thrombophlebitis). Caution is urged in prolonged use (two to four years) without periods in which alternative methods are used. Diabetics may suffer special problems. Individuals prone to migraine headache, epilepsy, or asthma may have additional attacks. But for the vast majority, the "pill" has so far proved to be effective and harmless.

THROMBOPHLEBITIS. *An inflammed vein* (G. thrombos, *a clot;* phlebos, *vein;* itis, *inflammation*).

Vasectomy

A vasectomy is an operation in which the vas deferens are cut. This is generally done by entering the scrotal sac and singling out the spermatic cord. The spermatic cord carries the spermatic artery, spermatic veins, lymphatic vessels, and the vas deferens within a fibrous connective tissue. The vas deferens is cut, and care is taken to be certain that the ends will not join each other again. The success of the operation in producing sterility is determined by examining a subsequent sample of ejaculum. Since only the vas deferens are cut, there is neither a diminution in volume of the ejaculate during orgasm, nor a reduction in sexual sensory sensations.

VASECTOMY. *An operation to remove a segment of the vas deferens* (L. vas, *vessel;* G. ektome, *to excise*).

Tubal Ligation

Tubal ligation is a procedure in which the oviducts are closed by ligation or cautery. The operation is more difficult than a vasectomy but if it is done after a delivery the tubes are readily accessible and a local or spinal anesthetic is used. As is the case with vasectomy, tubal ligation does not deprive a woman hormonally and therefore there should be no reduction in libido or change in physiology.

LIGATE. *To constrict a vessel with a thread* (*ligature*) (L. ligo, *to bind*).

LIBIDO. *Conscious or unconscious sexual desire* (L.).

SUMMARY

1. In the male reproductive system the hypothalamus, anterior pituitary gland, and testes collaborate to produce sperm and testosterone; and a set of tubules, glands, and a penis are present for coition and ejaculation.
2. Sperm are produced in seminiferous tubules, where they are nurtured by Sertoli cells. Testosterone, which promotes the development of male characteristics, including the male gonads, is produced in Leydig cells, which are located between the seminiferous tubules.
3. The female reproductive system consists of: (a) a hypothalamus, anterior pituitary gland, and ovaries, which collaborate to produce eggs,

estrogen, and progesterone; (b) a pair of fallopian tubes in which an egg cell may be fertilized; (c) a uterus, in which a fetus develops; (d) the posterior pituitary gland, from which oxytocin is released to stimulate the smooth muscle of the uterus during childbirth and of the mammary glands during breastfeeding; and (e) accessory sexual structures, including the external genitalia, a vagina, and mammary glands.
4. Approximately 400 immature ovarian follicles complete a maturation process during 30 years of fertility. An egg is released from a follicle around the 14th day of the 28-day menstrual cycle, after which the follicle becomes a corpus luteum, which produces increasing amounts of estrogen and progesterone.
5. In the absence of pregnancy, the endometrial lining of the uterus, which becomes increasingly thick from days 5 to 28 of the menstrual cycle, is shed, beginning on day 1 of the next cycle.
6. Birth control may be affected through several procedures, including: (a) the use of an IUD, which stimulates the uterus to expel a fertilized egg; (b) the "pill," an oral contraceptive which prevents development of an egg; (c) tubal ligation, in which the fallopian tubes are closed; and (d) vasectomy, in which the male's vas deferens are severed to prevent the ejaculation of sperm.

REVIEW QUESTIONS

1. Describe the route by which sperm travels from the seminiferous tubules to the urethra. Name—in the order of their appearance—the glands that are present along this pathway.
2. What are the functions of the Leydig cells and Sertoli cells? Where are these cells located?
3. What is the functional difference between a gonadotropic hormone and a sex hormone?
4. List the components of the female reproductive system and tell what their functions are.
5. What are the separate functions of estrogen and progesterone?
6. Describe the menstrual cycle with respect to endometrial and hormonal changes.
7. Oxytocin has an effect upon the function of the uterus only at the end of pregnancy, but affects lactation only after pregnancy. Suggest a hypothesis to explain this.
8. How would you explain the operation or effectiveness of the "pill" in birth control? the IUD?
9. What hormonal changes, if any, result from tubal ligation? from a vasectomy?

The Structural and Support Systems

14 The Skin
15 The Skeletal System
16 The Joints
17 The Physiology of Muscle
18 The Functional Anatomy of Skeletal Muscle

PART FOUR

The Skin 14

THE EPIDERMIS
    Functional anatomy of the epidermis
        Rate of epidermal growth
    Role of the epidermis in insensible perspiration
THE DERMIS
    Structure and histology of the dermis
    Blood supply to the skin
        Constriction and dilation of the blood vessels
        Coloration of the skin by the blood
SWEAT GLANDS
    Structure and histology of sweat glands
    Physiological classification of sweat glands
    Mechanism of sweating
    Excretory function of the skin
    Regulation of sweating
    Sweating abnormalities
THE NAILS
HAIR
    Hair and its growth cycle
    Baldness
SEBACEOUS GLANDS
ABNORMALITIES OF SKIN
    Primary and secondary lesions
    Common skin disorders

The skin has the following three main functions: (1) it contains the sensory receptors through which the general sensory nervous system may operate; (2) in the presence of sunlight it initiates the conversion of certain dietary sterols into vitamin D; and (3) it helps other systems of the body to maintain a relatively constant internal environment. The first two functions were discussed earlier. This chapter is concerned with the third function. This function is achieved through the skin's role in thermoregulation; its role in protecting the body against invasion by harmful agents such as infectious

The Epidermis

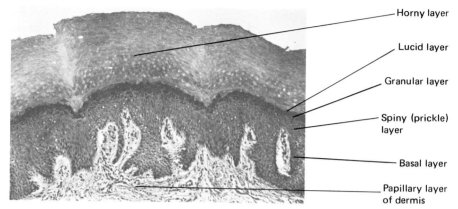

FIG. 14.1. Skin from the palm of the hand, showing five layers of epidermis.

microbes, chemicals, and ultraviolet sun rays; and its role in preventing excessive losses of water and electrolytes.

The skin consists of an external epidermal layer, an internal dermal layer, and four appendages—sweat glands, nails, hair, and sebaceous glands. Sections of skin from the general body surface and the palms are shown in Figures 14.1 and 14.2.

THE EPIDERMIS

EPIDERMIS. Outer skin (G.).

The epidermis has three major functions: (1) to manufacture a dead horny surface layer that helps in thermoregulation and in protecting the dermis against foreign invasion; (2) to manufacture melanin, which protects the body against ultraviolet sun rays; and (3) to give rise to the four appendages of the skin.

FIG. 14.2. Skin from the abdomen. The papillary and reticular layers are the outermost regions of the dermis.

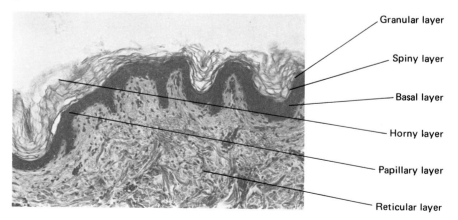

Functional Anatomy of the Epidermis

Four histologically-distinct layers are present in the epidermis throughout the body. A fifth layer is present in the palms and soles (Figure 14.1).

The lowermost layer—the *basal* or *germinal* layer—consists of basal cells and melanocytes. The *basal cells,* which appear columnar, cuboidal, or even squamous, depending on regional differences, give rise to the upper layers of epidermal cells and to the skin's appendages. The *melanocytes,* which are derived from the neural crests in the embryo, are distinguished from all other cells in the epidermis by their dendritic processes, much like those of astrocytes, which also are derived from the neural crest.

The function of melanocytes is to produce *melanin,* a pigment made partly from proteins and partly from the amino acid tyrosine under the influence of two enzymes (tyrosinase and peroxidase). In the presence of sunlight, tyrosinase and peroxidase operate more rapidly to provide larger quantities of melanin, which is secreted from the dendrites of the melanocytes and picked up by adjacent epithelial cells through the process of phagocytosis. Histologically, melanin is often seen to form a cap around the outer surface of the nuclei of the cells. It is believed that the melanin cap protects the nucleus against damaging radiation from the sun.

Studies have been conducted on dark and light skins to determine the basis for the differences in color. It turns out that all people, no matter what race, have the same number of melanocytes per unit area of skin. What makes the difference in shade is the amount of melanin produced by the melanocytes. Dark people have more melanin. The control mechanism for this is totally unknown.

What makes a difference in coloration, such as yellow versus red, is partly the quantity of melanin produced and perhaps also differences in the structure of the melanin. The structure of melanin, however, is unknown, other than to say it consists of derivatives of tyrosine which are complexed to proteins.

Immediately above the basal cell layer is the *spiny layer* or *prickle cell* layer. These names—spiny and prickle—are based on the appearance of filaments (prickles) that hold the epidermal cells to one another. In various skin diseases the prickle cells lose their filaments (prickles). As a result, the epidermis loses its cohesiveness, and blisters are formed. Jointly, the basal and spiny layers make up the *Malpighian* layer.

The next layer of the epidermis, moving toward the free surface of the skin, is the *granular layer,* an older form of the prickle cell layer. This layer retains its filamentous character and, in addition, produces keratohyalin granules.

As the granular cells age and are displaced toward the surface of the skin by newer cells, their granules coalesce, and the contents within the granules are transformed into *keratin,* a tough fibrous protein that protects the body against destructive chemicals, microorganisms, and physical forces.

In the palms and soles (Figure 14.1), the granular layer lies beneath a *lucid layer,* which in turn is covered by a *horny layer* or *corneum.* Else-

---

ASTROCYTES. *Star-shaped neuroglial cells* (G. astron, *star;* kytos, *cell*).

MELANIN. *Pigment in the skin* (G. melas, *black*).

MALPIGHI. *An Italian anatomist (1628–1694).*

KERATIN. *A hard proteinaceous substance found in such things as hair, nails, horns, and the free surface of the epidermis* (G. keras, *horn*).

CORNEUM. *The horny layer of the epidermis* (L. corneus, *horny*).

where in the body the granular layer is converted outright into a horny layer. Measuring about 100 to 150 micrometers in thickness over the general body surface, and 500 to 1300 micrometers thick on the palms and soles, the horny layer consists of cells that lack organelles and nearly everything else except keratin.

The cells of the horny layer are cemented to one another by a mucopolysaccharide (a nitrogenous carbohydrate) and — except for the uppermost layer or two of cells (out of 10 to 12 layers in the palms and soles, 3 to 5 layers elsewhere) — their intercellular spaces are occupied by water.

The horny layer is the skin's chief barrier against invasion, and also against losses of water and electrolytes, as can be demonstrated by stripping it off layer by layer with a strong adhesive tape. In addition, the horny layer screens out some of the sun's rays, thus reducing the load which melanin might otherwise have to absorb.

Rate of Epidermal Growth. In a normal person it takes about 4 weeks between the time a daughter cell is formed in the basal layer and the time it is shed from the horny layer. In certain diseases, however, this period is considerably reduced. In psoriasis, for example, the lifetime of a daughter cell is less than a week. The reduced lifetime is associated with an improperly structured epidermis which fails to function adequately.

PSORIASIS. *An itch-provoking disease in which silvery scaled pimples appear on the skin* (G.).

Role of the Epidermis in Insensible Perspiration

A certain amount of water — other than sweat — is lost from the body every moment of the day through the lungs and skin, and is referred to as *insensible perspiration*. The amount of this loss can be determined by weighing a person and measuring the person's intake of oxygen and output of carbon dioxide. After doing this, the insensible perspiration is calculated as the loss in body weight, plus the weight of carbon dioxide given off, minus the weight of oxygen taken in.

For an average normal adult the insensible perspiration amounts to about 30 to 50 ounces per day. Approximately 50% of it is exhaled from the lungs. The other 50% leaves through the skin.

It is important to realize that when scientists attempt to measure insensible perspiration, they do so under conditions in which an individual is not supposed to be sweating. That is, only insensible perspiration (invisible and intangible liquid) is meant to be measured. However, the palms, soles, and armpits of many individuals are sweating all the time. Nonetheless, only microscopic droplets are perspired, and since they do not form a film which is visible to the naked eye, no attempt can be made to correct for them in calculating the value of the insensible perspiration.

What is the significance of insensible perspiration, granting it is distinct from sweating? Very likely it is a mechanism which offsets normal tendencies to overheat. Normally, the human body operates at a core (rectal) temperature that is usually much higher internally (37°C, 98.6°F) than externally. Much of the time, however, there is a tendency for the internal temperature to exceed 37°C, especially when we are moderately active. However, such processes as the expulsion of moist warm air from the lungs, and the evaporation of water from the skin, counteract the tendency toward overheating. For each kilogram of water that is evaporated, 585 kilocalories of heat are lost. Hence, the daily loss of 900 to 1500 grams (milliliters) of water through insensible perspiration signifies a loss of about 530 to 880 kilocalories of heat. To show how dramatic this loss can be, if only one kilogram (liter) of water were to be lost immediately from the body in the form of evaporation (not kidney excretion), the internal temperature of a 150-pound person would drop 10°C.

In addition to offsetting the tendency to overheat, insensible perspiration may be helpful, insofar as the palms and soles are concerned, for providing friction in order to facilitate the handling of objects and performing of work. The palms and soles do not secrete an oily sebum, as does the skin elsewhere on the body, and in the absence of moisture on their surface, it would be more difficult to get a good grip on certain things. In this connection, the reader may reflect on a baseball batter's habit of spitting on his hands when called to bat.

THE DERMIS

The dermis is a dense irregular connective tissue with an abundant supply of blood and lymph vessels, nerve fibers, and fat cells (Figure 14.2). It performs several functions, among which are the following: (1) supporting the receptor elements of the general sensory nervous system; (2) supporting the appendages of the skin; (3) helping rid the body of certain invaders that pass through the epidermis, such as infectious bacteria, fungi, and viruses; and (4) helping in the process of thermoregulation.

DERMIS. *True skin; becomes leather when tanned (G.).*

In addition to these functions, the dermis—along with the hypodermis beneath it—is important in insulating the body (especially by virtue of its adipose tissue) and in protecting the underlying muscle and bone (by virtue of its collagen and elastic fibers as well as adipose tissue).

Structure and Histology of the Dermis

The dermis is composed of (a) collagenous, reticular, and elastic fibers which anchor various kinds of cells, such as fibroblasts, macrophages, mast cells, adipose cells, and smooth muscle cells; (b) appendages (sweat glands, sebaceous glands, nails, and hair); and (c) blood and lymph vessels. The fibers, in turn, are held adhesively within the organ by a ground substance which is produced by fibroblast cells in the presence of adequate levels of various vitamins, especially A and C.

FIBROBLASTS. *Elongated connective tissue cells that produce connective tissue fibers (L. fibra, fiber; G. blastos, germs).*

Two layers of tissue can be distinguished histologically in the dermis (Figure 14.2). Directly beneath the epidermis is the *papillary layer*. This

The Dermis

PAPILLA. *A small nipple-like or pimple-like structure* (L. *nipple*).

layer contains hair follicles, sebaceous glands, blood vessels, and nervous elements. Its name is derived from the fact that it displays papillae ("pimples") when the epidermis is stripped away from the skin.

Beneath the papillary layer is the *reticular layer*. It consists mainly of fibrous cells that interlace with one another and with the underlying connective tissue of the hypodermis, thus giving the appearance of a fibrous network (reticulum).

Blood Supply to the Skin

PLEXUS. *A network* (L.).

SUPERFICIAL. *On or near the surface, in contrast to deep* (L. *superficialis*).

The dermis receives its blood supply from a system of three arterial plexuses (networks). The deepest plexus lies at the level of the superficial fascia in the hypodermis. An intermediate plexus occurs between the hypodermis and dermis, and a superficial plexus is found just below the epidermis.

The arterial vessels furnish a nutritional supply of oxygenated blood to an extensive capillary network throughout the dermis. In turn, the capillaries drain into a system of veins.

Constriction and Dilation of the Blood Vessels. Adrenaline, noradrenaline, and angiotensin constrict the arterial vessels, lessening the amount of blood entering the skin. This happens on occasions when body heat is to be conserved. In addition to this mechanism of heat conservation, some of the arterials send their blood directly into veins, rather than through capillary networks. In bypassing the capillary networks, the blood retains heat within the body instead of permitting it to be released.

Acetylcholine, histamine, and various kinds of substances called *kinins* dilate the blood vessels of the skin to provide it with larger amounts of blood under certain conditions. For example, when there is need to cool the body, acetylcholine is called into action. When there is need to rid the skin of harmful invaders, histamine is released from nearby mast cells. Its presence causes capillary blood vessels to become dilated and leaky, making it easier for white blood cells to seep into tissue spaces occupied by the invaders. All too often, however, too much histamine is released and discomfort is felt, such as itching and pain from edema (fluid accumulation). Antihistamine drugs provide relief.

HEMOGLOBIN. *The red respiratory protein found in red blood cells* (G. haima, *blood;* globus, *sphere*).

Coloration of the Skin by the Blood. A close look at the blood vessels in skin reveals that some are reddish in color, while others are bluish. The red coloration signifies arterial passageways, which contain an oxygenated form of hemoglobin. Blue coloration denotes a venous vessel with deoxygenated hemoglobin.

Various shades of color may be seen in a light skinned person, depending on the circumstances. The skin may appear yellowish in a moment of fear, due to reflex actions in which the superficial blood vessels are closed off. In a moment of anger or emotion the skin may become reddened from increased arterial blood flow. On exposure to cold over a prolonged period, it may appear blue, due to stasis of blood in the capillaries, resulting in the deoxygenation of hemoglobin.

SWEAT GLANDS

Sweat glands (Figure 14.4) are exocrine (duct-bearing) glands derived from epidermal tissues. Two kinds of sweat glands are present in the body: eccrine and apocrine. The secretory cells of the *eccrine glands* pass a liquid (sweat) into a duct that leads to the surface of the skin. In contrast, the secretory cells of the *apocrine glands* (which are present mainly in the armpits and as modified glands in the mammary glands), pass a liquid plus a portion of the cell itself into the excretory duct (Figure 14.3).

FIG. 14.3. *Cells of eccrine glands (left) and apocrine glands (right).*

ECCRINE. *Exocrine glands that secrete sweat which is chemically different from apocrine sweat.*

APOCRINE. *Exocrine glands whose secretions include the tip region of the cell itself (G. apokrini, to separate).*

MYOEPITHELIAL CELLS. *Contractile cells in glands (G. mys, muscle).*

Structure and Histology of Sweat Glands

All sweat glands, regardless of type or location, are made of a secretory portion, called a glomerulus, and an excretory duct.

The *glomerulus* is a highly convoluted single tubule (Figure 14.4) made of cuboidal cells resting on a basement membrane. Between occasional cells and basement membrane there is a smooth muscle cell (a myoepithelial cell) which, upon contraction, facilitates the discharge of sweat from the glomerulus.

The *excretory ducts* of the sweat glands are made of two layers of cuboidal cells—a basal layer and a surface layer. In the case of eccrine glands, this tubule passes upward in corkscrew fashion to the surface of the epidermis. In contrast, the excretory duct of the apocrine gland terminates at a hair follicle, which carries the apocrine secretions to the surface of the skin. Close observation of the skin during sweating reveals the location of the pores of the active sweat glands.

In addition to the eccrine and apocrine sweat glands, the body contains modified sweat glands in the external ear canal. These glands produce *cerumen*, a waxy pigmented secretion which is more like sebum than sweat.

CERUMEN. *The soft, brownish ear wax secreted by the ceruminous glands of the external ear canal (L. cera, wax).*

Physiological Classification of Sweat Glands

From the viewpoint of function and purpose, there are three classes of sweat glands in the body. The first class, present in the palms and soles, responds mainly to mental or emotional stimuli and hardly at all to thermal stimuli except in extreme conditions such as after excessive exercise or when a heat stroke is imminent. It is interesting to note that although the palms and soles yield the largest percentage (66%) of insensible perspiration (when the body is not actively sweating), they yield the smallest percentage of sweat when the body is responding to overheating. A *heat stroke* occurs when not enough heat is lost from the body through evaporation (insensible perspiration and sweating) to compensate for the heat gained by the body from its surroundings. During a heat stroke sweating decreases or disappears. In addition, giddiness, irritability, headache, drowsiness, throbbing, labored breathing, a rapid and weak pulse, and stupor may occur. The

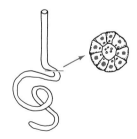

FIG. 14.4. *Sweat gland. Cells of the glomerulus (right) pass sweat into an excretory duct.*

internal body core (rectal) temperature generally rises from 37°C to 39°C to 41°C, and unless the body cools off the individual may enter a coma or die. Thus the proper functioning of the sweat glands is of crucial importance to our physical health.

The *mental-emotional sweat glands* are under the control of a sweat center, which lies in the premotor area of the brain. Through this center the sweat glands of the palms and soles are put into operation right away when a person is subjected to mental or emotional stress.

The second class of sweat glands is present over the entire body except the nail beds, soles, palms, armpits, and possibly the forehead. These sweat glands are under the control of a sweat center in the hypothalamus, but when there is extensive damage to the upper spinal cord they often become reflexly operated through centers all along the spinal cord below the level of injury. They respond to thermal stimuli and not to mental or emotional stimuli unless the latter are mixed with aggressive or violent physical actions on the part of the individual.

Unlike the mental-emotional sweat glands, the *thermal sweat glands* do not respond immediately to heat stimuli. Rather, there is a latent period—as for example when an individual is engaged briefly in exercise or is subject to a short period of hot sun. During this latency the threshold levels of nerve discharge are being approached and thereafter the autonomic nerves fire to induce sweating.

The glands of the third class respond to both mental-emotional stimuli and thermal stimuli. These glands occur in the armpits of all people (except in rare congenital cases in which sweat glands are absent all over the body), and in the forehead of certain people.

Mechanism of Sweating

GLOMERULUS. *A ball of capillaries (in the kidneys) or a ball of secretory cells (in the sweat glands)* (L.).

Based on microscopic measurements on histological preparations, it has been estimated that in each person, the secretory portions (glomeruli) of all the sweat glands can hold about 17 milliliters (one-half ounce) of sweat at any given moment and the excretory portions can hold an equal volume. Ordinarily, in the absence of mental, emotional, or thermal stimuli the sweat glands do not release their contents, except perhaps as insensible perspiration.

During moments of vigorous sweating, however, the sweat glands can secrete as much as 2 to 3 liters per hour under strenuous conditions, and up to about 10 to 12 liters in one day. How is it possible to achieve such productivity?

To begin with, the human body is well endowed with sweat glands. The number varies considerably from one individual to another and from one region to another, but it is generally in the order of 200 to 400 per square inch on the thigh, leg, cheek, and back of the trunk; about 700 to 1000 on the face and front of the trunk, and about 2000 on the palms and soles. Altogether, there are about 3 million sweat glands in the body. This number neither increases nor decreases after the age of about 30 months.

Secondly, the skin receives far more blood than is needed for nutritional purposes. For every square micrometer of surface area of sweat gland there is an adjacent surface area of 3 square micrometers of blood capillaries.

Excretory Function of the Skin

The skin is not a major excretory organ under resting conditions, but it can be an important one during exercise and work. In either case, whether at rest or at work, the skin excretes only about 1% of the body's waste carbon dioxide. The rest is excreted by the lungs.

A different situation holds true for nitrogen. Under resting conditions only a percent or two of the total nitrogen may be excreted, but for an ordinary person engaged in moderate activity, the skin may excrete up to about 9% of the total waste nitrogen. More than 95% of this material is ammonia (NH_3) and urea (NH_2CONH_2). During strenuous exercise or work, even greater percentages may be eliminated.

Regulation of Sweating

All sweat glands are innervated by the sympathetic nervous system. However, unlike the nerve fibers of this system, which innervate smooth muscle or cardiac muscle, the neurotransmitter that is liberated into the sweat glands is acetylcholine, not noradrenaline or adrenaline.

In addition to a cholinergic sympathetic innervation, the glossopharyngeal and trigeminal cranial nerves (Table 5.4) provide some individuals with a cholinergic parasympathetic supply to sweat glands in the face, particularly in the region of the forehead.

CHOLINERGIC. *Refers to synapses in which acetylcholine is liberated* (G. chole, *bile;* ergon, *work*).

In general, when thermal sweating occurs, it happens over the entire general surface. There may be regional differences with respect to rate of glandular secretion and quantity secreted, but there is normally no such thing as copious production of sweat on the arms and no sweat whatever on the legs.

The process of sweating is triggered reflexly to offset the possibility of any remarkable increase in body temperature. It is interesting in this regard that the reflex action passes more favorably along nerve pathways which lead to sweat glands whose pores are unobstructed. Thus, when an individual who is sweating in response to heat lies on one side of his body, the sweat glands on the other side take over the bulk of the task of secreting sweat and cooling the body.

Sweating Abnormalities

Excessive sweating (*hyperhidrosis*) may be congenital or acquired, localized to one region of the body (such as the armpits) or widespread through-

HIDROSIS. *Sweating* (G.).

The Nails

ANHYDROSIS. *A sweating deficiency* (G.).

out the body. Acquired hyperhidrosis may be due to damage in the central nervous system, an emotional disorder, or some unknown cause.

Anhydrosis—an excessive decrease in sweating ability—may be congenital or acquired, localized or widespread. The congenital case is usually due to a lack of sweat glands, whereas the acquired case may be caused by damage to the nervous system.

THE NAILS

Beginning around the ninth week of embryonic life, a portion of the epidermis on the dorsal side of the end of each toe and finger begins to fold under and move backward and laterally toward the tip of the digit. By this process, a *nail fold* is formed.

As you can see in Figure 14.5, the nail fold consists of a roof, a floor, and lateral walls. It is generally agreed that the bulk of the nail, if not all of it, is formed from the floor. The floor begins posteriorly where it meets with the roof and it terminates anteriorly in the vicinity of the *lunula*, the half-moon region which is usually visible on the thumb and may appear on other fingers as well.

The nail itself is made of layers of flattened, keratinized cells that have lost their organelles, much like the horny layer of the epidermis. The hardness of the nail is attributed to a parallel arrangement of keratin fibrils between the cells and also to the fact that the cells are densely and strongly stuck to one another.

The nail is fastened firmly to its nail bed by modified prickle cells. When the nail is so severely injured that it becomes separated from its nail bed, the nail is usually sloughed off and a new nail begins to grow in its place. As in embryonic life, the new nail develops from the floor of the nail fold, and it takes about 5 to 6 months to complete its development.

FIG. 14.5. *The nail.*

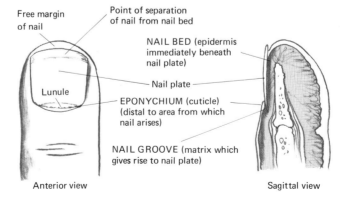

HAIR

A single hair is just one of several parts of an organ that consists of a hair follicle, the hair itself, one or more sebaceous glands, and a smooth muscle known as an *arrector pilorum* (Figure 14.6). The first three structures are discussed below. Nothing more will be said of the arrector pili other than that they respond to autonomic nerve stimulation in moments of emotion or when the skin is chilled, and in doing so they cause the hairs to stand erect and give rise to goose bumps.

ARRECTOR PILORUM. *A bundle of smooth muscle which erects a hair and creates a goose bump (L. arrigo, erect; pilus, hair).*

The *hair follicle* consists of an outer dermal sheath and an inner epidermal sheath. Its main function is to produce a hair. However, in order to carry out this function the follicle must be stimulated by an underlying structure called a *dermal papilla*.

The dermal papilla contains blood vessels that furnish oxygen, other nutrients, and hormones to the follicle. In addition, it contains nerve fibers for receiving sensory information and various phagocytes to protect it against foreign invasion.

If dermal papillae are destroyed, as happens in a third-degree burn, hair follicles cannot develop and the area becomes bald. On the other hand, if the hair follicles are destroyed but the dermal papillae are left intact, new hair will regenerate. Thus, in plastic surgery, for example, where skin may be grafted from the thigh to the forearm, the transplanted skin will lose its hair following surgery, but in time it will develop new hair—provided the dermal papillae of the forearm are functionally sound. So too, new skin will appear on the thigh and it also will grow hair—provided the dermal papillae of the thigh are functioning properly.

The *hair shaft* itself, looked at histologically in cross section, has a thin outer *cuticle* of horny cells, a thick pigmented *cortex* of spindle-shaped

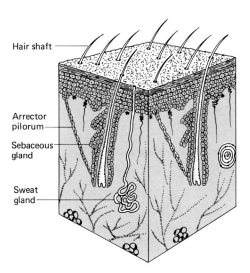

FIG. 14.6. *Structure of hair.*

cells, and a *medulla*. The cells of the medulla are nucleated near the hair follicle, but the nucleus is lost when these cells are displaced distally as the shaft grows out of the skin. The melanin pigment of the cortex cells ceases to be produced in some people as they age. The hair becomes completely transparent instead, and appears gray because of the pigmented hairs that are present nearby.

Hair and Its Growth Cycle

Hair has a definite growth phase, after which it enters a resting phase. At any one time many hair follicles all over the body are in the growth phase, while the others are in the resting phase. The life span of a hair varies depending on factors which are not known. On the average, an eyelash lives from 3 to 4 months, a scalp hair from 3 to 4 years.

Late in the resting phase of the growth cycle, the hair follicle separates from its firm attachment to the dermal papilla. Within the follicle, the root of the hair shaft swells up and its cells die, though they remain in the follicle. As a new growth phase is entered, a new hair pushes past the old one. As a result, two hair shafts protrude from the epidermis until the dead hair is shed.

The growth cycle may be interrupted by high fever, major surgery, or some other traumatic event, at which time hair follicles throughout the body may abruptly enter a resting phase. Two to four months later a new cycle begins and many hairs are shed as new hairs displace the old resting hairs. In general, recovery from this situation is complete.

Baldness

SEBORRHEA. *A skin disorder in which there is an excess of sebum* (L. sebum, *fat*; G. rhoia, *a flow*).

Baldness (*alopecia*) may arise for congenital reasons or in response to some traumatic situation (see above). Other causes include a destructive disease (for example, leprosy), a skin disease (for example, seborrheic dermatitis), a serious neural disorder, senility, or a long-term illness.

Most often, however, baldness is inherited. When baldness arises through heredity, it appears more often in males than in females, and the male sex hormone, testosterone, is needed to bring it out. However, female sex hormones are ineffective in treating it. For that matter, no treatment is available for this type of baldness, other than the grafting of skin from one part of the body—skin which includes the dermal papillae—to the hairless (glabrous) regions.

SEBACEOUS GLANDS

One or more sebaceous glands are always found associated with each hair of the body (Figure 14.6). They are absent in the soles and palms, as are hairs, but they are present in certain tissues which do not have hair. The latter include the outer surface of the lips, the glans penis, the inner surface of the prepuce, the labia minora of the vulva, the nipple of the breasts, and the tarsal (Meibomian) glands of the eyelids.

Sebaceous glands consist of several lobules of flattened, stratified squamouslike epithelial cells. These cells secrete their product (*sebum*) into a short duct which enters the upper part of a hair follicle. In tissues that lack hairs, the excretory duct of the gland passes the sebum directly to the surface of the skin.

Sebum is an oily material containing fatty acids, neutral fats, cholesterol, waxes, and various other fatty materials. It lubricates the hairs and skin and is believed to be capable of destroying certain microorganisms, such as bacteria and fungi.

In contrast to sweat, which is secreted from sweat glands in eccrine or apocrine fashion, the secretions of the sebaceous gland cells are *holocrine* (Figure 14.7). That is, the entire cell, along with its oily products, is discharged into the excretory duct of the gland.

FIG. 14.7. *Holocrine mode of secretion.*

SEBUM. *A fatty material secreted by sebaceous glands as a lubricant for hair (L.).*

HOLOCRINE. *A secretion which consists of altered cells of the gland itself, as in sebaceous secretions (G. holos, all; krino, to separate).*

ABNORMALITIES OF SKIN

The branch of science that deals primarily with problems of skin is called *dermatology*. It is beyond the scope of this text to cover even a single problem in detail, but by describing the basic characteristics of several common problems, the reader may gain a better understanding of normal skin.

Primary and Secondary Lesions

A *primary lesion* is a lesion, or change in the skin, that is formed as a skin disease progresses to its height. A *secondary lesion* is one that develops as a consequence of the primary lesion.

The common primary skin lesions are macules, papules, vesicles, and scales. A *macule* is a change in coloration. It may be due, for example, to deposits of pigment (freckles), dilation of capillary muscles (capillary nevus), or escape of blood from blood vessels into the skin (*petechiae* are minute hemorrhagic spots of pinpoint to pinhead size; *purpura* are larger hemorrhagic spots).

A *papule* is a small (1- to 5-millimeter) elevation (Figure 14.8). A large papule is called a *nodule*.

A *vesicle* is a blister formed by an accumulation of fluid in the epidermis, usually associated with the disintegration of epidermal cells in the vicinity. A *bulla* is a large vesicle. A *wheal* (hive) is an accumulation of fluid caused by edema. A *pustule* is a blister which contains pus. (*Pus* is a product of inflammation consisting of liquid, white blood cells, tissue debris, and living and dead microbes.)

MACULA. *A flat discolored spot. Maculas occur in the skin and are also found microanatomically in such places as the inner ear and kidneys (L.).*

FIG. 14.8. *Papule—a solid elevation of the dermal surface.*

254
Abnormalities of Skin

Scales are usually an abnormality in the formation of keratin.

Common secondary lesions include *fissures* (small cracks through the epidermis to the dermis); *ulcers* (destruction of a local region of the epidermis and part of the underlying dermis); and *scars* (replacements of epithelial, connective, muscular, or nerve tissue with fibers of connective tissue).

Common Skin Disorders

Skin disorders vary greatly in appearance, cause, and severity. *Dermatitis* is a generic term which can be applied to any skin disease in which pain, flaring, and swelling are seen. *Urticaria* is any skin disease in which itch-provoking wheals are formed. Known also as hives or nettle-rash, the disease may be due to an allergic reaction, infection, or psychic stimulus. The itching may be due to histamine (from mast cells), acetylcholine (from motor nerve endings), and other chemical substances (from other connective tissue elements).

Whereas itching in urticaria is associated with wheals, the itching of *pruritus* may occur independently of a skin lesion. Or it may arise from aging and degenerating tissues or from diseased tissues associated with such disorders as hemorrhoids, ulcers, nodules, or dermatitis.

Known to the layman as a birthmark or strawberry mark, a *nevus* is a localized developmental defect (a bluish-red macule) of the skin. It may appear at any time in life, though it occurs most often at or before birth.

A staphylococcal inflammation of a sebaceous or cerumen gland is called a *boil* or *furuncle*. An aggregation of boils is called a *carbuncle*. A well-established boil appears as a red, itching nodule and rapidly develops into a hard, cone-shaped projection surrounded by a red collar. As the body's defense mechanisms continue to muster an attack on the causative bacterium, a pustule (Figure 14.9) forms at the tip of the nodule. This pustule eventually breaks, some of the pus escapes, and the remainder follows after a period of time or is forced out by manipulation.

Cold sores, zoster, and warts are the three most common viral skin problems, aside from certain common immunological diseases such as measles and chicken pox. Cold sores are due to *Herpes simplex*, zoster to *Herpes zoster*, and warts (verrucae) to unidentified viruses. The *Herpes simplex* virus induces the formation of vesicles within the prickle layer of the epidermis in response to certain provocations, such as the common cold, menstruation, and exposure to cold or sunlight. Some people are susceptible, others are not. The tissues most commonly affected are the lips (herpes labialis); interior of the nose, tongue, mucosa of the cheeks, gums, and tonsils (herpes febrilis); glans penis and prepuce in males and the labia majora or minora in females (herpes genitalis).

Herpes zoster, known also as shingles, is caused by what is now known as the varicella-zoster virus, the virus which also causes chicken pox. It is believed that chicken pox is a primary disease, and only after having had this disease is it possible to acquire zoster. For reasons unknown, the virus

URTICARIA is commonly seen as hives, as a rash, or as an eruption of itching wheals (L. urtica, *a nettle*).

HEMORRHOIDS. Swollen "piles" of veins near the anus, which are likely to bleed and produce pain (G. haima, *blood;* rhoia, *a flow*).

ZOSTER. Herpes zoster or shingles (G. *a girdle*).

FIG. 14.9. *Pustule.*

sometimes remains inactive for periods up to many years in the tissues of people who have had chicken pox. It then becomes active, destroying one or more dorsal root ganglia, creating pain, depriving an individual of sensory ability within the afflicted region, and sometimes resulting in partial paralysis in local regions of the body. The virus passes along the nerve fibers into innervated regions of the skin where it causes blistering within the prickle cell layer. The vesicles and bullae that arise occur most often in the chest or abdomen and less often in the face, where they appear to girdle the infected region.

Warts (verrucae) are localized circumscribed overgrowths of the prickle cell layer, resulting in a thickening of the horny layer. The granular layer is absent in places or drastically altered in appearance. In time, the wart invades the dermis, which results in a defensive reaction on the part of white blood cells and other cells of the connective tissue. For this reason, warts generally disappear "spontaneously" within six months after they appear.

Fungal diseases of the skin include ringworm (tinea), actinomycosis, blastomycosis, sporotrichosis, and mycetoma. Scales, vesicles, pustules, macules, and various other characteristics appear, depending upon the species of fungus, the site of infestation, and the response of the skin. The fungi thrive on keratin, which they are able to digest, thereby causing fissures. Children are especially susceptible, but after puberty they become resistant, which suggests a chemical change in the sebum that lubricates the skin.

Eczema refers to a group of diseases in which redness of the skin and papules are the principal initial characteristics. Later, vesicles (or pustules —infected vesicles), scales, and fissures appear. In general, all forms of eczema are forms of dermatitis (though not all forms of dermatitis are cases of eczema). Frequently, however, a case of eczema is referred to as a special kind of dermatitis. Thus, *infantile eczema* and *atopic dermatitis* are used interchangeably for a skin rash that arises in certain infant allergies, as for example, when the baby is allergic to cow's milk. *Contact allergy* and *exfoliative dermatitis* are forms of eczema in which an individual is hypersensitive to substances that come into contact with the skin, as for example, formalin, carbolic acid, or the sap from poison ivy, poison oak, or poison sumac.

ECZEMA. *An eruption of the skin* (G. ekzema).

Psoriasis is a common, chronic dermatitis in which the rate of proliferation of basal cells is much increased, the prickle layer is greatly thickened, the granular layer is absent, and the horny layer consists of nucleated, incompletely keratinized cells, resulting in slightly raised, well-marginated, dry, reddish, papules covered by silvery scales. Inflammatory conditions are present in the papillary layer, and pin-point pitting of the nails or loss of translucency and thickening of the nail plates are almost always present. The cause of the disease is unknown, but it is linked with hereditary factors, a cold damp climate, and psychological problems.

Acne is an inflammation of the sebaceous glands and hair follicles, characterized by red papules or pustules on the skin. The problem may arise at puberty when increased amounts of the male sex hormone (testosterone)

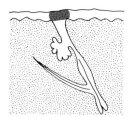

FIG. 14.10. *Blackhead.*

HEMANGIOMA. *Strawberry mark* (G. haima, *blood;* angeion, *vessel;* oma, *tumor*).

METASTASIS. *Movement of diseased tissue from one part of the body to another, as may occur in mumps, cancer, and bacterial diseases* (G. meta, *in the midst of;* stasis, *a placing*).

Summary

appear in both males and females and the sebaceous glands increase in size. Acne may also arise any time later in life. Regardless of when it appears, large quantities of sebum are produced and, for reasons unknown, the sebum becomes pastelike instead of oily and is forced toward the dermis, instead of being released onto the surface of the skin through a hair follicle. At this point tiny *whiteheads* appear, formed of white blood cells attracted from the dermis. Later, they often become *blackheads*, the discoloration probably caused by dirt which seeps in through the follicle or by oxidation of the sebaceous secretion (Figure 14.10). The whiteheads or blackheads pick up bacteria and become "pimples." The epidermis is red and sore (inflamed), then turns yellowish-white as white blood cells and bacteria form pus. Scarring occurs if excessive damage is done when the pimple erupts.

Papules, nodules, and pustules are *tumors*, as are cysts, which are fluid filled; corns; warts; moles (pigmented nevi); and hemangiomas (for example, strawberry marks). Most skin tumors are benign (harmless), but some are cancerous (malignant).

Three kinds of *skin cancer* are relatively common: basal cell carcinoma, epidermal cell cancer, and melanoma. In *basal cell carcinoma* the diseased cells invade the dermis – a feat which normal basal cells are unable to do. Fortunately, however, the malignant basal cells do not enter a blood vessel and metastasize (move to another tissue). Thus, the malignancy is localized. It can, however, produce an ulcer – a so-called rodent ulcer – and the cells can invade nearby cartilage and bone, causing serious destruction and possible further malignancy.

In *epidermal cell cancer* (epithelioma or squamous cell carcinoma) the cells enter blood and lymph vessels, which then carry them to other organs or distant regions of the skin where they may proliferate, forming a malignancy or neoplasia.

Melanoma, a cancer of melanocytes, is the most dangerous of skin cancers. The melanoma cells are very efficient in entering the dermis and metastasizing to other parts of the body.

SUMMARY

1. The skin consists of: (a) an epidermis, which protects the underlying dermis and gives rise to sweat glands, sebaceous glands, hair, and nails; and (b) a dermis, which harbors blood vessels, nerve fibers, lymph vessels, and various body defense cells.
2. The epidermis produces two important substances: melanin, which protects the skin against harmful radiations; and keratin, which protects the skin against abrasion and other forms of injury.
3. The escape of "insensible perspiration" is important in temperature regulation and in providing friction to the palms for gripping objects.
4. Sweat glands are simple tubules whose proximal secretory portion is

highly convoluted and whose distal excretory portion is a straight duct which leads either to the skin (eccrine gland) or to a hair follicle (apocrine gland). Eccrine gland cells secrete sweat only, while apocrine gland cells secrete sweat and the apical region of the cell.
5. Sweat glands in the palms and soles respond mainly to mental-emotional stimuli; sweat glands elsewhere in the body respond mainly to thermal stimuli; and sweat glands in the armpits and forehead respond to a combination of these stimuli.
6. As much as 2 or 3 liters of sweat can be secreted per hour in response to sympathetic cholinergic stimuli even though, collectively, the three million sweat glands can harbor only about 17 milliliters of liquid.
7. A nail is made of parallel plates of keratinized cells which are held to a nail bed by prickle cells. In a severe injury, the nail is shed and it takes 5 to 6 months to grow a new one.
8. Hair consists of: (a) a hair follicle from which the hair shaft grows; (b) a hair shaft, which serves mainly an esthetic function; (c) one or more sebaceous glands, which oil the hair and skin and perhaps protect the skin against certain microorganisms; and (d) a smooth muscle, called an arrector pilorum, which, upon contraction, raises the hair shaft and creates a goose bump.

REVIEW QUESTIONS

1. Describe the structure of the epidermis and specify the functions of its various layers.
2. Are there any functional distinctions between the papillary and reticular layers of the dermis?
3. How does insensible perspiration differ from sweating?
4. What kind of tissue are sweat glands made from? Describe the structure of a sweat gland.
5. Where are the sweat glands that respond mainly to mental or emotional stimuli? mainly to thermal stimuli? to a combination of these stimuli?
6. What is unique about the fact that sweat glands respond to sympathetic cholinergic stimuli? See Table 8.1 for assistance in answering this question.
7. Describe the functions of the hair follicles, the sebaceous glands, and the arrector pilori?
8. Describe one or more histological abnormalities in the following skin diseases: eczema, psoriasis, wart, acne, cold sore, shingles.

15 The Skeletal System

THE SKELETON
    Classification of bones
    Terms used in describing irregular surfaces on bones
THE AXIAL SKELETON
    The skull: special features
        Location and specific features of the skull bones
        The hyoid — A suspended skull bone
        Six more bones from the cranium
    The face
    The spinal, or vertebral, column
        Regional variations in the spinal column
        Disorders of the vertebral column
    The thoracic cage
THE APPENDICULAR SKELETON
    The shoulder girdle
    The upper appendages
        The arm
        The forearm
        The hand
    The pelvic girdle
    The lower appendages
        The thigh
        The leg
        The foot
THE STRUCTURE, FUNCTION, AND MAINTENANCE OF BONE
    The bone wall
        The periosteum
        The mineral layer
        The endosteum
    Bone marrow
    Blood circulation in bone
DEVELOPMENT OF BONE
    Intramembranous bone development
    Intracartilaginous bone development
        Formation of Haversian systems

Intracartilaginous development of short bones
Intracartilaginous development of irregular bones
Repair of fractured bone
DISORDERS AND DISEASES OF BONE

The skeletal system is composed of bones held together by ligaments. The skeletal system performs five services for the body: (1) protection, by enclosing the vital organs of the body; (2) aid to movement, by allowing the skeletal muscle to execute movements; (3) support, by serving as a framework to which tendons and fascia are attached, enabling skeletal muscles, viscera, and skin to obtain a holdfast; (4) storage, of calcium and phosphorus; and (5) production of blood cells.

The storage function of the skeletal system was discussed in Chapter 12 and the production of red blood cells is described in Chapter 29. In this chapter we will focus on the anatomy of the skeletal system and the physiology of bone tissue.

THE SKELETON

At birth there are about 275 bones in the body. In the course of development, three to five vertebrae fuse to become the sacrum, which is located in the pelvic region of the spinal cord. Four other vertebrae also fuse together, becoming the coccyx (tailbone). In addition, many of the skull bones fuse. Thus, in the course of development the number of bones is reduced.

In the adult, the skeleton consists of 206 named bones and a variable number of unnamed *sesamoid bones* (Figure 15.1). The latter, most of which resemble sesame seeds, are small, rounded bones. They develop in the capsules of certain joints or in tendons, where they serve to provide support or to reduce friction. Two outstanding sesamoids are the kneecap (patella) and one of the wrist bones (the pisiform).

SESAMOID. *Shaped like a sesame seed (G. sesamoeides).*

PATELLA. *A large sesamoid bone commonly called the kneecap (L. small plate).*

Classification of Bones

Bones may be classified as long, short, flat, irregular, or sesamoid. The *long bones* are those of the limbs, except the wrist, ankle, and kneecap. They consist of a central long shaft (*diaphysis*) between two or more end-regions (*epiphyses*) whose surface areas are jointed to (articulate with) other bones.

DIAPHYSIS. *The shaft of a long bone (G. dia, through; physis, growing).*

The so-called *short bones* are in the wrists and ankles. They are made of a spongy core within an outer shell of compact bone.

The *flat bones* include the ribs and many of the skull bones. They consist

PISIFORM. (L. *pisum*, *pea*; *forma*, *appearance*).

of two *plates* of compact mineral between which is sandwiched a spongy layer, technically called *diploe*.

All the remaining name-bearing bones are classified as *irregular*, except the patella and pisiform, which are sesamoid bones.

Terms Used in Describing Irregular Surfaces on Bones

All bones have distinct irregular surfaces which serve special functions. The following terms are used in describing some of the irregularities. The definitions which accompany the terms will be helpful to you in studying the figures in this chapter. The first group of terms describes *processes* (elevations and projections):

Condyle – a slightly rounded projection for articulation with another bone;

Crest – a ridge to which muscle is attached;

Head – extends from a constricted portion (neck) into a joint;

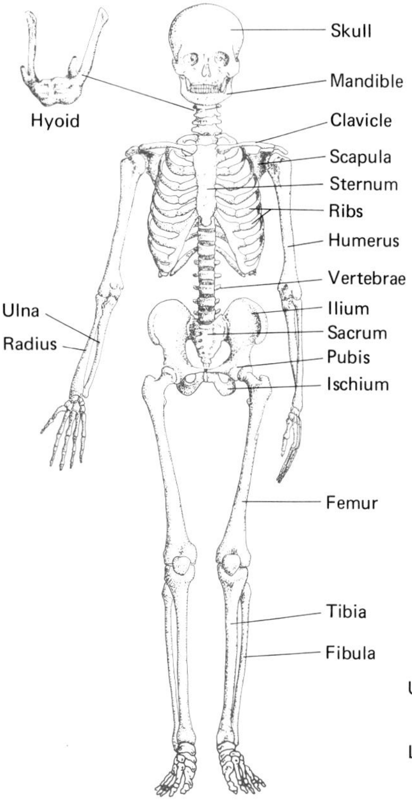

AXIAL SKELETON

		Number of bones
Skull:	Cranium	8
	Face	14
	Auditory ossicles	6
	Hyoid (throat)	1
Verebrae		26
Thorax:	Sternum	1
	Ribs	24

APPENDICULAR SKELETON

		Number of bones
Upper:	shoulder girdle	4
	arms and hands	60
Lower:	pelvic girdle	2
	legs and feet	60

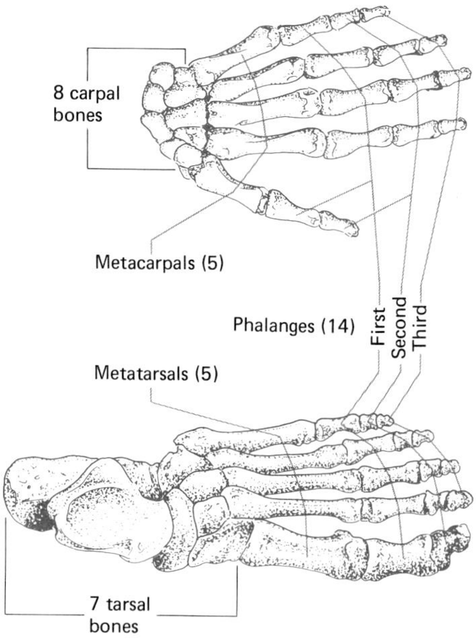

FIG. 15.1. *The skeleton.*

Pedicle—a stem, stalk, or constriction;
Ramus—a thin process which forms an angle with the main body;
Spine—a relatively sharp or pointed projection for attachment of muscle;
Trochanter—a large process to which muscle is attached;
Trochlea—a process shaped like a pulley;
Tubercle—a small, rounded projection for attachment of muscle;
Tuberosity—a large roughened projection for attachment of muscle.

The second group of anatomical terms is employed to distinguish among several different types of cavities, openings, grooves, and depressions:
Alveolus—a deep pit or socket;
Facet—a small, flat depression;
Foramen—a hole for the passage of other structures;
Fossa—a depression or concavity;
Meatus—a short canal or tube-shaped opening, as, for example, the opening that extends through the temporal skull bones from the inner ear to the outer ear.

THE AXIAL SKELETON

For convenience, the skeleton is said to be divided into two parts: an *axial skeleton* and an *appendicular skeleton* (Figure 15.1). The axial skeleton comprises the skull, the spinal column, and the thoracic cage.

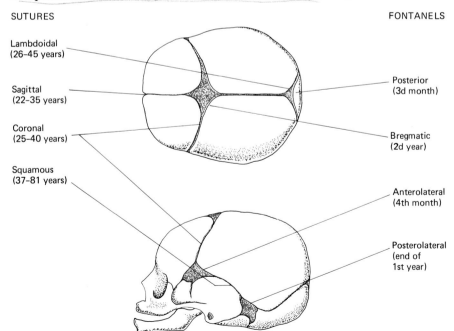

FIG. 15.2. *Sutures and fontanels. The sutures and fontanels become fully ossified at the different ages indicated. The sutures may not close fully until quite late in life.*

The Skull: Special Features

FONTANEL. *A membranous region between the angles of the cranial bones in infancy (Fr. fontaine, fountain).*

SUTURE. *(L. sutura, seam).*

Fontanels, sutures, sinuses, mastoid cells, and cavities are special features of the skull. *Fontanels* are membranous spaces that occur in certain cranial joints during fetal life and through early infancy. The names, location, and approximate ages at which they become fully ossified are shown in Figure 15.2. Fontanels are important in childbirth because they allow the baby's head to be compressed, thus facilitating delivery.

Sutures are joints between two cranial bones. With advancing years they ossify completely and thus become obscure. There are about 20 sutures, four of which are shown in Figure 15.2, along with the approximate beginning and peak years for their obliteration. These four sutures illustrate the range of ages over which all of the sutures begin to close. From this knowledge a specialist may be able to determine the age of a skull. Premature ossification of the sutures may result in microcephaly, a condition in which the skull has a capacity of less than about 1.4 liters. Individuals with this condition are usually mentally retarded due to restrictive brain development.

SINUS. *A hollow, cavity, or channel (L.).*

Sinuses are spaces within the bones adjacent to and surrounding the nasal cavity. Mucous secretions are continuously being passed from the lining cells of these spaces into the nasal and orbital cavities. A cold infection can cause a blockage and inflammation (*sinusitis*) in this area, thus altering the voice and causing discomfort.

Mastoid cells are numerous small intercommunicating openings within

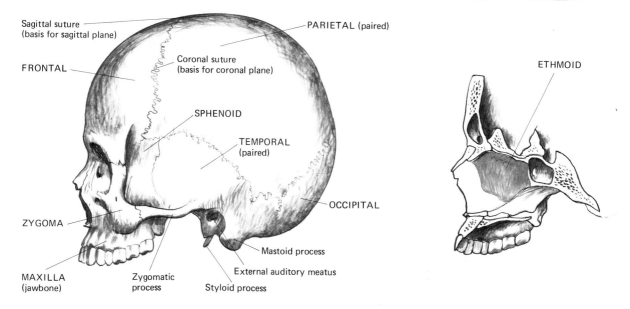

FIG. 15.3. *Seven of the eight cranial bones. The single ethmoid bone (Fig. 15.5) lies within the structure of the skull.*

The Skeletal System

the temporal bone. Portions of the inner ear are embedded in this bone. Excessive fluid from the inner ear passes into the middle ear by way of the mastoid cells. From the middle ear, the secretions pass by way of the eustachian tube into the pharynx. An inflammation of the mucosal lining of the mastoid cells is called *mastoiditis*.

Cavities are large openings within the skull. The *nasal cavity* harbors the nose, the *oral cavity* forms the mouth, the *orbits* hold the eyes, and the *cranial cavity* encloses the brain.

Location and Specific Features of the Skull Bones. The skull is usually discussed in terms of its two major components: the cranium (which encloses the brain), and the face. The *cranium* is made up of eight cranial bones, one frontal, two parietal, two temporal, one ethmoid, one occipital, and one sphenoid (Figure 15.3).

The *frontal bone* forms the forehead of the skull. Its supraorbital margins —which can be felt through the skin of the eyebrows—protect the eyes. Within the bone, directly over the orbits, is the *frontal sinus*, which opens into the nasal cavity. In certain cases of sinus colds, the mucosal lining of this sinus may become inflamed, rendering mild to severe discomfort.

The paired *parietal bones* constitute the walls of the cranial vault. (See Figure 15.3.) The paired *temporal bones* form the sides and part of the base of the skull. Feel them. Each one encloses an ear and articulates with the lower jaw (Figure 15.4). They were given their name from the Latin *tempora* (= time) because it is usually in the temporal region of the head that hair first turns gray.

PARIETAL. *Pertaining to the wall of a cavity (L. paries, wall).*

Three other features of the temporal bones are noteworthy. (1) A portion of each bone near the ear surrounds an *auditory meatus*, a canal which passes from the inner to the outer ear. (2) Behind and below the auditory meatus, the temporal bone contains numerous mastoid cells. (3) A third portion—at the base of the skull—contains the inner ear for hearing and the labyrinth with proprioceptive receptors for balance (Figure 15.4).

The *ethmoid* bone lies mainly within the skull cavity—buried away from

ETHMOID. *Resembling a sieve (G. ethmos, sieve; eidos, resembles).*

FIG. 15.4. *The right temporal bone. Lateral view.*

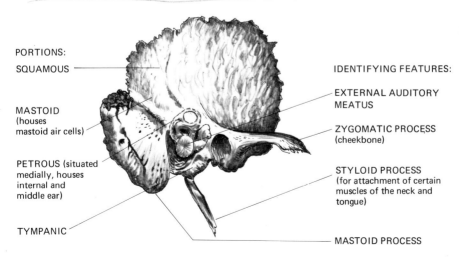

PORTIONS:
SQUAMOUS

MASTOID (houses mastoid air cells)

PETROUS (situated medially, houses internal and middle ear)

TYMPANIC

IDENTIFYING FEATURES:

EXTERNAL AUDITORY MEATUS

ZYGOMATIC PROCESS (cheekbone)

STYLOID PROCESS (for attachment of certain muscles of the neck and tongue)

MASTOID PROCESS

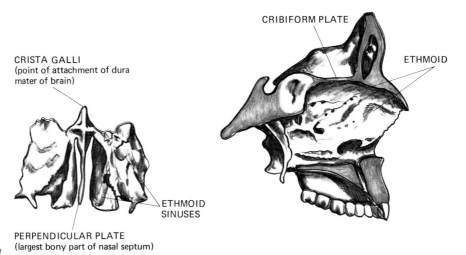

FIG. 15.5. *The ethmoid bone. Right-hand figure shows where the ethmoid is located.*

CRIBRIFORM. *Perforated* (L. cribrum, *a sieve;* forma, *form*).

NARIS. *Nostril* (L.).

view for the most part—between the cranial (or brain) cavity and the nasal cavities (Figure 15.5). The *cribriform plate* of the ethmoid bone forms the roof of the nasal cavity and part of the cranium. Axons from the olfactory receptors pass through the cribriform plate on their way to the brain. A front (anterior) view of the ethmoid bone reveals a thin median plate. Known as the *perpendicular plate* or *septum*, this plate separates the *nares*, or nostrils, from one another. Lateral masses on each side of the ethmoid bone contain ethmoid sinuses, which open into the nasal cavity.

Two final landmarks worth mentioning in conjunction with the ethmoid bone are the *superior* and *middle conchae*. Shaped like scrolls, these bony processes project medially, toward the nasal cavity from the lateral masses.

FIG. 15.6. *Base of the occipital bone.*

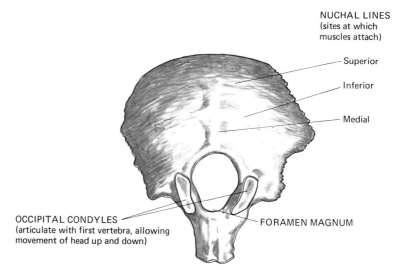

They serve to increase the surface area of the cavity for the purpose of warming incoming air. The efficiency of the warming process is increased by the mucosal lining of the nasal cavity, which has an extensive network of blood vessels.

The *occipital bone* forms the floor and lower rear wall of the cranium (Figure 15.6). Its principal landmark is the *foramen magnum*—a large hole through which the spinal cord passes into the brain. Anterior (in front of) to this opening are two laterally situated prominences (occipital condyles). These protuberances articulate with the first vertebra—the atlas—and allow the head to move freely.

The *sphenoid bone*, shaped like a sphinx, lies anterior to the temporal bone, in line with the orbits and nasal cavity (Figure 15.7). The sphenoid sinus, within the sphenoid bone, opens into the nasal cavity. The upper surface of this bone has a large depression—the *sella turcica* (Turk's saddle)—which encloses and protects the pituitary gland.

OCCIPITAL. *Region at the back of the head (L. occiput, cap).*

FORAMEN MAGNUM. *Literally, a large opening (L.).*

SPHENOID. *Wedge-shaped (G. sphen, wedge; eidos, appearance).*

The Hyoid—A Suspended Skull Bone. The *hyoid* bone lies in the neck, just beneath the chin. It is suspended by ligaments from the styloid processes of the temporal bone of the cranium. Its function is the attachment of most of the anterior muscles in the neck. Since it is not articulated with any other bone, the hyoid may be classified as a sesamoid bone.

HYOID. *U-shaped (G. hyoides).*

Six More Bones from the Cranium. Three more pairs of bones remain to be mentioned in conjunction with the skull. Situated in each middle ear, within the temporal bone, are the malleus, incus, and stapes. The *malleus* conveys sound waves from the tympanic membrane (eardrum) to the incus. The *incus* transmits the sound waves to the stapes, and the *stapes* transmits the sound waves to the oval window of the inner ear (Figure 9.1). These auditory ossicles thus serve to convey sound energy from the tympanic membrane to the inner ear.

MALLEUS, INCUS, and **STAPES.** *Latin words for hammer, anvil, and step.*

The Face

The face contains 14 facial bones (Figure 15.8). Seven of these bones can be felt with the fingers as *superficial bones*. The other seven are *deep bones*, which cannot be felt.

FIG. 15.7. View of superior surface of the sphenoid bone.

FOSSA OF SELLA TURCICA (seats pituitary gland)

OPENINGS FOR NERVES AND BLOOD VESSELS
- Optic foramen
- Foramen ovale
- Foramen spinosum

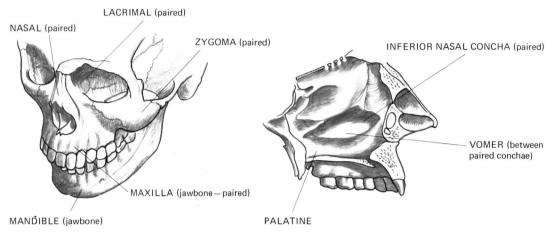

FIG. 15.8. *The fourteen facial bones.*

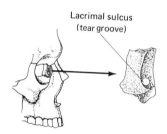

FIG. 15.9. *Lateral view of the right lacrimal bone.*

CONCHA. A shell-shaped structure, as occurs in the nose or outer ear (L. a shell).

FIG. 15.10. *The right palatine and maxillary bones (medial view).*

The *deep facial bones* are two lacrimals, two palatines, one vomer, and two inferior conchae. The *lacrimal bones* form part of the median wall of each orbit (Figure 15.9). A groove runs in a superior-inferior direction along each bone adjacent to the nose. Termed a *lacrimal sulcus*, this groove supports the lacrimal canal for the movement of tears from an eye to a nasal passageway.

The *palatine bones* form part of the lateral walls and floor of the nasal cavity, part of the roof of the mouth, the floor of the orbits, and the rear (posterior) portion of the hard palate (Figure 15.10). The *vomer* forms the posterior and inferior portion of the nasal septum (Figure 15.11).

The paired *inferior conchae* extend along the lateral walls of the nasal cavities (Figure 15.12). Each of these two bones—like the paired middle and superior conchae—serves to increase the surface area of the mucosal lining in the nasal cavity. However, the inferior conchae are bony, scroll-like plates, separate from the ethmoid bone.

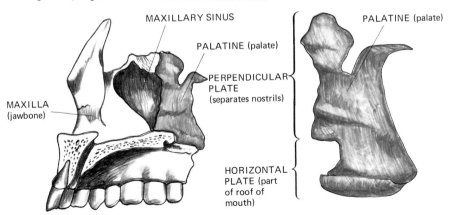

The Skeletal System

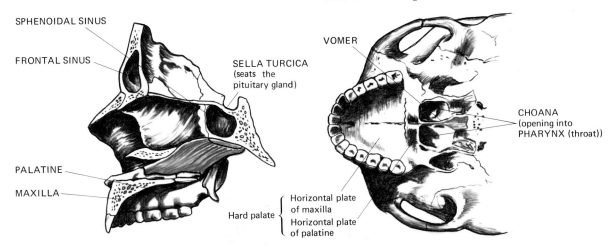

FIG. 15.11. *The vomer divides the internal nares (choanae) into right and left halves, and it forms the inferior part of the nasal septum.*

The *superficial facial bones* are two nasal, two zygomatic, two maxillary and one mandibular. The *nasal bones* lie side by side, uppermost on the median line of the nose (feel them). From them extends the cartilaginous flexible inferior region of the nose.

The *zygomatic bones* form the prominences of the cheek and part of the lateral and inferior walls of the orbits. Feel them by placing your fingers beneath the eyes.

The *maxillary bones* fuse with one another medially to form the upper jaw (feel them). They form the bulk of the hard palate in the roof of the mouth (Figure 15.13). In addition, they constitute part of the walls of the orbital and nasal cavities. A faulty union of the maxillae may result in a

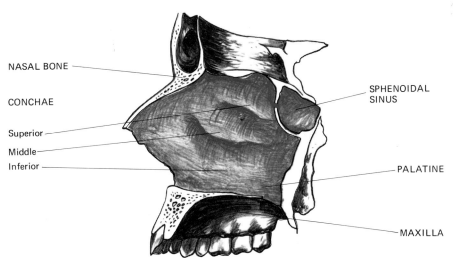

FIG. 15.12. *The conchae.*

The Axial Skeleton

FIG. 15.13. One of the paired maxillary bones, which make up the upper jaw. The alveolar process is the arch that holds the upper teeth.

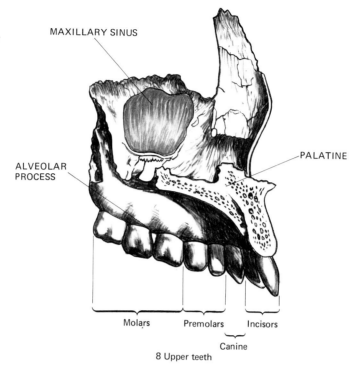

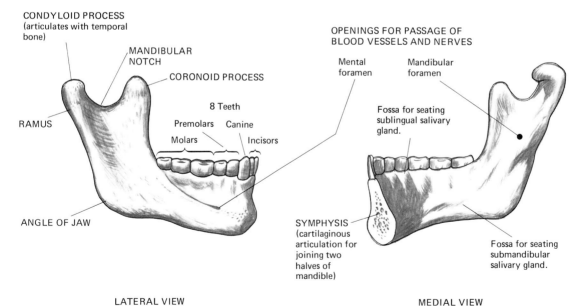

FIG. 15.14. The mandible.

The Skeletal System

congenital deformity known as a *cleft palate*. The cleft may be a mere splitting of the uvula, a more severe form, such as a slit on one side of the palate, or a dramatic absence of the palate. Modern surgery can often remedy some or most of the difficulties which otherwise arise with regard to speech and feeding.

Two other features of the maxillae are noteworthy. (1) The inferior border, called the *alveolar process*, is pocked with pits (alveoli), which form sockets around the roots of the upper teeth. (2) Present in each maxilla, beneath the orbits, is a maxillary sinus, which opens into the nasal cavity. Note that this is the fourth time sinuses have been mentioned in connection with bones, the others being the frontal, ethmoid, and sphenoid sinuses. Collectively, these four sinuses comprise the *paranasal sinuses*.

The *mandible* is the lower jawbone (Figure 15.14). Its paired condyles articulate with the temporal bone of the skull. The condyles can be felt by placing a finger just in front of the opening of each ear and moving the lower jaw.

MANDIBLE. *The lower jaw* (L. mandibula, *a jaw*).

On the upper ridge of the mandible are 16 sockets which enclose the roots of the lower teeth. Two foramina (openings) are also prominent on the mandible (Figure 15.14). The mental foramen—located at the chin—is a passageway for nerves and blood vessels to the chin. The mandibular foramen—located on the inner side of the angle (ramus) of the jaw bone—serves the same function for the lower teeth.

The Spinal, or Vertebral, Column

The adult spinal column consists of 26 bony segments (vertebrae): seven cervical (neck), 12 thoracic (chest), five lumbar (lower back). There are also one sacral and one coccygeal (tail) bone, each composed of multiple segments (Figure 15.15).

Each vertebra, except the first and last, has an anteriorly-located *vertebral body* (Figure 15.16). These vertebral bodies lie directly in line with one another and are separated from each other by an *intervertebral disc*. The vertebral bodies serve to bear weight, as is the case in maintaining a sitting or standing posture. The discs, made of a fibrous elastic cartilage, cushion the vertebrae and act as shock absorbers.

Each disc contains a great deal of water (ranging from about 90% to 70% from early life to old age). The water allows the disc to bulge toward the rear when a person bends forward, and to the front when bending backward. The bulge shifts from side to side when the spinal cord bends to opposite sides of the body, and it assumes a uniformly flat shape when the body is erect.

A pair of stalks, called *pedicles*, arise laterally from each vertebral body (Figure 15.16). The pedicles of one vertebra lie next to those of another. They are so arranged that the pedicles of two adjacent vertebrae form an opening through which the spinal nerves emerge.

In addition to a vertebral body and pedicles, each vertebra except the coccyx has a vertebral foramen, through which the spinal cord passes, and various kinds of processes. The processes serve one or more of three functions: to limit the kinds of movements the spine can make, to anchor other bones (ribs, pelvis, or shoulder girdle), and to anchor muscles and tendons.

Regional Variations in the Vertebral Column. The *cervical vertebrae* are built for flexibility. In addition to a vertebral foramen, they have a pair of openings (transverse foramen) for the passage of the vertebral arteries into the brain.

The first cervical vertebra—the atlas—supports the head (Figure 15.17). Its construction (with lateral articular facets) allows the occipital condyles

ATLAS. *In Greek mythology, a giant god who supported the heavens on his shoulders.*

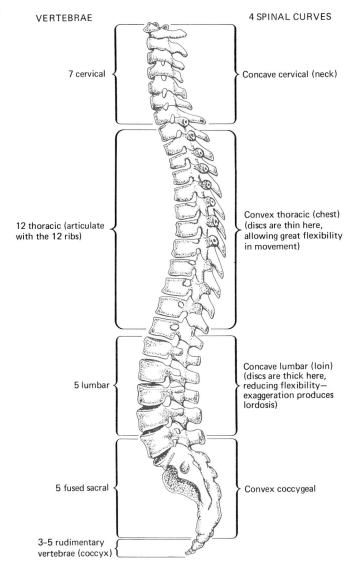

FIG. 15.15. *The vertebral column.*

VERTEBRAE

7 cervical

12 thoracic (articulate with the 12 ribs)

5 lumbar

5 fused sacral

3–5 rudimentary vertebrae (coccyx)

4 SPINAL CURVES

Concave cervical (neck)

Convex thoracic (chest) (discs are thin here, allowing great flexibility in movement)

Concave lumbar (loin) (discs are thick here, reducing flexibility— exaggeration produces lordosis)

Convex coccygeal

The Skeletal System

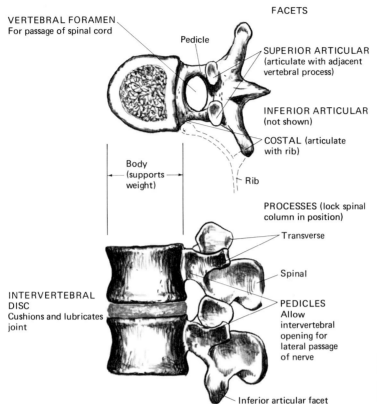

FIG. 15.16. *Thoracic vertebra, depicting the general features of vertebrae.*

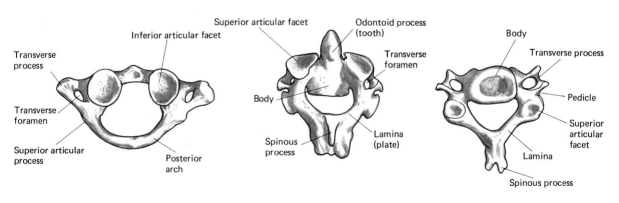

FIG. 15.17. *Cervical vertebrae — the three types.*

(rounded projections) of the occipital skull bone to rock back and forth when the head is raised or lowered.

The second cervical vertebra—the axis—has a unique *odontoid process* (Figure 15.17). This process, and the remaining construction of the vertebral body, allows the axis and skull to rotate on the spinal column.

The unique characteristic of the 12 *thoracic vertebrae* is the presence of smooth facets (small depressions) on the vertebral body and transverse processes. These facets are the sites at which the ribs are joined (Figure 15.16).

The main characteristic of the five *lumbar vertebrae* is their bulk (Figure 15.18). This property lends itself to the demands imposed upon the lower back for bearing the weight of the upper body.

The *sacrum*, located at the lower end of the spinal column, is formed by the fusing of three, four, or five vertebrae during development (Figure 15.19). It is wedged between the two hip bones. As such it forms an important integral part of the pelvic girdle (see below).

The *coccyx* is usually a fusion of four rudimentary vertebrae; it is a vestige of the ancestral tail (Figure 15.19).

Disorders of the Vertebral Column. When viewed from the side, the spinal column has four distinct curves: cervical, thoracic, lumbar, and sacral (see Figure 15.15). The convex thoracic and sacral curves are present at birth, giving the newborn a C-shape. Early in life, when the baby develops to the point where it raises its head, the curve of the seven cervical vertebrae at the top of the spine reverses itself and becomes concave. Further adjustments are made beginning around age one, when the baby stands and starts walking. The coccygeal and sacral vertebrae fuse, and the lumbar curve also becomes concave.

Improper development of the vertebral column can result in *curvature disorders:* kyphosis, lordosis, or scoliosis. *Kyphosis* (hunchback), is an exaggeration of the thoracic curve. *Lordosis* (swayback) is an exaggeration of the lumbar curve. *Scoliosis* (a crookedness) is an abnormal lateral curvature of the spine that may be caused by uneven musculature on the two sides of the spinal column.

SACRUM. *The name given to the sacral bone because it was believed to escape decay and become the basis of the resurrected body (L. cacer, sacred).*

COCCYX. *(G. kokkyx, a cuckoo, whose bill resembles the coccyx in form).*

FIG. 15.18. *Lumbar vertebrae.*

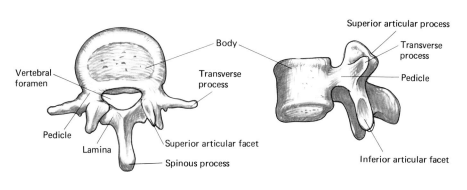

The Skeletal System

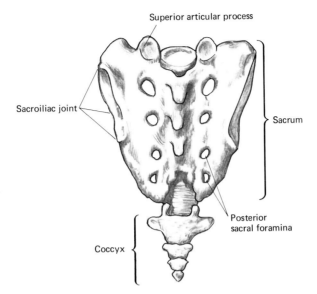

FIG. 15.19. *Posterior view of the sacrum and coccyx.*

A relatively common disorder of the vertebral column is what is frequently called a *slipped, ruptured,* or *herniated disc.* The intervertebral disc is made of a fibrous elastic cartilage with a high percentage of water. In young people it contains about 90% water and is very compressible; in older people it has about 70% water and it is not very compressible. This transition—from high to low water content—is part of the normal aging process, but when it occurs at age 20 or 30 it is usually due to a degenerative disc disease.

No matter how old a person is when the disc's water content drops, the condition predisposes the individual to a "slipped disc" (Figure 15.20). Alternatively—or simultaneously—weak tendons, ligaments, or muscles may be predisposing conditions.

In either case, a weakening arises in the outer ring (*annulus*) of the disc, or in the longitudinal ligaments that ordinarily help keep the discs in place.

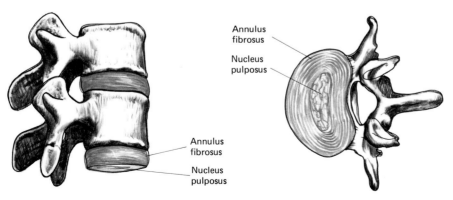

FIG. 15.20. *The intervertebral disc.*

SCIATICA. Pain in the hip, usually due to an inflamed sciatic nerve (L. sciaticus).

LUMBAGO. Pain associated with the tendons that are attached to vertebrae in the lumbar region (L. lumbus, loin).

Next, the inner, or central, region of the disc—the *nucleus pulposus*—bulges through the weakening. This allows the vertebrae to lower down upon one another, often pinching one or more nerves and thus causing pain.

In about 50% of the cases, the slip occurs between the fifth lumbar and first sacral vertebrae (where the sciatic nerve originates) giving rise to *sciatica*. In another 40% of the cases, the slip occurs between the fourth and fifth lumbar vertebrae, creating a case of *lumbago*. In the remaining 10% of cases, the person suffers both lumbago and sciatica.

To what extent are lumbago and sciatica due to disc degeneration and to what extent are they caused by a weakening of the longitudinal ligaments? In about 80% of the cases there is no skeletal disorder, and the discs appear to be normal. Very commonly the problem rests in muscles, ligaments, or tendons that are weakened from a sprain, lack of exercise, poor posture, or excessive straining.

The Thoracic Cage

The thoracic cage is composed of 12 pairs of ribs, 12 thoracic vertebrae, and the sternum (breastbone) (Figure 15.21). The first seven pairs of ribs are called *true ribs*—they unite anteriorly through their costal cartilage with the sternum. Ribs eight to 12 are called *false ribs*, because they are not attached to the sternum. Instead, they are united with each other and with rib seven. The bottom two ribs (11 and 12) are called *floating ribs*—they have no anterior attachment.

Each rib has three components: a head, a tubercle (a small, rounded

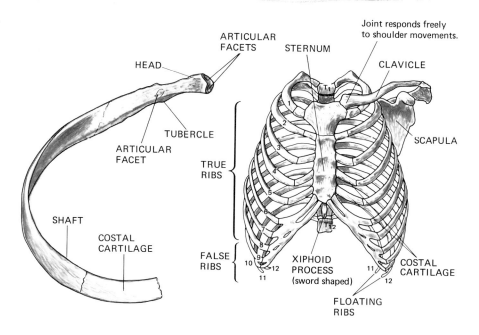

FIG. 15.21. *The thoracic cage. The right scapula (shoulder) and clavicle (collarbone) are not shown.*

projection), and a shaft. Facets (depressions) on the head and tubercle articulate with the thoracic vertebrae, which were mentioned earlier. The shaft is the remainder of the rib.

In the course of diagnosing and following up on certain blood diseases, such as leukemia, samples of bone marrow are removed from the sternum with a syringe. This procedure is known as a *sternal puncture*.

STERNUM. *The long, flat breastbone (G. sternon, the chest).*

THE APPENDICULAR SKELETON

The appendicular skeleton is composed of an upper extremity (the shoulder girdle and its paired appendages), and a lower extremity (the pelvic girdle and its paired appendages).

The Shoulder Girdle

This structure consists of two scapulas (shoulder blades) and two clavicles (collar bones) (Figure 15.22). It serves two functions: to attach the bones of the upper extremity to the axial skeleton, and to provide surfaces for the attachment of muscle.

Each *clavicle* is connected at one end to the sternum (breastbone); at its other end it is attached to the scapula. The *scapula* lies within the body wall in the upper rear region of the thoracic cage. Its glenoid fossa articulates with the arm bone.

CLAVICLE. *Collar bone (L. a small key).*

The Upper Appendages

The bones of the upper extremity make up the arm, forearm, and hand.

The Arm. A single bone, the *humerus*, is present in the upper arm—the arm proper (Figure 15.23). The head of the humerus articulates with the scapula. Just distal to this head is an anatomical neck. Slightly below this anatomical neck—below the point where the bulk of muscles are attached—is a surgical neck. The term *surgical neck* denotes an area most frequently subject to fracture.

HUMERUS. *The bone of the upper arm (L. shoulder).*

Another important area of the humerus is the "funny bone." The "funny bone" is actually a region, just below the medial epicondyle, along which a large nerve passes. When pinched, this nerve gives rise to tingling sensations in the forearm and hand.

EPICONDYLE. *A projection on or above another projection of bone (G. epi, upon; kondylos, a knuckle).*

The Forearm. Two bones, the *radius* and *ulna*, make up the forearm (Figure 15.23). The radius lies on the lateral side of the forearm and the ulna is on the medial side. The ulna can be felt posteriorly from the elbow to wrist. When your palm is held up (supinated) the radius and ulna are

RADIUS. *The lateral and shorter of the two bones of the forearm (L. spoke of a wheel).*

ULNA. The medial and larger of the two bones of the forearm (L. elbow, arm).

PHALANGES. The bones of the fingers, thumbs, and toes (G. phalangos).

parallel to one another. When your palm is held down (pronated) the two bones cross each other near the wrist. One of the most common fractures of the forearm, a fracture of the lower third of the radius, is caused by falling on the palms.

The Hand. The hand contains eight carpals, five metacarpals, and 14 phalanges. The *carpals* are bound close to one another. They are usually referred to as the wrist. The wrist is also the joints that are formed between the radius and ulna at one end, and the proximal carpals at the other end (Figure 15.24).

Distal to the carpals are five *metacarpals*, commonly called the palm. The rounded distal ends of metacarpals constitute the knuckles. Distal to the metacarpals are the 14 *phalanges*—two for the thumb and three for each finger.

FIG. 15.22. *The shoulder girdle.*

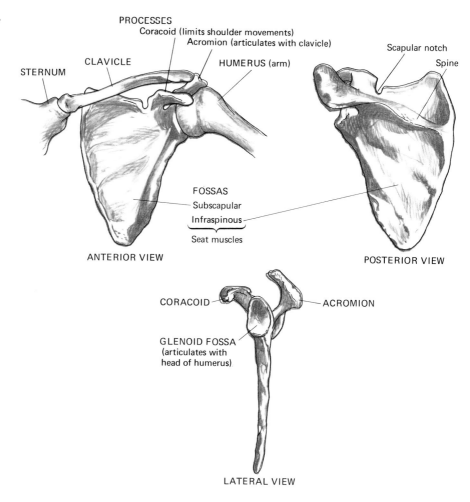

The Skeletal System

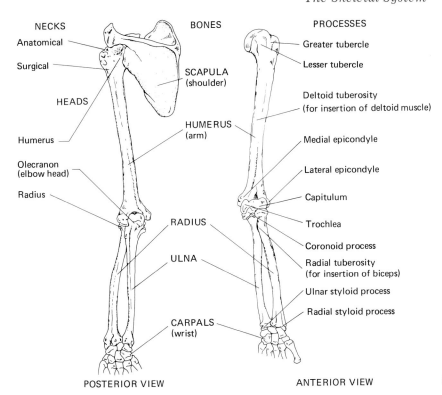

FIG. 15.23. *The arm and forearm.*

The Pelvic Girdle

The pelvic girdle is formed from four bones: two hip bones, a sacrum, which is sandwiched between the hip bones, and a coccyx. The sacrum and coccyx are described above.

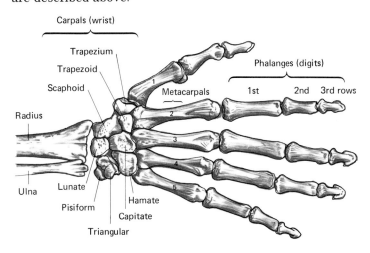

FIG. 15.24. *The right hand and wrist. Note that the carpals are named more or less according to shape.*

ILIUM. *The groin or flank (L.).*

SYMPHYSIS. *A union, such as the pubic symphysis between the pubic bones (G., a growing together).*

FIG. 15.25. *The pelvic girdle.*

Each hip bone (*os coxae;* innominate bone) is made up at birth of three parts: ilium, ischium, and pubis (Figure 15.25). In the course of development these bones fuse, forming an *acetabulum* — a socket which accepts the head of the femur (thigh bone) for articulation.

The *ilium* forms the uppermost region of the hip and articulates with the sacrum. Damage at this articulation — the *sacroiliac* — is usually very painful.

The *ischium,* which is the lowest part of the three-part hip bone, is also the strongest segment. It helps support the weight of the trunk when one sits down.

The *pubis* forms the anterior region of the hip bone. Each pubis unites medially with the other through a cartilaginous bridge known as the *pubic symphysis.*

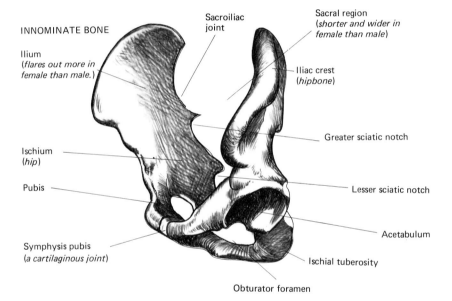

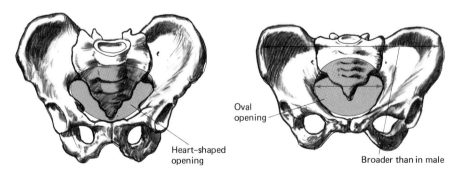

Functionally, the pelvic girdle serves three major purposes: it supports the trunk upon the thighs while standing, it permits sitting, and it affords protection to the urinary bladder, the ovaries, oviducts, and uterus (in females), and the terminal end of the large intestine.

The Lower Appendages

This region of the appendicular skeleton is made up of the thigh, knee, leg, and foot.

The Thigh. The thigh is built upon the *femur*, a large heavy bone which connects the hip bone to the tibia in the leg (Figure 15.26). The area formed around the connection of these two bones, known as the *knee* or knee joint, is protected by the *patella* (kneecap). The patella is the largest sesamoid bone in the body. It is embedded in the tendon of the large thigh muscle (quadriceps).

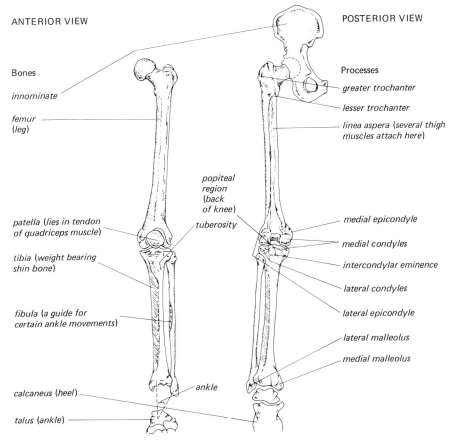

FIG. 15.26. *The thigh and leg.*

TIBIA. The medial and larger of the two bones of the leg (L. shinbone).

FIBULA. The lateral and smaller of the two bones of the leg (L. a clasp).

TARSALS. The bones which form the sole of the foot (G. tarsos, sole).

The Leg. The region of the body properly known as the leg contains two bones: tibia and fibula (Figure 15.26). The *tibia* (shinbone) transmits the weight of the trunk from the hip bone to the femur. Its laterally located parallel companion, the *fibula*, does not support weight. Rather, the fibula facilitates certain skeletal movements at the ankle. Forceful twisting movements of the ankle (joint between leg and foot) often result in fracture of the fibula a few inches above the ankle.

The Foot. Seven tarsal bones, five metatarsals, and 14 phalanges make up each foot (Figure 15.27). Together with ligaments, muscles, and tendons, these bones give rise to a *longitudinal arch* that runs from heel to toes and a *metatarsal arch* that runs from the lateral to medial side of the foot. Weakening of this structure may result in fallen arches (flat feet).

THE STRUCTURE, FUNCTION, AND MAINTENANCE OF BONE

A fully formed bone consists of two portions: an outer wall and an inner marrow.

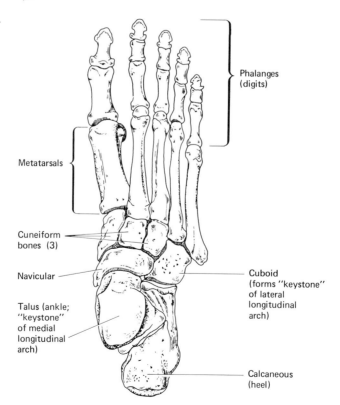

FIG. 15.27. *The right foot.*

The Skeletal System

The Bone Wall

The outer wall of bone, seen from the outside in, consists of a periosteum, a mineral layer, and an endosteum (Figure 15.28).

The Periosteum. The *periosteum* is a connective tissue that is fibrous and richly supplied with blood vessels. It has several essential functions. One is to anchor ligaments, tendons, and fascia. A second is to transmit blood vessels into the mineral layer. Another function is to proliferate primitive bone cells (osteoblasts) during the first 21 years, enabling the bone to grow in length and breadth. The periosteum also carries out this function during adulthood in the event of a bone fracture. Osteoblasts are occasionally seen in the periosteum of a normal adult. However, it is only after a fracture—during healing and regeneration—that they appear in large numbers.

PERIOSTEUM. *The thick fibrous membrane that surrounds most of the surface of a bone (G. periosteon, around the bone).*

OSTEOBLASTS. *Bone-forming cells (G. osteon, bone; blastos, germ).*

The Mineral Layer. The mineral, or bony, layer is made up either of compact bone or of spongy bone that is enclosed within compact bone of variable thickness.

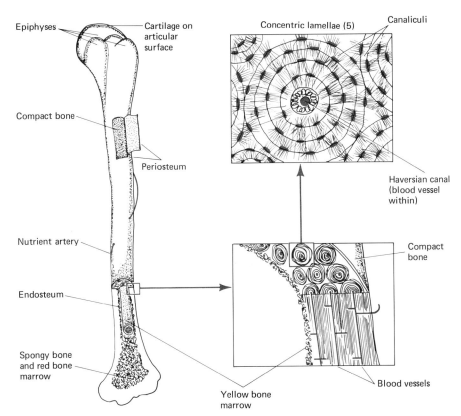

FIG. 15.28. *A long bone.*

Compact bone is an orderly arrangement of bony sheets and tiny canals. It arises mainly after birth.

There are basically two types of bony sheets. Type I lies adjacent to the periosteum and endosteum. In the diaphysis of long bones, the bony sheets are arranged in cylinders between the inner boundary of the periosteum and the outer boundary of marrow. These bony sheets are perforated obliquely or at a right angle with *Volkmann's canals*. These bony canals carry nerve fibers, blood vessels, and lymphatic vessels from the periosteum and marrow to the interspaces of the bony plates.

The Type II bony sheet is present in *Haversian systems*. These systems run along the main axis or axes of a bone. They consist of a central canal and about four to 20 concentric rings of bony sheets. Each sheet is about 3 to 7 micrometers in thickness. The central canal, usually about 50 micrometers in diameter, carries nerve fibers, blood vessels, and lymphatic vessels to bone cells. The bone cells (osteocytes) occupy tiny openings (lacunae) which occur between or within the concentric rings. A single osteocyte is found in each lacuna. The osteocyte extends relatively long cytoplasmic processes through tiny canals (canaliculi) into neighboring concentric rings.

Spongy bone is not as highly organized as compact bone. It is spongy in appearance, it lacks Haversian systems, and its interspaces are filled with bone marrow.

The Endosteum. The endosteum is a connective tissue that lines the cavities of the marrow and the canals of compact bone. Like the periosteum, it is important in proliferating bone cells during development or – in the case of an adult – following a fracture.

VOLKMANN. *German surgeon (1830–1889).*

OSTEOCYTE. *A mature bone cell (G. osteon, bone; kytos, cell).*

ENDOSTEUM. *The membrane which lines the surface of bone within the marrow cavity (G. endon, within; osteon, bone).*

Bone Marrow

Marrow is protoplasmic material that occupies the inner cavities of bones. It receives nutrients from one or more arteries which enter the bone at an oblique angle (Figure 15.29). It sends its products – red blood cells, white blood cells, and blood platelets – into the circulation by way of veins.

The bone marrow of the entire body weighs twice as much as the body's heaviest visceral organ, the liver. Bone marrow consists of: (a) a supporting framework of reticular phagocytic stem (blast) cells, (b) sinuses, and (c) freely-moving cells within the sinuses.

Until middle childhood, the marrow of all bones appears red. Throughout this period, all of the available marrow is required to provide blood cells for the developing child. Thereafter, increasing amounts of red marrow are converted to yellow marrow (adipose tissue). By the end of adolescence most of the marrow is yellow. Under stress, however, whenever red cells are needed (for example, in certain cases of anemia), the yellow marrow may change back to red marrow to produce the needed blood cells. Otherwise, in the normal adult, the red marrow occurs mainly in the sternum, ribs, vertebrae, and cranium.

Blood Circulation in Bone

In a long bone, one or more arteries perforates the shaft (diaphysis) and ends (epiphyses) of the bone, and enters the marrow cavity. For bones in general, numerous capillarylike arterioles feed into the bone through its periosteum (Figure 15.29).

The arterioles pass through the Volkmann's and Haversian canals into the marrow. Here they circulate through small capillaries and larger capillaries (sinusoids). From these spaces the blood enters small veins (venules). Some of the veins retrace the route of the arteries to the outside, but most veins leave the bone at extremities, at joints.

HAVERS. *A 17th-century English anatomist.*

DEVELOPMENT OF BONE

Most of the bones of the body develop from structures that are initially made of cartilage. This process is called *intracartilaginous development of bone*.

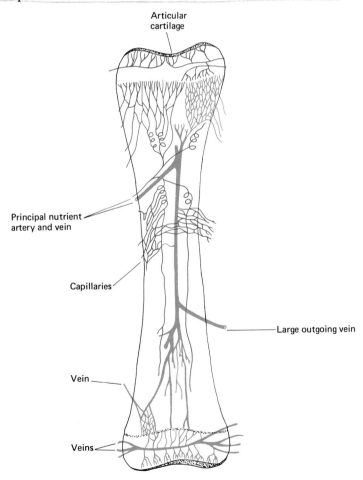

FIG. 15.29. *Blood vessels in bone. Capillaries are shown only in a small area, although they are found in great numbers throughout the bone.*

Development of Bone

In contrast, the flat bones of the cranial dome, portions of several other cranial bones, part of the mandible, and some of the bones of the face develop directly from primitive connective tissue, without the intervention of cartilage. This process is called *intramembranous development of bone*.

Intramembranous Bone Development

In this process, primitive bone cells, called *osteoblasts*, aggregate in various scattered bone-forming (ossification) centers. The earliest centers appear around the eighth week of fetal life. Within these centers the osteoblasts secrete a *matrix* of ground substance and collagen fibers. A rich supply of blood to this area brings with it calcium and phosphate. These minerals are incorporated into the matrix, which is then said to be *mineralized*. When deposition ceases, the osteoblast produces a capsule around itself and is now known as an *osteocyte*. Ultimately all of the "membrane-bounded" space which is destined to be a bone becomes *ossified* in this way.

By the time of birth, these "membrane" bones are well organized with outer layers of periosteum sandwiching a layer of spongy bone. Gradually, the periosteal layers deposit bony sheets. In the flat bones these sheets are made of compact bone, and they include Haversian systems. They form the so-called *tables*, whereas the spongy bone they enclose is called *diploe*.

DIPLOE. *Spongy bone between two layers of compact bone in the flat cranial bones* (G. double).

Intracartilaginous Bone Development

Intracartilaginous bone development, with respect to the *long bones*, is shown in Figures 15.30 and 15.31. As Figure 15.30 shows, the bone is cartilaginous at first; it is made primarily of cartilaginous cells (*chondrocytes*) and collagen fibers enclosed within an outer membrane, the *perichondrium*. Around the second month of fetal life, a *primary ossification center* arises internally—midway between the two ends of the cartilaginous bone. Here the cartilaginous cells begin to differentiate. They become larger and begin to mineralize calcium phosphate into their matrix.

At the same time, the inner layer of the perichondrial membrane undergoes differentiation. Two things now begin to take place: (1) concentric sheets of osteoblasts are laid down all along the periphery; and (2) a collar of spongy bone is formed around the periphery of the middle third of the shaft. This step of the process is essentially the same as intramembranous bone formation. Now the perichondrium (outer cartilage) is known as the *periosteum* (outer bone).

CARTILAGE. *A tough nonvascular connective tissue made of cartilage cells, fibers, and a ground substance* (L. cartilago, gristle).

PERICHONDRIUM. *A fibrous membrane which covers cartilage* (G. peri, beyond; chondros, cartilage).

Blood vessels from the periosteum begin to pass toward the primary ossification center by way of gaps in the spongy bone collar. These blood vessels are accompanied by chondroblasts, which at this point are known as osteoblasts. Simultaneously, cells that were formerly cartilaginous begin to differentiate into bone cells within the primary ossification center. However, these bone cells meet with an early fate, since nutrients cannot diffuse into their vicinity through the mineralized matrix. As these bone cells die, their mineral matrix weakens and begins to crumble.

Blood vessels from the spongy bone collar then pass into gaps which arise in the crackled matrix. Using remnants of the matrix as a scaffold, the osteoblasts begin to deposit calcium phosphate all about themselves. They become osteocytes, but in this case, they are in touch with blood vessels and receive the nutrients and oxygen necessary for survival.

As the blood vessels continue inward, the primary ossification center splits in two. Each half then appears to move toward one of the two ends of the developing bone. With continued proliferation of cartilage cells toward each end, the bone grows longer.

Another operation comes into the picture about this time: the *endosteum* appears. This germinative layer proliferates osteoblasts that tend to further

MATRIX. *Intercellular cement plus connective tissue fibers.*

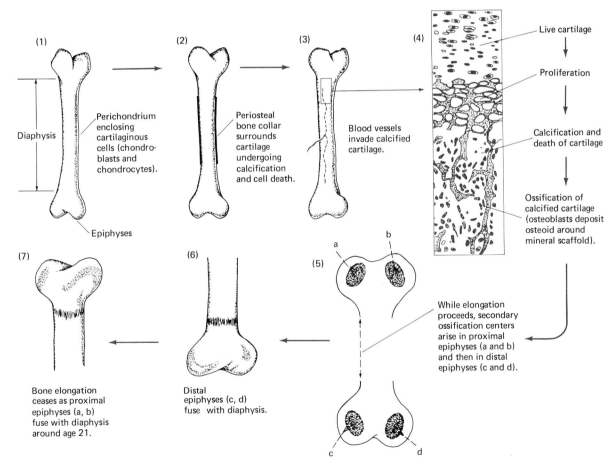

FIG. 15.30. *Intracartilaginous bone development. The elongation shown in the detailed drawing (4) is upward; cartilage is found in the growing end of the bone.*

OSTEOCLASTS.
Multinucleated cells which function in the absorption and removal of bone tissue (G. osteon, bone; klastos, broken).

increase the mineral layer of the wall. At the same time, *osteoclasts* appear, principally in the endosteum. It is believed by many scientists that these cells destroy the mineral matter near the marrow. Others believe that these cells, which are multinucleated, arise as a consequence of a process in which the mineral matter near the marrow is dissolved. No matter which interpretation is correct, the result is the same: the marrow cavity enlarges. More space is thus provided for the production of blood cells, and ultimately a lighter bone is formed.

Not all of the peripheral cartilage is ossified. Where the bone is to articulate with another bone, some cartilage remains as articular cartilage.

Just before birth, but mostly afterward, secondary ossification centers arise in the epiphyses. In essence, the same development process occurs as in the primary ossification center: (1) cartilage cells proliferate, differentiate, mineralize, and die; (2) the mineral weakens and crumbles, bringing about the formation of cavities; (3) the cavities are invaded by blood vessels from the periphery; and (4) the weak mineral is resorbed and new mineral is deposited by well-nourished bone cells. However, instead of cartilaginous cells proliferating along straight lines along the axis, they spread out peripherally in all directions. Thus, instead of elongation, we get enlargement in all directions.

Also—very importantly—a plate is left between each epiphysis and diaphysis. This *epiphyseal plate* gives rise to late elongation of the bones, from some time in infancy to about age 21, when growth stops. Cartilage cells proliferate at the distal end and are replaced by bone at the proximal end. These operations balance one another, so the epiphyseal plate remains constant in thickness. At the end of growth—which is dictated genetically

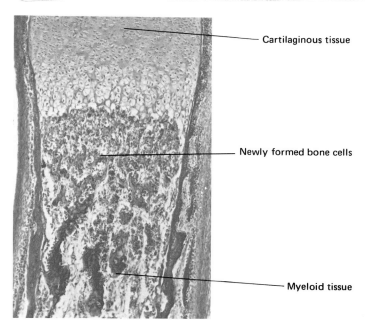

FIG. 15.31. Bone elongation in the finger of an infant. The bone is growing toward the top.

and carried out under hormonal control (chiefly by growth hormone)—the plate becomes entirely bony. Further elongation of the bone is no longer possible.

Throughout the development of bone—and throughout life, as stresses are imposed from accidents, disease, or mere use, or as calcium phosphate is added or withdrawn to meet metabolic needs—bone remodeling occurs. Old Haversian systems break down and new ones are added.

Formation of Haversian Systems. Haversian systems arise throughout life in various canals, tunnels, and grooves. They form themselves around blood vessels. At first, osteoblasts arrange themselves around the periphery of the tunnel. As these cells mature into osteocytes, another layer of osteoblasts comes into the picture, reducing the lumen further. Ultimately the entire lumen is occupied, except the central Haversian canal, which carries blood vessels, lymphatic capillaries, and nerve fibers.

Intracartilaginous Development of Short Bones. The intracartilaginous development of short bones is similar to that of long bones, except that only a primary, and not a secondary, ossification center arises. Spongy bone spreads in all directions, and when development is complete the external surface, except at points of articulation, is covered by compact bone. Later, much of the spongy bone is replaced by compact bone.

Intracartilaginous Development of Irregular Bones. For irregular bones, such as the vertebrae, several primary ossification centers arise through which the irregularities of contour may be preserved. In addition, several secondary ossification centers arise, usually where a bone is to make a bend or where it is to stop growing. The scapula (shoulder blade), for example, develops from two primary and seven secondary centers.

Repair of Fractured Bone

A certain amount of bleeding occurs whenever a bone breaks. A blood clot then forms (Figure 15.32). Thereafter, in the initial step of repair, the clot is invaded by *fibroblasts*. These cells differentiate into a cartilaginous growth, called a *callus*, which temporarily serves to bridge the fracture.

Next, osteoblasts appear in the periosteum and endosteum. Through their action a callus of spongy bone is formed. At about the same time, the cartilaginous callus becomes mineralized and then undergoes erosion. The callus of spongy bone is converted into compact bone and excessive bony material is dissolved.

FIBROBLASTS. *Cells that produce connective tissue fibers and differentiate into various other cells.*

DISORDERS AND DISEASES OF BONE

Bone tissue is subject to a variety of diseases and disorders. In *acromegaly*, the extremities of the body increase unduly in size, due to an excess of

Disorders and Diseases of Bone

ACROMEGALY. Enlargement of the head, face, hands, feet, and thorax due to excessive growth hormone (G. *akron, extremity;* megas, *large*).

growth hormone stemming from a pituitary tumor. Another disorder, *calcification*, is the deposition of calcium into tissue which ordinarily is not osteoid. It is often caused by excessive parathyroid hormone in the blood.

A *fracture* is a crack or break in a bone (Figure 15.33). In a *simple fracture* there is no external wound at the point of fracture. In a *greenstick fracture* there is a partial break with some splintering. In a *compound fracture* there is a wound through the surface of the skin, along with the fracture.

OSTEOMYELITIS. (G. osteon, *bone;* myelos, *marrow;* itis, *inflammation*).

Osteomyelitis is an inflammation of the bone marrow, the Haversian canals, and the periosteum. It is usually due to a staphylococcus, streptococcus, pneumococcus, or typhoid germ. *Osteodystrophy* is a disturbance in the growth of a bone. In the majority of cases the disease is acquired, not inherited. It is caused by a disorder in the endocrine system or by faulty cartilage and bone cells. In some cases the disease is congenital. Three of the more common acquired forms are: osteomalacia, Paget's disease, and osteitis fibrosa.

OSTEOMALACIA. A disease characterized by bone softening because the osteoid tissue has failed to mineralize sufficiently (G. osteon, *bone;* malakia, *softness*).

Osteomalacia, called *rickets* when it appears in infants and young children, is due to a deficiency in calcium, phosphorus, vitamin D, or any combination of these nutrients. Its major characteristic is an insufficient mineralization of osteoid. The bones become soft, resulting in a knock-kneed or bowlegged appearance. In rickets there is increased growth of osteoid at the epiphyseal plates, giving rise to bony swellings at these points.

Paget's disease (osteitis deformans), usually begins after age 40. At first there is excessive removal of calcium and phosphorus from bone, even though the level of these minerals in the blood is normal. The removal may be due to excessive amounts of the hydrolytic enzyme phosphatase. The

FIG. 15.32. *Repair of fractured bone.*

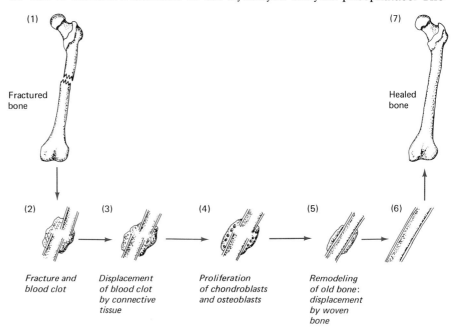

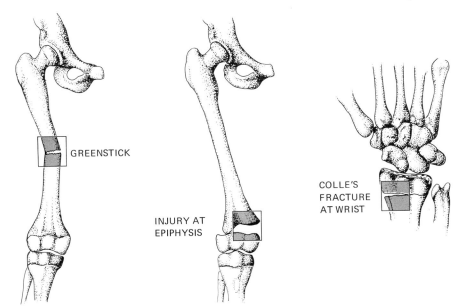

FIG. 15.33. *Two common children's fractures (left). Colle's fracture is common in adults.*

affected bones first become soft, later becoming hardened and thicker. Throughout this process the bone loses its normal shape and the individual becomes deformed.

In *osteitis fibrosa*, excessive parathyroid hormone brings about the removal of calcium phosphate from the bone, with accompanying weakening of the structure.

Osteoporosis is a reduction of bone matrix which can be traced to inadequate osteoblastic activity. It may be due to aging, deficiency of sex hormones, excessive amounts of hormones from the adrenal cortex, or poor nutrition. *Osteosclerosis* is a local increased density of bone. It is due to such things as chronic osteomyelitis, degenerative arthritis, and other factors, such as syphilis.

Bone tissue is also subject to tumors. *Multiple myeloma* is the most common malignant tumor arising in bone. It is characterized by cancerous plasma cells that develop in the marrow, erode adjacent regions of bone, and secrete abnormal antibodies (Bence Jones proteins) which can be found in the urine. *Osteosarcoma* is the second most common form of primary bone cancer. It is characterized by cancerous osteoblasts and osteocytes. More common than these primary forms of bone cancer are the secondary (metastatic) forms, which are derived most frequently from the breasts, lungs, prostate, thyroid, and kidneys.

OSTEOPOROSIS. *A condition in which existing bone is normally mineralized, but there is a reduction in quantity of bone (G. osteon, bone; poros, a pore).*

SUMMARY

1. Bones are classified as long, short, flat, irregular, or sesamoid.
2. The axial skeleton consists of the skull, spinal cord, and thoracic cage,

while the appendicular skeleton is made up of a shoulder girdle, a pelvic girdle and two upper and lower extremities.
3. The skull has fontanels, sutures, sinuses, and mastoid cells as special features which are not present elsewhere in the skeleton.
4. The cranium is formed from eight flat bones, two of which (the temporals) contain three additional (auditory) bones.
5. The neck harbors a suspended U-shaped bone, the hyoid, which supports the musculature of the throat.
6. The face contains 14 bones, seven of which are palpable while the other seven are too deeply situated to be felt.
7. The spinal column consists of 26 vertebrae: seven cervical, 12 thoracic, five lumbar, one sacral, and one coccygeal.
8. The vertebrae are separated from one another by intervertebral cartilaginous discs. When a disc "slips" from between the fifth lumbar and the first sacral vertebrae, it may give rise to sciatica.
9. The thoracic cage is composed of 12 thoracic vertebrae, the sternum, and 12 pairs of ribs, the lower five of which are "false ribs."
10. The shoulder girdle, which is made of two scapulas and two clavicles, attaches the humerus of the arms to the axial skeleton.
11. Each upper extremity is made of a shoulder girdle, a humerus (arm), a radius and ulna (forearm), eight carpals (wrist), five metacarpals (palm), and 14 phalanges.
12. The pelvic girdle, which is made of two hip bones, a sacrum, and a coccyx, attaches the femur of the thighs to the axial skeleton.
13. Each lower extremity is made of a pelvic girdle, a femur (thigh), fibula and tibia (leg), seven tarsal bones, five metatarsals, and 14 phalanges.
14. The wall of a bone is made of: (a) a periosteum, which brings blood vessels and nerves into the bone; (b) a mineral layer; and (c) an endosteum, which lines the mineral layer.
15. Certain bones develop within membranes through a process of mineralization, while other bones form from cartilage cells through a process in which cartilaginous cells proliferate, differentiate, mineralize, and die, after which they are replaced by bone cells.

REVIEW QUESTIONS

1. Can the hyoid bone be classified as a sesamoid? Explain.
2. What does the axial skeleton consist of? the appendicular skeleton?
3. Name at least three of the seven facial bones that can be felt. Where are they?
4. How are the 26 vertebrae distributed among the five regions of the spinal column?
5. Describe how a slipped disc may give rise to sciatica.
6. Define: shoulder girdle, upper extremity, pelvic girdle, lower extremity, diaphysis, epiphysis, Haversian system, Volkmann's canals.
7. Describe the structure of the wall of a long bone in the region of the diaphysis.
8. Why do chondroblasts die after they differentiate into bone-producing cells?
9. Name and describe the five most important bone diseases.

The Joints 16

STRUCTURAL MATERIALS OF JOINTS
    Ligaments
        Functions of ligaments
    Tendons
    Cartilage
        Functions of tendon and cartilage in joints
    Repair of ligament, tendon, and cartilage
        Providing nutrients to ligament, tendon, and cartilage
    Bursas and synovial membranes
CLASSIFICATION OF JOINTS
    Reference terms
    Joint movements
    Types of diarthrodial joints
        Distribution of the freely movable joints
        Strapping the joints together
DISORDERS AND DISEASES OF THE JOINTS

Each bone in the body, except the sesamoid bones and the hyoid bone, engages itself with at least one other bone to form a *joint*, or articulation. It is essential to understand the structure and movements of these joints if you wish to analyze specific skeletal movements, as physical education majors must; or if you wish to have a good appreciation of skeletal muscle actions, as all biologists must; or if you are to understand, as a medical practitioner must, why a sprain may sometimes be worse than a fracture.

Our approach to the study of joints will be first to describe the basic structural components of joints and then to classify the various types of joints. Next, we will say something about disorders and diseases of these important structures.

SESAMOID BONES. *Flat, somewhat ovoid bones, embedded in tendons in various regions of the body; The largest is the kneecap.*

STRUCTURAL MATERIALS OF JOINTS

Bones, ligament, cartilage, tendon, and bursa are the basic structural materials present at most joints.

Structural Materials of Joints

Ligaments

At every joint in the body one bone is connected to another bone by ligaments. In some joints, two or more bones may be connected to a third bone, as is in the wrist and ankle, where up to four bones are attached to a fifth bone.

A *ligament* is a dense connective tissue consisting predominantly of fibers and fibroblast cells within a ground substance. In most ligaments, collagenous fibers predominate, though elastic fibers are abundant in certain ligaments, such as those which run along the vertebral column.

Functions of Ligaments. For a long time the functions of ligaments were believed to be principally structural: to connect bone to bone; to restrict the movements of a joint; and to lend some support to the skeleton in bearing a load.

Through relatively recent research, however, it has become clear that ligaments perform at least two important physiological functions. First, they assume a load-bearing function, which, prior to around 1960, was usually attributed solely to muscle. In assuming this function, ligaments make it possible for the muscles to save a great deal of energy which they would otherwise have to expend. Second, ligaments—and very likely tendons and cartilage in a joint—signal fatigue when excessive work is done. The sensation of fatigue induces an individual to rest before any harm can be done to a muscle or joint.

How were these physiological functions deduced? Through the use of the electromyograph, a tool that makes it possible to determine whether a particular muscle is actively contracting when the body is assuming a particular posture or is involved in a certain kind of movement. The contractions are detected by recording electrical signals that come from specific muscles being tested. In making a record, anywhere from one to twelve electrodes are inserted into living muscle through the skin, or are placed upon the surface of the body directly above the selected muscles.

It will suffice to provide one example of electromyographic findings with respect to the two physiological functions of ligaments mentioned above. In support of the idea that ligaments spare muscles and energy, it has been found that there is reduced activity in shoulder and elbow muscles when one is carrying a heavy load, though there is considerable activity initially when the load is being lifted and placed in position. This means that after the initial lifting, ligaments must take over the weight-bearing function of muscle.

In support of the idea that ligaments signal fatigue, thus reducing the possibility of joint damage, it has been found that when a person hangs by the hands from a horizontal bar, reduced levels of electrical activity are measurable after a short period from muscles which enable one to assume this position. Nevertheless, although there is little measurable electrical activity—which signifies that the muscles are not undergoing contractile activities—the individual is able to hang for two or three minutes. Thereafter, fatigue sets in and the person is forced to release his grip. This fatigue

---

LIGAMENT. *Band of fibrous connective tissue which connects the articular ends of bones and sometimes envelops them in a capsule* (L. ligamentum, band, tie).

ELECTROMYOGRAPH. *Record of electric responses from muscles* (G. mys, *muscle;* gramma, *letter*).

is now attributed to pain (not to overworked muscles) which is believed to originate from the ligaments and perhaps other cartilaginous structures of the joints.

Tendons

Tendon, like ligament, is a tough connective tissue made up of fibers and fibroblast cells held together in a ground substance. It usually appears as a cord or band of fibrous material. At its point of attachment to muscle, the tendon often flares out, forming a sheet, known as an *aponeurosis*.

Tendons most commonly connect muscle to bone, though they also connect muscle to ligament, muscle to muscle, or muscle to fascia. In fact, in these latter three cases, tendons perform the connective function of fascia. When the connection occurs between muscle and ligament, the tendon and muscle—along with the ligaments—impose certain restrictions upon movements that the joint might otherwise make.

TENDON. *Connective tissue which connects muscle to bone, cartilage, ligament, or fascia (L. tendere, to stretch).*

GROUND SUBSTANCE. *Cellular secretory material which, together with fibers, forms the matrix of connective tissue.*

Cartilage

Cartilage is constructed like a tendon or a ligament, differing chiefly in having specialized cells (chrondocytes) in small cavities (lacunae), and having the potential to give rise to bone.

Functions of Tendon and Cartilage in Joints. There are two basic functions of tendon in joints: to connect muscle with bone or ligament, and to impose restrictions on movement.

Somewhat different functions are assumed by cartilage, three of which are general enough to merit singling out. First, cartilage is present as the articulating surface of bones in all of the freely movable (diarthrodial) joints of the body. As a substance which is far more elastic than bone, cartilage thus helps the joint withstand pressure and move more easily.

Second, cartilage is present as an elastic connection between certain immovable joints, such as those between the ribs and sternum; or at the pubic symphysis, where the pelvic bones are united in front of the body. Third, cartilage forms the bulk of intervertebral discs, and thus serves to support the spinal column and cushion various physical stresses encountered in daily life.

CARTILAGE. *Connective tissue which contains cartilage cells in cavities within its matrix (L. cartilago, wickerwork).*

Repair of Ligament, Tendon, and Cartilage

When a ligament or tendon becomes torn, as may happen in an athletic contest, each is able to repair itself. Fibroblasts begin to proliferate and new fibers, properly oriented, are produced. In time, the new fibers thicken and the injured material is resorbed.

Cartilage, in contrast to ligament and tendons, is not able to repair itself.

FIBROBLASTS. *Embryonic cells with the potential of differentiating into connective tissue cells.*

294
Structural Materials of Joints

CHONDROBLAST. *An embryonic cell which can mature into a cartilage cell or bone cell* (G. chondros, *cartilage;* blastos, *germ*).

This is because the adult cells are so highly specialized that they cannot divide. Chondroblasts from outside the cartilage—from a connective tissue covering (perichondrium)—come in to repair the injury. They invade the injured region, proliferate, and fill in the gap. A new matrix is thus deposited, fibers are produced and laid down in the matrix, and the chondroblasts mature into chondrocytes, each in its own lacuna. On occasion this process fails and a permanent fibrous scar tissue is developed instead. Sometimes this tissue becomes calcified pathologically into a bony substance.

Providing Nutrients to Ligament, Tendon, and Cartilage. As we saw in Chapter 3, ligaments, tendons, and cartilage contain very few cells relative to other tissues in the body (except bone). Hence, these tissues do not require large amounts of nutrients nor do they need a copious flow of blood to carry away their wastes. Most of their nutrients are supplied through diffusion from neighboring tissues. Most of their wastes are passed by way of diffusion toward other kinds of nearby tissues, such as a bursa or synovial membrane.

Bursas and Synovial Membranes

Bursas and synovial membranes are important components of our system of joints. Each of these structures must be clearly visualized in order to appreciate why it is possible for joints to operate as smoothly as they do.

BURSA. *A small fluid-filled sac located between certain parts of the body that move upon one another* (G. byrsa, *wineskin*).

A *bursa* is a closed fibrous sac. The bursas are located at points of friction in the skeletal system (Figure 16.1). Each is lined with a synovial membrane which contains synovial fluid (joint oil). The function of the bursa is to reduce friction during the movement of a joint, tendon, or bone.

Although there are more than 75 different bursas in the body, each one

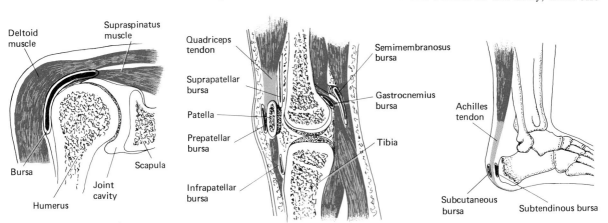

FIG. 16.1. *The various types of bursas can be seen in the shoulder joint (left), ankle, and knee.*

can be classified into one of four groups: a *submuscular bursa* is found over a bony prominence beneath a muscle; a *subtendinous bursa* is found over a bone, ligament, cartilage, or other hard tissue and beneath a tendon; a *subcutaneous bursa* is found over a bony prominence beneath the skin; an *articular bursa* is found between bones in a joint. Certain bursas may be classified into more than one of the preceding groups. The various types of bursas are illustrated in Figure 16.1.

A *synovial membrane* is part of a joint capsule, which in turn is present over any joint that is classified below as being freely movable (*diarthrodial*).

All freely movable joints—such as those in the shoulder, hip, ankle, elbow, knee, and between the digits (*phalanges*) of the fingers and feet—contain a cavity, or space, between the articulating surfaces (Figure 16.2). The cavity is present due to the arrangement of ligaments, which keep the articulating surfaces of the bones slightly apart from one another. Surrounding the cavity and joint ligaments is the joint capsule.

The joint capsule, in essence, has two layers. The outer layer is continuous with the periosteum (outer layer) of the two bones that make up the joint. The inner layer of the capsule—the synovial layer—lines the joint cavity and exudes a synovial fluid into the cavity. This fluid, which originates from blood plasma within the joint capsule, lubricates the joint, protects it against shock, and provides the ligaments and the articulating cartilaginous surfaces of the bones with nutrients.

SYNOVIAL *pertains to any movable joint and to the joint fluid (L.).*

PERIOSTEUM. *Fibrous membrane which covers bone along most of its surface (G. peri, above; osteo, bone).*

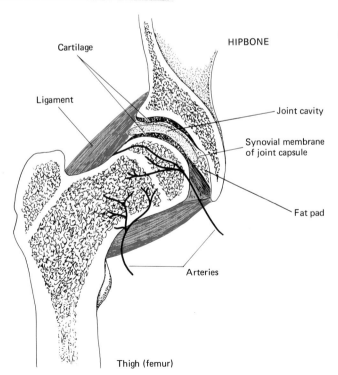

FIG. 16.2. *The hip joint, showing a joint cavity.*

Structural Materials of Joints

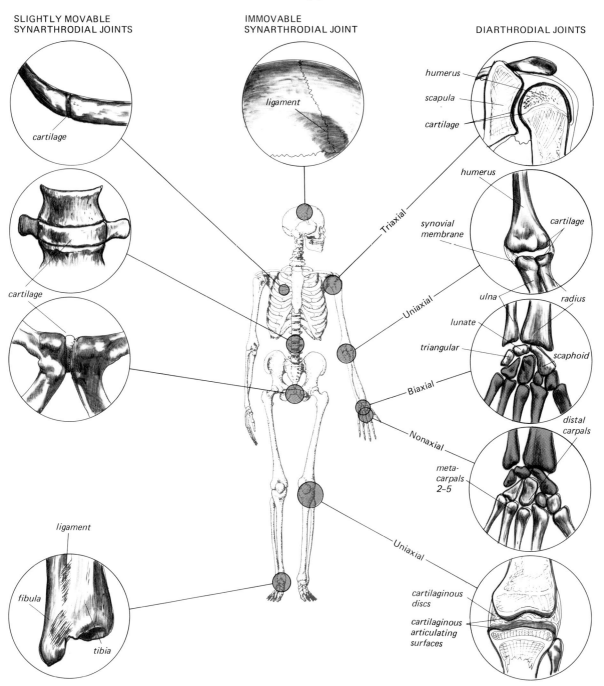

FIG. 16.3. Types of joints.

CLASSIFICATION OF JOINTS

Although there are many joints in the body, they can all be classified into two groups (Figure 16.3).

Synarthrodial joints have no cavity in the joint. Instead, the bones are united firmly by either cartilage or ligament. These joints may be classified further as either strictly immovable or slightly movable.

The *strictly immovable* joints are found in the fully ossified, mature cranium where ligament holds the bones together. They are also found in a few other regions, such as the symphysis pubis, where the pelvic bones are joined anteriorly along the midline of the body by cartilage (Figure 16.3).

Examples of *slightly movable* synarthrodial joints are found in the thorax, where ribs are joined to the sternum by the costal (rib) cartilages; in the leg, near the ankle, where the fibula is held to the tibia by ligaments; and between the vertebrae of the spinal column.

The second large class of joints, *diarthrodial joints*, have a separation, or articular cavity, between the two bones in the joint. Unlike synarthrodial joints, diarthrodial joints are freely movable and they are encased in a ligamentous capsule that is lined with a synovial membrane that secretes a lubricant (synovial fluid). Hence they are often called *synovial joints*.

The diarthrodial class of joints can be subdivided into four groups, depending on the number of axes through which movement is permitted.

Before proceeding into this classification, something will be said about the kinds of movements that our joints can make. The following discussion of terms should help you to understand these movements.

Reference Terms

A *plane* is a surface whose points can be connected to one another by straight lines which lie wholly within the surface. There are three anatomical planes (Figure 16.4). The *sagittal plane* extends from the front (anterior) of the body to the back (posterior). The *median sagittal plane* (the midsagittal plane) divides the body into right and left halves. The *coronal plane* extends from one side of the body to the other side. It divides the body into an anterior (front) and a posterior (rear) portion. The *transverse plane* divides the body into an upper (cranial) and a lower (caudal) portion.

An *axis* is a line along, around, or about which some part of the body may move. There are three types of anatomical axes (Figure 16.4). *Coronal axes* extend from one side of the body to the other side. *Sagittal axes* extend from the front to the back of the body. *Longitudinal axes* extend along the length of the body.

Joint Movements

Flexion is the movement that decreases the angle between a joint which moves around a coronal axis. This movement occurs in the anterior direc-

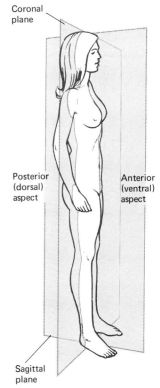

DIARTHRODIAL. *A freely movable joint (G. articulation).*

FIG. 16.4. *The planes and directions used to describe movements of the human body.*

Classification of Joints

tion for joints of the head, neck, upper extremity, trunk, and hips. In contrast, it occurs in the posterior direction for joints of the knee, ankle, foot, and toes (Figure 16.5).

Extension is the opposite of flexion, with one exception: in the case of the ankle—when the toes are raised toward the shin—the movement is referred to either as extension or, more commonly, as *dorsiflexion*.

DORSAL. *Of, or toward, the back* (L. dorsum, *back*).

Abduction is a movement within a coronal or transverse plane. The movement occurs away from the mid-sagittal plane of the body for all parts of the body except the fingers and toes. For the fingers, abduction refers to coronal movements of the first, second, fourth, and fifth fingers away from the third. For the toes, abduction refers to coronal movements of the first, third, fourth, and fifth toes from the second. *Adduction* is the opposite of abduction.

Rotation is a movement about a longitudinal axis for all parts of the body except the scapula, which may rotate around a sagittal axis or may be tilted (slightly rotated) around a coronal axis. To fully appreciate the rotational

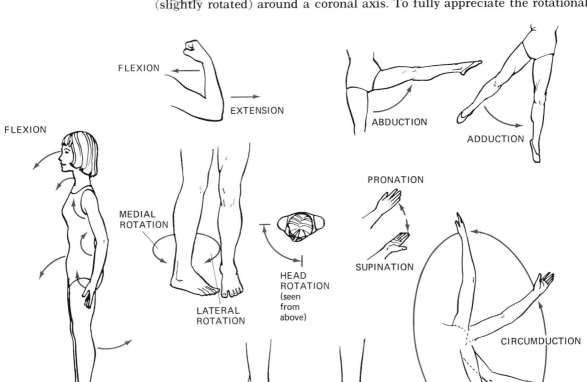

FIG. 16.5. *Flexion and other movements of the major body joints.*

movements, you should try to visualize (or actually carry out) the following types of movements:

(a) *Medial rotation.* Turn an arm or leg toward the midsagittal plane.

(b) *Lateral rotation.* Turn an arm or leg away from the midsagittal plane.

(c) *Head rotation.* Turn the head to the left or right to achieve left or right rotation respectively.

(d) *Thoracic and pelvic rotations.* Allow the thorax or pelvis to come further ahead along the left side than the right side to achieve clockwise rotation. Allow the left side to lag behind the right side to achieve counterclockwise rotation.

(e) *Scapular rotations.* To experience upward or lateral rotation, lift an arm sideways, allowing the scapula to rotate in the coronal plane. Return the arm and thus achieve medial or downward rotation. Roll your shoulders forward to experience tilting of the scapula around a coronal axis.

(f) *Supination.* Rotate the forearm, causing the palm of the hand to face anteriorly.

(g) *Pronation.* Rotate the forearm, causing the palm of the hand to face posteriorly (Figure 16.6).

Inversion is a combination of rotation and adduction of the foot, whereby the foot is swung medially. *Eversion* is a combination of rotation and abduction of the foot, whereby the foot is swung laterally.

Circumduction results from a combination of flexion, extension, abduction, and adduction, wherein the distal end of a part of the body (head, hand, or arm, for example) is moved along a circular route.

SCAPULA. *Shoulder blade (L.).*

SUPINATE. *To lie or place on the back, face upward, or to turn the hand palm upward (L. supinus, thrown backward).*

PRONATE. *To lie or place face down, or to turn the palm of the hand downward (L. pronus, leaning forward).*

Types of Diarthrodial Joints

And now we proceed to a discussion of the four major types of diarthrodial joints: nonaxial, uniaxial, biaxial, and triaxial.

The jointed surfaces of *nonaxial joints* are irregular in shape and allow limited movement in a planar, vibrating, or sliding sense, rather than along a definite axis of rotation. Examples are the carpal joints of the wrist (Figure 16.3) and the tarsal joints of the foot

Uniaxial joints are divided into two subtypes, according to the movements permitted by the joint: the hinge joint and the pivot joint. The *hinge joint*, which permits flexion and extension only, consists of a convex articulating surface which fits into a concavity of the other articulating surface. Examples of these joints are found between the ulna and humerus at the elbow and the femur and tibia at the knee. The *pivot joint*, which permits rotational movements only, consists of two articulating surfaces which allows one surface to roll over the other. Examples of this type of joint are found between the radius and ulna at the elbow and between the atlas and axis of the spinal column.

In a *biaxial joint*, a convex articulating surface fits into an oval concavity, allowing forward and backward movements (as in flexion and extension) and side-to-side movements (as in abduction and adduction). These move-

CARPALS. *Wrist bones.*

TARSALS. *Ankle bones.*

ATLAS. *The first vertebra, just beneath the skull.*

AXIS. *The second vertebra, upon which the atlas may rotate.*

HUMERUS. *The arm bone.*

FEMUR. *The thigh bone.*

ments, when performed sequentially—as for example with a pointed index finger at the metacarpal-first phalanx joint—give rise to circumduction.

Known commonly as ball-and-socket joints, *triaxial joints* occur in the shoulder (between the humerus and the glenoid fossa of the scapula) and in the hip joint (between the femur and acetabulum of the pelvic girdle). They are the most versatile joints in the body, permitting flexion, extension, abduction, adduction, and rotation.

Distribution of the Freely Movable Joints. Figure 16.3 shows what type of joint is present in specific regions of the skeleton. Table 16.1 is a listing of the joints, and it provides a description of the specific movements which are possible at a given joint.

Strapping the Joints Together. All movable joints of the body are held together by ligaments, as shown for the major joints of the body in Figures 16.7 to 16.13. In examining these figures the reader may note that the indi-

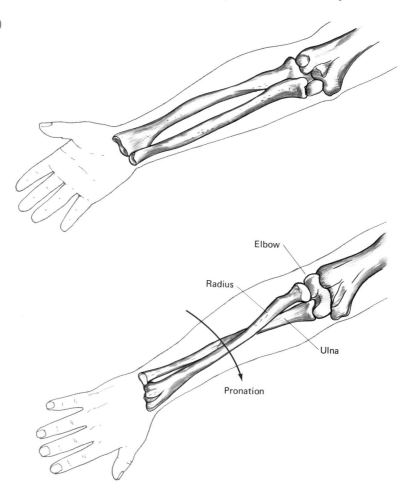

FIG. 16.6. *Supination* (left) *and* pronation (right) *are special forms of rotation.*

TABLE 16.1 The Freely Movable Joints of the Body

Region of Body	Joint	Type of Joint
Head	Condyloid process of mandible (jaw) and mandibular fossa of temporal bone	Uniaxial
Head/neck	Occipital condyles of cranium with articular facets of atlas	Uniaxial, hinge
Neck	Articular facets of atlas and axis, and condyloid process of axis	Uniaxial, pivot
Spinal Cord	Articular facets (along with cartilaginous articulations)	Biaxial—lateral flexion (movement in coronal plane) is also permitted
Shoulders	Glenoid fossa of scapula and head of humerus	Triaxial
Elbows	Head of ulna and capitulum of humerus	Uniaxial, hinge
Forearms	Proximal and distal joints of radius and ulna	Uniaxial, pivot, allowing supination and pronation
Wrists	Radius and ulna of arm and scaphoid, lunate, and triangular of hand	Biaxial
Hands	Carpal-metacarpals 2–5	Nonaxial
	Carpal-metacarpal of thumb	Triaxial
	Metacarpal-phalangeals 2–5	Triaxial
	Metacarpal-phalangeal of thumb	Uniaxial, hinge
	Interphalangeals of thumb and fingers	Uniaxial, hinge
Hips	Head of femur and acetabulum of pelvis	Triaxial
Knees	Condyles of femur with condyles of tibia	Uniaxial, hinge
Ankles	Tibia, fibula, and talus	Uniaxial, hinge
Feet	Talus and calcaneus	Nonaxial
	Ankle and tarsal joints, and tarsal-metatarsals	Give rise to combination of uniaxial, hinge and pivot movements, producing flexion, extension, adduction, abduction, inversion, and eversion
	Metatarsal and phalanges	Triaxial
	Interphalangeal	Uniaxial, hinge

FIG. 16.7. *Ligaments of the upper extremity.*

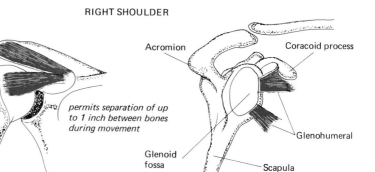

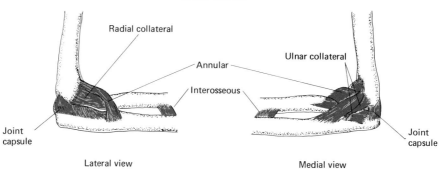

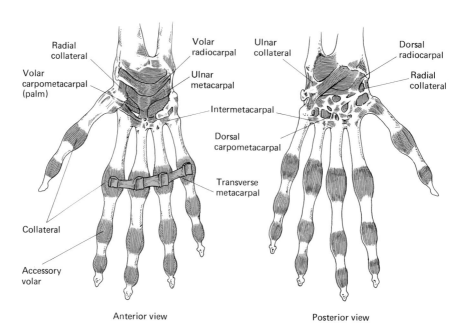

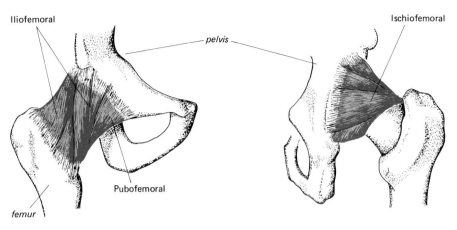

FIG. 16.8. Ligaments of the right hip.

vidual ligaments are named either for the bones they connect together, or else on the basis of the direction in which they run. Thus, the volar carpometacarpal ligament appears on the volar (= palmar) aspect of the wrist across the carpal and metacarpal joints (Figure 16.9). Similarly, collateral ligaments run along one side of a particular joint (Figures 16.7 and 16.13 through 16.16).

Some further explanation may be helpful with regard to the strapping together of the vertebral column (Figures 16.12 to 16.16). Here, five kinds of

CARPOMETACARPAL.
Joint between wrist and metacarpal bones.

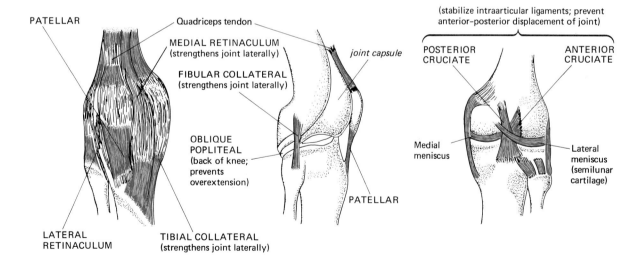

FIG. 16.9. Ligaments of the right knee.

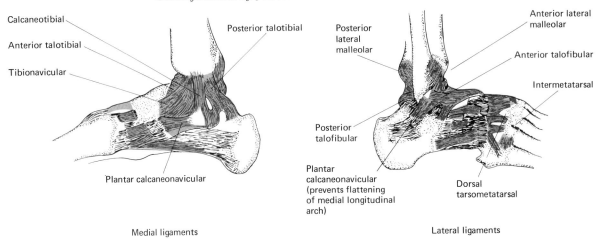

FIG. 16.10. Ankle ligaments. The medial ligaments may be sprained if the foot is twisted outward, the lateral ligaments if the foot is twisted inward.

ligaments are involved: (1) the *anterior* and *posterior longitudinal ligaments* connect the anterior and posterior aspects of the vertebral bodies together, from axis to sacrum; (2) the *ligamentum flavum* runs along the pedicles of the column, fastening together the laminas of the vertebrae—the broad bony extensions from the pedicles which project posteriorly and medially to form the spinal processes; (3) the *supraspinous ligament* con-

FIG. 16.11. Plantar ligaments—the ligaments of the foot.

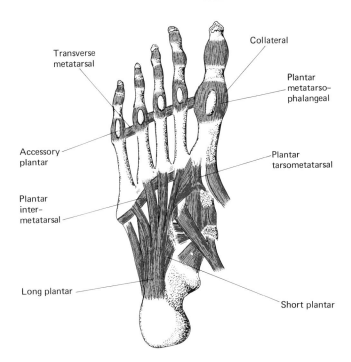

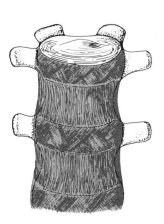

FIG. 16.12. *Anterior longitudinal ligaments of the spine.*

nects the dorsal tips of the spinous processes from the seventh cervical vertebra to the sacrum, all along the trunk; (4) additional strength is provided between adjacent vertebrae by the *interspinous ligaments,* which pass from the lateral surfaces of the spinal processes to the lamina of their adjacent vertebrae; (5) finally, there is the *ligamentum nuchae*—a ligament which is fastened from the base of the skull to the seventh cervical vertebra. This ligament contributes to the support of the head.

DISORDERS AND DISEASES OF THE JOINTS

FIG. 16.13. *Posterior longitudinal ligaments of the spine.*

Arthritis literally means an inflammation of an articulation. In common usage, it is synonymous with the older term *rheumatism.* There are several types of arthritis, and not all of them are actually inflammatory. *Infectious arthritis* is due to the syphilis spirochaete, gonorrhea bacteria, "strep" or streptococcus bacteria, or other infectious agent. *Metabolic arthritis,* usually noninflammatory, may occur in gout (due to faulty purine metabolism), rickets and osteomalacia (vitamin D deficiency), and scurvy (vitamin C deficiency).

Rheumatoid arthritis is usually due to an autoimmune reaction. It may accompany or result from rheumatic heart disease, in which some bacterium, fungus, or other organism invades the heart and other tissues of the body and changes the native characteristics of the tissue. A physical agent such as solar radiation may also bring about the alteration. The body's immune system responds by producing antibodies, which in turn react with the now-foreign tissue to dispose of it. When a joint becomes the target of the antibodies, an inflammatory reaction arises and in time the joint increases greatly in size.

Traumatic arthritis is due to a dislocation, strain, sprain, fracture, and the like. *Degenerative arthritis* is a condition in which degeneration occurs, as in rheumatoid arthritis, but for no known cause. In some forms of rheumatism, structures outside of the joint are affected. When the muscles are affected the condition is called *fibromyositis.*

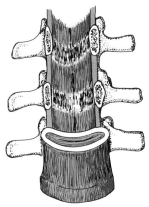

FIG. 16.14. *The ligamentum flavum.*

Most forms of arthritis can be cured. Thus (1) infectious arthritis may often be cured by getting rid of the causative organism; (2) metabolic arthritis may be overcome with vitamin D (for rickets and osteomalacia) or vitamin C (for scurvy); (3) gouty arthritis may be treated with colchicine—and predisposition to gout (high uric acid level in the blood) may be offset by Benemid; (4) rheumatoid arthritis may be alleviated with glucocorticoids; and (5) traumatic arthritis may disappear spontaneously some time after its formation. Rheumatic discomfort can often be reduced by assuming proper postures when standing, sitting, walking, and the like; by the application of heat; and by exercise.

A bunion (Figure 16.17) is an enlargement of the large toe joint (metatarsophalangeal joint) and thickening of the joint capsule, usually due to pressure from tight shoes which push the large toe toward the other toes.

Bursitis is an inflammation of a bursa. Commonly affected are the

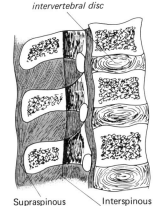

FIG. 16.15. *The spinal ligaments in sagittal view.*

Disorders and Diseases of the Joints

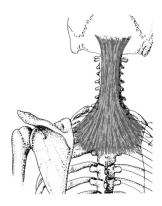

FIG. 16.16. The ligamentum nuchae.

BENEMID. Brand name for probenecid, a drug used to promote the excretion of uric acid from the blood by the kidneys.

patellar bursas (housemaid's knee) and the olecranon bursa (tennis elbow). A *dislocation* is a displacement of the ends of the bones in a joint.

Osteoarthrosis is a noninflammatory joint disease commonly known as osteoarthritis (though it is not truly an *-itis*, or inflammation), senescent arthritis (though it appears in young as well as old), menopausal arthritis (though it occurs in young men and women), and by several other terms. Osteoarthrosis is hereditary, and occurs more frequently in females than in males. It is characterized by joints (fingers, hips, knees) which become stiffer, more swollen, and increasingly painful in the course of time.

Although hope is available to sufferers of osteoarthrosis, nothing can be done at present to reverse the tissue changes. Some palliative measures (giving relief without curing) for osteoarthrosis include aspirin for pain, exercise for strengthening the muscles, use of corrective supports (such as corsets and properly fitted shoes), weight reduction (relief of extra burdensome work), surgical removal of water cysts from certain joints (elbow, "water on the knee"), and surgical removal of degenerated tissue from an afflicted joint.

A *sprain* is a tearing of muscles, ligaments, or tendons (Figure 16.2).

A *strain* is an unusual or excessive exertion that causes stretching of muscles and ligaments but no real damage. *Subluxation* is a partial joint disruption, and *synovitis* is an inflammation of a joint capsule.

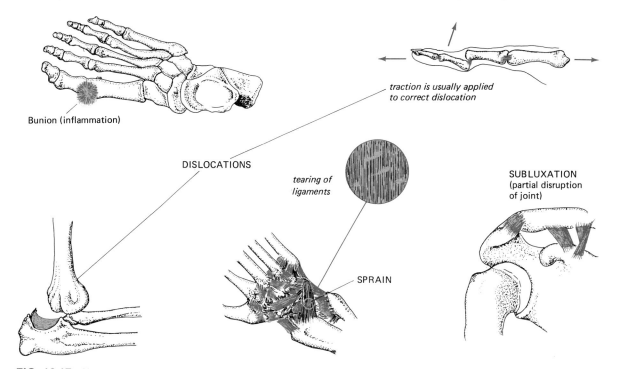

FIG. 16.17. Abnormalities of the joints. A sprain (detail) can involve tearing of muscle or tendon, as well as of ligament.

SUMMARY

1. Ligament is a dense connective tissue which connects bone to bone. It restricts the movements of the joints and supports the skeleton in carrying a load.
2. Tendon is a dense connective tissue which connects muscle to bone, to ligament, to another muscle, or to fascia.
3. Cartilage is constructed like a tendon or ligament, but it has chondrocytes, located within lacunae, instead of fibroblast cells. Joint cartilage is present as the articular surface in freely movable joints, as an elastic connection between immovable joints, and as the main tissue in intervertebral discs.
4. A bursa is a closed fibrous sac located at points of friction in the skeletal system, such as over a bony prominence beneath the skin.
5. A freely movable joint is enclosed in a joint capsule, which secretes a lubricant, called synovial fluid, into the capsule.
6. Joints are classified as movable (diarthrodial) or immovable (synarthrodial).
7. Freely movable joints allow the body, head, and appendages to move along three kinds of axes in three kinds of planes. The basic movements are flexion, extension, abduction, adduction, and rotation.
8. Movable joints may be classified as nonaxial, uniaxial (hinge and pivot), biaxial, and triaxial.

REVIEW QUESTIONS

1. What are the functional differences between ligament and tendon?
2. How does cartilage differ histologically from ligament and tendon?
3. Describe a bursa. Describe four types of body region in which a bursa may be located.
4. What is the difference between a bursa and a joint capsule?
5. Define: flexion, extension, abduction, adduction, pronation, supination, rotation.
6. Are cats and dogs able to pronate and supinate their forelimbs?
7. Give an example of the following types of movable joints: nonaxial, uniaxial, biaxial, triaxial.
8. Describe the following joint disorders and diseases: osteoarthrosis, rheumatoid arthritis, gout, bunion, bursitis, dislocation, sprain, strain.

17 The Physiology of Muscle

PROPERTIES OF MUSCLE TISSUE
    Responsiveness to stimuli
    Contractility
    Extensibility
    Elasticity
SKELETAL MUSCLE
    Structure of skeletal muscle cells
        The sarcoplasmic reticulum and bioelectricity
        The sarcosomes and energy production
    Contraction and relaxation of skeletal muscle
        Types of muscle contractions
    Skeletal muscle disorders
        Fasciculations
        Cramps
        Tetany
        Abnormal muscle weakness
SMOOTH MUSCLE
    Description and location of smooth muscle
    The mechanism of smooth muscle contraction
    Classification of smooth muscle
    Regulation of smooth muscle activity by nerves and hormones
CARDIAC MUSCLE
    Differences between skeletal and cardiac muscle
RESPONSIVENESS OF MUSCLE TO PHYSICAL TRAINING
    Classification of exercise

The movements of tissues, organs, limbs, and the body are due to alternating contractions and relaxations of muscle. There are three types of muscles in the body: smooth, cardiac, and skeletal. Altogether, these tissues make up about 50 percent of our total body weight.

This chapter begins with a discussion of the properties shared by all three types of muscle. Thereafter we discuss each type of muscle separately.

PROPERTIES OF MUSCLE TISSUE

Among the characteristics which define muscle, four are perhaps most notable: responsiveness to stimuli, contractility, extensibility, and elasticity.

Responsiveness to Stimuli

Muscle cells are sensitive to electrical, mechanical, chemical, and thermal energy. In their natural setting, muscle cells ordinarily respond to nerve impulses. This means they react to chemicals that are liberated from the endings of the nerve fibers which innervate them. The reaction may be an inhibitory one, in the case of cardiac and smooth muscle, or it may be excitatory for any of the three types of muscle. Excitation brings about a contractile response; that is, a contraction. Inhibition reduces or precludes a contractile movement.

Mechanical and electrical energy also are effective in stimulating muscle. The application of mechanical energy—such as stretching a muscle—may bring about a contractile response. The application of an electric shock is also effective in bringing about a response. The application of electric current is used in laboratory experiments to stimulate a muscle contraction. As a medical tool, it is used to treat heart fibrillation. The shock stops the heart muscle from fibrillating so that when the muscle begins to contract again, the regular rhythm may resume.

Contractility

Excitatory stimuli cause muscle fibers to shorten to different extents—sometimes to about 15 percent of their normal resting length.

Extensibility

Muscle may also be stretched. The smooth muscle of the blood vessels, stomach, intestine, uterus, and urinary bladder undergo considerable stretching as these organs become filled. The cardiac muscle of the heart stretches considerably during filling (diastole). Skeletal muscles also are extensible. When a flexor muscle contracts, an opposing extensor muscle relaxes and partly stretches.

DIASTOLE. *Period of relaxation during which the heart fills with blood (G. dilation).*

Elasticity

The elastic nature of muscle allows a muscle cell to regain its original size and shape after having been stretched or contracted.

SKELETAL MUSCLE

Skeletal muscle is also called striated, striped, or voluntary muscle. It is called *voluntary* because the fibers are under the control of the central, rather than the autonomic, nervous system. It is termed *striated* or *striped* because the individual cells, when seen under an ordinary light microscope, exhibit transverse striations (Figure 17.1).

Structure of Skeletal Muscle Cells

The smallest cellular unit of skeletal muscle is a long *fiber*, which ranges in length from 1 to 300 millimeters and in thickness from 10 to 100 micrometers. Although the cell is fibrous in shape, it contains organelles that are commonly found in other kinds of cells. Usually, however, the prefix *sarco* is used in naming the organelles of the muscle cells.

The entire skeletal muscle cell is bounded within a plasma membrane, known also as a *sarcolemma*. Across this membrane there is a potential difference in voltage. When the cell is not being stimulated, the interior surface of the sarcolemma bears a negative charge of about −70 to −80 millivolts.

Bathed by the sarcoplasm of the cell, just beneath the sarcolemma along the length of the cell, are several nuclei (Figure 17.2). Along with the rest of the cell's protein-synthesizing machinery, these nuclei are particularly active in producing contractile proteins, especially when an individual is working or exercising.

SARCO-. (G. sarx, *flesh*).

SARCOPLASM. *Cytoplasm of a muscle cell* (G. sarx, *flesh;* plasma, *thing formed*).

FIG. 17.1. *Skeletal muscle.*

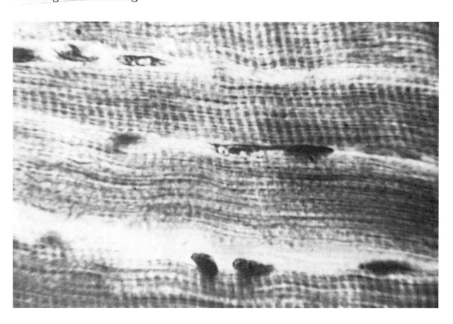

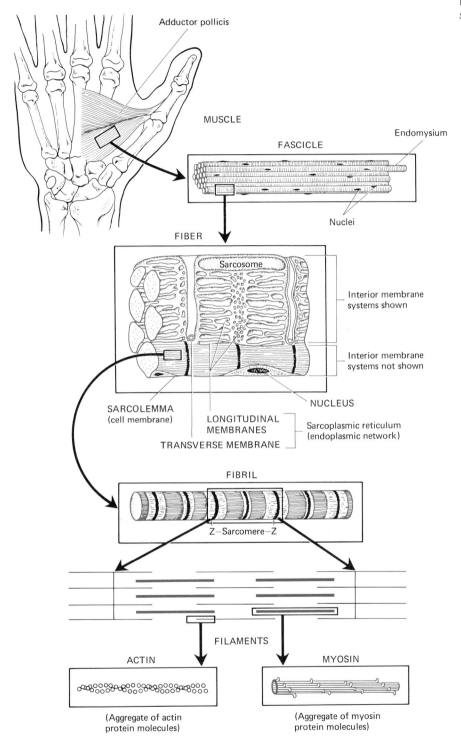

FIG. 17.2. *Structure of skeletal muscle.*

Skeletal Muscle

SARCOMERES. *The segments into which a muscle fibril is divided by Z discs* (G. sarx, *flesh;* meros, *part*).

The contractile proteins, *actin* and *myosin*, are assembled as *myofilaments*, which can be seen only through an electron microscope. These myofilaments are grouped into *myofibrils*, which can be seen with an ordinary microscope.

Each myofibril is made up of many repeating units, called *sarcomeres*. The sarcomeres are lined up end to end with one another, much like cars in a railroad train. When a muscle fiber contracts, all of the sarcomeres shorten successively, from the point where the fiber is stimulated—at the motor end plate—to each distal end of the fiber.

In Figure 17.2 you may note that the sarcomeres extend from one line—called a *Z-line*—to another Z-line. A Z-line is seen with an electron microscope as being denser than the material on either side of it. It is denser because one end of an actin myofilament is attached here to a portion of intracellular membrane. The other end of the actin myofilament—the free end—passes parallel to the longitudinal axis of the fiber toward the midline of the sarcomere. However, it does not quite reach the midline when the muscle fiber is relaxed (elongated). Nor does the actin myofilament that runs in line with it from the Z-line at the other end of the sarcomere. Consequently, a gap is left between the actin myofilaments which arise from the two Z-lines. This gap is bridged by the other type of contractile myofilament—the myosin myofilament. But the myosin myofilament does not run from one Z-line to the other Z-line. Rather, as you may see in the figure, the myosin myofilament merely straddles the gap between the actin myofilaments. During muscle contraction, the actin and myosin slide past one another and close the gaps.

The Sarcoplasmic Reticulum and Bioelectricity. In Figure 17.2, it can be seen that the myofibrils are enclosed in longitudinally-oriented membranes. At the Z-lines of the sarcomere, these membranes are in contact with transversely-oriented membranes. The transversely-oriented membranes appear as tubules (small tubes) which are continuous with the sarcolemma (the plasma membrane).

Collectively, these membranes form the *bioelectrical system* of the muscle cell. When the sarcolemma is excited by a stimulus that is strong enough to cause the generation of an action potential, the action potential travels from the point at which it arises (the motor end plate) to all other points on the sarcolemma surface. In the course of passing along the sarcolemma, the action potential is directed deeply into the thickness of the muscle fiber by way of the transversely-oriented membranes (*the T-system*). The action potential is also conveyed to all myofibrils within the muscle fiber by way of the longitudinally-oriented membranes (*the L-system*).

ENDOPLASMIC RETICULUM. *Network of membranes within a cell* (G. endon, *within;* plasma, *to form;* L. rete, *net*).

The longitudinally-oriented membrane system (the L-system) is formed from what is called the *endoplasmic reticulum* in other types of cells. In skeletal and cardiac muscle cells, the endoplasmic reticulum is commonly called the *sarcoplasmic reticulum*. The special arrangement of the sarcoplasmic reticulum within cardiac and skeletal muscle cells allows the cells to contract or relax quickly, so the cells can carry out their functions rapidly and repeatedly in a short period of time.

The Sarcosomes and Energy Production. Energy for the contraction of a muscle cell is provided directly by ATP, as described on page 316. Most of the cell's supply of ATP is produced in mitochondria, which are specifically known as *sarcosomes* in muscle.

SARCOSOMES. *Muscle mitochondria (G. sarx, flesh; somas, body).*

Although ATP is the immediate high energy metabolite for triggering a muscle contraction, other nutrients are needed as a source of energy for the production of ATP in muscle cells. In general, during most of the day, fatty acids serve as the principal bulk source of energy. Next to fatty acids, glycogen (animal starch) serves as a principal reservoir. Shortly after a meal, when the concentration of glucose in the blood is high, and in the event that strenuous muscular activity is needed, the blood glucose is catabolized in the muscle cell, providing a source of ATP. In moments of stress, when adrenaline and adrenal cortical hormones are released from the adrenal gland, glucose is made available through metabolic activities in the liver. Amino acids are also used to some extent as an energy source.

Regardless of whether ATP is formed in the muscle cell through the catabolism of fatty acids, glycogen, glucose, or amino acids, some of the ATP is used to produce *phosphocreatine*. This substance is kept on hand in case sudden bursts of muscular activity are needed. The normal conversion of the energy of fatty acids, glycogen, glucose, or amino acids into the energy of ATP is too slow for sudden bursts of activity. Phosphocreatine and ATP are produced enzymatically through a reaction which is readily reversible:

CATABOLISM. *A metabolic reaction in which a chemical is degraded and energy released (G. catabole, a breaking down).*

Creatine + ATP ⇌ Phosphocreatine + ADP

Contraction and Relaxation of Skeletal Muscle

Skeletal muscle does not normally contract unless it is stimulated by a nerve cell called a motor neuron. The cell body of the motor neuron is located in the spinal cord. The cell's main fibrous extension, which is called an axon, has numerous distal branches. The tip of each branch ends on a *motor end plate*. Each motor end plate is intimately joined to the plasma membrane of a skeletal muscle cell to form a *neuromuscular junction*. The muscle cell is said to be innervated. All of the muscle fibers that are innervated by a motor neuron, together with the motor neuron itself, constitute a *motor unit*. We will consider now what happens in a single muscle fiber when its motor neuron is not discharging action potentials. Next, we will see what happens when the motor neuron *is* discharging (Figure 17.3).

When a skeletal muscle fiber is at rest, it is electrically silent. That is, no action potentials are measurable when electrodes are placed on the plasma membrane of the fiber. There is, however, a resting potential. That is, the inner surface of the plasma membrane bears a negative charge, and the exterior surface has a positive charge.

When the muscle fiber is to be contracted, action potentials first pass along the axon of the motor neuron. At the motor end plate the action potential induces the release of acetylcholine, a neurotransmitter, from vesicles near the motor end plate. The acetylcholine diffuses across a small space between the motor end plate and the plasma membrane of the muscle cell. After doing so, the acetylcholine interacts with the plasma membrane. In consequence, positively-charged sodium ions rush into the muscle fiber. The negative charge that is present on the inner surface of the resting muscle fiber first becomes neutral, and then positive.

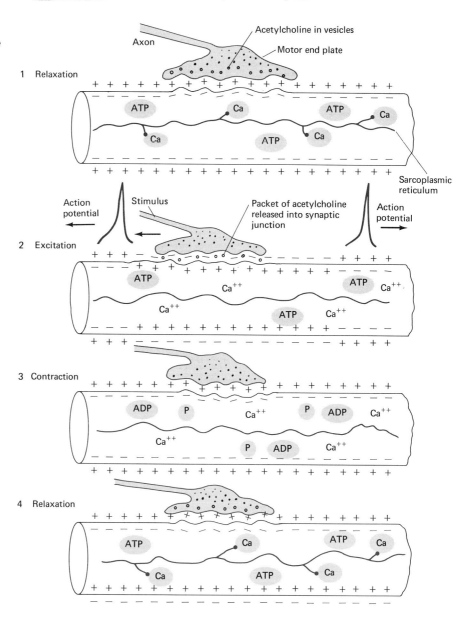

FIG. 17.3. Bioelectrical events in skeletal muscle contraction.

When a critical amount of neurotransmitter has been released—which happens when a sufficient number of action potentials travel along the nerve fiber—an action potential arises in the plasma membrane of the muscle fiber. The action potential may last for as little as 2 milliseconds or as long as 8 milliseconds. Regardless of duration, the action potential passes toward both ends of the muscle fiber. In doing so, it also passes into, and along, the entire surface of the transverse tubules. In addition, the energy of the action potential passes along the entire surface of the longitudinal membranes—that is, along the sarcoplasmic reticulum, which encloses the myofibrils.

As the action potential passes along the fiber calcium ions are released from the sarcolemma, from the transversely-oriented membranes, and from the longitudinally-oriented membranes. The calcium ions then interact with actin and myosin. To appreciate what happens as a result of this interaction, you must realize that the myosin filaments have so-called "cross bridges" which lock onto the actin filaments and hold them in place during periods of relaxation. Furthermore you must realize that *troponin*, another important molecule, is also present near the actin and myosin molecules. Let us now consider the interaction between actin, myosin, troponin, and calcium ions.

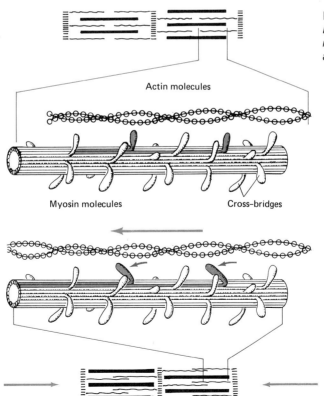

FIG. 17.4. *Cross-bridges between actin and myosin filaments bring about muscle contraction.*

In this interaction, the following events occur. First, ATP, the energy source for work done by the muscle fiber, is split by an enzyme that is part of the myosin molecule. Second, the protein *troponin*, which prevents the actin and myosin filaments from sliding past one another during relaxation, is changed so that it no longer holds the actin and myosin filaments in their stationary positions.

Third, the "cross bridges" of the myosin molecules, which have detached from their "locking" positions on the actin molecules, reach and out and attach to other portions of the actin molecules and pull the actin molecules toward a region halfway between the Z-lines of each sarcomere. When this happens the actin and myosin molecules appear to slide past one another and the muscle fiber contracts, or shortens. Fourth, the "cross bridges" may engage and disengage from the actin molecules several times during one contraction. With each engagement, the actin molecules are drawn closer to the center of the muscle fiber and the muscle fiber contracts further.

These reactions require ATP and calcium. When all of the necessary calcium has been released from the sarcoplasmic reticulum, the contraction ceases. As a final step, the calcium is withdrawn from the fluid surrounding the muscle fiber and is bound to the sarcoplasmic reticulum again. Magnesium is thought to be involved in this process of relaxation. Magnesium may be taken up by the sarcoplasmic reticulum when a contraction begins and released into the fluid surrounding the muscle fiber when relaxation begins.

Types of Muscle Contractions. The behavior of skeletal muscle contraction is often studied using one of the two experimental setups shown in Figure 17.5. In the setup on the right, the contractile tissue is allowed to shorten visibly upon being stimulated adequately with pulses of direct electrical current. The electricity may be applied through an electrode directly to the muscle fiber or indirectly to the muscle cell through a nerve fiber. This type of contraction—in which the muscle shortens visibly—is called an *isotonic contraction*. The tension which develops during contrac-

ISOTONIC. *Having equal tension (G. isos, equal; tonos, a stretching or tension).*

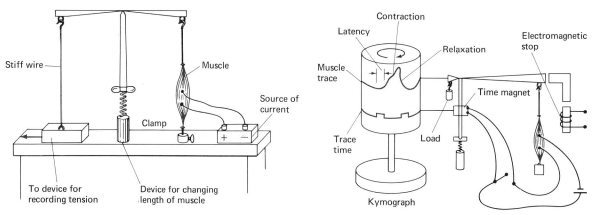

FIG. 17.5. *Experimental setups for studying isometric (left) and isotonic (right) muscle contractions.*

tion is spread equally throughout the muscle cells and their connective tissues.

In contrast, in the setup on the left the muscle is not allowed to shorten visibly. In this type of contraction, known as an *isometric contraction*, the myofibrils slide past one another, creating a state of tension, but the connective tissues, such as the endomysium, are not allowed to shorten. This is so because the muscle is required to raise an impossibly heavy load on the short arm of the lever. The tension that builds up in the muscle can be measured with a strain gauge, but although contractile activity is occurring within the muscle, no external *work* is done. That is, no mass is being moved over a distance. Nevertheless, in an isometric contraction heat is generated, measurable force (tension) is developed, and energy is consumed.

In the body, isotonic contractions are performed whenever a muscle shortens visibly and moves the part of the body in which it is inserted. Isometric contractions are performed when a muscle contracts but either the weight is too heavy to lift, or the person voluntarily decides not to allow the inserted limb to move.

The "isotonic" laboratory setup shown in Figure 17.5 can be used to demonstrate the *all-or-none law* for skeletal muscle contractions. This law says that if a muscle fiber is excited sufficiently, the fiber will contract throughout its entire length. That is, each sarcomere will shorten. If the muscle fiber is not excited enough, the muscle fiber will not contract at all.

The same laboratory setup can also be used to demonstrate tetanus. *Tetanus* is a sustained muscular contraction caused by a series of stimuli given so rapidly that the separate response to each stimulus fuses with the next. This is the type of contractile muscle activity our muscle fibers perform nearly every time they are used. This activity may be recorded graphically using a kymograph. Figure 17.6 shows how tetanus arises in the course of stimulating a muscle fiber repeatedly with brief pulses of electricity. When the muscle is stimulated very infrequently it responds with twitches. *Twitches* are contractile activities in which the duration of contraction roughly equals the duration of relaxation. With an increase in frequency of stimulation, the tension of contraction steadily grows larger and the duration of contraction exceeds the period of relaxation. Ultimately,

ISOMETRIC. *Not changing in dimension (G. isos, equal; metron, measure).*

ENDOMYSIUM. *Connective tissue sheath around a muscle fiber (G. endon, within; mys, muscle).*

TETANUS. *In physiology, a contracted state of muscle; in pathology, a disease, often fatal, caused by the toxins of certain bacteria (G. tetanos, tension).*

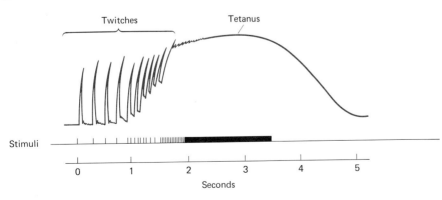

FIG. 17.6. *Activity of a muscle fiber as recorded on a kymograph. Note the effect of a sustained stimulus (color).*

Skeletal Muscle

at a sufficiently high frequency of stimulation, the muscle fiber enters a state of tetanus. In this state the muscle is contracting maximally, and although there is a tendency toward relaxation, no relaxation is actually achieved. The very brief period during which the fibers tend toward relaxation is called the *refractory period.* Stimulation of the muscle is ineffective during this period. The actin and myosin filaments slide away from one another, causing the muscle fibers to elongate ever so slightly, and calcium is taken back up into the sarcoplasmic reticulum. After this incomplete relaxation, the fibers return to the fully contracted state.

REFRACTORY. *Resistant to treatment or stimulation* (L. refractum, *to break up*).

Tetanic contractions may occur either isotonically or isometrically. They may also occur as a phasic contraction or a tonic contraction, depending upon the nature of the muscle. A *phasic contraction* is a very rapid contraction, as happens, for example, when the eyelids blink or when an arm is moved rapidly in reaching for an object. A *tonic contraction* is sustained over a longer period. Tonic contractions occur, for example, in the maintenance of posture.

PHASIC. *Brief* (G. phasis, *phase of the moon*).

In addition to being able to undergo active contractions of various sorts, normal skeletal muscle exhibits the property of tonicity. *Tonicity,* or *muscle tone,* refers to the belief that at all times some muscle fibers in a muscle are partially activated. Muscle tone is lost when a muscle's motor nerve fails to deliver "maintenance levels" of some unknown metabolite, perhaps acetylcholine, to the muscle. This may happen, for example, after a stroke or injury.

TONIC. *Of muscle tone or tension* (G. tonikos, *stretching*).

Skeletal Muscle Disorders

Skeletal muscle problems may be caused by a fault in the central nervous system, the peripheral nervous system, or the muscle itself. The topic of spastic and flaccid muscle paralysis due to injury in the central nervous system was covered in Chapter 7. Here, we will discuss several other common muscle problems.

Fasciculations. Fasciculations are involuntary contractions of groups (fascicles) of muscle fibers. The contractions can be seen right through the skin. The exact cause of this problem is not known other than to say it is accompanied by excessive nervous discharge. It occurs naturally in most of us at one time or another and is almost always nothing to be concerned about. It is frequently seen or felt as a local quiver; in certain nerve diseases fasciculation occurs commonly and serves as a telltale sign of the abnormality.

Cramps. Muscle *cramps* are also involuntary contractions. They are usually of sustained nature, often painful, and sometimes serious enough to be concerned about. They are associated most frequently with vigorous exercise, with sodium deficiency, with nighttime, with cold temperature, and with pregnancy. Cramps may also result from prolonged work in one's occupation (the pianist's fingers and forearms, the waiter's back and arm muscles, or the "writer's cramp" of the textbook editor).

Cramps may be relieved by massaging the affected muscle. The relief may be due to the fact that in massaging the muscle, the Golgi tendon receptor is activated. When this receptor is activated, afferent messages of inhibitory nature are transmitted to the spinal cord. In turn, alpha motor neurons discharge fewer action potentials to activate the muscle.

Tetany. Tetany is a disorder characterized by intermittent muscular contractions, often accompanied by sensations of burning, pricking, numbness, and pain. It is usually associated with a reduction of ionic calcium in the blood and other extracellular tissue fluids (interstitial fluid). Tetany may be caused by excessive deep breathing (hyperventilation), which tends to clear the body of abnormally large amounts of carbon dioxide. As a result, the pH of the blood increases (a condition known as alkalosis), and less ionic calcium is present. The nerves become hyperexcitable, and their discharge leads to the muscle spasms characteristic of tetany, as in lockjaw or in cases of abnormally-underactive parathyroid activity.

TETANY. *Abnormal muscle spasms, resulting usually from faulty calcium metabolism* (G. tetanos, *tension*).

ALKALOSIS. *A condition in which there is an abnormal increase in bicarbonate in the blood* (Arabic al-quili, *alkali;* G. osis, *condition*).

Abnormal Muscle Weakness. A muscle normally becomes weak or fatigued because of a depletion of neurotransmitter substance at the neuromuscular synapse; or an accumulation of an oxygen debt in the muscle itself; or perhaps inadequate nutrition, an imbalance of sodium, potassium, calcium, and magnesium, and the like.

Occasionally, however, an abnormal muscle weakness develops. The origin of such weakness may reside in the nervous system or in the muscle itself. One way of getting an overview of this subject is to consider the sites at which a disorder may originate, proceeding in a direction from the spinal cord — in which lie the cell bodies of the motor neurons — to the contractile units in the muscle fibers.

If there is a fault in the spinal cord, destruction of the cell bodies of motor neurons, as happens in certain forms of poliomyelitis, will result in denervation of the affected muscles. Whereas denervation of smooth and cardiac muscles is not followed by an atrophy (degeneration) of the muscles, denervation of skeletal muscle is. Skeletal muscle loses its tone and its striations. It suffers the same kind of losses when a peripheral nerve is cut. However, the peripheral nerve often regenerates, and if this happens within a reasonable period, usually one year, the muscle may regain its tone and function normally again.

POLIOMYELITIS. *A virus disease of nerve and muscle tissue* (G. polios, *gray;* mys, *muscle;* itis, *inflammation*).

The fault in abnormal muscle weakness may also lie at the nerve-muscle junction. A common muscular weakness, *myasthenia gravis*, results from an abnormality at the neuromuscular junction. This disease afflicts about 50,000 people in the United States, chiefly women over 30. It usually begins in the face and throat, and is a chronic, progressive disorder. Although any muscle or group of muscles can be affected, the facial, ocular, and shoulder girdle muscles are most commonly involved. Chewing, swallowing, and speech are often difficult.

Although the precise nature of the defect is uncertain, data have been gathered to suggest an abnormality in the motor end plate. Other data show that the muscle membrane beyond the motor end plate — the postsynaptic membrane — is inadequate in its capability for responding to acetylcholine.

In still other studies it has been shown that the motor end plate fails to depolarize, and at one time it was thought that this was due to a failure of the nerves to release acetylcholine; however, even the direct application of acetylcholine may fail to result in normal depolarization. Yet, in many cases, muscle strength returns when the patient receives a drug that works against acetylcholinesterase, the enzyme which ordinarily destroys acetylcholine in neuromuscular synapses. In some cases remission occurs; that is, the disease temporarily disappears.

MYASTHENIA GRAVIS. *A muscle disease in which the muscles become very weak, but do not atrophy* (G. mys, *muscle;* astheneia, *weakness*).

Myasthenia gravis in some people is related in some unknown way to an abnormal thymus. The removal of the thymus gland may or may not bring relief. Very recent work on myasthenia gravis has revealed that the disease is commonly due to a form of autoimmunity; this is discussed later in the book, in Chapter 32.

A muscle weakness may arise from an abnormal balance of potassium. The site of improper action in this case is the muscle membrane. Too much or too little potassium can trigger the weakness. Evidence of weakness lies in the fact that the muscles are not as responsive as normal muscle would be to electric shocks applied directly to the muscle.

Weakness is felt during rest periods following exercise. Where there are excessive levels of potassium in the blood the extra potassium is derived from the muscle cells. Excessive levels of potassium are also found in the urine. When an abnormally low level of potassium is present in the blood, an excessive amount of potassium is present in the muscle cells; an abnormally low amount of potassium is also found in the urine.

Muscle weakness may be caused by a fault beneath the muscle cell membrane. Here there is a failure in the process of coupling electrical excitation to muscle contraction. Following the passage of an action potential, calcium ordinarily is released from the sarcoplasmic reticulum. Thereafter actin and myosin begin to slide over one another. In the case of faulty calcium metabolism—as occurs, for example, in rickets and osteoporosis—the muscles may experience weakness. Vitamin D is able to correct the weakness in some cases. This suggests that the problem may reside in the transport of calcium across the plasma membrane of muscle fibers.

RICKETS and **OSTEOMALACIA.** *Bone diseases in which there is a deficiency of mineralization throughout the bone.*

Finally, muscle weakness may arise as a result of an abnormality within the myofibrils. Muscular dystrophy, for example, is an inborn abnormality characterized by, among other things, disintegration of the myofilaments.

SMOOTH MUSCLE

Smooth muscle differs histologically from skeletal and cardiac muscle in its lack of cross striations, as may be seen in Figure 17.7. It differs neurologically from skeletal muscle in being innervated by the autonomic nervous system.

Unlike skeletal muscle, few abnormalities are known for smooth muscle. The principal and most common disease—arteriosclerosis—will be described when we discuss problems of the circulatory system (see page 553).

Description and Location of Smooth Muscle

Smooth muscle cells are elliptical or oblong in shape. They are about 75 micrometers long and about 5 micrometers in diameter at the thickest point.

Smooth muscle cells may occur singly, in bundles, or in sheets. They occur *singly* adjacent to gland cells in the sweat glands, salivary glands, mammary glands, and other glands where, upon contraction, they facilitate the excretion of glandular secretions. They occur as *bundles* in the skin, where, upon contraction, they serve to raise a hair or a goose bump. And they occur in *sheets* in blood vessels and in the respiratory, urinary, digestive, and reproductive tracts.

In blood vessels the sheets of smooth muscle are arranged circularly, somewhat spirally. That is, the individual muscle fibers run somewhat obliquely along the circumference of the blood vessels. When these fibers shorten, they reduce the lumen of the blood vessel, causing the pressure of the blood to rise, but at the same time reducing the total volume of flow through the vessel.

In the digestive tract—which runs from the mouth to anus—one or two sheets of smooth muscle (depending on location) are arranged circularly and one sheet is arranged longitudinally (Figure 17.7). The circular sheet is arranged somewhat obliquely, as in blood vessels. In the longitudinal sheet, the smooth muscle fibers run along the long axis of the tract. When the circularly-arranged muscle contracts, the lumen is made smaller. When the longitudinal muscle contracts, the tract shortens along its long axis. Through the combination of these two movements the intestinal contents

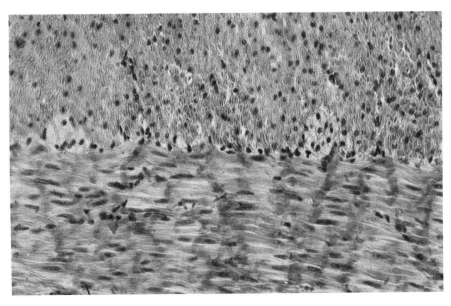

FIG. 17.7. *Smooth muscle. The muscle cells appear circular in cross section (top) and elliptical in longitudinal section (below); cell nuclei are visible as dark spots.*

are moved from the mouth to anus. This movement (peristalsis) is described further on pages 381 and 395.

In the respiratory, urinary, reproductive, and digestive tracts, where smooth muscle occurs as sheets, it is present as layers of circular and longitudinal muscle. In these tracts, as in the digestive tract, the smooth muscle serves primarily to propel the contents of the tract out of the body. Thus, mucous secretions are continuously being conveyed from the bronchi (air passages) of the lungs through the trachea (windpipe) to the mouth. From the mouth, the secretions are passed into the stomach during swallowing. So too, urine is continuously being transported from the kidneys into the bladder for storage by way of a muscularized tube (the ureter). Periodically, the bladder is emptied, whereupon the more distal tube of the excretory tract (the urethra) facilitates the ejection of this waste product from the body. And in the reproductive tract, smooth muscle facilitates the movements of the egg through the fallopian tube, of the sperm through the vas deferens, and of the baby from the uterus during childbirth.

The Mechanism of Smooth Muscle Contraction

When a smooth muscle cell is excited, action potentials of long duration — usually more than 20 milliseconds and upwards to five or ten seconds — pass along the cell membrane. In the wake of these potentials, the cell contracts slowly and thereafter relaxes again.

Very little is known about the basic mechanism of smooth muscle contraction. It may be presumed that actin and myosin, the principal contractile proteins, slide past one another, but there is no direct evidence of this action, as there is in skeletal muscle.

Sodium, potassium, calcium, and magnesium are needed in proper balance, but direct nervous stimulation is not as important as it is in skeletal muscle. Some cells are directly innervated whereas others are not. When an innervated cell is stimulated to contract, the action potentials that travel across its membrane serve also to excite neighboring smooth muscle cells, which may or may not be innervated. This excitatory mechanism, in which bioelectricity is transmitted directly from one cell to an adjacent cell without a chemical neurotransmitter serving as a mediator, is known as an *ephapse*, or electrical synapse.

The dependence of smooth muscle upon nerves differs from that of skeletal muscle fibers in two other important ways. First, whereas skeletal muscle fibers will not contract if their nerve supply is cut off, certain smooth muscles may do so, as described in the next section. Second, whereas skeletal muscle degenerates (atrophies) and becomes paralyzed when its nerves are destroyed, smooth muscle does not.

Classification of Smooth Muscle

Many smooth muscles may be classified into two groups, depending on how dependent the muscle is upon external nerves.

The first group includes muscles which do not function properly if their nerve supply is destroyed. These muscles are present in the eyeball (the circular and radial muscles of the pupil and the ciliary muscles used in conjunction with the lens). They are present also in the blood vessels throughout the body, the bronchi of the lungs, and the penis, clitoris, and accessory sex glands.

The second group of smooth muscles operates relatively independently of motor innervation. In general, most of these muscles are present in organs that may have to be distended over a prolonged period. They are present in the stomach, urinary bladder, and rectum, for example, which may be stretched considerably for several hours when food, urine, and feces, respectively, are present. They are present also in the uterus, which increases about tenfold in size during pregnancy.

Scattered diffusely throughout this second group of smooth muscle are *pacemaker cells*. When the muscle is stretched—as happens when food passes through the intestine or when the urinary bladder is being filled—tension develops to different extents in various regions of the tissue. The pacemaker cells respond to this tension by generating action potentials. These potentials pass through the muscle from cell to cell along paths of low resistance, due to the close proximity of adjacent cell membranes. In the wake of these action potentials, certain smooth muscle cells contract isotonically. They shorten visibly and serve to move food in the intestine, urine in the urinary tracts, semen in the vas deferens, and other materials in other organs. Other muscle cells respond with an isometric contraction. In this case, the tension diminishes and the cells relax after a very brief period, but they are now longer than they were before being stretched. This elasticity is so well developed in the smooth muscle of certain organs, such as the urinary bladder, stomach, and uterus, that it is possible to enlarge the organ to several times its original size.

Regulation of Smooth Muscle Activity by Nerves and Hormones

Unlike skeletal muscle fibers, smooth muscle fibers often lack direct innervation. Within bundles and sheets of smooth muscle, some of the cells may be innervated by sympathetic fibers, others by parasympathetic fibers, still others by both kinds of fibers. And the majority receive no innervation at all. The cells which lack innervation are usually tightly joined with other cells, one of which is directly innervated.

When the innervated cell is stimulated by a nerve cell, the effect of the stimulus—either inhibitory (hyperpolarizing) or excitatory (depolarizing)—is conveyed into the noninnervated cells along tight cell junctions, which are low in electrical resistance. In turn, the muscle relaxes or contracts, depending upon the combination of sympathetic and parasympathetic discharge.

The general overall effects of sympathetic and parasympathetic discharge into various smooth muscles throughout the body were given in Table 8.1. These effects are discussed again, separately, in conjunction with the various physiological systems in the following chapters.

Smooth muscle is subject to hormonal regulation as well as to neural control. The principal regulatory hormones include adrenaline, oxytocin, and angiotensin. These hormones and their actions are discussed elsewhere in the text.

CARDIAC MUSCLE

Cardiac muscle is a striated muscle found only in the heart (cardium). In the following text we describe cardiac muscle, its mechanism of action, and how it differs uniquely from skeletal muscle. Numerous pathologies are known to affect heart muscle; these will be discussed in Chapter 27, which covers disorders of the circulatory system.

Differences Between Skeletal and Cardiac Muscle

The cardiac muscle cell resembles the skeletal muscle cell in having actin and myosin myofibrils and in having a T-system and an L-system for rapid

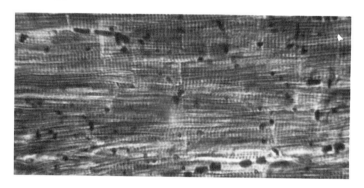
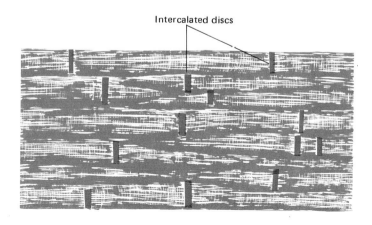

FIG. 17.8. Cardiac muscle, photomicrograph and drawing. Drawing shows the extensive interdigitation of the cells and the intercalated discs between cells.

propagation of action potentials from the plasma membrane to all of the myofibrils. But apart from these similarities, the two types of muscle cells and tissues differ fundamentally from one another.

To begin with, skeletal muscle cells have many nuclei whereas the cardiac muscle cell, like the smooth muscle cell, has a single nucleus. Second, cardiac muscle differs histologically from skeletal muscle in two significant ways. The cells of the heart are separated end-to-end from one another by *intercalated discs*. These are not separate structures, as the word "intercalated" might seem to suggest. Rather, the intercalated discs (see Figure 17.8) are pleats or folds of membrane, formed from the connecting ends of adjacent cells. These pleats interdigitate much the same way as one may interdigitate the left and right hands by locking the fingers of the left hand with those of the right hand. The pleats increase the surface area of the cell, and the interdigitations maximize the opportunity for transferring bioelectrical activity from one cell to another. Moreover, as may be seen in Figure 17.8, the discs do not extend in a straight line across the width of the fiber. Instead, they appear scattered throughout the tissue. This arrangement permits bioelectrical activity to move along pathways in such a manner as to allow the odd-shaped heart to contract, as a whole, in a three dimensional rather than a linear fashion. This behavioral feature of cardiac muscle is facilitated by another anatomical peculiarity of cardiac muscle; that is, heart muscle is an extensively branched, composite tissue. This also may be noted in Figure 17.8. Where branch points occur, there is virtually no delay in the transfer of bioelectrical activity from side-to-side in the muscle. Thus, the intercalated discs permit rapid contraction of the cardiac cells from end to end, and the branch points allow the excitation and contraction to proceed rapidly in a diagonal direction. In this way the heart muscle—which is a structure that is best described as being spiral in form—may contract in a coordinated manner to squeeze blood out of the heart's chambers in an orderly sequence.

A third difference between cardiac and skeletal muscle concerns the contractile trigger; that is, the stimulus that causes the individual cells to contract. Whereas skeletal muscle requires the stimulus of a nerve fiber, cardiac muscle can operate independently of the nervous system.

What do we mean by "independently"? To begin with, cardiac muscle contains five kinds of cells, the locations of which are shown in Figure 17.9. Atrial cells are located in the wall of the two upper chambers of the heart. Ventricular cells are located in the wall of the two lower chambers of the heart. Sinoatrial cells lie in the sinoatrial node, a piece of modified cardiac muscle in the wall of the upper anterior corner of the heart. Atrioventricular cells are located in the atrioventricular node. This knot of modified cardiac muscle cells lies in the wall at the junction of the upper and lower right side of the heart. Purkinje fibers stem from the atrioventricular node and pass downward to the tip (apex) of the heart and back upward again to the junction of the upper and lower chambers.

Each type of cell has a certain degree of automaticity. That is, these cells generate action potentials and they contract in the absence of external stimuli. Moreover, these automatic events occur rhythmically. This automatic rhythmic beat may be shown by placing samples of the cell types in

INTERCALATED DISCS. *Pleated and intertwined membranous endings of two adjacent heart muscle cells.*

ATRIAL CELLS. *Cells in the musculature of the two upper chambers of the heart (L. atrium, an entrance room).*

VENTRICULAR CELLS. *Cells in the musculature of the two lower chambers of the heart (L. ventriculum, a cavity or pouch).*

a dish of saline (salt) solution. The inherent rhythms of four of these different types of cells—examined at body temperature (37°C or 98.6°F)—are given in Figure 17.9.

In the same figure, note that each of the various types of cardiac muscle cells gives rise to action potentials which are similar in magnitude (about 100 millivolts) to those of nerve fibers, skeletal muscle fibers, and smooth muscle fibers, but are different in shape.

It should be emphasized that the action potentials shown in Figure 17.9 were recorded on an oscilloscope from individual heart cells. To obtain such data it is necessary to place one electrode on the muscle fiber and to keep a second electrode on the outside of the fiber. On page 499 it will be seen that

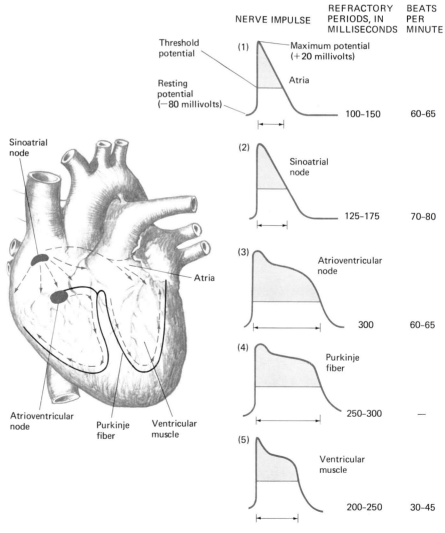

FIG. 17.9. Action potentials of the several types of heart muscle fibers. The action potentials move through the heart as shown by the arrows.

a very different kind of signal is obtained through electrocardiographic measurements from the heart as a whole.

The refractory period of the action potential for any given cardiac fiber is very long. The refractory period is the interval during which it is impossible to elicit another action potential from the fiber. It is a period during which the membrane of the fiber must reset itself if it is to be able to fire off another action potential.

The significance of a long refractory period is twofold. When the cardiac muscle cells are operating normally, the long refractory period ensures a definite period of absolute relaxation, during which the cell cannot be made to contract. Since the action potentials in heart cells progress in an orderly fashion from a point of origination (the SA node) to a point of termination (the cardiac muscle cells in the ventricles) during each heart cycle, the heart is assured a rhythmic beat in its overall pumping action. This is described in greater detail in Chapter 25.

Moreover, because of the long refractory period, it is not possible for heart muscle to incur an *oxygen debt*. Therefore the heart is able to work continuously. Because it is not possible to stimulate the cardiac muscle excessively into contractile states, it is impossible to drive the muscle biochemically to such an extent that the muscle cells begin to accumulate lactic acid. It will be seen on page 417 that skeletal muscle cells accumulate lactic acid when they are overworked. The accumulation of lactic acid signifies a relative deficiency of oxygen in the biochemical machinery of the cell. Without an adequate level of oxygen, it is not possible to convert lactic acid into carbon dioxide and water. Instead, the lactic acid lingers. The accumulation of lactic acid forces the skeletal muscle fiber to reduce its contractile activity or to stop working altogether, momentarily, until the oxygen store of the cell is replenished.

For skeletal muscle cells this is a desirable feature, inasmuch as it prevents excessive overwork and possible destruction of the cell. In the case of cardiac cells, an oxygen debt would be prohibitive inasmuch as it is necessary for the heart to supply a continuous stream of blood to the vital organs of the body. A stoppage of circulation, even momentarily, can result in fainting, stroke, or possibly death.

A fourth major difference between cardiac and skeletal muscle has to do with neural regulation. Like smooth muscle, cardiac muscle is innervated by nerve fibers from the autonomic nervous system. Although, as described above, the cardiac muscle cells can be operated in the absence of external nerves, they are nevertheless subject to regulation by the autonomic nervous system.

In general, when the sympathetic nerves discharge, adrenaline and noradrenaline are liberated into the cardiac muscle. These neurotransmitters decrease the refractory period of cardiac muscle cells. In doing so, they increase the frequency of the heartbeat. They also increase the force of cardiac contraction.

In contrast, when the parasympathetic fibers (the vagus fibers of the tenth cranial nerve) discharge, acetylcholine is liberated into the cardiac musculature in the upper chambers (atria) of the heart. There are scarcely any vagus fibers entering the lower chambers (ventricles) of the heart and there is, therefore, hardly any effect on the ventricles. The effect of acetylcholine in the atria is to increase the refractory period and to decrease the force of atrial contraction. In doing so, acetylcholine decreases the overall rate and force of cardiac contraction.

A fifth notable difference between skeletal and cardiac muscle concerns the ability of the cell to function in the absence of an external nerve supply. Whereas cardiac muscle cells do not cease to contract when their external nerve supply is cut off, skeletal muscle degenerates (atrophies) and becomes paralyzed. No one knows why, but in the absence of input from the central nervous system, skeletal muscle fails to perform its function.

ATROPHY. *A reduction in size* (G. atrophia, *want of food*).

RESPONSIVENESS OF MUSCLE TO PHYSICAL TRAINING

When skeletal and heart muscle fibers are not used, they become small and weak. When these muscle fibers are properly exercised, they increase in size and strength. The increase in size comes about through hypertrophy, not hyperplasia. That is, the individual muscle cells increase in size only, not in number.

The increase in heart muscle we are talking about is not the same as the cardiac hypertrophy, which will be discussed in Chapter 27 in connection with heart disease. Cardiac enlargement in connection with exercise may or may not be perceptible through X-ray procedures. However, enough evidence is on hand to show that exercise does increase the size of cardiac muscle as well as the size of skeletal muscle.

Classification of Exercise

Muscle exercise may be defined as any kind of activity through which one or more groups of muscles are strengthened. Regardless of how simple or complex the activity may be, muscle exercises may be grouped into four basic categories: isometric, isotonic, anaerobic, and aerobic.

In an *isometric exercise*, skeletal muscles are contracted from several seconds to about a minute at a time without producing a movement at the adjacent joint. Cupping the hands together and trying to pull one hand apart from the other is an example of an isometric exercise. Although this type of exercise is capable of increasing the size and strength of the few muscles that are involved, it has no significant effect on health because it does not demand an appreciable effort from the respiratory and circulatory system. Very little additional oxygen is consumed, the thorax is not expanded vigorously and the heart is not exercised. Isometric exercises are useful mainly to bedridden patients.

Isotonic exercises produce movements around skeletal joints, but, like isometric exercises, they do not place a demand upon smooth and cardiac muscle by way of increasing respiration and circulation. Isotonic exercises include calisthenics and moderate sports activities which are carried out for a short period such as croquet, miniature golf, and bowling.

Anaerobic exercises are muscular activities that are carried out so exhaustively that one is forced to rest after a very short period. A typical anaerobic exercise might be to run, cycle, or swim a short distance at top speed, when you are not otherwise in good physical condition. This type of exercise is called "anaerobic" because it causes an accumulation of lactic acid in the muscle cells as well as in the intercellular spaces of skeletal muscle tissue. As a result, an oxygen debt is said to arise. This means that the heart and overworked muscles, working together, were not efficient enough to bring about the total oxidation of the metabolites used by the skeletal muscle as its source of energy. The heart was unable to supply oxygen (by way of blood) rapidly enough, and the skeletal muscle fibers were not able to completely degrade the glucose from which they derive ATP for muscular contractions. While the oxygen debt persists, following the exercise, the muscles ache because the accumulated lactic acid and the overworked muscle fibers stimulate pain receptors. In time, after the lactic acid is removed and the muscle fibers recuperate, a healthy normal muscle tone is restored.

ANAEROBIC. *Oxygen poor.*

Aerobic exercises involve muscular activities that are carried out over an extensive period of time during which a demand is made upon the heart to circulate much more blood than is circulated at rest. Also, the thorax is called upon to oxygenate much more hemoglobin than when the person is at rest. A long vigorous walk, jogging or running, bicycling, and swimming are examples of aerobic exercises, as are certain vigorous sports, such as basketball and tennis, which include walking or running movements. When aerobic exercises are practiced five to seven days a week, training benefits are obtained that are not possible through isometric, isotonic, or anaerobic exercise alone. Physical endurance, improved breath control, an increase in the efficiency of heart action, greater muscular strength, a higher regularity in eliminating wastes, an ability to sleep better and awaken more energetically, and a reduction of psychological tension are the cardinal benefits of a program of aerobic exercise. Physical endurance, in particular, seems to improve sharply. (Physical endurance may be defined as the ability to perform an exercise which requires a heart rate of 150 beats per minute for about 15 to 20 minutes; the normal resting, adult heart rate is 70 beats per minute.)

AEROBIC. *Oxygen rich.*

Although it is possible to recognize four broad categories of exercise, most physical training programs actually include combinations of these categories. During a basketball game, for example, there are moments in which purely isometric or isotonic movements are executed, and there is a point of endurance which is often trespassed, leading to anaerobic exercises. For optimal development of muscle size and strength, however, as well as for general health, it is important to engage maximally in aerobic forms of exercise.

SUMMARY

1. Skeletal muscle cells are multinucleated fibers enclosed in a plasma membrane (sarcolemma), which is negatively charged on its interior surface. These fibers contract when action potentials pass along the sarcolemma and then deep into the cell along a transverse tubular membrane and then along a longitudinal tubular membrane that encloses a muscle fibril. In the course of contraction, filamentous molecules of actin and myosin slide past one another. This process occurs at the expense of ATP, which is produced mainly in mitochondria (sarcosomes).
2. Skeletal muscles may contract visibly (isotonically); or isometrically. In isometric contraction actin and myosin fibrils slide past one another but there is no visible external shortening of the muscle.
3. Muscles usually enter a state of tetanus when they contract. The tetanus may be rapid (phasic) as in running, or prolonged (tonic) as in standing.
4. Fasciculations (involuntary phasic contractions), cramps (involuntary spasmic contractions), tetany (e.g., involuntary contractions in response to deficient parathyroid hormone or calcium), and various abnormal muscle weaknesses are described. An abnormal muscle weakness may be caused by a disease in the spinal cord (polio), a disease at a neuromuscular junction (myasthenia gravis), a metabolic problem in the muscle membrane (potassium imbalance), a problem involving coupling of bioelectric excitation to mechanical contraction, or a disease of the muscle fibrils (muscular dystrophy).
5. Smooth muscle cells are elliptical, with a single nucleus. They may contract spontaneously or contract after being stimulated by an autonomic nerve fiber.
6. Cardiac muscle is made of singly nucleated extensively branched cells whose narrow ends may form intercalated discs with a branch of an adjacent cell. Cardiac muscle cells are able to contract in the complete absence of nerve cells, though they are normally regulated by excitatory and inhibitory stimuli from the autonomic nervous system. At the end of contraction, a relatively long period of time, known as the refractory period, must pass before the cell can be stimulated into another contraction phase. Because of this long refractory period, the heart cannot normally overwork nor can it accumulate lactic acid, which might prevent the heart from working adequately to satisfy the body's need for oxygen.

REVIEW QUESTIONS

1. Describe the structure of a skeletal muscle cell.
2. What biological events occur when a skeletal muscle fiber contracts?
3. How does an isotonic contraction differ from an isometric contraction?

4. Give an example of a phasic tetanus and a tonic tetanus.
5. Define: fasciculation, cramp, tetany, spastic paralysis, flaccid paralysis, myasthenia gravis, muscular dystrophy.
6. Describe the structure of a smooth muscle cell.
7. Where are smooth muscle cells located? How is smooth muscle activated into a contractile state?
8. In what two principal ways does cardiac muscle differ anatomically from skeletal muscle?
9. In what two ways is the relationship of nerve to muscle different between skeletal and cardiac muscle?
10. What is the adaptive significance of a long refractory period?

18 The Functional Anatomy of Skeletal Muscle

STRUCTURE OF SKELETAL MUSCLE
    Attachments of muscle
    Origin and insertion of muscle
FUNCTIONS OF SKELETAL MUSCLES
    Muscles of the head
    Muscles of the neck
    Muscles that move the scapulas
    Muscles that move the shoulders and arms
    Operation of the elbow
    Operation of the forearms and wrists
    Operation of the hands, fingers, and thumbs
    Muscles that operate the trunk
        Flexion, extension, and rotation of the trunk compression of the abdomen
        Muscles of respiration
        Muscles that support the abdominal-pelvic visceral
        Special muscles of defecation, urination, and erection
    Muscles that operate the hips and thighs
    Muscles that operate the knees
    Muscles that operate the ankles
    Muscles that operate the feet and toes

In Chapter 15 we examined the skeleton, paying special attention to the functions played by various architectural features of certain bones. Subsequently—in the chapter on joints—it was pointed out that certain joints are movable, and we learned that the directionality of movement is limited by the structure (uniaxial, biaxial, or triaxial) of the joint and by the arrangement of the ligaments, tendons, and cartilaginous membranes around the joint. In this chapter we will discuss the muscles which move the joints. We will also say something about muscles which move structures other than bones—structures such as the eyelids, the eyes, and the skin of the face.

  We will first discuss the gross structure of a muscle, how the muscle is

STRUCTURE OF SKELETAL MUSCLE

Each fiber of skeletal muscle is enclosed in a connective tissue sheath called the *endomysium*. Within this endomysium are blood and nerve fibers which nourish and stimulate the muscle cell.

The endomysium of one muscle fiber is connected to the endomysium of several neighboring muscle fibers, giving rise to an assemblage, or group, of fibers known as a *fascicle* (cable). Each fascicle is enclosed in a connective tissue sheath called a *perimysium*. This sheath carries larger blood vessels and nerve fibers from which the smaller nerve fibers and capillaries of the endomysium are derived.

FASCICLE. *A bundle of nerve, muscle, or tendon fibers* (L. fasciculus, *cable*).

Fascicles, in turn, are grouped into a *muscle*. Each muscle is separated from other muscles and tissues by another sheath of connective tissue—an *epimysium*—which carries nerves and blood vessels to the muscle as a whole. It is always this whole muscle, a group of fascicles, that is assigned a particular name (for example, sartorius, biceps, or gastrocnemius).

Muscles are separated from one another by deep fascia—a tough fibrous connective tissue that anchors muscle to bone or cartilage. It is continuous with the superficial fascia which connects skin to underlying structures.

SUPERFICIAL. *At or near the surface.*

Attachments of Muscle

Skeletal muscles are attached directly or indirectly to bone at one end, and to bone or another tissue at the other end. The direct attachment is made by way of fascia or perimysium to bone. The indirect attachment is made by way of a tendon or aponeurosis to bone. A *tendon* is a dense round, or flat cord of collagenous inelastic fibers. An *aponeurosis* is a broad, flat sheet of inelastic collagenous fibers—in essence an expanded tendon.

Origin and Insertion of Muscle

In order to perform work, a muscle must be attached to at least two points: the origin(s) and insertion(s). The *origin* is a relatively fixed point of attachment. That is, the object to which a muscle is attached at its origin usually does not move when the muscle contracts. The origin of most muscles is bone; the muscle is attached to the bone either directly by way of fascia or indirectly by way of tendon. The origin of other muscles is by way of fascia to another muscle. The *insertion* is an attachment to a movable part of the body, such as another bone, an eyelid, an eye, the tongue, the corners of the mouth, or the lips.

Structure of Skeletal Muscle

In general, when a muscle contracts, its insertion is brought nearer to its origin. There are exceptions, however. The principal exceptions are those muscles that are arranged circularly, as for example, the muscles which surround the orbits of the eyes, the mouth, and the anal opening. The circularly arranged muscles around the orbits and mouth give rise to facial expressions. The muscle around the anal opening regulates elimination. Each of these muscles contracts in a circular sense rather than having the insertion brought nearer to the origin.

Another exception to the generalization that a muscle's insertion is brought nearer to its origin upon contracting has to do with the *use* of a particular muscle. This exception can be described best with an example: When the biceps brachii is flexed, its insertion in the forearm is brought closer to its origin in the shoulder. In certain actions, however, such as climbing a rope, the forearm is fixed, and the origin (shoulder) is brought closer to the insertion (forearm).

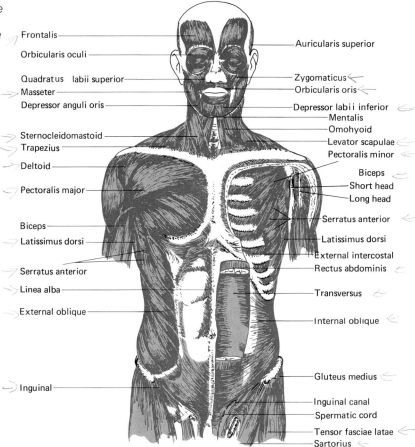

FIG. 18.1. *Muscles of the upper region of the body, anterior view and (opposite page) posterior view.*

335
The Functional Anatomy of Skeletal Muscle

FUNCTIONS OF SKELETAL MUSCLES

There are about 650 skeletal muscles in the body. The exact number depends on whether or not you count a certain fascicle as a single muscle or more than one muscle. This chapter will cover about 150 of these muscles. We will actually cover about 330 muscles in all, since many of the muscles covered occur either as a pair, or as a set of three, four, or more muscles. If we were to do so by simply listing and describing the muscles, you would rapidly lose interest. However, functional anatomy can be a very exciting topic, especially since there is nothing about it which you cannot relate personally to yourself.

In particular, at least 70 of the muscles covered in this chapter can be located externally—either by palpation (feeling) or by sight. These muscles are shown in Figures 18.1, 18.2, and 18.3. In these figures you may note

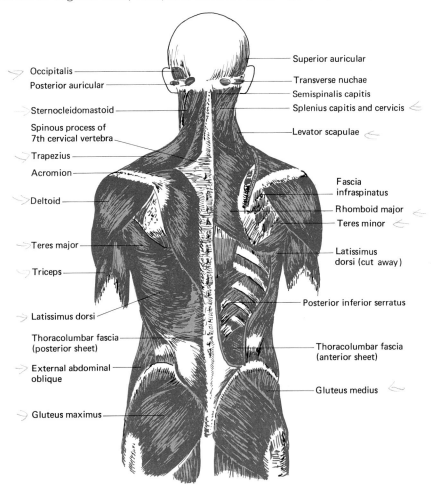

Functions of Skeletal Muscles

that many other muscles cannot be felt since they are located beneath the superficial muscles.

To further encourage you to approach a study of muscle with enthusiasm, we will use a technique in which the prime movers of a given joint are figuratively dissected out in most of the remaining figures of this chapter. Through this technique—coupled with a regional-functional approach—it should be possible to obtain a fundamental understanding of the basic mechanics of human motions.

It should help you as you begin to learn the names of the major muscles if you remember that each muscle is named on the basis of one or more of

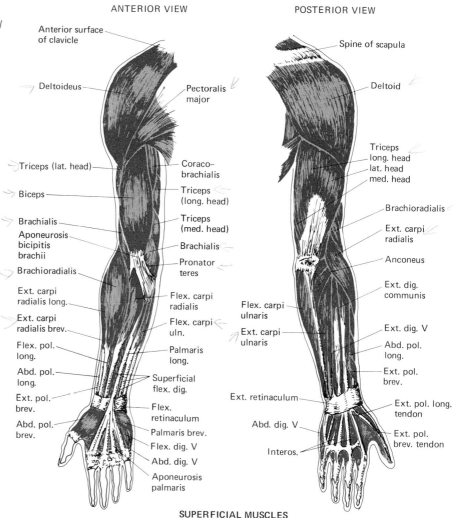

FIG. 18.2. *The superficial muscles and (opposite page) the deep muscles of the right extremity. The digiti and pollicis muscles move the fingers and thumb, respectively.*

The Functional Anatomy of Skeletal Muscle

the following: (1) shape (for example, trapezius, deltoid, and rhomboid); (2) position (palmaris); (3) size (gluteus maximus); (4) number of parts comprising the origin or head (biceps and triceps); (5) location of origin and insertion; or (6) action (for example, the erector spinae).

For further clarity, there is a diagram (or description) within many of the figures to illustrate the action(s) of a particular muscle or group of muscles. Finally, there is an alphabetical list, in Table 18.2 at the end of this chapter, of all of the muscles described in this text. Given in this table is the origin, insertion, and primary action of each of the body's major muscles.

Our procedure for covering the various muscles will be to work from the

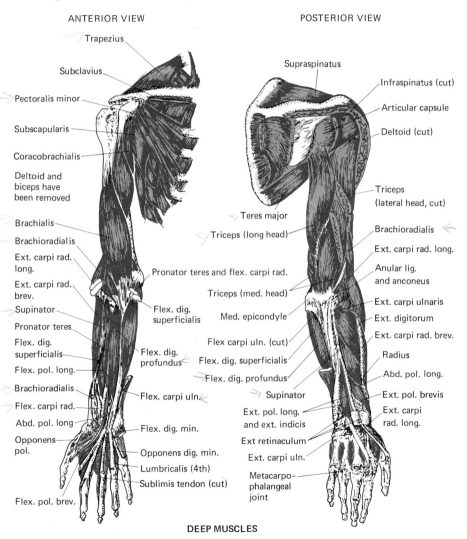

DEEP MUSCLES

338
Functions of Skeletal Muscles

face to the toes, in the order suggested by Table 18.1, which briefly lists the fundamental movements that are possible around a particular joint or in a given body region. (The parenthetical instructions in Table 18.1 indicate how to execute the particular action.)

In reading this chapter and examining the figures, it is important for the reader to realize that not all possible movements are covered, nor is any muscular action fully described from the point of view of bringing all of the participating muscles into the picture. In fact, even specialists in this field are not in full agreement regarding the exact role of certain muscles in cer-

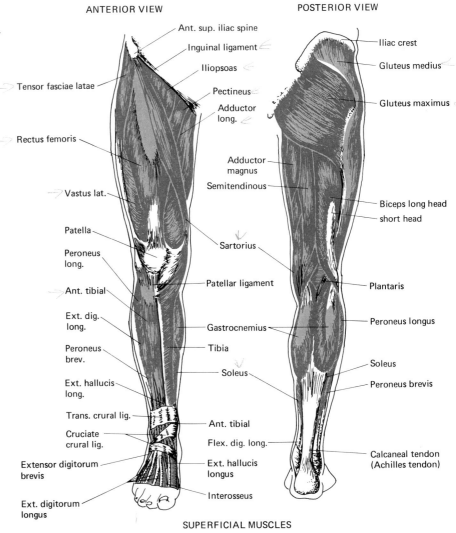

FIG. 18.3. Muscles of the lower appendages; superficial layer and (opposite page) deep layer.

The Functional Anatomy of Skeletal Muscle

tain movements. The student who wishes to know much more about this topic may refer to textbooks on kinesiology, the study of body movements.

KINESIOLOGY. *The science of muscle movements* (G. kinema, *movement;* logos, *study of*).

Muscles of the Head

The muscles of the head are used for facial expressions; mastication; tongue movements; operating the pharynx; and moving the soft palate. *The muscles of facial expressions* (Figures 18.4 and 18.5) are small

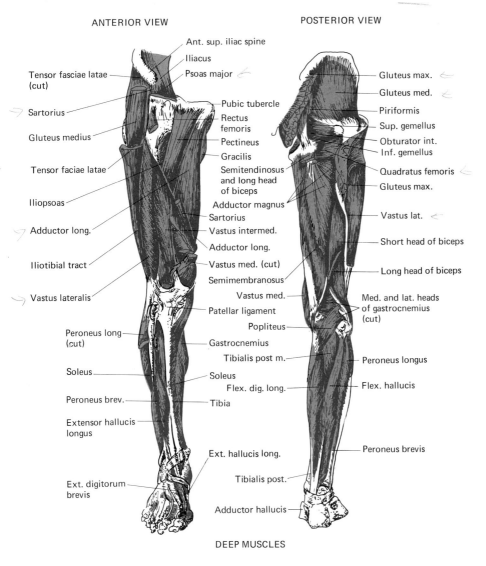

Functions of Skeletal Muscles

TABLE 18.1. Fundamental Movements of the Major Body Regions and Joints

Region	Fundamental Actions
Head	Facial expressions
	Movements of the eyeballs and upper eyelids
	Mastication (chewing)
	Tongue movements
	Swallowing
Neck	Flexion (move your head toward your chest)
	Extension (move your head backward)
	Rotation (turn your head to the left or right)
Scapulas	Adduction (move your scapulas toward one another); abduction (move scapulas away from each other)
	Elevation (shrug your shoulders up); depression (lower the shoulders)
	Adduction and downward rotation (move scapulas toward one another and downward)
	Abduction and upward rotation (move scapulas away from each other and upward)
Shoulders and Arms	Flexion (swing an arm up, in front of the body, along the side of your body—in a sagittal plane)
	Extension (swing an arm up, behind you, along the side of your body—in a sagittal plane)
	Vertical abduction (swing an arm up from the side of your body—in a coronal plane—to a horizontal level with the shoulders); vertical adduction (reverse the preceding action)
	Horizontal abduction (swing an arm horizontally from the side of the body to the front of the body in a transverse plane); horizontal adduction (reverse the preceding action)
	Medial rotation (rotate an arm medially)
	Lateral rotation (rotate an arm laterally)
Elbows and Forearms	Flexion (bring a hand to its shoulder)
	Extension (reverse the preceding movement)
	Supination (turn a palm face up)
	Pronation (turn a palm face down)
Wrists and Hands	Flexion (bend a palm toward its wrist)
	Extension (bend a palm away from its wrist)
Fingers	Flexion (bend the fingers at their various joints into the palm)
	Extension (bend the fingers away from the palm)
	Adduction (draw your fingers together)
	Abduction (spread your fingers apart)

Table 18.1 (continued)

Region	Fundamental Actions
Thumbs	Flexion (fold a thumb)
	Extension (open a thumb)
	Abduction (spread a thumb away from its index finger, but keep the digits in a horizontal plane throughout)
	Adduction (draw a thumb toward its index finger, keeping all of the digits in a coronal plane throughout)
	Opposition (draw a thumb toward any of the fingers in the same hand)
Trunk	Flexion (raise your back while lying down in the supine position—on your back)
	Extension (raise your back while lying down in the prone position—on your stomach)
	Lateral flexion and extension (bend the trunk laterally and return it medially)
	Rotation (twist your trunk to one side or the other)
	Compression of the abdomen
	Respiratory movements
	Special visceral functions
Hips and Thighs	Flexion (raise a thigh forward)
	Extension (send a thigh backward)
	Abduction (raise a thigh laterally)
	Adduction (bring a thigh from a lateral to a medial position)
	Lateral rotation (turn a thigh outward—away from the midsagittal plane—around a longitudinal axis)
	Medial rotation (reverse the preceding movement)
Knees and Legs	Flexion (bring a leg up behind you)
	Extension (raise a leg after first crossing it over the opposite knee while sitting down)
Ankles	Flexion (extend a foot)
Feet	Inversion (turn a foot inward—medially)
	Inversion and dorsiflexion (turn your foot in and up)
	Eversion (turn a foot laterally)
Toes	Flexion (turn the toes upward)
	Extension (turn the toes downward)
	Abduction (spread the toes apart laterally)
	Adduction (squeeze the toes together medially)

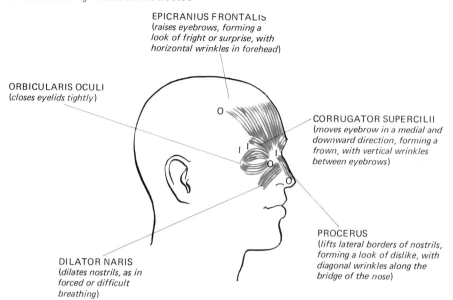

FIG. 18.4. Upper facial muscles.

muscles located mainly around the mouth, eyes, nose, and forehead. Some of them have their origin in bone. Others have their origin in an aponeurosis. Many of them have their insertion in soft tissues, such as the skin of the eyebrows, the corner or angle of the mouth, or the lips. A group of muscles

FIG. 18.5. Lower facial muscles.

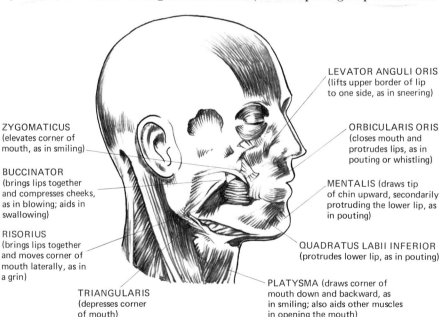

around the mouth can change a smile to a sneer. The ring of muscles around the eye can produce a wink or a squint. A sheet of muscle in the forehead allows us to raise our eyebrows.

The muscles of the upper eyelids and eyes permit us to draw the upper eyelid over the eyeball, and to turn the eyeball in various directions as indicated in Figure 18.6.

The muscles of mastication (Figure 18.7) include the powerful temporalis, which runs from the cranium to the lower jaw (mandible), and the masseter, which runs from the cheekbone to the lower jaw.

Tongue movements are possible because of a group of four muscles within the tongue. Collectively, they can make the tongue high in the center and lower along the edges, or vice versa. They can raise or lower the tongue and send it into or out of the mouth. With the aid of cheek muscles they can form a chute along which food is conveyed into the esophagus.

The muscles of swallowing — shown in Figure 18.8 — are located in the cranium, beneath and around the tongue, and in the wall of the throat. The muscles of the cranium, in conjunction with the soft palate, prevent food from re-entering the mouth — and from entering the nose — once swallowing is initiated. The intrinsic muscles of the tongue — and the external muscles associated with the tongue — serve to pass a bolus of food into the throat (pharynx). And the muscles of the wall of the pharynx — upon contracting — send the bolus on its way to the esophagus.

Additional muscles involved in swallowing — known as hyoid or strap muscles — are present in the interior of the throat (Figure 18.9). Some of these muscles serve to elevate the larynx and pharynx, preventing food from entering the lungs by way of the larynx. Others move the floor of the mouth, or help open the mouth, or help coordinate the movement of other companion hyoid muscles.

MASTICATE. *To chew* (L. masticare).

ESOPHAGUS. *The gullet* (G. oisein, *to carry;* phagein, *to eat*).

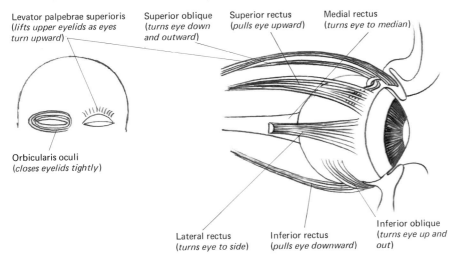

FIG. 18.6. *External eye muscles and muscles of the eyelids.*

Functions of Skeletal Muscles

Muscles of the Neck

The muscles of the neck move the head and the first two ribs.

Movements of the head are shown in Figure 18.10. Flexion, extension, and rotation are the primary movements of the neck that result in movements of the head. When the paired sternocleidomastoids contract simultaneously and independently of other muscles, the head is flexed. When only the posterior fascicles of the sternocleidomastoids contract with other muscles of the neck, the head is extended. When only one of the sternocleidomastoids contracts, the head is turned to the opposite side. In this situation the working muscle can be seen through the skin as a cord running from the back of the head to the breastbone (sternum) and collar bones (clavicles).

Entirely different muscles are involved in extending the neck. As may be seen in Figure 18.11, most of them originate in the occipital bone or cervical vertebrae and are inserted in thoracic vertebrae.

Elevation of the first two ribs (Figure 18.12) is carried out by muscles located in the neck beneath the sternocleidomastoids. Known as scalene muscles, they are especially important in bringing about an enlargement of the chest (thorax) during the inspiratory phase of breathing (inhalation). In addition, they assist the sternocleidomastoids in flexing, extending, and rotating the neck and head.

Muscles that Move the Scapulas

The scapulas, together with the clavicles (collar bones), constitute the shoulder girdle. Because of the joint between the clavicles and sternum (breastbone), each half of the shoulder girdle is able to move anteriorly and posteriorly around a longitudinal axis; cranially and caudally around a sagittal axis; and rotationally around a coronal axis. These movements are due mainly to three large muscles—the trapezius, rhomboideus, and serratus anterior—as shown in Figures 18.13 to 18.15.

Muscles that Move the Shoulders and Arms

The *shoulders* are defined as the two regions of the body where a scapula joins with a clavicle and humerus and is covered by the deltoid muscle. The *arm*, basically, is the humerus and its fleshy coverings.

The shoulders can be moved in many different ways, most of which are shown in Figures 18.16 to 18.20. In spite of this versatility, only about a dozen muscles are involved as prime movers, with several other muscles as minor assistants. The separate actions are carried out by single muscles or combinations of these muscles (for example, extension by the teres major and rotation by the latissimus dorsi and subscapularis, Figure 18.17.) Also,

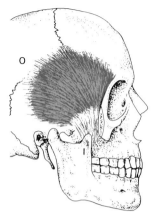

Temporalis (*closes jaw; elevates mandible*)

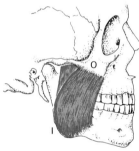

Masseter (*closes jaw*)

Pterygoideus lateralis (*assists in opening mouth; protracts mandible*)

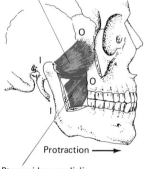

Protraction ⟶

Pterygoideus medialis (*closes and protracts jaw*)

FIG. 18.7. *Muscles of mastication.*

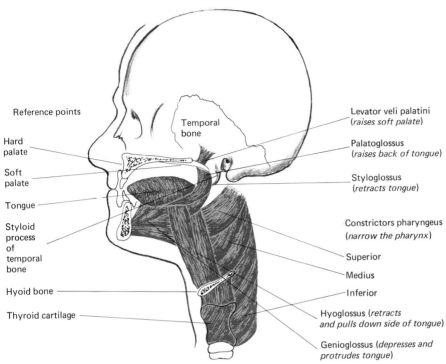

FIG. 18.8. *Muscles of swallowing.*

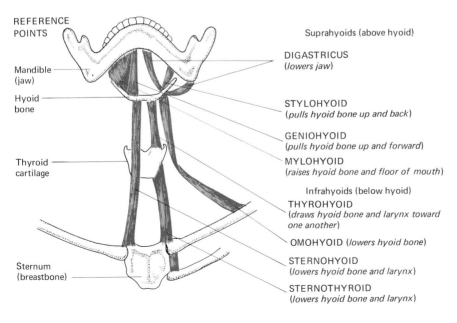

FIG. 18.9. *The strap muscles.*

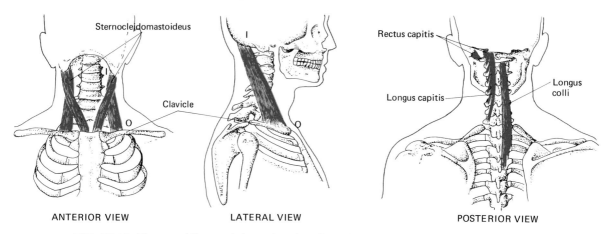

FIG. 18.10. Flexors of the neck (*muscles that draw the head downward toward the chest*).

a single muscle may carry out more than one function, for example, pectoralis major adducts the arm, flexes the shoulder, and rotates the arm inward—toward the body.

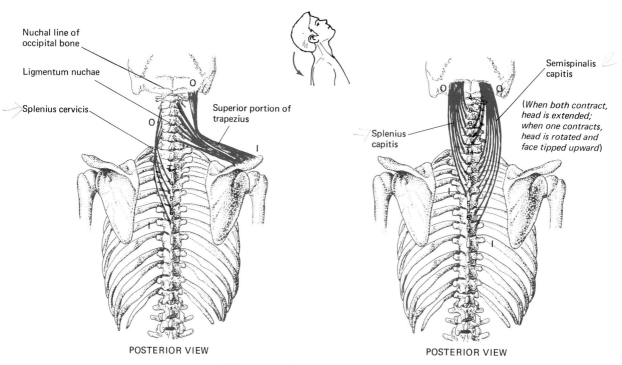

FIG. 18.11. Extensors of the neck.

Operation of the Elbow

As may be seen in Figure 18.21, only five muscles are involved at the elbow joint. They bring about flexion or extension of the forearm—the only movements permitted by the elbow.

Operation of the Forearms and Wrists

Pronation (turning the palm upward) and supination (turning the back of the hand upward) are the principal movements of the forearm, as may be

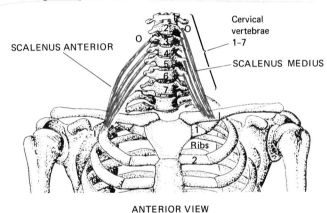

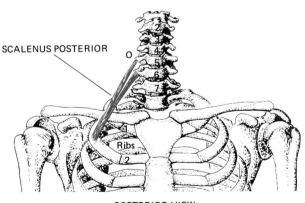

FIG. 18.12. Scalene muscles; together, the scalenes raise the first and second ribs during inspiration.

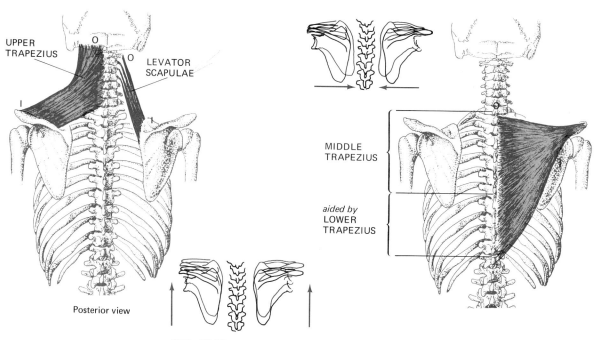

FIG. 18.13. *Scapular adduction and elevation.*

seen in Figure 18.22. Flexion and extension are executed at the elbow; pronation and supination at the forearm.

Operation of the Hands, Fingers, and Thumbs

The following are the principal movements of the hand: opening (extending) or closing (flexing) the hand or one or more of its fingers; spreading

FIG. 18.14. *Scapular adduction and downward rotation.*

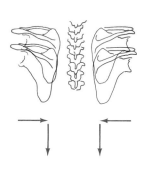
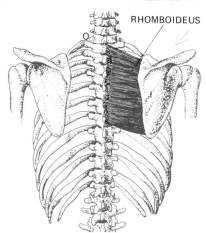

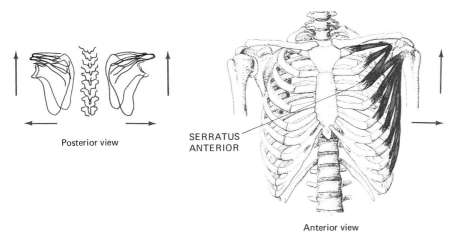

FIG. 18.15. *Scapular abduction and upward rotation.*

apart (abducting) or bringing together (adducting) two or more fingers; and opposing the thumb to one of the other fingers.

Extension and flexion are accomplished through a series of muscles that originate in the forearm (Figures 18.23 to 18.26). These muscles insert into tendons which, in turn, tunnel through sheaths that pass from the wrist, through the hands, into the fingers. The activities of these muscles

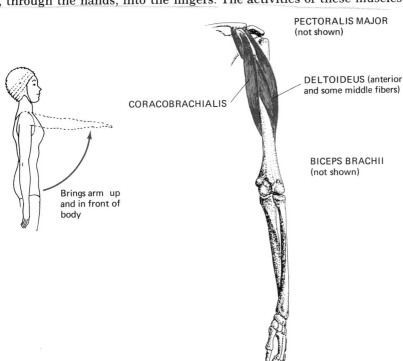

FIG. 18.16. *Muscles used in shoulder flexion; front view of right shoulder.*

FIG. 18.17. Muscles used in shoulder extension; lateral view of right shoulder.

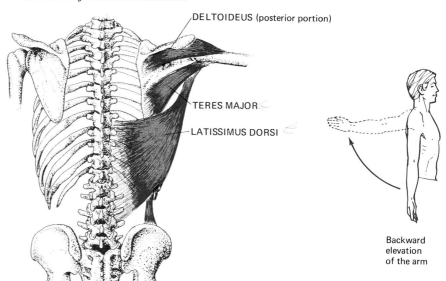

may be felt by placing the left hand on the right forearm and moving the right fingers. The sheaths provide protection to the tendons and keep them close to the bones they move.

The other movements of the hand—abduction, adduction, and opposition of the little fingers and thumb—are illustrated in Figures 18.27, 18.28, and 18.29.

FIG. 18.18. Muscles used in vertical abduction of shoulder.

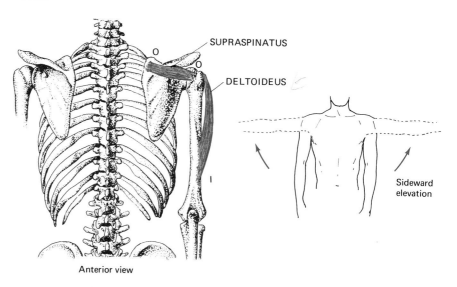

Muscles that Operate the Trunk

Under this heading are included: (1) muscles involved in the flexion, extension, and rotation of the trunk; (2) the muscle that compresses the abdomen; (3) muscles involved in respiration; (4) muscles that support the abdominal-pelvic viscera; and (5) muscles involved in defecation, urination, and erection of the penis.

Flexion, Extension, and Rotation of the Trunk. These movements are carried out by moving the spinal column of the trunk. Flexion (forward bending) of the spinal column is due mainly to the rectus abdominis muscle, which runs from the fifth pair of ribs to the symphysis pubis at the base of the pelvis (Figure 18.30).

Extension (backward bending) of the spinal column is achieved by contracting — bilaterally — the erector spinae and quadratus lumborum (Figure 18.31). The erector spinae is a long mass of muscle extending from the

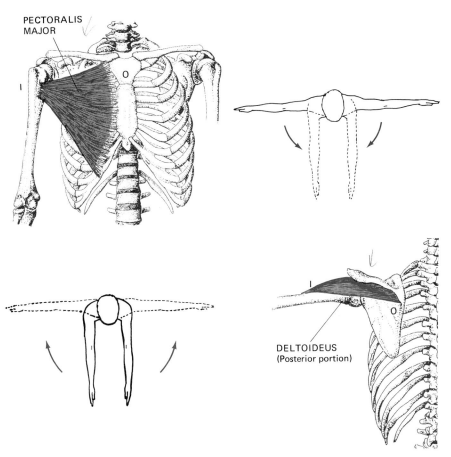

FIG. 18.19. *Muscles used (above) in horizontal flexion and (below) in horizontal extension of shoulder.*

UNILATERAL. *One-sided.*

skull to the tailbone (coccyx) along the spinal column. The quadratus lumborum forms part of the back wall of the abdomen (Figure 18.31). Unilateral contraction of the erector spinae and quadratus lumborum causes the trunk to bend toward the side along which the contraction occurs.

Rotation of the trunk is due mainly to two sets of obliquely oriented muscles, as shown in Figure 18.32.

Compression of the Abdomen. Are you interested in having a better appreciation of that portion of your body which so easily goes to fat when you are physically inactive? A re-examination of Figure 18.30 will show that only the rectus abdominis is present as musculature along the median face of the belly. The other abdominal muscles discussed so far—the oblique muscles—lie laterally to the rectus abdominis. To complete the picture take note in

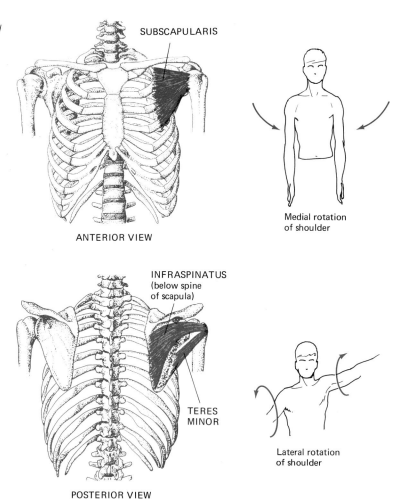

FIG. 18.20. *Muscles used in lateral and medial rotation of shoulder; anterior and top views.*

Figure 18.30 of the transversus abdominis. Now proceed to Figure 18.33, which shows this muscle more fully. Its main function is to compress the abdomen, as happens, for example, in eliminating waste products from the bowels or bladder, in vomiting, and in childbirth.

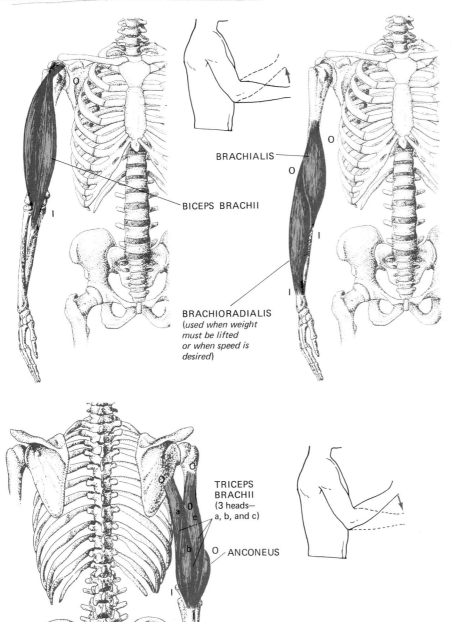

FIG. 18.21. *Elbow flexion and extension. Above: two front views of right arm; below: rear view of right arm.*

Functions of Skeletal Muscles

Muscles of Respiration. The muscles of respiration are located as shown in Figures 18.12 and 18.34, and in Figure 28.4 of Chapter 28. The muscles of respiration include (1) those which are essential in normal breathing – the scalenes, the diaphragm, and the external intercostals; and (2) those which assist in breathing – including the muscles of the vocal cords, the sternocleidomastoids in the neck, the pectorals and serratus anterior of the chest, and the trapezius. The role of the muscles that are essential for ordinary breathing is discussed in Chapter 28. The assisting muscles are used in forceful breathing, sneezing, sighing, shouting, singing, and similar activities.

The diaphragm is a dome-shaped structure composed of muscle and tendons (Figure 18.34). The tendons are attached to the ribs, spinal column, and lower tip of the breastbone. The upper convex portion of the diaphragm forms the floor of the thoracic cavity; the lower concave portion forms the roof of the abdominal cavity. Three large openings in the diaphragm are passageways for the esophagus, aorta, and the inferior vena cava, together with branches of the vagus nerve. Smaller openings are for the splanchnic nerves and hemiazygous vein.

The external intercostals are located between the ribs – obliquely oriented – where they serve to elevate the ribs. Together with the diaphragm and scalenes, the external intercostals bring about an enlargement of the thoracic cage. In doing so they create a vacuum which allows air to rush into the lungs.

Muscles that Support the Abdominal-Pelvic Viscera. The viscera are supported by muscles that reside primarily in the front wall of the abdomen, the inner surface of the pelvic girdle, the muscles of the pelvic diaphragm, and the muscles of the perineum.

The abdominal support of the abdominal-pelvic viscera stems from the external and internal oblique abdominal muscles, the transverse abdominal muscle, the rectus abdominis, the quadratus lumborum, and the psoas muscles. The latter two muscles form the posterior wall of the abdominal cavity (Figures 18.31 and 18.36), whereas the former muscles lie in the anterior and lateral walls of the abdomen (Figure 18.30).

Of particular interest with respect to the anterior abdominal musculature is the inguinal ligament of the oblique external muscle. This ligament can be felt through the skin. It runs diagonally from the crest of the hip bone to the pubis, along the thigh. In the vicinity of the pubic symphysis it partially supports the inguinal canal.

Ordinarily the inguinal canal carries a spermatic cord or a ligament of the uterus, depending on sex. Straining, which may occur during weight lifting, childbirth, or forceful elimination of fecal wastes, may cause a *hernia* (a break) to appear in the tissues which partly close the opening of the inguinal canal. Many kinds of hernias may arise in the body, but this particular type, an *indirect inguinal hernia,* is quite common. Because of the hernia, peritoneal and visceral tissues, particularly intestinal tissues, may protrude abnormally into the inguinal canal. Should these tissues become strangulated, as may happen in the 1.5-inch inguinal canal, they may fail to properly exchange oxygen, carbon dioxide, nutrients, and wastes,

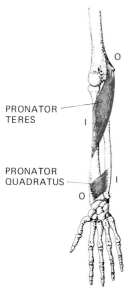

FIG. 18.22. *Muscles used in supination (a) and pronation (b) of the forearm (turning the palm up or down).*

The Functional Anatomy of Skeletal Muscle

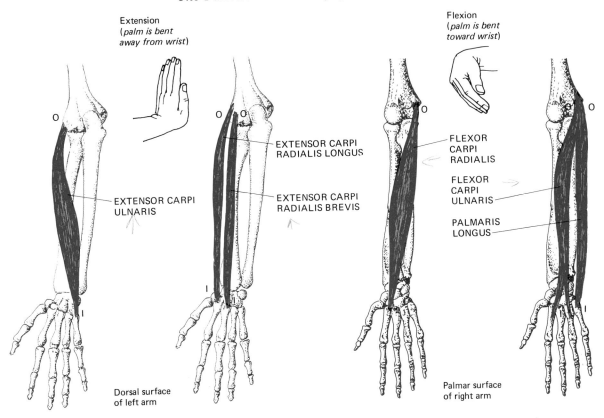

FIG. 18.23. Wrist flexion and extension. The palmaris longus is absent in many people.

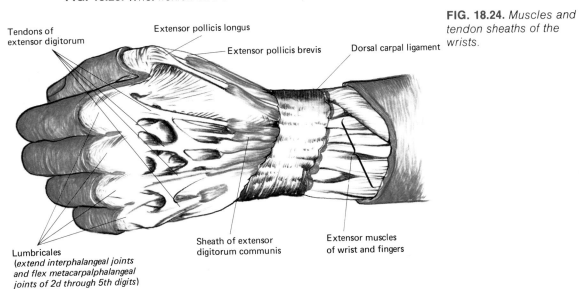

FIG. 18.24. Muscles and tendon sheaths of the wrists.

INGUINAL. *The groin* (L. *inguinalis*).

and the individual may consequently need an emergency operation. (In a *direct inguinal hernia* the visceral tissues pass directly through the abdominal wall medial to the inguinal canal, along the edge of the rectus abdominis into the body wall or scrotum.)

The pelvic viscera are given muscular support in the inner region of the pelvic girdle by the iliacus and the obturator internus (Figures 18.36 and 18.39). These muscles form the walls of the pelvic cavity. The iliacus flexes and rotates the thigh medially, whereas the internal obturator extends and rotates the thigh laterally.

Special Muscles of Defecation, Urination, and Erection. Muscular support of the pelvic viscera in the region of the pelvic diaphragm is provided by the levator ani (Figure 18.35) and the coccygeus (not illustrated). The levator ani plays a prominent role in raising the rectum during defecation.

Muscles of the perineum form the most inferior musculature of the pelvis (Figure 18.35). The perineum is the area between the thighs, from the

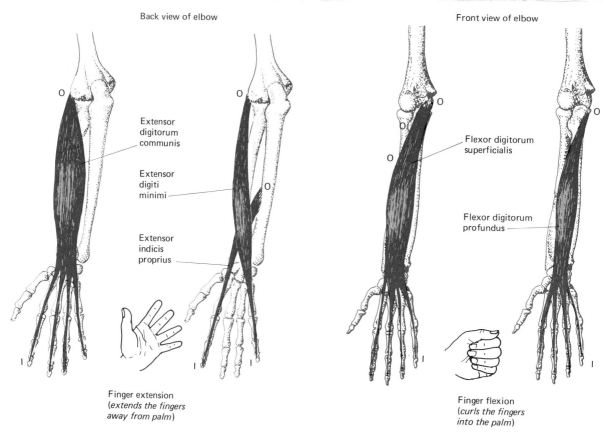

FIG. 18.25. *Finger extension and flexion.*

The Functional Anatomy of Skeletal Muscle

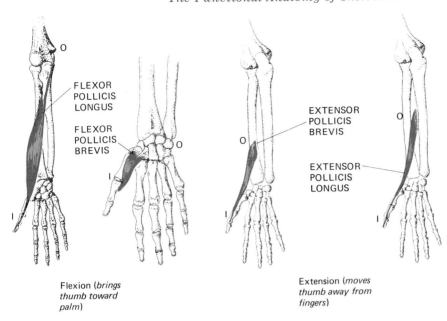

Flexion (brings thumb toward palm)

Extension (moves thumb away from fingers)

FIG. 18.26. *Muscles used in flexion and extension of the thumb: right hand, from front (left) and back (right).*

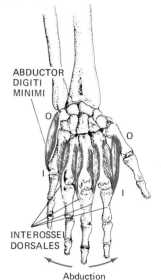

Abduction

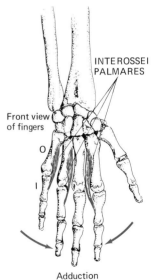

Adduction

FIG. 18.27. *Muscles of finger adduction and abduction.*

coccyx to the pubis, below the pelvic diaphragm. Its principal muscles are concerned with defecation, urination, and erection, as follows. Voluntary closure of the anal canal is accomplished through the external anal sphincter (Figure 18.35). Voluntary control of urination is due mainly to the external urethral sphincter, which constricts the initial length of the urinary duct (urethra) from the urinary bladder. This function is aided in the male by the bulbocavernosus muscle. The bulbocavernosus in the male also compresses the erectile tissue of the penis in bringing about an erection. In the female, the bulbocavernosus closes the vaginal opening and assists in erecting the clitoris. In both the male and female the ischiocavernosus muscle is also involved in the erection process.

Muscles that Operate the Hips and Thighs

By definition, a hip is that part of the body which surrounds and includes the joint formed by each thigh bone (femur) and the pelvis. The term "hip" is also used to refer to the fleshy part of the upper thigh. The musculature of the hip is easier to understand through illustrations than through verbal descriptions. See Figures 18.36 to 18.40.

Functions of Skeletal Muscles

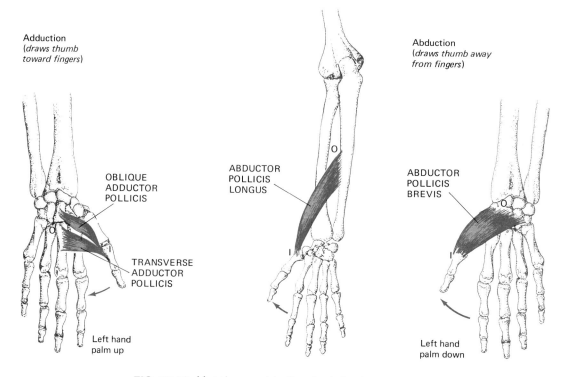

FIG. 18.28. *Muscles used in thumb abduction and adduction.*

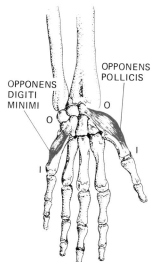

FIG. 18.29. *Muscles used in opposing the thumb and little finger.*

Muscles that Operate the Knees

Flexion and extension are the only movements normally allowed by the knee joint. As may be seen in Figure 18.41, a wide variety of thigh muscles are used in these movements.

The muscles which extend the knee—the rectus femoris and three vasti muscles—are known collectively as the quadriceps. They are located in front of the thigh bone, where they form one of the largest muscles of the body. As may be seen in Figure 18.41, the quadriceps has four separate origins, but only a single insertion—a common tendon which passes over the knee into the larger leg bone (tibia). The quadriceps is needed for maintaining balance when standing and it is an important locomotory muscle.

The principal muscles responsible for knee flexion—the biceps femoris, semitendinosus, and semimembranosus—are referred to collectively as the "hamstrings." They are located in the rear of the thigh, where they are used to direct the leg backward, as in walking or swimming, or to aid in hip move-

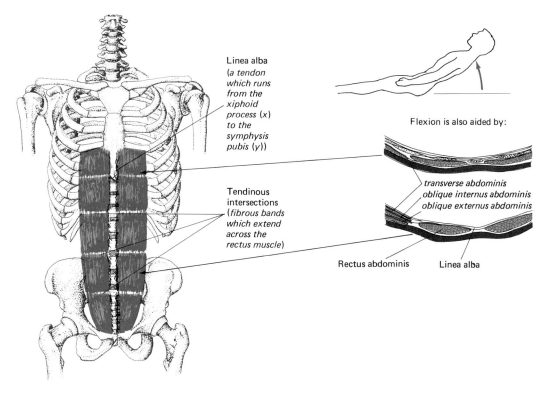

FIG. 18.30. Trunk flexion; the iliopsoas muscle (not shown) also functions in this movement.

ments involving adduction, rotation, and extension. Their flexor activities are continuously coordinated with the extensor activities of the quadriceps, permitting them to act as a brake for what would otherwise be jerky knee extensor movements.

One more muscle should be mentioned in conjunction with knee movements—the sartorius. Known as the "tailor" muscle because it allows one to sit with ankles crossed and knees widespread as tailors did in years long past, the sartorius passes diagonally across the front and median surfaces of the thigh. It flexes the leg at the knee, and the thigh at the hip. In addition, it rotates the thigh.

Muscles that Operate the Ankles

The ankle is the joint in which the tibia and fibula of the leg articulate with the talus of the foot. Flexion, dorsiflexion, inversion, and eversion, are the only movements normally allowed to this joint. As may be seen in Figure

18.42, two very large muscles (gastrocnemius and soleus), along with two lesser (peroneal) muscles, are involved in flexion. Fewer and different muscles are articulated in dorsal flexion of the ankle (Figure 18.43), in which the toes are brought toward the shin. Inversion and eversion are shown in Figure 18.44.

Muscles that Operate the Feet and Toes

This topic is best understood by referring to Figure 18.44 and Figure 18.45, and to Table 18.2, taking note of the origin, insertion, and action of each muscle illustrated. In examining the Table, bear in mind the similarities between the lumbricales and interossei in the hands and feet. The interossei of the feet are not illustrated, but they occupy a similar position to those of the hands (Figure 18.27).

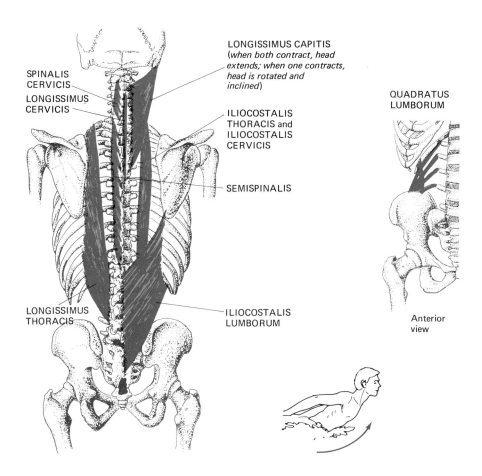

FIG. 18.31. *Muscles of trunk extension.*

The Functional Anatomy of Skeletal Muscle

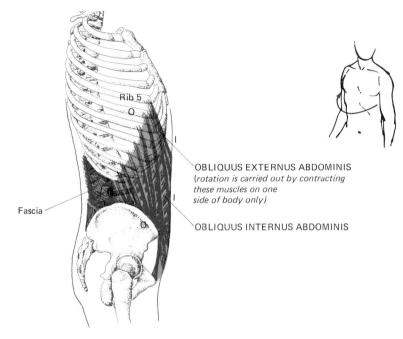

FIG. 18.32. *Muscles of trunk rotation.*

OBLIQUUS EXTERNUS ABDOMINIS
(*rotation is carried out by contracting these muscles on one side of body only*)

OBLIQUUS INTERNUS ABDOMINIS

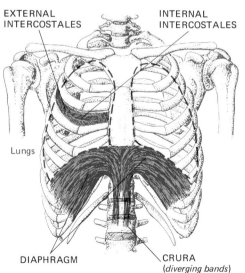

FIG. 18.34. *Thoracic muscles of respiration. There are eleven pairs each of external and internal intercostal muscles (not shown), which function in inhaling and exhaling, respectively.*

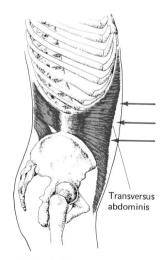

FIG. 18.33. *Muscles used in compression of the abdomen.*

Functions of Skeletal Muscles

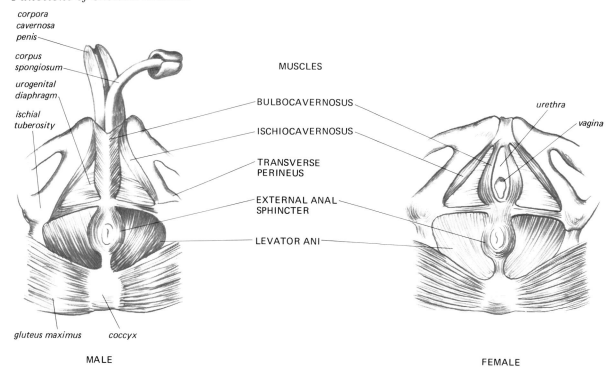

FIG. 18.35. *Muscles that support the abdominal-pelvic viscera.*

FIG. 18.36. *Hip flexion.*

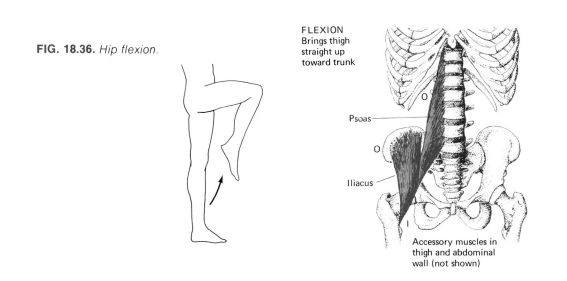

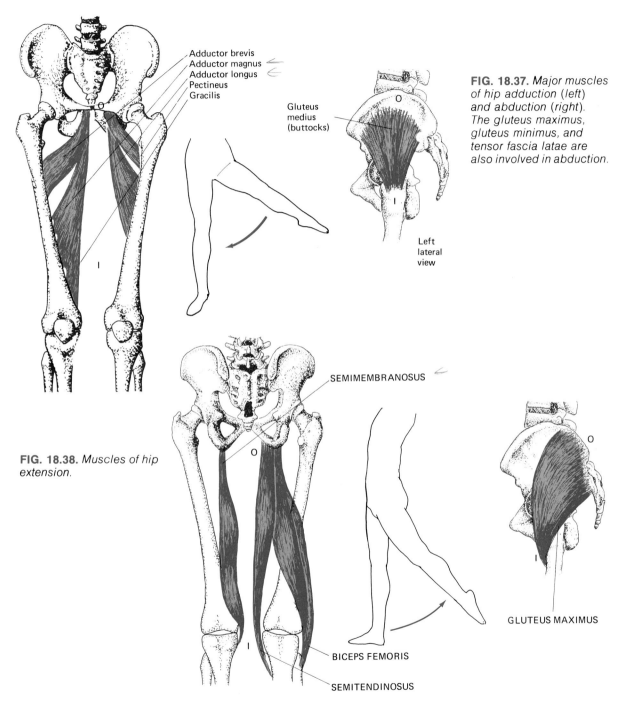

FIG. 18.37. Major muscles of hip adduction (left) and abduction (right). The gluteus maximus, gluteus minimus, and tensor fascia latae are also involved in abduction.

FIG. 18.38. Muscles of hip extension.

Functions of Skeletal Muscles

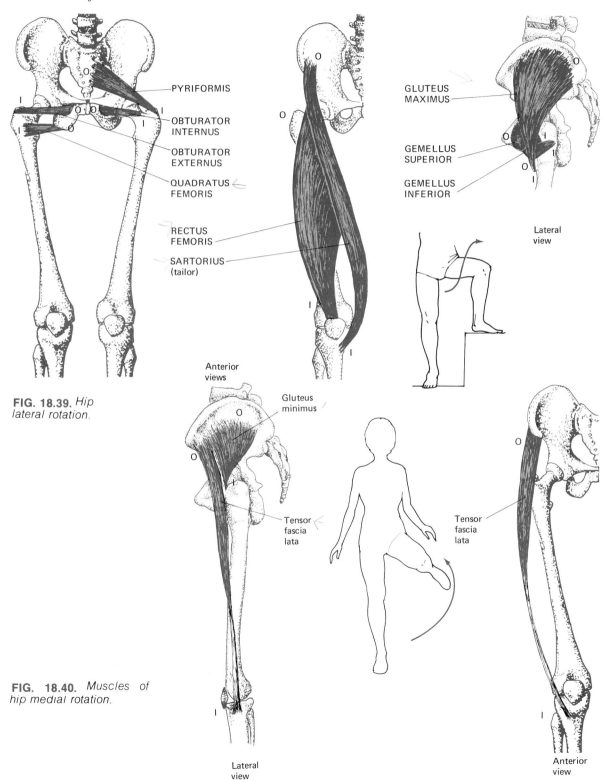

FIG. 18.39. Hip lateral rotation.

FIG. 18.40. Muscles of hip medial rotation.

The Functional Anatomy of Skeletal Muscle

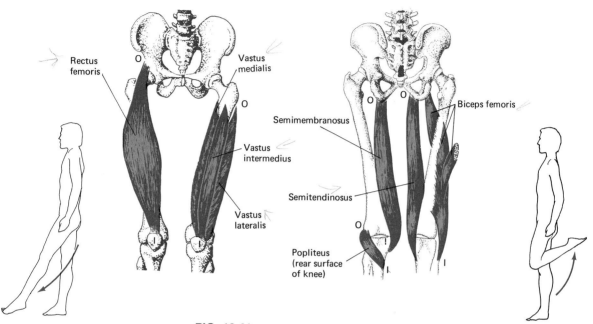

FIG. 18.41. Knee flexion and extension.

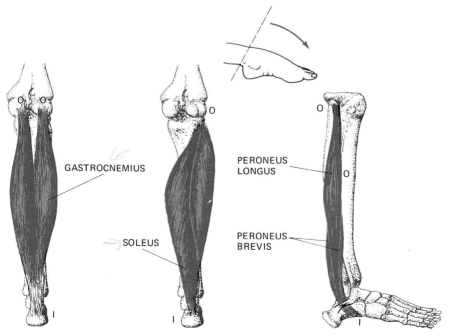

FIG. 18.42. Ankle flexion (plantar flexion).

366
Functions of Skeletal Muscles

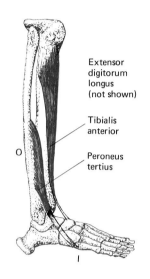

FIG. 18.43. *Ankle dorsiflexion.*

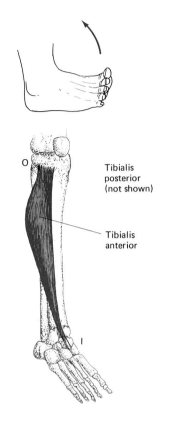

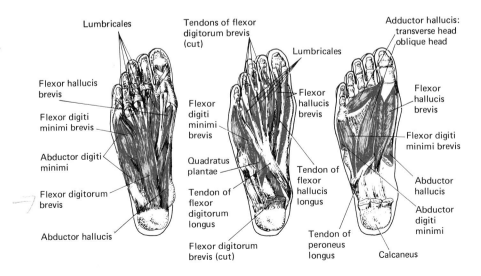

FIG. 18.45. *Muscles of the right foot. From the left: the first, second, and third muscle layers.*

TABLE 18.2. Skeletal Muscles

Name	Location	Origin	Insertion	Principal Action(s)
Abductor digiti minimi manus	Hand	Pisiform	Phalanx 1 of little finger	Abducts little finger
Abductor pollicis brevis	Hand	Trapezium and scaphoid	Phalanx 1	Abducts thumb
Abductor pollicis longus	Wrist	Radius and ulna	Metacarpal 1	Abducts thumb
Adductors brevis and longus	Thigh	Pubis	Linea aspera	Adducts thigh
Adductor magnus	Thigh	Pubis and ischium	Linea aspera	Adducts thigh
Adductor pollicis	Hand	Metacarpals 2 and 3, trapezoid, and capitate	Phalanx 1 of thumb	Adducts thumb
Anconeus	Elbow	Humerus	Ulna	Extends forearm
Biceps brachii	Arm	Scapula	Radius	Flexes and supinates forearm
Biceps femoris	Thigh	Ischium and linea aspera	Fibula	Extends thigh and flexes leg
Brachialis	Arm	Humerus	Ulna	Flexes forearm
Brachioradialis	Forearm	Humerus	Radius	Flexes forearm
Buccinator	Face	Maxillae and mandible	Orbicularis oris	Compresses cheek
Bulbocavernosus	Pelvis	Perineum	Male penis and urethra	Constricts urethra
	Pelvis	Perineum	Female clitoris	Constricts urethra
Coccygeus	Pelvis	Ischium	Sacrum and coccyx	Constricts vagina and erects clitoris Supports pelvic floor
Constrictors pharyngeus	Neck	Various cartilages, ligaments, and hyoid	Pharynx	Narrows pharynx in swallowing
Coracobrachialis	Arm	Scapula	Humerus	Raises arm anteriorly
Corrugator supercilii	Forehead	Orbicularis oculi	Skin above orbit	Frowning
Deltoideus	Shoulder	Scapula and clavicle	Humerus	Abducts, flexes, extends and rotates arm laterally and medially
Diaphragm	Junction of thorax and abdomen	Ribs 7–12, lumbar vertebrae and xiphoid	Crura	Enlarges thorax
Digastricus	Neck	Temporal and mandible	Hyoid	Moves hyoid up and forward and opens mouth
Dilator naris	Nose	Maxilla	Wall of nostril	Dilates nostrils
Epicranius frontalis	Forehead	Muscles above orbit	Skin of eyebrow and root of nose	Raises eyebrows

Functions of Skeletal Muscles

TABLE 18.2 (*continued*)

Name	Location	Origin	Insertion	Principal Action(s)
Extensor carpi radialis brevis, longus, and ulnaris	Forearm	Humerus	Metacarpal 3, 2, and 5, respectively	Extend and abduct hand
Extensor digiti minimi	Forearm	Humerus	Phalanx 1 of small finger	Extends small finger
Extensor digitorum brevis	Foot	Calcaneus	4 tendons of extensor longus and phalanx 1 of large toe	Extends toes
Extensor digitorum communis	Forearm	Humerus	Phalanges 2 and 3 of fingers	Extends fingers
Extensor digitorum longus	Leg	Tibia and fibula	Phalanges 2 and 3 of toes 2–5	Extends toes 2–5; dorsiflexes ankle
Extensor hallucis longus	Leg	Fibula	Large toe	Extends large toe
Extensor indicis	Forearm	Ulna	Phalanx 1 of index finger	Extends index finger
Extensor pollicis brevis	Forearm	Radius	Phalanx 1 of thumb	Extends thumb
Extensor pollicis longus	Forearm	Ulna	Phalanx 2 of thumb	Extends thumb
Flexor carpi radialis	Forearm	Humerus	Metacarpal 2	Flexes and abducts hand
Flexor carpi ulnaris	Forearm	Humerus and ulna	Metacarpal 5 and pisiform	Flexes and adducts hand
Flexor digitorum brevis	Sole	Calcaneus and plantar aponeurosis	Phalanges of toes 2–5	Flexes toes 2–5
Flexor digitorum longus	Leg	Tibia	Phalanges 2–5	Flexes toes 2–5
Flexor digitorum profundus	Forearm	Ulna	Distal phalanges	Flexes fingers
Flexor digitorum superficialis	Forearm	Humerus, ulna, and radius	Phalange 2 of fingers	Flexes fingers
Flexor hallucis brevis	Sole	All 3 cuneiforms and tibialis posterior	Proximal phalanx of large toe	Flexes large toe
Flexor hallucis longus	Leg	Fibula	Phalanges of large toe	Flexes large toe
Flexor pollicis brevis	Thumb	Distal carpals and transverse carpal ligament	Phalanx 1 of thumb	Flexes and adducts thumb
Flexor pollicis longus	Thumb	Radius and ulna	Distal phalanx of thumb	Flexes thumb
Gastrocnemius	Leg	Femur	Calcaneus	Extends foot (flexes ankle)
Gemellus	Pelvis	Ischium	Femur	Rotates thigh laterally

TABLE 18.2 (*continued*)

Name	Location	Origin	Insertion	Principal Action(s)
Genioglossus	Neck	Mandible	Lingual fascia	Depresses and protrudes tongue
Geniohyoideus	Neck	Mandible	Hyoid	Moves hyoid up and forward
Gluteus maximus	Buttocks	Ilium	Femur	Extends and laterally rotates thigh
Gluteus medius	Buttocks	Ilium	Femur	Abducts and medially rotates thigh
Gluteus minimus	Buttocks	Ilium	Femur	Abducts and medially rotates thigh
Gracilis	Thigh	Symphysis pubis	Tibia	Adducts thigh
Hyoglossus	Neck	Hyoid	Tongue	Retracts and pulls down side of tongue
Iliacus	Pelvis	Ilium and sacrum	Femur	Flexes and medially rotates thigh
Iliocostalis	Back of body	Ilium and ribs	Ribs and cervical transverse processes	Extends spine
Infraspinatus	Shoulder	Scapula	Humerus	Rotates arm laterally
Intercostales, external	Thorax	Ribs	Ribs	Enlarge thorax
Intercostales, internal	Thorax	Ribs	Ribs	Contract thorax
Interossei dorsales manus	Hand, 4 muscles	Sides of metacarpals	First phalanges fingers 2, 3, 4	1st adducts finger 3; 2d, 3d, 4th muscles abduct fingers 2, 3, 4
Interossei dorsales pedis	Foot, 4 muscles	Sides of metatarsals	First phalanges toes 2, 3, 4	1st adducts toe 2; 2d, 3d, 4th muscles abduct toes 2, 3, 4
Interossei palmares	Hand, 3 muscles	Metacarpals 2, 4, 5	Finger 2, 3, 5	Adduct fingers
Ischiocavernosus	Pelvis	Ischium	Penis or clitoris	Maintains penis or clitoris erect
Latissimus dorsi	Back of body	Ribs 10–12, spinous processes T6–L5 and ilium	Humerus	Extension, adduction, and medial rotation of arm
Levator ani	Pelvis	Pubis and ilium	Coccyx	Draws anus upward in defecation
Levator anguli oris	Face	Maxilla	Orbicularis oris and corners of mouth	Raises angle of mouth
Levator palpebrae	Face	Sphenoid	Eyelid	Raises upper eyelid
Levator scapulae	Neck	Transverse processes C1–C4	Scapula	Elevates scapula
Levator veli palatini	Mouth	Temporal bone and eustachian tube	Soft palate	Raises soft palate

Functions of Skeletal Muscles

TABLE 18.2 (*continued*)

Name	Location	Origin	Insertion	Principal Action(s)
Longissimi	Back of body	Vertebrae	Skull, vertebrae and ribs	Extends spine; bends it laterally
Longus capitis	Neck	Transverse processes C3–C6	Occipital bone	Flexes neck
Longus collis	Neck	Transverse processes and bodies C3–T7	Atlas and cervical bodies	Flexes and rotates neck
Lumbricales manus	Hand, 4 muscles	Tendons of flexor digitorum profundus	Phalanx 1 and extensor tendon of fingers 2–5	Flex 1st phalanges and extend 2d and 3d
Lumbricales pedis	Foot, 4 muscles	Tendons of flexor digitorum longus	Phalanx 1 and tensor tendon of toes 2–5	Flex 1st phalanges and extend 2d and 3d
Masseter	Face	Maxilla and zygomatic arch	Mandible	Elevates mandible
Mentalis	Chin	Mandible	Skin of chin and lower lip	Wrinkles chin and protrudes lip
Mylohyoideus	Mouth	Mandible	Hyoid	Elevates hyoid and floor of mouth
Obliquus externus abdominis	Abdominal wall	Ribs 5–12	Ilium, inguinal ligament and sheath of rectus	Bilateral contraction compresses abdomen; unilateral contraction rotates trunk.
Obliquus inferior	Eyeball	Maxilla	Sclera	Turns eye upward and outward
Obliquus internus abdominis	Abdominal wall	Ilium and lumbar fascia	Ribs 10–12 and sheath of rectus	Bilateral contraction compresses abdomen; unilateral contraction rotates trunk.
Obliquus superior	Eyeball	Optic foramen	Sclera	Turns eye downward and outward
Obturator externus and internus	Pelvis	Obturator foramen	Humerus	Rotates thigh laterally
Omohyoideus	Shoulder	Scapula	Hyoid	Lowers hyoid
Opponens digiti minimi	Hand	Hamate	Metacarpal 5	Draws little finger toward palm
Opponens pollicis	Hand	Trapezium and transverse carpal ligament	Metacarpal 1	Opposes thumb to other fingers
Orbicularis oculi	Around eyes	Orbits	Skin of eyelids	Closes eyelids
Orbicularis oris	Around mouth	Skin of lips and fibers of adjacent muscles	Corner of mouth	Closes lips; produces puckering

TABLE 18.2 (*continued*)

Name	Location	Origin	Insertion	Principal Action(s)
Palatoglossus	Mouth	Soft palate	Tongue	Raises back of tongue and narrows fauces
Palmaris longus	Forearm	Humerus	Palm	Flexes wrist (hand)
Pectineus	Thigh	Pubis	Femur	Adducts and flexes thigh
Pectoralis major	Thorax	Clavicle, sternum, ribs 1–7	Humerus	Flexes, adducts, and medially rotates arm
Pectoralis minor	Thorax	Ribs 3–5	Scapula	Depresses shoulder
Peroneus brevis	Leg	Fibula	Metatarsus 5	Everts and extends foot (dorsiflexes ankle)
Peroneus longus	Leg	Fibula	Metatarsus 1 and cuneiform 1	Everts and extends foot (dorsiflexes ankle)
Peroneus tertius	Ankle	Fibula	Metatarsus 5	Everts foot and dorsiflexes ankle
Piriformis	Pelvis	Sacrum	Humerus	Rotates thigh laterally
Plantaris	Thigh	Linea aspera	Calcaneus	Extends ankle (foot)
Platysma	Neck	Fascia of deltoideus and pectoralis major	Skin of lower face	Draws corner of mouth down and back
Popliteus	Calf	Femur	Tibia	Flexes knee (leg) and medially rotates hip (thigh)
Procerus	Face	Nose	Epicranius frontalis	Assists epicranius frontalis
Pronator quadratus	Forearm	Ulna	Radius	Pronates forearm
Pronator teres	Elbow	Humerus and ulna	Radius	Pronates forearm
Psoas	Pelvis	Lumbar transverse processes	Femur	Flexes hip (thigh)
Pterygoideus lateralis	Face	Sphenoid	Mandible	Protracts mandible
Pterygoideus medialis	Face	Sphenoid and maxilla	Mandible	Raises mandible
Quadratus femoris	Pelvis	Ischium	Femur	Rotates hip (thigh) laterally
Quadratus labii inferior	Face	Mandible	Skin of lower lip	Protrudes lower lip
Quadratus lumborum	Pelvis	Ilium and lower lumbar vertebrae	Upper lumbar vertebrae and rib 12	Flexes lumbar spine
Rectus abdominis	Abdomen	Pubis	Ribs 5–7	Flexes spine
Rectus capitis anterior	Neck	Atlas	Occipital bone	Turns and inclines head anteriorly

Functions of Skeletal Muscles

TABLE 18.2 (continued)

Name	Location	Origin	Insertion	Principal Action(s)
Rectus capitis posterior	Neck	Atlas	Occipital bone	Turns and inclines head posteriorly
Rectus femoris	Thigh	Ilium and acetabulum	Patella	Rotates hip (thigh) laterally
Rectus inferior	Eyeball	Optic foramen	Sclera	Moves eye downward
Rectus lateralis	Eyeball	Optic foramen	Sclera	Turns eye laterally
Rectus medialis	Eyeball	Optic foramen	Sclera	Rotates eye medially
Rectus superior	Eyeball	Optic foramen and sheath of optic nerve	Sclera	Rotates eye upward
Rhomboideus	Back of body	Spinous processes T1-T5	Scapula	Adducts scapula
Risorius	Face	Buccinator	Lateral skin of mouth	Draws skin around mouth laterally
Sartorius	Thigh	Ilium	Tibia	Flexes hip (thigh) and knee (leg); rotates thigh laterally and knee medially
Scaleni	Neck	Cervical transverse processes	Ribs 1 and 2	Elevate ribs and rotate neck
Semimembranosus	Thigh	Ischium	Tibia	Extends hip (thigh) and flexes knee (leg); rotates them medially
Semispinalis	Back of body	Vertebral processes	Occipital bone and vertebral processes	Extends and rotates spine
Semitendinosus	Thigh	Ischium	Tibia	Extends hip (thigh) and flexes knee (leg)
Serratus anterior	Thorax	Ribs 1–9	Scapula	Abducts and elevates scapula
Serratus posterior	Back of body	Thoracic and lumbar spinous processes	Ribs	Increases size of thorax
Soleus	Leg	Tibia and fibula	Calcaneus	Extends ankle (foot)
Sphincter ani externus	Anus	Coccyx	Perineum	Constricts anal opening
Splenius capitis and cervicis	Neck	Spinous processes of lower cervical and upper thoracic vertebrae	Nuchae and cervical transverse processes	Extends and rotates neck
Sternocleidomastoideus	Neck	Sternum and clavicle	Temporal bone	Flexes and rotates neck
Sternohyoideus	Neck	Sternum	Hyoid	Depresses hyoid and larynx

TABLE 18.2 (*continued*)

Name	Location	Origin	Insertion	Principal Action(s)
Sternothyroideus	Neck	Sternum	Thyroid cartilage	Depresses hyoid and larynx
Styloglossus	Mouth	Styloid process of temporal bone	Tongue	Retracts tongue
Stylohyoideus	Neck	Styloid process of temporal bone	Hyoid	Elevates and retracts hyoid
Subclavius	Thorax	Rib 1	Clavicle	Depresses clavicle
Subscapularis	Shoulder	Scapula	Humerus	Rotates arm medially
Supinator	Elbow	Humerus and ulna	Radius	Supinates forearm
Supraspinatus	Shoulder	Scapula	Humerus	Abducts arm vertically
Temporalis	Skull	Temporal bone	Mandible	Elevates mandible
Tensor fasciae latae	Pelvis	Ilium	Fascia lata	Abducts, flexes, and medially rotates hip (thigh)
Teres major	Shoulder	Scapula	Humerus	Adducts, extends, and medially rotates arm
Teres minor	Shoulder	Scapula	Humerus	Rotates arm laterally
Thyrohyoideus	Neck	Thyroid cartilage	Hyoid	Elevates larynx
Tibialis anterior	Leg	Tibia	Cuneiform 1 and metatarsus 1	Dorsiflexes and inverts foot
Tibialis posterior	Leg	Tibia and fibula	All 3 cuneiforms, calcaneus and navicular	Inverts foot
Transversus abdominis	Abdomen	Ribs 7–12 ilium and lumbar fascia	Pubis, linea alba and xiphoid	Compresses abdomen
Transversus perinei	Pelvis	Ischium	Median raphe	Supports perineal region
Trapezius	Back of body	Superior nucha, ligamentum nuchae and spinal processes C7–T12	Clavicle and scapula	Elevates, adducts, and depresses scapula
Triangularis	Face	Mandible	Skin of lower lip	Depresses corner of mouth
Triceps brachii	Arm	Humerus and scapula	Ulna	Extends elbow (forearm)
Vasti	Thigh	Femur and linea aspera	Patella	Extends knee (leg)
Zygomaticus	Face	Zygomatic bone	Corners of mouth	Draws skin around mouth up and back

SUMMARY

1. Skeletal muscle fibers, groups (fascicles) of fibers, and groups of fascicles (a muscle) are enclosed in increasingly thick connective tissue sheaths that carry correspondingly larger blood vessels and nerve fibers.
2. Muscles are attached at their origin to bone, and at their insertion to bone or some other tissue. The origin is a fixed point of attachment, and the insertion is the point of attachment which moves when the muscle contracts. Some muscles have more than one origin and/or more than one insertion.
3. About 150 pairs of muscle are described in terms of their position, origin(s), insertion(s), and main function(s).

REVIEW QUESTIONS

1. Describe the arrangement of a muscle from the level of whole muscle to muscle fiber.
2. Define muscle origin and muscle insertion.
3. Does each of the facial muscles have a distinct origin and insertion?
4. Where are the strap muscles?
5. What muscles are used in breathing?
6. What are the principal muscles of the shoulders?
7. What muscles allow you to nod your head? to turn your hands up? to accomplish a butterfly swim stroke? to jump a hurdle?
8. Name the muscles of the anterior abdominal wall.
9. What muscles make up the hamstrings? the quadriceps?
10. Name the two major leg muscles used in walking.

Digestion, Metabolism, and Nutrition

19 Digestion and Elimination
20 Absorption of Nutrients and Elimination of Cell Wastes
21 Principles of Metabolism
22 The Principal Organs of Metabolism
23 Caloric Requirements
24 Nutritional Requirements

PART FIVE

Digestion and Elimination

COMPONENTS OF THE DIGESTIVE SYSTEM
FUNCTIONAL ANATOMY OF THE DIGESTIVE TRACT
    The mucosa
    The submucosa
    The muscularis externa
    The adventitia
PREPARATORY DIGESTIVE ACTIVITIES IN THE MOUTH
THE PHARYNX, OR THROAT
    Swallowing
        Difficulties in swallowing
PASSAGE OF FOOD FROM THE ESOPHAGUS INTO THE STOMACH
    Two cardiac sphincter problems
    Vomiting
THE STOMACH
    Digestive activities and other activities of the stomach
    Regulating the gastric glands
    Regulating the emptying of the stomach
        Failure of the pyloric sphincter to open
    Ulcers
THE PANCREAS
    The role of the pancreas in digestion
THE LIVER AND THE GALLBLADDER
    Functional anatomy of the liver
    The role of the liver and gallbladder in digestion
    Common problems associated with the liver
    Two common problems of the gallbladder
THE SMALL INTESTINE
    Digestive activities in the small intestine
    Diverticula, diverticulosis, and diverticulitis
THE LARGE INTESTINE
    Functions of the large intestine
REGULATING THE OUTPUT OF THE DIGESTIVE JUICES
ELIMINATION
    Problems in elimination

Components of the Digestive System

DIGEST. To decompose, to hydrolyze (L. digere, to separate).

Digestion is the process of changing food into absorbable form. The digestion of solid food begins in the mouth with the process of chewing. The process continues in the stomach and small intestine, where enzymes degrade the food into its constituent molecules. When the chemical processes of digestion are complete, food has been converted into various kinds of metabolites including simple sugars, fats, amino acids, and the components of nucleic acids (Figure 19.1). These relatively small and simple metabolites can be absorbed by the cells of the body. Minute quantities of small peptides are also absorbed. Following their absorption, the metabolites undergo various metabolic reactions. Through these metabolic reactions, which take place at the cellular level, we satisfy our nutritional needs.

In this chapter we will discuss the digestion of food, the elimination of undigestible foods, and certain common abnormalities of the digestive system. Our discussion will proceed sequentially from the mouth to the anus. In the next chapter, we will discuss how nutrients are absorbed by cells. In the four succeeding chapters we will consider the topics of metabolism and nutrition.

COMPONENTS OF THE DIGESTIVE SYSTEM

The digestive system is composed of a digestive tract and four accessory glands: the salivary glands, the pancreatic glands, the liver, and the gall-

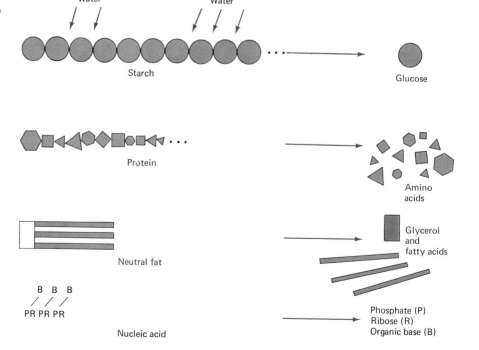

FIG. 19.1. *Digestion (hydrolysis) of food into absorbable nutrients.*

bladder (Figure 19.2). The digestive tract begins in the mouth, which contains the teeth and tongue for chewing and swallowing. The remainder of the digestive tract includes the pharynx, esophagus, stomach, small intestine, large intestine, and anus.

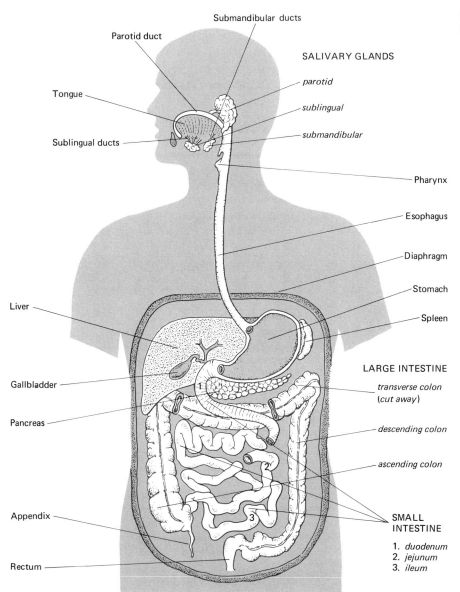

FIG. 19.2. *The digestive system.*

FUNCTIONAL ANATOMY OF THE DIGESTIVE TRACT

The digestive tract, from the esophagus to the rectum, is a tubular structure. When looked at in cross-section, the digestive tract has a four-layered wall. The innermost layer, which surrounds the lumen (opening) of the digestive tract, is called the *mucosa*. Proceeding away from the lumen and toward the body cavity, the remaining layers are the *submucosa*, the *muscularis externa*, and the *adventitia* or *serosa* (Figure 19.3).

The Mucosa

The *mucosa* has four major histological components: epithelium, basement membrane, lamina propria, and muscularis mucosa. The *epithelium* lines the lumen of the entire digestive tract. It secretes a lubricant called *mucus*

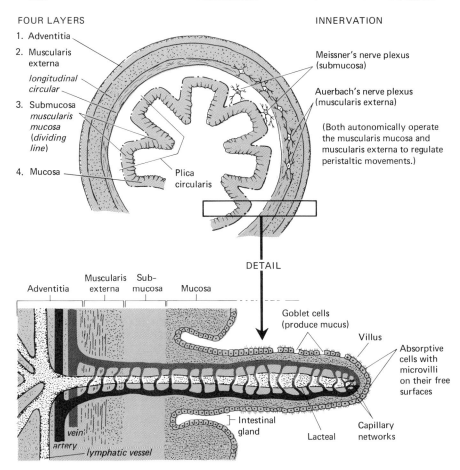

FIG. 19.3. *Representative cross-section of the small intestine.*

and it absorbs the final products of digestion. In the stomach, the epithelium plays an additional role—it secretes digestive enzymes. In the small intestine, the epithelium plays yet another role—it secretes an alkaline juice which neutralizes the acidity of stomach effluents.

The *basement membrane* anchors the epithelium to the lamina propria. It is produced by the epithelium. The *lamina propria* is a loose connective tissue. It harbors blood vessels and nerve fibers, and it contains patches of white blood cells that protect the body from invasion by microbes and other foreign material in the food. In addition, the lamina propria physically supports the muscularis mucosa.

The *muscularis mucosa* is a region of smooth muscle cells. It is prominent in the esophagus, moderate in the stomach and small intestine, and sparse in the large intestine. The major function of the muscularis mucosa is to aid the digestive tract in its action of mixing and transporting the intestinal contents.

MUCUS. *A viscid liquid consisting of water, inorganic salts, epithelial cells, etc. held in suspension by glycoproteins (L.).*

BASEMENT MEMBRANE. *A protein-carbohydrate complex secreted by epithelial cells to anchor them to connective tissue.*

The Submucosa

The *submucosa* lies beneath the mucosa. It is a loose connective tissue. Blood vessels and Meissner's nerve plexus are contained within it. The blood vessels carry away absorbed nutrients from the digestive tract and also furnish nutrients to the digestive tract. The nerve fibers within Meissner's plexus stimulate muscle fibers within the tract and in doing so keep the digestive tract in a continuous state of movement.

PLEXUS. *A network of interlacing blood vessels or nerves (L. a network).*

The Muscularis Externa

The *muscularis externa* is composed of a circular inner band of smooth muscle fibers and a longitudinal outer band of smooth muscle fibers. Auerbach's nerve plexus of the autonomic nervous system lies in a connective tissue between these bands of muscle fibers. Parasympathetic nerve fibers from this plexus release acetylcholine into the wall of the digestive tract, while sympathetic nerve fibers release noradrenaline. Acetylcholine stimulates the contractile activities of the muscularis mucosa and muscularis externa, while noradrenaline exerts an inhibitory effect on these muscles. The main function of the muscularis externa is to produce wavelike movements called *peristalsis*. These movements propel food and wastes from the esophagus to the rectum.

The Adventitia

The *adventitia* or *serosa* is the outermost wall of the digestive tract. It is a connective tissue. Its external surface may be freely exposed to the thoracic or abdominal cavity or it may be attached to some organ within the abdomi-

ADVENTITIA. *The external covering of an organ (L. adventicius, foreign).*

The Pharynx, or Throat

SEROSA. *A serous membrane* (L. serum; watery).

nal cavity. Where it is free, it is called an *adventitia* and where it is attached, it is called a *serosa*.

The serosa may be attached to mesentery (in the case of the intestine), or omentum (in the case of the stomach). The *omentum* is a layer of peritoneum which anchors the stomach to the body wall and to other abdominal organs. The *mesentery* is a layer of peritoneum which anchors the intestines and other abdominal organs to one another and to the body wall. Regardless of whether the adventitia is free or attached, its main function is to pass blood vessels, lymphatic vessels, and nerves into the inner walls of the digestive tract.

Regional differences occur here and there among the four walls of the digestive tract. We shall describe some of these differences as we discuss the separate regions of the tract.

PREPARATORY DIGESTIVE ACTIVITIES IN THE MOUTH

The mouth holds the food while the jaws and teeth masticate and reduce it to particles that can be swallowed. When solid food is eaten, the salivary juices moisten the food and lubricate the mouth. The salivary juices come from three kinds of glands (Figure 19.2). It is often said that salivary enzymes start the breakdown of fats and carbohydrates. However, the enzymatic action is very weak since food is kept within the mouth for only a very short time and the acidic conditions of the stomach, which are encountered within a few seconds after swallowing, prevent the salivary enzymes from continuing their catalytic actions. You can easily evaluate the effectiveness of salivary enzymes by determining how long you must keep a cracker or other starchy food in the mouth before you can taste the sweetness of sugar.

PEROXIDASE. *An enzyme that uses hydrogen peroxide to oxidize a reactant.*

An important role of saliva, aside from lubricating the mouth and moistening solid foods, is its destruction of microbes by a salivary enzyme called *peroxidase*. After the microorganisms are destroyed, the salivary digestive enzymes initiate the process of digesting the microbes. The destruction and digestion of microbes by salivary enzymes is important at all times, and perhaps especially at night, when swallowing is not periodically used (as it is during the day) to sweep microbes into the gut.

THE PHARYNX, OR THROAT

Food which is to be swallowed must first pass into the pharynx. If you look into a mirror with your mouth wide open you will see where this structure begins. Situated within the mouth are two arches. The base of one arch is tilted forward and the other backward, heading toward the throat. A bell-shaped structure, called a *uvula*, appears to dangle from the apex of the arches. Between the arches, which are technically known as the *pillars of fauces*, and beyond these pillars, lie the *tonsils*. The anterior pillars mark the end of the mouth and the beginning of the pharynx.

The pharynx, in short, is the rearmost wall of the oral opening. It runs upward into the nose and downward into the esophagus. Seven tubes or cavities—the mouth, trachea (via the larynx), esophagus, nostrils, and eustachean tubes—open into the pharynx. The tonsils, which are richly endowed with lymphocytes and macrophage cells, protect these tubes against foreign invaders.

Swallowing

When food is to be swallowed, it stimulates nerve endings that are located at the back of the tongue and pharynx. Afferent messages are then sent along cranial nerves 9 and 10 to the medulla oblongata, from which efferent messages are directed along cranial nerves 9 to 12. The reflex action that follows brings about swallowing (Figure 19.4).

First, the tongue is drawn upward and back to shove food from the mouth into the pharynx. Next, the soft palate is drawn backward so that the uvula, together with the soft palate, blocks the pharyngeal passageway into the nose. Third, the larynx is raised by the pharynx into a position beneath a cartilaginous structure called the *epiglottis*. This movement blocks the pharyngeal passageway into the trachea. Breathing is suspended momen-

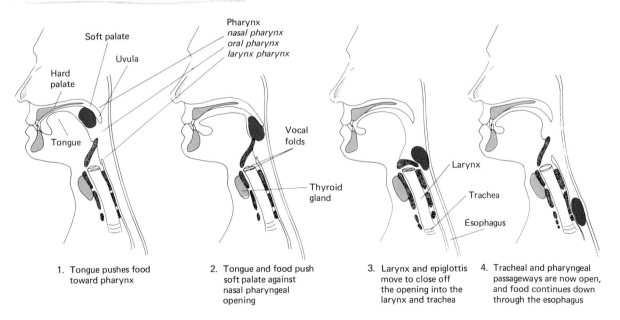

FIG. 19.4. *Swallowing.*

Passage of Food from the Esophagus into the Stomach

tarily, and the musculature of the esophagus contracts and relaxes rhythmically, creating peristaltic waves that drive the food, within two or three seconds, into the stomach.

Difficulties in Swallowing. Dysphagia, or difficulty in swallowing, may arise from a number of causes. Damage to the medulla oblongata (which provides nerves to the pharynx and esophagus) or to the ninth, tenth, eleventh, or twelfth cranial nerves brings about a *neurological dysphagia.* Myasthenia gravis, a muscle disease, may paralyze the tongue and make swallowing difficult, forcing an individual to lean the head backwards in order to get the food into the esophagus. Cancer or inflamed tissues in the pharynx or esophagus bring about an *obstructive dysphagia.* Psychological disturbances also may cause difficulty in swallowing.

PASSAGE OF FOOD FROM THE ESOPHAGUS INTO THE STOMACH

SPHINCTER. *A ring-shaped muscle that closes or opens a circular passageway by contracting or relaxing (G. sphinkter, band).*

The lowest inch or two of esophagus lies below the diaphragm, in the abdominal cavity. When food or saliva is not being swallowed, the musculature of this region is in a state of considerable contractile activity. The result of this contractile activity is the creation of a "physiological sphincter." An actual or anatomical *sphincter* is a ringlike band of muscle that surrounds a tube in the body and serves to close it. Examples of anatomical sphincters include the pyloric sphincter (Figure 19.5), the anal sphincters (Figure 18.35), and the sphincters of the urinary bladder. The "sphincter" in the lower esophagus ordinarily prevents the regurgitation of stomach materials.

When food enters the esophagus, the physiological sphincter—technically known as the *cardiac sphincter*—opens reflexly. Food is thus allowed to enter the stomach.

Two Cardiac Sphincter Problems

ACHALASIA. *Inability to relax (G. a, without; chalasis, relaxation).*

In abnormal situations the cardiac sphincter reflex fails: the sphincter does not relax and food is prevented from entering the stomach. This is known as *achalasia of the cardia.* During a meal, food accumulates in the esophagus until its weight overcomes the resistance of the sphincter. In the interim, an afflicted individual experiences the discomfort of dysphagia, and the regurgitation of food into the trachea and mouth.

Achalasia of the cardia is a complex neuromuscular problem. It may be due to loss of ganglion cells in Meissner's and Auerbach's nerve plexuses, or to diminished activity of these plexuses. Certain drugs, such as octyl nitrite, relax the sphincter and may be useful in some cases if taken before a meal. In severe cases, the sphincter may have to be cut by a surgeon.

A second relatively common cardiac sphincter problem arises from a *hiatus hernia*. "Hiatus" means a gap or space, and "hernia" means a protrusion. Ordinarily in the diaphragm there is a small hiatus through which pass the esophagus and the two vagus nerves. In some people, however, this opening becomes abnormally large. For such people an excessive rise in intra-abdominal pressure (as happens in stooping, bending, or lying down) will bring about herniation, whereby the upper part of the stomach passes through the esophageal hiatus into the thorax. In such a case the acidic gastric juice may enter the esophagus, and the pain of heartburn is felt back of the breastbone. In normal individuals the rise in intra-abdominal pressure does not push the stomach through the hiatus. Instead it causes the cardiac sphincter to seal itself tightly, thus preventing gastric juices from passing into the esophagus.

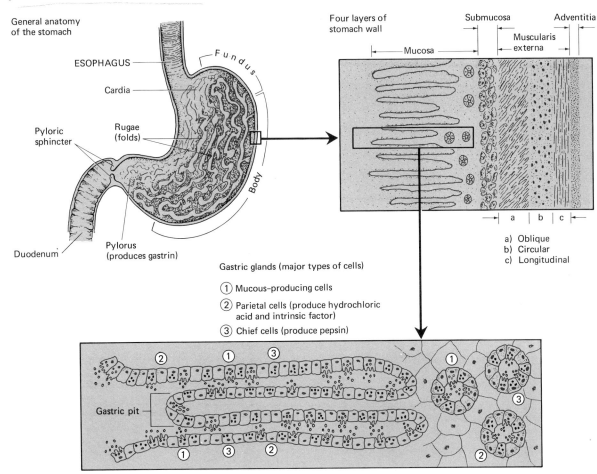

FIG. 19.5. *The stomach.*

Vomiting

Vomiting, in effect, is a reversal of swallowing, though it is a forceful and almost uncontrollable act. It is a reflex response to any one of various stimuli, including psychological factors (fear, anxiety, disgust); stimulation of nerve endings of the utricle in the ear (motion sickness); stimulation of pain-sensing nerves (nociceptors) in any organ of the body by drugs, poisons, or metabolic disturbances; or mechanical stimulation of the pharynx and esophagus, as happens, for example, when a finger is inserted into the throat.

In response to any one of these stimuli, the vomiting center in the medulla oblongata sends out efferent messages. Some of these messages are carried out by autonomic nerves, others by spinal nerves. The cardiac sphincter and the stomach become extremely relaxed, while the muscles of the abdominal wall and diaphragm contract forcefully. In consequence, the stomach contents are expelled (vomited) or an individual with an empty stomach experiences a dry heave (retching).

MEDULLA OBLONGATA. An extension of the spinal cord into the brain, forming the initial region of the brain stem.

THE STOMACH

The stomach is a digestive sac with a capacity of about 1 to 1.5 liters. When empty, the wall of this organ is thrown into folds, called *rugae*. The rugae unfold as the organ becomes distended with food.

The stomach can be divided into three anatomical regions (Figure 19.5) — cardia, body, and pylorus. The *cardia*, which was mentioned above in conjunction with the cardiac sphincter, is the region where the esophagus joins the stomach. It merges with the *body* of the stomach, which makes up the bulk of this organ and contains about 20 million gastric glands in its mucosa. The *pylorus*, which is the terminal region of the stomach, does not contain gastric glands. It produces *gastrin*, a hormone which stimulates gastric glands to produce and secrete a digestive juice. It also contains the *pyloric sphincter*, a band of smooth muscle which allows or prevents emptying of the stomach into the intestine.

In Figure 19.5 you may see a fourth anatomical region of the stomach. Called the *fundus* (which simply means a part that is far removed from the opening), its functions are identical to those of the body of the stomach.

PYLORUS. Circular opening leading from the stomach to the duodenum (G. pyloros, *gate keeper*).

Digestive Activities and Other Activities of the Stomach

The mucosa of the stomach has an endocrine function, an exocrine function, a function in absorption, and a unique nutritional function.

The *endocrine function* of the stomach is to produce and release gastrin, a hormone which promotes the activities of the stomach's gastric glands. Gastrin is produced in the pylorus and is released into the bloodstream in response to stimuli which arise as acidified food (*chyme*) enters the pyloric region of the stomach.

The *exocrine function* of the stomach is the secretion of a *gastric juice* from the gastric glands. This juice contains hydrochloric acid, which is secreted from parietal cells, digestive enzymes, which are secreted from chief cells, and mucus, which is secreted from mucous cells (Figure 19.2).

PARIETAL. Pertaining to a wall (L. paries).

Hydrochloric acid permits the digestive enzymes to catalyze their reactions optimally. The main digestive enzymes are *pepsin*, which breaks down proteins; *rennin*, which is present only in babies for breaking down protein (especially the casein protein of milk); and *gastric* lipase, which works upon small fats such as those which are abundant in dairy products. The mucous secretions protect the stomach wall from the erosive actions of hydrochloric acid.

Hydrochloric acid, as such, is not stored within the gastric glands. Rather, it is produced indirectly in a way which is not known exactly, but can be plausibly explained (on the basis of indirect evidence) as the result of two chemical reactions. These reactions are shown in Figure 19.6.

For the most part, the cells lining the inner surface of the stomach are not involved in the *absorption* of any appreciable amount of nutrients. Nonetheless, certain amounts of alcohol, glucose, aspirin, and a number of other substances pass into the bloodstream within the mucosa.

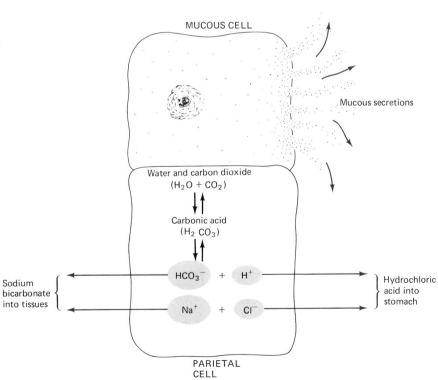

FIG. 19.6. *Production of hydrochloric acid by a parietal cell and of protective mucus by a mucous cell.*

The special and unique *nutritional function* of the stomach is to produce *intrinsic factor*. This substance is a protein necessary for the absorption of vitamin B_{12}. In its absence or deficiency—as sometimes happens for genetic or pathological reasons (severe ulcers, cancer, autoimmunity)—it is not possible to absorb enough vitamin B_{12} from the intestinal canal. As a result, *pernicious anemia* occurs. This is a chronic, progressive anemia, the chief characteristics of which are a decreased red blood cell count and a wide range of nervous disorders. Pernicious anemia may be corrected by administering vitamin B_{12} intramuscularly or subcutaneously.

PERNICIOUS. *Highly destructive; fatal (L. perniciosus).*

Regulating the Gastric Glands

The secretion of the gastric juice is regulated reflexly and hormonally. Various kinds of stimuli trigger the reflex pathway. Among these stimuli are psychic factors (thought, sight, or smell of food), the taste of food, and the distention of the stomach by food. Afferent messages then travel to the medulla oblongata in the brain. From here, efferent messages are sent through the tenth cranial (vagus) nerve to the gastric glands, establishing the *cephalic phase of gastric secretion*. The cephalic phase can be abolished by distaste, distress, disgust, or other adverse psychological stimuli.

VAGI. *The paired tenth cranial nerves. They "wander" all over the body, innervating most of the viscera (L. wandering).*

The endocrine control of gastric secretion, which is a slower and longer lasting control than the reflex type, is mediated by two hormones: gastrin and enterogastrone.

Gastrin is released from the pylorus of the stomach in response to the presence of food. It stimulates the release of gastric juices during what is referred to as the *gastric phase of digestion*. This phase is enhanced by vagus nerve discharge.

Enterogastrone is released from the small intestine in response to *chyme* (the acidic stomach contents) during what is called the *intestinal phase of gastric secretion*. Enterogastrone inhibits the release of gastric juices near the end of the gastric phase of digestion (two to four hours after a meal). Enterogastrone also stimulates the closure of the pyloric sphincter at the junction of the stomach and small intestine, thus prolonging the retention of food within the stomach.

Regulating the Emptying of the Stomach

Periodically, the pyloric sphincter relaxes and chyme is passed out of the stomach into the small intestine. This action occurs with increasing frequency following the first half hour after eating. During the first half hour, the pyloric sphincter is mainly in a contracted (closed) state, an effect attributed mainly to enterogastrone.

Failure of the Pyloric Sphincter to Open. Occasionally, a baby is born with a pyloric sphincter which fails to open. This defect is transmitted genetically. Milk cannot be passed into the stomach, vomiting and dehydration occur, and the life of the baby is at stake. Drugs may be helpful in some cases, but surgery is usually necessary.

Ulcers

An *ulcer* is a lesion or sore, on the surface of the skin or mucous membrane. It may arise anywhere on the surface of the body or within a canal which communicates from the interior to exterior of the body—such as the excretory, respiratory, reproductive, and digestive tracts.

The most common ulcer is the *peptic ulcer*, which is caused by the action of gastric juice. The peptic ulcer may appear in the lower end of the esophagus or in the initial region of the lower intestine, but it occurs most commonly near or on the lesser curvature of the stomach. For reasons which are not known, peptic ulcers occur more often in individuals with type O blood, than in persons with type A, B, or AB blood.

An ulcer may heal by itself, especially in cases where the cause of ulceration, such as emotional stress, is relieved. An example of a helpful diet is one which is low in fat and provides protein in the form of liquid (for example, milk) rather than fiber (meat).

When ulceration is severe, surgery is necessary. All surgical procedures are aimed at reducing acid secretion—either by removing the source of gastrin (the pylorus), or by cutting the branches of the tenth cranial nerve to the stomach.

THE PANCREAS

The pancreas, a tongue-shaped organ, is attached along its wider end (called a head) to the curve of the first portion of the small intestine (the duodenum). The body of the pancreas—which is separated from its head by a constriction (neck)—passes upward to the left of the body, behind the stomach. The tail of the pancreas is partly attached by connective tissues to the spleen (Figure 19.2).

The pancreas is covered on its outside by a loose connective tissue. This tissue invades the organ at intervals, dividing it into lobules. The interior of each lobule is made up of two kinds of glands: exocrine and endocrine.

The *exocrine glands*—which make up the bulk of the organ—are or-

PANCREAS. (G. pan, *all*; kreas, *flesh*).

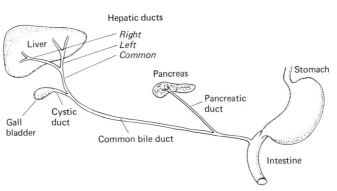

FIG. 19.7. *Pancreatic and hepatic ducts.*

ganized as compound tubular-alveolar structures. The ducts of these glands merge into a main excretory duct, called the *pancreatic duct*. En route to the duodenum, the pancreatic duct is joined by the *common bile duct* of the gallbladder. Together, these ducts enter the wall of the duodenum (Figure 19.7). The entrance into the duodenum is regulated by the *sphincter of Oddi*. This one-way valve prevents intestinal materials from passing into the pancreas, liver, or gallbladder.

The *endocrine glands* — known also as the *islets of Langerhans* — comprise a minor component of the pancreas. These glands secrete insulin and glucagon, two important hormones.

The Role of the Pancreas in Digestion

The pancreas secretes a juice which contains (1) bicarbonate for neutralizing the acidity of the stomach effluents (chyme), and (2) digestive enzymes. Chief among the enzymes, from the viewpoint of breaking down the bulk of food, are: *lipase*, which hydrolyzes fats; *nucleases*, which break down nucleic acids, such as DNA and RNA; *amylase*, which splits starch into maltose; and *proteases*, which continue the work started by pepsin in the stomach.

The proteases are produced and stored in the pancreas as inactive forms called *chymotrypsinogen* and *trypsinogen*. As inactive forms of enzymes, they are unable to catalyze chemical reactions through which proteins would be broken down. They are thus normally incapable of digesting away the organ (pancreas) in which they are stored. Nevertheless there is always a possibility they may become active while within the pancreas. To offset this possibility, the pancreas is endowed with several inhibitors of proteases — inhibitors that are very effective against chymotrypsin and trypsin.

Activation of chymotrypsinogen and trypsinogen occurs in the small intestine. Here an intestinally-derived enzyme, called *enterokinase*, converts trypsinogen to *trypsin*. Trypsin, in turn, converts additional quantities of trypsinogen to trypsin, and it converts chymotrypsinogen into *chymotrypsin*.

Jointly, the enzymes of the pancreas participate in the further breakdown of proteins, and they initiate the major action of breaking down carbohydrates, fats, and nucleic acids.

THE LIVER AND THE GALLBLADDER

The *liver* is an important organ in metabolism (Chapter 22) and body defense mechanisms (Chapter 32). As far as digestion is concerned, its only real important function is to provide the small intestine with bile, which is needed for the efficient hydrolysis and absorption of fats.

The *gallbladder* is a small contractile sac located directly beneath the liver. Stored within this sac is a variable amount of bile which the gallbladder receives from the liver (Figure 19.2). The bile is concentrated ten- to twenty-fold after entering the gallbladder. During a meal, when fatty materials enter the small intestine, the gallbladder contracts and sends some of its bile to the small intestine (Figure 19.7).

BILE. *An alkaline juice which contains bile salts, bile pigments, cholesterol, lecithin fats, and mucus* (L. bilis).

Functional Anatomy of the Liver

The liver weighs about three pounds in an average adult, and is the largest visceral organ. It is located in the upper abdomen, where it is kept in position by the peritoneum and by the pressure of its neighboring abdominal organs (Figure 19.2).

The liver is unequally divided into four lobes, and each lobe is subdivided into numerous *lobules*. Each lobule has a centrally-located vein. From this "central vein" the liver cells radiate outward—as *liver cords*—to a many-sided perimeter. Situated periodically along the perimeter are *portal areas*—areas which harbor a *portal vein*, a branch of the hepatic artery, and a bile duct. See Figure 19.8).

In order to appreciate liver function it is necessary to understand the relationship of the central vein and portal area to the liver cord. Each liver

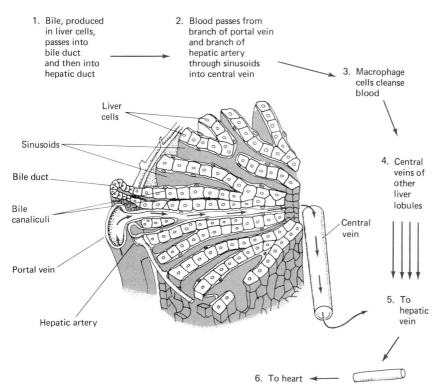

FIG. 19.8. *Cross-section of a single functional unit (lobule) of the liver.*

cord is two cells thick along its length, which extends from the central vein to the perimeter of a lobule. Sandwiched between these cells, along the length of the cord, is a *canaliculus* – a small canal. This canal serves to transport bile, which is produced by the liver cells. The bile is carried from the central vein to the portal area. From the portal area, the bile is carried by a *bile duct* to the *hepatic duct* which leads to the outside of the liver. From the hepatic duct the bile may be sent into the gallbladder, where it is concentrated and stored until needed, or it may be sent directly into the small intestine (Figure 19.7).

Whereas a canaliculus is sandwiched between the cells of a liver cord, the cord itself is sandwiched between two *sinusoids* (large capillary spaces). The sinusoids are lined by reticuloendothelial cells called *Kupffer cells*, which phagocytize foreign matter. The sinusoids also convey blood from the portal area to the central vein, which in turn conveys the blood to the heart.

KUPFFER CELLS. *Fixed macrophages that line the sinusoids (large blood capillaries) of the liver.*

The blood entering the liver is of two sorts. The blood coming from the branch of the hepatic artery is fully oxygenated, since it comes directly from the heart. The blood from the portal vein is partially oxygenated, since it is derived from the pancreas, spleen, stomach, and intestine (Figure 25.19). The hepatic artery furnishes oxygen to the liver while the portal vein furnishes nutrients from the digestive tract, hormones from the pancreas and small intestine, and red blood cells from the spleen.

The Role of the Liver and Gallbladder in Digestion

BILE SALTS. *Sodium salts of acid steroids, produced in the liver and stored, highly concentrated, in the gallbladder.*

The liver produces *bile salts*, which may be sent to the gallbladder for temporary storage, or to the duodenum to facilitate the digestion and absorption of fats. In the absence of a gallbladder, the liver must send the bile salts directly to the intestine, as needed.

Within the intestine the bile salts perform two functions. Together with lipase, which comes from the pancreas, bile salts bring about the digestion of fats. Bile salts do not digest the fats themselves; this is the role of lipase. Rather, bile salts render the fats more soluble in water, and in doing so, facilitate the action of lipase. The second function of bile salts, which is carried out without the need of lipase, is to facilitate the absorption of neutral fats and fatty acids into the epithelial cells of the intestinal mucosa.

In addition to concentrating bile salts, the gallbladder concentrates the breakdown products of hemoglobin. It receives these products from the liver, which in turn receives aged and broken down forms of hemoglobin from red blood cells that are destroyed in the macrophage system, especially in the spleen. Periodically, the musculature of the gallbladder contracts and sends the breakdown products into the intestinal tract, where they impart color to the stools and are themselves thus eliminated. In the absence of a gallbladder, the waste products of hemoglobin are discharged from the liver directly to the small intestine.

Common Problems Associated With the Liver

Hepatitis is a common inflammation of the liver. It is caused by a virus, which converts the epithelial liver cells into a fibrous scar tissue called cirrhosis. *Liver cirrhosis* may arise independently of hepatitis, as is common in chronic alcoholism, where a person fails to receive proper nourishment.

In addition to cirrhosis, an individual with hepatitis may develop *jaundice*. In an individual with jaundice, the whites of the eyes, the skin, and the body fluids become yellowish. Jaundice, however, is a symptom, not a disease. It signifies an excess of *bilirubin* and *biliverdin* — two of several breakdown products of hemoglobin — in the blood.

The three most common forms of jaundice are called hepatic jaundice, obstructive jaundice, and hemolytic jaundice. *Hepatic jaundice* arises in certain liver diseases, such as hepatitis or cancer. *Obstructive jaundice* arises whenever the liver canaliculi or bile ducts are blocked, as may happen in liver cirrhosis, cancer, bacterial infections, or when bile stones are formed. *Hemolytic jaundice* arises when there is excessive breakdown of red blood cells in certain forms of anemia.

CIRRHOSIS. *A disease in which parenchymal (epithelial) tissue degenerates and is replaced by fibrous connective tissue* (G. kirrhos, *orange-yellow;* osis, *condition*).

JAUNDICE. *A disease in which the whites of the eye, the skin, and the urine become abnormally yellow due to bile pigments in the blood* (L. galbinus, *yellow*).

Two Common Problems of the Gallbladder

Cholelithiasis (gallstones) and *Cholecystitis* (inflammation of the gallbladder) are two common problems of the gallbladder.

Gallstones may be formed from cholesterol and calcium, or from bilirubin (a breakdown product of hemoglobin) and calcium, but very commonly gallstones are mixtures of cholesterol, bilirubin, calcium, and other materials. Gallstones are twice as common in females than males and they are favored by age (over 40), multiple pregnancies, obesity, a high level of cholesterol in the blood, and diabetes.

Gallstones may block bile ducts, giving rise to jaundice. Upon enlarging, gallstones may irritate the gallbladder, causing considerable pain. In addition, gallstones may predispose an individual to cholecystitis, an inflammation of the gallbladder or cystic duct. Cholecystitis usually requires surgical removal of the gallbladder.

CHOLELITHIASIS. *Production of gallstones* (G. chole, *bile or gall;* lithos, *stone*).

THE SMALL INTESTINE

The small intestine is about 20 feet long. It is divided into the duodenum, jejunum, and ileum. The *duodenum* is about 10 inches long. Its submucosa contains goblet cells and *Brunner's glands,* exocrine glands whose function is not known, but which may secrete mucus. The duodenum is perforated near the stomach by the *pancreatic duct* and the *common bile*

BRUNNER (1653–1727) *described the duodenal glands which bear his name.*

The Small Intestine

These ducts bring pancreatic and liver secretions into the small intestine (Figure 19.7).

The *jejunum*, which is about 8 feet long, is not appreciably different from the *ileum*, which is about 12 feet long. These two regions of the small intestine differ primarily in their location. The jejunum is located in the upper abdomen and the ileum is in the lower abdomen.

Digestive Activities in the Small Intestine

In order to appreciate the functions of the small intestine, you must have a clear idea of its microanatomy. You may develop this mental picture from Figure 19.3 as we go along.

The presence of circular or spiral folds of the mucosa, called *plicae circulares*, is a unique characteristic of the small intestine. These folds, which can be seen with the naked eye, are not collapsible, as are the rugae of the stomach. Nipplelike structures, called *villi*, are present all over the plicae circulares as well as over the surface of the small intestine which is free of plicae. These villi, which project about one millimeter into the lumen of the small intestine, are not found elsewhere in the digestive tract.

Each villus is lined by thousands of columnar cells and goblet cells. The *goblet cells* secrete a mucus that lubricates the exposed surface of the mucosa. The *columnar cells* contain digestive enzymes in their plasma membrane. These enzymes catalyze the final step in carbohydrate and protein digestion. This action is one of the special functions of the small intestine. Through the actions of these enzymes, disaccharides and dipeptides are cleaved into monosaccharides and amino acids. Among the enzymes that are present as part of the cell membrane of the columnar epithelial cells are: *maltase*, which splits maltose into two glucose molecules; *sucrase*, which splits sucrose into glucose and fructose; and *dipeptidase*, which splits dipeptides into two amino acids.

Another special function of the small intestine is to absorb the products of digestion. Most of the products are absorbed into the columnar epithelial cells, while some products pass between these cells or into the goblet cells. This process is discussed in detail in the following chapter.

A third special function of the small intestine concerns the *intestinal glands*. These glands, which number about 200 million, are about half as long as the villi. They run parallel to, and beneath, the villi in the lamina propria; their secretory products are discharged into the intestinal lumen from pores between the bases of the villi (Figure 19.3).

The function of the intestinal glands is not to produce digestive enzymes, as is the case for the columnar epithelial cells of the villi. Rather, they share with the pancreas the task of neutralizing the acidity of stomach effluents. They accomplish this by secreting an alkaline juice in response to being stimulated by a hormone called *pancreozymin*. (Pancreozymin is now known to be identical to cholycystokinin. At one time these names were applied to what were believed to be two different substances.)

Within the neutralized fluid that is present in the intestinal lumen the

GOBLET CELLS. *Vase-shaped cells that produce and secrete mucus.*

PANCREOZYMIN. *A small intestine hormone that stimulates the secretion of pancreatic enzymes* (G. pankreas, *flesh;* zyme, *leaven*).

small intestine carries out a third function. It incubates the digestive enzymes from the pancreas, the bile which comes from the liver and the chyme (effluent) which comes from the stomach. It incubates these materials for a period of two to four hours. During this time disaccharides and dipeptides are produced, along with free fatty acids which are derived from neutral fats.

A fourth function of the small intestine is to propel forward the chyme which it receives from the stomach. It accomplishes this by virtue of its peristaltic activities. The peristaltic actions of the entire digestive tract are due mainly to the actions of the muscularis externa and to a lesser extent the muscularis mucosa. These muscles are in a continuous state of tonus (partial contraction). The tonus is so great that shortly after death (when tonus no longer exists) the length of the intestine increases two to three fold.

The tonus of the digestive tract is due mainly to nerve cells in Auerbach's and Meissner's plexuses. It is also due to stimulation by external nerves that enter the tract from the autonomic nervous systems. When the external nerves are cut, the frequency and strength of contractility are reduced to a minor extent. Local anesthesia, on the other hand, in which the nerve cells of Auerbach's and Meissner's plexuses are directly affected, reduces peristaltic activity significantly. Local anesthesia is commonly used in abdominal operations to facilitate surgery (Figure 19.9).

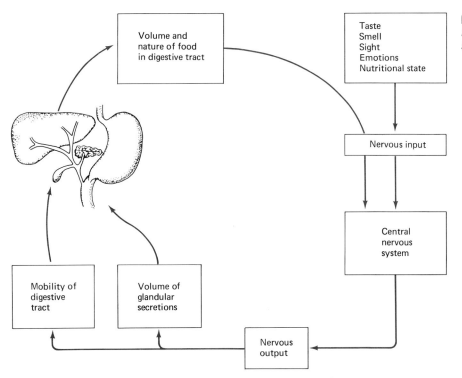

FIG. 19.9. *Nervous regulation of the digestive system.*

One final function of the small intestine which is worthy of mention is its role in adjusting the ratio of sodium to potassium in body fluids. The small intestine shares this function with the remainder of the digestive tract and with the kidneys and sweat glands. The adjustment of the sodium/potassium ratio is achieved through the action of *aldosterone*, a hormone from the adrenal cortex. The main effect of aldosterone is to bring more sodium into the body and send excess potassium out of the body.

Diverticula, Diverticulosis, and Diverticulitis

For reasons not fully known, local regions of the intestinal wall sometimes become weak. There is increasing evidence that the fault lies partly in the wall muscle. In the weakened areas, pressures arising within (or outside) the wall cause an outpocketing (or inpocketing). These pockets are called *diverticula*. In the event that they are not infected, the condition is called *diverticulosis*.

When an infection sets in, a diverticulum becomes inflamed, and the condition is now known as *diverticulitis;* it often becomes necessary to surgically remove the diseased tissue.

THE LARGE INTESTINE

CECUM. *The large blind pouch in which the large intestine begins* (L. caecum, blind).

The large intestine is about five feet long. Unlike the stomach and small intestine, it has no folds and no villi. It begins with a blind sac, called a *cecum*, below which extends the blind, nipplelike structure called the *appendix* (Figure 19.2). From the cecum, the next section of the large intestine, the *colon*, moves up until it meets the liver. It then moves across the abdomen until it meets the spleen, and then down to the left hipbone, where it curves into the middle of the pelvic cavity to become the *rectum*.

Functions of the Large Intestine

During a period of about six hours the undigested food and unabsorbed products of digestion pass through much of the colon. They then linger for a variable period, ranging up to days, in the descending colon and rectum. During its journey from cecum to rectum, the material in the colon is shifted about by peristaltic waves. Some of these waves convey the intestinal contents toward the rectum, others send the contents back toward the cecum. In the course of these back-and-forth moves, most of the water and some salts are reabsorbed into the bloodstream. The absorption of water and salts is the chief function of the large intestine. A minor function which is attributed to it by some scientists is the synthesis of vitamins by microorganisms and the absorption of these vitamins into the bloodstream.

REGULATING THE OUTPUT OF THE DIGESTIVE JUICES

Previously we discussed the hormones (gastrin and enterogastrone) and nerves (a pair of vagus nerves) which are involved in regulating the digestive activities of the stomach. Now we will say something about the regulation of the digestive functions of the pancreas, intestinal glands, liver, and gallbladder. All of the juices that enter the intestinal lumen are released from their sources through nervous or hormonal signals.

The nerve signals are sent from the medulla oblongata by way of the tenth cranial nerve to the intestinal glands, pancreas, liver, and gallbladder in response to the sight, thought, taste, and smell of food.

Two hormones and one special enzyme regulate the digestive activities of the tissues we are considering here. The first, *pancreozymin*, is released from the mucosa of the small intestine in response to chyme. Pancreozymin then enters the blood, circulates throughout the body, and interacts with the pancreas and gallbladder. It stimulates the release of digestive enzymes from the pancreas and causes the gallbladder to contract and release bile (Figure 19.10).

The second hormone, *secretin*, is also released from the intestinal mucosa in response to chyme. Its target tissues, however, are the liver and pancreas. It stimulates the liver to produce bile and causes the pancreas to release an alkaline juice.

The special enzyme, called *enterokinase*, is also produced in the mucosa of the small intestine. In response to chyme, it is released into the intestinal

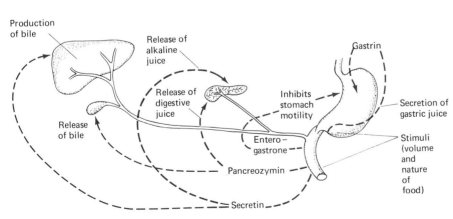

FIG. 19.10. *Hormonal regulation of the digestive system. Arrows indicate route of hormones in bloodstream.*

lumen where it catalyzes the conversion of inactive forms of trypsin and chymotrypsin into full-fledged enzymes.

ELIMINATION

The need to eliminate feces is triggered by the distension of the rectal wall by feces. Stretch receptors become excited, sensations are passed to the level of consciousness in the cerebral cortex, and a decision is made to empty the lower colon. A complex reflex action is put into order. To understand this reflex we must consider a few anatomical facts.

Within the last inch or two of the rectum, in what is called the *anal canal*, there is an enlarged smooth muscle in the muscularis mucosa. This muscle, which is called the *internal sphincter*, is under the control of the autonomic nervous system. Outside the internal sphincter is an *external sphincter*, which is composed of skeletal muscle and is under conscious control of the central nervous system.

When elimination is called for, the autonomic nervous system sends action potentials into the colon, rectum, and internal sphincter. The colon and rectum contract while the internal sphincter relaxes. During these actions the diaphragm usually descends into the abdomen, causing a rise in pressure within the abdomen. Elimination follows. At any point, however, the pudendal nerves of the central nervous system may effect the conscious function of contracting the external sphincter, thus preventing elimination.

Problems in Elimination

The most common problems of elimination are diarrhea and constipation.

DIARRHEA. (G. diarrhoia, a *flowing through*).

Diarrhea is a passage of excessive amounts of soft or fluid stools. It can be caused either by the failure of the ileum and colon to absorb sufficient water or by the premature discharge of the intestinal contents. Failure of the ileum and colon to absorb enough water may be due to excessive osmotic pressure in the intestinal lumen. Failure to absorb sufficient water may also occur after operations in which portions of the intestine are removed.

LAXATIVE. An agent that relieves constipation; a cathartic; a purgative.

Excessive osmotic pressure in the intestinal lumen is created intentionally with certain laxatives, as for example, magnesium sulfate. Excessive osmotic pressure may also develop unintentionally, as for example, when an individual is not able to digest milk sugar or when, because of certain diseases, such as an inflammation of the intestinal wall—enteritis, it is not possible to reabsorb enough sodium and other minerals that pass into the intestine in the pancreatic and intestinal juices.

Premature discharge of the intestinal contents is usually due to massive peristaltic waves in response to certain nerve stimuli. The stimuli may be rooted in anxieties; disease of the thyroid gland (thyrotoxicosis); or diseases of some other organ (such as a liver tumor), in which excessive amounts of nerve-stimulating chemicals, such as serotonin, are liberated. Other causes include food poisoning, food allergy, or irritation of the digestive tract by certain foods. In some cases diarrhea occurs for reasons totally unknown.

Constipation is the opposite of diarrhea. Here, bowel movements are infrequent and incomplete, and the stools are usually dry. Often the problem arises when attention is not given to the need to eliminate. In the event that the need is not overwhelmingly urgent, an individual may decide to hold off. The stool then becomes drier and smaller and less pressure is placed upon the rectum. But the stool remains, forming a plug and making it more difficult to eliminate newer stools. Should it become a practice to ignore the need to eliminate, an individual may find him or herself chronically constipated.

CONSTIPATION. (L. *constipare,* to press closely together).

Constipation may also arise from Hirschsprung's disease or spastic colon. In both of these situations stools are prevented from passing out of the anus. In *Hirschsprung's disease* — which is a congenital abnormality — the problem is due to abnormal relaxation of the colon and rectum and contraction of the internal sphincter. The abnormality rests in a deficiency of parasympathetic nerves to the colon, rectum, and anus. Paradoxically, in *spastic colon* — an abnormal response to certain foods, emotional stress, or drugs — there is excessive discharge of parasympathetic nerves to the colon, rectum, and anus. In this case the colon and rectum are abnormally contracted; their lumen is closed. In both cases — Hirschsprung's disease and spastic colon — the constipation can be relieved with spinal anesthetics.

SPASTIC. *Pertaining to spasmic contractions characterized by exaggerated reflexes and muscle tension.*

In some individuals, constipation occurs for reasons which are not understood. Some people do not have a need to eliminate once a day or even once every two or three days. When the need finally arises, it may be very difficult to eliminate. The situation may be a chronic one, perhaps even life long, and perhaps for such people it is a "normal" condition.

SUMMARY

1. The digestive system is composed of a digestive tract, the salivary glands, a pancreas, a liver, and a gallbladder.
2. The wall of the digestive tract is lined along its lumen by a mucous membrane which is connected, through a connective tissue, to smooth muscle. The tract is lined on its outside by a serous membrane which, in the abdominal cavity, is continuous with the visceral peritoneum.
3. The chief functions of saliva are: (a) to lubricate the mouth; (b) to facilitate swallowing by moistening solid foods; and (c) to destroy microorganisms.
4. Swallowing is a complex reflex in which the tongue draws food back toward the throat, the soft palate blocks the pharyngeal passage to

the nose, the epiglottis closes the laryngeal passage to the lungs, and peristaltic waves send the food or liquid into the stomach.
5. The stomach performs an endocrine function (secretes gastrin), an exocrine function (secretes acid, enzymes, and mucus), an absorptive function (absorbs, for example, alcohol, glucose, aspirin) and a nutritive function (produces intrinsic factor).
6. The pancreas secretes digestive enzymes and a bicarbonate fluid into the small intestine.
7. The gallbladder stores and secretes bile, which it receives from the liver. Bile works together with lipase to break down fats, and it works by itself to help the small intestine absorb fats.
8. Hepatitis is a liver disease whose common symptoms include cirrhosis and jaundice. Cholelithiasis (stones) and cholecystitis (inflammation) are gallbladder problems commonly encountered beginning around the fifth decade of life.
9. The small intestine secretes an alkaline fluid which, together with an alkaline fluid from the pancreas, neutralizes the acidic contents of the stomach. The small intestine also contains digestive enzymes in its microvilli for carrying out the final hydrolytic actions of digestion, after which the epithelial cells absorb the products of digestion.
10. The vagus nerve, gastrin, enterogasterone, pancreozymin, secretin, and enterokinase regulate the digestive activities.
11. The large intestine functions mainly to remove the water from the digestive wastes of the small intestine.
12. Elimination is a complex autonomic reflex subject to voluntary control.

REVIEW QUESTIONS

1. What is the digestive system composed of?
2. Describe the four walls or layers that make up the wall of the digestive tract.
3. What are the functions of saliva?
4. Define: peristalsis, pharynx, uvula, epiglottis, dysphagia, achalasia, hiatus hernia, gastric ulcer, hepatitis, cholelithiasis, cholecystitis, diverticulitis.
5. Describe the endocrine, exocrine, absorptive, and nutritive functions of the stomach.
6. What is the function of the liver in digestion? of the gallbladder?
7. Where are the digestive enzymes of the small intestine located?
8. How is the acidity of the stomach effluent neutralized in the small intestine?
9. What are the roles of the following substances in regulating the digestive system: vagus nerve, enterogastrone, pancreozymin, secretin, and enterokinase?
10. What is the major role of the large intestine?
11. Describe ways in which diarrhea and constipation may arise.

Absorption of Nutrients and Elimination of Cell Wastes

INTESTINAL ABSORPTION
    Food circulatory routes
        Screening the intestinally absorbed materials
        Why are there two food circulatory routes?
MECHANISMS OF CELLULAR TRANSPORT
    Diffusion
    Osmosis
    Endocytosis and exocytosis
    Facilitated diffusion
    Active transport
    Emulsification
FAULTY INTESTINAL ABSORPTION
    Faulty digestion
    Faulty pancreas and deficient bile
    Defective intestinal enzymes
    Defective transport mechanisms
    Anatomical abnormalities
    Histological abnormalities

All physiologically active cells must absorb nutrients and eliminate wastes. The act of passing materials into and out of a cell through the plasma membrane is called *cellular transport*.

In the preceding chapter we discussed the digestion of food. In this chapter we will consider, first, where the products of digestion go after entering intestinal cells. Next we will discuss the necessity of sending intestinally absorbed materials to the liver or lymph nodes prior to sending them into other organs of the body. Thereafter we explain why there are two routes, rather than one, through which intestinally absorbed materials are sent to most of the organs of the body. At this point, you should have a clear picture of the general process of intestinal absorption. To make this picture more meaningful we will discuss the mechanisms by which nutrients and wastes are transported across the plasma membrane of body cells. In conclusion, we will describe the major causes of intestinal malabsorption.

INTESTINAL ABSORPTION

VILLI. *Minute nipplelike projections from the intestinal mucosa.*

In thinking of the overall process of intestinal absorption it is helpful to visualize the absorptive structures of the small intestine. Imagine the canal of the small intestine, which is 6 meters long and 1 centimeter in diameter. The canal's internal surface falls into folds (plicae circulares) of mucous membrane. Visualize 4 million nipple-shaped projections, called *villi*, which protrude 1.0 millimeter from the mucous membrane into the intestinal canal. There are about 3 trillion cells lining the free surface of the villi. These cells are replaced every two to seven days. Each cell has about 3000 *microvilli* which protrude about 1 micrometer from the villi into the intestinal lumen.

The function of these projections is to provide the intestinal wall with an enormously increased absorptive surface. Altogether, the small intestine has a surface area of about 4500 square meters, or about 1.25 acres. If there were no microvilli, the surface area would be about 9 square meters and if there were no villi, the total surface area would be about 0.5 square meters.

Food becomes trapped upon and between the microvilli, where it lingers for a period of about two to four hours. During this period the food undergoes intestinal digestion and the nutritional products of digestion are absorbed. Most of the products are absorbed through cellular transport into the epithelial cells that line the villi. The remaining products are absorbed through intercellular spaces that lie between the cells in each villus. The products that are absorbed into epithelial cells through cellular transport pass back out of the cell through another surface of the cell into connective tissues. Here they reunite with the products that were absorbed through intercellular spaces. Next, within the core of a villus, the absorbed material passes into either a blood capillary or a lymph capillary.

Food Circulatory Routes

PORTAL VEIN. *The vein which leads partly deoxygenated blood from the tissues of the gut, pancreas, and spleen into the liver.*

If the absorbed material enters a blood capillary within the core of a villus, it is conveyed into the portal vein. The *portal vein* carries most of the products of digestion, including simple sugars, amino acids, minerals, and water-soluble vitamins, to the liver (Figure 20.1).

If the absorbed material enters a lymph capillary (a *lacteal*), it is conveyed into a series of increasingly large lymph vessels, which empty into great veins that enter the heart (Figure 20.2). The lacteals pick up only a small proportion of the products of digestion mentioned in the preceding paragraph. Mainly, they pick up fats and fat-soluble vitamins from the intestinal canal.

Although some of the intestinally absorbed material is metabolized directly by the intestinal epithelial cells, most of it passes on, unaltered, into one of the two circulatory routes described above. Most of the non-fatty material is taken directly into the liver by way of the portal system

Absorption of Nutrients and Elimination of Cell Wastes

while most of the fatty material does not enter the liver or other organs of the body until it has passed through the lymphatic system.

Screening the Intestinally Absorbed Materials. Why do intestinally absorbed materials pass into either the liver or the lymphatic system before passing into other organs of the body such as the heart and brain? Because foreign materials that cannot be used by the body—materials which may be toxic—must be screened out. Consider what happens along each of the food circulatory routes.

As the products of digestion travel along the route from the lacteal of a villus to the heart, all of the fatty lymphatic material passes through one or more lymph nodes. Within these nodes certain foreign materials, such as artificial food additives, are destroyed and disposed of by macrophages through the process of phagocytosis.

In the portal system, foreign material is screened in two different ways, depending on whether the material is particulate or molecular. The particulate material is devoured by Kupffer cells, macrophages that lie along the wall of the sinusoids of the liver (Figure 19.8). The nonparticulate molecular materials usually are converted into special kinds of salts within the

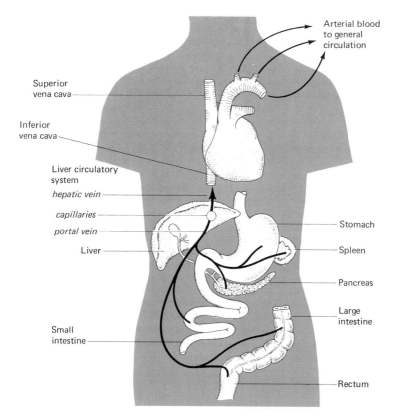

FIG. 20.1. *Portal system for circulation of nutrients. The heavy line represents the venous circuit.*

Intestinal Absorption

liver cells. These salts, which are mainly glucuronates, sulfates, and amino acid derivatives, are excreted from the body by the kidneys, as described in Chapter 31.

Why Are There Two Food Circulatory Routes? The reason for sending absorbed materials through two different routes into the general circulation has to do with the need to have energy available over long periods of time between meals. Nearly all of the intestinally absorbed carbohydrates and amino acids pass directly into the liver, while most of the fats pass slowly through the lymphatic system before they enter the blood itself. The carbohydrates provide an immediate source of energy to most of the tissues of the body and they are about the only kinds of metabolites which can be used by nerve cells for energy.

Most amino acids can easily be converted into glucose, thus serving as another immediate source of energy. In addition, they are needed to reconstruct the proteinaceous components of newly dividing cells. Fats, on the other hand, are useful over the long run between meals as a source of energy to tissues, especially skeletal muscle.

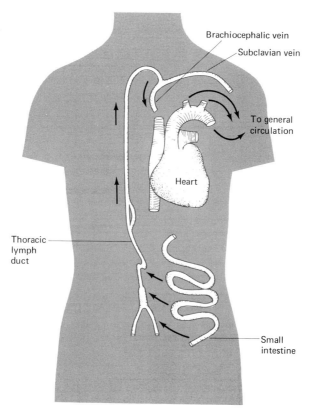

FIG. 20.2. *Lymphatic system for circulation of nutrients.*

Absorption of Nutrients and Elimination of Cell Wastes

MECHANISMS OF CELLULAR TRANSPORT

Nearly all of the digestible nutritional materials of the diet are absorbed. Normally, this is possible because food remains in the intestine for a sufficient period of time (two to four hours) and there is an adequate surface area (more than an acre) of intestine.

One might ask whether there are any substances in the diet which are absorbable, but not digestible. All of the vitamins and minerals are, to be sure. None of these substances needs to be digested—cleaved into smaller units—to become suitable for absorption.

One might also ask whether there are any digestible materials that might be absorbed without first being digested. The answer, again, is yes; there are three kinds in particular. Small amounts of dipeptides and large amounts of neutral fats are absorbed without being digested—this happens after each meal, assuming fats and proteins are present in the diet.

It is also possible that special kinds of proteins might be absorbed without being digested. Specifically, antibodies might be absorbed by babies who are breastfed. The babies might thus acquire specific immunity to certain diseases. This belief, however, is based upon observations made on cattle. As for children and adults, there is no evidence for intestinal absorption of antibodies. In fact, very shortly after birth we begin to produce our own antibodies in response to antigenic stimulation, as described in Chapter 32. Only infants therefore, might have need of an outside source of antibodies.

ANTIBODY. *A protein synthesized by lymphocytes in response to antigens which the antibodies subsequently dispose of.*

Aside from the possibility that antibody proteins may be absorbed by the intestine in nursing babies, there is no evidence that other kinds of proteins can be absorbed, except, possibly, intrinsic factor (page 388). This protein is produced in the stomach mucosa. It is needed if vitamin B_{12} is to be absorbed, thus preventing pernicious anemia. There is some evidence that intrinsic factor is absorbed through a process called phagocytosis.

The process of phagocytosis is just one of six mechanisms of cellular transport. The others are diffusion, osmosis, facilitated diffusion, active transport, and emulsification.

Diffusion

Diffusion is the directional movement of a substance from a region of high concentration to a region of low concentration. We have already mentioned that neurotransmitters diffuse from nerve endings to target cells through an intervening extracellular fluid in the course of a synapse. Inside living cells, innumerable substances are continuously diffusing from one kind of organelle, where they are produced, to other kinds of organelle, where they are consumed or modified. When we get above the level of cell organelles and begin to think of whole cells, tissues, and organs, diffusion is concerned mainly with the transport of oxygen and carbon dioxide. Most other substances are transported into or out of cells by other processes.

DIFFUSION. *A process in which particles move from a region of higher to one of lower concentration (L. diffusum, to spread out).*

Mechanisms of Cellular Transport

Osmosis

OSMOSIS. *The passage of a solvent through a semipermeable membrane, from a dilute to a concentrated solution* (G. osmos, *a pushing*).

Osmosis is a directional movement of water across a semipermeable membrane from a region of high concentration to a region of low concentration. This process was described fully in Chapter 1.

Osmosis is very important in the large intestine. Here, most of the water content of the feces is absorbed into the intestinal mucosa. If water is not reabsorbed, diarrhea occurs. Repeated bouts of diarrhea can lead to dehydration, and people suffering severe prolonged diarrhea should increase their fluid intake.

Endocytosis and Exocytosis

VESICLE. *A small saclike receptacle.*

PHAGO-. (G. phagein, *to eat*).

EMEIO-. (G. emetos, *vomit;* eidos, *appearance*).

Endocytosis and *exocytosis* are processes in which substances are transported into and out of cells, respectively, by way of *vesicles*. These processes can be observed directly through an electron microscope. When a particulate substance is transported by way of endocytosis the process is commonly called *phagocytosis*. When a liquid is transported in this manner the process is commonly called *pinocytosis*. On the other hand, when either a solid particle or a liquid substance is transported by way of exocytosis, the process is sometimes called *emeiocytosis*.

How are endocytosis and exocytosis brought about? To begin with, a substance which is to be "eaten," "drunk," or "vomited" stimulates the plasma membrane (Figure 20.3). By "stimulation" we mean that there is an interaction between the substance and certain "recognition" molecules in the cell membrane (Chapter 2). Following this interaction, the substance is engulfed into a *vesicle*, which is formed from the plasma membrane at the point of recognition. The vesicle with the engulfed substance detaches itself from the cell membrane and moves into the cell (endocytosis) or out (exocytosis). No hole is left where this process takes place, because the remainder of the cell membrane comes together.

Endocytosis and exocytosis occur commonly in the cells lining the mucous and serous membranes of the body. Mucous membranes line all of the canals which lead from the interior to the exterior of the body, such as the canal of the digestive tract. Serous membranes line the inner cavities of the body, such as the thoracic and abdominal cavities. Endocytosis and exocytosis are also prominent in all of the cells of the macrophage, or reticuloendothelial, system. This system, which extends throughout all the tissues of the body, protects the body against invasion by foreign substances.

Facilitated Diffusion

Facilitated diffusion is a special biological process in which materials are transported into or out of cells by specialized "carrier-molecules" in

Absorption of Nutrients and Elimination of Cell Wastes

the plasma membrane. In the discussion that follows, we will refer to the transportable material as "passenger-molecules." You may visualize the carrier-molecules as vehicles which take passenger molecules across a cell membrane.

There are several kinds of evidence for the existence of carrier-molecules. First, carrier-molecules are stereospecific. That is, they can selectively discriminate between two passenger-molecules that are identical to one another in every way except in being like left and right hands to one another. Carrier-molecules treat these mirror image molecules in different ways. For example, the sugar molecules D-glucose and L-glucose are identical to one another in every way except one. They have the same melting point, solubility, chemical reactivity, and other physical and chemical properties. The difference between them is that they differ in their three-dimensional configuration; they are mirror images of each other. This distinction, slight as it is, can be discerned with polarized light. It can also be discerned by the carrier-molecules within the cell membrane. Thus, D-sugars, which can be metabolized, are transported faster than L-sugars, which cannot be used by the body. So too, L-amino acids, which are metabolizable amino acids, are transported more rapidly than D-amino acids,

STEREO-. *Three dimensional (G. solid).*

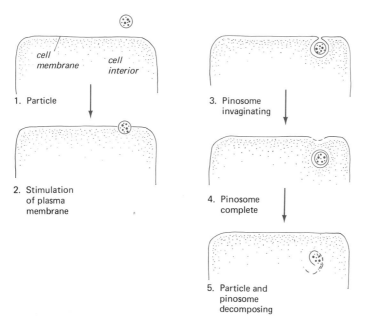

FIG. 20.3. *Transport of a fluid particle into a cell by pinocytosis.*

Mechanisms of Cellular Transport

which, like L-sugars, are of no dietary value. Some L-sugars and D-amino acids are not transported at all.

A second kind of evidence which demonstrates the existence of carrier-molecules has to do with Q_{10}. Q_{10} is a ratio of the rate at which a process occurs at one temperature to the rate at which the process occurs when the temperature is ten centigrade degrees lower. For example, Q_{10} equals the rate of a process at 20°C over its rate at 10°C. Facilitated diffusion has a Q_{10} of 2 or more. This means that the process occurs two or more times faster when the temperature is increased by ten degrees centigrade. In ordinary diffusion, the Q_{10} ranges from about 1.2 to 1.5. A Q_{10} of 2 or more signifies the making and breaking of chemical bonds. This in turn suggests the presence of the specialized carrier-molecules.

A third line of evidence for the existence of carrier-molecules is that they can be "poisoned" with very low concentrations of metabolic inhibitors, such as iodoacetate or mercury poisons. This does not happen in ordinary diffusion. If, for example, you gently layer a spoonful of 10% sugar into a cup of coffee, without stirring, the process of ordinary diffusion will not be hindered by the presence of 0.1% mercuric chloride. If the same concentration of this poison were present in the digestion tract, where facilitated diffusion occurs, the process of facilitated diffusion would be completely blocked.

One final piece of evidence of facilitated diffusion has to do with the concept of *saturation kinetics*. The following statements describe this concept: (1) The rate of transport is low when there are few passenger-molecules relative to the number of carrier-molecules in a system. (2) The rate increases when the number of passenger-molecules increases, even though temperature is kept constant. This is so because more carrier-molecules become occupied by passenger-molecules at one time. (3) As the concentration of passenger-molecules increases, the rate at which these molecules are transported also increases, but only to a limiting point. At this point, the saturation point, all of the carrier-molecules are occupied. Hence, the rate at which the passenger-molecules can be transported increases no further.

Active Transport

Active transport is a process in which a substance can be transported "uphill," from a region of low concentration to one of high concentration. It is carried out at the expense of ATP, a molecule which, when split, furnishes the energy for doing work. We do not know how this process operates, only that it does. We also know that if we try to operate an active transport system in an atmosphere of nitrogen (anaerobically), the system will not respond for long. The failure to respond favorably is due to the severe reduction in the supply of ATP, a supply which is brought about mainly in an atmosphere which has oxygen.

Active transport resembles facilitated diffusion in many ways: there is a

Absorption of Nutrients and Elimination of Cell Wastes

Q_{10} of 2 or more; the process is halted by poison; the process is stereospecific; and saturation kinetics can be observed.

Active transport differs from facilitated diffusion in two ways: (1) In active transport molecules can be transported uphill—from regions of low concentration to high concentration—or ionic molecules can be transported against an electrical gradient. For example, a negatively charged ion can be transported from a region where the charge is less negative (e.g., -50 millivolts) to a region where it is even more negative (e.g., -70 millivolts). (2) Active transport requires ATP as a source of energy.

Almost all of the simple sugars that have been studied, all of the natural amino acids and minerals, and certain vitamins (such as thiamine) are transported either actively into the intestinal cells, or through the process of facilitated diffusion. It is currently believed that the products of digestion are brought into the epithelial cells of the gut by way of active transport and that these same products are sent out of the epithelial cells, into a blood capillary or a lacteal, by way of facilitated diffusion.

ATP *(adenosine triphosphate) releases its energy when it is hydrolyzed. A phosphate group is released and* **ADP** *(adenosine diphosphate) is formed.*

Emulsification

Fats are hardly soluble, if at all, in water. If they are to be absorbed into a living cell they must be brought into solution. That is, they must be solvated by water. This is necessary if they are to pass through the aqueous envelope that lies over the surface of the cell they are to enter. The solution to the problem of insolubility calls for an emulsifier. An *emulsifier* is a substance which disperses the fat molecules and water molecules among each other, making them homogeneously distributed among themselves.

There are two major kinds of emulsifiers in the body: phospholipids and bile salts. The *phospholipids* operate as emulsifiers mainly within cell membranes. The *bile salts* operate in the intestinal juice of the small intestine.

Bile salts are steroids which are partly soluble in water and partly soluble in fat. They can remain in aqueous solution and attach themselves to fat molecules. In doing so they form an emulsion or colloidal suspension. The entire suspension—a bile salt-fat-water complex—can pass through the layer of water that lies over the surface of the cells lining the intestinal canal. After passing through this layer the complex comes into direct contact with the cell membrane. The cell membrane is itself an emulsion of protein, water, and fats (including phospholipids). The "food emulsion" diffuses readily into the cell through the "membrane emulsion."

In summary, fat transport is a process in which water and fat are brought into solution with one another by a solubilizer, or emulsifier. The resulting complex, an emulsion, is able to come into close contact with the plasma membrane, and then move into the cell itself. (A common example of an emulsion is mayonnaise. To make this dressing, two insoluble substances, oil and vinegar, are emulsified with egg yolk, a substance rich in phospholipids and the steroid cholesterol.)

EMULSION. *One liquid dispersed throughout the body of a second liquid (L. emulgere, to milk out).*

EMULSIFIER. *Agent used to assist in maintaining an emulsion.*

PHOSPHOLIPID. *A fat which has phosphorus within its chemical structure.*

BILE SALTS. *Steroids which are produced from cholesterol and are used in the gut to emulsify or solubilize fat.*

FAULTY INTESTINAL ABSORPTION

Improper intestinal absorption of digested materials may be caused by a number of factors.

Faulty Digestion

A digestive problem may be caused by an intestinal parasite, such as a tapeworm, food poisoning, an allergic response to certain foods, or some other biological or physical problem.

Often, on the other hand, the problem is caused by an emotional disturbance. This can lead to an excessive secretion of gastric juice (*hyperacidity*), intestinal muscle spasms which interfere with peristalsis (*spastic constipation*), or excessive muscular activity which results in failure to keep food in the stomach (vomiting) or intestine (diarrhea). Chronic emotional stress can induce the formation of ulcers.

SPASM. *Involuntary skeletal muscle contraction* (G. spasmos, *pull*).

Faulty Pancreas and Deficient Bile

Inadequate fat absorption, with or without faulty absorption of carbohydrates and proteins, may be caused by a deficiency of bile or pancreatic lipase. Ordinarily, an individual absorbs most of the dietary fat. Even when the intake of fat is as high as 8 or 9 ounces a day, about 95% of it is absorbed.

Faulty fat absorption can be diagnosed by placing an individual on a daily dietary intake of 4 ounces of fat. On this diet not more than a quarter ounce of fat should appear in the stools. Larger quantities signify either faulty fat absorption (bile deficiency), faulty fat digestion (lipase deficiency), or both. Faulty fat absorption may be the result of obstructive jaundice (blockage in the ductwork from the liver to the intestine), a disease of the pancreas (tumor or pancreatitis), or obstruction of the gallbladder by stones (cholelithiasis).

Defective Intestinal Enzymes

Occasionally, faulty absorption is caused by an enzyme deficiency. An example is lactase deficiency. Many adults either lack or are deficient in this enzyme. When such persons drink an excessive amount of milk, which might amount to only a cupful, the milk sugar (lactose) is not split into its components (glucose and galactose). Consequently, the sugar is not absorbed and abdominal pain, bloating, or diarrhea may occur.

Defective Transport Mechanisms

Some children are born without the carrier-molecule that transports glucose and galactose across the cell membrane. In such children these sugars remain in the intestinal canal after they are ingested. Their presence sets up a high osmotic pressure which draws water into the intestine from the body. Glucose and galactose also become metabolized by intestinal bacteria which convert them into organic acids such as lactic acid and propionic acid. The resulting acidic environment stimulates nerve endings, bringing about vigorous intestinal movements and the expulsion of the highly fluid feces (diarrhea). Unless this situation is recognized early, and fructose is substituted for dietary glucose and galactose, severe disturbances arise, ultimately leading to death.

Other kinds of carrier-molecule deficiencies are known. Most of these deficiencies involve carriers for amino acids. Two of the most common defects are (1) Hartnup's disease, in which the neutral amino acids, such as alanine and serine, are not properly absorbed, and (2) cystinuria, in which the dibasic amino acids, such as lysine and cystine, are improperly absorbed.

Anatomical Abnormalities

Often, for unknown reasons, the intestines develop anatomical abnormalities. These abnormalities include *constrictions* (strictures), *intussusceptions* (one portion of the intestine becomes telescoped within an adjacent section), inpocketings or outpocketings (*diverticula*), or a *fistula* (abnormal passageway from one region of the intestine to another or to some other visceral organ, such as the stomach). Surgery is usually required to correct these abnormalities.

Histological Abnormalities

Celiac disease in children and idiopathic steatorrhea in adults are diseases characterized by an abnormally structured mucosa in the small intestine as well as abnormalities elsewhere in the body.

Celiac disease is due to a congenital deficiency of an enzyme which reacts with gluten—a protein in wheat, barley, and rye flour. The cells of the intestinal mucosa become pathologically sensitive to gluten and gradually deteriorate. As a result, digestion and absorption are impaired, and large white fatty, frothy stools—with undigested meat fibers and undigested carbohydrates—are defecated.

In some children, celiac disease may be accompanied by *cystic fibrosis*. In other children cystic fibrosis arises in the apparent absence of a sensitivity to gluten. In either case, large cysts appear within the pancreas, and the normal polyhedral cells of this gland become fibrous and incapable of producing digestive enzymes. The pancreas thus undergoes a degeneration,

CELIAC. *Pertaining to the abdomen* (G. koilia, *belly*).

CYSTIC FIBROSIS. *A congenital disease of children, characterized by fibrosis of epithelial tissue, especially the pancreas* (G. kystis, *bladder, cyst, or sac*; fibrosis, *fibrous*).

IDIOPATHIC. (Of a disease) of an unknown cause (G. idios, one's own; pathos, disease).

as do gland cells in many other regions of the body, such as sweat glands and salivary glands.

Idiopathic steatorrhea, as its name suggests, is due to unknown causes. As with celiac disease, large quantities of undigested fat are present in the stools. In addition to the idiopathic form of steatorrhea, faulty absorption of fat (steatorrhea) may be due to a pancreatic or bile deficiency.

SUMMARY

1. The passing of materials into and out of cells is called cellular transport.
2. Intestinal absorption consists of cellular transport across the epithelial cells of the intestine followed by cellular transport at the opposite end of the epithelial cells into either a blood vessel or lymph vessel.
3. Most fatty nutrients are first transported through lymph nodes before entering the blood, while nonfatty nutrients are clarified of foreign material by way of the liver before entering other organs of the body.
4. Cellular transport is accomplished through several different mechanisms. These are diffusion; osmosis; facilitated diffusion; active transport; endocytosis or exocytosis; and solubilization or emulsification.
5. Faulty intestinal absorption may be caused by a problem in the intestine, the pancreas, or liver.

REVIEW QUESTIONS

1. Describe the route followed by a food particle in its passage from the lumen of the intestine to the bloodstream or lymph stream.
2. What is the hepatic portal system?
3. Why are all of the products of digestion passed into the hepatic portal system or the lymphatic system before they enter the general circulation?
4. What are the six primary mechanisms of cellular transport?
5. How is diffusion related to osmosis?
6. What is the difference between diffusion and facilitated diffusion? between facilitated diffusion and active transport?
7. What is an emulsion? Give an example of an emulsion.
8. Under what conditions might intestinal absorption be faulty due to a liver disease? a pancreatic disease? an intestinal disease?

Principles of Metabolism

CARBOHYDRATES
    Types of carbohydrates
    Carbohydrate metabolism
PROTEINS
    Types of proteins
        "Biological value"
        "Chemical score"
    Protein metabolism
FATS
    Types of fats
    Fat metabolism
        Beta oxidation
ACETATE
VITAMINS
MINERALS

Nutrition is the study of food requirements necessary for growth, activity, reproduction, lactation, and maintenance. Metabolism refers to the chemical changes through which nutrition is effected. *Metabolism* is the sum of all chemical changes which take place as food is converted to protoplasm, energy, and waste materials.

Like many other physiological systems, the metabolic system is exceedingly complex. A full appreciation of this system requires us to cover the following topics: (1) nutrients and the chemical changes (metabolism) they undergo; (2) the principal organs of metabolism and their regulation; (3) the caloric requirements of nutrition; and (4) nutritional aspects of the diet. The first of these topics is covered in this chapter, and the others are covered in the following three chapters.

Carbohydrates, proteins, and fats are the major, or bulk nutrients of the diet. Vitamins and minerals are minor nutrients. Water also is a nutrient but nothing will be said about its nutritional role. This chapter covers the bulk nutrients and the minor nutrients. We will describe them, classify them, and say something about their metabolism and their metabolic importance.

413

CARBOHYDRATES

Carbohydrates, as their name suggests, are compounds made of hydrated carbon atoms. Each carbon atom, on the average, carries the equivalent of one water molecule.

Types of Carbohydrates

Carbohydrates appear in food to a minor extent as monosaccharides (one-sugar units), to a larger extent as disaccharides (two-sugar units), and to a major extent as polysaccharides (polymers of several hundred to many thousands of sugar units).

Glucose is the most common monosaccharide in nature, followed by fructose and galactose. Glucose is by far the dominant sugar in blood, and it is practically the only metabolite that can be absorbed and used by nerve cells.

Sucrose (table sugar) is the most common disaccharide, followed by *lactose* (milk sugar). When these disaccharides are digested, sucrose yields glucose plus fructose while lactose yields glucose plus galactose.

Cellulose is the most abundant polysaccharide in nature. It is not digestible by humans and therefore is not available for metabolism. Its only nutritional value is to provide roughage and bulk to facilitate bowel movements.

Starch, the second most abundant polysaccharide, is completely digestible, due mainly to two enzymes, *diastase* from the pancreas and *maltase* from the epithelial cells of the small intestine. Diastase catalyzes the cleavage of maltose (disaccharide) units from starch and maltase catalyzes the cleavage of glucose (monosaccharide) units from maltose.

GLUCOSE. *Sometimes called dextrose; the sole building block of starch, glycogen, and cellulose.*

FRUCTOSE. *Sometimes called levulose; a major component of honey.*

Carbohydrate Metabolism

Food contains various kinds of carbohydrates, including starch, sucrose, and ribose, but when carbohydrate metabolism is complete, the sugar that enters the bloodstream in the largest quantity is glucose.

The relation of glucose to other kinds of nutrients in metabolism is shown in Figure 21.1. In this figure you will see that glucose is converted into metabolites of many kinds. We will discuss thirteen of them. In doing so we shall illustrate certain physiological principles, and at the same time point out the importance of each metabolite. As you read about these metabolites you should refer back to Figure 21.1. The number beside each metabolite in Figure 21.1 corresponds to the order in which the metabolites are discussed in the following pages. This will show you more clearly how each substance is related to the others.

INSULIN, *a pancreatic hormone, conserves nutrients in the body and limits the increase of blood sugar levels.*

1. *Glucose-6-phosphate.* Living cells synthesize glucose-6-phosphate at the expense of ATP. To form this compound, energy and a phosphate group are transferred from the ATP molecule to glucose. This conversion is facilitated by insulin.

Principles of Metabolism

Glucose-6-phosphate is energetically richer than glucose and it is an ionic compound with negative electric charges. On account of its higher energy content, glucose-6-phosphate is more reactive than glucose. It is immediately converted into one or another of several possible products, as shown in Figure 21.1. Because of its negative ionic charge, glucose-6-phosphate is repelled by the negative charges on the interior surface of the cell membrane. Its probability of escaping from the cell—unlike the probability of glucose escaping—is nearly zero.

2. *Glycogen.* The body's only storage polysaccharide is glycogen, a substance that chemically resembles starch. It is produced from glucose-6-phosphate in many organs of the body, such as the skin, brain, and kidneys. The bulk of it is produced and stored in the liver (3 to 4 ounces, following a meal) and in cardiac and skeletal muscle (8 to 10 ounces, most of the time).

Almost all of the glycogen in the liver is used up during an overnight fast. As may be seen in Figure 21.1, glycogen can be reconverted into glucose within the liver, and this carbohydrate can now be circulated in the blood to needy tissues throughout the body—especially to nerve tissues. With regard to skeletal and cardiac muscle, there is no enzyme present for reconverting glycogen into glucose. Instead, all of the glucose units from the glycogen store pass mainly into the Krebs cycle (see Figure 21.1), where the glucose is oxidized to provide energy (ATP) to the muscle cell.

Glycogen, unlike fat, is a temporary, short-lived source of energy. Once it

KREBS. *An English biochemist (1900–*

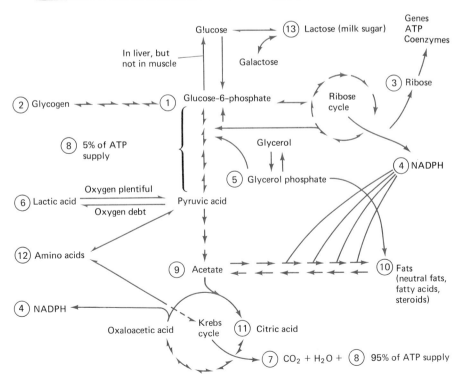

FIG. 21.1. *General metabolism. The number beside each metabolite corresponds to the order in which the metabolites are discussed in the text.*

Carbohydrates

is depleted—assuming an individual is fasting—it is necessary to rely on protein or fat as a source of energy.

3. *Ribose.* Ribose, a monosaccharide, is needed if the cell is to produce ATP, RNA, DNA, and certain coenzymes. ATP, the currency of physiological energy, was described in Chapter 1. RNA and DNA, which are important in protein synthesis and in cell division, were discussed in Chapter 2. *Coenzymes* are substances that work with certain enzymes to catalyze chemical reactions in living cells. Figure 21.2 shows where many of the vitamins—as coenzymes—perform a metabolic function. Several of these coenzymes contain ribose as an integral part of their structure.

NADPH. *The reduced form of nicotinamide adenine dinucleotide phosphate, a derivative of vitamin B_1.*

NADP. *The oxidized form of NADPH.*

4. *NADPH.* NADPH is a coenzyme derived from vitamin B_1. The most important fact about NADPH is that it is absolutely necessary for producing fats from carbohydrates and proteins. When the diet is rich in carbohydrates, most of our NADPH is produced from NADP in the course of making ribose from glucose (Figure 21.1). When the diet is poor in carbohydrate and rich in protein, most of the NADPH is probably produced from NADP in the course of making pyruvic acid from malic acid (an intermediate product of the Krebs cycle, Figure 21.1).

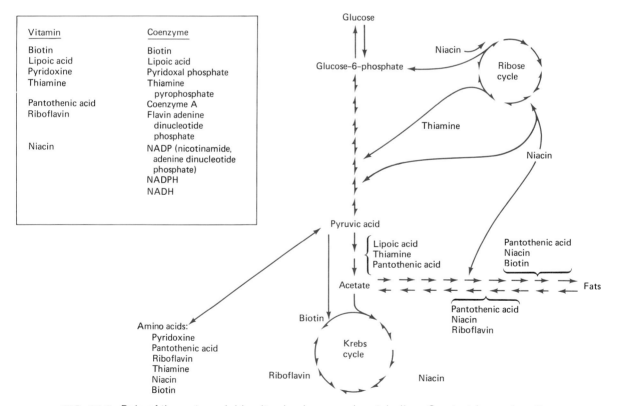

FIG. 21.2. *Role of the water-soluble vitamins in general metabolism. See text for explanation.*

Principles of Metabolism

5. *Glycerol Phosphate.* Glycerol phosphate, or just glycerol alone, may be used by liver cells to produce neutral fats. *Neutral fats*, known also as *triglycerides*, are the main storage fats in adipose cells. In adipose cells, however, only glycerol phosphate (and not glycerol) can be used to synthesize neutral fats. This is an important nutritional fact since it points out the importance of carbohydrate metabolism to the process of fat storage in adipose cells. You may see how this comes about by studying Figure 21.1.

NEUTRAL FAT. *A fat which has no acidic (H^+) components.*

6. *Lactic Acid.* In the presence of adequate amounts of oxygen, lactic acid is converted into pyruvic acid and then into acetate. Acetate can be channeled into various metabolic pathways (Figure 21.1).

In the absence of adequate amounts of oxygen, however, lactic acid accumulates in the body. It accumulates because there is not enough oxygen to oxidize the carbon and hydrogen atoms of pyruvic acid. The accumulation of lactic acid occurs during anxiety, wherein larger than normal amounts of lactic acid appear in the blood. Lactic acid also accumulates in overworked skeletal muscles, as often happens during strenuous physical activity. Usually, the affected muscles feel painful and stiff. The pain is due to the stimulation of nociceptors by the accumulating acid, and the stiffness is due to the difficulty actin and myosin have in sliding past one another. The muscles are said to have incurred an *oxygen debt*. After a period of time—which can be reduced by heat and massage—an adequate amount of oxygen will be brought into the muscle to allow lactic acid to pass on into the terminal steps of oxidation (Krebs cycle, Figure 21.1). In contrast to skeletal muscle, heart muscle cannot incur an oxygen debt. The reasons are given in Chapter 17.

NOCICEPTOR. *A pain receptor.*

OXYGEN DEBT. *The accumulation of lactic acid in skeletal muscles owing to an inefficient coupling of work with metabolism.*

The ability to incur an oxygen debt has a survival value. Without an oxygen debt, the muscles might be overexerted without forewarning. They might become irreversibly damaged. With an oxygen debt, on the other hand, the individual is virtually forced to rest.

7. *Carbon Dioxide and Water.* The production of carbon dioxide and water correlated with oxygen consumption, and therefore with energy production. More than 95% of these products are derived through the terminal steps of oxidation in the mitochondria, where 95% of the body's intake of oxygen is consumed. The consumption of 1 liter of oxygen is equivalent to the liberation of 5 Calories of energy from food.

8. *ATP.* In the course of consuming the bulk of oxygen and producing the bulk of carbon dioxide and water, organic molecules pass through the *Krebs cycle*, which is also known as the *citric acid cycle* or *tricarboxylic acid (TCA) cycle*. The reader is urged at this point to note where this activity occurs with respect to overall metabolism (Figure 21.1). It is through the enzymatic activities of Krebs cycle, which occur in the mitochondria of cells, that more than 95% of the ATP is produced. Smaller quantities are produced earlier in carbohydrate metabolism, in the steps between glucose and pyruvic acid.

9. *Acetate.* Prior to its entrance into Krebs cycle, glucose is converted into acetate. Figure 21.1 shows that in carbohydrate metabolism acetate is

Proteins

derived directly from pyruvic acid. Acetate is a two-carbon molecule which condenses with a four-carbon oxaloacetate molecule as it enters the Krebs cycle. The product of this condensation is citric acid, a six-carbon molecule.

10. *Fats.* In general, two classes of fats may be produced from glucose when excessive carbohydrates are eaten. The central molecule in this conversion is acetate (Figure 21.1).

Neutral Fats. Eight to 10 acetate molecules usually condense to form fatty acid molecules composed of 16 to 20 carbon atoms. Three fatty acid molecules can unite with glycerol to produce a neutral fat. This union is due to *esterification*, a reaction carried out between an alcohol and an acid. Esterification is the reverse of hydrolysis. To form a neutral fat, three H atoms are removed from glycerol, and an OH group is removed from each of three fatty acids to form three ester bonds.

HYDROLYSIS. *Splitting a chemical bond with water.*

Steroids. Instead of being used to produce fatty acids and neutral fats, acetate molecules may be assembled into cholesterol. This steroid, in turn, can be used by liver cells to produce bile salts; by the adrenal cortex to produce aldosterone and cortisol; and by the gonads to produce sex hormones (Figure 21.1).

11. *Citric Acid.* As indicated above, citric acid is the product of a reaction in which acetate combines with oxaloacetate. When excessive amounts of carbohydrate are eaten, large quantities of citric acid build up within certain cells, especially the liver cells, and as a result, the metabolic machinery turns toward the synthesis of fats from carbohydrates.

12. *Amino Acids.* Fourteen of the 22 natural amino acids are called "*nonessential amino acids.*" They can be produced in the body from fats (or carbohydrates) and a source of ammonia; they are therefore not essential to the diet. The remaining 8 amino acids are "*essential.*" That is, they must be included in the diet in order to meet the nutritional needs of the body. These essential amino acids are isoleucine, leucine, lysine, methionine, phenylalanine, threonine, tryptophan, and valine.

13. *Lactose.* The sugar molecule lactose is produced in the mammary glands from glucose and galactose. It is produced in increasing amounts during pregnancy; its production continues if breast-feeding is practiced.

PROTEINS

PROTEIN. *A polymer made of amino acids* (G. proteios, first in rank).

Proteins are polymers made from amino acids. Amino acids are organic molecules that contain at least one carboxylic acid group (COO^-H^+) and one amino group (NH_2). When the carboxylic group on one amino acid is attached to the amino group on another amino acid a water molecule is split off in the process and a peptide bond is formed. During digestion, water is reintroduced and the peptide bond is split. This process is illustrated on page 19.

Although many kinds of amino acid are found in nature, there are only 22 so-called "natural" amino acids. Each of these contains carbon, oxygen, hydrogen, and nitrogen. Three of the natural amino acids also contain sulfur (S).

Types of Proteins

In the section on carbohydrates we described various types of carbohydrates, including glucose, starch, and cellulose. Similarly, in the section on fats we will discuss various types of fats. But with proteins, from a nutritional viewpoint, the situation is different. Here, we can only talk about the nutritional value of a protein. This is usually done in terms of the "biological value" and the "chemical score" of a particular source of protein, such as egg protein, meat protein, and vegetable protein.

"Biological Value." The *biological value* of a protein indicates the protein's capacity for maintaining an individual in a proper nitrogen balance. Three situations may be considered in this connection. (1) In general, while growing, individuals tend to be on a nitrogen *positive balance;* that is, more nitrogen is taken in than is given out. The same relationship holds true for well-nourished athletes who are in training. (2) For most healthy normal adults, the amount of nitrogen absorbed is about equal to the amount excreted; the individual is on a nitrogen *neutral balance.* (3) In cases of protein malnutrition, more nitrogen is excreted than is absorbed; the individual is on a nitrogen *negative balance.*

In practice, the biological value of a protein is assessed by feeding a particular protein (eggs, for example) to a population of test subjects, and measuring how much nitrogen is absorbed and how much is retained. Suppose, for example, that 100 grams of nitrogen are consumed as protein. If 10 grams of nitrogen are then found in the feces, 90 grams of nitrogen were absorbed by the intestine. If 10 grams are found in the urine (in addition to the 10 grams in the feces), 80 grams were retained by the body. Now we set up a formula:

$$\frac{\text{Nitrogen}}{\text{assimilated}} = \frac{\text{grams nitrogen retained by the body}}{\text{grams nitrogen absorbed into the intestine}} \times 100$$

For the example given here, 89% of the nitrogen was assimilated.

Some proteins are assimilated better than others. Egg protein is considered the "perfect protein" from this viewpoint. It is assigned a biological value of 100. Most proteins are less effective insofar as nitrogen retention is concerned. This may be seen in Table 21.1. In general, animal proteins have a higher biological value for humans than plant proteins.

"Chemical Score." Some proteins are rich in essential amino acids while others are relatively poor. Gelatin, for example, has no tryptophan. Casein,

ASSIMILATION or **ANABOLISM.** *The process of transforming food into protoplasm or energy* (*L.* assimilatio, *to make similar*).

TABLE 21.1. Nutritional Values of Proteins.[1]

	Chemical Score	Biological Value
CHEESE	100	90
MILK (COW)	80	80
EGG	100	100
PORK TENDERLOIN	90	
FISH (LEAN)	70	95
OATS	80	90
RYE	80	60
RICE	70	70
WHITE FLOUR	50	50
WHEAT GERM	60	90
SOY FLOUR	70	70
POTATO	60	80
SWEET POTATO	80	
BEANS	50	45
PEAS	60	70
SPINACH	70	65

1. *Protein requirements.* Food and Agricultural Organisation, United Nations, Committee on Protein Requirements. FAO Nutritional Studies No. 16 (Rome, 1957).

CASEIN. *A protein obtained from milk by treating milk with renin or acids (L. caseus, cheese).*

or milk protein, is low in sulfur-containing essential amino acids. Egg is rich in all of the essential amino acids. The capacity of a protein to provide essential amino acids is measured by its ability to support growth or maintain health. One way to provide an index to this capacity is to assign the protein a *"chemical score."*

The chemical score of a given food measures the extent to which it furnishes its least abundant essential amino acid. A protein may be deficient in more than one essential amino acid (as determined by its failure to support growth or to maintain health), but the chemical score is based upon the amino acid which is *most* limiting in this respect.

Potato protein, for example, is deficient in four amino acids, but it is most limiting with respect to tryptophan. The chemical score for potato varies from one crop to another, but it has an average chemical score of about 72. This value is based on what the Food and Agricultural Organization of the United Nations has established as a baseline value for the amount of tryptophan that potatoes should contain if they are to support growth and maintain health.

Table 21.1 lists the chemical scores of a variety of common protein sources. From this table it may be noted that egg has the highest chemical score, and that animal proteins are superior to plant proteins.

Protein Metabolism

Protein metabolism begins with the digestion of proteins and absorption of amino acids into the villi of the intestine. From the villi, the amino acids pass into the blood and then into the liver, where they enter one of three pathways (Figure 21.3).

Some of the amino acids are synthesized into albumin, globulins, enzymes, or other kinds of proteins. Other amino acids do not enter liver cells. Instead, they pass on to other tissues in the body, where they are used for synthesizing hormones, neurotransmitters, enzymes, or structural proteins for cells. Other amino acids, through regulatory mechanisms which are totally unknown to us, are treated by the body as excess material. They are stripped of their nitrogen, which is channeled (as ammonia, NH_3) into the formation of urea ($CO(NH_2)_2$), our principal nitrogenous waste product. The carbon skeletons of these amino acids are converted into glucose or fatty acids, or they are channeled into the Krebs cycle to produce ATP.

The process of converting amino acids into glucose is called *gluconeogenesis* (Figure 21.1). All amino acids, except leucine, are known to be gluconeogenic. Glycerol, glycerol phosphate, and lactic acid are the only other kinds of common metabolites which can be converted into glucose through the process of gluconeogenesis.

Gluconeogenesis is extremely important in providing glucose to the nervous system, and to the lens of the eye, whenever liver glycogen is exhausted. Liver glycogen may be exhausted at the approach of dawn, following an overnight fasting period, or it may be exhausted during the daytime, several hours after a meal, especially if a person has engaged in vigorous

GLUCONEOGENESIS. *The process of producing glucose from amino acids, glycerol, or lactic acid* (G. glykos, sweet; neo, new; genesis, to make).

FIG. 21.3. *The pathway of protein metabolism.*

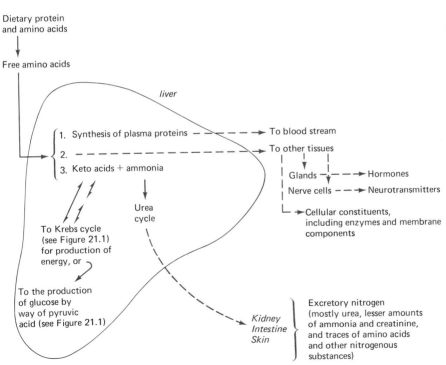

physical activity. Since the nervous system and lens can use only glucose as an energy source, gluconeogenesis is an important process.

FATS

Fats (lipids) are metabolites composed predominantly of carbon and hydrogen atoms. Fats come in a variety of shapes and sizes. Most of them carry an occasional oxygen atom, and a few of them (phospholipids) carry phosphorus and nitrogen as well.

Types of Fats

FAT. *A class of combustible, ether-soluble, water-insoluble organic molecules.*

Six classes of fats are found throughout the body: fatty acids, neutral fats, phospholipids, sphingolipids, steroids and prostaglandins.

Fatty Acids. Fatty acids bear COO^-H^+ on one end of the molecule. The hydrogen ion can dissociate ($COO^-H^+ \rightleftharpoons COO^- + H^+$) and it is this ion itself which is the acid.

FATTY ACID. *A fat with an acid (H^+) component.*

A fatty acid may be saturated, monounsaturated, or polyunsaturated. In a *saturated fatty acid*, such as palmitic acid, no additional hydrogen atoms can be attached to the molecule's carbon atoms. Saturated fatty acids are useful in the diet mainly as a source of energy. In a *monounsaturated fatty acid*, such as oleic acid, one hydrogen atom can be added to each of two interior neighboring carbon atoms. Monounsaturated fatty acids also are useful in the diet mainly as a source of energy. In a *polyunsaturated fatty acid*, two or more pairs of carbon atoms can be hydrogenated. Linolenic and linoleic acids are very important dietary polyunsaturated fatty acids. They are called "essential" fatty acids because they must be eaten; they cannot be produced in the body from other metabolites to meet physiological requirements.

Neutral Fats. When the COO^-H^+ group of a fatty acid unites with an OH group of a glycerol molecule, the acidity of the fatty acid is neutralized. Glycerol is able to combine in this way with a total of three fatty acids. When this happens, the product is called a *triglyceride*, or *neutral fat*. Approximately 98% of all lipids eaten belong in this category of fats.

TRIGLYCERIDES. *Neutral fats, made of three fatty acids and glycerol.*

Phospholipids. Lecithin is typical of the phospholipids. These fats resemble neutral fats but they contain phosphorus and nitrogen. Phospholipids are especially prominent in cell membranes, where they aid in the transport of metabolites into and out of the cell.

Sphingolipids. The sphingolipids are phospholipids that consist of a fatty acid, a nitrogenous alcohol (*sphingosine*), a sugar, and a phosphorus-containing substance. Sphingolipids are found in membranes. Their importance is known because of *lipid storage diseases* which involve problems in sphingolipid metabolism. Ten lipid storage diseases are known (Gaucher's disease and Niemann-Pick disease, for example). All of these diseases are inherited. They afflict the nervous system and other organs, and cause degeneration of the kidney, liver, spleen, skin, and other organs. Today these diseases can be accurately diagnosed and treated with various forms of therapy.

Steroids. Cholesterol, bile salts, sex hormones, and adrenal cortical hormones are steroids, or cyclic fats. Certain dietary steroids are the precursors of vitamin D.

Prostaglandins. Prostaglandins, which are derivatives of fatty acids, were so named because they were first isolated from the prostate gland as a component of semen. Although prostaglandins are highly concentrated in the prostate gland, they are present in every tissue of the body.

PROSTAGLANDIN. *A fat which stimulates smooth muscle.*

Two major classes of prostaglandins are known. These classes, which are based on differences in chemical structure, are called PGE and PGF. Class *PGE* prostaglandins predominate in semen and they are known to have a powerful inhibitory influence on the contraction of the nonpregnant uterus. This finding suggests that PGE prostaglandins play a role in relaxing the uterus to facilitate the movement of sperm into the fallopian tubes. Class *PGF* prostaglandins predominate in menstrual fluid where they stimulate contractions of the uterine muscle. This contractile action suggests that PGF prostaglandins help expel the menses. During menstruation the uterine content of PGF prostaglandins reaches its highest level.

Interestingly, both PGE and PGF prostaglandins stimulate the contraction of the smooth muscle of a pregnant uterus. This suggests a role of prostaglandins in childbirth. It also suggests that production and release of prostaglandins may be the basis for the operation of the intrauterine device (IUD) in birth control. The possibility of using prostaglandins directly as a birth control device is currently undergoing research.

In addition to affecting the uterine muscle, class PGE prostaglandins dilate the bronchi in the lungs, while prostaglandins of class PGF constrict the bronchi. This information may be useful someday for the treatment of asthma. Also, certain prostaglandins inhibit stomach gastric secretions, and have been used with some success for reducing distress and pain caused by stomach and duodenal ulcers. Finally, certain prostaglandins produce pain in an inflammation, pain which can often be reduced by aspirin (Chapter 31).

ASTHMA. *A constriction of the chest characterized by difficult breathing, coughing, and wheezing (G., a panting).*

Fat Metabolism

Small fatty acids, such as the butyric acid ($CH_3CH_2CH_2COOH$) commonly found in dairy products, pass into the bloodstream of the intestinal tract and

CHYLOMICRON. *A microscopic globule of fat and protein appearing in the blood after a meal.*

are taken directly to the liver. Larger fatty acids, neutral fats, and other lipids pass by way of the lymph stream into the blood. They enter the lymph and blood as *chylomicrons*—globules that contain the fats complexed with protein.

The fats are taken out of the chylomicrons by adipose cells and liver cells (Figure 21.4). In these tissues the fats are cleaved, yielding three fatty acids (and one glycerol molecule) per molecule of neutral fat.

Fatty acids ultimately enter one of the three metabolic pathways indicated in Figure 21.4. They may be sent to tissues throughout the body as a source of energy. They may be used by any cell of the body to form membranes or other structures. Or they may be degraded by the process of beta oxidation into acetate units.

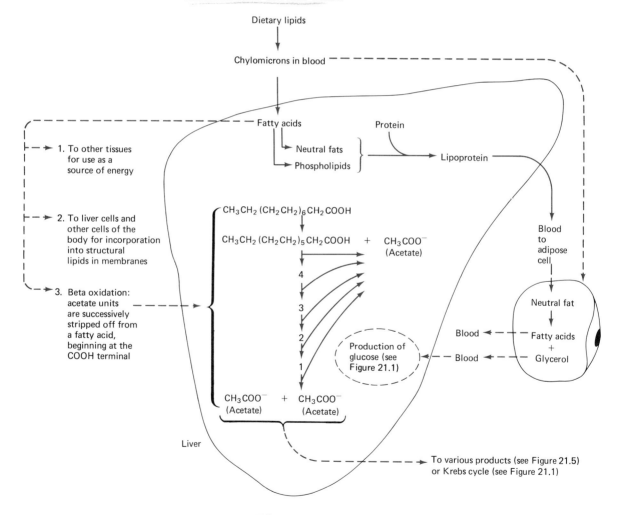

FIG. 21.4. *Fat metabolism.*

Principles of Metabolism

Beta Oxidation. In beta oxidation, a specialized metabolic process, a fatty acid is cleaved—two carbon atoms at a time—beginning at the COO^-H^+ end, as shown in Figure 21.4. (The COO^-H^+ carbon is called the alpha carbon; the carbon next to it is the beta carbon—hence the name beta oxidation.)

As a result of beta oxidation, a fatty acid may be converted completely into acetate. A 16-carbon fatty acid (such as palmitic acid) yields 8 acetate fragments.

BETA OXIDATION. *A metabolic process in which fatty acids are converted into acetate.*

ACETATE

Acetate, which actually is found in the cell as *acetyl coenzyme A*—a very complex derivative—is an extremely important compound in the economy of the living cell. Figure 21.1 shows that acetate is produced not only from fatty acids but from protein and carbohydrate as well. Above all else, the survival of any living cell is dependent upon a source of energy. What better safeguard could nature have provided than the ATP-generating system of the Krebs cycle, into which all three classes of the bulk foodstuffs can be directed? (See Figure 21.1).

ACETYL COENZYME A. *A coenzyme made from acetate and a vitamin B derivative.*

Approximately 95% of our energy is ultimately derived from ATP that is generated by combusting acetate through the Krebs cycle. It has been estimated that approximately 70% of the carbon atoms from carbohydrates and glycerol, 50% of the carbon atoms from amino acids, and 100% of the carbon atoms of fatty acids become acetate in the course of metabolism.

In addition to its importance as the mediator for the conversion of carbohydrate, protein, and fat into energy, acetate serves as an important building block for the synthesis of numerous kinds of metabolites, such as cholesterol, the sex hormones, the hormones of the adrenal cortex, bile salts, and fatty acids (Figure 21.5).

Although acetate can be derived from fatty acids, and fatty acids can be produced from acetate, entirely different enzymes and metabolic pathways

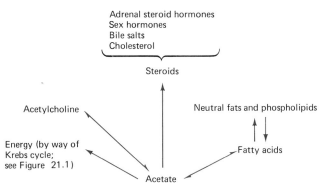

FIG. 21.5. *Metabolic products of acetate.*

are involved in each of these processes. In general, the synthesis of fatty acids is stepped up in the presence of excess dietary carbohydrates and insulin, whereas beta oxidation is promoted when the dietary level of carbohydrate is reduced.

VITAMINS

The word *vitamin* was coined from "vital" and "amine," in reference to the vital need for certain nitrogenous (amine) substances in the diet. (Inasmuch as vitamins A, C, D, E, and K are not nitrogenous, it is perhaps more helpful—and certainly more accurate—to think of vitamins as *vit*al substances needed at some *min*imum level.)

There are two major classes of vitamins: the water-soluble and the fat-soluble. Vitamins A, D, E, and K are fat-soluble vitamins. All other known vitamins are water-soluble. Animal foods (meat, milk, eggs, and fish) are generally richer in fat-soluble vitamins than plant foods. Plant foods are richer in water-soluble vitamins.

Table 21.2 lists the vitamins and some of their metabolic functions. Figure 21.2 shows where some of the water-soluble vitamins carry out their metabolic functions. In all cases, the vitamins act as coenzymes; that is, they act together with enzymes to catalyze metabolic reactions. The fat-soluble vitamins work together with various components of cell membranes to bring about special effects such as transport of calcium (vitamin D), reception of light (vitamin A), synthesis of certain blood-clotting factors (vitamin K), and possibly prevention of the oxidations which lead to aging (vitamin E).

Some vitamins behave as coenzymes without having to be modified. Biotin and vitamin C fall into this class. Most vitamins, however, must be modified by enzyme-catalyzed reactions before they can function as coenzymes. Thus, with respect to Figure 21.2, wherever niacin appears, either nicotinamide adenine dinucleotide or nicotinamide adenine dinucleotide phosphate (NAD or NADP) is the coenzyme. In certain reactions either one of these coenzymes can work effectively. In other reactions only one of them can be used. Thiamine pyrophosphate, coenzyme A, and pyridoxal phosphate are the coenzyme forms of thiamine, pantothenic acid, and vitamin B_6, respectively.

MINERALS

A mineral is any inorganic substance from the earth's crust. In our bodies, minerals perform either a structural or a regulatory function. The *structural function* is exemplified best by the incorporation of calcium and phosphate into the skeleton. The *regulatory function* of minerals is exemplified by: (1) the roles played by sodium, chloride, and potassium in the propagation of impulses along nerve and muscle membranes; (2) the role of calcium in the muscle contraction and magnesium in muscle relaxation; (3) the action of iron, as part of hemoglobin, in carrying oxygen

TABLE 21.2. Principal metabolic functions of vitamins.

Water-soluble Vitamins	Metabolic Functions
Thiamine	Metabolism of carbohydrates, especially the conversion of pyruvic acid to acetate and the combustion of sugars from the ribose cycle
Riboflavin	Involved in metabolism—especially combustion—of carbohydrates and amino acids
Niacin	Metabolism of carbohydrates, amino acids, and fats; involved in more than 100 different kinds of metabolic reactions
Pyridoxine	Amino acid metabolism
Biotin	Metabolism of amino acids and pyruvic acid; synthesis of fatty acids
Pantothenic acid	Metabolism of carbohydrates, amino acids, and fatty acids; a component of coenzyme A and therefore an activator of acetate (acetyl coenzyme A)
Folic acid	Metabolism of nucleic acids and certain amino acids
Vitamin B_{12}	Metabolism of nucleic acids and certain amino acids
Ascorbic acid (vitamin C)	Overall metabolism; promotes oxidation of fatty acids and certain amino acids; also aids in formation of certain kinds of proteins, such as collagen
Fat-soluble vitamins	Metabolic Functions
Vitamin A	Important in synthesizing mucopolysaccharides (nitrogen-containing sugars) which are found in connective tissues; also used in producing light-receiving sensory portion of the visual receptor cells
Vitamin D	Enhances absorption of calcium (and phosphate) in various cells
Vitamin E	Protects unsaturated fatty acids and vitamin A against being excessively oxidized
Vitamin K	Essential for production of certain blood coagulation factors and in the production of ATP in the Krebs cycle

Review Questions

through the circulatory system; (4) various minerals (zinc, copper, iron, magnesium, potassium, molybdenum) that act as coenzymes in catalyzing enzyme reactions; and (5) the functioning of minerals, in general, in maintaining a proper water balance between intracellular and extracellular fluids.

Five minerals—calcium, phosphorus, iron, magnesium, and fluoride—are of general dietary importance. Recommended daily allowances of these minerals are given in Table 24.3, and a discussion of each of these minerals is presented in Chapter 24.

SUMMARY

1. The focal point of carbohydrate metabolism is glucose, which can be converted to glycogen, ribose, lactic acid, acetate, fats, carbon dioxide, water, and many other metabolites.
2. Proteins are degraded by way of amino acids, which are stripped of ammonia, at which point they can be converted into glucose, fat, carbon dioxide, water, and many other metabolites. All of the proteins can be ranked in terms of their ability to maintain nitrogen balance or to provide essential amino acids.
3. Fats are degraded mainly by way of beta oxidation, which yields acetate fragments.
4. Acetate, which is produced from carbohydrates, fats, and proteins, can be oxidized to produce ATP. Or it can be used to synthesize cholesterol, steroid hormones, and bile salts.
5. Water-soluble vitamins are important as coenzymes in metabolic reactions.
6. Minerals perform structural and/or regulatory functions in the body.

REVIEW QUESTIONS

1. List at least eight substances that are metabolically derived from glucose. Give one important fact about each of these substances.
2. What are proteins made of?
3. Define chemical score and biological value with respect to protein metabolism.
4. Differentiate carbohydrates, fats, and proteins on the basis of their chemical composition.
5. Define: fatty acid, neutral fat, phospholipid, prostaglandin.
6. What is lecithin?
7. Describe beta oxidation.
8. Name at least five important substances that are produced in the body from acetate.
9. Why are water-soluble vitamins important in general metabolism?
10. Name the four fat-soluble vitamins and give a function of each.

The Principal Organs of Metabolism   22

PRINCIPAL ORGANS OF METABOLISM
    The liver
        Interconversion of carbohydrates, fats, and proteins
        Removal of excess nitrogen from the body
        Detoxication
        Destruction of worn-out hemoglobin
        Storage
        Synthesis
    Muscle
    Adipose
    Kidneys
    Central nervous system
HORMONAL REGULATION OF METABOLISM
    Blood levels of glucose
        The oral glucose tolerance test
        Hypoglycemia and hyperglycemia
    Blood levels of proteins
    Blood levels of fats
THE PRINCIPAL REGULATORY HORMONES OF METABOLISM
    Insulin
    Glucagon
    Adrenaline
    Growth hormone
    Glucocorticoids
    Thyroxine
TRANSLATION OF HORMONAL MESSAGES INTO ACTION

In the preceding chapter we discussed the main classes of nutrients and their metabolism. In this chapter we will describe the specific roles of the major organs of metabolism. We will also discuss how these organs are regulated in order to maintain normal levels of carbohydrates, fats, and proteins in the blood.

Principal Organs of Metabolism

PRINCIPAL ORGANS OF METABOLISM

Although each of the body's organs is directly involved in metabolism, five organs are particularly worthy of singling out for their special roles. These organs are the liver, muscles, adipose tissue, kidneys, and the central nervous system.

The Liver

The liver, which is described in Chapter 19, is the central effector of metabolism. It is especially noteworthy for the following six functions:

Interconversion of Carbohydrates, Fats, and Proteins. Although many kinds of cells can interconvert carbohydrates, fats, and proteins, the liver carries out this process on a grand scale in accord with bodily needs. If, for example, the blood sugar (glucose) level is lower than normal, the liver produces additional amounts of glucose from amino acids, glycerol, and lactic acid. On the other hand, if certain nonessential amino acids are needed, they can be synthesized in the liver from fats or carbohydrates. Finally, if the blood sugar level is adequate and excessive protein or carbohydrate is eaten, the carbon framework of the surplus protein and carbohydrate molecules can be converted by the liver into fat molecules.

Removal of Excess Nitrogen from the Body. There is no provision in living cells for the storage of excess nitrogenous compounds, such as amino acids. Liver cells strip ammonia (NH_3) from excessive nitrogenous compounds and convert the ammonia into *urea* ($CO(NH_2)_2$). The urea is sent by way of the bloodstream to the kidneys, from where it is excreted in the urine.

UREA. *The major nitrogen-containing waste of the body.*

Detoxication. Many kinds of foreign substances, such as drugs and pollutants — substances which cannot be used as a source of energy or matter for protoplasm — are eliminated soon after entering the body. Most of these foreign substances are first modified in liver cells. They are partially oxidized or reduced or they undergo some other chemical modification. Some foreign materials do not have to undergo preliminary modification. In either case, the foreign substance is usually converted by the liver into a salt which can be eliminated by the gut or skin, but mainly by the kidneys. This topic is discussed further in Chapter 31.

Destruction of Worn-out Hemoglobin. Red blood cells have an average life span of 120 days in our body. When they become worn-out they are destroyed by macrophage cells throughout the body, including the Kupffer cells of the liver. This topic is discussed further in Chapter 29.

Storage. The liver stores varying amounts of carbohydrates, proteins, and fats, most of the body's fat-soluble vitamins (A,D,E,K), and various metals, particularly iron and copper.

Synthesis. The liver synthesizes a wide variety of substances needed by the body. The principal substances are glycogen, albumin, globulins, prothrombin, fibrinogen, bile salts, heme, glycerol phosphate, and neutral fats.

(1) *Glycogen.* Shortly after a meal, when excessive amounts of glucose and amino acids begin to appear in the blood, the liver begins to convert these metabolites into glycogen. About two hours after a meal, liver glycogen becomes the principal source of glucose, which is sent to needy cells throughout the body.

In addition to producing glycogen, the liver is important as a clearinghouse for large loads of glucose. It is able to do this because it possesses a special kind of *hexokinase*—an enzyme which converts glucose to glucose-6-phosphate (Figure 21.1). Although hexokinases are widespread throughout the body, and various forms may be found in a cell, only the liver cell has a type called *glucokinase*.

In contrast to other kinds of hexokinases, glucokinase has an extremely high binding affinity and reaction velocity for glucose, which explains its ability to convert large quantities of glucose into glycogen. Glucokinase disappears when the diet is low in carbohydrates or when a person is fasting, and it appears when large amounts of glucose enter the blood from the intestine. When glucokinase is not present, the type of hexokinase found in most of the cells of the body initiates the conversion of blood glucose into liver glucose-6-phosphate.

(2) *Albumin.* Approximately 10 grams of this glycoprotein are produced daily in liver cells. It is sent into the bloodstream as a serum protein, where it is maintained at a relatively constant concentration of 4 grams per 100 milliliters of plasma. Each albumin molecule remains for about 3 weeks in the bloodstream, where it serves at least four functions: as a reserve protein in the event of nutritional need; as a buffer of pH; as a vehicle for conveying some otherwise-insoluble fatty acids and steroids through the bloodstream; and as an osmotic agent which promotes an exchange of wastes and nutrients in capillary beds and facilitates a balance of fluid between the intracellular and extracellular spaces. (The role of albumin as an osmotic agent is discussed further in Chapter 26.)

(3) *Globulins.* Three classes of globulins (specialized glycoproteins) are present in blood plasma: alpha globulins, beta globulins, and gamma globulins. Each class can be separated from the others through a process called *electrophoresis*, which is shown in Figure 32.4.

The *alpha* and *beta globulins* are produced by the liver. Each of these classes contains a variety of proteins which perform various functions in the body. *Transferrin*, for example, is a beta globulin which transports iron to needy tissues. (This process is discussed in Chapter 24.) *Ceruloplasmin* is an alpha globulin whose metabolic function is not understood. Its absence, however, causes the liver to degenerate (Wilson's disease). *Angiotensin* and several other constrictors and dilators of blood vessels are also carried in the alpha or beta globulin fractions of the plasma proteins.

The *gamma globulin* proteins, in contrast to alpha and beta globulins,

HEXOKINASE. *An enzyme that catalyzes the attachment of a phosphate group on a six-carbon (hexo) sugar, such as glucose, galactose, or fructose.*

GLOBULINS. *Serum proteins which differ from albumin (also a serum protein) in being less soluble and larger in molecular weight.*

ANGIOTENSIN. *A small protein that induces constriction of blood vessels.*

are produced mainly by lymphocytes and plasma cells. Most of the proteins which are present in the gamma globulin fraction of blood are antibodies (Chapter 32).

(4) *Prothrombin and fibrinogen.* These proteins are used in blood clotting. They are secreted by liver cells into the bloodstream, where they are kept at relatively constant levels, always ready for clotting should the need arise (Chapter 31).

(5) *Bile salts.* Bile salts (Chapters 19 and 20) are steroids used in the digestion and absorption of fats. They are produced in liver cells from cholesterol.

(6) *Heme.* An important component of the blood, heme contains nitrogen and iron. It is produced in young red blood cells (reticulocytes), and it is then combined with a protein (globin) to form *hemoglobin* (Chapter 29). Heme is also produced in liver cells, after which it is sent through the blood to reticulocytes in red bone marrow.

(7) *Glycerol phosphate and neutral fats.* Varying amounts of fatty acids are picked up by the liver from time to time. Some of these fatty acids are used as a source of energy by the liver. Others are incorporated into the liver's protoplasm and still others are treated as excess.

The excess fatty acids are combined with glycerol phosphate, a metabolite produced from glucose (Figure 21.1). Through a series of enzyme reactions, the phosphate ion is stripped from glycerol phosphate and a molecule of neutral fat is produced. A portion of this neutral fat is retained by the liver (normal liver contains about 4% fat by weight) and the remainder is sent into the bloodstream, where some of it is picked up by needy tissues and the rest is taken up for storage by adipose cells.

HEMOGLOBIN. *A protein which carries oxygen in the blood* (G. haima, *blood;* L. globus, *ball*).

RETICULOCYTES. *Young red blood cells which, upon being stained, display an inner network of membranes* (L. reticulum, *net;* G. kytos, *cell*).

Muscle

Muscle tissue is the body's chief user of energy. Through its incessant tonal activity — whether it is being employed actively in movements or not — heat is generated. The body thus operates at a high temperature (about 37°C), permitting enzymatic activities, impulse propagation, muscle contraction, and other activities to proceed at higher speeds than would be possible at lower temperatures.

Some of the energy used by muscle is derived from glycogen (Chapter 21). Another large fraction is derived from fatty acids, by way of beta oxidation, which cleaves acetate fragments from the fatty acid molecule (Chapter 21). The acetate molecule thus produced may pass into the Krebs cycle, where its energy content is used to produce about 10 molecules of ATP.

Adipose

Fat cells and sheets of adipose tissue, scattered widely throughout the body, constitute, collectively, an organ. The functions of this organ in metabolism and nutrition are twofold: to store potential energy in the form of neutral fats, and to serve as a waste bed in which excessive amounts of dietary carbohydrates, proteins, and fats are stored in the form of neutral fats. A normal amount of neutral fat in adipose tissue is beneficial, while too much is detrimental.

When blood sugar levels drop below normal, and the glycogen reserve in the liver and elsewhere is considerably reduced, neutral fats are mobilized from their storage sites in fat cells. They are released into the blood and carried to other cells, such as liver or muscle cells, where they may be used as needed.

On the other hand, when there is an excess of neutral fats in the blood, an abundance of chylomicrons or lipoproteins or an excess of glucose in the blood, the fat cells begin to store fats. Let us consider these three situations separately.

In the first case, when there is an excess of fats in the bloodstream, the adipose cell simply picks up a certain quantity of fat for storage (Figure 22.1).

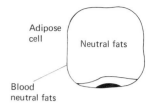

FIG. 22.1. *Direct absorption of neutral fats into a fat cell.*

In the second case, where there is an abundance of chylomicrons and lipoproteins in the blood, the fat cell mobilizes its lipoprotein lipase. *Lipoprotein lipase*, an enzyme, comes out of the fat cell and passes into a blood capillary, where it splits fatty acids out of the chylomicrons and lipoproteins. The mobilization of lipoprotein lipase and the cleavage of chylomicrons and lipoproteins increase in proportion to the amount of glucose and insulin present in the blood. Later in this chapter we will see how high levels of insulin and glucose in the blood are related to diabetes and obesity. At this point it will be helpful to know why high levels of insulin and glucose lead to *obesity*. To begin with, insulin stimulates adipose cells to absorb the fatty acids which were released from the chylomicrons and lipoproteins by lipoprotein lipase. Second, insulin stimulates adipose cells to absorb the excess glucose that is present in the blood. Third, insulin stimulates the conversion of this glucose into glycerol phosphate. This happens within the adipose cell where it is now possible to condense the glycerol phosphate with three fatty acids to form a neutral fat (Figure 22.2).

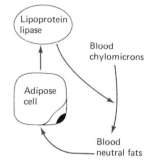

FIG. 22.2. *Absorption of neutral fats from chylomicrons.*

In the third case, where excess glucose is in the blood, insulin again comes into the picture. Through its influence, the adipose cell produces a large amount of glycerol phosphate, NADPH, and acetate. These substances are produced from glucose (Figure 21.1). In turn, the NADPH and acetate are used to produce fatty acids. The fatty acids combine with glycerol phosphate, as in the preceding case, to form neutral fats which the adipose cell then stores (Figure 22.3).

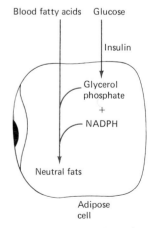

FIG. 22.3. *Formation of neutral fats in fat cell from blood glucose and blood fatty acids.*

Kidneys

The kidneys have four main functions in metabolism and nutrition. First, they rid the body of excessive nitrogen by excreting urea, which is produced from ammonia in the liver. Second, they excrete excess water or prevent the loss of water when necessary. Third, they regulate the mineral levels in the blood. Fourth, they produce ammonia (NH_3) from glutamine (an amino acid). The ammonia traps excessive acid (H^+) and forms ammonium ions (NH_4^+). The ammonium ions are excreted, and in this way the body rids itself of excess amounts of ammonia and acid. (The functions of the kidneys are described more fully in Chapter 30.)

Central Nervous System

HYPOTHALAMUS. *A region of the brain from which many autonomic motor actions arise.*

The central nervous system is important in metabolism for at least three reasons. First, it contains an appetite-satiety center, in the hypothalamus of the brain. This center receives signals which indicate a need or desire for food (appetite). It also receives signals that tell us that a sufficient amount of food has been eaten (satiety).

The appetite-satiety center receives some of its information from the periphery of the body. For example, nerve signals are sent to this center from the stomach and intestine to indicate how much food and liquid have been taken in. Metabolic "signals" are also sent to the appetite-satiety center. These signals are simply the levels of various metabolites in the blood which passes through the center. In addition, the appetite-satiety center receives psychological input from the cerebral cortex, perhaps by way of the limbic system (Chapter 11). This information concerns food and beverage preferences and quantities. All of this information—the peripheral nerve and metabolic signals and the psychological stimuli—is integrated in the brain. With the passage of time, as this information continues to be integrated, a person either will continue to have an appetite or will feel satiated.

LIMBIC LOBE. *The cerebral cortex bordering on several sensory, motor, and emotional regions of the brain.*

The second important feature about the central nervous system concerns its regulation of several kinds of metabolic hormones, including growth hormone, thyroxine, glucocorticoids, adrenaline, and acetylcholine. Here we will say something about acetylcholine, leaving a discussion of the other hormones for later.

Acetylcholine is released from postganglionic fibers when the parasympathetic nervous system is stimulated. This happens, for example, during a meal and shortly thereafter. In addition to stimulating digestive activities, acetylcholine promotes the release of insulin from the pancreas. The primary role of insulin is to facilitate the transfer of metabolites, such as glucose, amino acids, and fatty acids, from the bloodstream into the cells. There they are converted into glycogen, proteins, and fats.

The third important feature of the central nervous system is its need for about 25% of the body's total energy requirement when the body is in a resting state. When an individual is on a balanced diet, about 60% of the sugar in the blood is normally used by the brain and spinal cord. In periods of starvation, or in situations where an individual is on a very low carbohydrate diet, the brain becomes more dependent upon acetate and acetoacetate as sources of energy.

How is it possible for the central nervous system to take up more than two thirds of the blood sugar when it, in fact, makes up less than 5% of the body weight? To answer this question let us consider two situations—the period immediately following a meal (the absorptive period) and the period between meals (the postabsorptive period).

During the *absorptive period* metabolites are entering the blood from the intestine. Increasing amounts of metabolites are absorbed during the first hour or two, after which diminishing amounts are taken in. During the first hour or so the blood sugar level rises. In response to this rise the pancreas releases insulin. In response to the stimulatory effect of insulin, most of the cells throughout the body take in blood sugar.

Now let us consider the *postabsorptive period*. Beginning two to three hours after a meal, blood sugar levels are relatively low (normal) and the pancreas secretes very little insulin. Because of this relative absence of insulin, very little blood sugar is allowed to enter any kind of a cell except cells in brain tissue. The cells that do not have access to glucose, such as gland and muscle cells, rely upon fatty acids as a source of energy. The fatty acids are received by these cells from the liver and adipose by way of the bloodstream.

Brain tissue, in contrast to other kinds of tissue, absorbs sugar readily whether or not insulin is present. The nerve and glial tissue of the brain must absorb glucose because it is not able to use amino acids and fatty acids as a source of energy. Since blood insulin levels are low most of the time, and because this tissue is able to absorb large amounts of glucose in the absence of insulin, the majority of sugar molecules in the blood are absorbed by brain or nerve tissue. The brain or nerve tissue can do this even though it makes up less than 5% of the body weight.

HORMONAL REGULATION OF METABOLISM

We have already seen how the endocrine system regulates the metabolism of calcium, phosphate, sodium, potassium, and iodine (Chapter 12). In this chapter we will discuss how hormones regulate the metabolism of carbohydrates, proteins, and fats. We shall examine the roles of the following hormones: insulin and glucagon (from the pancreas), adrenaline (from the adrenal medulla), growth hormone (from the pituitary gland), glucocorticoids (cortisol and cortisone from the adrenal cortex), and thyroxine (from the thyroid gland). As an introduction to this topic we will discuss the normal blood levels of glucose, proteins, and fats.

Blood Levels of Glucose

Glucose is released into the blood from liver cells at the rate of 2 to 3.5 milligrams per kilogram body weight per minute. This amounts to an average of about 300 grams (10 ounces) of glucose per day in a person weighing 70 kilograms (150 pounds).

GLUCONEOGENESIS. The process of producing glucose from amino acids, glycerol, or lactic acid (G. *glykos, sweet;* neo, *new;* genesis, *to produce*).

In an individual at rest, the breakdown of glycogen accounts for about 75% of the liver's output of glucose. An additional 20% of the glucose output is derived from lactic acid by way of gluconeogenesis, and the remaining 5% comes from amino acids, also by way of gluconeogenesis.

In an individual at rest, nervous tissue accounts for the consumption of about 60% (150 grams) of the blood glucose. Muscle consumes about 20%, and the other tissues of the body, including the blood cells, account for the remaining 20%.

Beginning about three hours after a meal, and continuing until the person eats again, the blood sugar level is maintained normally at 60 to 100 milligrams per 100 milliliters of venous blood. When the blood sugar levels fall below 50 mg per 100 ml, the individual is said to be *hypoglycemic*. When blood sugar levels rise above 130 mg per 100 ml, the person is said to be *hyperglycemic*.

Whenever a question is raised about an individual's metabolic management of blood sugar (for example, in suspected cases of diabetes mellitus or in certain cases of obesity), a *glucose tolerance test* is usually given. This test provides information on blood sugar levels at various times after ingesting a measured amount of glucose. The change in blood sugar levels over time provides information about the rate at which blood sugar is used up.

The Oral Glucose Tolerance Test. Although glucose tolerance tests can be made by administering a load of glucose intravenously, they are usually made by giving glucose orally; that is, by mouth. For at least three days before the test the person is placed on a diet which includes at least 300 grams of carbohydrate per day. Total fasting is required from shortly after supper time on the day before the test until the test is over. The test is started by withdrawing a sample of blood from the person early in the morning. The person is then asked to drink a glucose solution which contains about one gram of glucose per pound of ideal body weight. Additional blood samples are obtained half an hour, one hour, two hours, three hours,

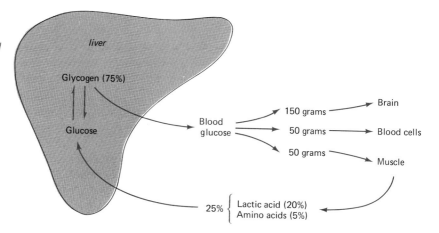

FIG. 22.4. Approximate adult quantities and distribution of glucose and related substances in the circulatory system.

and four hours later. Additional blood samples may be taken if the results of the test suggest that the person has diabetes or hypoglycemia.

Figure 22.5 shows representative curves for glucose tolerance tests for normal children and adults, diabetic children and adults, and obese adults. In these graphs you should note the following important general findings. (1) Normal children are less "tolerant" of glucose in their blood than normal adults. That is, the blood sugar is not allowed to rise to a peak in children as it does in the adult. (2) Peak levels of glucose are found half an hour to one and a half hours after drinking the glucose solution. In normal adults there is usually less than 150 milligrams of glucose per 100 milliliters of blood at the peak. (3) Also, higher levels of glucose and insulin appear in the blood of obese adults than in normal adults, causing the metabolic machinery to produce fat. In regard to this third point it should not be inferred that a person becomes obese because of a high output of insulin. On the contrary, when obesity is overcome, as may happen through proper dieting, blood insulin drops to a normal level.

OBESITY. *An excess of weight inconsistent with good health* (L. obesitas, fatness).

Hypoglycemia and Hyperglycemia. Abnormally low blood sugar (*hypoglycemia*) may be due to any one of several different causes, some of which are listed in the table below:

Induced causes of hypoglycemia	Endocrine causes of hypoglycemia
Starvation	Overactivity of pancreas (from tumor or other cause)
Excessive muscular exercise	
Chronic alcoholism	Underactive pituitary (insufficient ACTH, growth hormone, etc.)
Overdose of insulin	
Liver damage	Underactive adrenal cortex (inadequate production of glucocorticoids)

When hypoglycemia reaches a point where the normal physiology is severely disturbed, the following symptoms appear: confusion, tremors, reduced body temperature, cold sweat, rapid heartbeat, hallucinations, convulsions, and coma.

A person may enter a coma for any one of various reasons, among which are hyperglycemia or hypoglycemia. When a diabetic person becomes comatose, a doctor must know the cause before he can treat the patient. A diabetic who goes into a coma because of hyperglycemia needs insulin. A diabetic going into a coma because of hypoglycemia needs glucose.

A different curve than those shown in Figure 22.5 is seen in hypoglycemia. In a hypoglycemic person, very low concentrations of glucose (50 mg per 100 ml of blood) are found just before taking the glucose solution, and very low levels are found again after the peak is reached.

Diabetic hyperglycemia and hypoglycemia may be distinguished as follows. In diabetic hyperglycemia the person goes into *acidosis*, a state in which the blood pH becomes abnormally low and the following symptoms

HYPERVENTILATION.
Excessively deep breathing, which may be voluntary or may arise involuntarily from rising levels of CO_2 in the blood.

appear: hyperventilation, low blood pressure, rapid heartbeat, dehydration, and a slow onset of coma. In diabetic hypoglycemia, which is caused by the administration of too much insulin (insulin shock), the onset of coma is rapid and there is neither hyperventilation nor dehydration.

When the cause of a diabetic coma is unknown, it is best to draw blood, rapidly measure the level of glucose and ketones through simple procedures, and then administer glucose or insulin, depending on the situation.

Blood Levels of Proteins

Protein circulates in the plasma blood at a relatively constant level of 7 grams per 100 milliliters. About 60% of this protein is albumin. Most of the remainder consists of globulins, and a small portion is composed of clotting proteins and enzymes.

KWASHIORKOR. *A protein deficiency disease (Ghanaian; literally, "red boy").*

MARASMUS. *A gradual wasting of the body due to deficiencies in protein and calories (G., a withering).*

From a nutritional viewpoint, albumin is the only important plasma protein because it can be drawn upon during prolonged fasting. In protein starvation (*kwashiorkor*), or in cases where both protein and caloric intake are inadequate (*marasmus*), the blood level of albumin is greatly reduced, while the globulin level remains relatively unchanged, as does the level of clotting proteins. (Kwashiorkor in children is characterized by a pot belly, a moon face, scanty fat, and edema. Marasmus in children is characterized by an old man's face, no fat, no edema, and extreme underweight.)

With regard to amino acids, the blood level is about 30 milligrams per

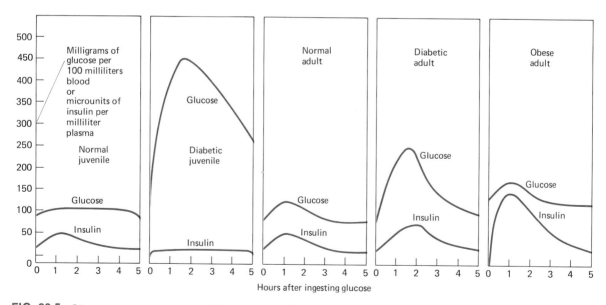

FIG. 22.5. *Glucose tolerance curves. "Diabetic adult" refers to maturity-onset diabetes, which usually appears after age 40. (Juvenile data based on Parker et al., 1968; adult data based on Perley and Kipnis, 1966.)*

100 milliliters of plasma in newborn babies, about half as high in children up to about puberty, and (again) about 30 mg per 100 ml in adults. Since plasma accounts for about 50% of the volume of blood, normal levels of amino acids in the blood are about the same as the levels of blood sugar (60 to 100 mg per 100 ml blood). The concentration of amino acids is usually lowest in the early morning and highest around midafternoon. The low value in the morning may be the consequence of a peak output of cortisol and cortisone from the adrenal cortex. These hormones stimulate gluconeogenesis, the process through which amino acids, lactic acid, and glycerol are converted into glucose. Gluconeogenesis is necessary early in the morning because the supply of glycogen in the liver has been largely depleted during the night.

Blood Levels of Fats

Shortly after a meal, fats enter the lacteals of the intestinal villi, where they combine with protein to form what are known as *chylomicrons*. Foreign materials are screened out of the chylomicrons in lymph nodes. The chylomicrons then enter the large veins which empty into the heart. The chylomicrons are then sent into the circulatory system where they are acted upon by the lipoprotein lipase, which is present in various kinds of cells throughout the body. The bulk of the chylomicrons are probably split apart by a lipoprotein lipase of liver, skeletal muscle, and adipose cells. Chylomicrons are spheres with an average diameter of 0.5 micrometers. They contain about 81% neutral fats, 10% cholesterol, 7% phospholipid, and 2% protein (globulin). They are slowly removed from the blood over a period of about four to eight hours after a meal, and they are not detectable in the blood after an overnight fast.

After an overnight fast, the total quantity of lipids in blood is about 300 to 800 milligrams per 100 ml serum. About 45% of this lipid is phospholipid, another 35% is neutral fats, 15% is cholesterol, and 5% is free fatty acids. Since only 5% of the total lipid is in the form of free fatty acids, the level of this lipid in the blood is about 27 mg per 100 ml. You may compare this figure with fasting blood levels of glucose (70 mg per 100 ml) and amino acids (30 mg per 100 ml), but in doing so you should keep three points in mind. First, fatty acids contain more than twice as many calories per unit weight (9 Calories per gram) than carbohydrates (4 Calories per gram) or protein (4 Calories per gram). Second, more of our dietary intake is routinely stored as fat than as glycogen, and there is virtually no provision for protein storage. Excess fat is stored in fat cells, but even fasting blood itself contains much more fat (300 to 800 mg per 100 ml serum) than glucose (70 mg per 100 ml) or amino acids (30 mg per 100 ml). Third, fatty acids in blood have a very short half-life of two to three minutes. This means that fatty acids are very rapidly consumed by most kinds of cells throughout the body (though not by nerve tissue). It has been estimated that fatty acids supply between 50% and 90% of the total energy requirement of the body in

CHYLOMICRON. *A very small globule of fat (0.5 to 1.5 micrometer diameter) that appears in the lymph and blood after a meal (G. chylos, juice; mikros, small).*

HALF-LIFE. *The time required for half of any given substance to lose its identity, and thus its "life."*

The Principal Regulatory Hormones of Metabolism

the fasting state. It is mainly only right after a meal that glucose and amino acids are readily used as sources of energy.

The fatty acids are carried in the bloodstream as components of phospholipids and neutral fats. The phospholipids and neutral fats are present in chylomicrons as we mentioned earlier, but they are also present in lipoproteins. However, the bulk of the lipids that are transported in the blood and given up to tissues in need are not the phospholipids or neutral fats. Rather, the bulk lipids are the free fatty acids that are carried through the bloodstream by albumin molecules. The albumin molecules combine with the fatty acids as albumin-fatty acid complexes. Upwards to 25 grams of fatty acids may be transported in the blood in one hour. If this quantity of fat were to be fully combusted it would yield about 250 Calories—more than enough to meet the basal caloric needs of even a very large person.

THE PRINCIPAL REGULATORY HORMONES OF METABOLISM

Six kinds of hormones are generally involved in regulating the metabolism of carbohydrate, protein, and fat. Their primary roles are indicated in Figure 22.6, which you should refer to as you read this section.

Insulin

INSULIN. A "nutrient-conservation" hormone.

The chief function of *insulin* is to mobilize the metabolic machinery in the direction of storing carbohydrates, proteins, and fats into most of the various kinds of cells throughout the body. It affects the liver, adipose, and

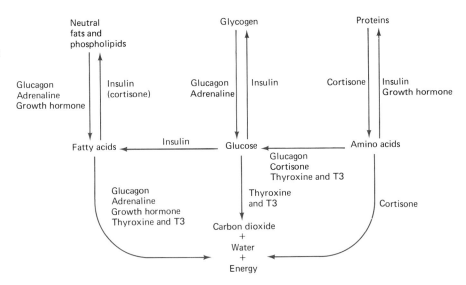

FIG. 22.6. Hormonal regulation of fatty acids, glucose, and amino acids in the blood.

kidney cells principally, muscle to a lesser extent, and nerve tissue not at all. Insulin may accurately be called a "nutrient conservation" hormone.

Small quantities of insulin are usually present in the blood, and significantly larger quantities are present after a meal, depending on how much glucose is absorbed from the intestinal canal (Figure 22.5). The pancreas is continuously active in secreting insulin, and the liver and other receptor tissues are continuously active in consuming this hormone. In a single passage of blood through the liver about 40% to 50% of its insulin content is used up. The half-life of insulin is only about 10 minutes.

In addition to promoting the entry of carbohydrates, proteins, and fats into various kinds of cells, insulin facilitates the following metabolic actions. (1) It stimulates glycogen synthesis in cells throughout the body, especially in liver and muscle cells. (2) It stimulates fat synthesis in adipose cells. (3) It antagonizes the action of glucocorticoids, and in doing so it cuts down on gluconeogenesis; insulin thus promotes the conservation of protein in cells throughout the body (Figure 22.7).

The primary regulatory role of insulin is to keep the blood sugar level low (Figure 22.8). As increasing amounts of carbohydrate are taken into the body, correspondingly higher amounts of insulin appear in the blood. In turn, larger quantities of glycerol phosphate and fatty acids are produced, giving rise to larger quantities of neutral fats for storage in adipose cells.

Glucagon

The main function of *glucagon* is to bring the blood levels of glucose and fatty acids up to a normal level whenever these metabolites tend to diminish. Accordingly, glucagon may be called a "nutrient-stabilizing hormone." In Figure 22.6 you may see that all of the actions of glucagon are antagonistic to those of insulin. For this reason it would be appropriate to dub glucagon the "anti-insulin hormone."

Glucagon is secreted into the blood whenever the blood level of glucose drops or the blood level of amino acids rises. Glucagon may react with

GLUCAGON. An "anti-insulin" hormone.

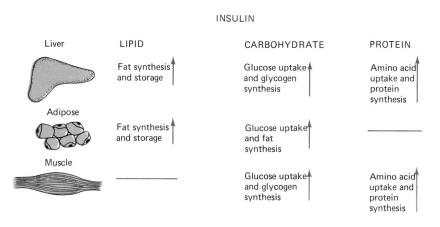

FIG. 22.7. Influence of insulin on metabolism in different tissues. Vertical arrows signify an increase.

The Principal Regulatory Hormones of Metabolism

various cells throughout the body, but its principal target organs are liver and adipose. It stimulates the liver cells to break down glycogen into glucose and to reduce the blood level of amino acids by converting them into glucose (gluconeogenesis). Through its interaction with fat cells, glucagon promotes the breakdown of neutral fats.

About 0.1 to 1.0 milligrams of glucagon are secreted from the pancreas each day. As is the case for insulin, the half-life of glucagon is very short, about 15 to 20 minutes.

Adrenaline

ADRENALINE. *A "quick-energy" hormone.*

Figure 22.6 shows that the actions of adrenaline are essentially the same as those of glucagon. Each of these hormones produces energy by initiating

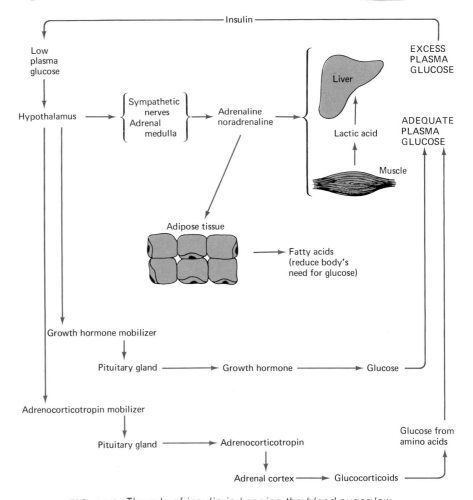

FIG. 22.8. *The role of insulin in keeping the blood sugar low.*

the combustion of carbohydrates and fatty acids. Although low levels of each of these hormones are present in the blood at all times, increasingly larger quantities of glucagon are secreted in response to metabolic stimuli (low sugar and high amino acid levels in blood); and increasingly larger quantities of adrenaline are secreted in response to emotional or mental stimuli from the brain. Adrenaline may very aptly be called the "quick-energy hormone."

Growth Hormone

The main function of growth hormone is to promote growth. It is critically important in bone development, in the enlargement of the mammary glands during pregnancy, and in the replacement of millions of body cells daily in the bloodstream, skin, intestinal mucosa, and other body tissues. Growth hormone is also important in the enlargement of muscle cells as a result of physical activity.

In carrying out its main function, growth hormone spares proteins from being broken down as an energy source and promotes the breakdown of fats instead. Accordingly, growth hormone may be called the "protein-sparing hormone." It appears in increasingly large amounts during periods of prolonged fasting, though it also appears between meals when carbohydrate levels in the blood begin to wane. The net effect of growth hormone is to reduce the use of carbohydrates and protein and to force fatty acids out of fat cells for use by other kinds of cells.

GROWTH HORMONE. A "protein-conservation" hormone.

Glucocorticoids

Glucocorticoids, such as cortisol and cortisone, are primarily "protein-utilizing hormones." They are released into the blood at times when glucose levels are low. They are present at their highest levels in the morning, when the glycogen and glucose stores are depleted. On being released from the adrenal cortex they promote the process of gluconeogenesis, in which amino acids, glycerol, and lactic acid are converted into glucose.

In addition to their gluconeogenic action, cortisone and cortisol promote the production of fat. They do this indirectly by stimulating the release of insulin in response to the glucose which appears in the blood as a result of their gluconeogenic action. As indicated previously, one of insulin's functions is to stimulate the production of fat.

GLUCOCORTICOIDS. "Protein-utilizing" hormones.

CORTISONE and **CORTISOL** are glucocorticoids.

Thyroxine

Thyroxine is best described as a "heat-generating hormone." This description is very appropriate because the production of body heat drops about 40% when the thyroid gland is removed.

Figure 22.6 shows that thyroxine promotes the oxidation of glucose, amino acids, and fatty acids. This effect is achieved by increasing the rate

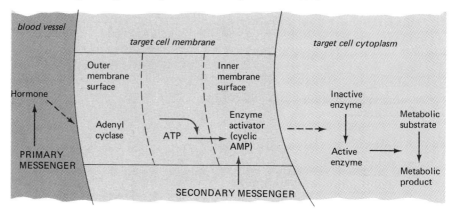

FIG. 22.9. The role of cyclic AMP in metabolism.

THYROXINE. A "heat-producing" thyroid hormone.

of oxidation of acetate (which is derived from glucose, amino acids, and fatty acids), in a desirable but seemingly inefficient way. Instead of coupling the release of energy from the oxidation of acetate with the production of ATP, thyroxine brings about the release of acetate's energy as heat. It is this heat that brings the temperature of the body up to the normal 37°C. At this temperature enzymes operate faster than they would at most environmental temperatures. Also, impulses are propagated at higher speeds along nerve and muscle fibers, the muscles contract more rapidly, and glands produce and release their secretory products in larger quantity.

TABLE 22.1. Metabolic Systems Activated by the "Secondary Messenger," Cyclic AMP[a]

Primary Messengers (Hormones)	Target Tissue	Metabolic Response in Presence of cAMP
Adrenaline and noradrenaline	Liver and muscle	Production of glucose from glycogen
	Adipose cells	Oxidation of fats
Glucocorticoids	Adipose cells	Oxidation of fats
Glucagon	Adipose cells	Oxidation of fats
	Liver	Production of glucose from glycogen
	Pancreas	Release of glucagon
Adrenocorticotropin	Adrenal cortex	Production of steroid hormones
Luteinizing hormone	Corpus luteum	Production of sex hormones
Thyrotropin	Thyroid gland	Production of thyroxine and T3
Antidiuretic hormone	Kidney	Reabsorption of water
Parathyroid hormone	Kidney	Excretion of phosphate
	Bone	Transfer of calcium into blood

[a] Cyclic AMP (cAMP) is also involved in a large variety of other physiological processes besides those listed here.

The Principal Organs of Metabolism

TRANSLATION OF HORMONAL MESSAGES INTO ACTION

Hormones may be considered *primary messengers*. Upon being released from cells in one kind of tissue they pass on to another kind of tissue, where they bring about an effect. The effect, however, is often achieved through a *secondary messenger* known as cyclic AMP.

As may be seen in Figure 22.9, *cyclic AMP* is produced enzymatically from ATP within the membrane of a cell which is stimulated by a hormone. In turn, cyclic AMP activates an inactive enzyme which carries out a metabolic process, such as breaking down glycogen into glucose.

Many metabolic processes involve the action of cyclic AMP. Some of these processes are listed in Table 22.1.

CYCLIC AMP. *A compound produced from ATP.*

SUMMARY

1. The liver is especially important in metabolism for the following functions: (a) storage of glycogen, iron, copper, fat-soluble vitamins, and other metabolites; (b) interconversion of carbohydrates, fats, and proteins; (c) removal of waste nitrogen from the blood; (d) conversion of foreign compounds into excretable salts; (e) destruction of worn-out hemoglobin; and (f) synthesis of glycogen, albumin, globulins, clotting proteins, bile salts, and other compounds.
2. Muscle is the body's chief consumer of energy, while adipose serves either as a reservoir of energy or as a depot for the collection of excessive carbon in the diet.
3. The kidneys are important in metabolism in that they rid the body of excess amounts of nitrogen, water, minerals, and hydrogen ions.
4. The central nervous system contains an appetite-satiety center through which we are prompted to eat or to abstain from eating. The central nervous system is also important in metabolism because it consumes 60% of the blood sugar and it regulates adrenaline, growth hormone, and acetylcholine, as well as the output of glucocorticoids and thyroxine.
5. When the fasting level of blood sugar rises above 130 mg per 100 ml, a person is hyperglycemic; when the level is below 50 mg per 100 ml, the person is hypoglycemic. Hyperglycemia and hypoglycemia are determined through a glucose tolerance text.
6. Fats absorbed in a meal are circulated as chylomicrons, which provide the body with fatty acids and other fats for a period of several hours. The bulk of the body's energy, however, is derived from fatty acids that are carried in the blood by albumin.
7. The principal endocrines of metabolism are insulin (a "nutrient-conservation hormone"), glucagon (an "anti-insulin hormone"), adrenaline

(a "quick-energy hormone"), growth hormone (a "protein-sparing hormone"), cortisol ("a protein-utilizing hormone"), and thyroxine (a "heat-generating hormone").

REVIEW QUESTIONS

1. List six general functions of the liver in metabolism.
2. Of what importance is muscle in general metabolism?
3. What happens to fats that appear in the blood after a meal?
4. Why are the kidneys important in general metabolism?
5. Describe at least three unique features of the central nervous system which are involved in general metabolism.
6. Define the terms *hyperglycemia* and *hypoglycemia*. Describe how these conditions can be diagnosed through a glucose tolerance test.
7. Indicate the kinds of synonyms that can be applied to the following hormones to denote their major function: insulin; glucagon; adrenaline; growth hormone; cortisol; thyroxine.
8. Tell how each of the hormones in question 7, above, carries out its main metabolic function.
9. At what concentrations are glucose, amino acids, fatty acids, and fats present in the blood after an overnight fast?

Caloric Requirements

CALORIES AND HOW THEY ARE MEASURED
    Measuring the caloric content of food elements
        The caloric value of foods
    Measuring the caloric consumption of humans
CALORIC NEEDS OF AN INDIVIDUAL
    Basal metabolism
        Surface area
        Weight
        Age
        Sex
        Nutritional status
        Health
    Specific dynamic effect
    Activity metabolism
AN APPROXIMATE DETERMINATION OF DAILY CALORIC NEEDS
REGULATION OF FOOD INTAKE
THE ESSENCE OF DESIGNS FOR LOSING WEIGHT
    The low-carbohydrate diet
    Evaluating the low-carbohydrate diets

In the preceding two chapters we discussed nutrients and their metabolism, and the organs of metabolism and their regulation. In this chapter the following topics are covered: (1) calories, and how they are measured; (2) caloric needs of an individual; (3) regulation of food intake; and (4) the essence of designs for losing weight.

CALORIES AND HOW THEY ARE MEASURED

A *calorie* is a unit of heat. It is defined as the quantity of heat needed to raise the temperature of 1 gram of water 1 degree centigrade, from 15°C to 16°C.

Calories and How They Are Measured

CALORIE. Amount of heat needed to raise 1 gram of water 1 degree centigrade (C).

KILOCALORIE. Amount of heat needed to raise 1000 grams of water 1 degree centigrade (C.).

It would be impractical in nutritional work to deal with calories, since this unit is extremely small as far as human nutrition is concerned. Kilocalories are used instead, and are designated by the word Calories, spelled with a capital C. A *kilocalorie* is the quantity of heat required to raise the temperature of 1000 grams of water 1 degree centigrade.

To understand dietary energy relationships you must know something about the caloric content of food elements (carbohydrates, proteins, fats); the caloric value of foods (meat, milk, beans, etc.); and the caloric requirements of individuals.

Measuring the Caloric Content of Food Elements

The caloric content of food elements is measured with a bomb calorimeter. A *bomb calorimeter* consists of two parts (Figure 23.1). One part, called the bomb, is a heavy steel chamber in which a sample of organic material is combusted. To do this, the sample is placed on a nickel or platinum dish. A fine platinum wire is placed in contact with the sample. The bomb is then filled with oxygen at a pressure of about 25 atmospheres and is placed into the second part of the calorimeter, a water bath contained in a thickly insulated box. The weight of the water in the water bath is known precisely. A thermometer is immersed in the water bath, which is completely sealed from the room outside the box. Electric current is passed through the platinum wire, causing the organic material to combust (burn) in an explosive manner. The heat given off during combustion is transferred to the water.

The caloric value of the sample is determined from the weight of the water and its change in temperature upon combustion. If, for example, the combustion of 1 gram of organic material produces an increase of 4°C in 1000 grams (milliliters) of pure water, the substance is said to have a caloric value of 4 Calories (kilocalories).

The caloric value of many kinds of carbohydrates, fats, and proteins have been measured through this procedure. The *caloric content* of carbohydrates ranges from about 3.6 to 4.1 Calories per gram. Fat contains 8.4 to

FIG. 23.1. *Bomb calorimeter.*

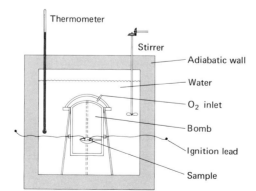

Caloric Requirements

9.4 Calories per gram. The caloric content of proteins ranges from about 4.5 Calories per gram for certain vegetable proteins to 5.8 Calories per gram for meat and eggs.

The *average* combustion values of a gram of carbohydrate, fat, and protein are 4, 9, and 5.2 Calories, respectively. However, these values are based on bomb calorimetry, in which total combustion occurs. That is, all of the carbon is converted to carbon dioxide, all of the hydrogen to water, and all of the nitrogen and sulfur to oxidized forms such as nitrous oxide (NO) and sulfur dioxide (SO_2). It turns out that nitrogen, which accounts for about 16% of the weight of protein, is eliminated from the human body mainly as urea (NH_2CONH_2), and partly as uric acid ($C_6H_4O_3N_4$) and ammonia (NH_3). In other words, protein is not completely combusted in the human body. Taking into account the nitrogen that is excreted in an incompletely combusted form, a value of 4 Calories is assigned as the caloric value of a gram of protein.

UREA, (NH_2CONH_2). *A neutral nitrogenous compound.*

In summary, approximately 4 Calories of energy are derived from the combustion of one gram of either carbohydrate or protein. More than twice as much energy (9 Calories) is derived through the combustion of 1 gram of fat.

The Caloric Value of Foods. In assigning a caloric value to a particular foodstuff—such as a pudding, ice cream, beer, or a vegetable—two kinds of data are necessary: the percentage of carbohydrate, fat, protein, and water, the latter of which has a caloric value of zero; and the digestibility of the ingredients. The first of these points needs no additional comment. The second one does.

DIGESTION. *Conversion of food into an assimilable form* (L. *digestio, to separate*).

Let us consider three examples. (1) Cellulose, among carbohydrates, is virtually indigestible by humans. Hence it has a caloric value of zero, insofar as human nutrition is concerned. (2) The carbohydrate of certain grades of wheat is about 95% digestible. Other grades are about 90% digestible. (3) The protein of milk is about 100% digestible. The protein of peas, beans, and nuts is about 80% digestible.

Knowing the percentage composition of a given foodstuff and the caloric contents of its digestible carbohydrate, fat, and protein, it is possible to assign the foodstuff a caloric value per unit weight or volume. No attention is paid to nucleic acids and vitamins, both of which are found in relatively low concentrations in food. Table 24.4 lists some common foods and their caloric values.

Measuring the Caloric Consumption of Humans

The caloric expenditures of a resting or active human being can be measured by direct or indirect calorimetry. *Direct calorimetry* is similar to bomb calorimetry, except that the heat measured is that given off from a human body rather than from a "bomb." The heat passes into an air space within a closed room. It is then picked up by water passing slowly through a long tube that snakes back and forth along an extensive area of the ceiling and walls. A calculation is made of the volume of water and the increase in its temperature in order to determine how many calories were expended.

Indirect calorimetry is based on the principle that the caloric expenditure of an individual is related to oxygen consumption, carbon dioxide output, and nitrogen excretion in the urine. Accordingly, a special apparatus is used for measuring oxygen consumption and carbon dioxide production, and the quantity of nitrogen in the urine is measured chemically.

The exact procedure is described in a reference listed at the end of this text. In practice, indirect calorimetry is based on the following principles:

MOLE or GRAM MOLECULE. *An amount of an element or compound equal in numerical value to the molecular weight of the substance. One mole of glucose is 180 grams.*

(1) A pure carbohydrate, when combusted, releases 1 mole of carbon dioxide for every mole of oxygen consumed. That is, the *respiratory quotient*. (R.Q.), CO_2/O_2, equals 1.0. When 1 mole of glucose is combusted, for example, energy is released along with 6 moles of carbon dioxide and 6 moles of water. The equation of this reaction is:

$$C_6H_{12}O_6 + 6\ O_2 \rightarrow 6\ CO_2 + 6\ H_2O + 672\ \textbf{Calories (180 grams)}$$

From this equation it may be noted that the consumption of 1 mole of oxygen is associated with the release of $672/6 (=112)$ Calories. Since 1 mole of oxygen is 22.4 liters, you can see that we get 5 Calories of energy for every liter of oxygen used in combusting carbohydrates. That is, $112/22.4 = 5$.

(2) When pure fats are combusted, an R.Q. of about 0.7 is obtained. For the neutral fat palmitin, for example:

R.Q. *Respiratory quotient — volume of carbon dioxide respired over volume of oxygen inspired.*

$$C_{51}H_{98}O_6 + 72.5\ O_2 \rightarrow 51\ CO_2 + 49\ H_2O + 7657\ \textbf{Calories (806 grams)}$$

from this equation it may be noted that the consumption of 1 mole of oxygen is associated with the release of $7657/72.5 (=103)$ Calories. Here you can see that we get a little less than 5 Calories of energy for every liter of oxygen used in combusting fats. That is, $103/22.4 = 4.75$.

The two preceding equations show that in order to determine how much oxygen is needed for complete combustion of any kind of carbohydrate or fat it is necessary to assume that there are 2 oxygen atoms for each carbon atom and 1 oxygen atom for each pair of hydrogen atoms in the combustible molecule.

A comparison of the two preceding equations shows that approximately the same amount of energy is liberated from foods per unit of oxygen consumption. This is so regardless of whether carbohydrate or fat is combusted. You can see, however, that more energy is derived from 1 gram of fat than from 1 gram of carbohydrate.

(3) When mixtures of fats and carbohydrates are combusted we obtain respiratory quotients between 0.7 and 1.0. For such mixtures there is a corresponding release of 4.7 to 5.0 Calories per liter of oxygen consumed. From the correspondence that exists between R.Q. and Calories per oxygen consumed it is possible to determine how much carbohydrate and how much fat are combusted during a test period.

(4) The exact respiratory quotient for protein cannot be calculated because the nitrogen is not oxidized. Since most of the nitrogen in urine comes from protein, however, we can estimate how much protein was combusted. To make this estimate we first measure the urinary nitrogen. Each gram of

Caloric Requirements

urinary nitrogen represents 6.25 grams of protein, since protein contains, on the average, 16% nitrogen (6.25 × 16 = 100). It has been established experimentally that about 6 liters of oxygen are consumed in breaking down 6.25 grams of protein, or about 1 liter of oxygen for each gram of protein that is used. Since the consumption of 1 liter of oxygen means we have obtained about 5 Calories of energy, we can assume that we have obtained 5 Calories of energy for each gram of waste nitrogen that is found in the urine.

EACH GRAM OF PROTEIN contains .16 g nitrogen; thus there will be 6.25 g protein for 1 g N.

Based on the preceding four points, it is possible to determine how many Calories are consumed by an individual engaged in any kind of activity.

TABLE 23.1. Energy Cost of Various Activities

Activity	Calories Expended, Average[a]
Sleep	100
Seated, studying	150
Standing	200
Slow walk	300
Brisk walk	450
Running	800

Walking	General Conditions	Energy Cost (Cal/hr) by Body Weight (pounds)		
		100 lbs.	150 lbs.	200 lbs.
Slow walk	Level ground	100	150	200
Slow walk	Slightly uphill	150	250	300
Brisk walk	Level ground	150	200	250
Brisk walk	Slightly uphill	200	300	400

Other Activities	Activity Metabolism (Cal/hr/kilogram of body weight)[b]
Bicycling (fast)	7.5
Bicycling (slow)	2.5
Dancing	4–6
Driving car	1.0
Eating	0.5
Gardening	4.0
Ironing, typing	1.0
Running	7.0
Tennis	5.0
Walking (3 mph)	2.0
Walking (4 mph)	3.5
Walking (5 mph)	4.0

[a] Quantities are expressed as percentages of the energy cost of sleeping, arbitrarily reckoned here as 100 Calories.

[b] Activity metabolism = energy cost, excluding value of basal metabolism and specific dynamic effect.

Calaric Needs of An Individual

Very careful measurements have been made over many years on numerous individuals engaged in various sorts of activities, ranging from sleep to strenuous work and exercise. It is through such measurements and related approaches (direct calorimetry) that tables, such as Table 23.1, are established.

CALORIC NEEDS OF AN INDIVIDUAL

The caloric expenditure of an individual can be broken down into three components: basal metabolism, specific dynamic effect, and activity metabolism.

Basal Metabolism

BASAL METABOLISM. *Amount of energy expended under wakeful, but completely restful, comfortable conditions.*

The basal component of overall metabolism represents the lowest level of physiological activity in the waking state. It is measured in terms of Calories per hour. It signifies the lowest (basal) level of cell activity associated with nervous activity, muscle tone, respiration, circulation, excretion, and the like.

AMBIENT. *About, around, or surrounding (L. ambi, around).*

A proper measurement of the basal metabolic rate requires that a person be comfortable, well rested, and free of emotional stress. In addition, the ambient (surrounding) temperature must be neither too warm nor too cold.

TABLE 23.2. Average Basal Metabolic Rates[a]

Age (years)	Basal Metabolic Rate (kilocalories per m^2 of body surface per hour)	
	Males	Females
3	60	54
5	56	53
7	52	50
9	49	46
11	46	43
13	44	40
15	43	38
18	41	36
22	39	35
27	38	35
34	37	35
40	36	34
55	35	33
65	34	32
70	33	31

[a] Based on Boothby, *Handbook of Biological Data*, 1956. W. S. Spector, editor.

Caloric Requirements

Also, the test should be performed long after any meals—usually after an overnight fast.

Table 23.2 shows that the basal metabolic rate varies with age, sex, and surface area. The basal metabolic rate is related to six primary conditions: body surface area, weight, age, sex, nutritional status, and health.

Surface Area. The machinery of the body is operated at 37°C, whereas the surrounding temperature is usually 15 to 20 degrees lower. The body is thus a radiator of heat. In general, the heat radiated by the body is directly proportional to surface area.

You may determine your surface area from Figure 23.2. This value, in

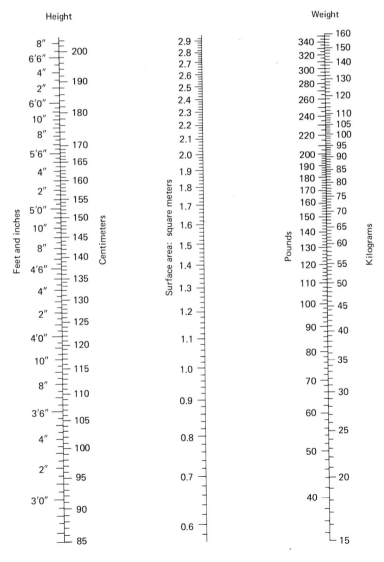

FIG. 23.2. *A nomogram used to determine body surface area. Use a ruler to line up the figures for your weight and height, and read your surface area in square meters. (From Boothby et al., 1936.)*

conjunction with Table 23.2, will allow you to determine the normal basal rate for your age. The basal rates listed in Table 23.2 are average values that were derived from thousands of careful measurements made in American and British laboratories.

Weight. In general, the surface area of an individual is closely related to weight. Heavier people, in comparison to lighter people, must circulate more blood, breathe more air, operate a heavier musculature, and provide more fuel to the body at the basal level. For adults the basal metabolism is generally about 0.5 Calories per pound per hour.

Age. The metabolic rate drops continuously from infancy to old age. This may be seen in Table 23.2.

Sex. Males have a higher metabolic rate, per surface area, than females. This is because males have a proportionally higher quantity of muscle. In this connection, males also carry a large number of red blood cells per unit volume of blood. They may thus furnish larger quantities of oxygen (per unit weight of tissue) to satisfy the higher basal metabolic demand.

Nutritional Status. Prolonged starvation decreases the basal metabolic rate. Scientific studies on this topic—with World War II conscientious objectors as subjects—have shown that the basal metabolic rate may drop as much as 30% after several months of starvation. This drop is partly correlated with a drop in weight—particularly muscle weight—which in turn imposes a smaller demand upon the circulatory, respiratory, excretory, and metabolic systems.

Health. Fever, pneumonia, hyperthyroidism, and various other disorders increase the basal metabolic rate. Hypothyroidism, on the other hand, decreases it.

Specific Dynamic Effect

The specific dynamic effect is the component of overall metabolism which accounts for the consumption of oxygen, above the basal amount, when an individual is tested after feeding rather than fasting. Under otherwise similar conditions, the fed individual consumes additional oxygen to provide energy for the absorption of food, the production of additional urea, and the interconversions (of carbohydrate, fat, and protein) which are necessary to achieve a metabolic balance. On a pure protein diet the specific dynamic effect constitutes a significant fraction of the total caloric intake (about 30%), whereas with a pure fat or carbohydrate diet it amounts to only about 5%. For a balanced diet, its value is about 10%.

Activity Metabolism

In addition to the calories used for basal metabolism and for the specific dynamic effect, calories are needed to provide energy for physical activities.

Caloric Requirements

The number of calories required for any specific activity varies considerably from one individual to another and from one situation to another. Among a host of variables, the energy required depends upon how much work is done, how fast it is being done, and how efficiently it is carried out. In addition, energy demands depend upon whether the ambient temperature is very high or low. High temperatures often lead to the additional physiological activity of sweating, and low temperatures lead to shivering. Both of these activities require additional energy. Table 23.1 lists the calories expended by people engaged in activities of various kinds.

AN APPROXIMATE DETERMINATION OF DAILY CALORIC NEEDS

On the average, each of us requires a certain intake of calories if we are to maintain a relatively constant weight. In this connection it is important to realize that one pound of human flesh, on the average, is equivalent to 3500 Calories. This number of Calories must be taken in, or refused, if an individual is to gain or lose a pound. As a rule of thumb, 1 dry ounce of either carbohydrate or protein yields 115 Calories, and 1 ounce of fat yields 260 Calories.

Your average daily caloric requirement can be determined in three steps. First, determine what your normal weight should be. Assume a normal weight of 100 pounds if you are 5 feet tall. Add or subtract 5 pounds respectively for each inch above or below 5 feet. Add or subtract 4 or 6 pounds instead of 5, if your body build is small-boned or heavy-boned, respectively.

Next, multiply the value obtained above by a number which takes age and sex into consideration. If you are in the 15 to 29 age bracket, multiply by 19 if you are a male, and by 16 if you are a female. If you are in the 30 to 45 age bracket, multiply by 18 or 15, respectively. If you are over 45 years old multiply by 16 or 13.

Finally, make an adjustment to take into consideration the amount of physical activity you engage in. Subtract 10% of the product obtained in step 2 if you are sedentary. Or add 5, 10, 15, 20, or 25%, depending on how much more active than normal you are. The average activity level of a normal person based on observations of a large population of people is as follows:

        8-9 hours sleeping and lying down
        5-6 hours sitting down activities
        6-8 hours standing up activities
        1-3 hours walking activities
        1-3 hours strenuous work or athletics

Having carried out the above steps, you now have a value of your approximate daily caloric need. Data are provided in Table 24.4 on the caloric value of common foods. In looking over these data, keep in mind that the caloric content of foodstuffs is highly variable from one source to another. Even eggs may vary considerably in composition, depending upon the diet of the chicken. If chickens are fed a commercial poultry feed, for example, and an examination is made of the fatty acid content in the eggs, about 35% of the fatty acids are of the saturated type, 50% are monounsaturated, and 15% are polysaturated. In contrast, if a polyunsaturated oil is substituted for the saturated fats of poultry feed, about 35% of the fatty acids in the eggs are of the saturated type, 15% are monounsaturated and 35% are polyunsaturated.

REGULATION OF FOOD INTAKE

EMACIATION. *Extreme loss of flesh* (L. emaciare, *to make thin*).

OBESITY. *Generalized weight excess* (L. obesitas, *fatness*).

SUBSCAPULAR REGION. *The area of the back beneath the shoulder blades.*

Emaciation is defined as an extreme loss of flesh. The individual is excessively lean (has an abnormally small amount of adipose). Many individuals throughout the world suffer from this problem, although in the economically rich countries obesity is more prevalent.

Obesity is defined as adiposity in excess of what is consistent with good health. In general, a person is obese when excessive amounts of adipose hang from the arms and are present around the abdomen and in the subscapular region. Obesity is always accompanied by overweight, though the converse may not be true. An individual may be overweight because of exercise. Certain weight lifters, for example, are overweight but not obese. Overweight may also be caused by edema.

Looked at as simply as possible, there are perhaps five major factors involved in determining how much food or how many calories an individual will consume: genetic predisposition; early eating habits; endocrine imbalances; physiological cues; and psychological or cultural influences.

Genetic Predisposition. Some individuals have a genetic tendency toward obesity or emaciation. Obesity in children is about ten times more common in the offspring of fat parents than in the children of thin parents. So too, emaciation is most common in children whose parents are very thin.

It is conceivable that the "satiety-center" of the brain in an obese individual (Chapter 22) requires an excessive amount of sensory input—and the "satiety-center" in the lean individual requires a subnormal amount of input—from the following sources of stimulation: nutrient levels in the blood; stomach and intestinal fullness and motility; psychological factors; and other unknown physiological factors. If this is true, it would partially explain the greater desire for food in obese individuals and the lower-than-normal desire for food in thin people.

In addition, there is some evidence that obese persons have digestive and metabolic systems that may be more efficient than those of lean persons in obtaining and retaining larger quantities of calories from a given amount of food.

Early Eating Habits. In experiments with rats it has been shown that there is a large increase in the number of fat cells when an animal is injected intravenously with excessive amounts of glucose before 16 weeks of age. In contrast, when rats are force-fed in this way sometime after 16 weeks of age, the result is an increase in the size of fat cells (hypertrophy) rather than an increase in number (hyperplasia). In studies on humans it has been found that fat cells are more abundant and larger in obese people than in thin people. It is very probable that people who eat large quantities of food early in life induce in themselves the formation of a large number of fat cells. There is, after all, a limit to the amount of food that can be excreted as waste carbon dioxide or stored in a fat cell. When a baby or infant, or perhaps an older child, takes in an excessive amount of food, the excessive carbon can be stored only as fat, calling for additional fat cells.

Hormonal Factors. An imbalance of hormones in the blood may lead to an excessive appetite or a loss of appetite. A person with excessive thyroxine, for example, may feel the need to eat more than normal amounts of food. Because of the excessive levels of thyroxine, however, more than a normal percentage of the caloric intake is converted into heat. The consequence of this inefficiency in metabolism usually leads to a weight loss. So too, the metabolic system of an individual with an underactive thyroid gland is inefficient in combusting foods. This type of person tends to gain weight on a low caloric intake; accordingly, the person may not want to eat a normal quantity of food.

In some women, hormonal imbalances occur during pregnancy. Excessive amounts of sex hormones are often produced, along with excessive amounts of insulin, leading to a craving for food and obesity. So too, males who undergo total castration, and therefore incur a deficiency in testosterone, tend to increase their appetite and become obese.

Physiological Cues. The urge to eat or to stop eating is triggered, in part, by certain kinds of stimuli which the appetite-satiety center receives from the periphery of the body. One type of stimulus comes from the stomach in the form of nerve impulses that are generated in response to stomach motility (movement). An increase in stomach motility is generally associated with an increase in appetite, and a decrease in stomach motility is associated with satiety. Nerve impulses are also sent to the appetite-satiety center in response to stomach distension. Increases and decreases in stomach distension are also associated with satiety and appetite, respectively. Another type of stimulus that is directed to the appetite-satiety center is the difference in glucose concentration in the arterial and venous blood. A large difference means there is a considerable amount of glucose in arterial blood; a small difference means the blood glucose level is very low. Large arterial-venous blood glucose differences are associated with satiety, while small differences are linked with appetite.

INSULIN, *a pancreatic hormone, promotes the retention of fats, carbohydrates, and proteins in the body and prevents blood sugar levels from rising to abnormally high levels.*

Psychological and Cultural Influences. It is noteworthy that obesity does not occur in nature except when an animal may be preparing itself for hibernation, migration, or some other special activity during which fasting will take place. It is also noteworthy that obesity is not very prevalent among certain ethnic groups, such as Orientals. Culture indeed plays a role in determining the amounts and kinds of foods we eat.

But it is perhaps correct to say that psychological factors play a predominant influence in regulating food intake. So strong are these influences that a person may be led by emotional factors alone to a 50% reduction in weight (as in certain cases of anorexia nervosa) or a twofold or greater increase in weight (extreme obesity).

How do psychological factors influence appetite or satiety? No one knows, though it is well established that "appetite" and "satiety" centers exist in the hypothalamus of the brain, and that higher regulatory centers exist in the cerebral cortex. Thus, destruction of the floor of the hypothalamus leads to obesity, whereas destruction of the lateral walls leads to emaciation. Damage to both sides of the floor of the thalamus, certain regions of the amygdala, the hippocampus, or the frontal lobes may lead to obesity. These regions of the brain are influenced by events which occur within our bodies and by external stimuli, such as the sight and smell of food.

Although nothing is known about the mechanism through which appetite and satiety operate, two general observations about obese individuals have been made. First, some of the metabolic features of obese people are strikingly different from those of lean individuals. Often there is an excess of insulin in the blood, a higher level of blood sugar (a higher glucose tolerance) shortly after a meal, a predisposition to excessively low blood sugar levels (hypoglycemia) between meals, and a much higher incidence of adult-onset diabetes, (see Chapter 22).

Second, where psychological tests have been made, obese individuals have been shown to be less responsive to internal (physiological) cues and more responsive to external (psychological) cues. For example, in an experiment where speeded-up clocks were used, obese subjects "felt hungry" when dinner time showed up on the clock, whereas lean subjects expressed hunger when their physiological state "led them" to a need for food (at their usual dinner time).

HYPOTHALAMUS. *A region of the brain from which many autonomic nervous activities arise.*

AMYGDALA and **HIPPOCAMPUS.** *Regions of the brain closely associated with smelling, eating, and emotions.*

THE ESSENCE OF DESIGNS FOR LOSING WEIGHT

Until recently, all weight-reducing diets were based on exercise and a decrease in caloric intake. For any individual who is willing to comply, this approach is the best. It has stood the test of time, and it certainly is adequate and suitable from a physiological standpoint.

The Low-carbohydrate Diet

Most individuals who tend to be overweight are not consistently willing to exercise or to reduce their caloric intake. Because of this, and based on our current knowledge of metabolism and nutrition, several versions of a low-carbohydrate diet have been developed by various people, including John Yudkins, Irwin Stillman, and Robert Atkins. Some of these diets specify a low-carbohydrate regime (30 to 60 grams per day), while others allow no carbohydrates for an extended initial weight-reducing period that is followed by a permanent low-carbohydrate diet.

What are the premises upon which low-carbohydrate weight-reduction diets are based? Let us consider the question using essentially the arguments advanced by Robert Atkins, as well as those provided by several others.

A historical basis for an argument, to begin with: cavemen ate mainly meat, fruits, and leaves, taking in very small quantities of carbohydrate, perhaps only 15 grams (60 Calories) a day. But about 7,000 years ago man began to cultivate grains, thus adding to the carbohydrate intake. And then, only recently, man started eating sugar. In 1750, the average American consumed 4 pounds of sugar per year. In 1840, as man began to refine sugar for the first time, about 20 pounds were consumed per person per year. Currently, the average American consumes more than 100 pounds per year, much of it in syrups, candies, carbonated beverages, baked goods, and other kinds of sweets. It is argued, therefore, that the composition of man's diet has changed considerably in recent times, and that this change toward a greater consumption of carbohydrates is the cause of obesity in many cases.

Next we come to several arguments which proponents of low-carbohydrate diets claim to be the scientific basis for their viewpoint. Let us consider what these arguments are and how sound they are. We will state the main thesis of each argument and then make a comment which you can reflect upon in view of what you now know about metabolism.

Argument: carbohydrates are absorbed and combusted more rapidly than fats and proteins. Therefore, there is a general tendency to crave food more frequently when carbohydrates exceed a certain caloric level in the diet than when they are below a certain level.

Comment: there may be considerable merit in this argument, but no scientific evidence is yet at hand.

Argument: carbohydrates stimulate the release of insulin from the pancreas and, in doing so, rapidly induce a feeling of hunger. If the carbohydrate intake is reduced to near zero, little or no insulin will be released, hunger will not be felt as often or as intensely, and additional weight — in the form of fat — will not be produced.

Comment: this argument is plausible, though it has never been proved scientifically. It is known — as the argument says — that insulin is released from the pancreas in proportion to the level of sugar in the blood, and it is known that insulin promotes the conversion of carbohydrates into fat.

Argument: excessive amounts of protein, which *are* allowed in the low-carbohydrate, weight-reducing diet, are converted into glucose, acetate, and ketone bodies. These substances will satisfy the metabolic needs of the central nervous system.

Comment: you may recall that the oxygen uptake of the central nervous system accounts for about 25% of the body's resting metabolism, and that the brain is sustained mainly at the expense of glucose. It may be added here that a number of results from experiments dealing with the physiology of starvation suggest that the brain becomes somewhat more dependent upon acetoacetate and acetate as an energy source when the availability of blood sugar is considerably reduced.

The Essence of Designs for Losing Weight

KETONE BODIES.
Acetoacetic acid and acetone.

Argument: when dietary carbohydrates are significantly reduced, the anterior pituitary gland releases a "fat-mobilizing" hormone. Also called ketogenic hormone, this endocrine secretion stimulates the release of fats from adipose cells. The fat travels to liver cells and other cells where it is rapidly broken down into acetate. The acetate provides the brain with whatever energy it cannot obtain because of low carbohydrate intake.

Comment: the "fat-mobilizing" hormone was initially described in a scientific journal in 1960, but its existence has never been confirmed by other investigators. This hormone was said to be released from the pituitary gland only when the level of glucose in blood reached a critically low level, which can happen when no carbohydrates are eaten for several days. Today it is known that the "fat-mobilizing" hormone is not a new, or even a single, hormone. Rather, it is a combination of growth hormone, adrenaline, noradrenaline, glucocorticoids, glucagon, and perhaps thyroxine.

Argument: excessive amounts of dietary fats and protein are converted to ketone bodies when dietary carbohydrate is restricted to the critically-low level at which "fat-mobilizing" hormone is secreted. There is no need, therefore, to reduce the dietary intake of fat and protein.

Comment: to begin with, ketone bodies are produced enzymatically when excessive amounts of acetate accumulate in a cell (Figure 23.3). Two acetic acid molecules (CH_3COOH) condense to form acetoacetic acid ($CH_3\underline{CO}CH_2COOH$) plus water ($H_2O$). Through another enzyme reaction, acetoacetic acid is converted to acetone ($CH_3\underline{CO}CH_3$). (Note: the underlined $\underline{CO}$ groups in the preceding formulas are the ketone groups which give acetone and acetoacetic acid the name of "ketone bodies.")

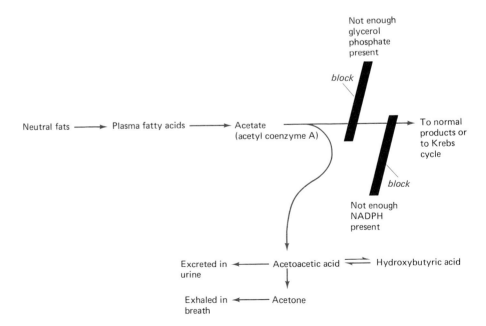

FIG. 23.3. Two possible blocks which force acetate into the production of excessive ketone bodies.

Acetoacetic acid is an excretable acid, and acetone is an exhalable organic solvent that smells like nail polish remover. In an untreated diabetic these substances accumulate to the point where they induce nausea, dizzy spells, vomiting, and fainting. In addition, acetoacetic acid, by virtue of being an acid, can reduce the pH of the blood and thus create a case of acidosis. The reduction of pH induces rapid and deeper breathing (hyperventilation), which results in the loss of carbon dioxide. This loss tends to restore the normal pH, inasmuch as less carbonic acid is formed ($H_2O + CO_2 \rightleftharpoons H_2CO_3$). At the same time, however, the depletion of carbon dioxide results in a reduction of bicarbonate ions (HCO_3^-) in the blood, since bicarbonate also is derived from carbon dioxide ($H_2O + CO_2 \rightleftharpoons H_2CO_3 \rightleftharpoons H^+ + HCO_3^-$). Excessive losses of bicarbonate ultimately lead to convulsions and coma.

ACIDOSIS. *A condition in which there is an excess of carbon dioxide and a deficiency of bicarbonate ions in the blood.*

Although the production of ketone bodies causes acidosis in a diabetic who is not being treated with insulin, it does not appear to do so in normal individuals on a low-carbohydrate or zero-carbohydrate diet. However, it is difficult to say what subtle ill effects may be caused by the continued production of ketone bodies over long periods of time.

Also, regardless of what effect excessive ketones may have, there is a real fallacy to the notion that the renal (kidney) excretion of acetoacetic acid (and a related derivative called beta-hydroxybutyric acid), and the exhalation of acetone can lead to a significant reduction in weight. At best, 100 grams of these substances might be lost in one day. Inasmuch as 1 gram of acetoacetic acid is equivalent to about 5 kilocalories, one may lose 500 kilocalories (1/7th of a pound) at best. On a weekly basis this could lead to the loss of 1 pound. However, the proponents of low-carbohydrate diets do not recommend a reduction in caloric intake, and therefore the loss of one pound through ketone excretion may easily be offset by excessive food intake.

Evaluating the Low-carbohydrate Diets

If you were familiar with the claims of advocates of low-carbohydrate diets before reading this text, you might wonder whether low-carbohydrate diets lead to any real and substantial loss of weight.

To begin with, regardless of what kind of diet an individual is on, there is generally a rapid initial loss of weight. Lean tissue tends to be lost first and rapidly; fat tissue later and more slowly. Why is this so?

It turns out that one ounce of lean tissue is equivalent to about 25 to 50 kilocalories. An individual does not have to experience a considerable reduction in caloric intake, therefore, to realize a measurable loss of weight. The loss of 1 ounce of fat, on the other hand, is equivalent to the loss of about 250 kilocalories. Large losses of this magnitude can have a significant effect on the size and appearance of an individual. Why should there be such a wide difference in the caloric value of lean versus fat tissue? Because lean tissue contains a great deal of water, whereas fat contains only trace quantities of water. Why, then, is lean tissue lost before fat? Because carbohydrates and protein are burned more quickly and easily than fat, and in fact, must be utilized if fat is to be combusted.

How then can one explain significant weight losses in individuals who remain on low-carbohydrate diets over a long period? It is believed that such individuals have an intense desire to achieve their goal and that the high protein-fat intake turns off their appetite before they take in an excess of calories (it is hard to make up with fat and protein the 50% of caloric intake which is normally derived from carbohydrates). Also, individuals who remain on low-carbohydrate diets over a long period discipline themselves to stick to their diet. Of the people who are inclined toward obesity however, only about 10% are successful in losing weight, and maintaining the weight loss, with any reducing diet.

SUMMARY

1. The caloric value of a food is determined on the basis of its quantity of digestible carbohydrates, proteins, and fats.
2. The human caloric values of digestible carbohydrate, protein, and fat are 4, 4, and 9 kilocalories (Calories), respectively.
3. A person's caloric needs can be broken down into three components: basal metabolic rate, specific dynamic effect, and activity metabolism.
4. Basal metabolism refers to the minimum number of Calories required to operate the body under the most relaxed conditions. The basal metabolic rate increases with weight and surface area, it decreases with age, and it is higher in males than in females.
5. The specific dynamic effect refers to the caloric component that is in excess of basal needs, and is needed to provide energy for the digestion and absorption of nutrients and for the arrangement of a proper balance among carbohydrates, proteins, and fats in the liver and other tissues.
6. Activity metabolism refers to the Calories required—in excess of the basal rate and the specific dynamic effect—to carry out physical activities.
7. Food intake is regulated by genetic factors, early eating habits, hormonal factors, physiological cues, psychological factors, and cultural influences.
8. The most reliable long-term weight-losing design rests upon a reduction of caloric intake, an adequate exercise program, and a minimum intake of sugar.

REVIEW QUESTIONS

1. On what basis (or bases) are caloric values assigned to a given foodstuff?
2. What are the average caloric values of carbohydrates, protein, and fats in human metabolism?

3. Define: basal metabolic rate, specific dynamic effect, activity metabolism.
4. What is the relationship between the magnitude of basal metabolic rate and each of the following: weight, surface area, age, sex?
5. Name six general factors that influence or regulate food intake.
6. List at least three statements used by proponents of "low-carbohydrate" diets in support of their claims for the effectiveness of these diets. Argue for or against these statements.
7. Where in the body is the appetite–satiety regulatory station located?
8. What kinds of physiological cues might be used by the appetite–satiety station to regulate food intake?
9. How might excessive eating in infancy or at puberty dispose a person to obesity later in life?

24 Nutritional Requirements

CARBOHYDRATES
    Essential carbohydrates
    Storage of carbohydrates
    Carbohydrates: facts and fallacies
PROTEINS
    Essential proteins
    Storage of proteins
    Proteins: facts and fallacies
FATS
    Essential fats
    Storage of fats
    Fats: facts and fallacies
VITAMINS
    Essential vitamins
    Storage of vitamins
    Some facts, fallacies, and uncertainties about vitamins
MINERALS
    Calcium
    Phosphorus
    Iron
        Absorption and excretion of iron
        Distribution of iron in the body
        Sources of metabolic iron
    Magnesium
    Iodine
STANDARDS FOR A NUTRITIOUS DIET
    Nutritional characteristics of seven classes of foods
    Richest sources of nutrients per caloric intake
    Alcoholic beverages
    Milk and cheese

Nutritional Requirements

In the last three chapters we discussed metabolism, its regulation, and our caloric requirements. In this chapter, we will take up topics that concern the quality and composition of a diet. We begin with a discussion of carbohydrates, proteins, fats, and vitamins. For each of these classes of nutrients we will say something about their functions, their essentiality or non-essentiality in the diet, and their storage in the body. Certain required dietary minerals will also be discussed. In conclusion, we will discuss the qualities of a nutritious diet.

NUTRITION. *The sum of the metabolic processes in the growth, maintenance, and repair of the body* (L. nutrire, *to nourish*).

CARBOHYDRATES

The principal functions of carbohydrates may be summarized in four points. First, they serve as one of several sources of energy to most tissues of the body. Second, they serve as the main and almost only source of energy for the nervous system. Third, carbohydrates stimulate the release of insulin, whereby carbohydrates, fats, and proteins may be maintained, or conserved, in the body (as opposed to being excreted wastefully, as happens in untreated diabetics). Fourth, carbohydrates are needed for the synthesis of other kinds of materials, such as proteins and fats.

Essential Carbohydrates

No single carbohydrate is essential in the diet. When a nutritionist says that a substance is *"essential,"* she means that the substance must be provided in the diet because the body cannot synthesize it fast enough (if at all) to satisfy a physiological requirement. Glucose, fructose, and all other carbohydrates in the body can be synthesized entirely from fats and proteins, as needed. No single kind of carbohydrate has to be taken into the body as food.

Storage of Carbohydrates

A certain amount of carbohydrate is stored in the body in an inactive form (glycogen) as an energy reservoir. The principal deposits of glycogen are in the liver and the skeletal muscle, though additional amounts of glycogen are present in the skin and other organs.

Approximately 3 to 4 ounces of glycogen are stored in the liver of a 150-pound individual following a meal. This storehouse is depleted usually after about 4 hours of activity, or after an overnight fast, during sleep. As shown in Figure 21.1, the liver cells contain an enzyme which brings about the release of glucose from the glycogen stores. The glucose is sent through the bloodstream to needy tissues throughout the body.

GLYCOGEN. *Animal starch* (G. glykys, *sweet;* gen, *to make*).

Approximately 8 to 10 ounces of glycogen are present in the skeletal muscle of a 150-pound individual at most times. During moments of work, skeletal muscle glycogen tends to be reduced; during moments of rest, skeletal muscle glycogen is replenished. Skeletal muscle, unlike liver, does not have an enzyme for producing glucose from glycogen (Figure 21.1). Since it does not have this enzyme, skeletal muscle must convert its glycogen into metabolites which produce energy or can be used for the synthesis of fat.

Carbohydrates: Facts and Fallacies

Theory: Carbohydrates are superior to fats and proteins as a source of quick energy.

Comment: Carbohydrates are the only kind of metabolite that the brain can use as a source of energy, and we perceive all of our actions through the brain. Furthermore carbohydrates provide energy faster to the brain than proteins or the glycerol of fats, both of which must first be converted to glucose. It is very likely, therefore, that we merely think carbohydrates are superior to fats and proteins for a quick energy pickup. The other tissues of the body can use carbohydrates, fats, or proteins as a source of energy. Skeletal and cardiac muscles, which are the tissues most closely linked with energy requirements, derive their energy directly from ATP, and secondarily from phosphocreatine, as described earlier in Chapter 17. ATP and phosphocreatine production is dependent upon the metabolism of acetate, which is derived from carbohydrates, proteins, and fats (Figure 21.5). The conversions involve a variety of vitamins and hormones, and the sequential events are not understood. If carbohydrate is superior to fat and protein as a quick energy pickup the reasons are probably that fats in the lymphatic system travel to muscle cells more slowly than carbohydrates in the bloodstream, while proteins and fats are not as rapidly digested and absorbed as carbohydrates.

Theory: Carbohydrates are not as fattening as fats.

Comment: This theory, quite simply, is false. A gram of carbohydrate is only 40% as caloric as a gram of fat, but as we have seen in the preceding chapter, a gram of carbohydrate releases about as much energy as a gram of fat per unit volume of oxygen consumption. The point is, all excess stores of carbohydrate, beyond whatever energy an individual requires in a given moment, are converted into fat.

PROTEINS

Protein is needed to construct cells, the ground substance of connective tissues, plasma proteins, enzymes, certain kinds of hormones, neurotrans-

---

ATP (adenosine triphosphate) *releases energy when its terminal phosphate group is cleaved by hydrolysis.*

mitters, contractile protein in muscle cells, and numerous other kinds of protoplasmic material.

Protein is also used as a source of energy. In many tissues of the body some of the amino acids which make up protein can be modified through various enzyme reactions, after which they can enter the Krebs cycle for the production of ATP. In the liver, amino acids can be modified and made into glucose. The glucose can be stored as glycogen or can be sent into the bloodstream to be used by other cells of the body.

PROTEIN. *A polymer made of amino acids* (G. proteios, first in rank).

Essential Proteins

No single type of protein is essential in the diet. However, eight kinds of amino acids are essential (see page 418).

In general, a healthful diet should contain about 20% or more protein on a caloric basis. This percentage amounts to about 2 grams of protein per kilogram of body weight. You might compare this figure with some of the data given in Table 24.1, "Recommended Daily Dietary Allowances." In Table 24.1 you should also note that additional quantities of protein are needed during pregnancy, lactation, and menstruation. (In another table, Table 24.4, you may note that a wide variety of foods would provide between 15 and 20 percent of the needed protein on a caloric basis.) Excessive amounts of dietary protein are not disadvantageous to health since they can be combusted or converted into carbohydrates or fats.

AMINO ACID. *A compound which contains an amino* (NH_2) *group and an acid* (H^+).

Storage of Proteins

Proteins and other nitrogenous compounds are not stored as reserve products in the body. Under ordinary circumstances the body excretes all of the nitrogen it cannot use. When inadequate amounts of protein are eaten, however, the body may degrade some of its own skeletal muscle proteins. In protein starvation diseases such as in kwashiorkor or marasmus, which we described earlier, the body uses some of the albumin which it normally circulates in the blood. A reduction in blood albumin usually leads to edema and circulatory disturbances.

Proteins: Facts and Fallacies

Theory: A worker needs more protein than a nonworker.
  Comment: False. A worker needs more energy, not more protein. Like a nonworker's body, the worker's body operates on the principle of nitrogen balance.
  Theory: A heavy load of protein may damage the kidneys.
  Comment: False, except when the kidneys are already diseased. By itself, a heavy intake of protein cannot harm the kidneys. It is true that greater

TABLE 24.1. Recommended Daily Dietary Allowances[a]

	Age (years)	Weight (lbs.)	Height (inches)	kcal	Protein (gm)	Fat-soluble Vitamins			Water-soluble Vitamins			
						Vitamin A Activity (IU)	Vitamin D (IU)	Vitamin E Activity (IU)	Ascorbic Acid (mg)	Folic Acid (mg)	Niacin[b] (mg)	
Children	1–6	26–42	32–43	1100–1600	25–30	2000–2500	400	10	40	0.2	8–11	
	6–10	51–62	48–52	2000–2200	35–40	3500	400	15	40	0.3	13–15	
Males	10–14	77–95	55–59	2500–2700	45–50	4500–5000	400	20	40–45	0.4	17–18	
	22	130–147	67–69	3000–2800	60	5000	400	25–30	55–60	0.4	20–18	
	22–55	154	69	2800	65	5000	—	30	60	0.4	18	
	55–75+	154	67	2400	65	5000	—	30	60	0.4	14	
Females	10–14	77–97	56–61	2250–2300	50	4500–5000	400	20	40–45	0.4	15	
	14–22	114–128	62–64	2400–2000	55	5000	400	25	50–55	0.4	16–13	
	22–55	128	64	2000–1850	55	5000	—	25	55	0.4	13	
	55–75+	128	62	1700	55	5000	—	25	55	0.4	13	
Females (pregnancy)				+200	65	6000	400	30	60	0.8	15	
Females (lactation)				+1000	75	8000	400	30	60	0.5	20	

Nutritional Requirements

	Age (years)	Water-soluble Vitamins (continued)				Minerals				
		Ribo-flavin (mg)	Thia-mine (mg)	Vita-min B_6 (mg)	Vita-min B_{12} (µg)	Cal-cium (g)	Phos-phorus (g)	Io-dine (µg)	Iron (mg)	Magne-sium (mg)
Children	1–6	0.6–0.9	0.6–0.8	0.5–0.9	2–4	0.8	0.7	55–80	15–10	100–200
	6–10	1.1–1.2	1.0–1.1	1.0–1.2	4–5	0.9–1.0	0.9–1.0	100–110	10	250
Males	10–14	1.3–1.4	1.3–1.4	1.4–1.6	5	1.2–1.4	1.2–1.4	125–135	10–18	300–350
	22	1.5–1.6	1.5–1.4	1.8–2.0	5	1.4–0.8	1.4–0.8	150–140	18–10	400
	22–55	1.7	1.4	2.0	5	0.8	0.8	140	10	350
	55–75+	1.7	1.2	2.0	6	0.8	0.8	110	10	350
Females	10–14	1.3–1.4	1.1–1.2	1.4–1.6	5	1.2–1.3	1.2–1.3	110–115	18	300–350
	14–22	1.4–1.5	1.2–1.0	1.8–2.0	5	1.3–0.8	1.3–0.8	120–100	18	350
	22–55	1.5	1.0	2.0	5	0.8	0.8	100–90	18	300
	55–75+	1.5	1.0	2.0	6	0.8	0.8	80	10	300
Females (pregnancy)		1.8	+0.1	2.5	8	+0.4	+0.4	125	18	450
Females (lactation)		2.0	+0.5	2.5	6	+0.5	+0.5	150	18	450

[a] Data from National Academy of Sciences–National Research Council, Food and Nutrition Board (1968), for normal healthy individuals in the United States.

[b] 1 mg of niacin also may be obtained from the metabolism of 60 mg of the amino acid tryptophan.

quantities of urea are formed within the liver and excreted by the kidneys when excessive amounts of protein are eaten. However, the stepped up metabolic and physiological functions of the liver and kidneys do not damage these organs.

Theory: Vegetarians derive a greater benefit from their diet than meat eaters.

Comment: False. Animal proteins are superior to plant proteins for purposes of human nutrition (Table 21.1).

FATS

FAT. *A class of combustible, ether-soluble, water-insoluble organic molecules.*

LINOLEIC and LINOLENIC ACIDS. *Polyunsaturated fatty acids.*

With the exception of the "essential" fatty acids, most of the fats in the body can be made from carbohydrates or proteins. Dietary fats are important as a source of the essential fatty acids (linoleic and linolenic acids). In addition, dietary fats provide energy, and they add flavor to food. (You may imagine how little flavor butter would have if it did not contain butyric acid, or how tasteless steak would be if the fat were somehow stripped from each muscle fiber.)

Fats are very important in the body. They are found in myelin sheaths around nerve fibers, in endomysial sheaths around muscle fibers, in fat cells as reserve fats, in the bloodstream as reserve fats (lipoproteins, albumin-lipid complexes), in the liver and gallbladder as bile salts, and in various organs as steroid hormones (cortisol, testosterone, vitamin D_3).

Essential Fats

Two of the polyunsaturated fats (linoleic and linolenic acids) are "essential" fatty acids. However, there is probably no known case of an individual suffering a dietary deficiency of one or both of these substances. As long as enough calories are ingested, the "essential" fatty acids are present in sufficient quantity to meet physiological needs. The "essentiality" of these fatty acids is based upon observations of laboratory rats that were placed on diets deficient in linolenic and linoleic acids. Unhealthy, scaly tails and sterility were telltale signs of the deficiency.

In a well-balanced diet about 2% and 3% of the caloric intake of an adult or a child, respectively, should consist of essential fatty acids. It is noteworthy that the essential fatty acids of breast milk provide about 4% to 5% of an infant's caloric needs. In Table 24.2 it may be seen that dairy products and coconuts contain large quantities of short-chain, saturated fatty acids, and that most plant foods contain larger percentages of polyunsaturated fatty acids than animal foods.

The essential fatty acids and the phosphorus-containing fats, such as lecithin, are used in constructing cell membranes. It is considered possible

TABLE 24.2. Percentage of Fatty Acids in Various Cooking Oils and Fats

Fatty Acid	Soy Bean Oil	Corn Oil	Coconut Oil	Olive Oil	Peanut Oil	Safflower Oil	Butter	Cow's Milk[a]	Human Milk[a]
Saturated, small chain (fewer than 14 carbon atoms)	0	0	60	0	0	0	11	8	6
Saturated, large chain (more than 12 carbon atoms)	13	15	31	9	14	7	52	57	38
Monounsaturated	29	51	7	85	56	18	31	26	41
Polyunsaturated	58	34	2	6	30	75	6	9	15

[a] Included for comparison.

that certain nervous disorders, such as neuritis, may be linked with faulty metabolism of nerve cell membranes.

In the subcutaneous layer of the body wall, fats play a role in insulation, preventing losses of heat. In and around tissues, fats serve to shield muscles, nerves, and glands against damage during physical encounters.

NEURITIS. *Lesion or disease in a nerve* (G. neuron; itis, *inflammation*).

Storage of Fats

All excessive dietary carbon—whether in the form of fat, carbohydrate, or protein—is stored in adipose cells. When adipose cells are "empty" they are not much larger than most other cells of the connective tissues—about 70 micrometers in diameter. When excessive amounts of organic nutrients are eaten, however, the cytoplasmic compartment of the fat cell can enlarge enormously, achieving a diameter of 130 micrometers or more.

Fats: Facts and Fallacies

Theory: Fat is more fattening than carbohydrate.

Comment: False. On a diet low in carbohydrate and high in fat, some of the excess fat is stored as such and a small part is excreted in the form of ketone bodies (Chapter 23). On a diet low in fat and high in carbohydrate, the excess carbohydrate is converted to fat.

Theory: High cholesterol levels in the blood come from a high dietary intake of cholesterol.

CHOLESTEROL. *A fatty alcohol* (G. chole, *bile*; stereos, *solid*; ol, *for alcohol*).

Comment: This is probably false. Many individuals eat large quantities of cholesterol, but nevertheless have a lower level of cholesterol in their blood than those who take in very little cholesterol. The relationship of cholesterol to heart disease is a controversial issue. When it is discussed, however, some important facts are usually ignored:

(1) In general, the intestinal capacity for absorbing cholesterol is less than about 0.5 grams per day. An intake of massive amounts of cholesterol (such as 10 grams, which is equivalent to three dozen eggs) may increase the uptake, but only up to about 1 gram.

(2) Whereas only 0.5 to 1 gram of cholesterol is intestinally absorbed, about 0.7 to 1.2 grams of the blood cholesterol is produced daily in the body, by the liver and the intestinal mucosa. It has been learned that dietary cholesterol suppresses the production of cholesterol in the liver, and that bile acids (derivatives of cholesterol, which pass from the liver to the intestinal tract) suppress the production of cholesterol in the intestinal mucosa. It is believed that tissues elsewhere in the body, although capable of producing cholesterol, do not secrete cholesterol into the bloodstream.

(3) About 1 to 3 grams of cholesterol are excreted daily in the bile, by way of the intestinal tract.

From these facts you may see that abnormally high levels of blood cholesterol are not caused by a diet rich in cholesterol, unless the metabolic mechanisms for regulating cholesterol levels are faulty. If, for example, excessive levels of cholesterol are absorbed from the gut, an abnormally high concentration of cholesterol may appear in the blood. Even if excessive amounts of cholesterol are intestinally absorbed, however, the blood level will be normal, provided the liver responds appropriately by reducing its rate of cholesterol production.

The level of cholesterol in the blood depends upon a host of factors (genetic, dietary, metabolic, and hormonal) about which literally nothing is understood. Only two general statements appear to hold true in the cholesterol controversy. First, obesity and/or a tendency toward hardening of the arteries are correlated with high levels of blood cholesterol. Second, an abnormally high blood cholesterol level may be reduced by a diet that is high in polyunsaturated fats relative to saturated fats; low in sugar; and low in calories.

VITAMINS

Table 24.3 lists the vitamins, their primary functions, and one or more symptoms of a dietary vitamin deficiency.

Essential Vitamins

Until 1966, the United States Food and Drug Administration listed Minimum Daily Requirements for vitamins and certain other nutrients. In 1966, the same bureau proposed that its listing of Minimum Daily Requirements be replaced by Recommended Daily Allowances, as set forth by the National Research Council of the National Academy of Sciences.

The listing of the Recommended Daily Allowances is given in Table 24.3. In examining this list you should keep in mind that the "allowances" are average values. No one knows what is best for an individual from time to time. The "allowances" are based on experimental observations over many years on a large number of Americans following a wide range of dietary programs.

Also in Table 24.3 you may note that certain vitamin requirements are specified in terms of *international units* (I.U.), as well as milligrams. Why are units used instead of weight? Because the international unit is a measure of the biological effectiveness of a vitamin, rather than a measure of its amount. This is important because some vitamins may be present in foods in multiple form. The different forms have a similar effect, but they are of different physiological strength. Vitamin E, for example, occurs in at least three forms: alpha, beta, and gamma. So too, vitamin D is present in different forms, as we pointed out in Chapter 12. A substance is said to have a certain number of "units" of vitamin activity when it brings about a certain amount of biological response. A *unit* of vitamin E, for example, is the total amount of vitamin E needed to enable a young female rat—with sterility

TABLE 24.3. Vitamins: Requirements, Functions, and Effects of Deficiency

Vitamin	Minimum Daily Requirement	Chief Functions	Primary Symptoms of Deficiency
Ascorbic acid	30 mg	General metabolism, especially in connective tissue	Improper deposition of collagen into connective tissue (including bone) and of dentine into teeth; from 5th to 11th month pain and tenderness in the extremities, disinclination to move, drawing up of legs, scorbutic rosary on chest; anemia
Niacin (can be synthesized from the amino acid tryptophan)	20 mg	General metabolism	Pellagra (tough skin); dermatitis (skin disease), dementia (mental deterioration) and diarrhea; weakness, irritability, loss of appetite (anorexia), insomnia, abdominal pains, burning sensations (deficiencies arise only in deficiency of both nicotinamide and tryptophan)
Riboflavin	2.5 mg	General metabolism	Dermatitis (skin disease); increase of blood vessels in cornea of eye; degeneration of the nervous system and liver; impaired reproduction; loss of hair; cataract of lens of eye
Pantothenic acid	10 mg	General metabolism	Burning feet syndrome; numbness and tingling
Pyridoxine	1.5 mg	General metabolism, especially of proteins	Acrodynia (soles and palms are extremely sensitive to pressure); degeneration of myelin in peripheral nerves; anemia; dermatitis
Thiamine	1.5 mg	General metabolism, especially of carbohydrates	Beriberi: accumulation of lactic and pyruvic acid in blood and tissues (onset of deficiency is delayed by a diet rich in fat) Wet form: edema, congestive heart failure; unusually slow heart at rest, unusually fast on exertion Dry form: multiple neuritis, progressive muscle weakness First reported symptom is usually loss of appetite, anxiety, irritability, and exhaustion

Vitamin	Minimum Daily Requirement	Chief Functions	Primary Symptoms of Deficiency
Folic acid	1 mg	General metabolism, especially of nucleic acids and protein	Anemia; degeneration in nervous system
B-12 (requires "intrinsic factor" from stomach for proper absorption)	1 microgram (mostly synthesized by bacteria in gut)	Synthesis of nucleic acids	Pernicious anemia; nerve damage
A	0.5 mg (4000 units)	An integral part of rods and cones for vision	Night blindness; loss of tear secretion; keratinization of cornea; scaling of skin
D	0.01 mg (400 units)	Induces the synthesis of an intestinal protein which picks up calcium from the intestinal tract; may be required for uptake of calcium in other cells	Rickets in children; osteomalacia in adults; both diseases characterized by a faulty deposition of calcium in the organic matrix of bone
E	Not known	Not known	Anemia due to faulty metabolism in bone marrow
K	1 mg (mostly synthesized by bacteria in gut)	Synthesis of several blood clotting factors	Tendency to bleed easily

induced by vitamin deficiency—to bear live young. A unit of vitamin A corresponds to the biological response obtained from 0.3 micrograms of pure vitamin A or 0.6 micrograms of pure beta carotene. A unit of vitamin D is that amount which is equally effective in preventing rickets as 25 nanograms of pure 1,25-dihydroxy vitamin D_3 (the active form of vitamin D, referred to in Chapter 12).

In Table 24.3 you may note that there are no allowances for pantothenic acid or for vitamins B_{12} and K. Pantothenic acid is present in nearly all foods. A deficiency of this vitamin usually occurs only in starvation diets or inadequate diets; pantothenic acid deficiency is common in chronic alcoholism. Vitamins B_{12} and K are usually synthesized in adequate quantity by bacteria in the lower intestine.

Storage of Vitamins

The fat-soluble vitamins are stored to some extent in the liver and to some extent in fat tissues. The water-soluble vitamins are generally converted into coenzymes and are used as such on a daily basis. Unlike carbohydrates, proteins, and fats, which are energy sources, vitamins are strictly regulatory agents. Thus the body has few safeguard features for their maintenance and retention.

Some Facts, Fallacies, and Uncertainties About Vitamins

Theory: Excessive vitamin A or vitamin D causes toxicity.

Comment: An excessive intake (100 fold or more) of vitamin A may be toxic. The symptoms of toxicity include drowsiness, headache, loss of appetite, vomiting, and bone demineralization. An excessive intake of vitamin D also may be toxic. The symptoms of toxicity are similar to those of excessive vitamin A. Although excessive levels of vitamins A and D are toxic, there are no known cases of toxicity among people who eat foods that are available commercially. Toxicity is produced only when people take supplemental vitamin pills excessively.

Theory: Excessive dietary water-soluble vitamins are excreted.

Comment: Excessive water-soluble vitamins generally do not produce toxic effects. Indeed, large doses of niacin and pantothenic acid as well as perhaps other water-soluble vitamins are used singly or in combinations for treating certain forms of mental illness.

For the most part, excessive levels of the water-soluble vitamins are in fact excreted, being flushed out of the kidneys without being metabolized. Or they are metabolized into inactive substances which are then excreted. Vitamin C may be converted into oxalic acid, vitamin B_6 into pyridoxic acid, and niacin into methylnicotinamide.

Theory: Proper metabolism hinges on a balance of vitamins.

OXALIC ACID, METHYLNICOTINAMIDE and **PYRIDOXIC ACID** *exist as ionic species in the bloodstream. Their electric charges prevent them from entering cells, and they are excreted in the urine.*

Comment: At this point you are urged to re-examine Figure 21.2. From this figure it should be clear that there is considerable overlap—and certain distinctions or uniquenesses—in the functions of individual vitamins. Both pantothenic acid and niacin, for example, are needed to synthesize or degrade fatty acids. In addition, however, whereas biotin is needed for the synthesis of fatty acids, riboflavin is required for their breakdown.

A re-examination of Table 24.2 at this point will also help you appreciate the interplay among vitamins in the overall scheme of metabolism. A study of column three, for example, will show that anemias may arise from a deficiency of various vitamins. Similarly, skin diseases may arise when any one of several vitamins is below the minimum requirement. Anemia, skin diseases, and other vitamin-deficiency diseases are merely outward signs of complex internal (cellular) problems.

Theory: Large quantities of certain vitamins are especially beneficial to one or another aspect of health.

Comment: This viewpoint has been expressed by a number of authors. Linus Pauling, for example, suggests (in *Vitamin C and the Common Cold*) that for most people, 1 to 2 grams of vitamin C may be beneficial on a daily basis; that the optimum intake may range from 250 milligrams to 10 grams, depending upon the individual; and that levels even higher than 10 grams may be needed when a common cold develops. Despite Pauling's suggestions, the Recommended Daily Allowance is much lower—60 milligrams.

Pauling advanced a number of biological arguments favoring his point of view. Four of these arguments can be used by way of illustration: (1) Gorillas, like humans, are unable to synthesize vitamin C from glucose. However, they eat fruits and vegetables, which provide them with what would be equivalent to a daily intake of 4 grams of vitamin C for an average human. (2) Rats, like most mammals, are able to produce vitamin C from glucose. They do this on a daily basis at a level corresponding to about 4 grams for an average human. (3) A 2500 Calorie diet of fresh raw fruits and vegetables contains about 2.5 grams of vitamin C. Pauling suggests that the caveman ate large quantities of fruits and vegetables, that we have not had a sufficient lapse in our evolution to evolve any special mechanisms for overriding our primeval needs, and that even if we ate 2500 Calories worth of fruits and vegetables, the cooking of these foods in metal containers would bring about large losses of vitamin C. (4) Pauling admonishes the dogma of the nutritionists who, in his viewpoint, in the course of suggesting daily vitamin "allowances," imply that, "If you don't eat vitamin C, you will come down with scurvy. Therefore, if you don't come down with scurvy, you're eating enough vitamin C."

From the preceding, you may note the logic of Pauling's arguments. Nevertheless, failure to consider the complexity of the human metabolic machinery, to ignore the possibility of special adaptive features which may have evolved over the past few thousand years, and to concentrate on a single vitamin while ignoring the others, casts some doubt upon the validity of his arguments.

Some food faddists recommend large amounts of vitamin E, with or with-

LECITHIN. Any of a group of phospholipids containing the alcohol choline (G. lekithos, *egg yolk*).

out vitamin C, as an antiaging substance, and also as a substance that may stimulate the libido (sex drive).

Lecithin (not actually a vitamin, but a family of phospholipids) has been recommended as a remedy for nervous tension and as an aid to softer, smoother skin.

It is perhaps fair to say that none of the claims for excessive amounts of vitamins has a firm basis in scientific fact. Indeed, the many self-regulating systems within the body, as discussed in this text, rule against the possibility that massive doses of vitamins may be helpful. What no one knows, however, is the difference between what is an excessive amount of vitamins and what is an optimum amount for a particular individual.

A careful analysis of claims made by advocates of megadoses (extremely large doses) of vitamins shows that the claims are arbitrary, that they fail to account for individual metabolic differences (of which we know virtually nothing), and that they are usually geared toward individuals with physical or psychological problems of unknown cause.

Nevertheless, a considerable amount of logical and plausible reasoning may be mustered in support of a claim for a particular vitamin—although such arguments invariably neglect many aspects of metabolism, such as the role of the various other vitamins in the metabolic machinery. In view of the recent rash of popular books on the subject, it is important, and even necessary, for a student to understand the biological principles involved if he or she is to be able to judge the validity of various authors' claims. Our discussion should have sufficed to point out the plausibility—but also the inadequacy, incompleteness, or lack of validity—of some of the claims made for particular vitamins or combinations of vitamins.

MINERALS

Most of the minerals needed in the body are adequately available in most kinds of diets. When the composition of an individual's diet is improperly balanced, or when someone is on a weight-reducing diet, the diet may become deficient in calcium, phosphorus, iron, magnesium, iodine, or any combination of these essential minerals (see Table 24.4).

Calcium

Calcium is needed for proper growth of bones and teeth, for blood coagulation, and for the proper functioning of cell membranes. Calcium is so important—to all cells of the body—that it is kept in the blood at a constant level of 9 to 11 milligrams per 100 milliliters. It is maintained at this level through the combined actions of parathyroid hormone, calcitonin, and vitamin D, as was described in Chapter 12.

Phosphorus

Phosphorus is needed for proper growth of bones and teeth. It is also necessary in all areas of metabolism, where it is an integral part of numerous metabolites.

Phosphorus is widespread in foods. It occurs mainly as calcium phosphate salts and in nucleic acids. Like calcium, phosphorus is maintained at a constant level in the blood, at a level of about 3 milligrams per 100 milliliters.

Iron

Iron is present in hemoglobin and in various enzymes. When a person develops an iron deficiency—regardless of whether the deficiency is due to an inadequate intake of iron, faulty absorption of iron, improper iron metabolism, or excessive losses of iron—the individual becomes anemic.

Absorption and Excretion of Iron. Iron is absorbed into epithelial cells of the small intestine. After entering an intestinal cell, it may be conveyed into the bloodstream, or kept as storage material in the intestinal cell.

When iron moves into the bloodstream from an intestinal cell it is picked up by *transferrin*, a beta globulin. This protein carries the iron through all the tissues of the body. As the blood passes through certain tissues in need of iron, such as red bone marrow or the liver, the iron is released from transferrin, and the transferrin molecule becomes available for the uptake of additional iron from the intestinal epithelial cells.

When iron is stored in an intestinal epithelial cell, it becomes locked into a protein called *ferritin*. Ferritin releases the iron into the bloodstream whenever a metabolic need arises. In the event that a need does not arise within the lifetime of the intestinal epithelial cell—which is about 2 to 5 days—the iron-ferritin complex is cast into the lumen of the intestine, as part of the sloughed-off intestinal cell. In this way the intestine prevents the accumulation of excessive amounts of iron in the body.

TRANSFERRIN. *A protein capable of combining with two atoms of iron, serving to transport iron in the blood (L. trans, to pass; ferrum, iron).*

FERRITIN. *An iron-protein storage form of iron within tissues (L. ferrum, iron).*

Distribution of Iron in the Body. In men, about 2.2 grams of iron are distributed in 25 trillion red blood cells. In women, these figures are 1.7 grams of iron and 22 trillion red blood cells. Another 1.0 and 0.4 grams, respectively, are present elsewhere, particularly in the liver. Still a third quantity of iron—about 0.5 grams—is present in red blood cells that are continuously degenerating in the reticuloendothelial system.

Sources of Metabolic Iron. Normal adults need about 20 milligrams of iron daily. About 75% of this quantity is derived from red blood cells that are degenerating in the reticuloendothelial system. Another 20% is derived from stores that are present in the liver and elsewhere in the body. Only about 5% of the needed iron is obtained from the diet.

An ordinary diet contains about 15 milligrams of iron per 2500 Calories of food. About 7% of this iron (1 mg) is absorbed. Larger quantities may be absorbed when special needs arise, as for example in certain cases of anemia, during menstruation, and in persons who lose excessive amounts of blood because of gastric ulcers or hemorrhoids. As much as 30% of the dietary iron, or about 5 mg, can be absorbed from the ordinary diet. Since 25 mg to 125 mg of iron are lost in menstruation, and even higher amounts are lost in individuals with severe ulcers or hemorrhoids, the upper limit of efficiency in absorption (30%) will not always provide a sufficient amount of iron, even over a long period. For most women, and for most persons with mild ulceration or hemorrhoids, the difference can be made up by drawing iron from the ferritin stores in the intestinal epithelium. For some individuals, however, it may be necessary to take supplemental iron tablets. You may note the higher Recommended Daily Allowance for iron by women during pregnancy and lactation (Table 24.1).

HEMORRHOIDS. *Enlarged and twisted veins in the rectum, anus, or esophagus* (G. haimorrois, *a vein liable to discharge blood*).

Magnesium

Magnesium is needed for the proper performance of muscle and nerve cells and for certain enzyme activities. Approximately 300 mg of magnesium are required daily to replace magnesium lost in the urine, sweat, and feces. A deficiency may lead to depression, muscular weakness, dizziness, and other problems.

Iodine

About 25 mg of iodine are present in the adult body. About half — 10 mg — is present in the thyroid gland and the remainder is diffused throughout the rest of the body, particularly as thyroxine. About 12 micrograms of iodine are needed daily in infancy. Upwards to about 120 micrograms are needed during growth, and decreasing amounts are required during aging (Table 24.1). Iodine deficiencies induce metabolic problems, as indicated in Chapter 12.

THYROXINE. *A hormone from the thyroid gland; increases the metabolic rate.*

STANDARDS FOR A NUTRITIOUS DIET

A balanced diet for a normal person should contain — on a caloric basis — about 20% protein, 30% fat, and 50% carbohydrate, with sucrose (table sugar) kept at a very low level. In addition, a nutritionally sound diet should have neither an excess nor a deficiency of calories, an ample quantity of fruits and vegetables (which are rich in minerals and water-soluble vitamins), and meat, fish, and dairy products (which are rich in fat-soluble vitamins as well as in certain water-soluble vitamins).

Another way of looking at a well-balanced diet is in terms of its caloric foods, its body-building foods, and its protective foods. *Caloric foods* are relatively high in carbohydrate and/or fat. They include such items as oils and fats, sugars and syrups, baked goods, potatoes, peas, beans, and nuts. The *body-building foods* are relatively high in protein and certain vitamins. They include such foods as meat, fish, eggs, and dairy foods, excluding butter. Fruits and vegetables are mainly *protective foods*, because of their high content of minerals and water-soluble vitamins.

In a balanced diet about half of the caloric intake should be composed of caloric foods. The body-building foods should make up about 40% of the diet, and the remaining 10% should consist of protective foods.

Table 24.4 is a compilation of common foods. The data are based on portions that would very likely be consumed by an adult at one mealtime, as, for example, an 8-ounce glass of milk or four slices of bacon. In examining this table, however, you should realize that the data are approximate; the figures are rounded off from the most extensive compilation currently available—the Agriculture Handbook Number 8, U.S. Government Printing Office, Washington, D.C. You should also keep in mind that the table is not specific with regard to the items listed (for example, chuck versus loin of beef), or to the quality (for example, choice grade versus good grade of meat), or to the age of certain food items (fresh versus frozen vegetables).

In spite of its limitations, a number of interesting observations can be drawn from Table 24.4. Our comments on these observations will take into account, first, the general qualities of the various classes of foods listed in the table, and, second, the food items which have the highest concentration (per calorie of food) of a specific nutrient. The data on foods which are very rich in some way are indicated by a light blue tone. The data for foods with extremely low concentrations of a particular nutrient are indicated by by light shading. At the end of this chapter, there is a discussion of milk, cheese, and alcoholic beverages—food items which are perhaps most often discussed and misunderstood in conjunction with nutrition.

Nutritional Characteristics of Seven Classes of Foods

Dairy Foods. The outstanding feature of dairy foods is their high content of calcium and riboflavin.

Meat, Fish, and Shellfish. The outstanding feature of this class—on a per serving basis—is its very high content of calories, protein, phosphorus, iron, vitamin A, thiamin, riboflavin, and niacin, and fat—although most fish usually contains little or no fat. Carbohydrate and vitamin C are present in negligible amounts, if at all.

Fruits. Except for an occasional exception, such as the avocado, fruits contain relatively low quantities of protein and fat per serving, and a high quantity of carbohydrate. They are most notable for their high content of niacin and vitamin C.

Colored Vegetables. The red, yellow, green, and other colored vegetables are high in vitamin A, niacin, and vitamin C. In addition, some of the deep green vegetables, such as broccoli and spinach, are rich in iron.

TABLE 24.4. Nutritive Value of Various Foods (Edible Parts Only). (Light tone over the figures in the table indicates extremely low concentration of a particular nutrient; colored tone indicates high concentration.)

	Measure	Calories	Protein (grams)	Fat (grams)	Carbohydrate (grams)	Calcium (milligrams)	Phosphorus (milligrams)	Iron (micrograms)	Magnesium (milligrams)	Vitamin A (international units)	Thiamin (micrograms)	Riboflavin (micrograms)	Niacin (micrograms)	Vitamin C (milligrams)
DAIRY FOODS														
Milk	1 cup	160	9	9	12	330	260	100	30	400	80	500	300	2
Skim milk or buttermilk	1 cup	90	9	—	12	330	260	100	30	10	90	500	300	2
Yogurt (skim milk)	1 cup	125	8	4	12	300	220	100	30	200	100	400	200	2
Cheese (cheddar, Swiss, American)	2 cubic inch	125	8	10	—	250	250	350	30	400	60	120	—	0
Cottage cheese (uncreamed)	1/2 cup	85	17	—	3	90	170	400	—	10	30	30	100	0
Eggs	2 medium	120	9	9	1/2	50	150	1600	9	900	75	250	75	0
Ice cream (regular)	1/2 pint	240	5	13	25	150	125	100	30	450	40	200	100	1
MEAT, FSH, SEAFOOD														
Bacon, fried	4 slices	170	2	18	0	4	80	1000	30	0	150	100	1700	—
Beef, roasted	6 oz.	600	50	50	0	20	250	6000	45	80	80	400	8000	—
Pork, broiled	6 oz.	800	40	70	0	20	500	6000	40	0	800	400	11000	—
Lamb, cooked	6 oz.	700	50	55	0	18	400	3000	35	0	200	400	10000	—
Chicken, roasted	6 oz.	600	50	40	0	20	440	4000	45	1800	100	400	14000	—
Haddock, fried	6 oz.	340	40	15	0	80	500	2500	45	—	80	150	6000	2
Sardines	3 oz.	160	20	9	0	370	200	2500	45	200	20	150	4500	—
Clams, lobster, shrimp (raw)	6 oz.	180	40	3	2	120	350	3000	50	—	160	300	200	—
FRUITS														
Apple (1)	5 oz.	70	—	—	18	8	10	400	15	100	40	20	100	5
Banana (1)	5 oz.	90	1	—	21	9	30	700	50	210	50	60	700	10
Grapefruit (1/2)	5 oz.	80	1	—	20	30	30	700	4	100	70	30	300	65
Orange (2 1/2")	5 oz.	80	1	—	20	70	30	700	15	350	150	70	700	80
Peach (2 2/2")	5 oz.	65	1	—	15	15	30	700	15	2300	30	70	1500	10
Watermelon (1" × 8")	7 oz.	60	1	—	14	15	25	1000	35	1200	60	60	300	15
Avocado	5 oz.	200	3	18	7	12	60	750	80	300	12	200	200	15
Dates	5 oz.	450	2	1	110	95	80	4500	100	80	120	140	3300	0

Nutritional Requirements

Food	Serving													
COLORED VEGETABLES														
Beans, green	1 cup	35	2	—	7	60	45	800	30	600	90	100	500	20
Broccoli	1 cup	40	4	—	6	130	90	1100	25	3000	120	300	1100	130
Brussels sprouts	1 cup	50	4	—	8	150	90	1700	30	16000	120	220	1100	130
Carrots	1 cup	60	2	—	13	50	50	1000	25	100	60	90	900	12
Lettuce	3 oz.	15	1	—	3	40	30	2000	10	100	60	60	300	8
Spinach	1 cup	35	4	—	5	140	15	4000	80	12000	150	300	900	70
Squash, summer	1 cup	30	2	—	6	52	45	600	15	600	75	140	1600	30
Tomato (2½")	1	30	1	1	7	18	40	600	15	1200	90	50	1000	30
Squash, winter	1 cup	130	4	1	30	57	110	1600	15	8600	100	270	1400	28
Corn	1 cup	70	3	—	15	2	70	500	50	300	90	80	1000	7
Sweet potato	1 cup	170	2	1	40	47	85	1000	30	11500	130	90	90	25
Peas	1 cup	80	4	0.5	12	25	100	2000	25	650	400	150	3000	30
WHITE VEGETABLES														
Cauliflower	1 cup	25	2	—	5	25	50	1100	25	60	100	100	700	80
Celery	3 stalks	15	1	—	4	40	30	300	20	200	30	30	300	10
Cucumber	1.5 × 6"	25	1	—	5	30	45	1500	10	250	50	70	300	18
Onion (white)	3"	60	2	—	13	45	60	900	10	60	45	60	300	15
Radishes	1 cup	20	1	—	4	50	50	1000	15	10	30	30	300	30
Potato	1 cup	90	2	—	20	10	50	600	20	—	100	40	1000	2
FATS AND OILS														
Butter or margarine	⅓ × 1 × 1"	36	—	4	—	1	1	0	1	170	0	0	0	0
Salad or cooking oil	1 tbsp.	125	0	14	0	0	0	0	—	—	—	0	0	0
CEREAL AND GRAIN PRODUCTS														
Bread	1 slice	70	2	1	13	20	25	600	10	—	60	50	60	—
Plain cake, no icing	2 × 8" slice	550	7	18	90	100	125	500		150	40	150	400	0
Saltine crackers	8	90	2	2	16	8	20	400		0	—	—	—	—
Roll, hamburger or hot dog	1.5 oz.	110	3	2	21	30	50	800		—	110	70	400	0
Pound cake	½ × 4 × 8" 1 slice,	140	2	9	14	6	25	200		80	10	30	900	0
Doughnut	1 oz.	120	1	6	15	13	30	400		30	50	50	10	—
Spaghetti, cooked	1 cup	150	5	—	32	11	75	1300	30	0	200	110	1500	0
Rice, white, cooked	1 cup	220	4	—	50	21	55	1800		0	230	20	2100	0
Corn flakes	1 cup	90	2	—	21	4	10	400		0	110	20	500	—
Pancakes, plain	3 cakes	180	5	6	27	180	200	900		120	120	180	600	—
Oatmeal, cooked	½ cup	60	3	1	9	10	90	1100			150	30	200	0

TABLE 24.4. (Continued)

	Measure	Calories	Protein (grams)	Fat (grams)	Carbohydrate (grams)	Calcium (milligrams)	Phosphorus (milligrams)	Iron (micrograms)	Magnesium (milligrams)	Vitamin A (international units)	Thiamin (micrograms)	Riboflavin (micrograms)	Niacin (micrograms)	Vitamin C (milligrams)
BEVERAGES														
Liquor, 80 proof (gin, rum, vodka, whiskey)	2 fl. oz.	130	—	—	—	—	—	—	—	—	—	—	—	—
Wine, table	3½ fl. oz.	85	—	0	4	9	40	400	15	—	—	10	—	—
Beer	12 fl. oz.	150	1	0	14	18	100	—	—	—	100	2200	100	—
Cola, ginger ale, root beer	12 fl. oz.	145	0	0	37	—	—	—	0	0	0	0	0	1
Tea, coffee	8 fl. oz.	2	—	—	—	—	—	—	—	—	—	—	—	3
NUTS AND SNACKS														0
Mixed (almonds, cashews, peanuts, pecans, walnuts)	½ cup	630	20	55	20	220	500	5000	300	0	75	100	2500	—
Coconut	½ cup	360	3	35	10	15	100	2000	100	0	50	20	500	—
Popcorn, plain	1 cup	25	1	—	5	1	20	200	3	—	—	10	100	1
Candy, chocolate, plain	1 oz.	145	2	9	16	65	70	300	—	80	20	100	100	0
Honey	1 tbsp.	65	—	0	17	1	1	100	0	0	—	10	100	—
Jellies	1 tbsp.	50	—	—	13	4	—	300	0	2	—	10	250	—
Table sugar	1 tbsp.	40	0	0	10	0	—	—	—	0	0	0	0	0

Unlike the preceding classes of food, colored vegetables (and white vegetables) are especially desirable in weight-losing diets because they have very few calories per serving as compared to other classes of food. The weight-watching individual must therefore consume much larger quantities of vegetables in order to take in as many calories as would be obtained from a smaller portion of, say, meat. This, in turn, means an individual is very likely to spend much more time at the meal table. During this time, the smell, taste, and visual receptors become accommodated to the smell, taste, and sight of food, and the stomach becomes distended sufficiently to turn off the appetite center before excessive calories are eaten.

White Vegetables. Although low in calories, the white vegetables are not nearly as nutritionally rich as the colored vegetables. There are exceptions, of course, as indicated in Table 24.4.

Fats and Oils (liquid fats). Animal fats have very little nutritional value, although they are a rich source of energy. In contrast, plant fats have a high caloric value and are also a rich source of essential fats and polyunsaturated fats.

Cereals and Grain Products. Cereals and grains are the seeds of domestic grasses. Sometimes they are eaten after only moderate alteration, such as boiling (rice and oatmeal). More often, they are milled and mixed with vitamins and other ingredients, after which they may be subject to further modification by the cook before mealtime. In general, they are well-balanced in terms of their proportions of carbohydrate, fat, and protein, and many of them are highly caloric per serving.

Richest Sources of Nutrients Per Caloric Intake

Proteins. As classes of food, meat, fish, and dairy foods generally offer the highest protein content per calorie consumed. Uncreamed cottage cheese, skim milk, and buttermilk are especially prominent in this regard, though certain vegetables, such as broccoli, Brussels sprouts, and spinach are also rich.

Fats. Nuts, meat, and whole milk offer the highest fat content per calorie intake, whereas fruits and vegetables (with certain exceptions, such as avocado) are very poor.

Carbohydrates. Milk and fruit contribute richly to the carbohydrate content of a mixed diet, though the richest sources per caloric intake are the primary sugar foods: table sugar, honey, jellies, candy, and the like.

Calcium. Dairy products in general are the most highly concentrated source of calcium. Specific members of various other food classes, such as sardines, broccoli, Brussels sprouts, and spinach, are also rich in calcium.

Phosphorus. Dairy products and meat are the richest sources of phosphorus.

Standards for a Nutritious Diet

Iron. Although meat, fish, and shellfish are rich sources of iron, certain vegetables, such as peas and spinach, eggs, and some other foods are as bountiful or more so, as shown in Table 24.4.

Magnesium. Magnesium appears to be uniformly distributed among many foods in the various classes shown in Table 24.4, though some food items, such as dates and coconut, are somewhat richer than others. In general, fruits and vegetables have the greatest quantity on a "per calorie" basis.

Vitamins. The colored vegetables are the best sources of three vitamins: thiamin, riboflavin, and vitamin C. White vegetables and fruits are also rich sources of vitamin C. Milk and yogurt, however, contain the highest quantity of riboflavin per calorie of food, whereas meat is the richest source of niacin.

Alcoholic Beverages

Alcohol has a caloric content of 7 Calories per gram. Its concentration in a beverage is expressed either as percent (volume of alcohol per total volume of beverage) or as *proof*; 2 proof is equal to 1%.

Regardless of what kind of alcoholic beverage is consumed, the alcohol itself is metabolized, almost entirely in the liver, at a constant rate. In general, it is broken down at the rate of 1 ounce (of 90 proof liquor) per hour. Beer (6 proof alcohol; 3% by volume) is broken down at the rate of about 8 ounces per hour. The exact rate of breakdown depends on how fast two specific enzymes (alcohol dehydrogenase and a microsomal oxidase) are able to work. No matter which enzyme is employed, acetaldehyde is produced from the alcohol molecule, and it, in turn, is converted into acetate.

Derived coenzyme forms of niacin are required for the conversion of alcohol into acetate. However, neither increased amounts of niacin, nor derived coenzyme forms (NAD, NADP), administered orally or intravenously, have been found to hasten the process of converting alcohol into acetate. Nor, for that matter, is there anything which is known to be capable of speeding up this process—not fresh air, cold temperature, exercise, caffeine, or food. The plain fact is that alcohol is metabolized at a constant rate, in a closed system, at 37°C, during which time, depending on its concentration in blood, it may produce (at different times) euphoria (a mood of well-being), mild sedation, or depression of the central nervous system.

Do alcoholic beverages provide any nutritional value? Whiskey and other hard drinks do not—except for calories—though beers, ales, and wines do, as shown in Table 24.4.

Milk and Cheese

For the past 50 years, there has been controversy regarding the benefit of human milk versus cow's milk. A comparison of these two food items shows that, from a strictly nutritional viewpoint, neither appears to be superior in every respect.

The fat content of cow's milk is reduced from about 5% to about 0.2% when whole milk is converted to skim milk. At the same time, large amounts of vitamin A and D are lost (which can be returned to the milk artificially), but the calcium, phosphorus, and protein content is not significantly diminished.

Evaporated milk is cow's milk from which half the water has been removed. Condensed milk is evaporated milk to which more milk sugar (lactose) is added.

Cheese is made from milk, which may or may not be allowed to sour first. Thereafter, rennet is added, causing a clot to appear in the liquid—now called *whey*. The clot, or *curd*, contains many kinds of nutrients, including most of the milk's 4% protein, and varying but usually large amounts of fat. The curd is separated from the whey and placed in a cool room to ripen. During ripening, bacterial fermentation takes place. Cow's milk is used for most cheeses, sheep's milk is used for Roquefort, and goat's milk is used for Norwegian cheeses.

From a nutritional viewpoint (see Table 24.4), cheese ranks favorably with milk as a source of high quality protein. It may in fact be more desirable than milk, in view of its lower carbohydrate content.

SUMMARY

1. The principal special functions of dietary carbohydrates are that they serve as the main source of energy to the central nervous system and they stimulate the release of insulin, which operates to conserve nutrients in the body.
2. No carbohydrate is essential to the diet.
3. Carbohydrates are stored in the body as glycogen. Excessive carbohydrates are converted into fat.
4. Carbohydrates may be superior to fats and proteins as a quick source of energy because they reach the liver sooner after being eaten.
5. Carbohydrates are as fattening as fats, on a caloric basis.
6. The principal special functions of dietary protein are that they serve as building material for the formation of protoplasm and they can be used to produce glucose when needed.
7. No protein is essential to the diet, but eight kinds of amino acids are essential.
8. Protein is not stored in the body over the long period as a reserve nutrient.
9. A worker does not need more protein than a nonworker; excessive proteins are not injurious to the kidneys; and vegetarians do not derive a greater benefit from their diet than people who eat meat.
10. The principal essential functions of dietary fats are that they provide certain necessary components for constructing cell membranes and they are a rich source of fat-soluble vitamins.

11. Linoleic and linolenic acids are essential fatty acids.
12. Excessive fats are stored in adipose cells.
13. High cholesterol levels in the blood do not necessarily arise because of a high dietary cholesterol intake.
14. An excessive intake of vitamins A or D can cause toxicity, whereas excess water-soluble vitamins are generally excreted in the urine.
15. A proper balance of small quantities of vitamins in a well-balanced diet is healthful, whereas the value of megadoses of one or two vitamins is generally advantageous only in certain medical cases.
16. Calcium, phosphorus, iron, magnesium, and iodine are the minerals most likely to be deficient in a diet.
17. For a normal person, a balanced diet should contain about 20% protein, 30% fat, and 50% carbohydrate on a caloric basis, with sucrose kept at a very low level.
18. A well-balanced diet should contain certain amounts of caloric foods, body-building foods, and protective foods.

REVIEW QUESTIONS

1. What are the principal special functions of carbohydrates in the diet?
2. What carbohydrates are essential in the diet?
3. Tell why carbohydrates are (or are not) superior to fats and proteins as a source of quick energy. Defend your conclusion.
4. What are the principal special functions of proteins in the diet?
5. Are any proteins essential to the diet? Explain.
6. What are the principal special functions of fats in the diet?
7. Name two essential fatty acids.
8. Argue the following, pro or con: High cholesterol levels in the blood can be reduced by reducing cholesterol intake.
9. What vitamins, taken in excess, are toxic?
10. What five minerals, if any, are likely to be deficient in a person's diet?
11. What percentages of protein, fat, and carbohydrate are considered desirable in a balanced diet?

Transport Systems: Circulation, Respiration, and Excretion

PART SIX

25 Circulatory System: Functional Anatomy and Electrocardiography

26 Dynamic Characteristics of the Circulatory System

27 Functional Disorders of the Circulatory System

28 The Respiratory System

29 The Red Blood Cell System

30 The Excretory System

Circulatory System: Functional Anatomy and Electrocardiography

BODY FLUIDS
FUNCTIONAL ANATOMY OF THE HEART
    The four chambers of the heart
    The four valves of the heart
    The three-layered wall of the heart
    The bioelectric system of the heart
        Influence of the nervous system on the myogenic system
THE CARDIAC CYCLE
ELECTROCARDIOGRAPHY
    Rate of heartbeat
    Normal rhythm, arrhythmia, and dysrhythmia
    The P wave
    The P-R interval
    The QRS complex
    The ST segment
    The T wave
FUNCTIONAL ANATOMY OF BLOOD AND LYMPH VESSELS
    General structure of blood and lymph vessels
    The elastic arteries
    The distributive arteries
    The arterioles, capillaries, and venules
    The medium veins
    The large veins
    Functional anatomy of the lymphatic system
CIRCULATORY ROUTES
    Pulmonary circuit
    Systemic circuit
        Circuit through the neck and head
        Circuit through the upper extremity
        Thoracic circuits
        The coronary circuit
        Abdominal circuits
        Circuit through the lower extremity

Body Fluids

The *circulatory system*, known also as the *cardiovascular system*, is the vehicle through which most of our living cells receive their nutrients and dispose of their wastes. The circulatory system consists of the heart, the blood vessels, and the lymphatic vessels.

The main task of the cardiovascular system is to circulate fluids, and we therefore begin this chapter with a description of the various kinds of fluid in the body. Certain anatomical and electrical features of the heart are described next. It then becomes possible for us to discuss the *cardiac cycle* – the cycle of mechanical and electrical events that result in the pumping action of the heart. With the cardiac cycle as background, the stage is set for a look at *electrocardiography* – the science that provides insight into whether the heart's electrical system is functioning properly. In the event of malfunction, the electrocardiograph is often useful in locating the difficulty.

Next we turn to the second major anatomical component of the cardiovascular system – the *vasculature*. The various kinds of blood vessels and lympathic vessels are described from an anatomical and functional point of view. These descriptions are followed by brief discussions of certain key circulatory routes through which blood and lymph are conveyed.

VASCULAR. *Pertaining to, consisting of, or provided with, vessels (L. vasculum, small vessel).*

BODY FLUIDS

Nearly 70% of the body's weight is fluid; in an average adult this fluid amounts to about 40 liters. Slightly more than half of this volume occurs within living cells as *intracellular fluid*. The remainder is found outside of cells as *extracellular fluid*. The principal chemical components of these fluids are shown in Figure 30.8, where you may see how these two fluids differ from one another.

Extracellular fluid may be subdivided into three classes: interstitial fluid, blood, and transcellular fluid.

Interstitial fluid, which makes up about three fourths of the extracellular fluid, is present in two compartments of the body. It is present mainly in the space between the trillions of cells of the body, where it provides a suitable environment for the cells. It is also present as a lymphatic fluid in the lymphatic vessels. When interstitial fluid accumulates in abnormally large quantities, edema (swelling) occurs in the tissues. When this happens, the body fluids are said to be imbalanced.

Approximately 5 liters of *blood* are present in an average adult. About half of it is extracellular fluid, known as *plasma*. The other half is composed of various blood cells, whose liquid content is intracellular fluid.

Transcellular fluid is a term applied to a variety of special fluids that are found outside of cells, but cannot be classified as interstitial fluid or blood. These fluids include digestive juice, the synovial fluid between movable joints, the fluids in the eyeballs, and the cerebrospinal fluid.

BLOOD. *Plasma plus blood cells.*

INTERSTITIAL. *Pertaining to the extracellular spaces, or to elements situated between cells (L. inter, between; sistere, to place or situate).*

Circulatory System: Functional Anatomy and Electrocardiography

FUNCTIONAL ANATOMY OF THE HEART

The *heart* is a richly vascularized double pump. Its main function is to provide the force that is needed to move the blood through the circulatory system.

The heart is suspended and supported within the thoracic cavity in a *pericardial cavity* (Figure 4.5). The inner, visceral wall of the sac envelopes the heart and the outer, parietal wall of the sac is attached, through fascia, to the diaphragm and breastbone (sternum).

The heart is made up of the following parts: four chambers that are partitioned by an elastic skeleton; four valves; a three-layered wall; and a bioelectrical system.

PARIETAL. *Refers to the outer (versus the visceral) wall of an anatomical cavity, such as the abdominal or pericardial cavity (L. paries, wall).*

The Four Chambers of the Heart

The heart's skeleton is a tough fibrous tissue with four valve rings (Figure 25.2, next page). Located medially in this skeleton is the *septum,* which separates the right (pulmonary) pump from the left (systemic) pump.

Each of the heart's two pumps consists of two chambers (Figure 25.1). The upper chambers, called *atria,* serve mainly to collect blood from the veins and to force this blood into the lower chambers, called *ventricles*.

The *left atrium* sends oxygenated blood into the *left ventricle* which, in turn, pumps the oxygenated blood through the systemic circuit to all regions of the body. The *right atrium* sends deoxygenated blood into the *right ventricle,* from which it is pumped into the pulmonary circuit of the lungs.

SEPTUM. *A partition between two spaces (L. fence).*

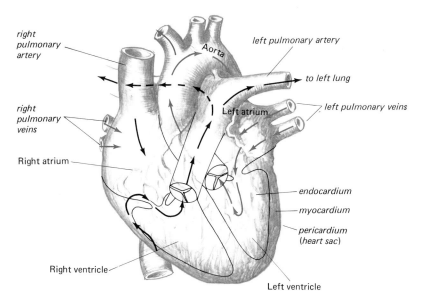

FIG. 25.1. Chambers of the heart and routes of circulation.

Functional Anatomy of the Heart

The *systemic circuit* is used to deliver nutrients, including oxygen, to all tissues of the body. The *pulmonary circuit* is used to rid the body of waste carbon dioxide and to reoxygenate the blood. Both of these circuits are described further on page 516.

The Four Valves of the Heart

Blood enters the atria freely whenever the walls of these chambers are not contracting. That is, there are no valves between the incoming blood vessels and the chambers of the atria.

But to get blood into the ventricles—and to get the blood out of the ventricles into the systemic and pulmonary circuits—four valves must be opened.

Each valve is seated in a fibrous ring which constitutes a part of the heart's skeleton. Two of the valves are located between the atria and ventricles. The other two valves are between the ventricles and outgoing blood vessels (arteries).

The valve rings which are located between the atria and ventricles house the two *atrio-ventricular (AV) valves*—the tricuspid valve and the bicuspid (or mitral) valve. The *tricuspid valve* allows venous blood to pass from the right atrium to the right ventricle. The *bicuspid valve* allows arterial blood to pass from the left atrium to the left ventricle.

The valve rings which lie at the origin of the aorta and the pulmonary artery house the three-cusped *semilunar* (aortic and pulmonary) *valves*. The *pulmonary valve* opens into the pulmonary artery to discharge venous (deoxygenated) blood from the right side of the heart into the lungs. The *aortic valve* opens into the aorta to allow arterial (oxygenated) blood from the left side of the heart into the systemic circulation.

TRICUSPID. (L. tri, *three;* cuspis, *point*).

SEMILUNAR. *Crescent-shaped.*

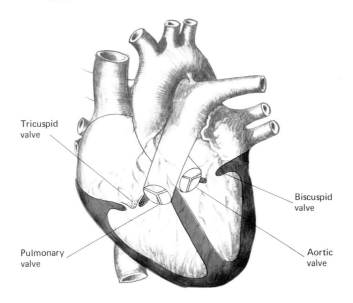

FIG. 25.2. *Valves of the heart.*

The Three-Layered Wall of the Heart

The valves comprise part of the innermost wall of the heart, called the *endocardium*. The free surface of the endocardium faces (lines) the four chambers of the heart. It is lined with a simple squamous epithelium. This epithelium, which is specifically called an *endothelium*, is continuous with the endothelium of all of the blood vessels in the body.

The remainder of the endocardium is composed mainly of connective tissue, as is the *epicardium* – the outermost wall of the heart.

Sandwiched between the endocardium and epicardium is the *myocardium*, the thickest layer of the heart. In essence, the myocardium is a spiraling muscle whose thickness varies from one chamber to the next. It is thin in the atria, where it serves to pump blood against a low resistance and over a short distance into the ventricles. It is thicker in the right ventricle, where it pumps blood against a higher resistance and over a longer distance, through the pulmonary circuit. It is thickest in the left ventricle, which has to pump blood against a very high resistance and over a very long course through the systemic circuit.

ENDOTHELIUM. *Cellular tissue which lines the serous cavities, blood vessels, and blood passageways throughout the body (G. endos, within; thele, nipple).*

The Bioelectric System of the Heart

Ordinarily, when a cardiac muscle cell is not in a state of contraction, it bears a negative electrical charge on the inner surface of its external membrane and a positive charge on its external surface. In order to bring about a cardiac contraction, this distribution of charge must be reversed to a certain extent. If the reversal proceeds far enough – to what is called the *threshold voltage* – an action potential arises. This potential travels from one end of the myocardial cell to the other. In its wake, calcium ions are released from attachment sites that are situated all along internally-located membranes. In response to the release of calcium, actin and myosin – the contractile proteins of muscle – slide past one another, giving rise to what we see as a contraction. (The details of this process were discussed more fully in Chapter 17.)

Three questions must now be raised in conjunction with the heart's bioelectrical system. How do action potentials arise in the heart? What regulates the sequence of action potentials so that the atria and ventricles will beat (contract) in an orderly manner? How is the normal sequence speeded up or slowed down in accord with bodily needs?

The formation of action potentials in the heart is a spontaneous act that occurs independently of the nervous system. Because of this, the heart is said to be operated myogenically rather than neurogenically. This phenomenon occurs also in certain kinds of smooth muscle cells. It occurs through mechanisms no one knows anything about. The fact that it does happen in cardiac muscle cells independently of nervous input is proved by the observation that the embryonic heart exhibits action potentials and begins to beat before it receives its nerve supply.

MYOGENIC *activity is generated from within muscle rather than by way of nerves (G. mys, muscle; gen, beginning).*

The sequencing of action potentials through the heart is regulated by the *pacemaker system*. This system consists of two clusters of modified

Functional Anatomy of the Heart

HIS. *A Swiss anatomist and embryologist (1831–1904) who introduced the terms dendrite, neuroblast, and spongioblast.*

PURKINJE, *a Bohemian physiologist (1787–1869), was one of the first to section and study tissues by using the microtome.*

cardiac muscle cells and numerous fibers arising from these cell clusters (Figure 25.3). One cluster of cells is located in a knot of tissue, called the *sinoatrial node (SA node)*, in the upper right myocardium of the right atrium. The other cluster is the *atrioventricular node (AV node)*. This node is located at the junction of the right atrium and right ventricle. This group gives rise to a third group which passes as a bundle of fibers (the *bundle of His*) along the midline of the heart between the two ventricles. Numerous myogenic muscle fibers, called *Purkinje fibers*, branch out from the bundle of His, spread themselves along the endocardium, and pass into the myocardium. The branches of the Purkinje fibers sweep down in fan-like fashion toward the apex (bottom tip) of the heart and then course backward again toward the junction of the atria and ventricles, as shown in Figure 25.3.

How does the pacemaker system bring about a sequencing of action potentials in the heart so that the atria will contract before the ventricles? First, action potentials arise spontaneously and rhythmically in the SA node. These action potentials cause a contraction of the atrial musculature, beginning in the region of the SA node and spreading out in all directions throughout the atria.

Second, while the atrial muscle is contracting, action potentials from the SA node spread toward the AV node, which itself becomes activated. Action potentials are then directed from the AV node to the bundle of His and then into the most distal reaches of the Purkinje fibers. In the wake of these action potentials, the ventricles contract, after which the modified cardiac muscle cells in the SA node give rise to another spontaneous action potential, bringing about another automatic sequence of bioelectrical actions.

FIG. 25.3. *The electrical conduction system of the heart; arrows show flow of electric impulse in heart.*

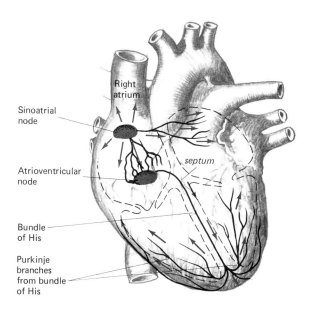

Influence of the Nervous System on the Myogenic System. The automatic beat of the heart can be altered by the nervous system. Some of these alterations are brought on strictly through the autonomic nervous system, as happens, for example, when the body shifts from rest to exercise. Other alterations are effected through the autonomic nervous system, but are initiated by the central nervous system, as happens, for example, during periods of emotional stress or fright.

The nervous system exerts its influence on the myogenic system by way of parasympathetic and sympathetic nerve fibers:

(1) The parasympathetic fibers stem from the vagus nerve (the tenth cranial nerve). They are found in the SA and AV nodes and in the atria, but not in the ventricles. When they are discharged, they release acetylcholine. This neurotransmitter reduces the heartbeat by increasing the refractory period of the atrial tissues. It also reduces the extent to which the ventricles may fill up. This is done by reducing the force of atrial contraction.

(2) The sympathetic nervous system, in contrast to the parasympathetic system, sends nerve fibers into the ventricles as well as the atria. When these fibers are discharged, noradrenaline and adrenaline are released from their tips. These neurotransmitters cause the heart to beat faster by reducing the refractory period of the cardiac action potentials. In addition, they cause the heart to contract more forcefully.

Altogether, the electrical activity of each region of the heart is coordinated in a manner whereby a *cardiac impulse* originates in the SA node and terminates in the cardiac musculature of the ventricles. This impulse has a characteristic electrical form which may vary slightly and predictably, depending on whether the sympathetic or parasympathetic nervous system is very active.

The characteristic, complex electrical form of the cardiac impulse can be converted into a visual graphic form known as the *electrocardiogram* (or ECG or EKG). We shall discuss this cardiac impulse, but first it is necessary to summarize what has been said so far about the electrical and mechanical activities of the heart. We will do this by describing the cardiac cycle.

REFRACTORY. *Resisting stimulation (L. refractus, to turn back).*

CARDIAC IMPULSE. *The electrical signal that originates in the heart and is measured as an electrocardiogram.*

THE CARDIAC CYCLE

This cycle of electrical and mechanical activities is shown in Figure 25.4. It consists of three rhythmically recurring events known as diastole, atrial systole, and ventricular systole.

Diastole is the filling phase of the cardiac cycle. It begins with the closing of the aortic and pulmonary valves. The closing of these valves can be heard through a stethoscope as the sound "DUB." In this phase, only the atria are filled first, because the valves between the atria and ventricles—the AV valves—are closed. Later, the AV valves open and the ventricles fill up.

DIASTOLE. *The phase of the cardiac cycle in which the heart relaxes (G. diastole, to dilate or expand).*

The Cardiac Cycle

SYSTOLE. *The phase of the cardiac cycle in which the heart contracts (G. systolo, a contracting).*

Diastole is terminated when a volley of action potentials arises in the SA node and atrial myocardium. Now atrial systole begins.

In *atrial systole*, the atrial musculature contracts. In doing so, blood is forced out of the atria into the ventricles. The ventricles, which were previously filled (during diastole), now become "overfilled." Their walls become stretched and, while this is happening, the cardiac impulse is momentarily arrested at the AV node. This arrest, or delay, virtually guarantees that the atria will have enough time to empty their contents into the ventricles before the ventricles themselves undergo a contraction.

Atrial systole is terminated when a volley of action potentials arises in the AV node and passes into the bundle of His and the Purkinje branches of this bundle. Now ventricular systole begins.

During *ventricular systole*, the ventricular myocardium contracts, giving rise to an increase in pressure within the ventricular chambers. As a result of this rise in pressure the AV valves close, and the sound "LUB"

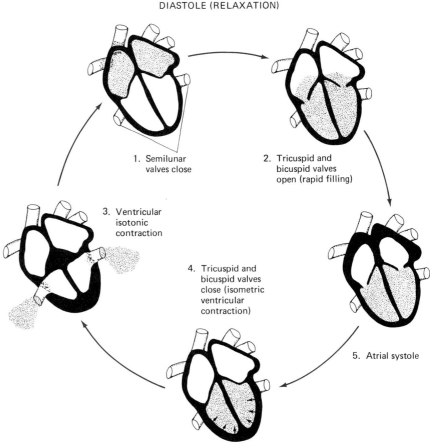

FIG. 25.4. *The cardiac cycle. Through the stethoscope, the sound "dub" is heart at point 1, the sound "lub" at point 4.*

can be heard through a stethoscope. At first, the ventricles contract isometrically—that is, without undergoing a visible reduction in size. Then they contract isotonically. In doing so, the ventricular chamber is reduced in size and blood is forcefully ejected against the pulmonary and aortic valves, which now open to release some of the blood into the pulmonary artery and aorta.

ISOTONIC *muscle contraction occurs when a muscle shortens visibly* (G. *isos, equal;* tonos, *accent*).

One important additional effect of ventricular contraction is that the atria enlarge and a vacuum is created within them. This vacuum facilitates refilling the heart in the initial phase of a new diastole.

One important note should be added at this point with regard to the route of the cardiac impulse through the ventricles. Following the delay at the AV node, the cardiac impulse continues on from the AV node to the apex of the ventricles by way of the bundle of His in the septum of the heart. At the apex of the ventricles, some of the Purkinje fibers terminate, and their action potential induces an action potential in the ventricular myocardium. Other Purkinje fibers are longer and bend back toward the base of the ventricles—heading toward the atrial-ventricular junction. All along this pathway some of these Purkinje fibers terminate, and in doing so they direct their action potential into the ventricular myocardium.

By having a sequence of action potentials flowing from the apex to the base of the ventricles, a normally healthy individual is virtually guaranteed that the ventricles will begin their contraction at the apex and conclude at the atria. In this way the ventricles are able to squeeze blood out of their chambers, from the bottom to the top.

Having completed its route from the SA node to the base of the ventricular muscle, the old cardiac impulse dies out and a new one arises in the SA node.

Each cardiac impulse is a complex of the action potentials which arise in all of the cardiac cells. As such, each cardiac impulse has a complex wave form. The individual deflections of this wave form are designated alphabetically—beginning midway through the alphabet—as P, Q, R, S, T (Figure 25.5).

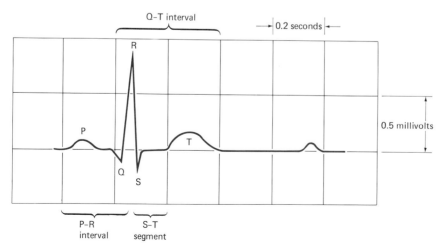

FIG. 25.5. *The electrocardiogram.*

DEPOLARIZATION.
Lessening the difference in electric charge that normally exists across an unstimulated cell membrane.

The P wave represents depolarization of the atrial tissue. It immediately precedes atrial systole. The QRS waves represent depolarization of the ventricles. They immediately precede ventricular systole. The T wave represents repolarization (resetting) of the ventricular tissues.

The PQRST waves should not be confused with the action potentials themselves. Action potentials in cardiac cells were discussed on page 326 and in Figure 17.9. Reference to Figure 25.5 will show that the cardiac impulse is the result of the specific kinds of action potentials which are exhibited by the various kinds of cardiac cells. From a practical point of view, it is important to know that the normal shape of the cardiac impulse is due to the differences in refractory period of the various kinds of cardiac cells and the distances and directions in which the action potentials travel.

The reader may recall that the *refractory period* is the interval during which a second action potential cannot arise along a cell membrane. During this interval the physical and chemical characteristics which were disrupted in the cell membrane in giving rise to the first action potential must now be re-established. In the event that the refractory period is abnormally shortened or abnormally prolonged, an abnormal cardiac impulse will follow.

ELECTROCARDIOGRAPHY

The cardiac impulse can be recorded with a sensitive electronic device (the *electrocardiograph*) through electrodes which are placed upon the surface of the body. Due to losses of electric current in passing from the heart to the surface of the body, the electrocardiograph receives considerably less voltage (about 1 to 2 millivolts) than actually exists across the individual cardiac cells (about 100 millivolts). The time span in which the cardiac impulse occurs, however, is the same. By placing the electrodes in the right places, it is possible to determine whether the electrical activity of the heart is normal. This activity is recorded mechanically. The changes in voltage are registered on the y axis of a graph, while time is denoted on the x axis (Figure 25.5).

It is important to realize that the graphic record – the electrocardiogram – does not predict the fate of the heart. A person with a normal ECG may suffer a fatal heart attack within minutes of the recording, while a person with a very abnormal ECG may live many years thereafter. Nevertheless, electrocardiography is a powerful diagnostic tool in determining heart problems and in checking the effects of various medical treatment procedures.

Two different methods are used in electrocardiography. One is known as the pattern method. The other is called the vector method.

Using the *pattern method*, the examiner studies the ECG to see if it corresponds with either a normal pattern, or a pattern associated with any one of numerous kinds of heart disease. Using the *vector method*, the

Circulatory System: Functional Anatomy and Electrocardiography

examiner determines whether the direction and strength of various electrical components of the cardiac impulse are normal. We will return to these methods, but first a word on electrocardiographic abnormalities.

The cardiac abnormalities that can be identified through electrocardiography fall into two large classes: rhythm problems, the major ones of which are discussed below; and problems in which a regular rhythm of some sort may or may not occur, but some component of the cardiac impulse is distorted. The distortion may be seen as a change in direction (for example, inversion of P wave) or magnitude (for example, enlarged R wave).

Among the second class of problems are such diseases as *myocardial infarction*, in which some blood-deprived and scarred dead tissue appears following a heart attack; *ventricular hypertrophy*, in which the size of one or both ventricles increases beyond normal, often due to some valvular disease; blockage of cardiac impulse conduction in the bundle of His;

INFARCTION. A portion of tissue that is dying because it has been deprived of blood (L. infarctus, to stuff).

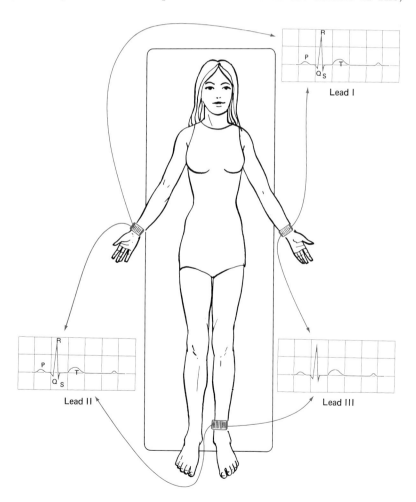

FIG. 25.6. Standard limb leads in electrocardiography.

EMBOLISM. The obstruction of a vein or artery by an embolus—a foreign body (G. *en, in;* ballein, *to throw*).

disease of the pericardium; and *pulmonary embolism,* in which tissue or blood clots obstruct the arterial passageway into the lungs, thus increasing the work of the right ventricle in pumping blood into the lungs.

The first class of problems—the rhythm problems—can be diagnosed only through the pattern method of interpreting electrocardiograms. This method, along with various rhythm problems, is described below.

The second class of problems listed above can be diagnosed in most cases also through the pattern method. However, the vector method holds certain advantages over the pattern method in these cases. Unfortunately, techniques used in the vector method cannot be described here for lack of space. We will just say that the vector method offers a more basic approach to electrocardiography and requires less recall of wave patterns. This is an advantage, inasmuch as the interpretation of electrocardiograms is based on special diagrams of the sort one studies in physics with regard to forces.

Regardless of whether the pattern method or vector method is used, a certain number of *leads* are used in recording. A lead is a particular placement of electrodes upon the skin. Figures 25.6 and 25.7 illustrate three *extremity leads* and six *chest leads,* often referred to as *precordial* leads. These nine leads are used routinely in the pattern method of interpretation. (Three additional leads are used in the vector method). The electrocardiograms shown in the text which follows are based on patterns that were obtained with lead II.

PRECORDIAL. The anterior surface of the lower part of the chest (L. prae *before;* cor, *heart*).

In the pattern method of interpreting electrocardiograms, the following features are studied: (1) rate; (2) rhythm; (3) the nature of the P wave;

FIG. 25.7. *Precordial (chest) leads in electrocardiography. These are used in combination with those shown in Figure 25.6.*

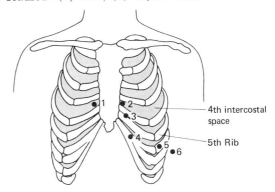

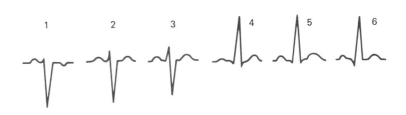

Circulatory System: Functional Anatomy and Electrocardiography

(4) the P-R interval; (5) the nature of the QRS complex; (6) the slope of the ST segment; and (7) the nature of the T wave.

Depending on the characteristics of the above features, an individual may be diagnosed as having a normal or abnormal electrocardiogram. When the ECG is abnormal, it is usually possible to determine precisely what part of the heart is affected.

It is beyond the scope of this text to cover each of the very many known abnormalities. Only the most common abnormalities are singled out here. They will introduce the reader to the pattern method of interpreting electrocardiograms, and they will indicate how a given pattern explains what part or parts of the heart are abnormal.

Rate of Heartbeat

A normal adult who is awake, but resting, has a heart rate of about 70 beats per minute. Lower heart rates—around 60 per minute—are normal in well-trained athletes. In these individuals the heart works more efficiently in providing adequate volumes of blood to the body.

In most other people, a rate of 60 or less is called *bradycardia*. Bradycardia may be caused by increased parasympathetic (vagal) stimulation, by heart disease, or by an infectious illness (such as influenza); or the cause may be idiopathic (unknown).

A heart rate which exceeds 100 in an adult at rest is called *tachycardia*. Rapid beating of this sort may be caused by thyroid disease (exophthalmos), some form of heart disease, or some unknown cause.

When tachycardia increases to well above 200, the atria and/or ventricles are in a state of *flutter or fibrillation*. In these cases, the heart ceases to contract rhythmically.

EXOPHTHALMOS, a thyroid disorder, causes abnormal protrusion of the eyeballs (G. ex, *out;* ophthalmos, *eye*).

Normal Rhythm, Arrhythmia, and Dysrhythmia

The PQRST waves of the normal cardiac impulse recur with a normal periodicity, as shown here for an average young adult:

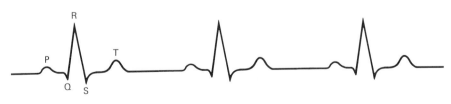

This normal periodicity is known as a *sinus rhythm*. In contrast to a normal sinus rhythm, a heart may display a dysrhythmia or an arrhythmia. *Sinus arrhythmia*, one type of arrhythmia, is exemplified here:

SINUS RHYTHM. Normal cardiac rhythm proceeding from the sinoatrial node.

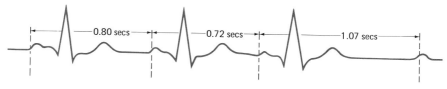

It is common in childhood and it tends to disappear with aging, but it may be seen in normal adults. It is often, but not always, related to the breathing cycle. A faster rhythm may occur during inspiration, a slower rhythm during expiration. Since a normal individual draws about 12 to 15 breaths of air each minute, and the heart beats about 70 times per minute, the slow and fast rhythms may occur two to three beats apart from one another.

FIG. 25.8. Several electrocardiograms, showing different cardiac conditions.

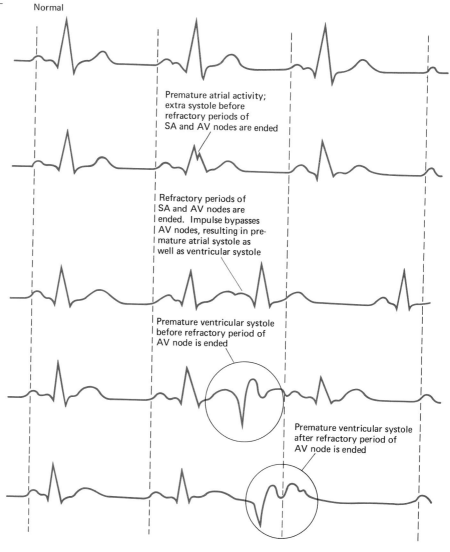

Circulatory System: Functional Anatomy and Electrocardiography

Dysrhythmias are arrhythmias in which extra systoles occur. For reasons that are not too well understood, electrical activity may arise *ectopically*; that is, some electrical activity may be generated out of phase.

The ectopic activity may be initiated in either the atria, as shown in the second and third of the patterns in Figure 25.8, or it may be initiated in the ventricles, as shown in the fourth and fifth patterns. Also, the activity may arise *before* the refractory period of the SA and AV nodes is terminated (as indicated in patterns 2 and 4) or *after* the refractory period of the AV node is over (patterns 3 and 5). The consequence of each of these actions is indicated in the graphs.

ECTOPIC heartbeat originates outside of the sinoatrial node (G. ek, out of; topos, place).

Extra systoles may or may not be something to be concerned about. Much depends upon the medical history of the individual and how often the extra systoles arise.

In the case of *AV nodal rhythm* or *dysrhythmia*, the activity is rhythmic, but it arises in the AV node rather than in the SA node. What is bad about it is that the cardiac impulse passes upward into the atria and downward into the ventricles more or less simultaneously. The following ECG exemplifies one form of nodal rhythm:

NODAL RHYTHM. The cardiac rhythm when the heart is controlled by the atrioventricular node.

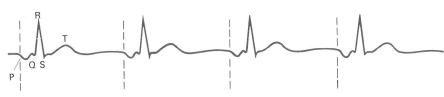

Atrial flutter and *atrial fibrillation* are extreme forms of tachychardia. In atrial flutter, the atria beat very weakly at a rate of about 250 to 300 times per minute. Furthermore, there may be two, three, four or more atrial systoles (P waves) for one ventricular systole (QRS complex) as shown here:

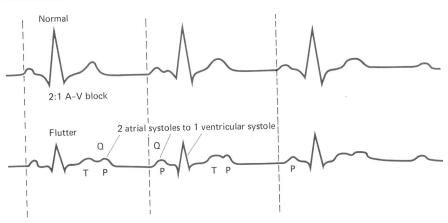

In *atrial fibrillation*, the atrial walls vibrate about 400 times or more per minute. Moreover, the fibrillatory waves (P waves) occur so rapidly and are so fine that they escape detection with a normal electrocardiograph. In this case, only the QRS-T complex is seen, as shown here:

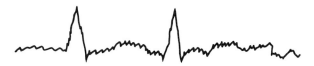

Atrial flutter or fibrillation may arise in coronary artery disease, mitral valve disease, diseased thyroid (thyrotoxicosis), and in other diseases. Regardless of cause, the chief concern is that each atrial flutter (contraction) is weak, more so as the frequency of beat increases. As a result, the ventricles usually do not become adequately filled. When the frequency of fibrillation reaches a critical point, the atria become totally useless as a pump in filling the ventricles. That is, they cannot execute an atrial systole.

Atrial flutter or fibrillation often may be relieved with digitalis or quinidine. The exact mechanism of *digitalis* action is not understood, but two effects are known. First, it allows a longer period for atrial repolarization, and thus cuts down on the frequency of atrial flutter. Second, it lengthens the time in which conduction is delayed at the AV node. Some physicians believe this action serves to prevent weak volleys of action potentials from passing into the ventricles. Strong coordinated signals, on the other hand, are allowed to continue along. For most problems of heart disease, digitalis is an extremely beneficial drug. In some cases, however, and especially in cases where an overdose may be erroneously administered, very deleterious effects may follow.

Quinidine lengthens the refractory period of myocardial cells in general. This gives the SA node a chance to regain control of the rhythm, to initiate a normal sinus rhythm, and thus to cut down on atrial flutter.

Ventricular flutter and *ventricular fibrillation* are two forms of dysrhythmia. Like atrial flutter and atrial fibrillation, they are extreme forms of tachycardia. Ventricular flutter occurs when the heart beats more than 200 times per minute and the ECG shows that the P wave, QRS complex, and T wave are not distinct. It rarely lasts more than a few seconds to a minute, after which it either reverts to some other slower rhythm or accelerates into ventricular fibrillation:

DIGITALIS. *A family of drugs that are obtained from the foxglove plant; used in certain heart diseases to increase the efficiency of the heart.*

QUINIDINE. *An alkaloid drug used to correct faulty heart rhythms.*

Ventricular flutter

Ventricular fibrillation

Circulatory System: Functional Anatomy and Electrocardiography

In ventricular fibrillation—which may be triggered by atrial flutter (or atrial fibrillation)—the heart appears as masses of contracting muscle areas interspersed among masses of relaxing muscle. It looks somewhat like a pail full of wiggling fly larvae. Random multiple electrical waves rise and fall haphazardly, here and there. The individual fibrils of the muscle cells contract and relax exceedingly fast, but the cells, and the muscle as a whole, fail to contract and relax in a coordinated manner.

Ventricular fibrillation is much more serious than atrial fibrillation. Whereas atrial fibrillation cancels out the value of the atria as pumps, ventricular fibrillation is an immediate threat to life, and emergency treatment must be administered. Emergency treatment calls for the use of electric shock treatment to defibrillate the heart.

It is probable that ventricular fibrillation accounts for most of the "sudden deaths" which arise during a heart attack. Sudden death is a technical medical term which refers to death that occurs within one hour of the onset of any abnormal health symptom.

The P Wave

The P wave, like the other waves of the ECG, has certain normal characteristics in shape and size, as shown in the normal traces above. When there is a radical change, certain cardiac problems may be diagnosed, as shown in the following list:

Characteristics of trace	Example of probable abnormality
Inversion of wave	AV nodal rhythm
Increased amplitude	Atrial enlargement
Increased width	Left atrial enlargement
Notched wave	Mitral valve disease
Pointed wave	Right atrial disorder
Absence of wave	SA block

The P-R Interval

The P-R interval is measured from the beginning of the P wave to the beginning of the Q wave. It represents the spread of atrial depolarization from the SA node to the ventricles, plus the delay of the impulse in the AV node. Its normal duration is about 0.12 to 0.2 seconds.

An abnormally short interval often indicates an *AV nodal rhythm*—a rhythm which originates in the AV node instead of the SA node.

An abnormally long P-R interval usually denotes a *block* (interruption of the electrical system) between the atria and ventricles. The block may be due to heart disease, thyroid disease, an infectious disease, digitalis medication, or some idiopathic cause. Atrioventricular blocks are classified as

HEART BLOCK. *A condition in which the cardiac impulse does not pass unobstructed from the sinoatrial node to the ventricles.*

first-degree, second-degree, or third-degree AV heart blocks. In first-degree AV heart block, the P-R interval is merely prolonged. This same feature, plus periodic skipping of heart beats, characterizes second-degree AV heart block. Skipping occurs when a P wave fails altogether to cross the atrial-ventricular junction to depolarize the ventricles:

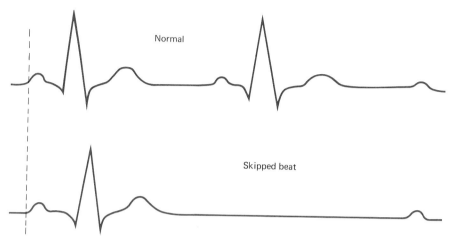

In third-degree AV heart block the conduction path from the atria to ventricles is completely interrupted. In consequence, the myogenic tissue of the atria and ventricles set up their own pacemaker rhythms, independently of each other, as indicated here:

Heart blocks may be treated with an artificial pacemaker, an electronic device that is implanted by a physician.

The QRS Complex

The normal duration of the QRS complex is about 0.04 to 0.10 seconds in an adult. An abnormally longer period may denote a block in the bundle of His. An abnormally larger complex may suggest that the heart has become *hypertrophied* (overgrown) to compensate for improperly working valves.

The ST Segment

The ST segment of the ECG follows the QRS complex and precedes the T wave. It usually slopes gently upward as it gives rise to the T wave. This gentle upward slope is usually seen regardless of whether an individual does or does not have a heart problem. However, if the unhealthy individual is tested during strenuous exercise, the slope may be abnormally depressed, as shown here:

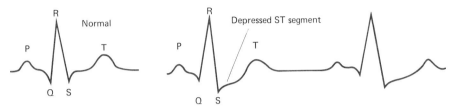

In fact, one of the most reliable signs of cardiac ischemia, a disease of the cardiac arteries in which insufficient blood is provided to the myocardium, is the horizontal depression of the ST segment, with or without inversion of the T wave, when an ECG is taken during or after physical exertion.

The T Wave

The T wave represents repolarization (resetting) of the ventricles. When it appears distorted, inverted, diminished, or exaggerated, a faulty repolarization is supected.

FUNCTIONAL ANATOMY OF BLOOD AND LYMPH VESSELS

After leaving the heart, blood passes through a series of vessels. This series is usually arranged in the following order: elastic arteries, distributive arteries, arterioles, capillaries, venules, medium veins, and large veins. The blood then re-enters the heart. However, in the course of passing through the capillaries a net amount of fluid leaks out into the interstitial spaces. Most of this fluid is returned to the bloodstream by way of the lymphatic system (Figure 25.9).

General Structure of Blood and Lymph Vessels

All blood and lymph vessels, with the exception of capillaries, are seen to have three walls when viewed in cross-section with an ordinary microscope. (See Figure 25.10.) Capillaries (Figure 25.11) are made up of only the first of these walls.

The innermost wall—the *intima*—is essentially an epithelial lining on a basement membrane, within a connective tissue covering. The epithelial lining is known also as an *endothelium*. It is continuous throughout the entire circulatory system.

The middle wall—the *media*—is rich in smooth muscle cells and/or elastic fibers, depending on the particular kind of vessel. The outermost wall—the *externa*—is composed chiefly of a connective tissue rich in elastic and collagenous fibers.

The Elastic Arteries

ARTERY. *A vessel that conveys blood away from the heart* (G. arteria, windpipe).

The *elastic arteries* are the ones that come directly from the heart (the aorta and pulmonary arteries), the ones that branch off the aorta to supply the head and upper extremity (the vertebrals and common carotids), and the arteries that branch off the aorta to supply the lower extremity (the common iliacs). The location of these arteries may be seen in Figure 25.12.

The distinctive anatomical feature of the elastic arteries is that their media is rich in elastic fibers. During ventricular systole, blood is forced from the heart into these arteries, causing them to swell up. In becoming swollen, they convert a portion of the heart's force of contraction into potential energy. During diastole—when the heart is relaxing—the elastic arteries recoil. In doing so, the potential energy is converted to kinetic

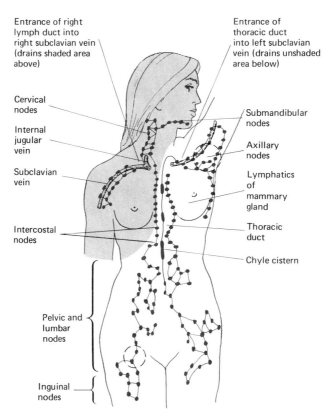

FIG. 25.9. *The lymphatic system. The circles indicate frequent regions of secondary cancer growth due to metastases (movement of cancer cells from other primary sites of cancer).*

energy, which serves to drive the blood into the capillary networks throughout the diastolic phase of the cardiac cycle.

The Distributive Arteries

The *distributive arteries* are those that branch directly from the elastic arteries to supply blood to the organs of the body. Their distinctive anatomical feature is a predominance of smooth muscle in their medial wall. For this reason they are also known as *muscular arteries*.

Unlike the elastic arteries, the distributive arteries do not simply expand and recoil in response to the systolic and diastolic phases of the cardiac cycle. Rather, due to a rich sympathetic (and sparse parasympathetic) innervation, the distributive arteries can be constricted to reduce the flow of blood into an organ, or dilated to allow more blood into an organ.

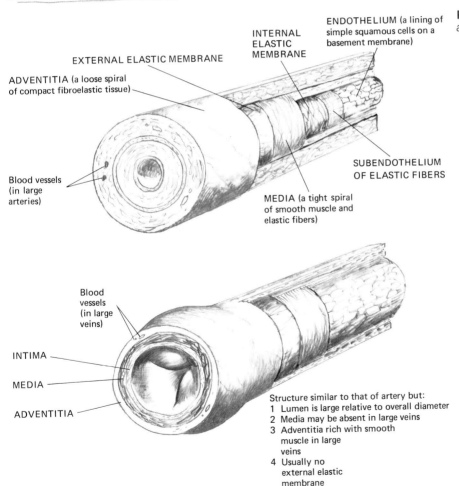

FIG. 25.10. *Structure of arteries and veins.*

FIG. 25.11. *Structure of capillaries.*

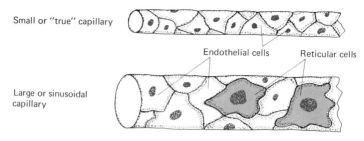

The Arterioles, Capillaries, and Venules

ARTERIOLE. *A very small artery (G.).*

After leaving the distributive arteries, blood passes through *arterioles*. These are small arteries which branch extensively and get narrower and narrower. This is a gradual and progressive process in which the arterioles

FIG. 25.12. *The systemic arterial circuit.*

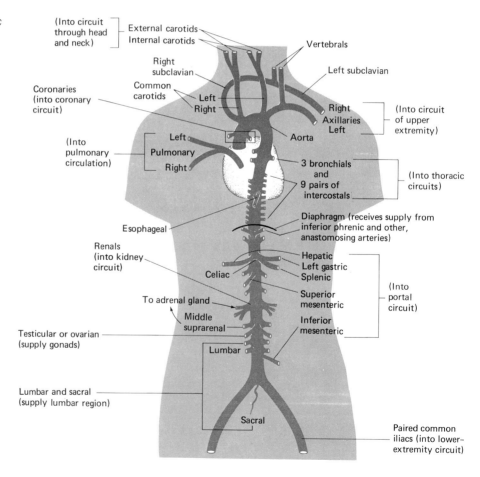

eventually lose all of their media and externa. At this point the arterioles are known as *capillaries*.

In actuality, blood capillaries are tissue spaces which are continuous with one another. They are lined with a thin endothelium (tissue covering) of squamous cells. They lie between and are continuous with arterioles and venules.

In some organs—such as the liver, spleen, lymph nodes, and bone marrow—the capillary spaces between the arteries and veins are lined partly by endothelial cells and partly by *reticular*, or *macrophage*, *cells*. In these cases the capillaries are referred to as *sinusoids*.

Sinusoids are generally larger in diameter (10 to 40 micrometers) than capillaries (7 to 14 micrometers). In addition, they perform two principal functions rather than one. They allow an exchange of nutrients and wastes with the tissues through which they pass, and they serve to cleanse the blood of certain impurities. Ordinary capillaries perform the first of these functions, but not the second. The cleansing of blood impurities is accomplished through *phagocytosis* by the reticular cells.

After passing through capillaries or sinusoids, the blood usually enters a *venule*, a vessel which has a media and externa as well as an intima. From here the blood passes into the medium veins.

CAPILLARY. *Any tube with a fine, hairlike bore (L. capillus, hair).*

ENDOTHELIAL CELLS *line the lumen, or bore, of serous cavities, blood vessels, lymphatic vessels, and the heart (G. endos, within;* thele, *nipple).*

The Medium Veins

Whereas distributive arteries bring blood into an organ, *medium veins* take blood out of the organ. These two kinds of vessels, when seen side by side in a histological preparation, differ principally from one another in that the vein has a much larger bore (lumen) and considerably thinner walls. In addition, the medium veins have leaflike flaps projecting from the intima into the lumen. These flaps form one-way valves which serve to prevent a backflow of blood toward the capillaries. Some of their locations may be seen in arm veins if you push the blood in these veins towards the hand. A bulge or swelling develops at the nearest valve, where the blood backs up against the valve.

VEIN. *A vessel that carries blood out of capillaries and toward the heart (L.* vena *blood vessel).*

The Large Veins

Large veins drain larger regions of the body. Among these are the following (Figure 25.13): (a) the common iliac veins, which drain the lower extremity; (b) the portal vein, which drains the digestive system; (c) the jugular veins, which drain the head and brain, (d) the axillary veins, which drain the upper extremity; and (e) the pulmonary veins, which send "arterial blood" into the left atrium. Except for the pulmonary veins, all of these large veins ultimately empty into the heart by way of the *superior* and *inferior vena cavae*, the largest of the large veins.

Unlike medium veins, the large veins have no valves and have a very thin media. They are often called *capacity vessels*, because collectively they

JUGULAR. *Pertains to the neck above the collarbone (L.* jugulum, *collarbone).*

VENA CAVA. *Either of two large veins which enter the right atrium of the heart (L.* vena, *vein;* cava, *hollow).*

VASA VASORUM. Literally, "vessels within a vessel" (L. vas, vessel).

may contain up to about two thirds of the body's blood at one time. From Figure 25.10, it may be seen that, like elastic arteries and some of the larger distributive arteries and medium veins, their walls receive a blood supply of their own. This supply is furnished through so-called *vasa vasora*.

Functional Anatomy of the Lymphatic System

Excess interstitial fluids are returned to the bloodstream by way of the *lymphatic system*. This system includes lymphatic capillaries, lymph vessels, lymph ducts, and lymph nodes.

Lymph capillaries resemble blood capillaries in that they are tissue spaces lined by a thin squamous endothelium. They differ, however, in that one end of each lymph capillary ends blindly between cells rather than being connected (like the capillaries) to other circulatory vessels. That is, lymph capillaries are not interposed between afferent (incoming) and efferent (outgoing) vessels; the entire lymph system is efferent. Excessive interstitial fluids are picked up all along the length of the lymph capillary and are conveyed into lymph vessels that carry the fluid in only one direction: toward the heart.

The lymph vessels are similar to medium veins in several ways. They have a large lumen relative to vessel diameter, one-way valves, and an intima, media, and externa. Unlike veins, however, lymph vessels carry *lymph*—interstitial fluid—not blood. Lymph differs from blood primarily

LYMPH. *Interstitial fluid in the lymphatic vessels* (G. Lympha, *water goddess*).

FIG. 25.13. *The systemic venous circuit.*

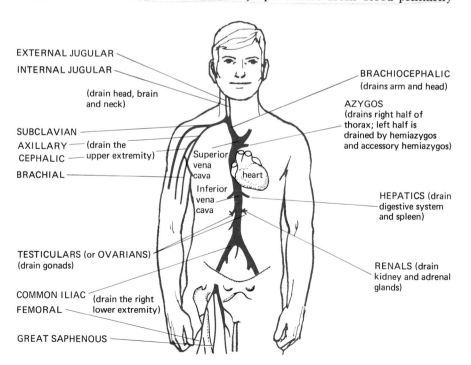

in having a lower concentration of protein (2% versus 7%) and in having only two kinds of "blood" cells. It contains lymphocytes and monocytes, but no red blood cells, granulocytes, or platelets.

Lymphatic vessels pass out of organs and into one of two large lymphatic ducts, as shown in Figure 25.9. The *thoracic duct* begins just below the diaphragm, against the rear wall of the abdomen. Its lymphatic contents are derived from the lower half of the body, the left upper limb, and the left halves of the head and thorax. The duct itself starts as a bag, called the *chyle cistern*, about two inches long, adjacent to the first and second lumbar vertebrae. From here it runs up the back wall, alongside the aorta. It passes through the diaphragm and goes into the left subclavian vein in the lower region of the neck.

The *right lymphatic duct* receives lymph from the right arm and the right halves of the head and thorax. It drains into the right subclavian vein. Semilunar one-way valves are present at the junction to prevent blood from flowing from the vein into the duct.

Approximately 500 to 700 small bean-shaped glandular bodies (1 to 25 millimeters long) lie along the route of the larger lymph vessels and ducts. Called *lymph nodes*, these organs furnish antibodies, lymphocytes, and other cells to the blood. In addition, they dispose of foreign materials, including antigens.

In the above discussion, blood was said to pass from arteries to capillaries to veins. Now it is appropriate to mention several exceptions to this plan:

(1) In the skin and in the mucosa of the digestive tract, blood may pass directly from an arteriole to a venule, bypassing a capillary bed. This connection, an *arteriovenous anastomosis*, is used when there is little need for blood in a capillary bed at a particular moment. During vigorous exercise, for example, a number of anastomoses may be closed, thus allowing relatively larger amounts of blood into a region of tissue, such as the skin of the hand. In periods of rest, the anastomoses open up and less blood passes into the region.

(2) In the glomerulus of the kidney (Chapter 30), blood passes from an arteriole to a capillary network and then to another arteriole and capillary network before entering the venules. In passing through the first capillary network a portion of the blood is filtered out and cleansed by the kidney tubules, which pass most of the cleansed filtrate into the second capillary network, where it is reabsorbed into the bloodstream.

(3) In the digestive system, and in the hypothalamic-anterior pituitary region, there are *portal systems*, in which blood from a capillary network enters a vein and then passes into another capillary network and another vein before returning to the heart.

LYMPHOCYTES and **MONOCYTES** *are the two kinds of agranulocytes present in the blood.*

CHYLE CISTERN. *A saclike structure at the beginning of the thoracic duct, located at the level of the twelfth thoracic vertebra (L. cisterna, a reservoir; chylos, a milky white emulsion).*

ARTERIOVENOUS ANASTOMOSIS. *A direct passage from an artery to a vein (G. anastomosis, an opening).*

CIRCULATORY ROUTES

Approximately 260 named arteries and many more unnamed arterioles carry blood from the heart to various tissues throughout the body. About

Circulatory Routes

240 named veins and many more unnamed venules return the blood to the heart.

The entire circulatory system can be divided into two major circuits — the pulmonary and the systemic. The pulmonary circuit receives all the blood from the right ventricle. The systemic circuit receives all the blood from the left ventricle.

Pulmonary Circuit

PULMONARY. (L. *pulmonarius, of the lungs*).

The *pulmonary circuit* begins with the pulmonary artery (Figure 25.14). This artery divides into a *right pulmonary artery* and a *left pulmonary artery*. Both of these arteries branch extensively (about 15 times) in the course of passing through the right and left lungs, respectively. The branches send dark, unoxygenated blood into the capillary spaces that surround the air pockets (alveoli) in the lungs. Here the blood picks up oxygen and gives up carbon dioxide. The newly-oxygenated blood is then carried to the left atrium of the heart by way of four *pulmonary veins*.

Systemic Circuit

SYSTEMIC. *Of or affecting the entire bodily system* (G. systema, *to place together*).

The *systemic circuit* includes all the circuits through the body except the pulmonary circuit. For convenience, it can be subdivided into a number of regional circuits. Here we shall not be concerned with any of the special circuits, such as those which extend through the thyroid glands, the pituitary gland, the eyes, or the cerebellum of the brain. Rather, attention will be focused on the following general circuits: (1) neck and head; (2) upper extremity; (3) thorax; (4) abdomen; and (5) lower extremity.

FIG. 25.14. *The pulmonary circuit.*

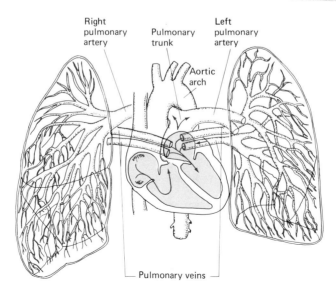

Circulatory System: Functional Anatomy and Electrocardiography

Not all of the circuits in each of these regions can be discussed. Nor can complete circuits—going from the heart and winding up back at the heart—be discussed in some cases. In particular, we must leave out numerous *anastomoses*—connections from one artery, which feeds one tissue, to another artery, which feeds another tissue. Important anastomoses occur at several of the joints, in the circle of Willis (Figure 25.15, below), and—to a lesser extent—within the coronary circuit. Anastomoses may serve to protect the body in the event that a major arterial route is obstructed.

All the blood entering the systemic circuit is monitored within the arch of the aorta with respect to blood pressure, oxygen content, and content of carbon dioxide. The blood pressure is sensed by pressure receptors within the aortic wall in what is called the *aortic sinus*. The oxygen and carbon dioxide content are checked by chemoreceptors in regions of the aortic wall.

AORTIC SINUS. *One of several pouchlike dilations of the aorta in the vicinity of the semilunar valves; each one contains pressure receptor nerve fibers.*

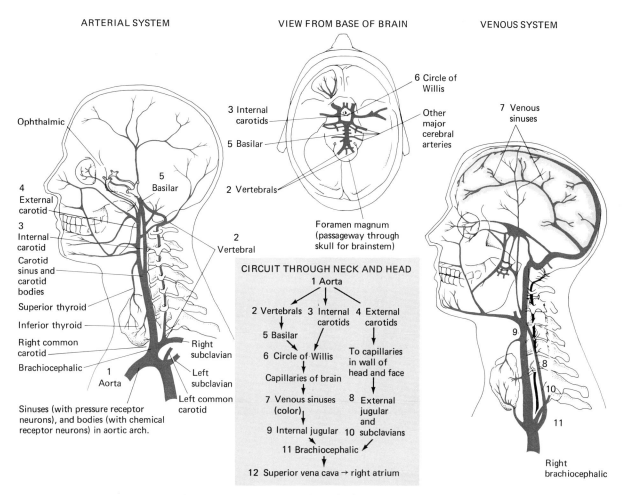

FIG. 25.15. *Blood circuit through the neck and head.*

Circulatory Routes

AORTIC BODY. *Masses of epithelial-like tissue which are present in the aortic arch, respond to changes in blood levels of oxygen and carbon dioxide, and convey information to the brain by way of the vagus nerve.*

CAROTID ARTERY *receives its name from the Greek karoun—to bring about a stupor or sleep—because compression of this artery induces unconsciousness.*

CAROTID SINUS. *Pressure receptor for signaling blood pressure.*

CAROTID BODY. *Chemoreceptor for signaling low oxygen, high carbon dioxide.*

These regions are known collectively as the *aortic body*. When the blood pressure is too high or low, or the carbon dioxide content is too high, corrective reflex actions are executed as described below and in Chapter 26.

You are urged at this point to examine Figures 25.12 and 25.13. They show where each of the following regional circuits originates and terminates.

Circuit Through the Neck and Head. This circuit is shown in Figure 25.15. It originates in the arch of the aorta and terminates in the superior vena cava. Two features are especially noteworthy about it. First, at the branching point of the external and internal carotid arteries—within the neck—is a slight dilation. Known as the *carotid sinus*, it contains pressure receptors which monitor the pressure of the blood to the brain. When the pressure is too high, a nerve reflex action is called upon to reduce it. When the blood pressure is too low, reflex actions are mustered up to increase it. Through these self-regulating mechanisms, which are described more fully beginning on page 541, the brain is ordinarily assured an adequate supply of oxygen and nutrients.

Also in the vicinity of the bifurcation (branching) of each internal and external carotid artery is a *carotid body*. Within this mass of tissue are nerve endings which monitor the carbon dioxide and oxygen content of the blood. When the carbon dioxide content is too high, or the oxygen content is too low, reflex actions are initiated to correct the situation. This topic is discussed further on page 546.

Another very important aspect of the circuit through the neck and head

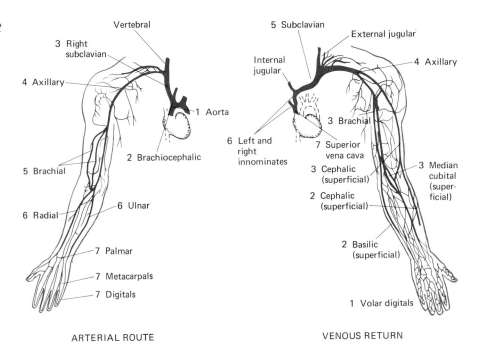

FIG. 25.16. *Blood circuit through the upper extremity.*

Circulatory System: Functional Anatomy and Electrocardiography

concerns the *circle of Willis*. This circle is formed from several arteries, as shown in Figure 25.15.

The arteries from the circle of Willis branch repeatedly and ensure an adequate exchange of nutrients and wastes in the brain. The arrangement of the brain arteries as the circle of Willis carries with it a survival value. Should one of the arteries become obstructed, the brain may receive an adequate circulation from other arteries in the circle.

Circuit Through the Upper Extremity. This circuit is shown in Figure 25.16. Two important points are especially important with regard to certain veins in this circuit. First note the location of the subclavian veins. These veins receive the two major lymph ducts of the body. The left subclavian receives the left thoracic duct; the right subclavian receives the right lymphatic duct. Now note that two kinds of veins are present in the upper extremity: *superficial veins* and *deep veins*. The superficial veins (for example, the cephalic and basilic veins) lie in the superficial fascia directly beneath the skin. They are veins which are easy to get at for removing samples of blood, for blood transfusions, and for intravenous feeding. The deep veins lie side by side with more deeply located arteries.

Thoracic Circuits. The thoracic circuits begin with either a visceral or a parietal branch of the subclavian artery or the aorta, as shown in Figure

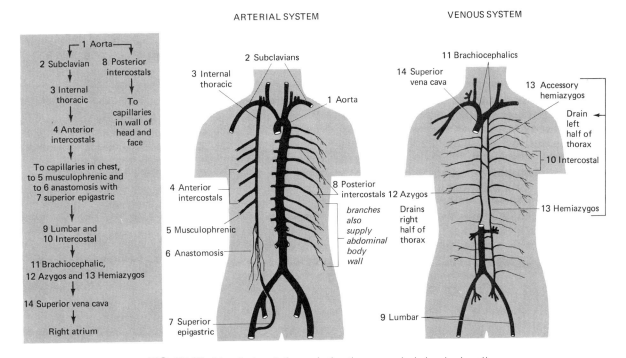

FIG. 25.17. *Blood circuit through the thorax and abdominal wall.*

Circulatory Routes

25.17. The visceral branches pass into the thoracic cavity to supply the soft organs. The parietal branches pass outward from the thoracic cavity to supply the body wall.

The *parietal branches* of the thoracic aorta are as follows: one pair of superior phrenic arteries, which supply the upper surface of the diaphragm; and one pair of subcostal arteries which, together with nine pairs of intercostal arteries, supply the thoracic body wall. The intercostal arteries pass laterally into the thoracic wall, carrying blood to the chest muscle, the pleural membranes around the lungs, the vertebrae, and the meningeal membranes that enclose the spinal cord.

The *visceral branches* of the thoracic aorta are as follows: a pair of mediastinal arteries, which furnish blood to the mediastinum (the central division wall of the thoracic cavity, which envelopes all of the thoracic viscera except the lungs); two or three pericardial arteries, which supply the pericardial sac that encloses the heart; four or five esophageal arteries, which pass into the esophagus; one right and two left bronchial arteries, which supply the lung tissues and are independent of the pulmonary circulation; and a pair of coronary arteries, which feed the coronary circuit.

The Coronary Circuit. This circuit begins with paired *coronary arteries* that stem from the ascending aorta (see Figure 25.18). The arteries branch into numerous capillaries in the wall of the heart. The capillaries drain into veins. Some of these veins empty directly into any one of the chambers of

PHRENIC. *Pertaining to the diaphragm* (G. *phren,* *midriff*).

COSTAL. *Pertaining to the ribs* (L. *costa, rib*).

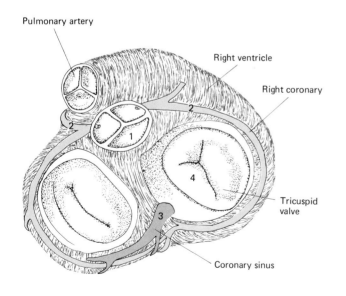
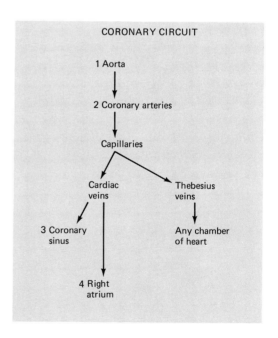

Superior view of transverse section

FIG. 25.18. *Coronary blood circuit.*

Circulatory System: Functional Anatomy and Electrocardiography

the heart. Others converge into a *coronary sinus* (hollow) which empties into the right atrium.

Abdominal Circuits. The abdominal circuits, like the thoracic circuits, begin with either a visceral or a parietal branch of the aorta. The *parietal branches* of the abdominal aorta and the structures they supply are as follows: a pair of inferior phrenic arteries, which supply the lower surface of the diaphragm; and four or five pairs of lumbar arteries, which supply the muscles of the back and abdominal wall.

The *visceral branches* of the abdominal aorta and the structures they supply are as follows: a pair of testicular or ovarian arteries, which supply the male or female gonads, respectively; a pair of middle suprarenal arteries, which supply the adrenal gland; a pair of renal arteries, which supply the kidneys; a single middle sacral artery, which supplies the lower lumbar vertebrae, the sacrum, and the coccyx; and several additional arteries which will now be described in greater detail.

The *circuit through the digestive organs and spleen* is shown in Figure 25.19. It may be noted that the blood of this circuit is derived from three aortic branches: the celiac artery, the superior mesenteric artery, and the inferior mesenteric artery.

The latter two arteries—the mesenteric arteries—furnish oxygenated blood directly to the intestines. The celiac artery, in contrast, subdivides and sends oxygenated blood into four different organs through three different

CORONARY. *A term applied to structures that encircle an organ; the coronary arteries encircle, or "crown," the heart (L. coronarius, wreath).*

CELIAC *refers to the abdominal region (G. koilia, belly).*

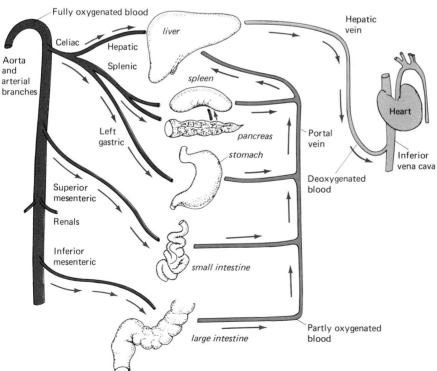

FIG. 25.19. *Hepatic–portal blood circuit.*

GLUCAGON. A pancreatic hormone which tends to elevate the blood sugar level.

arteries: (1) liver (hepatic artery); (2) spleen and pancreas (splenic artery); and (3) stomach (left gastric artery).

At this point, look back at Figure 25.19; you will note that whereas the liver receives oxygenated blood from the hepatic artery, it also receives partially-deoxygenated blood (about 10% to 30% deoxygenated) from the spleen, pancreas, stomach, and intestines. This blood passes into the sinusoids of the liver, by way of the portal vein.

The portal vein may bring a rich supply of one or more of the following materials into the liver: (a) blood cells from storage depots (so-called red pulp areas) in the spleen; (b) glucagon and insulin from the pancreas; and (c) nutrients and hormones from the intestine.

The blood cells may be used to adjust the *hematocrit* (the concentration of red blood cells in the plasma). Worn-out red blood cells that were not destroyed in the spleen may be destroyed here, in the liver. The other healthy blood cells pass on out of the liver, through the hepatic vein into the inferior vena cava and then into the right atrium, for general circulation.

Circuit Through the Lower Extremity. This circuit is shown in Figure 25.20. As in the upper extremity, there are deep and superficial veins in the

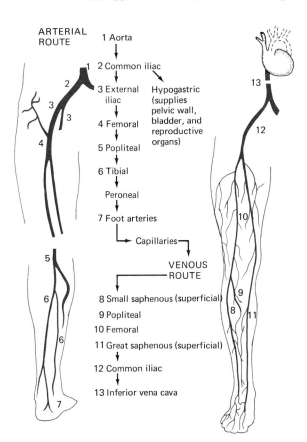

FIG. 25.20. *Blood circuit through the lower extremity.*

lower extremity. Two problems—two disorders of the venous system—may often occur in this circuit: *varicosities* and *phlebitis*.

Whereas the deep veins of the legs are suitably supported by fascia and muscle, the superficial veins (the saphenous veins) receive relatively little anatomical support. Undue pressure upon superficial veins may cause them to stretch, dilate, thin out, twist, and enlarge. The resulting *varicosity* may prevent the vein from carrying out its function in circulation. Varicosities also commonly occur in the rectum (hemorrhoids); and to a lesser extent in the scrotum (varicocele) and lower esophagus (esophageal varices). Vomiting of blood (hematemesis)—while sometimes indicative of a gastric ulcer—may also arise when an esophageal varix ruptures.

VARICOSITY. *A dilated, twisted, knotted blood vessel (L.* varix, *dilated vein).*

The blood in the vein may stagnate into an immobile pool. In severe cases, the varicosed vein may have to be stripped out of the flesh. In these cases, the remaining superficial veins, and all of the deep veins, take over the task of draining the lower extremity.

Some individuals, for reasons which are totally unknown, are susceptible to saphenous varicosities. Obese persons are often susceptible, inasmuch as their saphenous veins have to withstand considerable weight. Pregnant women are prone by virtue of pressure from the enlarging uterus upon the iliac veins, where these veins cross the pelvis.

Superficial leg veins are generally the most susceptible to inflammation. The inflamed veins generally bleed and are painful, a condition technically known as *phlebitis*. Blood clots (thrombi) may occur at the same time, creating a *thrombophlebitis*. Drugs, trauma, and strain may bring about this abnormality.

PHLEBITIS. *An inflammation in a vein (G.* phlebo, *vein;* itis, *inflammation).*

THROMBOPHLEBITIS. *Inflammation of a vein associated with a blood clot (G.* thrombos, *lump;* phlebo-, *vein;* itis, *inflammation).*

SUMMARY

1. About half of the body's fluid is intracellular and half is extracellular. Three fourths of the extracellular fluid is interstitial. The remainder is either blood, lymph, or transcellular fluid.
2. The heart, which is enclosed in a pericardial sac, has four chambers, four valves, a three-layered wall, and a bioelectric system of modified muscle. The action potentials of this system pass sequentially through an SA node, an AV node, the Bundle of His, and Purkinje branches. These action potentials arise spontaneously, as do action potentials that arise in the atrial and ventricular musculature subsequent to the discharge of the SA node and Purkinje branches, respectively.
3. The right atrium receives venous blood, which is pumped into the right ventricle and then into the lungs. The left atrium receives oxygenated blood, which is pumped into the left ventricle and then into the systemic circulation.
4. In the cardiac cycle, diastole (relaxation) begins with the closure of the aortic and pulmonary valves, producing the sound DUB. Ventricular systole (contraction) is signified by the closing of the AV valves, creating the sound LUB.

5. The pattern method of electrocardiography can be used to determine whether a heart is healthy. An analysis of rate, rhythm, P-R interval, slope of the ST segment, and the nature of the P, QRS, and T waves is used to diagnose certain forms of heart disease.
6. The wall of a blood or lymph vessel is made of three layers: intima, media, and externa. Capillaries lack a media and externa. Elastic arteries expand with blood during systole and they squeeze this blood into distributive arteries during diastole. Distributive arteries distribute blood to organs and regions of the body where small arteries then convey the blood into capillaries.
7. Venules receive deoxygenated blood from capillaries. This blood passes into medium-sized veins and then into large veins, which pass the blood into the heart by way of two vena cavae.
8. The lymph system consists of: (a) lymph capillaries, which pick up interstitial fluid; (b) lymph nodes, which screen out foreign matter; and (c) lymph vessels and ducts, which convey the lymph to the heart by way of the subclavian veins.
9. The pulmonary circuit is used to oxygenate venous blood and the systemic circuit is used to send oxygenated blood to regional circuits throughout the body.
10. The circle of Willis, an arrangement of arteries within the cerebral circuit, provides some insurance against death in the event of a minor brain hemorrhage by permitting blood to reach the brain via an alternate route.
11. The portal circuit functions to bring nutrients, hormones and blood cells from the digestive tract, pancreas, and spleen to the liver.
12. Superficial veins of the lower extremity are susceptible to varicosities and phlebitis.

REVIEW QUESTIONS

1. Describe the heart in terms of its four chambers, four valves, its three-layered wall, and its bioelectric system.
2. How is bioelectricity, in the form of the cardiac impulse, routed through the heart during a single cardiac cycle?
3. Describe the sequential flow of blood through the chambers of the heart.
4. Describe the sequence of contractile and valvular events that occur during a single cardiac cycle.
5. Tell what the P,Q,R,S, and T waves in an ECG denote with respect to the events that occur in a single cardiac cycle.
6. What two drugs are of primary value in treating a heart whose rhythm is faulty?
7. Define: intima, media, externa, pulmonary circuit, systemic circuit, circle of Willis, superficial veins, varicosity, phlebitis, portal circuit.
8. What special functions are carried out by elastic arteries? by distributive arteries? by small arteries?
9. Where does lymph originate? What is the course of lymph through the body, and where is it ultimately taken?
10. What are the special functions of the carotid and aortic bodies and sinuses?

Dynamic Characteristics of the Circulatory System

KINDS OF PRESSURE IN THE CIRCULATORY SYSTEM
    Hydrostatic pressure in the circulatory system
    Blood pressure in the circulatory system
        Measuring the systemic arterial blood pressure
        Values of blood pressure
    Osmotic pressure in the body
RESISTANCE ENCOUNTERED BY BLOOD IN ITS FLOW
EXCHANGE OF NUTRIENTS AND WASTES
RETURN OF VENOUS BLOOD TO THE HEART
VARYING THE CARDIAC OUTPUT TO MEET THE BODY'S NEED FOR OXYGEN
SHIFTING THE BLOOD FROM INACTIVE TO ACTIVE REGIONS OF THE BODY
THE PULSE—AN INDICATOR OF GENERAL CONDITIONS IN THE CIRCULATORY SYSTEM
REGULATION OF THE CIRCULATORY SYSTEM
    Autoregulation of the heart
        Myogenic contraction
        Starling's law of the heart
    Nervous control of the circulatory system
        Sympathetic influences upon the circulatory system
        Parasympathetic influences upon the circulatory system
REFLEX REGULATION OF THE CIRCULATORY SYSTEM
    Central input
    Peripheral input
        Stretch receptors
        Pressure receptors
        Chemical receptors

The major overall function of the circulatory system is to transport nutrients (including oxygen) to the tissues, and to transport waste products from the tissues to the excretory organs—the lungs, kidneys, gut, and skin. This two-

Kinds of Pressure in the Circulatory System

fold function is achieved in healthy individuals primarily by adjusting the cardiac output (the volume of blood per ventricle per minute), and secondarily by shifting blood from relatively inactive to active tissues of the body.

But before getting into these primary and secondary adjustments it is necessary to discuss a variety of other topics. The various sources of pressure must be discussed first: hydrostatic pressure, arterial blood pressure, and osmotic pressure. The arterial blood pressure drops continuously as various kinds of resistance are encountered by the blood in its flow through the blood vessels. The various forms of resistance are discussed, accordingly. Because of the resistance encountered in the capillaries, and because the capillary beds, collectively, are like a large ocean receiving a supply of blood from many small rivers (the arterioles), blood flows very slowly—slower than in the arteries and veins. This provides enough time for an efficient exchange of nutrients and wastes. However, because of the reduction in pressure encountered in these capillary beds, there is scarcely any driving force of the heart left to deliver venous blood to the heart. Various mechanisms other than arterial pressure are employed instead, as described below.

With the preceding as background, we can then discuss the fundamental relationship between oxygen consumption (one's energy requirements), cardiac output (one aspect of the cardiovascular system's response to energy requirements), and the shifting of blood from inactive to active regions of the body (another aspect of the cardiovascular system's response to energy requirements).

In concluding this chapter, we give a brief description of the pulse—the mechanical wave which accompanies each heartbeat—with an explanation of how an evaluation of this wave can be useful in determining the general condition of the cardiovascular system, and a brief discussion of the mechanisms that regulate the circulatory system.

KINDS OF PRESSURE IN THE CIRCULATORY SYSTEM

MASS. *The quantity of matter in a body* (G. maza, cake).

Pressure is defined as any kind of force which acts against a resistance. *Force* (F) may be defined as mass (m) times acceleration (a). That is, $F = ma$.

Any object has *mass*. When the object lies outside of the earth's field of gravity, it is a *weightless mass*, a mass which cannot be accelerated toward earth by the earth's gravitational field. When the object enters the earth's field of gravity it becomes accelerated toward the earth by the force of the earth's gravity. And when the object comes to rest at sea level, its mass can be measured accurately and expressed as *weight*. In other words, at sea level, mass (m) times acceleration due to the force of gravity (a), equals weight (w). That is, $ma = w$.

From the preceding discussion it may be noted that force equals mass times acceleration (at any level of atmosphere), and force equals weight (at sea level).

Going back now to the definition of pressure (any kind of force which acts against a resistance) it can be seen that when weight is held against any form of resistance, pressure is being exerted. Air molecules, for ex-

Dynamic Characteristics of the Circulatory System

ample, exert air pressure upon our body due to the force of gravity. A person who stands upon your body is exerting pressure upon your body. Blood (mass), accelerated by the force of the heart's contraction, encounters resistance in the blood vessels, giving rise to pressure.

Three kinds of pressure are important with respect to the circulatory system: hydrostatic pressure, blood pressure, and osmotic pressure.

Hydrostatic Pressure in the Circulatory System

Hydrostatic pressure (P) is equal to the height (h) of a stationary column of liquid, multiplied by the density (d) of the liquid. That is, $P = hd$. With regard to the circulatory system, the density of blood is about the same as the density of water. The density of water—that is, its weight per unit volume (grams per milliliter)—is defined as one.

Hydrostatic pressure is an important factor in raising arterial blood from the heart to the brain and in raising venous blood into the heart from positions below the level of the heart. Hence, for venous blood to return from the feet to the heart, in an individual six feet tall, a hydrostatic pressure of about four feet must be overcome. So too, in order to drive blood toward the brain, the force of cardiac contraction must overcome a hydrostatic pressure of about two feet.

DENSITY. *The quantity of material present in a unit volume of some material.*

HYDROSTATIC PRESSURE *is that exerted by a column of a liquid (G. hydro, water; stat, to stop).*

Blood Pressure in the Circulatory System

The contraction of the heart provides the force that is needed to create blood pressure. The vasculature creates the resistance against which the force acts. The blood is the object which has mass, and is being accelerated.

To appreciate the meaning of the above paragraph, consider what happens during each ventricular systole and diastole. During *ventricular* systole, a certain amount of blood is ejected from the heart into the large elastic arteries. These arteries cannot accommodate the ejected blood without expanding themselves. They resist the force of cardiac contraction, and thus this cardiac force is presented to the arteries as pressure.

How much pressure is generated during this action? The answer to this question varies, depending upon whether you are considering the systemic circuit or the pulmonary circuit. In the systemic circuit of a normal, average, healthy adult at rest, the pressure is about 120 millimeters mercury. What does this figure—120 mm Hg—mean? It means the force of the heart is able to lift a column of mercury about 5 inches upward—into the aorta (Figure 1.2). Inasmuch as mercury is 13.6 times heavier than water, it also means the force of cardiac contraction is sufficient to lift a column of blood about 70 inches high. Clearly, this is a sufficient force to raise blood to the highest regions of the brain.

In the pulmonary circuit, the systolic pressure is about 25 millimeters mercury (Figure 1.2). Why is this pressure so much lower than the systemic systolic pressure? Mainly because the force of cardiac contraction in the right ventricle is much lower than in the left ventricle. A larger force is not

VASCULATURE. *The blood vessels, lymph vessels, and the heart (L. vasculum, small vessel).*

SYSTOLE. *The contractile phase of the cardiac cycle (G. to contract).*

Kinds of Pressure in the Circulatory System

required because the resistance of the arterial and capillary passageways is considerably lower in the pulmonary circuit than in the systemic circuit. In addition, a smaller force of contraction is sufficient in the right ventricle because the height to which it must pump blood is lower than for the left ventricle.

During *diastole*, the larger elastic arteries recoil. In doing so, they provide a force which continues to accelerate the blood onward toward the capillary beds. Along the way, resistance is encountered (within the arterial passageways and in the capillaries) and a diastolic pressure is generated.

What is the average minimum diastolic pressure in a resting adult? In a normal healthy adult it is about 80 millimeters mercury in the systemic circuit; about 2 mm Hg in the pulmonary circuit.

Measuring the Systemic Arterial Blood Pressure. The common way of measuring the systemic arterial blood pressure is with an inflatable cuff and a *sphygmomanometer* (Figure 26.1). The cuff is wrapped around the arm above the elbow, with the elbow kept bent and at the level of the heart. The cuff is inflated to a point where it cuts off the flow of blood through the radial artery. The cutoff occurs when sound cannot be heard through a *stethoscope* which is placed over the radial artery. Some of the air in the cuff is then slowly released. As soon as a sound can be heard through the stethoscope—the sound of the flowing blood—the pressure is read from the sphygmomanometer. This reading is called the *systolic pressure*—the pressure in the cuff which is barely below the pressure needed by the heart to drive blood through the artery. Now additional air is released from the cuff. At a point where the sound of blood flow changes from turbulent to streamline (a sound that cannot be heard) another reading is taken. This reading is the *diastolic pressure*—the pressure exerted by a recoiling aorta during the relaxation phase of the cardiac cycle.

Values of Blood Pressure. The average normal blood pressure for an adult at rest is 120/80 (systolic/diastolic). However, lower or higher values are also normal, depending on the person's sex, age, and physical condition. Below age 35 it is generally lower in females than males, and above age 40 it rises faster in females than in males, probably due to hormonal changes at menopause.

When we stand, the systolic and diastolic pressure may rise a few millimeters mercury. The rise in diastolic pressure usually increases more rapidly than the rise in systolic pressure. This indicates a greater increase in resistance to flow of blood through the small arteries and the capillary beds (a greater peripheral resistance).

The greater increase in diastolic over systolic pressure, upon rising to a standing pressure, reflects a smaller pulse pressure. *Pulse pressure* is simply the difference between the systolic and diastolic pressure. A high pulse pressure, compared to a lower one, is correlated with a greater delivery of blood per minute (cardiac output). Upon getting up from a lying to standing position, you may experience a drop in pulse pressure and cardiac output. This is referred to as *orthostatic hypotension*. In exaggerated cases the drop in pulse pressure may cause fainting.

DIASTOLE. *The relaxation phase of the cardiac cycle (G. to dilate).*

SPHYGMOMANOMETER. *An instrument for measuring arterial blood pressure in the systemic circuit (G. sphygmos, pulse; manos, rare; metron, a measure).*

STETHOSCOPE. *An instrument for studying sounds coming from a body (G. stethos, breast; skopein, to examine).*

DIASTOLIC PRESSURE. *The residual pressure which remains in the arterial vessels after the heart has finished its contraction.*

ORTHOSTATIC. *Pertaining to, or caused by, upright standing (G. orthos, straight; statikos, causing to stand).*

The ratio of systolic to diastolic pressure also is an important indicator of the person's physical condition. A value of 130/85 may be just as healthful as 120/80. Blood pressure equal to or greater than 160/95 is a sure and dangerous sign of hypertension, a condition which is discussed in the following chapter. Blood pressure below normal is generally considered to be safe, certainly nothing to be concerned about, except in extreme cases.

Osmotic Pressure in the Body

Ordinarily, there is very little fluid between cells—in what is called the interstitial space. Nutrients and oxygen in the capillaries can diffuse

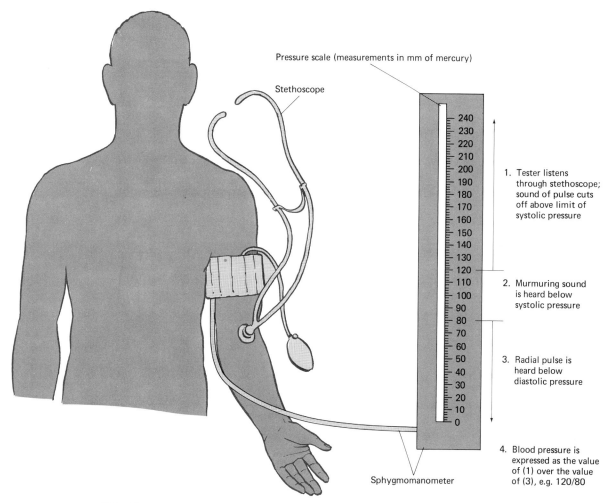

FIG. 26.1. *Method for determining blood pressure with a sphygmomanometer.*

Kinds of Pressure in the Circulatory System

quickly through this fluid to the cells. So too, wastes, including carbon dioxide, can diffuse rapidly through the interstitial fluid into a blood capillary. In certain diseases, however—such as in some allergies, in kidney disease, and in heart failure—excessive amounts of fluid accumulate in the interstitial space. This excessive accumulation leads to a swelling (edema) in the tissues.

EDEMA. *An accumulation of fluid in tissue spaces* (G. oidema, *swelling*).

Where does this liquid come from? From the blood. Why, then, doesn't edema ordinarily occur spontaneously? The answer to this question has to do with osmotic pressure.

In Chapter 1, osmotic pressure was described as a force that draws water across a semipermeable membrane from a region of high water concentration to a region of lower water concentration. It was discussed then, and shown in Figure 1.10, that two solutions with unequal concentrations would have unequal osmotic pressure. The solution with the higher solute concentration (and therefore the lower water concentration) has a higher osmotic pressure.

With regard to the circulatory system, the following osmotic situation—shown in Figure 26.2—prevails in a healthy individual and accounts for the absence of edema. (1) The concentration of protein in blood plasma is 7%, whereas the concentration in interstitial fluid (including lymph) is 2%. (All other solutes in these two fluids are present in equal concentrations.) Because plasma has a higher concentration of protein than lymph, water is drawn into the bloodstream from the interstitial spaces. (2) The concentration of solutes inside body cells is greater than the concentration outside of

INTERSTITIAL. *Pertains to the spaces, and matter, between cells* (L. inter, *between;* sistere, *to place*).

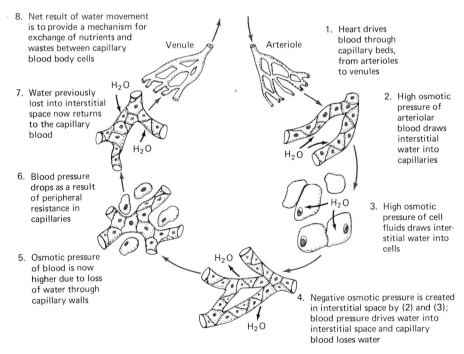

FIG. 26.2. *Role of osmosis in the circulatory system. Edema develops if water fails to return to venular capillaries at point (7).*

8. Net result of water movement is to provide a mechanism for exchange of nutrients and wastes between capillary blood body cells

7. Water previously lost into interstitial space now returns to the capillary blood

6. Blood pressure drops as a result of peripheral resistance in capillaries

5. Osmotic pressure of blood is now higher due to loss of water through capillary walls

1. Heart drives blood through capillary beds, from arterioles to venules

2. High osmotic pressure of arteriolar blood draws interstitial water into capillaries

3. High osmotic pressure of cell fluids draws interstitial water into cells

4. Negative osmotic pressure is created in interstitial space by (2) and (3); blood pressure drives water into interstitial space and capillary blood loses water

cells. Because of this, water moves into the cells from the interstitial fluid. (3) The osmotic attraction of water into the capillaries and cells leaves only a thin film of interstitial fluid around the cells. (4) The higher osmotic pressures in the capillaries and in the cell fluids creates a negative pressure in the interstitial space. When this vacuum pressure—about 5 millimeters mercury—fails to hold, and tends instead toward zero or a positive value, edema follows. That is, water moves from the blood plasma into the interstitial space, causing the tissue to swell.

RESISTANCE ENCOUNTERED BY BLOOD IN ITS FLOW

When blood is ejected from the heart it meets an *elastic resistance* in the large arteries. In the event that the arteries are inelastic, as happens in varying degrees of arteriosclerosis, a greater resistance is encountered and a higher blood pressure results.

In addition to elastic resistance, blood encounters a *peripheral resistance* in the circulatory system. By definition, the peripheral resistors are the smaller arteries and the capillary networks. *Arterial peripheral resistance* is related to the diameter of the passageway in the smaller arteries. These arteries may be made narrower (constricted) by the action of sympathetic nerves. As an artery becomes increasingly constricted, it offers more and more resistance to the flow of blood. As a result, the volume of blood which can flow through the artery into the capillaries is reduced. It has been demonstrated experimentally that a 20% reduction in diameter cuts the flow of blood in half. Looked at from the other side, a 20% increase in diameter doubles the flow of blood.

Capillary peripheral resistance arises because the blood encounters a considerable amount of frictional resistance as it passes through certain capillary beds. The reader should try to visualize what is happening. If we discount sinusoidal capillaries for the moment, the situation is this: the red blood cells, with an average diameter of 8.5 micrometers, must pass through capillaries, whose average diameter is about 8 micrometers. To do so, the biconcave cells take on the shape of a bullet. The cells and the plasma around them rub against the capillary walls (Figure 26.3). Through this action, friction reduces the velocity of flow. The reduction in flow allows more time for the exchange of nutrients and wastes. This is a desirable feature, inasmuch as the supply of nutrients, and the capacity for picking up wastes, both diminish increasingly as the blood gets further away from the arteries and closer to the veins.

The capillary peripheral resistance varies from time to time, depending on how many capillary networks are open. Not all of the capillary networks are open at all times in the body. In fact, in certain regions of the body there are direct cross-connections between arterioles and venules. Known as *arteriovenous anastomoses*, these cross-connections can be opened or closed by certain smooth muscle cells, which form a sphincter.

PERIPHERAL. *Lying to the outside of the central region of a body.*

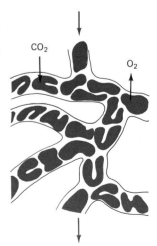

FIG. 26.3. *Movement of red blood cells in capillaries.*

When an arteriovenous anastomosis is closed, the arterial blood passes through the regular capillary bed before entering a venule. When the anastomosis is open, much of the arterial blood bypasses the capillaries and goes directly to the venules.

ARTERIOVENOUS ANASTOMOSIS. *A direct connection between an artery and a vein* (G. *anastomosis, an opening*).

Arteriovenous anastomoses are common in the skin, where they can allow a restriction of blood flow when heat has to be conserved or they can allow a copious flow when heat has to be discharged. Arteriovenous anastomoses are also common in the mucous membrane of the entire digestive tract, where they can allow a liberal flow of blood after a meal or a reduced flow in other circumstances, as for example, when the blood is needed by skeletal muscle for work or exercise.

In addition to elastic resistance and peripheral resistance, the circulatory system may become subjected to excessive metabolic resistance and/or excessive resistance due to an abnormally high number of red blood cells (polycythemia).

VISCOSITY. *Property of a liquid which makes it resist the tendency to flow* (L. viscidus, *sticky*).

Excessive *metabolic resistance* arises when the fat content of the blood becomes abnormally high. The additional fat increases the viscosity of the blood and makes it necessary for the heart to work harder to drive the blood through the circulatory system.

Excessive red blood cells (polycythemia) also impose a greater burden on the heart, making it more difficult to circulate the blood.

EXCHANGE OF NUTRIENTS AND WASTES

As blood travels through the capillary beds it slows down enormously, facilitating the process of diffusion—the movement of nutrients into the tissues and the movement of waste products out of the tissues and into the blood.

RETURN OF VENOUS BLOOD TO THE HEART

As blood passes beyond the capillary beds into venules, and then into the medium veins, the driving force of the heart dwindles, becoming entirely dissipated in the large veins. This may be seen in Figure 26.4, which shows not only that the blood pressure drops continuously from the heart to the large veins but also that the pressure alternates between zero and subzero near the opening into the right atrium.

We may now ask whether the force of the heart's contraction is sufficient to drive blood from the heart all the way back to the heart, encountering all the resistance it must meet along the way. A partial answer to this question depends upon what region of the body the blood is directed to and coming from. Another part of the answer depends upon whether you are standing still, moving about, or lying down.

In general, the body regions above the heart (especially the brain) drain their venous blood into the heart partly because of hydrostatic pressure—that is, by virtue of gravity—and partly through contraction of the musculature in the medium and small veins. These mechanisms are used in the standing and supine (lying down) positions, whether an individual is resting or active.

A very different situation is involved with regard to the return of venous blood from body regions below the heart. Let us use as an example the return of venous blood from the feet.

To begin with, we must consider the U-tube principle, which is shown in Figure 26.5. This principle states that the level of liquid in two limbs of a U-shaped tube will be the same. To a large extent this principle holds true. Much of the blood which is forced down the arteries and arterioles (the *arterials*), heading toward the feet, is raised to a higher level as it passes into the ascending veins. But the veins are large-capacity vessels that can store larger volumes of blood than the arterials. Hence, there is a tendency for veins to collect and pool the blood as it comes from the capillaries. This tendency is especially noticeable when one stands "at attention" for a long period, especially with the knees locked in position. With the knees thus thrust backward, some of the veins close. Because of this, and because of the hydrostatic pressure imposed upon the blood in the feet while standing

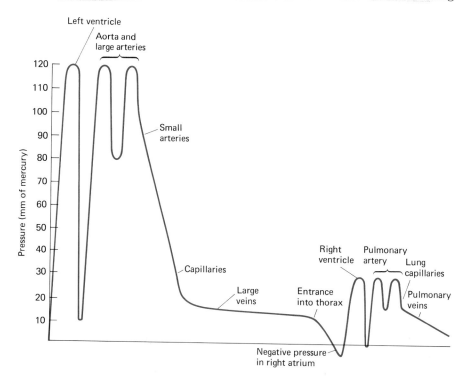

FIG. 26.4. *Blood pressure varies greatly throughout the circulatory system.*

at attention, the U-tube principle does not hold strictly, and, unless the individual shuffles his or her feet, or gets into a supine (lying down) position, not enough blood will be returned to the heart, the brain in turn will get less than its needed share, and fainting will follow.

How is it possible to get blood all the way back to the heart, from the feet—or even from the hands—when an individual is in the fully upright position? The U-tube principle contributes in large measure, but in addition the following four mechanisms are important. First is a *venous pump*. This pump consists of the skeletal and smooth muscle that is present around, or in the vicinity of, veins throughout the body. Due to the incessant tonic activity of this musculature—and especially that of the skeletal muscle during physical activity—the blood is squeezed through the veins toward the heart. The presence of one-way valves in some of these vessels prevents blood from moving away from the heart.

The second mechanism concerns *venular constrictions*. The contractions of the medium and small veins send blood upward, and the presence of one-way valves in some of these vessels prevents blood from moving downward. Third is an *atrial vacuum pump*. Immediately after ventricular systole a negative pressure develops in the heart (the negative pressure referred to above)—a pressure which sucks blood into the right atrium during diastole.

A *thoracic pump* is the fourth mechanism. During the inspirational phase of breathing a negative pressure is created within the thorax. This vacuum aids the circulatory system in bringing venous blood from the inferior and superior vena cavae into the heart.

VARYING THE CARDIAC OUTPUT TO MEET THE BODY'S NEED FOR OXYGEN

The quantity of blood which is forced out of the heart at any moment depends entirely on how much energy the body needs; that is, on how much

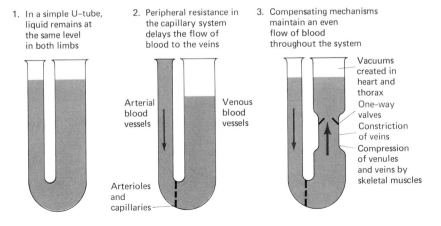

FIG. 26.5. *Using the U-tube analogy to understand blood flow. A variety of mechanisms are used to return venous blood to the heart.*

1. In a simple U-tube, liquid remains at the same level in both limbs
2. Peripheral resistance in the capillary system delays the flow of blood to the veins
3. Compensating mechanisms maintain an even flow of blood throughout the system

Dynamic Characteristics of the Circulatory System

oxygen is needed by the body. (You may recall that the amount of energy that is derived from the combustion of food is scientifically predictable on the basis of how much oxygen is being consumed). Above all else, life is sustained only at expense of energy—and energy is available to the body only in the presence of a supply of oxygen.

How is the output of blood from the heart related to oxygen consumption? The all-important answer to this question is found in the following formula, known as the *Fick equation:*

$$\frac{\text{milliliters } O_2 \text{ consumed}}{\text{minute}} = (A_{O_2} - V_{O_2}) \times (\text{cardiac output})$$

In this equation A_{O_2} represents milliliters of oxygen per 100 milliliters of arterial blood. V_{O_2} has the same meaning, but with respect to venous blood. Cardiac output, as indicated previously in this chapter, denotes the volume of blood sent out from either one of the two ventricles during a period of one minute.

In accord with the above equation, it is possible to measure cardiac output by knowing (a) how much oxygen an individual is consuming per unit of time, (b) the average oxygen content in the arterial blood, and (c) the average content of oxygen in the venous blood.

Let us assume an individual is using 250 milliliters of oxygen per minute. (Incidentally, this quantity corresponds to 1800 Calories per day—a normal basal metabolic rate for a young adult.)

Let us also assume that the arterial and venous contents of blood are 20 and 15 milliliters of oxygen per 100 milliliters of blood, respectively. Placing these three numbers into the Fick equation, we get:

$$\frac{250 \text{ ml } O_2}{\text{min}} = \left(\frac{20 \text{ ml } O_2}{100 \text{ ml arterial blood}} - \frac{15 \text{ ml } O_2}{100 \text{ ml venous blood}}\right)\left(\begin{array}{c}\text{cardiac}\\\text{output}\end{array}\right)$$

Solving this equation, we get a cardiac output of 5 liters per minute.

Although cardiac output is dictated by how much oxygen is needed, the numerical value of cardiac output is determined by three variables: frequency of heartbeat; volume of blood emitted per heartbeat (the stroke volume); and the value of $A_{O_2} - V_{O_2}$ (milliliters of oxygen extracted from 100 milliliters of blood as the blood passes through the capillary networks from the arteries to the veins).

Suppose, for example, that 250 milliliters of oxygen are required per minute by an individual. In accord with the example given above, this requirement can be satisfied by pumping 4 liters of blood per minute and extracting 25% of the arterial oxygen content (20/100–15/100). It can also be satisfied, however, by extracting less oxygen and pumping more blood, or extracting more oxygen and pumping less blood.

Alternatively, if we assume that 25% of the oxygen will be extracted from a cardiac output of 5 liters—in accord with the example—it is possible to pump this volume of blood by increasing the heartbeat and decreasing the stroke volume or increasing the stroke volume and decreasing the heartbeat.

Varying the Cardiac Output to Meet the Body's Need for Oxygen

By far the most efficient way of satisfying our need for oxygen is to extract as much oxygen as possible from the arterial blood and to increase the stroke volume. The extraction of a larger quantity of oxygen during each passage of blood from the arteries to veins, and the ejection of a larger volume of blood per heartbeat, reduces the work which is required of the heart in satisfying the body's need for oxygen. In this connection, as mentioned in the last chapter, the heartbeat of a well-conditioned athlete is lower than that of an ordinary person where equal amounts of oxygen are consumed. At rest, for example, the heart of an average person beats about 70 times per minute, compared to 60 beats in a highly trained athlete.

Oxygen consumption and the three determinants of cardiac output (oxygen extractability, heartbeat, and stroke volume) are related to a large number of variables. Several of the more important, conspicuous variables are age, sex, and physical condition. The relationships are shown in Figure 26.6.

From the graphs in Figures 26.6 and 26.7 you may see that nature has imposed upper limits on stroke volume (about 160 milliliters), heart rate (about 200 beats per minute), cardiac output (about 25 to 30 liters per minute), arterial-venous oxygen difference (about 16 milliliters oxygen per 100 milliliters of blood), and oxygen consumption (about 5 liters per minute).

The limitation of oxygen consumption stands supreme among the cardiovascular limits. In fact, the single most important difference between life and death—at least, as the cardiovascular system is concerned—is whether the oxygen requirements of vital processes can be met. If more oxygen is needed than can be provided, death is sure to follow. This is precisely what happens, in the long run, during a chronic period of heart failure.

STROKE VOLUME. *The volume of blood ejected by either ventricle during a single contraction of the heart.*

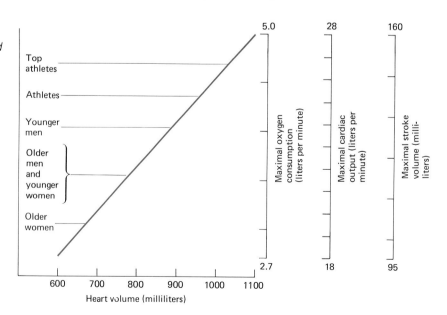

FIG. 26.6. *Dynamic cardiovascular relationships. (After Astrand and Rodahl, 1970.)*

Dynamic Characteristics of the Circulatory System

Clearly, the body's oxygen requirements will not be met when there is inadequate cardiac output due to a drop in stroke volume or rate of heartbeat (problems that often occur in conjunction with heart disorders). Similarly, heart failure or anemia may make it impossible to obtain oxygen by way of the blood.

In concluding this section, it is worthwhile to single out two other important relationships in conjunction with oxygen consumption (and therefore cardiac output). One such relationship concerns *oxygen consumption* and *body weight*. Oxygen consumption increases in proportion to an individual's body weight. This relationship has an important bearing on the work of the heart with respect to obesity. It shows that the heart in any individual must work harder and harder as more and more weight is added to the frame. This partly explains why it is dangerous to put on excessive weight when one's heart is diseased or failing.

The second relationship concerns *oxygen consumption, intensity of work,* and *quantity of work*. In Figure 26.8 it may be seen that oxygen consumption increases in proportion to the intensity of work done by an individual. Intensity of work is defined in terms of Calories of work per minute. This relationship may have an important bearing on heart attack. Often, a heart attack is precipitated not so much by how *much* work is done, but by how *intense* the work may be. A short period of severe exertion—as shoveling snow, a rapid dash in an emergency situation, or some other intense effort—may bring calamity, whereas long periods of work with low caloric expenditure per unit of time may be harmless.

In Figure 26.8 it may also be noted that the amount of work an individual can do before reaching a point of exhaustion is related to the intensity of work. *Exhaustion* may be defined as an inability to do work, no matter how

HEART FAILURE. *A combination of symptoms which result from failure of the heart to pump blood properly. Left heart failure gives rise to very difficult breathing, and right heart failure produces a dramatic edema, among other symptoms.*

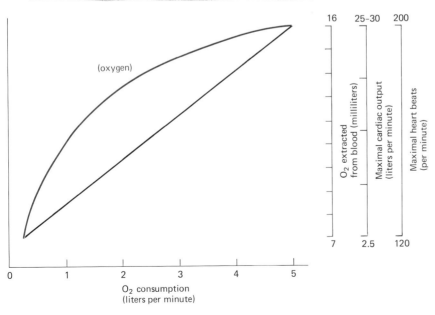

FIG. 26.7. *The limits of oxygen consumption remain relatively constant with increasing age.*

hard the person tries. The body simply refuses to respond in spite of one's conscious desire. From the relationship shown in the figure, it should be clear that an individual can only do a certain amount of work before exhaustion sets in. That work can be done very intensively in a short period of time, or less intensively over a longer period.

SHIFTING THE BLOOD FROM INACTIVE TO ACTIVE REGIONS OF THE BODY

Each organ in the body requires a certain amount of blood, depending on how intensively it is carrying out its functions. During heavy work and exercise, the heart and skeletal muscles need a greater proportion of blood than many other tissues in the body. After a meal, the digestive system requires more blood than various other tissues. Clearly it would be a great advantage to the heart—from an energy viewpoint—if it were possible to satisfy the varying needs of tissues without having to pump a great deal of blood around. This advantage is afforded by the ability to shift blood from relatively inactive to relatively active regions of the body.

How this is done is not known precisely. However, we do know that nerve reflex activities operate through the autonomic nervous system to constrict one or another distributive artery, thus reducing the blood supply to the organ served by the constricted artery. Because of this, it is possible for the body to get along with about 7% of its total weight as blood, whereas it might otherwise be necessary to have a greater percentage of the body's weight in the form of blood.

How is the blood distributed in the body under various conditions? A set of data pertinent to this topic is given in Table 26.1 with respect to an average person at rest and a well-trained athlete doing maximum exercise.

Before proceeding, it is necessary to define *maximum exercise*. An indi-

AUTONOMIC NERVOUS SYSTEM. *A division of the nervous system which operates at all times, whether we are awake or asleep.*

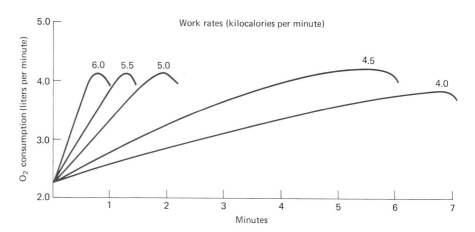

FIG. 26.8. Oxygen consumption in relation to intensity of work; O_2 consumption drops off sharply at the point of exhaustion. (After Astrand and Rodahl, 1970.)

vidual is said to be doing maximum exercise when oxygen consumption begins to drop, even though the individual might then take on a heavier work load (for which there should be a corresponding increase in oxygen consumption). This can be seen in Figure 26.8. Physical fatigue and exhaustion sets in at, or just beyond, the point of maximal exercise. Perhaps you may recall a personal experience of your own where maximal exercise was achieved.

Several interesting features may be noted in Table 26.1 with regard to regional distribution of blood during rest and exercise. First, it may be seen that the flow of blood through the brain remains relatively constant—regardless of whether an individual is resting or under severe physical exertion. This constancy also holds true when a person is under severe mental exertion.

Second, it may be seen that during exercise the largest fraction of the cardiac output is distributed to the skeletal muscles.

Third, it may be noted that the heart muscle (the myocardium) receives the second largest share of the cardiac output. In spite of this, however, the percentage of cardiac output to the heart remains approximately constant—about 4%.

Although the data shown in Table 26.1 are helpful in seeing what happens to the distribution of cardiac output when one goes from rest to maximal exercise, they fail to show what happens between these extremes. For example, the flow of blood in the skin changes greatly as one moves progressively from rest to vigorous exercise to maximal exercise. During rest, assuming the surroundings are comfortable, the skin receives a relatively small proportion of the cardiac output—about 10% of the 5 liters which are pumped out each minute. During submaximal exercise, the blood flow increases through the skin. About 12% to 15% of the cardiac output may now be flowing through. This increase is provided mainly for the purpose of dissipating body heat. But now, finally, if one pushes the work effort to the maximal point—an effort which might be sustained a minute or two, perhaps five or ten minutes—a smaller percentage is sent to the skin. The skin, in fact, becomes pallid as blood is diverted away from it and from in-

TABLE 26.1. Distribution of Cardiac Output During Rest and During Maximal Exercise

	Resting (output = 6 liters per minute)	Maximal Exercise (output = 25 liters per minute)
Heart	4%	4%
Brain	13%	3%
Skeletal muscle	21%	88%
Digestive system	24%	1%
Kidneys	19%	1%
Skin	9%	2.5%
Other	10%	0.5%

active muscles, and is sent instead to the skeletal and cardiac musculature, where the need is intense.

THE PULSE—AN INDICATOR OF GENERAL CONDITIONS IN THE CIRCULATORY SYSTEM

PULSE. Intermittent change in tension in an arterial wall (L. pulsus, a striking).

RADIAL ARTERY. An artery in the forearm (L. radius, lateral bone of forearm).

As the left ventricle contracts and sends blood into the aorta, the force of the blood causes a mechanical pressure wave—called a *pulse*—to be sent along the main arterials. In the case of arteries located close to the surface of the body the pulse can be felt (palpated) with a finger.

The pulse is most commonly evaluated by palpating the radial artery (see Figure 26.9). However, when this artery is not easily accessible, as may happen in certain accidents, when it is difficult to feel the pulse at this point, or when information is needed on regional blood circuits, other points may be tested, as shown in Figure 26.9.

Three valuable kinds of information may be obtained when a pulse is properly evaluated: the rate of heartbeat, the rhythm of heartbeat, and information about whether the heart is delivering a normal volume of blood.

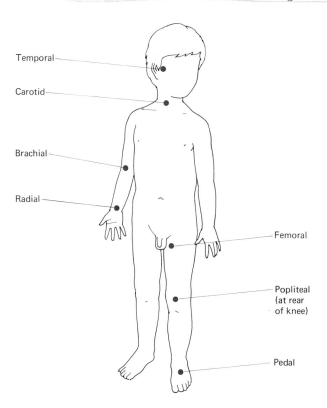

FIG. 26.9. Locations of commonly tested pulses. The popliteal pulse is located at the rear of the knee.

In the evaluation procedure, this latter point is referred to as the *pulse depth*.

Pulse Rate. The rate of pulsation equals the rate of heartbeat. A normal heart rate in a resting healthy adult is about 70 beats per minute.

The heartbeat increases in a variety of circumstances; including (1) exercise; (2) an increase in body temperature—as happens during vigorous exercise on a hot day, or when one is feverish; (3) when the blood volume is less than normal, as may occur following a hemorrhage or during shock; (4) in various forms of heart disease; and (5) when the electrical rhythm is stepped up considerably.

A reduced heartbeat may be found under a variety of other conditions, such as (1) in a trained athlete; (2) during convalescence from various infectious diseases, such as influenza; (3) in certain heart diseases; and (4) when the electrical activity is stepped down or blocked.

Pulse Rhythm. There is usually a regular, cyclic pattern in the beat of the heart. In certain disease states, however, (such as thyroiditis) or when the body is poisoned (as by excessive digitalis), the pattern may become slightly or markedly arrhythmic.

DIGITALIS. *A therapeutic mixture of compounds obtained from the dried leaf of foxglove, used to correct cardiac rhythm problems.*

Pulse Depth. In a healthy individual there is ordinarily a certain fullness of the pulse wave. We then speak of a strong pulse. This fullness can be greatly diminished under abnormal conditions, such as hemorrhage, shock, arteriosclerosis, and certain heart diseases. In these cases the pulse is said to be weak.

REGULATION OF THE CIRCULATORY SYSTEM

The circulatory system is regulated by the nervous system and by certain built-in features of its own. We begin this section with a discussion of the latter—the autoregulatory features of the heart.

Autoregulation of the Heart

AUTOREGULATION. *Self-regulation.*

The circulatory system has two important self-regulatory features: a myogenic operation and an ability to regulate the force of contraction in proportion to its blood volume.

Myogenic Contraction. You may recall that the word *myogenic* refers to a form of muscle activity in which contractions arise due to action potentials that originate in the muscle itself. The heart is a myogenic muscle. It is able to beat completely independently of the nervous system. In fact, it accomplishes its job in just this very way throughout periods of sleep and in periods of wakefulness when the body is inactive and the mind is relaxed. During these periods the heart normally beats at around 70 times a minute.

Regulation of the Circulatory System

Perhaps you may better appreciate the concept of myogenic contraction by considering heart transplants and pacemakers. A heart transplant patient receives a heart that is completely denervated. And yet, assuming all other factors are favorable, the heart may beat on and on for years, furnishing oxygenated, nutrient-rich blood to all of the tissues of the body. Pacemaker patients carry a transistorized electronic device (an artifical pacemaker) which stimulates and regulates the beat of the heart. Such individuals need this device because their own myogenic and neurogenic regulatory mechanisms are not completely dependable.

Starling's Law of the Heart. The ability of the heart to regulate its force of contraction in proportion to the degree of filling is called Starling's Law of the Heart. This law holds true whether or not a heart is innervated. When a small volume of blood is received during filling, the heart contracts moderately during systole. When a large volume is received, the heart contracts more forcefully.

The importance—and value—of the Starling relationship is twofold: (a) When a large volume of blood is received—as happens, for example, during exercise—a vigorous contraction ensures delivery of the blood to needy tissues and precludes the possibility of building up excess reserves that would tend to overfill the heart. (b) On the other hand, when a small volume of blood is received—as happens, for example, during periods of inactivity or sleep—the heart is not called upon to do excessive work in expelling the blood.

Starling's law merits serious practical consideration where heart transplants are concerned. Large individuals receiving a small heart will nevertheless require a certain minimal cardiac output in order to furnish their body with needed oxygen and nutrients. For such individuals, in spite of a lack of innervation, large volumes of blood enter a relatively small heart. The heart contracts more forcefully than with a smaller volume of blood, and of course, a greater amount of work is demanded of the heart.

Nervous Control of the Circulatory System

Although the heart is capable of beating myogenically—in the total absence of nerve stimulation—it is at all times subject to regulation by the autonomic nervous system. As a rule, the autonomic actions are executed reflexly, although they are always subject to actions we perform voluntarily, such as eating, sleeping, and walking.

In discussing the regulation of the circulatory system by the nervous system, it will be helpful to describe the separate actions of the sympathetic and parasympathetic nervous systems and then to discuss the major reflex actions they execute.

Sympathetic Influences Upon the Circulatory System. By far, the sympathetic system dominates the parasympathetic system with respect to circulation. It imposes itself upon the heart and upon the blood vessels.

PACEMAKER. *The sino-atrial node.*

INNERVATE. *To supply a tissue with nerves.*

Dynamic Characteristics of the Circulatory System

Effects on the Heart. When sympathetic nerves release adrenaline and noradrenaline into the heart, two different kinds of responses are produced: the heart contracts more forcefully and at a greater frequency.

To what extent can these two actions increase the cardiac output? To answer this question we must first consider the maximum amount of blood the heart can eject with each heartbeat and the maximum heartbeats per unit time.

Ordinarily, in a normal adult at rest, there are about 130 milliliters of blood in the heart immediately after diastole and just before systole. About 60 milliliters are ejected with each heartbeat. A larger quantity—as much as 120 milliliters—can be ejected by increasing the force of contraction. If we assume that the heart is beating 70 times a minute, then, by increasing the force of contraction, the amount of blood emitted can be increased from about 5 liters per minute to about 10 liters per minute. Even larger quantities than this can be emitted. For this to happen, larger volumes of blood must be taken into the heart during diastole.

Ordinarily, in a normal healthy resting adult, the heart beats about 70 times per minute. During vigorous exercise the beat may be increased to about 100 or 120 beats per minute and even higher. If we assume that the heart is able to fill up and nearly empty itself with each beat, and if we consider an increase in heartbeat from 70 beats per minute to 105 beats per minute, it follows that the cardiac output value of 10 liters can be increased by 50% to 15 liters. In well-trained athletes the cardiac output can be increased to as much as 25 liters ($6\frac{1}{2}$ gallons) per minute or even higher.

Effects on Blood Vessels. The sympathetic nervous system brings about two different kinds of responses from the blood vessels, depending upon whether or not the adrenal medulla is brought into action. (You may recall from Chapter 8 that the adrenal medulla is homologous to a postganglionic autonomic nerve.)

In the event that the secretory action of the adrenal medulla is *not* called into action and, further, if the remainder of the sympathetic nervous system is activated, the main neurotransmitter present in the smooth muscles throughout the body is noradrenaline.

Noradrenaline is a *vasoconstrictor*. That is, it brings about the contraction of smooth muscle in any blood vessel with which it comes in contact. In doing so, it causes the blood vessel to constrict. In consequence, the pressure of the blood within the vessel increases, and the blood itself flows more rapidly.

In the event that the adrenal medulla *is* activated—as happens during exercise; stress; in any situation where additional muscular energy is required (such as shivering); or in situations where it is necessary to reduce heat losses from the skin—the main neurotransmitter present in muscles throughout the body is adrenaline.

Adrenaline causes certain blood vessels to constrict and others to dilate. This is in striking contrast to noradrenaline, which always acts as a vasoconstrictor.

Why does adrenaline exert a twofold action? To begin with, it is called upon only during movements when the muscles of the body require more

ADRENAL MEDULLA. *The central or inner region of the adrenal gland. It discharges small quantities of noradrenaline and large quantities of adrenaline.*

NOR- *refers to the fact that adrenaline, but not noradrenaline, has a methyl (CH_3) radical as part of its structure (German,* nur ohne *radical, only without a radical).*

ADRENALINE. *A metabolite whose name and synonym* **(epinephrine)** *are based on the fact that large quantities of this chemical come from the adrenal gland* (L. ad, near; ren, kidney); (G. epi, upon; nephros, kidney).

VISCERA. *The soft internal organs* (L. viscus).

than their average share of the cardiac output. In carrying out its actions, adrenaline causes the arteries in the heart and skeletal muscles to dilate. This permits a larger volume of blood to flow with ease into the heart and skeletal muscle and thus allows a greater exchange of nutrients and wastes within these tissues.

In contrast to the arteries of the heart and skeletal muscles, the arteries of the skin and visceral organs constrict under the influence of adrenaline. The skin and viscera thus receive less than their average share of blood. This is an adaptive feature which permits the heart and skeletal muscle to have more oxygen and nutrients than would otherwise be the case. At the same time, the reduction of blood flow through the skin prevents losses of heat to the outside air. The heat is kept within the body instead. It raises the temperature within the body ever so slightly, but enough to allow a maximum speed in the execution of nerve impulse propagation, muscle contraction, and reflex actions.

Although sympathetic stimulation can increase both the frequency and force of heart contraction, it is important to realize that there are upper limits to the heartbeat and cardiac output. At slightly elevated heartbeats and moderately forceful contractions, the heart is able to deliver increased quantities of blood to needy tissues. But as the frequency of beat increases more and more, less and less time is available for filling during each diastole. So no matter how forceful a contraction may be, even at a low frequency of heartbeat, there is a limit to the amount of blood the ventricles can accommodate.

Parasympathetic Influences Upon the Circulatory System. The parasympathetic nervous system exerts a major effect upon the heart, but only a minor effect upon the blood vessels. It exerts its action by way of the vagus nerve, which releases acetylcholine from its nerve endings.

VAGUS. *Either of the paired tenth cranial nerves.*

Effects on the Heart. When acetylcholine is liberated into the heart, three kinds of responses follow: the frequency of heartbeat is reduced; the force of contraction decreases; and the length of the functional refractory period is increased. This threefold action is brought into play whenever it is necessary to counter the effects of the sympathetic nervous system. In this way the increased cardiac output, blood pressure, and heartbeat that is brought about by the sympathetic nervous system can be reduced toward normal resting values.

Effect on Blood Vessels. The parasympathetic nervous system exerts a dilatory effect when it is mustered into action. Its role in this regard is generally very minor. It is important in the penis, whereby blood may rush into cavernous spaces to bring about erection. It may be helpful in the heart, though there is little evidence or reason to believe it is important in this regard. It may also be helpful in certain viscera whose blood vessels are innervated by the vagus nerve. But this action also is of minor importance. Perhaps the only blood vessels which partake a benefit from the vagus nerve (aside from the penis), are the arterioles of the salivary glands. These arterioles dilate during a meal, providing additional nutrients to the glands when they are active in producing and secreting their products.

REFLEX REGULATION OF THE CIRCULATORY SYSTEM

Having said something about the autoregulatory features of the heart and the nervous influences upon the circulatory system, we can now consider the major reflex regulatory actions.

The highest regulatory centers for circulation are located in the medulla oblongata of the brain stem. Called the *cardio-inhibitory* and *cardio-acceleratory centers*, these regions of the brain receive central input from the cerebral cortex and peripheral input from sensory neurons located in the great veins (vena cavae), atria, aorta, and carotid arteries.

MEDULLA OBLONGATA. A 4-centimeter extension of the spinal cord into the brain stem; the medulla partly controls a number of visceral functions.

Central Input

The *central input* is largely psychological in nature. It consists of information which may arouse the spirit of fighting, the emotions of love, a feeling of relaxation, or the syndrome of anxiety. This syndrome, which affects about 5% of all Americans, is difficult to diagnose, inasmuch as it develops slowly, is episodic (comes and goes), and cannot be characterized discretely or distinctly from one individual to another. In general, people suffering anxiety have more of the following traits than those who are free of anxiety: palpitation (violent heartbeat which is felt in the chest); chest pain (possibly muscular pains; possibly angina pectoris); breathlessness; apprehensiveness; headache; paresthesias (abnormal spontaneous sensations such as burning, pricking, numbness); weakness; insomnia, unhappiness; syncope (fainting); and frequent urination. Large quantities of lactic acid are often found in the blood, and the adrenal medulla is active in secreting adrenaline.

ANGINA PECTORIS. Chest pain from oxygen deficiency in the heart (L. angina, quinsy, choking; pectoris, chest).

Peripheral Input

The *peripheral input*, in contrast to the central, is entirely chemical or physical in nature. It consists of information which indicates any one of the following: that the volume of blood entering the atria is larger than normal; that the blood pressure in the aorta or carotid arteries is too high; that the oxygen content of the blood which enters the aorta or carotids is too low; and that the carbon dioxide content is too high.

Stretch Receptors. The *stretch receptors*, which respond to large volumes of blood entering the atria, are located in the large veins (vena cavae) and the walls of the atria. These neurons convey their impulses through the tenth cranial nerve (vagus) to the medulla oblongata. They initiate the so-called *Bainbridge reflex*—a cardio-acceleratory reflex. Accordingly, when a large volume of venous blood enters the heart, the stretch receptors fire impulses into the medulla oblongata. As a result, efferent messages are directed to decrease the activity of the parasympathetic system. Through these actions a greater amount of venous blood is cleared from the heart.

Summary

AORTIC SINUS. *Pressure receptor for signaling blood pressure.*

CAROTID SINUS. *Pressure receptor for signaling blood pressure.*

Pressure Receptors. The *pressure receptors*, which respond to high blood pressure in the aorta and carotid arteries, are called the *aortic sinus* and *carotid sinus*. The aortic sinus lies within the arch of the aorta. The carotid sinus is present in the bifurcation (branching) of each common carotid artery—within the neck. See Figure 25.15.

The aortic sinus monitors the blood pressure of the systemic circulation. The carotid sinus monitors the blood pressure to the cerebral circulation. When the pressure tends to rise, a cardio-inhibitory reflex is initiated. That is, afferent messages are sent along the ninth and tenth cranial nerves into the medulla oblongata. As a result, the parasympathetic outflow is increased, the sympathetic outflow is decreased, the heart slows in its action, and the blood pressure drops.

AORTIC BODY. *Chemoreceptor for signaling low oxygen, high carbon dioxide.*

CAROTID BODY. *Chemoreceptor for signaling low oxygen, high carbon dioxide.*

Chemical Receptors. The *chemoreceptors* which respond to low oxygen or high carbon dioxide are located in the *aortic* and *carotid bodies*. Here—in the wall of the aortic arch and in the carotid sinus (Figure 25.15)—are aggregates of sensory nerve fibers. It is believed that these sensory cells respond to low concentrations of oxygen in the blood. Generally, however, low concentrations of oxygen are accompanied by a relatively high concentration of carbon dioxide. Most investigators believe that the chemoreceptors are more sensitive to high concentrations of carbon dioxide. It is known that high concentrations of carbon dioxide tend to reduce the pH of the blood. It tends to do so by forming carbonic acid ($H_2O + CO_2 \rightleftharpoons H_2CO_3$), which dissociates into bicarbonate and acid ($HCO_3^- + H^+$). When the body's buffering systems are unable to cope fully with the acid (H^+) that is produced, the normal blood pH of 7.4 drops slightly. This drop in pH, if severe enough (down to about 7.2) will upset various metabolic, neural, and muscular processes. It is believed that such serious drops in pH can be more harmful than a relative shortage of oxygen. No matter what the interpretation may be, it is important to realize that the brain consumes about 25% of the oxygen used by the body under basal conditions (when at rest).

Conceivably, the carotid bodies may be more sensitive to oxygen than to carbon dioxide, as the blood flows past them into the brain. In contrast, the aortic body may be more sensitive to carbon dioxide than to oxygen, as the blood flows past this body into the systemic circulation. Whatever the case may be, changes in the levels of oxygen and carbon dioxide are continuously monitored by the aortic and carotid bodies. In response to low oxygen or high carbon dioxide levels, the receptors in these bodies relay messages to the medulla oblongata to bring about an increase in circulation.

SUMMARY

1. A systemic arterial pressure of 120 mm Hg, in a normal adult at rest, is well in excess of what is needed to overcome hydrostatic pressure in sending blood into the brain.

Dynamic Characteristics of the Circulatory System

2. Due to a higher osmotic pressure in blood and in cells than in the space between cells, there is a negative interstitial pressure, and fluid is normally not allowed to accumulate between cells.
3. In the course of its flow, blood encounters: (a) an elastic resistance in large arteries; (b) an arterial resistance in medium and small arteries, which increases in proportion to vasoconstriction; and (c) a capillary resistance due to friction. The total resistance to flow increases when excess fat accumulates in the blood and when excess blood cells are present.
4. Blood is returned to the heart from veins by virtue of: (a) the U-tube principle; (b) a venous pump, which is effected by skeletal muscle contractions; (c) venular constrictions; (d) an atrial vacuum pump; and (e) a thoracic pump.
5. The amount of blood that is sent out of the heart each minute is ordinarily equal to the amount of oxygen consumed divided by the amount of oxygen extracted from the blood. This relation is expressed mathematically in the Fisk equation.
6. The most efficient way of satisfying our need for oxygen is to increase the volume of blood ejected per heartbeat and to extract a larger percentage of oxygen from the blood as it passes through capillary networks.
7. Another way in which the circulatory system responds to the body's need for oxygen is by shifting blood from less physically active tissues to regions that are more active.
8. The rate, rhythm, and depth of the pulse are valuable indicators of the condition of the circulatory system.
9. The circulatory system is regulated through: (a) autonomic features built into the bioelectric system of the heart; (b) Starling's law, which says that the force of the heart's contraction is proportional to degree of filling; (c) the sympathetic nervous system, which increases the force and frequency of cardiac contraction; (d) the parasympathetic system's converse effect; (e) the vasoconstrictor effect of noradrenaline on all arteries; the effect of adrenaline, which constricts the arteries of the skin and viscera but dilates the arteries of the heart and skeletal muscles; and (f) the pressure and chemical receptors in the aorta and carotid arteries, which initiate reflexes that maintain proper levels of pressure, oxygen, and carbon dioxide in the blood.

REVIEW QUESTIONS

1. Tell how edema arises when normal osmotic relationships between the capillaries and cells break down.
2. What kinds of factors contribute resistance to blood in its flow?
3. What mechanisms are used by the body to convey blood from the veins to the heart?

Review Questions

4. What information is contained in the Fick equation?
5. Considering that there is relatively little blood in the body, what are the principal mechanisms by which the body is provided with adequate amounts of oxygen during periods of severe work or exercise?
6. What characteristics of the pulse are valuable indicators of the condition of the circulatory system?
7. How are blood vessels and the heart regulated by the autonomic nervous system?
8. In what specific ways are adrenaline and noradrenaline different with respect to their constricting effect upon various kinds of arterial blood vessels?
9. Describe Starling's law of the heart.

Functional Disorders of the Circulatory System 27

SHOCK
FAINTING
DEHYDRATION
WATER INTOXICATION AND EDEMA
HYPERTENSION
ARTERIOSCLEROSIS AND ATHEROSCLEROSIS
HEART DISEASE
    Endocardial diseases
    Congenital heart diseases
    Hypertensive and arteriosclerotic heart diseases
    Coronary heart disease
    Pericardial diseases
    Myocardial diseases
HEART FAILURE

The circulatory system may suffer a variety of abnormalities. The major disorders, in terms of either frequency or severity, are discussed in this chapter.

Shock and *fainting* are reactions triggered by the autonomic nervous system. They are conditions in which the normal physiology of the circulatory system is disrupted for a relatively short period of time. *Dehydration, water intoxication,* and *edema* are body fluid imbalances which impose a burden on the heart.

Arteriosclerosis (hardening of the arteries) and *hypertension* (high blood pressure) also impose an excessive burden upon the heart. Arteriosclerosis is always accompanied by hypertension, though hypertension may arise in the absence of arteriosclerosis. Hypertension—alone or together with arteriosclerosis—may lead to a form of heart disease known as *hypertensive heart disease.*

Another form of heart disease, *coronary heart disease,* arises when the heart is deprived of an adequate supply of blood. When an individual experiences an acute deprivation—whereby some of the heart muscle cells die—the individual is said to have a *heart attack* or *myocardial infarction.*

Coronary heart disease, and all *other forms of heart disease*, reduce the work capacity of the heart, which in turn may lead to heart failure. Certain bodily abnormalities other than cardiac problems—such as kidney disease—also may lead to heart failure. *Heart failure* is any condition of the body in which the circulatory load exceeds the heart's capacity to accomplish its work.

Each of the italicized words, or topics, in the paragraph above represents a condition in which there is some disruption—which may be moderate or severe—of the circulatory system. These are discussed separately below, not only because they are subjects everyone should know something about but also because an understanding of these problems relates to the application of physiological principles.

SHOCK

SHOCK. *A circulatory disorder caused by injury or psychic disturbance; symptoms are reduced blood pressure, weak, rapid pulse, and mental confusion (French* chocquer, *to strike).*

Shock represents a severe disturbance of the autonomic nervous system. In actuality, it is a symptom of faulty blood circulation; it may or may not be accompanied by a loss of blood.

Shock may arise in response to an emotional disturbance, an allergic reaction, a burn or accident, a coronary occlusion (blockage of one or more arteries in the heart), or some other cause.

Shock is characterized by: (1) a pale, cold, sweaty skin—because the peripheral circulation is low and the peripheral veins are "empty"; (2) an increase in heartbeat—because the heart is trying to compensate for a low arterial pressure; (3) a weak and rapid pulse; (4) rapid breathing; (5) thirst; (6) a low level of consciousness; and (7) a certain degree of confusion.

FAINTING

FAINT. *To swoon and to fall unconscious (Old French* faindre, *to feign).*

SYNCOPE. *Fainting (G.* synkope, *sudden loss of strength).*

Shock may or may not be accompanied by *fainting*, depending on whether enough blood flows through the brain per unit time. Ordinarily the brain receives about 15% of the 5 liters of blood which the heart pumps each minute in a healthy resting adult of average size. When the quantity is reduced by about 60%, loss of consciousness (fainting, or syncope) follows.

Fainting (a sudden fall in systemic blood pressure) may be triggered by any one of a number of things. For some of these causes the resulting faint may be given a special descriptive term. *Postural syncope* is due to prolonged standing at attention, so that blood collects in the feet. Fainting forces the individual to lie down, and thus allows the blood pressure to return to normal.

Carotid sinus syncope may be brought on by emotions. It is a spontaneous faint attributed to overactivity of the pressure receptors in the carotid

Functional Disorders of the Circulatory System

sinus. *Micturition syncope* occurs in association with emptying the bladder. It is more common in men than women, perhaps because of posture. It is also called *psychomotor epilepsy* in view of its association with epileptics. *Vasovagal syncope* is attributed to overstimulation of the blood vessels by the vagus nerve, which causes them to relax and dilate, resulting in a fall in blood pressure.

Fainting may also be caused by a punch in the chin, testes, breast, or "pit of the stomach" (actually the *solar plexus* of autonomic nerves behind the stomach). It may be induced by a heart attack, shock, or, in the case of a diabetic, a violent coughing spell stimulated by alkalosis, which follows a fall in blood sugar after an overdose of insulin.

MICTURITION. *Urination (L. micturire, to make water).*

DEHYDRATION

In severe cases of water depletion, *dehydration* sets in. This may arise in a case of (1) excessive diarrhea; (2) sodium depletion due to hemorrhage, sweating, or failure of the kidney; and (3) in cases where water output greatly exceeds water intake, as for example in diabetes insipidus.

A very complex combination of events arises during dehydration. First there is loss of water from the extracellular space. In consequence, the extracellular osmotic pressure increases. Water is then drawn from the cells into the extracellular space. Now cellular dehydration sets in. The situation may be so severe that—in spite of the actions of antidiuretic hormone and aldosterone—water and sodium continue to be lost. In time, unless the situation is corrected, the following changes come into the picture: the hematocrit (the percentage of red blood cells in the blood) rises; the concentration of protein in plasma rises; the cardiac output becomes lower; the pulse pressure narrows (systolic pressure goes down, diastolic pressure goes up)—creating a state of hypotension which results in weakness, faint, and a rapid heartbeat; a selective vasoconstriction occurs within the skin—causing the skin to become cold and clammy; and as the amount of water in the body continues to drop, blood begins to flow more slowly through the kidneys—making it impossible for the kidneys to clear the blood of urea. This leads to high levels of urea in the blood (uremia) and coma.

ANTIDIURETIC HORMONE or **VASOPRESSIN.** *Posterior pituitary hormone that tends to prevent excessive losses of water into the urine (G. anti, against; diourein, to pass water).*

VASOCONSTRICTION. *Reduction of the lumen, or bore, of an arterial vessel (L. vas, vessel).*

WATER INTOXICATION AND EDEMA

Edema is a condition with the following characteristics: the blood albumin decreases in concentration; the venous pressure drops; the blood (arterial) pressure drops; and water accumulates excessively in the extracellular, interstitial fluid compartment. Edema may occur locally, or it may be widespread. It is especially common around the ankles, among the subcutaneous tissue of the buttock and in the loose folds beneath the eyes.

Water intoxication resembles edema. Here there is excessive interstitial fluid—as in edema—but in addition there is an excess of intracellular fluid. Also, unlike edema, it usually involves the whole body.

Water intoxication may arise in a case of trauma, or in hospital situations where intravenous feeding is carried out too rapidly. Headache, dizziness, mental confusion, perspiration, muscle fasciculations and convulsions are some of the symptoms which may be seen in a case of water intoxication.

In a more general sense, edema and/or water intoxication may develop whenever excessive quantities of water accumulate in the body. Excessive quantities may accumulate under any one of the following conditions:

(1) failure of the kidney to excrete sufficient amounts of sodium; (2) secretion of excessive antidiuretic hormone, as for example, from a pituitary tumor; (3) partial blockage of the aorta, usually because of atherosclerosis, giving rise to hypertension; (4) arteriosclerosis or some other disease of the renal arteries, giving rise to hypertension; (5) blockage of lymph vessels, due to an inflammation, leading to increased interstitial pressure; (6) reduction in the osmotic pressure of the blood due to a loss of protein in burns or kidney disease, leading to a decrease in interstitial pressure; and (7) release of larger than normal quantities of histamine from mast cells, platelets, and basophils, causing capillaries to become leaky.

The retention of excessive amounts of water is usually, if not always, accompanied by a retention of excessive amounts of sodium. The larger burden of fluid which is forced upon the heart causes it to work harder to circulate the blood. Overstretching the heart, however, leads to an enlargement of the heart itself (myocardial hypertrophy) as well as what may be a permanent dilation of the arteries. The cardiac output cannot be stepped up. When the edema extends to the lungs (pulmonary edema), normal gaseous exchanges of oxygen and carbon dioxide are prevented. Acidosis, headache, nausea and a wide range of other complications set in, often leading to convulsions and coma.

TRAUMA *usually signifies a severe mechanical, physical, or psychological injury* (G. traumatikos, *of wounds*).

FASCICULATION. *An involuntary contraction of a bundle of muscle fibers* (L. fasciculus, *a cable*).

HYPERTENSION

By definition, hypertension is an elevation of arterial blood pressure beyond what is normal for an individual. The normal, healthy average-sized adult usually has a blood pressure of 120/80 when the systolic/diastolic pressures are measured while in the seated, relaxed position. Higher values are normally expected as an individual ages—particularly so, perhaps, because a certain amount of arteriosclerosis inevitably creeps in.

As long as the ratio of systolic to diastolic pressure is around three to two, and provided the measured values do not exceed 135/90, the pressure might be considered perfectly normal. Systolic pressures somewhat in excess of

Functional Disorders of the Circulatory System

140 might be viewed as warning signs of potential circulatory problems. Diastolic pressures greater than 90 are even more foreboding, since they signify a high peripheral resistance—a situation in which a higher pressure (diastolic) is needed in order to drive the blood through arterioles into capillary beds.

Hypertension is usually rooted in one or more physiological systems other than the circulatory system itself. Four physiological systems are particularly prominent in this regard: the nervous system; the endocrine system; the excretory system; and the cardiovascular system. Emotional tension—working through the cerebral cortex and hypothalamus—brings about a greater degree of sympathetic activity which, among other things, has the effect of increasing the peripheral resistance to blood flow. Over a long period, constant excessive stimulation of the arteries and heart can bring about physical changes, and these may often be reflected as high blood pressure.

The endocrine contribution to hypertension stems mainly from aldosterone and renin. When abnormally large amounts of aldosterone are present, excessive amounts of sodium chloride are retained in the body. This tends to keep water in. An extra load of fluid is thus imposed on the heart and arteries—a load which calls for an increase in blood pressure.

When excessive amounts of renin are secreted, or when the kidney fails to excrete surplus amounts of sodium, *renin*—a hormone which is produced in the kidney—acts upon an inactive form of *angiotensin* in the blood (angiotensin I) to produce a different form, angiotensin II. Angiotensin II, a small protein, does two things: it constricts the smooth muscle of arteries, thus directly raising the blood pressure; and it reacts with the adrenal cortex, inducing a release of aldosterone, which—when released in excess—prevents sodium from being excreted. Excessive amounts of sodium then accumulate, and these, in turn, draw water from the urinary tubules back into the blood. Edema is thus promoted, and with it, hypertension.

RENIN. *An enzyme or hormone released from the kidney in response to low blood pressure.*

ANGIOTENSIN. *A small blood protein which increases blood pressure and stimulates the release of aldosterone from the adrenal cortex (G. angio, vessel; L. tensio, tension).*

The cardiovascular system's peculiar contribution to hypertension—aside from the fact that it mediates the contributions of the brain, adrenal cortex, and kidney—is its tendency toward *sclerosis* (hardening) of the arteries. In itself, hardening of the arteries does not allow the force of cardiac contraction to be dissipated during systole. The arterial walls remain somewhat inelastic. In order to provide a sufficient driving force thereafter during diastole, the heart is forced to contract more forcefully, thus setting up a higher blood pressure.

ARTERIOSCLEROSIS AND ATHEROSCLEROSIS

Arteriosclerosis, when used without qualification, means hardening of the arteries. Two types are commonly known: *atherosclerosis*, a disease of the intima, and *medial sclerosis*, a disease of the medial layer. Medial sclerosis is commonly known simply as arteriosclerosis.

Arteriosclerosis and Atherosclerosis

Atherosclerosis is the chief cause of death in the United States. It is a disease of large and medium arteries. It affects principally the aorta—which arises from the heart and furnishes blood to the entire body except the lungs; the coronary arteries—which supply the heart; the cerebrals—which supply the brain; the iliacs—which supply the abdomen; and the femorals—which supply the legs.

Three kinds of lesions are frequently seen in arteries afflicted with atherosclerosis: fatty streaks, fibrous plaques, and the "complicated lesion." *Fatty streaks* are mainly yellowish fat deposits, found for the most part within smooth muscle cells in the intima. *Fibrous plaques* are whitish elevations that protrude into the lumen of the artery. They consist chiefly of fat-containing smooth muscle cells which, together with collagen and elastic fibers, lie above an accumulation of extracellular fat and dead cell matter. *"Complicated lesions"* are believed to be fibrous plaques that have been modified by calcification and extensive cell death.

Until recently, it was believed that atherosclerosis was the result of a form of faulty metabolism, which caused large deposits of fat to be laid down within the intima. Now, as a result of intense study, we have a better understanding of the condition.

To begin with, atherosclerosis is initiated with the aging process, and it is believed to be prompted, hastened, or provoked by injuries that occur within the intima. In children, there are few muscle cells within the intima. In the course of aging—up to a point—increasingly larger numbers of muscle cells slowly develop, bringing about an increase in the thickness of the intima. In controlled experimental situations with monkeys, in which the intima is purposely scratched through the rasping actions of an inserted catheter, the smooth muscle cells proliferate more rapidly.

Next, it is important to appreciate that atherosclerosis begins here and there throughout the body—presumably at vulnerable spots, and certainly at select focal points where excessive numbers of smooth muscle cells appear. These cells begin to deposit fat within themselves or into the connective tissue spaces outside of themselves. In this connection it is also known that smooth muscle cells proliferate much more rapidly in the presence, rather than the absence, of *lipoproteins*, and especially in the presence of lipoproteins which contain large amounts of *cholesterol*.

It is believed that the atherosclerotic tissue first appears as fatty streaks of muscle cells. Further, since the smooth muscle cells produce and secrete elastic fibers, collagen, and mucopolysaccharides (carbohydrate polymers containing an amino sugar), it is likely that this secretory activity—along with the proliferation of smooth muscle cells and their production of fats—leads to the formation of fibrous plaques. Finally, the death of additional cells, coagulation of blood within the vicinity, and the calcification of the tissues leads to the "complicated lesion" of atherosclerosis.

As a consequence of atherosclerosis, the affected artery becomes hardened (sclerotic), loses its tonicity, and often sheds some of the sclerotic material. The shed material enters the lumen as an *embolus*—called a *thrombus* when it contains large quantities of blood elements—and can

---

PLAQUE. *A patch or flat plate (French, disc or plate).*

COLLAGEN. *A gelatinous fibrous substance found in connective tissue (G. kolla, glue; gen, to make).*

INTIMA. *The inner wall of a blood vessel (L. intimus, intimate).*

LIPOPROTEINS. *Complexes formed from lipids and globulin proteins in the blood.*

CHOLESTEROL. *A fatlike alcohol (G. chole, bile; stereos, solid).*

lead to coronary thrombosis (heart attack), cerebral thrombosis (stroke), or vascular occlusion (blockage of blood vessels).

Arteriosclerosis (medial sclerosis) involves the smaller arteries. It is seen as a deposition of calcium in the medial wall with little or no encroachment on the lumen. It is a frequent and perhaps general accompaniment to aging.

STROKE. *Usually refers to a sudden loss of a bodily function because of a hemorrhage or embolus in the brain.*

HEART DISEASE

Heart disease is a collective term for a wide variety of disorders, each of which—singly or in combination—reduces the heart's ability to adjust the cardiac output to meet the body's need for oxygen. The disease may affect the valves alone, the musculature alone, the electrical conductance system alone, the pericardium or endocardium alone, or any combination of these anatomical components.

The major forms of heart disease are: endocardial; congenital; hypertensive; coronary; pericardial; and myocardial. The order of this listing is strictly arbitrary and is not meant to signify an order of importance, an order of frequency, or any other arrangement. Moreover, many heart diseases can fall into more than one of these categories.

Two conditions that occur most commonly in heart disease—unless corrective measures are taken—are cardiac enlargement (cardiomegaly) and heart failure. (Heart failure is discussed separately at the end of this chapter.)

Cardiac enlargement occurs whenever the heart is unable to furnish blood efficiently to needy tissues. The heart may enlarge because of the stresses imposed upon it. Or it may enlarge as a means of compensating for its inefficiency. It may also enlarge for both of these reasons.

When the heart enlarges in response to stress, it appears as a dilated organ. When it enlarges for compensatory reasons, its musculature becomes thicker (hypertrophied). These features—dilation and hypertrophy—may occur singly or together.

Often the heart becomes dilated first. It then attempts to compensate for the lesser quantity of work it is able to do in its dilated condition. And in the course of compensation the heart musculature hypertrophies.

CARDIOMEGALY. *Pathological enlargement of the heart (G. kardio, heart; megas, large).*

Endocardial Diseases

Endocardial diseases may involve only the membrane which covers a valve or valves (valvular disease), or it may involve the general lining of the chambers of the heart (mural disease). The disease may be an inflammatory one (valvular endocarditis; mural endocarditis). Or it may not involve an inflammation (valvular endocardosis; mural endocardosis). An inflammation is said to occur when a tissue is painful, swollen (with edema), and reddened (due to leaky blood capillaries). Other characteristic symptoms may also be seen.

Endocardial disease may be due to bacteria (for example, in diphtheria, typhoid fever, or rheumatic heart disease); to syphilis; to a metabolic problem (such as a disease of the thyroid gland); to a congenital condition; or to some other cause. Regardless of the cause, the disease itself generally leads to valvular "incompetence"—which may or may not be accompanied by *stenosis*.

Incompetence refers to the failure of one or more of the four valves to close properly. In consequence, blood is allowed to flow backward. Mitral incompetence and aortic incompetence are the most common forms. Pulmonary incompetence and tricuspid incompetence occur less often. In mitral incompetence blood regurgitates into the left atrium during systole. In aortic incompetence blood regurgitates into the left ventricle during diastole.

Stenosis is a narrowing of any canal. With respect to the heart, it refers to the narrowing of the opening of a valve. The most commonly afflicted valves are the aortic and mitral valves.

Whenever incompetence or stenosis arises, the heart necessarily works harder in order to satisfy bodily requirements for blood. In time, one or more chambers may dilate (enlarge their opening). In an effort to compensate for the additional work requirement, the chamber hypertrophies (thickens). In this way—during mitral stenosis, for example—the left ventricle may force a sufficient quantity of blood into the lungs. So too, in aortic valve incompetence, a hypertrophied left ventricle may eject a sufficient quantity of blood into the systemic circulation.

But should the difficulty persist—assuming heart failure and death do not follow—the stenosis or incompetence may become so severe that it is no longer possible for the heart to compensate by adding on to the musculature (that is, by further hypertrophy).

In *uncompensated mitral valve stenosis* the regurgitating blood may become dammed up in the lungs, since it cannot pass from the left atrium into the left ventricle. The liver becomes congested, fluid accumulates in the abdominal cavity (ascites), edema is evident, the extremities become cold, and the cheeks become flushed. Long before these symptoms show up, however, it is possible to remedy the situation. In mitral stenosis the narrowed valve may be slit. Other surgical procedures may be used in the event of incompetence without stenosis.

Uncompensated aortic valve incompetence results in a general deficiency of oxygen in the systemic arteries. Dizziness, tinnitus (buzzing or ringing in the ears), and flashes of light may be experienced. Here too it is often possible to ward off the symptoms by using surgical procedures.

Congenital Heart Diseases

Any disease which is seen shortly after birth, and may be attributed to having been established while the fetus was developing in the uterus, is termed

STENOSIS. *Constriction or narrowing* (G.).

MITRAL VALVE. *The two-cusped valve between the left atrium and left ventricle of the heart* (G. mitra, *headband resembling a bishop's miter*).

ASCITES. *An accumulation of fluid in the abdominal cavity, associated most often with heart failure, cirrhosis of the liver, kidney failure, or protein starvation* (G. askos, *a bag*).

Functional Disorders of the Circulatory System

a *congenital disease*. Improperly formed valves, a perforation (opening) in the wall between the atria or ventricles, and any other abnormal anatomical or functional characteristic falls into the category of congenital heart disease (Figure 27.1). Often, it is possible—using open heart surgical procedures—to correct the abnormality.

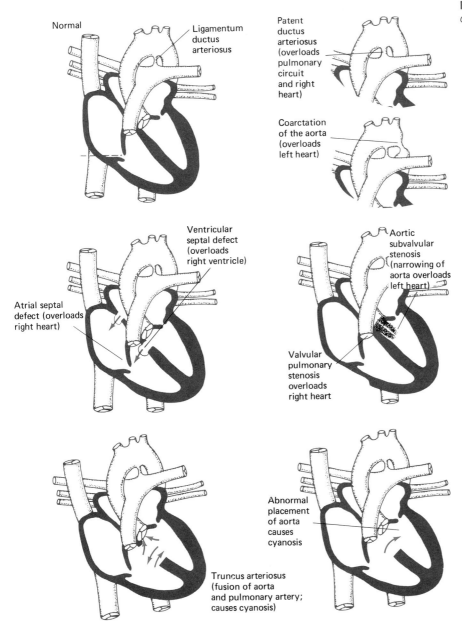

FIG. 27.1. *Examples of congenital heart defects.*

Hypertensive and Arteriosclerotic Heart Diseases

Abnormally high systemic blood pressure is a reflection of additional work which is being demanded of the heart in getting the blood pumped through the body. Similarly, an overly high pulmonary arterial blood pressure is a reflection of an excessive work burden which the heart assumes in getting blood through the lungs. In either of these cases—excessive systemic or pulmonary pressure—one or more chambers of the heart may dilate and/or hypertrophy in consequence.

Hypertension may be due to a number of different causes, as was discussed earlier. Arteriosclerosis, which was also described earlier, is just one of several possible causes of hypertension. In a very high percentage of cases of arteriosclerotic and hypertensive heart disease, the individual's blood cholesterol level is well above normal.

Coronary Heart Disease

ANEURYSM. A weakened area in the wall of the heart or an artery, causing the area to pulsate painfully with each contraction of the heart (G. ana, *up*; eurys, *broad*).

An excessive amount of blood pressure in the coronary arteries may cause the blood vessel to rupture and the blood to hemorrhage. Alternatively, even if the blood pressure is normal, there may be a weak spot (an *aneurysm*) in a coronary vessel—an aneurysm which can be broken easily, resulting in hemorrhage. In either of these cases, or in other cases where the coronary vessels are diseased—whereby blood clots (thrombi, emboli) or any other kind of clot (tissue or fatty emboli) block the circulatory passageway or restrict the circulatory passageway—the individual is said to have a *coronary disease*.

Coronary heart disease may make itself known suddenly and violently—in the form of a *heart attack*. Or it may manifest itself more gradually—in the form of *angina pectoris*.

Angina pectoris is a warning to an individual from an overtaxed heart. We might say, somewhat fancifully (but accurately) that the heart is crying for oxygen and begging for an easier way of life. The cry may be triggered while an individual is at rest, but usually an emotional or physical strain is needed. The angina is felt as a spasm of the coronary arterioles or the heart muscle itself. The afflicted individual "hears the cry" in the form of pain, which arises in the chest and radiates along the left arm to the small finger and ring finger.

The "attack" may be halted with nitroglycerine tablets, which are swallowed or placed beneath the tongue. Nitroglycerine induces a reduction in blood pressure and slows the flow of blood through the heart.

HEMORRHAGE. Loss of blood from blood vessels (G. haima, *blood*; rhegnynai, *to burst*).

INFARCTION. A region of dead tissue due to a loss of blood flow.

Angina pectoris differs from a heart attack in that there is no major noticeable aftereffect. A heart attack is often due to a coronary occlusion (blockage of coronary vessels by an embolus or thrombus) and it may be precipitated by the rupture of an aneurysm wherein hemorrhage follows, resulting in blood clotting. The aftereffect of a heart attack is a *myocardial infarct*—the death of those cardiac muscle cells which were seriously afflicted with an oxygen deficiency. The dead cardiac cells are eventually replaced by fibrous scar tissue, which can be identified histologically.

During the heart attack myoglobin is released from broken cardiac muscle cells into the urine, where its presence for a period of about 12 hours is a telltale sign of the heart attack. Myoglobin, an important respiratory protein, is described on page 572.

Angina pectoris is not the only disorder that causes chest pain. Pain in the chest may be due to any one of various causes, including sore ribs, sore muscles, inflammation of the pleural coverings of the lungs, damage to a spinal nerve under a rib, and myocardial heart disease. These sources of pain, unlike the pain of angina pectoris (reduced blood flow through the coronary arteries), fail to respond to nitroglycerine.

Pericardial Diseases

The pericardium is the outer connective tissue covering over the myocardium. Like the endocardium, this covering may become inflamed (pericarditis) or diseased in some other sense (pericardosis). In pericardosis, the pericardium may become infiltrated with fat or fibrous material. As in endocardial heart disease, the disease may spread into the musculature (myocardium).

PERICARDIUM. *The sac that encloses the heart and the roots of the large blood vessels* (L. peri, *around;* G. kardia, *heart*).

Myocardial Diseases

Myocardial disease affects the musculature and is not associated significantly with the preceding five kinds of heart disease. That is, the myocardium is diseased, but the arteries and valves are relatively healthy, there is no evidence of a congenital malformation, and the pulmonary and systemic blood pressure is near normal. Of course, as the reader might now correctly surmise, it may be very difficult for a physician to distinguish myocardial diseases from other forms of heart disease. How is this distinction made in many cases?

To begin with, a myocardial patient may not even have the symptoms of myocardial disease. Such patients merely have what are called *presenting signs* — signs which present themselves as being potential symptoms of heart disease. An X-ray picture may suggest cardiac enlargement. Or an ECG may appear abnormal.

An individual with a well-developed myocardial disease, however, usually has all, or most, of the following symptoms: (1) cardiac enlargement; (2) certain kinds of cardiac arrhythmia; (3) certain kinds of chest pain; (4) congestive heart failure; and (5) miscellaneous symptoms such as murmurs and syncope (fainting).

The cardiac enlargement of myocardial disease may be of the dilated or the hypertrophied type, or both. But in the case of myocardial disease, the dilation and hypertrophy may (and often do) occur although the valves are normal and there is virtually no systemic hypertension. On autopsy, the hypertrophied muscle usually displays an abnormal organization of the muscle bundles.

The cardiac arrhythmia which most often characterizes a myocardial

AUTOPSY. *An examination of the body after death* (G. autopsia, *seeing with one's own eyes*).

ARRHYTHMIA. *Literally, "without rhythm."*

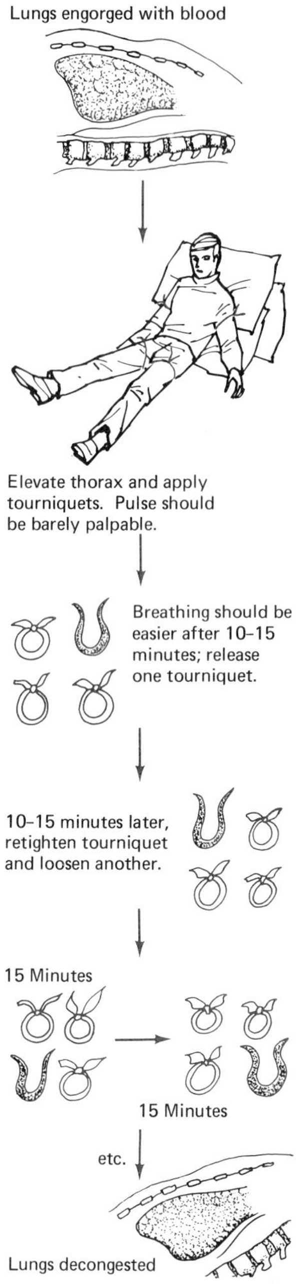

disease is one which is heard as a gallop rhythm, somewhat like the sound of a galloping horse. It is a rhythm with extra (abnormal) sounds. In addition, extrasystoles may occur, as well as short, violent (paroxysmal) periods of abnormally rapid heartbeating (tachycardia) or long, sustained periods of tachycardia. Whether tachycardia is occurring or not, the myocardial patient's pulse commonly is felt as a "pulsus alternans"—a pulse that is alternately strong and weak.

The chest pain of the myocardial patient is frequently different from the "classical" case of angina pectoris. In fact, one or two different kinds of pain may be experienced. The first kind—a pleuritic chest pain—is associated with an abnormal, frictional rubbing of the membranous connective tissues outside of the heart, but within the chest, against the chest wall. This rubbing produces a telltale sound which can be easily heard with a stethoscope.

The other type of myocardial chest pain is of a coronary nature. That is, like angina pectoris, it is associated with the heart itself. Unlike angina pectoris, however, the myocardial chest pain usually occurs at the end of a fatiguing experience, rather than during the experience. Also, it may last 30 to 60 minutes or longer, rather than a few seconds or minutes. Moreover, it is not completely relieved by nitroglycerine.

The congestive heart failure of myocardial disease is essentially similar in character to congestive heart failure which arises from other forms of heart disease. By congestion is meant an accumulation of blood which the heart fails to discharge. Usually the congestion occurs in the left heart. Less often it occurs in the right heart, and sometimes it occurs in both halves of the heart.

Left-sided congestive heart failure is associated with, or confirmed by, a gallop rhythm, pulsus alternans (alternating strong and weak pulsations), and a congestion of the lungs. Lung congestion occurs as a result of the backing up of blood which the left heart fails to pump into the systemic circulation. When this situation becomes critical the heart will cease to function. Figure 27.2 illustrates several emergency steps which must be taken if life is to be sustained, should the critical stage of left-sided congestive heart failure develop at a time when medical help is not available immediately.

Right-sided congestive heart failure is associated with a number of features, including the following: (a) an enlarged liver (hepatomegaly), as a result of the backing up of blood which the right heart fails to pump into the pulmonary circulation; (b) distension of the neck veins, also a result of the backing up of blood; and (c) peripheral edema—a collection of fluids in the interstitial spaces of the body, particularly in the face and extremities.

Figure 27.3 illustrates emergency treatment of an individual whose heart has stopped beating for reasons which are unknown at the time, but may be due to right or left heart failure (or both) or to a heart attack.

Myocardial diseases can be broken down into two major classes, primary and secondary.

Primary myocardial disease is also known as idiopathic myocardial disease, because the cause of this disease is unknown. It occurs approximately

FIG. 27.2. *Emergency treatment of left-sided heart failure.*

Functional Disorders of the Circulatory System

three times as frequently as does secondary myocardial disease. Often it occurs suddenly, without warning. It may occur, for example, in the last three months of pregnancy, or in individuals who have recently recovered from a viral illness or an illness associated with diarrhea, fever, a skin rash, or an inflammation of the pericardium or the pleural membranes.

Secondary myocardial disease may be caused by any one of a large number of different things, such as: (1) a general infection; (2) a nutritional disorder, such as kwashiorkor (a severe protein deficiency), a vitamin deficiency, certain forms of anemia, and deficiencies of potassium and magnesium; (3) endocrine disorders, such as hyperthyroidism, acromegaly, Cushing's syndrome, aldosteronism, and pheochromocytoma; (4) neurologic disorders (for example, muscular dystrophy); (5) collagen diseases — diseases in which the collagen of connective tissues is improperly developed or is destroyed — such as rheumatic arthritis, rheumatic fever, and scleroderma; and (6) physical causes, such as X-ray radiation, toxic conditions such as uremia and alcoholism, or trauma — as happens often following a severe auto accident.

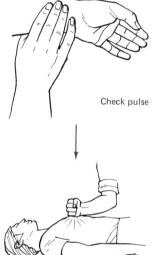

Check pulse

Thump chest 5-6 times; this may trigger a heartbeat

Clear mouth, blow air into lungs 12-15 times per minute

HEART FAILURE

Heart failure is defined as a condition in which the heart is unable to accomplish the work it is called upon to do. The fundamental physiological abnormality is a failure of the myocardium to contract forcefully enough to muster up an adequate blood pressure and/or cardiac output.

Heart failure is characterized by labored breathing, fatigue (even while at rest), and an abnormal, or pathological, retention of sodium and water by the kidney — a retention which results in systemic edema and/or pulmonary edema. In Table 26.1 the reader may note that a smaller percentage of cardiac output passes through the kidneys and a larger percentage is needed for the heart and skeletal muscles.

There are many and various causes of heart failure, including: (1) any chronic strain upon the heart, as for example, heart disease or obesity; (2) any chronic strain upon the lungs, as for example, emphysema; and (3) chronic kidney disease, wherein excessive amounts of body fluids are retained and are imposed as an excessive burden upon the heart.

Push sternum down one or two inches, 60 times per minute

FIG. 27.3. *Emergency treatment of heart arrest.*

What physiological characteristics break down as the heart tends toward a failing condition? To begin with, the cardiac output may be normal when an individual in the early stages of heart failure is at rest. Unfortunately, however, the peripheral resistance of the circulatory system is greater than normal. It should be recalled that peripheral resistance is the opposition offered by constricting arterioles and capillaries to the flow of blood. From the following formula it can be seen that if the cardiac output is normal or above normal, and if the peripheral resistance is increased, the arterial blood pressure must increase — if the cardiac output is to remain normal:

arterial pressure = cardiac output × peripheral resistance

Summary

It is important to realize that an increase in peripheral resistance is reflected as an increase in *diastolic* blood pressure. This is shown in the systolic/diastolic ratio when a measurement is made of the systemic blood pressure.

Because of the increased peripheral resistance—which is caused by the edema that characterizes heart failure—the heart must work harder. Unfortunately, because of this additional work, a greater amount of oxygen is needed. From the Fick equation

$$\frac{\text{ml } O_2}{\text{min}} = \frac{\text{ml blood}}{\text{min}} \times \left(\frac{\text{ml } O_2}{100 \text{ ml arterial blood}} - \frac{\text{ml } O_2}{100 \text{ ml venous blood}} \right)$$

CARDIAC OUTPUT. *The volume of blood ejected in one minute from either ventricle of the heart.*

it may be noted that an increase in oxygen consumption demands either that the cardiac output be increased or that more oxygen must be given by arterial blood to the tissues before the blood enters the veins. This was discussed on page 535.

You may also recall that there are upper limits to cardiac output and to the amount of oxygen that can be extracted from arterial blood as this blood passes through capillary spaces in the tissues. For a failing heart, these limits are reduced considerably.

As the limits in oxygen extraction and cardiac output are being reached, the arterial pressure may increase in accordance with the earlier equation (arterial pressure equals cardiac output times peripheral resistance). This increase in pressure may speed up the flow of blood to satisfy the needs of the body for oxygen. But, again, there is a limit to arterial pressure—a limit which imposes an excessive strain upon the heart. Unless corrections can be made blood will flow more slowly through the kidney. As a result, more sodium and water will be retained, urea will rise to a toxic level in the blood, and the heart will fail altogether; the result of heart failure, therefore, is often death.

SUMMARY

1. Shock represents a severe disturbance of the autonomic nervous system. It is characterized chiefly by a pale, cold, sweaty skin, a weak rapid pulse, and mental confusion.
2. Fainting occurs when the brain receives less than about 60% of its normal share of blood, which is usually about 750 milliliters per minute in an average adult.
3. Edema occurs when the normally negative interstitial fluid pressure rises to a value near zero. It is characterized by an accumulation of fluid between cells in tissues.
4. Hypertension, an elevation of systemic arterial pressure, is usually caused by some abnormality in the nervous, endocrine, excretory, and/or cardiovascular system.

5. Atherosclerosis, a form of arteriosclerosis (hardening of the arteries), is characterized by a faulty deposition of fibers and fats in the intima of the arteries.
6. Cardiac enlargement and heart failure are common features of long-standing heart disease.
7. Endocardial diseases usually involve one or more valves of the heart and/or the lining of the chambers. Valvular incompetence or stenosis are common characteristics of these diseases, which may arise congenitally, or be caused by an infectious agent, an abnormal thyroid gland, or something else.
8. In a congenital heart disease, the valves and/or attached vessels are improperly formed before birth.
9. Hypertensive or arteriosclerotic heart disease is caused by a chronically high blood pressure.
10. Coronary heart disease is caused by a faulty circulation of blood from the coronary arteries into the musculature of the heart. The most severe result of this disease, a heart attack, may be preceded over a long period by episodes of angina pectoris (chest pain). Angina can be relieved by nitroglycerine, but the disease itself generally calls for a different way of life.
11. Pericardial disease affects the outer connective tissue of the heart muscle.
12. Myocardial heart disease affects the muscle itself and usually results in left-sided congestive heart failure and/or (less often) right-sided congestive heart failure.
13. Heart failure is a disease in which the heart is unable to furnish enough oxygenated blood for the body's needs.

REVIEW QUESTIONS

1. Are there any characteristic distinctions—aside from unconsciousness—between shock and fainting?
2. Describe the osmotic alterations that bring about edema.
3. Define: hypertension, arteriosclerosis, cardiac hypertrophy, valvular incompetence, stenosis, congenital heart disease.
4. Write the word "hypertension" and encircle it with words and phrases that denote contributory or causal factors to it.
5. How is atherosclerosis distinguished as a particular form of arteriosclerosis? Describe how atherosclerosis comes about.
6. How does angina pectoris differ from a heart attack? What drug is helpful in relieving the cause of angina pectoris?
7. What organs of the body does blood back up into in left-sided congestive heart failure? in right-sided congestive heart failure?
8. Discuss heart failure in terms of cardiac output, oxygen extraction from blood, frequency of heartbeat, and oxygen requirements.

28 The Respiratory System

COMPONENTS OF THE RESPIRATORY SYSTEM
MECHANISM OF BREATHING
    Alterations in breathing
FUNCTIONS OF THE UPPER RESPIRATORY TRACT
FUNCTIONS OF THE LOWER RESPIRATORY TRACT
EXCHANGE OF OXYGEN AND CARBON DIOXIDE
BREATHING CAPACITY
LUNG DISEASES AND BREATHING PROBLEMS
CONDITIONS FOR NORMAL RESPIRATION
BLOOD BUFFERING

In the course of obtaining energy, a living cell consumes oxygen and gives off carbon dioxide and water. This process is called *internal respiration*.

Related to internal respiration is the process of *external respiration*, also called *ventilation* or *breathing*, the process in which we inspire and expire air.

During the inspiratory phase of breathing, air which is rich in oxygen and poor in carbon dioxide enters the lungs. Here, the oxygen is picked up by hemoglobin, an iron-containing protein in the red blood cells. After the newly-oxygenated blood leaves the lungs (the pulmonary circuit), it enters the systemic circuit. Here, as the red blood cells pass through the capillary beds, the hemoglobin gives up a certain amount of oxygen. At the same time, the carbon dioxide which is produced through internal respiration diffuses into the capillary blood. This waste product, upon arriving in the lungs, is sent out of the body during the expiratory phase of breathing.

This chapter is concerned with the respiratory system, which carries out the process of external respiration. We begin with a description of the components of this system and a discussion on the mechanism and regulation of breathing. These topics lead to discussions on how oxygen and carbon dioxide are exchanged in the capillaries. Next, we specify how the functioning of the breathing apparatus is evaluated in terms of the air capacity of the lungs and how rapidly one is able to breathe in and out. Difficulty in breathing may disturb the normal hydrogen ion concentration of the blood. Conversely, disturbances of the hydrogen ion concentration due to certain metabolic problems may be corrected by the respiratory system. Hence, the topic

565
The Respiratory System

of buffering is discussed. The following chapter discusses respiratory problems which may arise due to a blood abnormality.

COMPONENTS OF THE RESPIRATORY SYSTEM

The respiratory system consists of the following parts, which are shown in Figures 28.1 and 28.2: (1) central and peripheral neural structures through which the rate and depth of breathing are controlled; (2) the breathing musculature, through which the size and shape of the thorax (chest) are altered to bring air in or out; (3) an upper respiratory tract, outside of the thorax,

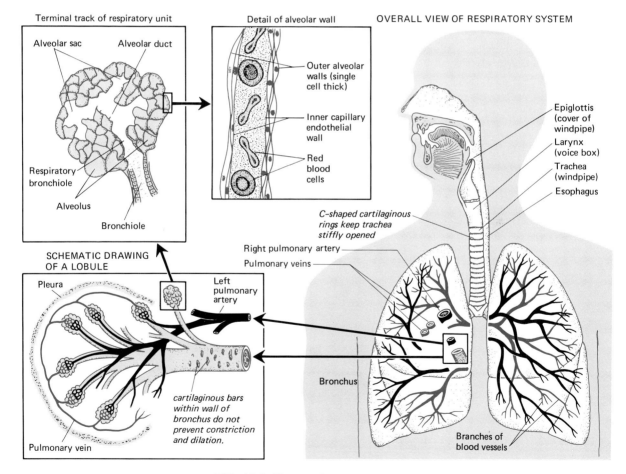

FIG. 28.1. *The respiratory system.*

Mechanism of Breathing

through which air is brought into and out of the lungs; and (4) a lower respiratory tract, mostly within the thorax, through which oxygen and carbon dioxide are exchanged.

The lower respiratory tract consists of the following parts, which are shown in Figure 28.1: (1) a trachea, for bringing air into the thorax; (2) bronchi (tracheal branches), which regulate how much air will enter or leave the lungs; and (3) alveoli, tiny air sacs in which ventilation takes place.

TRACHEA or **WINDPIPE**.
The cartilaginous tube which conveys air from the larynx to the bronchi (G. tracheia, *windpipe*).

MECHANISM OF BREATHING

The filling and emptying of the lungs is a passive process. It follows the active process of enlarging the thorax. The *thorax* is enclosed and bounded by (a) the tissues of the neck, (b) the ribs and their intervening costal cartilages and intercostal muscles, (c) the spinal column and sternum (breast-

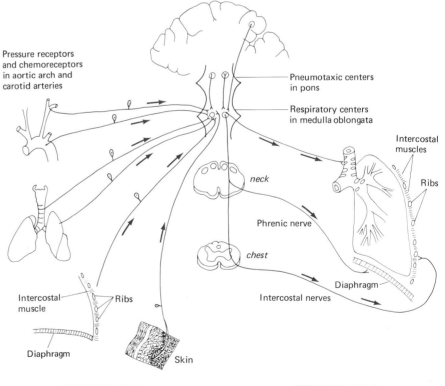

FIG. 28.2. *Regulation of breathing.*

AFFERENT PATHS   EFFERENT PATHS

bone) and (d) the diaphragm, a dome-like sheet of skeletal muscle that separates the thoracic cavity from the abdominal cavity below.

The inner wall of the thorax is lined by a *pleural membrane* which is composed of two layers. The outer, *parietal* layer runs along the entire inner surface of the thorax. The inner, *visceral* layer reflects itself inwardly to cover the lungs. Between the parietal and visceral layers is a thin, moist *pleural space* in which the pressure is relatively low — about 4 millimeters mercury below atmospheric pressure.

When air is to be taken into the lungs the scalene, external intercostal, and diaphragm muscles contract. These muscles contract as part of a complex reflex action which occurs through the medulla oblongata and pons (Figure 28.2). The contraction of the scalene and intercostal muscles causes the rib cage to enlarge. The diaphragm, which bulges upward after expiration, flattens down when it contracts, causing the thoracic cavity to enlarge. As the thorax enlarges, the pressure between the parietal and visceral pleura decreases to about −8 millimeters mercury (Figure 28.3). This drop in pressure, or vacuum, draws air into the lungs, filling them much like a balloon.

When air is to be released from the lungs (*expiration*), the muscles between the ribs and within the diaphragm relax and the internal intercostals contract, causing the thorax and lungs to decrease in size.

The scalene, intercostal, and diaphragm muscles are used for routine breathing. Other muscles are used for special breathing actions at one time or another (Figure 28.4).

PLEURA. Serous membranes that cover and envelop the lungs (G. pleura, rib).

INTERCOSTALS. The muscles between the ribs which help to enlarge the thoracic cavity (L. inter, between; costa, rib).

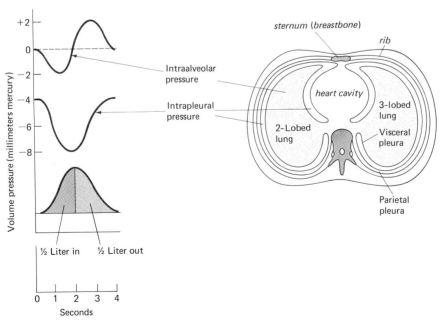

FIG. 28.3. Pressure-volume relations in normal breathing.

Alterations in Breathing

The rhythm, rate, and depth of breathing are subject to change depending on various physiological and psychological influences. In normal breathing the rhythm, rate, and depth of breathing are triggered in response to the levels of oxygen and carbon dioxide in the blood. Superimposed on this normal reflex action are other reflex activities and voluntary, conscious breathing efforts.

Among the other reflex activities are: sneezing; swallowing; coughing; hiccupping; reactions to internal or external temperature changes; and reactions to varying levels of oxygen, carbon dioxide, and blood pressure.

Sneezing arises in response to such things as nasal irritations, which may be due to odors or sudden temperature changes from the outside; virus infections; or tears from the nasolachrymal sacs on the inside of the nose. In sneezing, there is a deep inspiration followed by a forced expiration.

In *swallowing*, food contacts the pharynx and causes the glottis to close.

NASOLACHRYMAL. Pertains to the nose, tear glands, and tear ducts (L. nasus, *nose;* lacrima, *tear*).

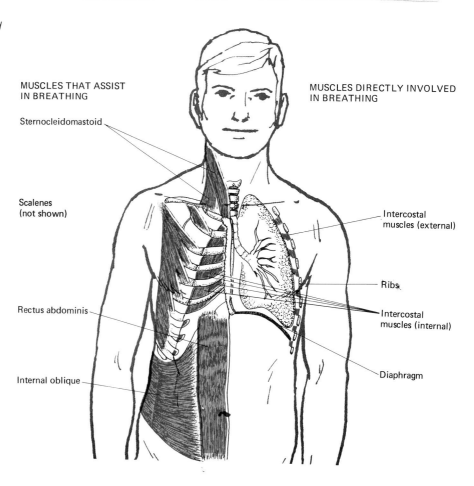

FIG. 28.4. Muscles used in breathing.

The glottis is the passageway through the larynx, a passageway which is constricted by contracting the musculature of the laryngeal wall. When the glottis closes, neither food nor air is allowed to enter the trachea, which itself always remains open. In short, breathing is momentarily suspended when we swallow.

In *coughing,* the glottis closes momentarily due to irritation of the larynx or trachea—perhaps by food that "went the wrong way" or by infectious respiratory agents. Air pressure builds up within the lungs, and the glottis then opens suddenly, releasing a blast of air, often carrying mucous material with it.

Hiccupping also involves the glottis. Due to stimulation of stretch receptors within the thoracic or abdominal wall, the diaphragm is induced to contract spasmodically, with the glottis closed. Hiccupping often may be stopped by stimulating certain nerve endings within the pharyngeal walls with one's finger, by holding one's breath as long as possible, or by breathing into and out of a paper bag. Holding one's breath and breathing into and out of a paper bag result in the accumulation of carbon dioxide and a partial depletion of oxygen within the body. Through nerve reflex pathways, these conditions turn off the involuntary mechanism of hiccupping.

An *increase in blood temperature or carbon dioxide,* or a *decrease in blood pressure or oxygen* reflexly trigger quicker, and/or deeper respirations. The converse situations reduce the rate and/or depth of breathing. In the case of temperature, the receptors are mainly in the hypothalamus. The receptors for detecting changes in oxygen (or carbon dioxide) and pressure in the blood are in the carotid and aortic bodies and sinuses, respectively (see Chapter 26).

Voluntary control of breathing can be superimposed upon most of the reflex activities. Holding of the breath, singing, speaking, laughing, crying, yawning, and a number of psychic factors, including fear, excitement, rage, and suspense call for interrupting or prolonging the inspiratory or expiratory phase of breathing as well as the depth of breathing.

GLOTTIS. *The opening between the vocal folds through which air passes into the trachea (G.).*

SPASM. *A sudden abnormal involuntary contraction (G. to wrench).*

FUNCTIONS OF THE UPPER RESPIRATORY TRACT

The upper respiratory tract includes the nose, the nasal and pharyngeal passageways, the larynx, and the trachea.

The *nose* is used passively, most of the time, for letting air into and out of the lungs during breathing. When an especially deep breath is to be taken, the musculature of the nose may be used to enlarge the nostrils, also called nares. When the nose is blocked, as happens sometimes when a person has a cold, the mouth is used for breathing.

The *nasal and pharyngeal passageways* serve to moisten, warm, and conduct the incoming air and to discharge the spent air. These passageways, which extend from the nostrils to the base of the throat, are lined by an epithelium which consists, for the most part, of ciliated columnar cells and goblet cells. Both of these kinds of cells secrete fluids onto their surfaces in the passageway of the throat. As inspired air sweeps by this fluid on its way to the lungs, certain foreign materials in it become trapped in the mucous secretions. The particles and secretions are carried toward the esophagus by the whiplike action of the cilia. Ultimately, they are swallowed or expelled from the mouth.

The *larynx* consists of a cartilaginous framework to which muscles and *vocal folds* (also called *vocal cords*) are attached. Air must pass through this structure during respiration, whether one is vocalizing or not. In the event that one is not vocalizing, the vocal folds are relaxed and there is a large air passageway. During vocalization, the muscles move the cartilages, and in doing so, the tension on the vocal folds is adjusted to bring about a certain quality of sound.

The *trachea* is a tube whose lumen is held rigidly open at all times by C-shaped cartilaginous rings. These rings may be felt by placing a finger on the neck.

The trachea is lined with ciliated cells and mucus-secreting goblet cells. The cilia beat continuously, in whiplike fashion, sending a stream of mucus from the lungs to the opening of the esophagus. The mucus moistens and warms the incoming air. At the same time, it traps some of the foreign particulate matter from the air. The trapped matter is conveyed into the esophagus and is thus prevented from entering the lungs.

CILIA. *Hairlike extensions from the epithelial cells lining the respiratory passageways. The cilia beat about 50,000 times per hour, sending fluid upward into the esophagus.*

VOCAL FOLDS. *Muscular folds in the passageway of the larynx. They can be made to vibrate, thus producing sounds.*

ESOPHAGUS or GULLET. *The tube which carries food from the pharynx to the stomach (G. oisophagos, gullet).*

FUNCTIONS OF THE LOWER RESPIRATORY TRACT

The trachea divides into two *bronchi*, one of which enters each lung. Each bronchus branches repeatedly—about nine times—into tubes of smaller diameter, called *bronchioles* (see Figure 28.1). In this process of transition, the C-shaped cartilaginous rings of the trachea phase out, and are replaced by two or three small bar-shaped cartilages. These bars lend structural support, preventing the tube from sagging and folding. In addition to the cartilaginous bars, the bronchi have a good deal of circularly-arranged smooth muscle. This musculature is controlled by the autonomic nervous system, which brings about two kinds of effects, depending on whether the sympathetic or parasympathetic nerves are being discharged.

During sympathetic stimulation, adrenaline and noradrenaline are released into the smooth musculature. These neurotransmitters inhibit the contractility of the bronchial muscles, resulting in a larger air passageway. This autonomic function allows a richer flow of air when needed—as for example, in exercise.

During parasympathetic stimulation, acetylcholine is released into the smooth musculature. This neurotransmitter imposes an excitatory effect, causing the muscles to contract and bringing about a narrowing of the passageway. The consequent restriction of airflow may be desirable in some situations, as for example, in cold air. Cold air carries more oxygen per unit

NEUROTRANSMITTER. *A chemical substance which is released from nerve fibers to stimulate their target cells.*

volume than warm air. Therefore, it is energetically favorable to reduce the total intake of air in cold weather. By taking in a smaller volume per breath, a sufficient quantity of oxygen is available—and at the same time less energy is consumed in warming up the air.

Alveolar ducts are passageways that are derived from the bronchioles (Figure 28.1). The transition from the bronchioles to alveolar ducts is one in which less and less smooth muscle and cartilage are found. Also, the wall becomes increasingly thin, ultimately giving rise to an *alveolar sac*. The alveolar sac is composed of a cluster of alveoli.

An *alveolus* is a hollow, berrylike structure (Figure 28.1). Its wall is formed mainly of extremely thin, squamous epithelial cells. Macrophages—called *dust cells*—comprise a part of the wall.

In Figure 28.1, you may note that the effectiveness in the exchange of gases is due in large part to the physical relationships of the alveoli to the blood capillaries. The alveoli have a relatively large hollow space and an extremely thin wall. Thus, there is an extensive surface area of alveolar wall exposed to the air space within the alveolar hollow. On the outer side of the alveolar wall—lying between adjacent alveoli—are capillary spaces. Note that these spaces are much wider than the thicknesses of two adjacent capillary walls put together.

MACROPHAGE. *A cell which devours unwanted materials in the tissue.*

EXCHANGE OF OXYGEN AND CARBON DIOXIDE

Approximately 5 liters of blood flow through the lungs every minute in an adult at rest. Harboring about 25 trillion red blood cells, this fluid flows over an enormous surface area, which is ventilated by air in the alveoli.

Without hemoglobin, 100 milliliters of blood would be able to hold about one milliter of oxygen. With hemoglobin, however, blood can carry about twenty times as much oxygen. This may be seen in Figure 28.5.

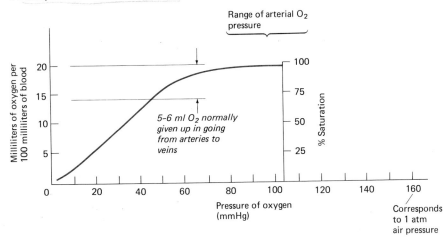

FIG. 28.5. *Amount of oxygen in blood, and percent saturation of blood in relation to the partial pressure of oxygen.*

Two other important facts may be gleaned from the same figure. One is that blood becomes saturated with oxygen even when the pressure of oxygen within the alveoli is less than 90 millimeters mercury. Normal oxygen pressure at one atmosphere in the air is about 160 millimeters mercury. (This pressure—one fifth of atmospheric pressure—is known as the *partial pressure* of oxygen. Oxygen constitutes one fifth of the total composition of air and exerts a corresponding "partial" pressure—one fifth that of the total air pressure.) Thus, there is a large safety factor which ensures adequate ventilation as one climbs to a higher elevation, where the air is rarefied, or as one encounters air—polluted air, for example—which is poorer than normal in oxygen content.

The other important fact is that when arterial blood passes through capillary beds and emerges in the veins, about 4 to 5 milliliters of oxygen are given up by the hemoglobin in each 100 milliliters of blood. This situation is generally true for an average healthy normal individual. Larger and smaller quantities are released during exercise and in heart failure, respectively.

There is a second respiratory protein in the body, called *myoglobin*, which has an even greater affinity for oxygen than does hemoglobin. Myoglobin is a protein found only within skeletal and cardiac muscle cells. It is one fourth the size of hemoglobin and it closely resembles one of the four monomers (units) which make up hemoglobin. Its greater affinity for oxygen means that in environments of low oxygen content, as in active muscle, myoglobin will remove oxygen from hemoglobin. The oxygen then becomes available within the muscle cell for oxidizing any of various metabolites.

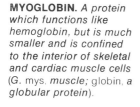

MYOGLOBIN. *A protein which functions like hemoglobin, but is much smaller and is confined to the interior of skeletal and cardiac muscle cells (G. mys, muscle; globin, a globular protein).*

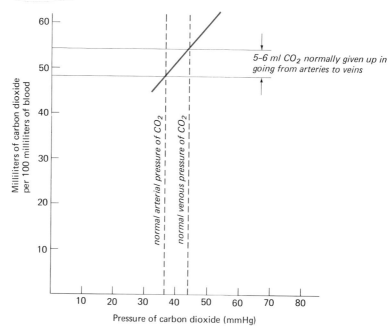

FIG. 28.6. *Amount of carbon dioxide in blood in relation to the partial pressure of carbon dioxide.*

A comparison of Figure 28.5 with Figure 28.6 will show that there is more than twice as much carbon dioxide as oxygen in arterial blood and about three times as much in venous blood. The seemingly excessive quantities of carbon dioxide are needed to provide a reservoir of bicarbonate for buffering the pH of body fluids. The difference—between two and three times as much carbon dioxide as oxygen in arterial and venous blood, respectively—reflects the fact that, in general, for every molecule of oxygen consumed in metabolism, one molecule of carbon dioxide is produced. In this connection, two additional facts are needed to complete the picture of oxygen and carbon dioxide exchange:

First, as oxygenated hemoglobin passes into a capillary it encounters an environment with an increased amount of carbon dioxide. The carbon dioxide reduced the pH in the vicinity of the oxygenated hemoglobin, and this causes hemoglobin to release some of its oxygen. The amount of oxygen released is directly proportional to the amount of carbon dioxide present.

Second, nonoxygenated hemoglobin holds on to more carbon dioxide than oxygenated hemoglobin (Figure 28.6). This fundamental property of hemoglobin enables it to rid the tissues efficiently of waste carbon dioxide. Otherwise there would be a tendency for carbon dioxide to accumulate. Carbon dioxide is, in fact, about twenty times more soluble in body fluids than is oxygen.

CARBON DIOXIDE. *When wetted by water, as happens within the body, carbon dioxide forms carbonic acid ($CO_2 + H_2O \rightarrow H_2CO_3$).*

BREATHING CAPACITY

During rest, about 0.5 liters of air move in and out of the lungs with each breath. Because of this alternating motion—which somewhat resembles

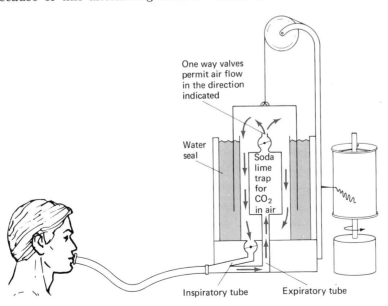

FIG. 28.7. *Apparatus for measuring changes in lung volume.*

TIDAL VOLUME. *The volume of air which comes in and out of the lungs during a single normal inspiration and expiration; about 0.5 liters.*

VITAL CAPACITY. *The total volume of inspiratory and expiratory air of the lungs.*

the tides of the ocean—the total volume exchanged in this way is called the *tidal volume*. An *inspiratory reserve* of 2.5 liters can be added by inhaling as deeply as possible. And an *expiratory reserve* of 2.5 liters can be sent out by exhaling as deeply as possible. On top of this total so-called *vital capacity* of 5.0 liters, there is a *residual volume* of about 1.5 liters which cannot be forced out, no matter how hard one tries.

The volume of air one may inspire or expire is just one of two important measures used in evaluating the performance of the respiratory system. The *rate* of inspiration and/or expiration is another important measure. Figure 28.7 illustrates how the rate may be measured.

LUNG DISEASES AND BREATHING PROBLEMS

A measure of *vital capacity* and the rate at which one inhales or exhales can be used to diagnose restrictive or obstructive lung diseases.

A *restrictive lung disease* is characterized by an inability to expand the chest properly. The neuromuscular apparatus may be at fault. Or there may be air within the thoracic cavity itself due to a puncture in the chest wall, or an opening within the lungs due to some disease. The resulting air in the chest (the *pneumothorax*) cancels out the normal negative pressure of the thorax or pleura and thus restricts the rate and depth of breathing.

ASTHMA. *A violent, painful breathing due to an inflammation of the bronchi and other air passageways (G. a panting).*

EMPHYSEMA. *Usually refers to a disease in which there is overdistension of the air sacs in the lungs; partial obstruction of the bronchi permits air to enter the air sacs but prevents it from leaving (G. emphysan, inflate or blow in).*

An *obstructive lung disease* is one in which air is obstructed in its passage to the alveoli. Examples of such diseases are asthma, emphysema, pneumonia, and pulmonary edema. In *asthma*, the trachea and/or bronchi are inflamed, usually due to an allergic reaction. As a rule, the asthmatic has greater difficulty breathing in than out. In *emphysema*, the alveolar walls lose their elasticity. The alveoli fail to collapse as much as they should during exhalation. Accordingly, they cannot take in as much air with the next breath. In general, the emphysema patient has greater difficulty breathing out than in. Emphysema often begins with a bronchial infection or inflammation, which, usually over the course of many years, gradually destroys the alveoli. *Pneumonia* may be due to a bacterium, a virus, a chemical pollutant or a physical agent such as asbestos. The characteristic response is an inflammation of the alveoli, which results in a pooling of liquid in spaces that should normally accommodate only air. In *pulmonary edema*, a characteristic of certain kinds of heart disease, the heart fails to move the blood adequately through the body and lungs. As in restrictive lung disease, the rate and/or capacity of breathing is abnormal.

CONDITIONS FOR NORMAL RESPIRATION

In order for the respiratory system to carry out its three major functions—oxygenation of venous blood, expulsion of waste carbon dioxide, and blood buffering—four conditions must be met simultaneously: (1) *Adequate ventilation*—a sufficient amount of air must enter and leave the lungs; (2) *Adequate perfusion*—blood must pass unobstructedly through the capillaries within the alveoli; (3) *Adequate alveolar permeability*—oxygen must be able to move into the blood and carbon dioxide into the air sacs; and (4) *Adequate exchange of oxygen and carbon dioxide* must be possible between the tissues and blood.

One or more of these four conditions will fail to hold in a case of obstructive or restrictive lung disease, in other disease situations (such as anemia or carbon monoxide poisoning), and in still other situations, such as drug intoxication or physical stress. Such a failure of the respiratory system generally leads to one or more of the following physiological defects: *hypoxemia*, in which oxygen is below the normal level in the blood; *hypercarbia*, in which carbon dioxide is above normal in the blood; *hyperpnea*, an increased rate or depth of breathing; and *dyspnea*, a difficult, uncomfortable, labored breathing. Dyspnea is present when the amount of maximum voluntary ventilation in one minute is less than 35% of normal. Normal maximum voluntary ventilation (10 liters) runs about twenty times the tidal volume (0.5 liters).

When hypoxemia is due to inadequate alveolar permeability, there is a sharp drop in arterial oxygen when the person exercises. It may be due to a variety of causes, including emphysema, tuberculosis, and pneumoconiosis (a destruction of alveolar elasticity, seen in coal miners, asbestos workers, and cotton workers). In this case, hypoxemia may not lead to hypercarbia. Why is this so? Because the amount of carbon dioxide produced is related to the amount of oxygen consumed. Accordingly, the afflicted person will tend to reduce his or her work load before hypercarbia sets in.

When hypoxemia is due to an inadequate flow of blood through the alveoli, it may lead to hypercarbia and labored breathing. However, whereas the labored breathing (hyperventilation) may correct hypercarbia, it can do nothing for hypoxemia. The reason for this is that the amount of blood which is able to flow through the lungs becomes fully saturated with oxygen, and no amount of hyperventilation can add more to it. On the other hand, the excessive amounts of carbon dioxide that might develop can be released through hyperventilation. Even a small decrease in the pressure of carbon dioxide within the lungs will bring about a large loss of carbon dioxide from hemoglobin, as shown by the steepness of the pressure-volume curve for carbon dioxide in Figure 28.7.

Hypoxemia that is due to inadequate ventilation may be corrected by hyperventilation. If it cannot be corrected by an increase in the rate or depth of breathing, it will lead to hypercarbia and eventually to inadequate ventilation (hypoventilation).

PERFUSE. *To pass fluid through spaces, as, for example, blood through capillaries in the alveolar walls (L. perfusio, to pour over).*

HYPOXEMIA. *An abnormally low level of oxygen in the blood (G. hypo, below; oxys, oxygen; haima, blood).*

HYPERCARBIA. *An abnormally high level of carbon dioxide in blood; also called hypercapnia (G. hyper, above; kapnos, smoke).*

DYSPNEA. *Difficult, painful, faulty breathing (G. dys, bad; pnoia, a breathing).*

HYPERVENTILATION or HYPERPNEA. *Forced breathing resulting in an increase in the rate or depth of respiration, or both; involuntary hyperventilation is a response to excessive carbon dioxide or deficient oxygen in the blood.*

BLOOD BUFFERING

Normal blood is very slightly alkaline, with a pH value of about 7.4. This pH is equivalent to having 40 nanograms (40×10^{-9} grams) of hydrogen ions

NANOGRAM. *One thousandth of one millionth of a gram (G. nanos, a dwarf).*

per liter of blood. The normal range is about 35 to 45 nanograms. In contrast, the interstitial fluid is much more acidic. It displays a normal range of 40 to 70 nanograms and the intracellular fluid may range from 50 to 1000 nanograms per liter.

In general, metabolic reactions tend to increase the acidity of body fluids. This is due mainly to the production of carbon dioxide, which goes into solution to produce carbonic acid ($CO_2 + H_2O \rightarrow H_2CO_3$). At the normal pH of blood (7.4), carbonic acid dissociates to form a strong acid (H^+) and a weak base (bicarbonate, HCO_3^-).

Ordinarily, about 300 liters of carbon dioxide are produced each day by the average person. This is equivalent to 15 grams of hydrogen ions. Normally only about 0.003×15 grams is disposed of by the kidney, and about 0.05 times 15 grams can be excreted maximally. The remaining 95% to 99% must be removed through exhalation.

Failure to rid the body of excessive hydrogen ions leads to a condition called *acidosis*. Excretion of too many hydrogen ions, on the other hand, leads to *alkalosis*.

ACIDOSIS. *An excess of acid in the blood.*

ALKALOSIS. *An excess of bicarbonate (deficiency of hydrogen ions) in the blood.*

Acidosis and alkalosis may have their basis in a metabolic or respiratory problem. *Respiratory acidosis* is characterized by a higher than normal concentration of carbon dioxide in the blood. It may be caused by inadequate ventilation and it may be compensated for by the kidney, through the excretion of hydrogen ions in the urine. *Metabolic acidosis* is characterized by the presence of increased hydrogen ions in the blood without an abnormally high concentration of carbon dioxide (H_2CO_3). It may arise through an oxygen debt, wherein excessive amounts of lactic acid are produced in skeletal muscle. Or it may arise when the kidney fails to excrete enough hydrogen ions. Ordinarily, metabolic acidosis is compensated by exhaling larger quantities of carbon dioxide.

OXYGEN DEBT. *Accumulation of lactic acid in tissues due to inefficient oxidation of metabolites.*

Respiratory alkalosis is characterized by deficient quantities of carbon dioxide in the blood. It may be caused by hyperventilation, as is often the case for untreated diabetics who are trying to compensate for a metabolic acidosis caused by the production of excessive acetoacetic and hydroxybutyric acids. The kidney is called upon to respond to respiratory alkalosis, wherein it sends out bicarbonate (HCO_3^-) and phosphate ions ($HPO_4^=$) which otherwise buffer hydrogen ions. In addition, the kidney acts to retain hydrogen ions.

Metabolic alkalosis is characterized by a decrease in the normal amount of hydrogen ions. It may be brought about by excessive vomiting or diarrhea. In both cases, there is a loss of gastric juice. The blood then becomes alkaline, and the kidney responds by excreting bicarbonate and phosphate ions and by retaining hydrogen ions, as in respiratory alkalosis.

SUMMARY

1. When the scalene, external intercostal, and diaphragm muscles contract, the thorax enlarges. In doing so, a thoracic vacuum is created and

air passes into the lungs. When the same muscles relax, air is expelled from the lungs.
2. Normal rhythmic breathing is regulated through the autonomic nervous system, as are such reflexes as coughing, sneezing, swallowing, and hiccupping. Voluntary control of breathing can be superimposed on autonomic reflex actions.
3. The upper respiratory tract includes the nose, nasal and pharyngeal passages, larynx, and trachea. This tract serves to conduct air into and out of the lungs, to moisten the air, and to produce sounds.
4. In the lower respiratory tract the bronchi and bronchioles can be constricted or dilated to reduce or increase the flow of air into and out of the alveolar sacs, where oxygen and carbon dioxide are exchanged.
5. Blood with hemoglobin carries about fifty times as much oxygen as it would without hemoglobin. Approximately similar volumes of oxygen are exchanged for carbon dioxide in the capillaries, though the total volume exchanged depends on how much activity a person is engaged in.
6. Myoglobin, a protein which is found in skeletal and cardiac muscle cells, has a greater affinity for oxygen than hemoglobin. Myoglobin serves to shuttle oxygen from hemoglobin to the metabolic machinery of the muscle cell. It also acts as a reservoir of oxygen for moments of intense need.
7. A half liter of air moves into and out of the lungs with each breath. About 12 to 15 breaths are taken each minute. During work, an inspiratory reserve of about 2.5 liters and an expiratory reserve of 2.5 liters can be superimposed on the tidal volume of 0.5 liters to allow additional ventilation, and the rate of breathing can be stepped up several fold.
8. In order to oxygenate venous blood adequately, to expel waste carbon dioxide sufficiently, and to buffer blood properly, the respiratory system must be adequately ventilated, it must be properly perfused with blood, its alveoli must be sufficiently permeable to allow an exchange of oxygen and carbon dioxide, and the blood must be able to take on carbon dioxide and give off oxygen to the tissues. Failure in one or more of these respects may lead to a subnormal level of oxygen in the blood (hypoxemia), and/or excess levels of carbon dioxide (hypercarbia) which leads to hyperpnea (hyperventilation) or dyspnea (labored, painful breathing).

REVIEW QUESTIONS

1. Name the muscles that are used to create a negative pressure in the lungs during inspiration. Describe their functioning.
2. Where are the highest nerve centers through which breathing is reflexly regulated?
3. List the anatomical components of the upper respiratory tract. What is the function of each component in breathing?

Review Questions

4. Specify a function for each component of the lower respiratory tract.
5. Draw the curve which relates partial pressure of oxygen to percent of oxygen on hemoglobin. Next, discuss what can be said if the curve were to shift to the right, as happens when the pH is reduced (that is, when the acid content increases).
6. What is myoglobin? Describe briefly the function of myoglobin in respiration.
7. Define: tidal volume, inspiratory reserve volume, expiratory reserve volume, vital capacity.
8. How is an obstructive lung disease distinguished from a restrictive lung disease?
9. What four demands must be adequately met by the respiratory system if venous blood is to be properly ventilated?
10. Define: hypoxemia, hypercarbia, hyperpnea, dyspnea, hyperventilation, acidosis, alkalosis.

The Red Blood Cell System 29

PRODUCTION OF RED BLOOD CELLS
ASSESSING THE PHYSIOLOGY OF RED BLOOD CELLS
    Anemias
        Anemias due to destruction of red blood cells
        Anemias due to insufficient production of red blood cells
        Anemias due to abnormal maturation of red blood cells
    Physiological adjustments to anemia
    Polycythemia

Blood is a specialized connective tissue in the circulatory system. It serves a variety of purposes. Its cellular components were described in Chapter 3. In several chapters since then we have discussed its functions in endocrinology, nutrition, buffering of pH, and the transport of oxygen and carbon dioxide. In the next chapter we will describe the function of blood in carrying waste products through the kidney, and in the last two chapters we will describe the important function of blood in body defense mechanisms. This function is performed mainly by the white blood cells and platelet cells.

This chapter focuses on the physiology of blood with respect to its red blood cells. We begin with a discussion on the production of red blood cells, then describe how to assess the normality of blood, and conclude with a discussion on the disorders caused by a deficiency or excess of red blood cells.

PRODUCTION OF RED BLOOD CELLS

Red blood cells are also known as *erythrocytes*, and their formation is called *erythropoiesis*. In fetal life, the red blood cells are formed in several organs, including the liver, spleen, and later, the bone marrow. After birth, the marrow of most bones takes over the job and still later, as relatively fewer red cells are needed, red blood cells are produced only in the marrow of the

MARROW. *The central or inner soft fatty material, as in bone.*

ribs, vertebrae, breastbone, collar bone, scapulas, pelvic bones, and cranium.

Whenever red bone marrow stops producing red blood cells, it becomes a fatty yellowish marrow. However, whenever the need for red blood cells arises, the yellowish marrow may become red marrow again and start producing red blood cells.

Within red marrow, the process of forming red blood cells takes about seven days. To begin with, a so-called "committed" stem cell or blast cell gives rise to a *pronormoblast*—a large cell which stains deep blue with Wright's stain when blood is smeared as a film on a slide (see Figure 29.1).

PRONORMOBLAST. *A young normoblast (G. pro, before; L. norma, rule; G. blastos, germ).*

The pronormoblast loses its nucleoli and becomes known as a *normoblast*. It differentiates (matures, changes, and develops) into a polychromatophilic cell—a cell which begins to make hemoglobin, and which, as its name implies, has an affinity for various colored dyes in Wright's stain. Red now appears, adding to the blue of the earlier (basophilic) normoblast. Still later, red appears as a dominant color in what is called an acidophilic (eosinophilic) normoblast, a cell in which the nucleus shrinks and then disappears.

EOSINOPHILIC. *Refers to cytoplasmic granules which pick up a red acidic stain called eosin (G. philos, love of).*

After seven days in the marrow, the red blood cell emerges into the circulating blood as a *reticulocyte*. Known by this name because certain dyes (such as methylene blue) can reveal a network (reticulum) of lingering RNA, the reticulocyte remains as it is for a day to a day and a half. It then acquires the characteristics of a fully mature red blood cell: a biconcave disc, about 8.5 micrometers in diameter.

After about 120 days, give or take 20 days, the red blood cell is phagocytized by a cell in the reticuloendothelial system—the system of macrophage cells that lie in the sinusoids (large capillaries) of the liver, lymph nodes, thymus gland, spleen, and numerous other tissues throughout the body.

ASSESSING THE PHYSIOLOGY OF RED BLOOD CELLS

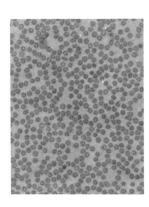

FIG. 29.1. *Red blood cells.*

Most people have a normal supply of red blood cells and hemoglobin. Sometimes, however, there is not enough hemoglobin, a condition called *anemia*, or too many red cells, a disorder called *polycythemia*. An assessment of blood to find out if it is normal, anemic, or polycythemic requires the following steps:

(1) A *blood count* with an electronic cell counter, to determine the number of red blood cells per microliter (mm^3). The normal values are 4.5 million and 5 million for women and men, respectively. Fewer than 3.9 and 4.4, respectively, signify anemia. More than 6 million signifies polycythemia.

(2) A determination of *hematocrit*—the percentage of blood occupied by red blood cells. This value is read from a graduated centrifuge tube, after sedimenting the cells through centrifugation. A normal hematocrit lies between 45% and 50%. Less than about 36% and 41% for women and men, respectively, signifies anemia.

(3) A determination of *hemoglobin concentration* is made through a process called *colorimetry*. A normal concentration is about 14 and 16 grams per 100 milliliters of blood for women and men, respectively. Quantities less than 12 and 14 grams, respectively, signify anemia.

(4) Study of a *blood smear* – prepared as shown in Figure 29.2. The size and shape of the red blood cells are measured and evaluated. The normal size in a smear is about 8 micrometers in diameter. Any abnormally small cells (microcytes) and large cells (macrocytes) are noted. The normal shape is a biconcave disc, appearing to have less color in the center than in the periphery. Abnormal shapes include spheres, which are seen in the disease, spherocytosis; ovals and teardrops, seen in megaloblastic (large blast cells in blood) forms of anemia; target cells, which seem to have a bull's-eye in the center of the disc and appear in certain anemias where hemoglobin is produced incorrectly (thalassemia, hemoglobin C, liver disease); siderocytes, in which nonhemoglobin iron collects in cells that are unable to produce hemoglobin properly in spite of having a sufficient amount of iron; ghost cells – broken cells that have lost their hemoglobin due to cell breakage (hemolysis); agglutinated cells – clumps of cells which aggregate due to a reaction with antibodies in the plasma; sickle cells (indicating sickle cell anemia); and other, variously distorted cells, often indicating hemolytic anemia.

(5) A *reticulocyte count* is also done in the course of evaluating a blood smear. It is expressed as the percent of red blood cells multiplied by the observed hematocrit and divided by 45, the normal hematocrit. A value of 1% is normal. Higher values usually mean that the bone marrow is responding to anemia. The response may arise as an adaptive feature of the body following a rapid loss of blood. Or it may arise after a therapeutic procedure where iron, folic acid, pyridoxine, or vitamin B_{12} is administered in case of a deficiency. Lower than normal values may be due to a disease in the marrow, as in aplastic anemia, or to some other factor which depresses erythropoiesis.

(6) A *measurement of bilirubin in the blood plasma* is carried out to get information on whether there has been excessive destruction of red blood cells – and therefore of hemoglobin – in the body.

Ordinarily, as red blood cells reach old age (senescence) they are selectively destroyed by the macrophage system. The globin (the protein of

COLORIMETER. *An instrument for measuring intensity and quality of color.*

SIDEROCYTE. *A red blood cell which contains iron in forms other than hemoglobin (G. sideros, iron; kytos, cell).*

SICKLE CELL ANEMIA. *A hereditary, chronic hemolytic anemia occurring chiefly in dark-skinned people; characterized by the red blood cells becoming sickle-shaped when hemoglobin is deoxygenated.*

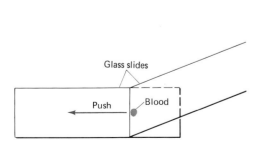
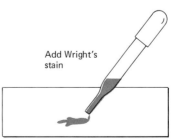

FIG. 29.2. *Preparation of a blood smear.*

hemoglobin) is broken down through hydrolysis and the resulting amino acids are used for metabolic needs. The iron is cleaved from the heme (the remainder of hemoglobin) and is either stored in the phagocytic cell which destroys the hemoglobin, or is sent outside the cell where it is picked up by transferrin, which carries it to the bone marrow, to be used again in erythropoiesis. The remainder of the heme unit, called porphyrin, is converted into biliverdin. Biliverdin is converted into *bilirubin* within macrophage cells.

BILIRUBIN. *The principal pigment of bile, formed by reducing biliverdin, which in turn is formed from broken down hemoglobin* (L. bilis, *bile;* ruber, *red*).

Bilirubin then enters the plasma where it attaches itself to albumin. When this complex reaches the liver, a metabolic reaction occurs in the liver cells. Through the reaction, bilirubin becomes conjugated with a sugar to form bilirubin glucuronide. In this form it enters the bile and is sent into the gut, where bacteria convert the material into urobilinogen. It is in this form that most of the bilirubin is eliminated. Small amounts of urobilinogen, however, are reabsorbed into the bloodstream instead. Now they pass into the kidney, and then into the urine and out of the body.

Normally, the liver and kidney, working together, can keep the bilirubin content of plasma below 200 micrograms per 100 milliliters, the bilirubin glucuronide below 600 micrograms per 100 ml, and the urobilinogen level below 4 milligrams per 24 hours in the urine.

When the macrophage system steps up its destruction to an abnormal rate (called extravascular hemolysis), the liver and kidney may not be able to handle the load. Consequently, up to twenty times as much bilirubin may be present in the plasma, attached to albumin. In addition, much larger quantities of urobilinogen appear in the urine. These larger quantities, incidentally, make the urine turn dark on standing.

(7) A determination is made of the *color* of the plasma after centrifuging the blood. Ordinarily, plasma is a pale straw color. When excessive amounts of red blood cells are destroyed within blood vessels (intravascular hemolysis), the color may become reddish to purple, or even brown.

HAPTOGLOBIN. *An alpha globulin which participates in the removal of hemoglobin breakdown products from the blood* (G. haptein, *to fasten;* L. globus, *ball*).

Normally, about 10% of the red blood cells break down within the blood vessels, rather than within reticuloendothelial cells (extravascular hemolysis). The hemoglobin of the hemolyzed cells becomes attached to a globulin protein in the plasma, a protein called *haptoglobin*. The hemoglobin-haptoglobin conjugate is then destroyed in the reticuloendothelial system.

When the intravascular destruction of red blood cells becomes abnormally high, more haptoglobin is broken down (as the hemoglobin-haptoglobin complex) than can be replaced. Therefore, the body's supply of haptoglobin becomes depleted.

METHEMOGLOBIN. *A form of hemoglobin in which the iron is present as the oxidized (ferric) rather than reduced (ferrous) form* (G. meta, *denoting change;* haima, *blood;* L. globus, *ball*).

As a result, hemoglobin appears in the plasma, as does methemoglobin. *Methemoglobin* is a form of hemoglobin in which the iron appears not normally, as ferrous (Fe^{++}) iron, but rather, as an oxidized ferric (Fe^{+++}) form. (Normal hemoglobin is always in the reduced, ferrous form. When it becomes oxygenated, it remains in the ferrous form and binds loosely to oxygen.)

Together, the hemoglobin and methemoglobin give the plasma a red color. In addition, the methemoglobin is subject to further breakdown—into ferriheme, which combines with albumin in the plasma and gives it a brownish appearance. Finally, because larger than normal quantities of red cells are broken down—with the release of more hemoglobin—the level of bilirubin rises in the blood and hemoglobin may appear in the urine (hemoglobinuria).

All of these changes—change in plasma color, increase in plasma bilirubin, and hemoglobin in the urine—are telltale signs of hemolytic anemia.

FERRIHEME. *An oxidized form of heme, a component of hemoglobin (L. ferrum, iron; G. haima, blood).*

Anemias

An individual is said to be anemic when there are less than normal values for one or more of the following measurements: red blood cells per microliter; hemoglobin per 100 milliliters of blood; or hematocrit. Anemia frequently shows up in an individual who is short of breath, fatigues easily, has pale skin and mucous membranes (like those in the wall of the pharynx) and experiences quivering, fluttering, or murmuring of the heart.

ANEMIA. *A deficiency of hemoglobin or red blood cells.*

There are more than a hundred different forms of anemia. Some forms arise congenitally, others as a result of a dietary or endocrine deficiency. Still others as a result of infection or exposure to radiation.

Based on their cause, all anemias may be placed into one or more of the following four classes: (1) sudden (acute) loss of blood, due to hemorrhage (menstruation, ulcers, injury); (2) excessive destruction of red blood cells; (3) insufficient production of red blood cells; and (4) abnormalities in the development of red blood cells. Each of these causes, except for the first, is described below.

Regardless of its cause, the picture of anemia which is revealed by a blood smear usually takes one of the following three forms. Less often, a combination of features is observed:

(1) Smaller than normal cells (microcytic) with less than normal coloration (hypochromic). The most common cause of this form of anemia is iron deficiency. Less common causes include a hereditary inability to synthesize one of the two polypeptide chains of hemoglobin (thalassemia) and an inability to synthesize hemoglobin properly, in the presence of sufficient iron, unless pyridoxine (vitamin B_6) is administered.

THALASSEMIA. *A hereditary hemolytic anemia which occurs principally among people in Mediterranean families (G. thalassa, sea).*

(2) Normally-colored (normochromic) and normally-sized (normocytic) cells with or without either abnormally shaped red blood cells (poikilocytes) or variously-colored large cells (polychromatophilic macrocytes), or both kinds of abnormal cells. These forms of anemia may arise from any of a very large variety of causes, including chronic infections, cancer, kidney disease, hormonal imbalances, and drugs.

(3) Large (macrocytic) cells in the form of ovals, teardrops, or circles. Various forms of macrocytic anemias are known. Again, various causes may give rise to these forms, including a genetic deficiency, a deficiency of vitamin B_{12} (needed for producing hemoglobin), or a deficiency of folic acid, which is required in protein synthesis.

In the following discussion, anemias are described in terms of some of their probable causes. It should be emphasized, however, that there is great

Assessing the Physiology of Red Blood Cells

overlap among the causes and forms of anemias. This discussion is intended, therefore, mainly as a guide or introduction to the subject.

Anemias Due to Destruction of Red Blood Cells. Anemias which are due to destruction of red blood cells in the body are called *hemolytic anemias.* Any one of several kinds of disturbances may lead to these diseases. Chief among them is an abnormally high rate of phagocytosis in the macrophage system, especially within the spleen. This situation is usually caused by autoimmunity; that is, the immunological system "sees" a "foreign" quality in the red blood cells and begins to destroy them. Aging, a drug, or some other agent may be responsible for making the red blood cells seem foreign in the first place. When excessive destruction of red blood cells occurs in the spleen, as may be determined through a range of tests, the removal of the spleen (splenectomy) often remedies the situation.

Other causes of hemolytic anemia are: (1) rupture of red blood cells, as in sickle cell anemia; (2) destruction of red blood cells which clot abnormally in blood vessels; (3) destruction of red blood cells by toxins from certain bacteria (for example, *Clostridium*), venom from snakes, and poisonous secretions of certain other animals; and (4) destruction of red blood cells which, because of a certain enzyme deficiency (for example, pyruvate kinase), are abnormally structured.

Anemias Due to Insufficient Production of Red Blood Cells. This type of anemia may be due to an iron deficiency, damage to red bone marrow, or a deficiency in the erythropoietin system.

The iron deficiency may come about as a result of insufficient iron in the diet, inadequate intestinal absorption of iron, chronic forms of hemorrhage (ulcers, menstruation), depletion of iron due to its transfer to a developing fetus in pregnancy, inflammation in the macrophage system, or some other cause. In children, iron deficiency anemia is almost always due to inadequate dietary intake. In adults it is more often due to losses of blood.

Anemia due to damage to the bone marrow is usually seen as a reduction not only in red blood cells but white blood cells and platelets as well. Where there is a general reduction in all of the various blood cells, along with less than normal numbers of cells (hypoplasia) in the marrow, the disease is called *aplastic anemia.*

Aplastic anemia may be congenital, or it may be caused by, among other things, chemicals, drugs, radiation, viruses, bacteria, autoimmunity, an inflammation in the pancreas, or cancer. It may also result from a pituitary or thyroid deficiency or a kidney disease.

Some forms of anemia are related to disturbances in the *erythropoietin system,* a hormone system about which little is so far known. At present it is believed that erythropoietin is produced in the kidney, and it is believed that it is stored there in an inactive form. When the blood level of oxygen in arterial blood drops below a critical level, erythropoietin is released into the blood. It is then activated by some blood-borne factor.

In ways which are not yet understood, erythropoietin stimulates the bone marrow to produce red blood cells. When erythropoietin in the blood is low—as happens in kidney disease, in certain endocrine problems (such as

---

PHAGOCYTOSIS. *Literally, "cell eating"* (G. phagein, *to eat;* kyto, *cell;* osis, *activity*).

AUTOIMMUNITY. *An immune reaction in which the body reacts to its own tissues as though they were foreign material; the body tissues, accordingly, are attacked and destroyed by lymphocytes or antibodies.*

ERYTHROPOIETIN. *A hormone which stimulates the production of red blood cells* (G. erythros, *red;* poiesis, *production*).

underactive thyroid or pituitary gland), and in some cases of malnutrition—the production of red blood cells also is reduced.

Anemias Due to Abnormal Maturation of Red Blood Cells. These anemias are usually seen as megaloblastic, microcytic, or combinations of the two. *Megaloblastic anemias*, such as those which arise due to a deficiency of vitamin B_{12} (pernicious anemia) or folic acid, are characterized by a predominance of large blast cells in the marrow relative to the normal blast cells (normoblasts). The circulating red cells are larger than normal, and often oval or irregularly shaped, and there is a larger content of hemoglobin per cell. However, the hematocrit is reduced, as is the cell count.

In *microcytic anemia*—which may be due to iron deficiency, an abnormality in the structuring of hemoglobin (such as thalassemia), or the heme unit of hemoglobin—many of the red cells appear smaller than normal, and less intensely red.

PERNICIOUS ANEMIA. *A deadly form of anemia, due to a deficiency of vitamin B_{12}, which brings about destruction of the central nervous system.*

Physiological Adjustments to Anemia

The body responds to anemia in a number of ways. Chief among these are: an increase in cardiac output; an increase in heartbeat; labored breathing (dyspnea) or orthopnea (labored breathing in any position except sitting or standing); an increase in erythropoietin; and a shift of the oxygen dissociation curve (Figure 28.5) to the right.

The fifth point deserves additional comment. It may be noticed in Figure 28.5 that a shift of the curve to the right means that more oxygen will be unloaded from hemoglobin as blood moves into tissues which have lower

ORTHOPNEA. *A condition in which it is necessary to sit up to breathe more easily (G. orthos, straight; pnoia, breathing).*

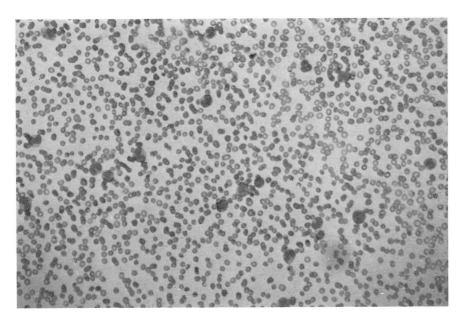

FIG. 29.3. Blood smear from a leukemia victim, showing a deficiency of red blood cells (anemia). White blood cells (which appear dark in the photograph) are present in unusually large numbers.

ambient quantities of oxygen. This shift can be ascertained analytically, upon examining an individual's blood. It is due most likely to a metabolic adjustment through which larger than usual quantities of DPG are produced in the course of metabolizing glucose within the red blood cells.

Polycythemia

POLYCYTHEMIA. *An excessive amount of red blood cells in the blood.*

Three conditions are found in polycythemia. First, there are more than about 6 million red blood cells per microliter of blood. The normal values are 4.5 and 5 million for women and men, respectively. Second, the hemoglobin level rises and remains consistently at about 18 grams per 100 milliliters of blood. The normal value is 14 for women, 16 for men. Third, the hematocrit reaches a value of about 55%. Normal is 45% to 50%.

In spite of these higher quantities, an individual with polycythemia characteristically has lower than normal levels of oxygen in the blood. This may be due to the sluggish movement of the blood through the circulatory system. Headache, dizziness, tinnitus, and thrombosis are common accompaniments of the disease.

There are two basic kinds of polycythemia, referred to as *primary* and *secondary*. *Primary polycythemia* usually arises in people over 40, and reaches a peak incidence between 55 and 60. It is characterized by an unregulated, highly proliferative bone marrow. The cause is unknown. A very wide array of disorders may result from this condition.

PHLEBOTOMY. *Removal of blood from a vein (G.* phlebs, *vein;* tome, *to cut).*

In some cases, the situation can be remedied with radiation. Alternatively, certain drugs are helpful, or bloodletting (phlebotomy) may also be practiced. In bloodletting an incision is made into a vein and blood is allowed to escape. This immediately reduces the red blood cell mass. It also reduces the availability of iron for erythropoiesis.

In *secondary polycythemia,* a wide variety of disorders—such as in the ventilation or perfusion of lungs or in the release of oxygen by hemoglobin—leads to excessive production of erythropoietin. This hormone stimulates the production of red blood cells. The distinction between primary and secondary polycythemia is that in the latter, a disorder leads to an increase in erythropoietin.

SUMMARY

1. Red blood cells are produced from stem cells over a period of about one week in bone marrow, after which an immature red blood cell, a reticulocyte, enters the circulation. After a day or two the reticulocyte becomes a mature red blood cell, which lives for about 120 days and is then destroyed in the reticuloendothelial system.
2. The normality of red blood cell physiology is evaluated by (a) taking a blood count (number of red blood cells per microliter); (b) obtaining a

hematocrit (volume percent of red blood cells in blood); (c) determining the hemoglobin concentration; (d) examining the shape, size, color, and appearance of red blood cells in a smear; (e) obtaining a reticulocyte count; (f) measuring the bilirubin in blood plasma; and (g) evaluating the color of plasma.
3. Any abnormality in the results of these tests can lead to a diagnosis of anemia, polycythemia, or other blood disease.
4. Anemia or polycythemia is present when a person is abnormally low or high, respectively, in any one of the following characteristics: blood count, hematocrit, hemoglobin concentration.

REVIEW QUESTIONS

1. Describe the life cycle of red blood cells.
2. What characteristics of red blood cells are tested or examined to determine whether the cells' physiology is normal?
3. Define: hematocrit, bilirubin, anemia, polycythemia, phlebotomy.
4. Give approximate normal values for the following: blood count, reticulocyte count, hemoglobin concentration, hematocrit.
5. List four kinds of anemia. What is the principal symptom of each?
6. What kinds of vitamin deficiencies are most often associated with anemia?
7. Why is it harmful to have an excess of red blood cells?

30 The Excretory System

THE RENAL EXCRETORY SYSTEM
    The kidneys
        The renal circulatory system
        The uriniferous tubule
    Remaining components of the excretory system
FORMATION OF URINE
    Osmolality
    Physiology of urine formation
    The countercurrent multiplier
    Forming a concentrated or a dilute urine
EVALUATION OF KIDNEY FUNCTION
    Glomerular filtration rate
    Clearance
ROLE OF THE KIDNEYS IN REGULATING THE COMPOSITION AND QUANTITY OF BODY FLUIDS
    Proper acid-base balance
        Chemical buffers
        Role of the lungs in buffering
        Role of the kidneys in buffering
    Balance needed for proper cell membrane behavior
        Sodium balance
        Potassium balance
        Calcium balance
        Magnesium balance
        Water balance
ENDOCRINE FUNCTIONS OF THE KIDNEYS
URINALYSIS AND KIDNEY PROBLEMS
    Urinalysis
        Appearance of urine
        pH of urine
        Specific gravity
        Urinary protein
        Urinary glucose
        Urinary ketones
        Microscopic particulate materials

Kidney disease
Acute glomerulonephritis
Chronic glomerulonephritis
Nephrotic syndrome

Excretion is the process in which undigested residues of food and the waste products of metabolism are passed out of the body. This process is carried out by the skin, intestine, lungs, and kidneys.

The skin usually plays a minor role in excretion, a role which is normally limited to the shedding of worn-out keratinized cells and hair. During periods of heavy exercise, or in hot environments, however, the skin may excrete significant amounts of lactic acid and urea. In fact, at high sweating rates enough urea is eliminated in sweat to introduce serious errors in nutritional studies on nitrogen balance when an investigator fails to take this outlet of urea into consideration.

The intestine's role in excretion, aside from eliminating undigested food residues, is limited mainly to the elimination of bile wastes, which are produced in the liver. These wastes are derived chiefly from the hemoglobin of worn-out red blood cells.

The role of the lungs in excretion was described in Chapter 28, where the discussion centered on the importance of the lungs in ridding the body of waste carbon dioxide—a volatile acid. The lungs also dispose of small quantities of other volatile materials, such as acetone and ammonia, both of which are derived in the course of normal metabolism.

The kidneys play a major role in excretion. It is largely through these organs that the blood is cleared of the following substances: (a) nitrogenous wastes, of which urea (NH_2CONH_2) is the major component, followed by ammonia (NH_3) and uric acid ($C_5H_4N_4O_3$); (b) ionic wastes, especially excessive inorganic ions (for example, sodium, potassium, chloride, and hydrogen), but also organic ions (such as lactate, pryuvate, acetoacetate, and acetate); and (c) excessive amounts of water.

Although the skin, intestine, and respiratory systems participate in excretion, and although each of these organs performs an important role with regard to one or more waste products, this chapter is concerned with the kidney. The kidney is important not only in eliminating wastes but also in buffering the pH of blood, regulating the ionic balance in the body, regulating the body's water balance, and in several endocrine functions.

We begin with an anatomical description of the renal (kidney) excretory system, followed by a description of how urine is formed. We then discuss a special mechanism through which urines of various concentration—from very dilute to very concentrated—may be formed. This mechanism falls under the heading of countercurrent multiplier hypothesis.

We then provide a brief introduction to the principal means by which kidney performance is evaluated. One section focuses on excretion. The next three sections concern the other kidney functions: buffering of blood

LACTIC ACID. *A three-carbon compound which accumulates in overworked skeletal muscle.*

UREA. *A neutral end product of nitrogen metabolism, formed from two molecules of ammonia and one molecule of carbon dioxide.*

URINE. *The fluid excreted by the kidneys (L.).*

pH; regulation of ionic balance; regulation of water balance; and endocrine activities. In concluding this chapter, a brief description is provided of urinalysis and kidney disease.

THE RENAL EXCRETORY SYSTEM

The renal excretory system consists of the following parts shown in Figure 30.1: (1) a pair of *kidneys*, which produce urine from certain components of blood; (2) a pair of *urinary pelves* and *ureters*, which lead the urine out of the kidneys; (3) a *urinary bladder*, in which urine is stored until excreted; and (4) a *urethra*—a duct through which the urine is discharged from the body.

PELVIS. *A basin or basin-shaped cavity (L.).*

The Kidneys

The kidneys are bean-shaped (lentiform) structures, about five inches long in the average adult (Figure 30.2). They are located in the upper abdomen, firmly attached by fascia, beneath the peritoneum, to the dorsal body wall.

Each kidney is enclosed in a thin capsule of connective tissue. Just beneath the capsule, along the convex surface of each kidney, is an area called the *cortex*. Heading into the central, inner core of the kidney is the

CORTEX. *The peripheral portion of an organ, just beneath a connective tissue capsule (L. bark).*

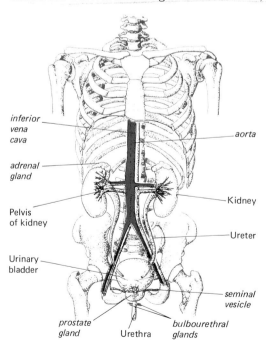

FIG. 30.1. *The excretory system.*

The Excretory System

medulla. At the concavity of each kidney is an indentation, known as the *hilum*. Here one can see a *renal artery*, which comes to the kidney from the aorta; a *renal vein*, which passes from the kidney to the inferior vena cava, and the *pelvis*, which carries urine from the kidney to the bladder by way of a ureter.

A microscopic examination of the cortex shows many globular structures, known as *glomeruli*. A liquid component of blood, known as an *ultrafiltrate*, is filtered out of these glomeruli. This liquid is usually devoid of blood cells and has very small quantities, if any, of large molecules, such as proteins. The ultrafiltrate thus differs from serum (blood plasma minus clotting factors) as well as from plasma.

A microscopic examination of the medulla shows many tubules, known as uriniferous tubules. Each tubule encloses a glomerulus at its proximal end, where the tubule receives the ultrafiltrate. At the distal end of each uriniferous tubule about 1% of the ultrafiltrate emerges as *urine*, which is conveyed through the pelvis and ureter to the bladder, for storage.

The Renal Circulatory System. A basic understanding of how a kidney works requires a clear picture of the route taken by blood into and out of the kidney and the route taken by the ultrafiltrate from a glomerulus into the pelvis. Blood enters each kidney from the aorta by way of a renal artery. The renal artery branches successively into the following arteries, as shown in Figure 30.3: (1) *interlobar arteries*, which lie between lobes of uriniferous tubules; (2) *arcuate arteries*, which run in the form of an arc, between the cortex and medulla; (3) *interlobular arteries*, which run between small lobes (lobules) of uriniferous tubules; and (4) *afferent arterioles*.

MEDULLA. *The central part of an organ (L.).*

HILUM or **HILUS.** *An opening or recess in an organ, usually for the entrance and exit of vessels.*

GLOMERULUS. *A small ball of blood capillaries in the kidney or a ball of gland cells in sweat glands (L. a small knot).*

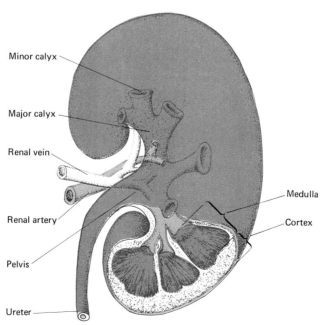

FIG. 30.2. Anatomy of the kidney.

The Renal Excretory System

AFFERENT. Leading toward (L.).

Each afferent arteriole—in heading toward the venous system—branches repeatedly into a capillary network. The numerous capillary branches fold back upon themselves to form the *glomerulus*. Within this glomerulus the many divergent capillary branches of the afferent arteriole gradually converge into fewer and larger outgoing branches which ultimately leave the glomerular knot as a single *efferent arteriole*.

EFFERENT. Leading away from (L.).

The blood carried by the efferent arteriole is thicker than the blood that came into the kidney from the heart. It is thicker because a liquid portion of it—without blood cells and plasma proteins—was filtered out of the glomerulus into the uriniferous tubule.

The thicker, slower-flowing blood of the efferent arteriole passes into one of two different kinds of capillary systems, depending on where the glomerulus lies within the cortex. If the glomerulus lies far from the medulla, in the outer cortical region, the efferent arteriole passes its blood into a capillary which snakes around in all directions within the cortex. Called a *peritubular plexus*, this capillary network serves mainly to exchange wastes and nutrients in the cortex of the kidney.

PERITUBULAR. Refers to haphazardly twisted blood capillaries around uriniferous tubules that are confined to the cortex of the kidney (G. peri, around).

If the glomerulus lies near the medulla, in most cases its efferent arteriole thins out as a capillary which is directed straight to the hilum, into the deepest region of the medulla. After that, it is reflected back again, as a linear tube, into the cortex. The capillary is thus shaped like a hairpin. Called a *vasum rectum*, the capillary serves to exchange nutrients and wastes in the medulla, and, in addition, is important as part of a mechanism through which urines of various concentrations may be formed.

VASA RECTA. Blood capillaries which run parallel to uriniferous tubules deep into the medulla of the kidney (L. vas, vessel; rectus, straight).

After passing through either of these capillary networks, the blood continues on into a series of veins. The veins bear names that correspond to the arteries of the renal circulation (Figure 30.3).

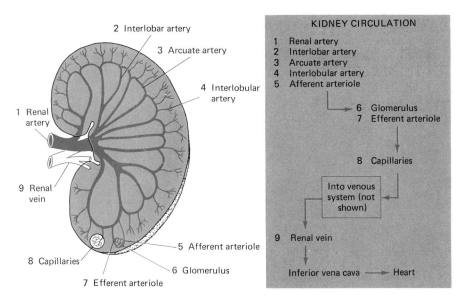

FIG. 30.3. *Renal circulation.*

The Uriniferous Tubule. The *uriniferous tubule* converts ultrafiltrate into urine, which it carries to the pelvis of the kidney. The uriniferous tubule, together with the glomerulus, is called a *nephron*, which is commonly referred to as the functional unit of the kidney. The uriniferous tubule consists of the following five components, which are shown in Figure 30.4:

(1) *Bowman's capsule.* This is a globular structure with an opening at one pole (vascular pole)—through which the afferent arteriole enters and the efferent arteriole exits; and a tubule—called the *proximal tubule*—at the opposite pole (the renal pole). Together with the glomerulus, Bowman's capsule is known as a *renal corpuscle* (Figure 30.5).

The capsule has a double wall. The inner, visceral wall is a very permeable layer through which the ultrafiltrate enters the uriniferous tubule. The outer, parietal wall unites with the visceral wall at the vascular pole, and becomes the proximal tubule at the renal pole. For all practical purposes, it is helpful to think of the outer parietal wall and the proximal tubule as being a funnel which receives a liquid (an ultrafiltrate) from the glomerulus, after the ultrafiltrate passes through the sievelike visceral layer.

NEPHRON. *A renal unit, consisting of a glomerulus, Bowman's capsule, and a uriniferous tubule (G. nephros, kidney).*

VASCULAR POLE. *The region of a renal corpuscle into which blood vessels enter and exit (L. vasculum, vessel).*

RENAL POLE. *The region of a renal corpuscle through which an ultrafiltrate passes into the uriniferous tubule (L. ren, kidney).*

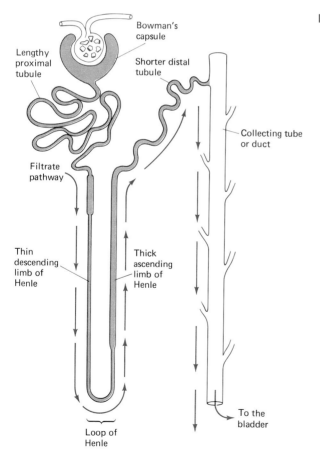

FIG. 30.4. *Schematic drawing of a uriniferous tubule.*

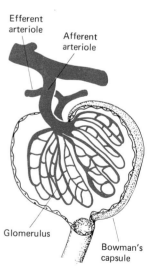

FIG. 30.5. *Renal corpuscle. Filtrate leaves through uriniferous tubule at bottom.*

PARS RECTUM. *Either of the two straight portions of the uriniferous tubule which are attached to the two convoluted portions of the tubule (L.* pars, *a part;* rectus, *straight).*

MITOCHONDRIA. *Organelles in which the bulk of a cell's directly usable source of energy (ATP) is formed (G.* mitos, *thread;* chondrion, *granule).*

HENLE *(1809-1855). A German anatomist among whose principal contributions was his early work on epithelial tissues.*

DISTAL. *Situated further from a reference point (L.* distans, *distant).*

PROXIMAL. *Situated nearer to a reference point (L.* proximus, *nearest).*

(2) *Proximal tubule.* This tubule lies mainly within the cortex, where it is present as a convoluted tubule, a tubule which snakes and twists back and forth. It then straightens out and passes into the medulla as a *pars rectum.*

The function of the proximal tubule is to absorb about 80% of the ultrafiltrate and, in doing so, to make minor but important changes in the composition of the ultrafiltrate. Nearly all of the sugar and many other desirable metabolites, such as amino acids, fatty acids, certain minerals, and vitamins are absorbed by the cuboidal cells which line the tubule.

These cuboidal cells are richly equipped for absorbing desirable metabolites. Their free surface within the lumen of the tubule is greatly increased in surface area by virtue of microvilli. And their metabolic machinery includes a wealthy supply of mitochondria, which provide the cell with the energy it needs to carry out the active transport of solutes from the ultrafiltrate into the cell.

After the material enters the cell it is passed out of the cell, at the perimeter of the tubule, into the interstitial fluid. From here it diffuses toward a capillary, where it is *reabsorbed* into the bloodstream.

(3) *Loop of Henle.* This loop is a three-part structure: the descending pars rectum; a thin hairpin loop with a long descending limb and a short ascending limb; and an ascending pars rectum.

Like the descending pars rectum, the wall of the ascending pars rectum is lined with cuboidal cells. In contrast, the wall of the thin hairpin loop is lined with thin, flattened squamous cells.

The descending limb of this loop passes deeply into the medulla. After passing as deeply toward the hilum as possible, it turns back upon itself, forming the hairpin loop. Shortly after turning back, the ascending limb thickens. Squamous cells give way to cuboidal cells and the tubule passes—straight as can be—as an ascending pars rectum into the cortex.

The function of the loop of Henle—particularly the cuboidal cells of the ascending limb—is to set up an osmotic gradient that will allow the kidney to produce a very concentrated urine.

(4) *Distal tubule.* This tubule, like the proximal tubule, is convoluted. However, it is much shorter than the proximal tubule and its cells are not equipped for absorbing large quantities of solute from the ultrafiltrate. Nevertheless, to a lesser extent, this tubule is committed to the same three functions. One of these functions—reabsorption—was described above.

Another function concerns *ion exchange.* Here, one ion from the ultrafiltrate may be exchanged for a different kind of ion from within the cell of the tubule. Magnesium, for example, might be exchanged for calcium, or iron for molybdenum. In this way the kidney can regulate the composition of certain ions in the body.

The third function—secretion—is important as a mechanism through which the kidney is able to excrete certain substances more rapidly than would be possible through filtration alone. What is the value of this mechanism?

To begin with, it should be realized that only a portion of the liquid fraction of the blood is filtered each time the blood passes through the kidney.

The remainder of the liquid fraction continues along into the capillaries that stem from the efferent arterioles—that is, into the vasa recta and peritubular plexuses.

It follows, therefore, that certain waste substances—which are destined to be excreted—are present in the blood of these capillary beds. These waste substances now pass out of the capillaries into the interstitial spaces. From here they diffuse into the vicinity of a proximal or distal tubule. The waste substance is picked up by the cells of the tubules, through which they are transported and thereafter *secreted* into the lumen of the tubules. From here they continue on, down to the urinary bladder.

(5) *Collecting tubules.* These structures, known also as collecting ducts, receive the ultrafiltrate from the distal convoluted tubule. A single collecting tubule drains numerous distal tubules.

The wall of the collecting tubule is lined by cells that are cuboidal to columnar in shape. Under the influence of two different hormones (antidiuretic hormone and aldosterone) these cells—and to some extent the cuboidal epithelial cells of the distal convoluted tubule—make all but the final adjustment in the concentration of water and solute in the ultrafiltrate.

ANTIDIURETIC HORMONE. *Known also as* vasopressin, *this hormone promotes the retention of water in the body* (G. anti, *against;* diourein, *to pass urine*).

ALDOSTERONE. *A hormone which promotes the retention of sodium in the body.*

Remaining Components of the Excretory System

From the collecting tubules, the filtrate—now called urine—passes successively through the pelvis, ureter, bladder, and urethra, before it leaves the body. In addition to these four components, the excretory system includes a neural component, which is shown in Figure 30.6 and discussed below.

The urinary pelvis. Within this funnel-shaped structure groups of 10 to 15 collecting tubules converge to form ducts of Bellini. Numerous ducts of Bellini, in turn, spill their urinary fluid into the ureter.

BELLINI (1643–1704). *An Italian anatomist who is best known for his discovery of the renal excretory ducts (1662).*

The ureter. This is a tubular structure extending from the pelvis of each kidney to the urinary bladder. A cross-sectional view of the ureter shows a lumen encircled by a mucous membrane (a mucosa). Sandwiched between the mucosa and the outermost wall of the ureter (the adventitia) are three layers of smooth muscle—two that are longitudinally oriented and a third that is circularly oriented.

The function of the ureter is to propel urine into the bladder. It does this through peristaltic action at the rate of about 75 milliliters (2.5 ounces) per hour.

The urinary bladder. This structure is a reservoir for a temporary storage of urine. Like the ureter and most of the urethra, its mucosa is lined with a transitional epithelium and it has a considerable amount of smooth muscle in its wall. Both of these tissues are capable of considerable expansion, which they undergo frequently, especially when one holds back on urination in spite of the need to do so.

TRANSITIONAL EPITHELIUM *appears to have fewer layers of cells when the ureter, urinary bladder, or urethra is distended with fluid.*

When the need to urinate arises, a complex neural reflex is activated, as is shown in Figure 30.6. Both the autonomic and voluntary nervous systems are called into action in adults, whereas in infants the voluntary system is

SPHINCTER. *A circular band of muscle around an opening of an organ (G. sphingein, to bind tightly).*

not yet fully developed. As a result of executing this complex reflex act, the smooth muscle of the bladder contracts, and the urine passes through a sphincter into the urethra.

FORMATION OF URINE

The human kidney is able to produce a urine that ranges in concentration from about 50 milliosmoles to 1400 milliosmoles. It accomplishes this through mechanisms that are described in the next section, but a full appreciation of this topic requires a thorough understanding of osmolality.

Osmolality

In Chapter 1 we described osmotic pressure and in Chapters 20 and 26 we described how osmotic pressure is involved in digestion and circulation. The process itself is one in which water diffuses across a semipermeable

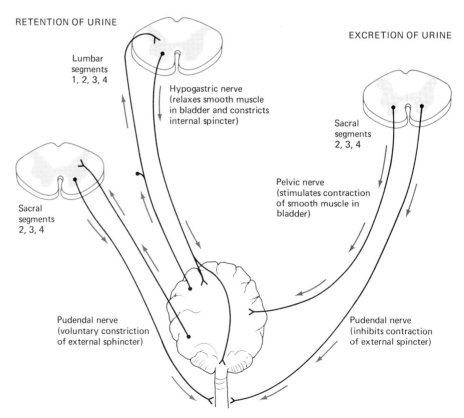

FIG. 30.6. *A highly schematic rendering of the regulation of urination by the nervous system.*

The Excretory System

membrane from a region of high water concentration (low solute concentration) to low water concentration (high solute concentration). The formation of urine with a higher concentration of solutes (mainly sodium chloride) than blood plasma is due to a mechanism which sets up and makes use of an osmotic gradient ranging from a osmolarity of about 300 milliosmoles in the cortex to 1200 milliosmoles deep within the medulla.

What is osmolality? It is a measure of osmotic pressure. One *osmole* is an amount of solute which, when added to a liter of pure water, will depress the freezing point of the water by 1.86°C. A milliosmole of solute in 1 liter of water will depress the freezing point by 1/1000th of 1.86°C. A 300 milliosmolal solution—as in the blood plasma and kidney cortex—freezes at (0.3) (−1.86°C), or −0.56°C.

How is osmolality related to osmotic pressure? A solution of 1 milliosmole per liter of water exerts an osmotic pressure of 0.025 atmospheres at body temperature. Since 1 atmosphere of pressure equals 760 millimeters mercury, an osmotic pressure of 1 milliosmole equals:

$$0.025 \text{ atmospheres} \times \frac{760 \text{ mm Hg}}{\text{atmosphere}} = 19 \text{ mm Hg}$$

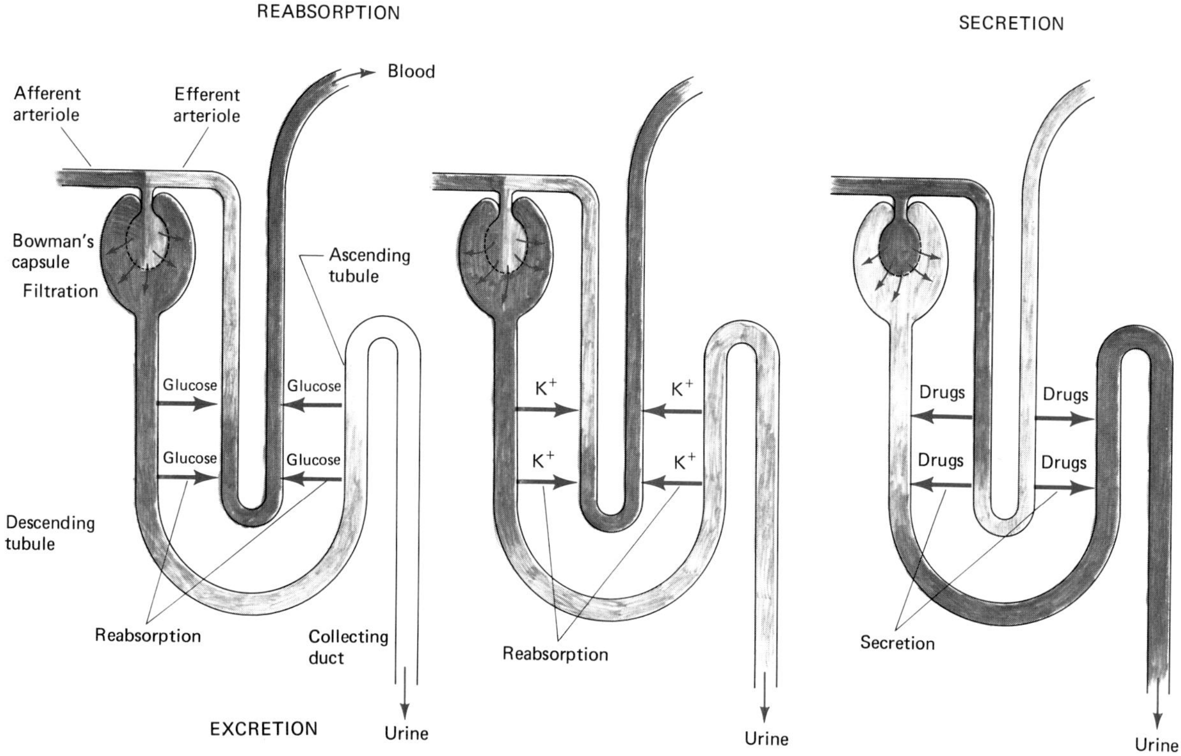

ISOOSMOTIC. *Refers to two solutions—such as an ultrafiltrate and blood plasma—which have an equal osmotic pressure (G. iso, equal; osmos, pushing).*

OSMOLALITY. *Refers to gram molecular weight quantities of solute added to one liter of water.*

FIG. 30.7. *The kidney is able to regulate reabsorption so as to produce a urine more concentrated in potassium and other metabolites than in glucose. Secretion is used for rapid excretion of some substances, as shown here for drugs.*

Formation of Urine

Thus, if a solution of 1 milliosmole were separated by a semipermeable membrane from a solution of pure water in a closed system, as shown in Figure 30.7 (on the previous page), a pressure of 19 mm Hg would be registered on a manometer above the milliosmolal solution at $+37°C$.

As indicated above, the milliosmolal solution will freeze at $-0.00186°C$. In practice, biologists measure and express osmotic pressure in terms of milliosmoles. It is easier to do this than to measure the osmotic pressure manometrically, as is idealized in Figure 30.7.

Physiology of Urine Formation

The process of forming urine begins with *filtration*. Eighty percent of the resulting ultrafiltrate is reabsorbed through the proximal tubule. Solutes and water are absorbed in approximately equal proportions here. Therefore, the osmolality of the ultrafiltrate remains relatively unchanged. The ultrafiltrate is approximately isoosmotic with blood plasma.

Although there is no substantial change in the concentration of ultrafiltrate as it passes through the proximal tubule, certain minor solutes are nevertheless scarcely reabsorbed and others not at all.

The extent to which a given substance is reabsorbed depends upon the substance's threshold. The cells that line the lumen of the tubules determine the threshold. Threshold refers to a concentration limit. When the concentration of a substance in the ultrafiltrate is below its threshold, it is reabsorbed. When it occurs at a higher concentration it is not taken into the cell of a tubule. Rather, it continues along as a component of urine.

THRESHOLD. *The point at which an action begins.*

Some substances are said to be *high threshold* substances. Glucose, sodium chloride, and amino acids fall into this category. These substances are almost entirely reabsorbed, even when present in the ultrafiltrate in fairly high concentration.

Potassium is a substance of *medium threshold*. A lower percentage of it is reabsorbed than substances of high threshold. Urea, uric acid, and phosphates are substances of *low threshold*. Relatively smaller amounts are reabsorbed.

Finally, some materials are said to have *no threshold*. Examples are sulfate and numerous kinds of drugs. These substances are filtered, secreted, and excreted. In short, they are disposed of almost in their entirety as they make a single journey through the kidney. They are simply eliminated—as though the kidney "knows" they should not be kept within the body.

After the isoosmotic ultrafiltrate passes into the pars recta of the proximal tubule, the countercurrent multiplier comes into the picture.

HYPEROSMOLAR. *Refers to a solution with a greater osmotic pressure than some other solution (G. hyper, above; osmos, pushing).*

The *countercurrent multiplier* is a mechanism which has been hypothesized to explain how a measurable concentration gradient of sodium chloride has come about within the kidney. That is, it attempts to explain the presence of a very high (hyperosmolar) concentration (1200 milliosmoles per liter) at the bottom of the loops deep within the medulla, isoosmolality in the convoluted regions of the proximal and distal tubules, and the sometimes very low concentration (hypoosmolality) in the distal convoluted tubule.

The Countercurrent Multiplier

The function of the countercurrent multiplier is to set the stage for forming a concentrated urine. After this happens, antidiuretic hormone and aldosterone come into the picture to determine the final osmolality of the urine. Most physiologists believe that the countercurrent multiplier works like this:

(1) An ultrafiltrate moves down—and then up—the hairpin loop. This constitutes a *countercurrent* (see Figure 30.8).

(2) In passing downward, water leaves the tubule. In doing so, the remaining ultrafiltrate becomes increasingly concentrated. The outward movement of water is a passive process, in response to a higher osmotic pressure outside the tubule.

(3) The higher osmotic pressure outside the tubule is produced by the active transport of sodium from the pars recta of the ascending limb. To establish this pressure, sodium is transported actively from the lumen into the cuboidal cells of the pars recta; the sodium is then passed out of the cells into the interstitial space outside of the tubule. Chloride follows it passively, neutralizing its charge. Water, however, remains within the tubule because the cells of the ascending tubule are impermeable to it.

(4) The concentration gradient of sodium chloride outside of the ascending limb and pars recta permits the establishment of a concentration gradient of water in the descending pars recta and limb of Henle. At the top of the pars recta water is drawn out osmotically. This renders the fluid within the tubule osmotically richer. But as the fluid continues its journey toward the bottom of the loop it encounters a region outside of the tubule which has an even greater osmotic pressure. Additional water leaves. This process continues throughout the journey to the bottom of the loop. In

PASSIVE PROCESS. *An event which does not require metabolic energy.*

ACTIVE TRANSPORT. *The movement of a solute across a membrane at the expense of ATP.*

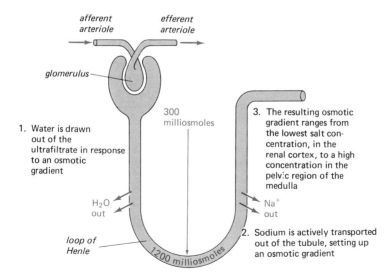

FIG. 30.8. *Creation of an osmotic gradient in the uriniferous tubules. Sodium is removed in the ascending thick limb, which is the only part of the tubule with the metabolic machinery for carrying on the process of active transport.*

consequence, the ultrafiltrate becomes increasingly concentrated, ultimately achieving a value of about 1200 milliosmoles per liter.

(5) At each level of the loop and pars recta there is an approximate similar difference in osmolality between the fluid within the ascending limb and that of the descending limb (Figure 30.8). However, the difference between the concentration of the fluid at the top and bottom of the loop is considerably greater. On this account the countercurrent effect at a given level is said to be multiplied along the length of the tube. Hence the term *multiplier* in countercurrent multiplier.

(6) A final component of the countercurrent multiplier hypothesis has to do with the *countercurrent exchanger*. In the absence of this exchanger, water and sodium chloride would accumulate indefinitely outside of the loop.

The countercurrent exchanger consists of the vasa recta. A countercurrent of blood flows through these hairpin capillaries. As the blood moves toward the bottom of the vasa recta, water is drawn out into regions of increasingly higher osmolality. At the same time, sodium moves into the bloodstream. The net effect is to create a gradient of increasing osmolality *within* the vasa recta toward the bottom of the capillary loops.

As the blood moves upward in the ascending limbs of the vasa recta, water is drawn in and sodium moves out.

Each of the four gradients—two within the vasa recta, two within the loop of the nephron—is maintained at a level whereby the effect of the countercurrent exchanger is to bring a net increase of water and sodium chloride into the bloodstream.

Forming a Concentrated or a Dilute Urine

An understanding of the mechanism for establishing the osmotic gradient is mainly of academic value. But an understanding of the significance of the gradient is tantamount to knowing how a hyperosmotic urine can be formed.

A hyperosmotic urine can be formed only in the presence of antidiuretic hormone. An absence of this hormone results in diabetes insipidus, a disease in which copious amounts of a very dilute urine are formed (Figure 30.9).

Antidiuretic hormone stimulates the cells that comprise the wall of the distal convoluted tubules and the collecting tubules. In doing so, these cells become permeable to water. That is, water is allowed to pass into and through these cells—from the lumen—with ease. However, there is perhaps scarcely any pressure *within* the tubules to drive the water into the permeable cells. Nonetheless, there *is* a driving force, and this force is the osmotic gradient within the kidney—outside of the collecting tubules.

In the cortical region, the water is drawn out of the collecting tubules by a relatively low osmotic pressure. As the urine progresses toward the medulla it becomes more concentrated, due to this loss of water. But then, too, in moving toward the hilum, the urine passes through regions of in-

DIABETES INSIPIDUS. *A disease characterized by excessive urination due to a deficiency of antidiuretic hormone* (G. diabetes, *flow;* insipidus, *tasteless*).

LUMEN. *The bore, or opening, in a tube* (L. light).

creasingly higher osmotic pressure—due to the gradient set up by the countercurrent multiplier. Accordingly, the urine surrenders increasingly more and more water, thus becoming more and more hyperosmotic. Finally, a point may be reached when the urine achieves an osmolality of 1400 milliosmoles. At this point the osmotic pressure within the kidney is insufficient to draw out additional water. The urine thus passes out at its highest concentration.

This highest osmolality can be achieved only when the antidiuretic hormone level is sufficiently high, and this situation can arise only in cases where the body fluid level is very low—and therefore the need to conserve water is greatest.

In the relative absence of antidiuretic hormone—and assuming the body needs to rid itself of excess water while at the same time maintaining sodium chloride—the following situation may hold. The cells of the distal convoluted tubules and collecting tubules become impermeable to the movement of water from the urine into the body. Thus, the osmotic gradient within the kidney is of no apparent value at the moment. As a consequence, an isoosmotic or slightly hypoosmotic urine tends to be formed.

To augment the tendency toward a dilute urine, and to conserve on sodium chloride, aldosterone is brought into the picture. This hormone, which is produced in the adrenal cortex, works specifically on the cells of the collecting tubules and perhaps the distal convoluted tubules. Its action is to bring sodium out of the urine back into the blood. Chloride follows the sodium as a means of neutralizing sodium's positive charge. In effect, the urine is made hypoosmotic. The osmolality ultimately achieved depends upon the relative amounts of antidiuretic hormone and aldosterone, which in turn depend upon the levels of water and sodium chloride in the body fluids.

ANTIDIURETIC HORMONE.
A hormone which is produced in the hypothalamus and stored in the posterior pituitary gland, and functions to retain water in the body.

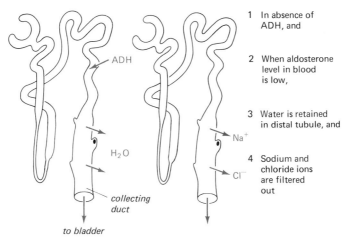

HIGHLY CONCENTRATED URINE

1. Antidiuretic hormone (ADH) makes collecting ducts more permeable to water

2. Water is drawn out of urine by osmotic gradient (Figure 30.8)

VERY DILUTE URINE

1. In absence of ADH, and

2. When aldosterone level in blood is low,

3. Water is retained in distal tubule, and

4. Sodium and chloride ions are filtered out

FIG. 30.9. *Concentration of urine is controlled in the collecting ducts by ADH from the pituitary and mineralcorticoids from the adrenal cortex.*

EVALUATION OF KIDNEY FUNCTION

Kidney function can be evaluated through a series of tests. Of the tests, two are particularly worthy of description, not only because of their value in assessing kidney function but also for their academic value in understanding kidney function.

GLOMERULAR FILTRATION RATE. *The rate at which an ultrafiltrate is formed.*

One test is the *glomerular filtration rate* (GFR). This test provides a quantitative value for the rate at which blood is filtered through the glomeruli. The value for a normal adult is about 125 milliliters per minute per 1.7 square meters of body surface. Lower values may signify partial to total destruction of the glomeruli.

CLEARANCE. *The maximum rate at which the kidneys can clear the blood of a particular substance.*

The second test is called the *clearance* (Cm) test. It provides a quantitative value for the rate at which the kidney is clearing the blood of any particular substance.

Glomerular Filtration Rate

The glomerular filtration rate (GFR) is derived from the relation: concentration in plasma times the volume of plasma filtered per minute equals the concentration in urine times the volume of urine excreted per min. The GFR is the volume of plasma filtered per minute. Note that the relation is similar to an expression used commonly in chemistry and sometimes in daily life:

$$C_1 V_1 = C_2 V_2$$

In this expression C_1 and C_2 represent two concentrations of a single substance and V_1 and V_2 are volumes of the substance. Thus,

$$GFR = \frac{(U)(V)}{P}$$

where U is mg of the filtered substance per milliliter urine, V is milliliters of urine per minute, and P is milligrams of filtered substance per milliliter of plasma. It should be noted that the value of GFR is given in terms of milliliters per minute.

By definition, GFR is the maximum rate at which the glomeruli, collectively, filter the plasma. In order to determine this rate, an appropriate marker must be used—a marker which (a) passes readily out of the glomerulus into Bowman's capsule, (b) does not influence the physiology of the kidney, (c) cannot be absorbed or secreted by the tubules, and (d) can be measured quantitatively in a valid, accurate, and relatively easy manner.

The choice marker for measuring GFR is *inulin*. Not to be confused with insulin, inulin is a polysaccharide made entirely of fructose. Its molecular weight is about 5000, and it can be obtained as a radioactive substance. As such it can be quantified with ease.

The Excretory System

An experimental determination of GFR calls for the intravenous injection of inulin, the voidance of urine at time zero, and the collection and analyses of blood and urine samples after about one or more hours. The concentration of inulin in the blood and urine are inserted into the above formula, as is the value for the volume of urine. The GFR is then calculated.

Clearance

The clearance (Cm) test is used to estimate the maximum rate at which plasma can be cleared of a given substance. It may be noted that nothing is said about how much material is cleared from the body in a given time. Rather, the volume of plasma cleared of a given substance is the matter of importance.

Proper kidney function may be determined by calculating the Cm for urea. The person is voided at zero time. Later, samples of blood and urine are analyzed and Cm is calculated from a formula which bears the same format as the formula for GFR:

$$Cm_{urea} = \frac{(U_{urea})(V)}{P_{urea}}$$

It is not necessary to administer urea in the same way that it is necessary to administer inulin to determine GFR. This is because the concentration of urea in urine, under normal conditions, is directly proportional to the concentration of urea in plasma. Urea is always present in blood and urine. Its concentration in blood is normally 15 to 40 milligrams per 100 milliliters when approximately 80 to 100 grams of protein are consumed per day. In a normal healthy individual, the rate at which urea is cleared is about 60% of the glomerular filtration rate (GFR). When the rate for the clearance of urea drops considerably, urea will begin to increase in the blood (assuming protein intake is not seriously restricted). When the value of blood-urea is abnormally high, a kidney problem is indicated. It might be said here, incidentally, that when the concentration of blood-urea is abnormally low, the liver may be faulty in converting ammonia (from protein breakdown) into urea.

UREA. *The major waste product of nitrogen metabolism.*

AMMONIA. *A toxic waste product of nitrogen metabolism, most of which is converted to the less toxic waste product urea.*

ROLE OF THE KIDNEYS IN REGULATING THE COMPOSITION AND QUANTITY OF BODY FLUIDS

Body fluids can be classified as intracellular and extracellular, depending on whether the fluid is within or outside the cells. Extracellular fluid can be further divided into interstitial fluid (which includes lymph) and blood plasma.

When the electrolytes (ions in solution) of the extracellular and intracellular fluids are "normally" distributed, as shown in Figure 30.10, and when the protein concentration in the blood plasma and lymph are about 7% and 2%, respectively, the body fluids are said to be properly balanced.

A proper balance is achieved through the interplay of several systems, chief of which are the circulatory, endocrine, and excretory systems. When the circulatory system begins to fail, an abnormal fluid balance develops. This may happen in heart disease; or when the endocrine system does not produce adequate amounts of aldosterone and antidiuretic hormone; or when the kidneys begin to fail, as in kidney disease.

A normal or proper fluid balance is necessary in order to achieve and maintain the following physiological functions: (1) a proper acid-base balance, so that the pH of blood is maintained around 7.4; (2) normal cell membrane behavior, allowing normal neuromuscular and glandular secretory activities, among others; (3) a balance of water which tends neither

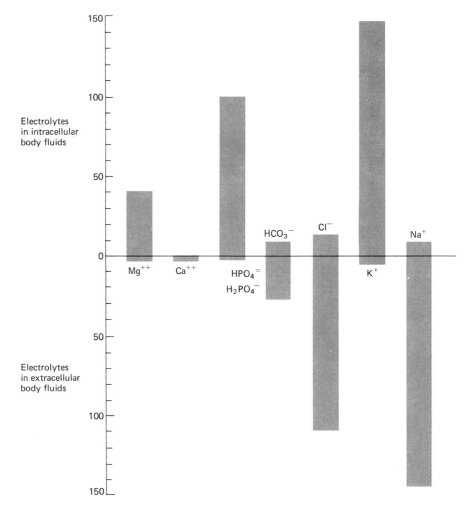

FIG. 30.10. Normal distribution of seven inorganic metabolites—magnesium, calcium, phosphate, bicarbonate, chloride, potassium, and sodium. Values are given in milliequivalents per liter of body fluid. (Milliequivalents express the concentration of positive or negative charges of a substance.)

The Excretory System

toward edema nor dehydration; and (4) a suitable exchange of nutrients and wastes across the walls of blood capillaries as described earlier.

Proper Acid–Base Balance

Acids and bases are produced continuously in cells during metabolism. Each of these substances tends to change the pH of the body fluids. The normal pH (power of the hydrogen ion) of blood is 7.4. A tendency toward lower pH values is called *acidosis*, and toward higher pH values, *alkalosis*.

Failure to maintain the blood pH near a value of 7.4 can bring about deleterious effects. Alkalosis, for example, can bring excessive potassium ions into muscle cells, making them weak. Acidosis can result in a potassium deficit, which may also weaken muscle, or induce abnormal heart behavior, as shown through electrocardiography.

To prevent acidosis or alkalosis, the body employs three mechanisms: chemical buffers, expulsion of carbon dioxide through the lungs, and ionic exchanges in the kidneys. The buffering system is concerned mainly with getting rid of excessive acids. Only one base—ammonia (NH_3)—is produced to any major extent in metabolism, and most of this base is converted into urea (NH_2CONH_2), a perfectly neutral substance, within the liver. The remaining bases—such as creatine and adenine—may produce a minor tendency toward alkalosis, but the main cause of alkalosis is the loss of excessive amounts of acids, as may happen with vomiting.

ACIDOSIS. *A condition in which the pH of blood decreases below about 7.35.*

ALKALOSIS. *A condition in which the pH of blood increases above about 7.45.*

ADENINE. *One of several organic bases found in certain metabolites such as RNA, DNA, and ATP.*

Chemical Buffers. **Chemical buffers** are substances which donate or accept hydrogen ions without allowing the pH of the solution to change. For example, if hydrogen ions are produced in a tissue due to metabolic reactions that produce lactic acid, the hydrogen ions may be picked up by sodium dibasic phosphate (Na_2HPO_4)—which then becomes sodium monobasic phosphate (NaH_2PO_4)—without a corresponding change in pH.

The body contains three major chemical buffer systems, shown here:

(1) The phosphates: $NaH_2PO_4 \underset{+H^+}{\overset{-H^+}{\rightleftharpoons}} Na_2HPO_4$

(2) The proteins. Here, the carboxyl and amino groups do the buffering:

$COOH \underset{+H^+}{\overset{-H^+}{\rightleftharpoons}} COO^- \qquad NH_2 \underset{-H^+}{\overset{+H^+}{\rightleftharpoons}} NH_3$

(3) The bicarbonates: $NaHCO_3 \underset{+H^+}{\overset{-H^+}{\rightleftharpoons}} Na_2CO_3$

DIBASIC PHOSPHATE. *A type of phosphoric acid (H_3PO_4) which has two basic ions, such as sodium or potassium (Na_2PO_4, $NaKPO_4$).*

The first of these systems—the *phosphate system*—works predominately within cells, where phosphate ions are present in high concentration. The second system—the *proteins*—functions chiefly within the blood, where soluble proteins are present within the plasma, and hemoglobin within the red blood cells. The third system—the *bicarbonates*—operates principally in the extracellular fluids, where it is formed from carbon dioxide and water, as follows: $CO_2 + H_2O \rightleftharpoons H_2CO_3$ (carbonic acid) $\rightleftharpoons H^+ + HCO_3^-$ (bicarbonate).

Although buffers are exceedingly important in offsetting a tendency

Role of the Kidneys

pH. *Values less than 7.0 signify acidity, and greater than 7.0 alkalinity.*

toward alkalosis or acidosis, they are limited in their capacity for accepting or donating hydrogen ions. That is, there is only a limited quantity of each buffering agent, whereas the body has a potential for producing larger quantities of acids and bases. For example, far more carbonic acid can be produced in one day than can be buffered by the above systems. All that we can really attribute to the chemical buffers is a temporary role in keeping the blood pH around 7.4 – a role which is played until the lungs and kidneys make the final adjustments.

Role of the Lungs in Buffering. This topic was discussed earlier, in Chapter 28. In summary, the principal role of the lungs in maintaining blood at pH 7.4 is to rid the body of the bulk of acid which is produced. That is, it clears the body of excessive carbon dioxide. Failure to do so would definitely lead toward acidosis, as shown in this equation:

$$CO_2 + H_2O \rightleftharpoons H_2CO_3 \rightleftharpoons H^+ + HCO_3^-$$

In addition to preventing the accumulation of large quantities of carbonic acid (H_2CO_3), the lungs are important, along with the bicarbonate buffer system, in excreting certain amounts of other acids. This can be illustrated with the following equation, which shows how the hydrogen ion of hydrochloric acid (H Cl) can be eliminated:

$$NaHCO_3 + HCl \rightleftharpoons NaCl + H_2CO_3 \rightleftharpoons NaCl + H_2O + CO_2 \uparrow$$

The arrow after carbon dioxide signifies the loss of this volatile acid as a gas.

Role of the Kidneys in Buffering. Just as the chemical buffers have a finite and temporary capacity for buffering, so too the lungs are limited in ridding the body of excessive acids because bicarbonate (HCO_3^-) is used up in the course of eliminating an acid. Inasmuch as there is only a finite amount of bicarbonate in the body, and inasmuch as this buffer is produced from carbon dioxide, a time comes when carbon dioxide must be retained to produce bicarbonate. However, in the process of producing bicarbonate, hydrogen ions (acid) are generated.

Thus it becomes necessary for the kidney to ultimately rid the body of excessive hydrogen ions *per se*. This is achieved largely through ion exchange. In essence, a series of simple inorganic reactions are carried out in or on the cells that comprise the tubules. These reactions fall into two major classes:

(1) Hydrogen is substituted mainly for sodium, and to a lesser extent for some other cation, such as potassium,

CATION. *An atom or molecule that loses one or more electrons and thus acquires one or more positive charges.*

$$Na_2HPO_4 + H_2CO_3 \rightarrow \underset{\text{(excreted)}}{NaH_2PO_4} + \underset{\text{(reabsorbed)}}{NaHCO_3}$$

whereby bicarbonate is conserved.

(2) Ammonia (NH_3) is used, allowing the kidney to rid the body of a waste product derived from protein. In addition, the body retains its cation reserves and at the same time conserves its bicarbonate reserve

CATION RESERVE. *The ions in the blood which are positively charged, such as sodium Na^+ or calcium Ca^{++}.*

$$Na_2SO_4 + 2H_2CO_3 + 2NH_3 \rightarrow \underset{\text{(excreted)}}{(NH_4)_2SO_4} + \underset{\text{(reabsorbed)}}{2NaHCO_3}$$

Balance Needed for Proper Cell Membrane Behavior

It is absolutely necessary to have a proper distribution of sodium, potassium, calcium, and magnesium, if cell membranes of nerve cells, muscle cells, secretory and absorptive epithelial cells, and numerous other kinds of cells are to operate properly.

These substances are balanced through ion exchange reactions—where one kind of positively or negatively charged ion in the ultrafiltrate is exchanged for correspondingly charged ion in the interstitial fluids.

Sodium Balance. Sodium chloride accounts for about 95% of the osmotic pressure in extracellular fluid. In normal individuals, it varies less than 1% in its concentration of about 140 milligrams per liter. This value is maintained by a delicate operation in which two hormones play major roles. These hormones are aldosterone and antidiuretic hormone.

Aldosterone—which is released from the cortex of the adrenal gland, probably in response to low sodium levels in the blood—reacts with the cells of the distal tubules and collecting tubules to keep sodium from being excreted.

ALDOSTERONE. *A hormone which functions to retain sodium in the body.*

Antidiuretic hormone—which is released from the posterior pituitary gland in response to high sodium levels in the blood—reacts with the cells of the distal tubules and collecting tubules to prevent water from being excreted. Exercise, emotion, barbiturates, morphine, and certain other drugs enhance the release of antidiuretic hormone, whereas alcoholic beverages inhibit its release.

In almost all cases, a lower than normal level of sodium in the body is due to *losses* of sodium, rather than inadequate intake. The loss may be due to sweating, diarrhea, vomiting, hemorrhage, or a kidney problem. It could bring about fatigue, sometimes muscular pain, often a disturbance in the circulation, dizziness, nausea, and vomiting.

Higher than normal levels of sodium are brought on secondarily to some other physiological problem, such as kidney disease or a deficiency of aldosterone. In turn, higher than normal levels of water are retained by the body, resulting in edema, which places an extra burden on the heart. Excessive levels of sodium are usually present in cases of heart failure.

EDEMA. *An accumulation of fluid in tissues* (G. oidema, *swelling*).

Potassium Balance. Potassium is maintained in the extracellular fluids at a relatively fixed concentration of about 40 milligrams per liter. It is maintained at this level—indirectly—through the action of aldosterone. As aldosterone reacts with the kidneys and the mucosa of the large intestine, sodium is brought into the bloodstream and potassium is excreted.

An abnormally low level of potassium in the body may be due to a reduction in intake—as happens during starvation or when the appetite is low over a long period. As a rule, individuals on a negative nitrogen balance lose about 3 mg of potassium for every gram of nitrogen excreted in excess of nitrogen intake. Excessive losses of potassium may arise also as a result of diarrhea, a kidney disorder, or excessive secretion of aldosterone.

Low levels of potassium, as well as high levels, lead to weakness of skeletal muscle. In addition, low levels of potassium often bring about faulty heart contractions—as shown through electrocardiography.

Role of the Kidneys

PARATHYROID HORMONE. *A chemical which increases blood calcium levels.*

Calcium Balance. Calcium is held in the extracellular fluid at a constant level of about 10 milligrams per liter. Parathyroid hormone, calcitonin, and vitamin D maintain the calcium at this level. The action of *parathyroid hormone* is directed primarily at three kinds of tissues. First, together with *vitamin D* it reacts with intestinal epithelial cells to bring calcium into the body. Second, it reacts with the kidney tubules to prevent the loss of calcium from the body. Third, it reacts with the bone cells (osteocytes or osteoclasts) to bring calcium into the blood, as needed.

Calcitonin reacts with osteoblasts and/or osteocytes to bring about the deposition of calcium in bone.

The effects of calcium deficiency on bone development were discussed in Chapters 12 and 15. Regardless of whether the deficiency results in rickets, osteomalacia, osteoporosis, or some other disease, it is accompanied by pain which may be relieved in a seated or lying position. Generally, there is also muscle weakness and there may also be mild to moderate tetany.

Excessive calcium in the blood may lead to a loss of appetite, vomiting, abdominal pains, excessive urination, thirst, and constipation.

Magnesium Balance. Magnesium is maintained at a constant level of about 1.6 milligrams per liter in the extracellular fluid. Nothing is known about its regulation except that it may share a transport vehicle with calcium in the cell membrane of the kidney tubules. In general, when large quantities of calcium are excreted, low levels of magnesium also are eliminated, and vice versa.

Excessive losses of magnesium may occur when the adrenal cortex sends out too much aldosterone or when the kidneys are working improperly. Cirrhosis of the liver may also cause excessive magnesium losses, as may the removal of a tumor from the parathyroid gland.

TREMORS. *Involuntary trembling, shaking, or quivering (L. tremor, to tremble).*

Somewhat nonspecific effects are caused by an excessive loss of magnesium. Among them are muscle tremors and weakness, convulsions, excessive calcium in the blood, and the production of calcium stones in the kidney.

Water Balance. The amount of water in the body is determined by the concentration and distribution of all the solutes between the intracellular and extracellular compartments, primarily by the extracellular concentration of sodium chloride.

An appropriate balance of water in the body is achieved mainly through the regulation of sodium. This ion—together with the anions which neutralize its charge (chiefly chloride)—accounts for about 95% of the total osmotic pressure in the extracellular fluid. In a normal person its level in the extracellular fluid does not vary much more than one percent from day to day. Its regulation is achieved through the action of aldosterone and antidiuretic hormone in the kidney as described above.

OSMORECEPTORS. *Nerve cells which are stimulated in proportion to the concentration of molecules dissolved in body fluids.*

As the body tends toward a deficiency of water, the concentration of electrolytes, particularly sodium chloride, increases. Through neural mechanisms of unknown nature, the sensation of *thirst* is aroused. In addition, osmoreceptors are stimulated within the hypothalamus and perhaps other regions of the brain. One result of such stimulation is the release of antidiuretic hormone from the posterior lobe of the pituitary gland. This hormone reacts with the distal tubules and collecting ducts of the kidney. As a result, less water is allowed to pass out of the body.

As the body tends toward an excess of water, its concentration of sodium chloride decreases. The sensation of thirst is not aroused and the supply of antidiuretic hormone to the kidney is either reduced or turned off. In addition, the adrenal cortex may be stimulated to release an extra amount of aldosterone. This hormone reacts with the kidney tubules to prevent the release of sodium ions, at the same time allowing the excretion of additional amounts of potassium ions. The relative lack of antidiuretic hormone, together with the presence of aldosterone, results in the loss of water and the retention of salt.

ENDOCRINE FUNCTIONS OF THE KIDNEYS

The kidneys are known to be the source of one hormone—renin. However, they may also be viewed as the producer of another hormone—active vitamin D, and they are now known to produce a third kind of hormone—erythropoietin. The latter hormone was discussed in Chapter 29, and active vitamin D was discussed in Chapter 12.

Renin is produced by a cluster of cells—a so-called *polar cushion*—which is present in the medial wall of the afferent arteriole, at a point where the afferent arteriole enters a glomerulus. See Figure 30.5.

MEDIAL WALL. *The middle or central wall.*

Next to the polar cushion is a sensory device, called a *macula densa*. It is seen as a cluster of crowded cells in the outer wall of the distal convoluted tubule. It occurs where the distal convoluted tubule lies between the afferent and efferent arterioles—at the entryway into Bowman's capsule.

MACULA DENSA. *Literally, a dense spot (L.).*

When blood enters the kidney at a pressure that is lower than normal, the approximately 1 million maculae densae sense the low pressure. In turn, they somehow stimulate the polar cushion to secrete renin.

Renin then passes into the blood, where it reacts with an inactive form of angiotensin, a small protein. As a result, an active form of angiotensin is produced.

Angiotensin has two functions. First, it reacts with the smooth muscle of arteries, causing them to constrict—and thus raise the blood pressure. The second function of angiotensin also is directed at raising the blood pressure, though a different mechanism is used. In this case, angiotensin stimulates the adrenal cortex to release aldosterone.

ANGIOTENSIN. *A blood protein that stimulates vasoconstriction and the release of aldosterone from the adrenal cortex, resulting in increased blood pressure.*

Aldosterone reacts with the distal convoluted tubule and collecting tubules, causing them to prevent the excretion of sodium. The resulting increase of sodium chloride in the body increases the circulatory system's peripheral resistance. This means the heart must now become more forceful to move a given volume of blood through the tissues. Hence, blood pressure rises.

URINALYSIS AND KIDNEY PROBLEMS

Whenever someone is suspected of having a kidney disease or some general body disorder, it is necessary to examine certain characteristics of the urine.

Urinalysis

> **URINALYSIS.** An examination of the physical and chemical properties of a sample of urine.

In a standard *urinalysis*, seven characteristics are examined: appearance, pH, specific gravity, protein, glucose, ketones, and microscopic particulate materials. In addition, specific other substances may be analyzed.

Appearance of Urine. Freshly voided urine ranges in color from yellow to dark amber, depending on the volume of water present. When abnormal constituents are present, a different color appears. Red or brown urine may be due to the presence of blood, abnormal breakdown products of hemoglobin, melanin, or certain drugs. The presence of bile produces a yellow foam when urine is shaken, and the sample may turn brown.

Fresh urine is normally a transparent liquid. It may become cloudy in time due to the gradual conversion of urea into ammonia. This conversion causes the urine to become alkaline, bringing about the precipitation of phosphates. When excess cloudiness occurs, or when freshly voided urine is cloudy, the cloudy appearance is usually caused by pus, bacteria, salts of uric acid or phosphoric acid, mucus, or proteins.

pH of Urine. Urine is usually slightly acidic, having a pH of about 6. The pH may rise above 7.5 in cases of alkalosis. In the event of fever or acidosis, the pH of urine can go to values below 5.

> **SPECIFIC GRAVITY.** Ratio of the mass of volume of a substance to an equal volume of a reference substance.

Specific Gravity. The weight of a known volume of fluid divided by the weight of the same volume of pure water is the specific gravity of the fluid. The specific gravity of urine ranges from 1.010 to 1.025. The lower value, 1.010, is also the specific gravity of plasma ultrafiltrate.

Specific gravity is important in determining kidney disease and in evaluating the progress of kidney disease. Specific gravity reflects the ability of the kidney to form a concentrated urine.

In a urine concentration test, the person is deprived of water for a long period, usually 16 hours or more. During this time, a normal person will concentrate urine to a specific gravity of 1.020 to 1.025.

Values below 1.020 appear in certain kidney diseases, such as glomerulonephritis, or in other kinds of diseases, such as diabetes insipidus (due to a deficiency of antidiuretic hormone). Unusually high values appear in fever, certain kidney diseases, and other diseases, such as diabetes mellitus.

Specific gravity tests, by themselves, are not absolutely indicative of kidney health or kidney disease.

Urinary Protein. Protein is normally not present in urine beyond a value of about 100 milligrams per day. The kind of protein which appears most often is albumin. *Albumin* is one of the smallest and most highly concentrated proteins in blood. When it appears in urine in excess of 150 milligrams per day, the condition is called *albuminuria*. Albuminuria, or proteinuria, may be due to severe muscle exertion, pregnancy, kidney disease, a disease in the pelvis, ureter, prostate gland, or some other cause. Special proteins, called Bence-Jones proteins, are present in the urine of about 50% of individuals suffering from multiple myeloma.

ALBUMINURIA. *The presence of serum albumin, globulin, and other proteins in the urine.*

Urinary Glucose. Glucose is absent from urine in healthy individuals except sometimes during a normal pregnancy. When glucose is present in urine (*glucosuria*), it may be accompanied by either a normal level of glucose in the blood or an abnormally high level of blood glucose (hyperglycemia). Glucosuria without hyperglycemia may be due to a subnormal kidney threshold for glucose. It may also be due to pregnancy, certain metabolic disorders, or toxic chemicals such as carbon monoxide, lead, and mercury.

When glucosuria occurs together with hyperglycemia, the cause may be diabetes mellitus, severe intracranial pressure, Cushing's syndrome, a tumor in the adrenal medulla (pheochromocytoma), abnormally high thyroid activity, certain types of anesthesia, such as ether, or some other factor.

CUSHING'S SYNDROME. *A condition of obesity and muscular weakness, caused by excessive levels of adrenal cortical hormones in the blood.*

Urinary Ketones. Ketone bodies (acetone and acetoacetic acid) may appear in the urine of individuals on extremely low carbohydrate diets. Ketone bodies appear also during starvation, in normal pregnancies, in dehydration, and in untreated cases of diabetes mellitus.

Microscopic Particulate Materials. When small particles of material are present in urine they can be located microscopically by examining the sediment of a centrifuged sample of urine. Red blood cells, white blood cells, casts, and crystals may be present. The presence of red blood cells may be due to a kidney disease, kidney stones, or a disease of the lower urinary tract. White blood cells in urine often suggest an infection somewhere within the urinary system. *Casts* are deposits of fats, waxes, epithelial cells, blood, and other materials, singly or in combination. The composition of casts is often helpful in diagnosing a kidney problem and in following the course of treatment of the problem. *Crystals* in the urine usually suggest a metabolic disorder. In acid urine, the crystals are most often salts of uric acid, cystine, or calcium oxalate. In alkaline urine, the crystals occur most often as phosphates.

CAST. *A mass of material that has taken the form of some cavity in which it has been molded, such as bronchial or intestinal casts, and (most commonly) renal casts (Old Norse,* kasta).

Kidney Disease

An *acute* kidney disease is one that arises suddenly, over a short period of time. A *chronic* disease develops progressively, over a long period of time. The disease may affect the basement membranes in the kidney, the glomeruli, the tubules, the pelvis, the ureter, or the bladder. A kidney disease may involve a combination of these structures or a single pyramid of neph-

CALYX. *A cuplike division of the pelvis of the kidney (G. kalyx, cup).*

rons which feed into a single minor calyx (Figure 30.2). The disease may be caused by an infectious agent (as in acute glomerulonephritis), high blood pressure (as in malignant hypertension), certain abnormalities in pregnancy, or any one of a number of other causes.

The three relatively common kidney diseases are acute glomerulonephritis, chronic glomerulonephritis, and the nephrotic syndrome. These kidney diseases, as well as any other, can ultimately bring about *renal failure*, in which the kidneys cannot excrete their load of wastes. In renal failure there is an elevation of urea in the blood (uremia), widespread edema, a failure to properly concentrate the urine, a failure to regulate blood pH, a failure to balance the body's electrolytes, and numerous other abnormalities.

RENAL FAILURE. *Failure of the kidneys to clear the body properly of toxic waste products.*

Acute Glomerulonephritis. Acute glomerulonephritis is usually caused by a strain of streptococcus bacteria. During the infection the color of the urine is usually red to brown, due to breakdown products of hemoglobin. Edema (especially in the upper eyelids), anemia, and proteinuria (0.5 to 3g per day) are also present.

GLOMERULONEPHRITIS. *An inflammation of the kidney affecting the glomeruli primarily (L. glomus, ball; G. nephros, kidney; itis, inflammation).*

The most characteristic feature is the presence of red blood cell casts in the urine. Casts of white blood cells may also be present. The renal function tests, such as urea clearance, indicate faulty performances in only about half the cases. Often, however, it is not possible to concentrate a urine, as is shown by a low specific gravity. Biopsies taken from patients usually reveal swollen glomeruli.

The disease appears more often in children than in adults. In children the mortality is only about 1%; complete recovery is usually the rule. In adults, about 25% to 50% of the patients develop chronic glomerulonephritis. Other adults recover completely and still others suffer renal failure, which, if left untreated, results in death.

Chronic Glomerulonephritis. Chronic glomerulonephritis may be preceded by a bacterial infection, but it usually cannot be related to some specific cause. The disease develops over a period of years. There are few abnormalities in the early years, though casts of red blood cells and epithelial cells are frequently present along with proteinuria. As the years go by, the kidneys gradually lose their ability to produce a concentrated urine. The symptoms of chronic glomerulonephritis usually do not show up until about 75% of the nephrons are destroyed. Ultimately renal failure sets in and one can generally find granular and waxy casts in the urine.

Nephrotic Syndrome. The *nephrotic syndrome* is characterized by high proteinuria (more than 3.5g per day), an abnormally low level of albumin in the blood, and a high level of cholesterol in the blood. The low level of albumin is tied in with the high level of cholesterol because the production of cholesterol in the liver is linked with production of albumin. An examination of a sample of kidney tissue (a biopsy) with the electron microscope commonly reveals abnormal basement membranes in the tubules. In many cases this abnormality may be seen with an ordinary microscope. As is generally characteristic of other kidney diseases, the urine may contain casts, elevated levels of protein, and blood cells.

BIOPSY. *An excised tissue which is examined to establish a diagnosis (G. bios, life; opsis, vision).*

SUMMARY

1. The renal excretory system consists of a pair of kidneys, a pair of urinary pelves and ureters, a urinary bladder, and a urethra.
2. Blood enters the kidney through the renal arteries. After passing through a series of decreasing arteries, a portion of the blood is filtered out of a knot of capillaries (glomerulus) into a capsule (Bowman's) and then through a uriniferous tubule. In passing through this tubule, 99% of the filtrate is reabsorbed into the blood and the remaining 1%, which carries waste products, passes into a pelvis. From the pelvis, the urine is carried peristaltically by a ureter to the bladder for storage.
3. A complex reflex, executed autonomously but subject to voluntary regulation, causes: (a) the bladder to contract and (b) the urinary sphincters to open, sending the urine out of the body by way of the urethra.
4. Concentrated urines are produced by virtue of a countercurrent multiplier and antidiuretic hormone. Dilute urines are due mainly to aldosterone, whose action brings sodium chloride into the bloodstream from the lumen of the uriniferous tubules.
5. Kidney performance can be evaluated by a variety of techniques including: (a) a test of the glomerular filtration rate, which tells how well the glomeruli are filtering the blood; and (b) a test for clearance, which tells how well the kidneys are clearing the blood of a toxic substance, such as urea.
6. Acid-base balance in the blood is achieved partly through ion exchange mechanisms in the kidneys. Hydrogen ions are conserved or excreted, depending on whether the blood is tending toward a condition of alkalosis or acidosis.
7. Sodium and potassium balance in the blood is achieved partly through the hormonal action of aldosterone in the kidney, whereby sodium is retained and potassium is excreted.
8. Calcium and magnesium balance in the blood is achieved partly through the actions of parathyroid hormone and vitamin D, whereby calcium is retained and magnesium is excreted.
9. The kidneys have endocrine cells which secrete: (a) renin, which increases the blood pressure by activating an inactive form of angiotensin in the blood, (b) an active form of vitamin D, which is formed from an inactive form, and (c) erythropoietin, a hormone that stimulates the production of red blood cells.
10. In a standard urinalysis a sample of urine is examined for appearance, pH, specific gravity, and the presence of protein, glucose, ketones, and particulate matter.

REVIEW QUESTIONS

1. List the anatomical components of the renal excretory system. What is the major function of each of these components?

Review Questions

2. Define: osmolality, renin, angiotensin, glomerulus, Bowman's capsule, uriniferous tubule, loop of Henle, collecting duct or tubule, macula densa, ureter, urethra.
3. Describe the blood circulatory route through the kidneys.
4. Explain how it is possible for the kidneys to produce a urine higher in osmolality than the blood plasma?
5. What are the functions of antidiuretic hormone and aldosterone in renal excretion?
6. What is signified or suggested by an abnormally low glomerular filtration rate?
7. What kind of information is provided by a clearance test?
8. Describe the role of the kidneys in buffering.
9. What hormones are produced in the kidneys? Describe their functions.
10. What kinds of information can be revealed about the kidneys through a urinalysis?
11. Define glomerulonephritis, the nephrotic syndrome, and renal failure.

The Mechanisms of Defense

31 Body Defense Systems: Nonimmunological
32 Body Defense Systems: Immunological

PART SEVEN

Body Defense Systems: Nonimmunological 31

THE BODY'S FIRST LINE OF DEFENSE
    The epidermis
    The mucosa
THE BODY'S SECOND LINE OF DEFENSE
    Steps in blood clotting
    Blood clotting problems
        Abnormal bleeding
        Vitamin K, coumarin drugs, and blood clotting
        The liver and clotting
        Platelets and clotting
        Hemophilia—a hereditary clotting disorder
THE BODY'S THIRD LINE OF DEFENSE
    Xenometabolism
    The interferon system
    The reticuloendothelial system
        The inflammatory response—helpful and adverse consequences
        The role of prostaglandins and aspirin in inflammation and anti-inflammation

The body is constantly subject to continuous challenge and assault by physical, chemical, or biological entities—such as insecticides, factory and automobile emissions, drugs, and microorganisms. Also, some cells and tissues, locally or widespread throughout the body, become changed due to wear and tear, genetic mutation, aging, disease, drugs, injury, and similar causes. Regardless of cause, altered cells and tissues challenge the body, assault it, or impose an adverse effect upon its normal physiological processes.

In response to this "challenge and assault," the body is endowed with several potential defense mechanisms. They are "potential" defense mechanisms because under one set of circumstances or another they may work adversely and bring destruction, rather than aid, to the body.

Altogether, the body must rely on two lines of defense against challenge

The Body's First Line of Defense

and assault: it must either evade would-be invaders, or — failing to do so — it must respond physiologically and dispose of the physical, chemical, or biological invader.

This chapter begins with a discussion of the body's first line of defense against invasion: the epidermis of the skin; and the epithelium of the mucosa, the membrane which lines body surfaces that are not covered by skin.

In the second section of this chapter we discuss the second line of defense — the blood clotting system — which is called into action in the event of injury to the skin or mucosa. Various kinds of blood clotting problems are also discussed at this point.

In the third section of this chapter we assume a potential invader is successful in entering the bloodstream and must be disposed of. Now the third line of defense is called into action. There are four units in this line, depending on the nature of the invader and the particular response which the body musters up: xenometabolism, the interferon system, the reticuloendothelial system, and the immunological system.

If the invading substance can be metabolized mainly within the liver, whereby it is converted into one of several special kinds of salts which can be excreted mainly by the kidney, the substance is called a *xenobiotic* and this aspect of the third line of defense may be called *xenometabolism*.

If the invading substance is a virus, it may succeed in destroying some cells which it invades, but before doing so these cells produce and release a substance called *interferon*. Interferon provides a form of protection to some of the nearby cells.

The *reticuloendothelial system* consists of tissues throughout the body that are capable of producing phagocytes which keep the body clean of worn-out tissue debris and foreign invaders, such as bacteria. This system may operate by itself, or together with the interferon system and/or immunological system.

The *immunological system*, together with the reticuloendothelial system, disposes of certain kinds of foreign invaders, and also sets up a "memory system" whereby the foreign invader is recognized the next time it enters the body. Thus, the foreign matter may be disposed of more rapidly. Often, in the course of aging, some cells acquire seemingly "foreign" properties; such cells may be attacked by the immunological system.

With regard to the third line of defense, we shall cover the first three systems in this chapter and reserve the immunological system for the following chapter. To conclude this chapter, there is a discussion on inflammation — a tissue reaction which often arises in conjunction with the third line of defense.

EPIDERMIS. *Five layers of epithelial cells which cover the dermis, or true skin.*

XENOMETABOLISM. *Metabolism of foreign compounds (G. xenos, stranger).*

PHAGOCYTE. *A cell that keeps tissues free of foreign material (G. phagein, to eat; kytos, cell).*

THE BODY'S FIRST LINE OF DEFENSE

All surfaces of the body are covered either by skin or mucosa. The *skin* is a two-layered organ consisting of an outer epidermis and an inner dermis.

Body Defense Systems: Nonimmunological

From the viewpoint of body defense, the epidermis provides a first line of defense, while the dermis is involved in the second and third lines of defense.

The Epidermis

The epidermis defends the body in at least three ways. Through its keratinized horny layer, the body is afforded some resistance to the destructive forces of acids, bases, organic solvents, fire, sharp objects, and other potentially harmful agents. Second, through the process of shedding cells of the horny layer, bacteria may be dislodged. Third, through the secretion of sebum, fungal growth may be prohibited. Ringworm, for example, is less common after puberty, when greater quantities of sebum, or perhaps a modified sebum are secreted. In addition, through the keratinized layer, along with melanin in the lower epidermal layers, the epidermis screens the body from the harmful rays of the sun.

KERATIN. *A tough protein derived from squamous cells of the epidermis as the cells lose their nucleus and begin to dry out.*

MELANIN. *A brown-black pigment which colors the skin and protects the skin against harmful rays from the sun.*

The Mucosa

The *mucosa* is a mucous membrane that covers the free surfaces of the body which are not covered by skin. It is found as the inner lining of the respiratory, digestive, reproductive, and excretory ducts. It is also present on the underside of the eyelids (as a conjunctiva) and in the ductwork of the ear.

As is the case with the skin, we consider only the epithelial lining of the mucosa as a first line of defense. The underlying connective tissues of the mucosa provide second and third lines of defense.

The epithelial lining of mucosa contains two basic types of cells: mucous cells and ciliated cells.

Mucous cells are specialized for producing mucus—a viscous, sticky material which contains *mucin,* a complex of carbohydrate and protein. This material is secreted from the plasma membrane, following its assembly within the Golgi apparatus. Outside the cell it serves to lubricate a passageway and, from the viewpoint of evasion—perhaps to trap microorganisms, other foreign matter, and dead tissue debris.

Ciliated cells display numerous small, hairlike extensions from the cell surface which border a canal. Each of these extensions, or *cilia,* beat back and forth in such a way as to drive mucous secretions—and would-be invaders—from the canal toward the outer opening in the body.

MUCOSA. *A wet membrane which lines all surfaces of the body that communicate with the air and are not covered by skin.*

THE BODY'S SECOND LINE OF DEFENSE

Whereas the skin and mucosa may ward off would-be foreign invaders of the body, occasions arise when these surfaces are damaged, thus allowing

The Body's Second Line of Defense

the invaders to move directly into the connective tissues at the site of the wound. To reduce the incidence of invasion and to prevent losses of body fluids, the blood clotting system is called into play.

Steps in Blood Clotting

Looked at simply, the overall process of blood clotting consists of seven steps (see Figure 31.1):

(1) *Constriction of arterioles which supply blood to the injured area.* This is a reflex action carried out by the autonomic nervous system. It reduces the flow of blood to the injured area.

(2) *Aggregation of blood platelets to form a hemostatic plug.* Ordinarily, platelets are kept apart from one another through electrostatic repulsive charges. When a blood vessel is injured, however, the platelets are attracted

AUTONOMIC NERVOUS SYSTEM. *That part of the nervous system which regulates the visceral organs.*

FIG. 31.1. *The mechanism of blood clotting. Heparin, an anticoagulant, works by blocking step (3c), the transformation of prothrombin to thrombin.*

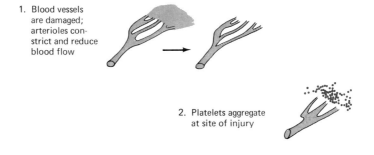
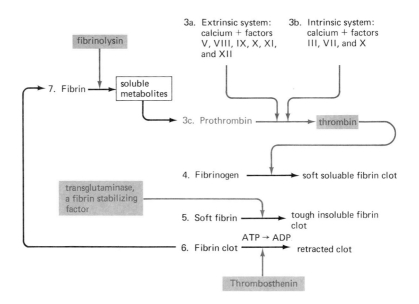

to one another in large numbers at the site of injury, resulting in blockage of the blood flow.

(3) *Conversion of prothrombin to thrombin.* Prothrombin, which is produced by the liver and carried in the blood, is an inactive form of *thrombin*, an enzyme. The conversion of prothrombin to thrombin is triggered by a sequence of chemical reactions which involve a number of so-called *clotting factors*, all of which, except calcium, are proteins. One or more clotting factors may be deficient in people who have a clotting problem.

From Figure 31.1 it may be seen that one series of clotting factors is present in blood platelet cells. This series constitutes the *intrinsic system* for initiating a clot.

The *extrinsic system* for initiating a clot is shown also in Figure 31.1. This system consists of clotting factors found outside of the blood vessels. A deficiency in either the intrinsic or the extrinsic system may result in abnormal bleeding.

(4) *Conversion of fibrinogen to fibrin by the enzymatic action of thrombin.* Fibrinogen, like prothrombin, is synthesized in the liver and stored in the blood.

(5) *Converting a soft fibrin clot into a tough fibrin clot.* This reaction is catalyzed by an enzyme (transglutaminase) which, if deficient, leaves the individual with a soft, vulnerable clot.

(6) *Retracting the clot.* In this step, another enzyme comes into the picture. Called *thrombosthenin*, it splits ATP into ADP and causes the clot to shrink, thus allowing the blood vessels to assume their normal size.

(7) *Fibrinolysis.* In this process, fibrin is destroyed by an enzyme, *fibrinolysin*, whereupon it becomes soluble again. This happens gradually, while the injured tissue is being repaired.

THROMBIN. *An enzyme that induces clotting by catalyzing the conversion of fibrinogen to fibrin (G. thrombos, clot).*

CLOTTING FACTORS. *Metabolites that are essential in forming a normal clot.*

FIBRIN. *The principal fibrous material in a blood clot.*

FIBRINOLYSIS. *Breakdown of a clot (L. fibra, fiber; G. lysis, tear apart).*

Blood Clotting Problems

Abnormal blood clotting may arise at any time in the lifetime of an individual. It may show up as prolonged bleeding, failure to form a clot, and/or internal hemorrhaging. One form of internal hemorrhaging is often seen in the skin as a *bruise*—a *purpura* which forms when one bumps against an object ever so slightly. Other forms of internal hemorrhaging may occur when blood vessels rupture within connective tissues elsewhere in the body, as for example, within a visceral organ.

Abnormal Bleeding. Failure to form a proper clot in a normal time period may be due to a number of different causes including: (1) vitamin K deficiency or an iatrogenic cause (use of coumarin and other drugs); (2) liver disease; (3) a platelet problem; and (4) a hereditary disorder.

Vitamin K, Coumarin Drugs, and Clotting. Vitamin K is needed in sufficient quantity for the synthesis of four of the clotting factors that are involved in the conversion of prothrombin to thrombin. Liver, meat, fish, and vegetable oils are good sources of this vitamin.

Coumarin drugs are anticoagulants which closely resemble vitamin K

PURPURA. *Usually, a hemorrhage in skin (L., purple).*

IATROGENIC. *Problems caused by medical treatment (G. iatros, physician; gen, to produce).*

in chemical structure. Coumarin anticoagulants were developed following the discovery of hemorrhagic outbreaks among cattle which had been feeding on improperly cured clover. Chemical analyses revealed the presence of dicoumarol in the clover. Because of coumarin's antivitamin K activity, coumarin anticoagulants are used beneficially in certain cases of coronary thrombosis and pulmonary embolism and against clots that may appear elsewhere in the body.

The Liver and Clotting. Two of the very important proteins used in a clot — fibrinogen and prothrombin — are produced by the liver. It follows, that liver disease, such as cirrhosis or hepatitis, may lead to a deficiency of these materials.

CIRRHOSIS. *Usually refers to the conversion of liver cells into functionless fibers.*

Platelets and Clotting. Normally there are $250,000 \pm 40,000$ platelets in a cubic millimeter (microliter) of blood. When this number drops to about less than 60,000, bleeding disorders usually arise. (However, blood clotting may be maintained properly by larger, young platelets at even lower platelet counts.)

Purple blotches may appear in the skin here and there, especially where pressure is applied. The symptoms of this problem are aptly termed *thrombocytopenic purpura*.

Thrombocytopenia (platelet deficiency), the most common cause of abnormal bleeding, may be due to: (1) a reduction in the production of platelets; (2) destruction of platelets; (3) exhausting the supply of platelets; (4) pooling of platelets in the spleen; or (5) a combination of these causes.

A *decrease in the production of platelets* — from the normal 35,000 per microliter per day — may be due to damage to the bone marrow, such as from radiation or an infection. It may also arise as a result of a deficiency of vitamin B_{12} or folic acid.

SPLEEN. *The largest lymphatic organ in the body, situated below the diaphragm on the left side* (G. splēn).

Where there is *destruction of platelets*, the process usually is initiated autoimmunologically, whereby the platelets acquire some foreign character and are attacked and destroyed by the body's immunological system. If the platelets are only moderately damaged they are removed, chiefly by the spleen. If they are more seriously damaged, the entire reticuloendothelial system — especially the liver — is involved in their removal. In this case, the removal of the spleen may be unwarranted.

Alternatively to destruction, a low platelet count may arise following *prolonged bleeding*, especially where internal clotting becomes extensive. In this case, the platelets become depleted of their clotting factors.

Ordinarily, about two thirds of the platelets are maintained in the circulatory system. The remainder are stored in lymphoid tissues, mainly the spleen. In certain diseases, such as liver cirrhosis, the spleen enlarges (splenomegaly) and more platelets are stored in the spleen. Fewer platelets are thus available, when needed, in the event of bleeding.

Two functional abnormalities may be involved in thrombocytopenic purpura: (1) a reduction in the rate at which fibrin is produced, due to a decrease in platelet clotting factors; and (2) failure of clot retraction due to a deficiency in one of the platelet factors (thrombosthenin).

Thrombocytopenia purpura may sometimes be corrected with drugs.

Cortisone, for example, may be administered with the hope that it may reduce the destruction of blood platelets by the spleen. Where drugs fail, the spleen may be removed.

Purpuras may be due to causes other than thrombocytopenia. There may be a normal number of platelets, but the platelets may be partly or totally lacking in one or more clotting factors. Also, the capillaries may be faulty in allowing platelets to adhere to them in order to allow a clot to form.

Hemophilia – a Hereditary Clotting Disorder. Hemophilia is a sex-linked recessive gene disorder in which one of two clotting factors (VIII or IX) is partly or totally lacking. The platelet count is usually normal, but the bleeding time, following an injury, is prolonged.

Several kinds of hemophilia are known. They vary with respect to the specific clotting factor deficiency (VIII or IX), and the severity of bleeding. About 25% to 50% of hemophiliacs bleed excessively only after a severe injury, 10% to 25% after a minor injury, and 1% to 10% bleed spontaneously.

It should be emphasized at this point, before passing on to the next topic, that the major problem faced by the hemophiliac is not external bleeding from a surface wound, as is commonly indicated in popular literature. Instead, the critical problem is internal hemorrhaging, which may easily be brought on by falling or bumping against another person or an object.

CORTISONE. *An adrenal cortical hormone which is often used as a drug to treat inflammation.*

HEMOPHILIA. *A sex-linked hereditary clotting disease in which there is prolonged bleeding* (G. haima, *blood;* phil, *love*).

THE BODY'S THIRD LINE OF DEFENSE

Neither the epithelial surface of the skin and mucosa, nor the blood clotting system are 100% effective in preventing foreign materials from entering the body. Bacteria, viruses, pollens, dust, drugs, and non-nutritive components of food, such as caffeine and artificial flavoring agents enter the bloodstream on one occasion or another. None of these substances can be tolerated within the body, and although one or more of these materials may accumulate to a certain point – such as tobacco residues in the lungs of a smoker – the body is generally able to dispose of them.

Xenometabolism

Xenometabolism is the metabolism of strange or foreign compounds, i.e., of xenobiotics. A *xenobiotic* may be defined as a substance that is neither incorporated into protoplasm nor used as a source of energy. In addition, a xenobiotic is usually not an antigen (as described on page 648) and it is often subject to a peculiar form of metabolism as described below. Drugs in general, gasoline, certain components of cosmetics and deodorants, DDT and other biocides, and a whole host of inorganic and organic pollutants and other synthetic products are examples of xenobiotics.

XENOMETABOLISM. *Metabolism of substances that are of no economic value to a cell* (G. xenos, *stranger*).

XENOBIOTIC. *A substance that furnishes neither energy nor a source of matter for protoplasm (G. xenos, foreign; bios, life).*

Xenobiotics may enter the body through the skin, as is the case with gasoline, dyes used in histological work, and preservatives used in fixing animals for anatomical studies (most commonly formaldehyde and carbolic acid). Xenobiotics may also enter the body through a mucosa, as is the case with aspirin, penicillin, and food additives in the intestinal tract and with ether, chloroform, and air pollutants in the respiratory system.

Once a xenobiotic penetrates the epithelium of the skin or mucosa, it may circulate through the blood to some other region of the skin or mucosa and leave the body unchanged.

CONJUGATED. *Refers to a complex formed between two compounds.*

More often, however, the xenobiotic becomes subject to one or more enzymatic reactions in the liver. In the end, the xenobiotic is conjugated with glucuronic acid, or sulfuric acid, or an amino acid. In all cases, the conjugated product is a salt which carries a negative or a positive charge.

The conjugated xenobiotic is discharged from a liver cell into the interstitial fluid. It may be sent along with the liver's bile juice into the intestine, from which it is eliminated, along with the feces. More often by far, however, it is sent by way of the bloodstream to the kidney.

Due to a negative or positive ionic charge on the conjugated xenobiotic, it is usually not possible for this substance to enter any cell along the way to the kidney. So too, within the kidney, the conjugated xenobiotic is "seen" as a substance that does not have to be conserved by the body, and accordingly, it is secreted.

PYRIDINE. *A colorless nitrogenous solvent.*

Although xenobiotics are converted into excretable products through metabolic processes within the liver, the product in all cases is not necessarily less toxic than the xenobiotic. On the contrary, in a few cases the product is more toxic. Pyridine, for example, which in itself is somewhat toxic, is converted into methyl pyridine, which is even more toxic. The liver simply has no way of discriminating one type of xenobiotic from another in terms of whether the xenobiotic or its salt-product will be more toxic. Rather, the liver cells merely process certain reactive groups on xenobiotic molecules through their xenometabolic machinery. The product is discharged from the liver cell into the blood, and although its ionic charge renders it a low probability of entering any cell along the way to the kidney, it may nevertheless interact with the cell membrane of nerve cells, or muscle cells, or some other kind of cell in a toxic way. Or it may enter a cell and interfere toxically with the cell's metabolic machinery.

The Interferon System

A major defense mechanism of the body against viruses—and possibly other microbes—is interferon (Figure 31.2). *Interferon* is a protein which is produced in cells upon being challenged by various viruses and other microorganisms.

INTERFERON. *A class of proteins that inhibits viral multiplication.*

RNA. *Ribonucleic acid, a necessary polymer for synthesis of proteins in biological cells.*

Interferon comes in various forms, perhaps depending upon the particular virus involved. It is believed that the messenger RNA that is needed for the production of interferon is normally inhibited by a repressor protein. When a virus enters a cell, the inhibition is relieved, and the cell begins to produce interferon.

Body Defense Systems: Nonimmunological

When any one of a variety of viruses (such as mumps, yellow fever, chicken pox, or the rhinovirus of colds) invades the body, interferon appears in the blood within a few hours. The interferon does not actually destroy the virus. Rather, it is believed to interact with the nuclei of uninfected cells. In doing so, it *interferes* with the virus' normal ability to take over the cell's nuclear apparatus for replicating its own self. That is, the function of interferon is to inhibit the multiplication of viruses which enter healthy cells. Interferon serves mainly to prevent the spread of a viral infection, and it may also have the same effect on other microbial infections.

VIRUS. *A microorganism, formed of RNA or DNA and protein, and incapable of multiplying itself except through the use of the protein synthesizing apparatus of a host cell which the virus invades.*

The Reticuloendothelial System

This system is comprised of phagocytes that occur in most connective tissues of the body (including the dermis of the skin and all but the epithelium of mucosal membranes). They are also found in the sinusoidal capillary spaces of the viscera.

There are two functions of the reticuloendothelial system: to initiate the process of rendering immunity, as described in the next chapter; and to engage in phagocytosis without calling the immunological system into action.

Phagocytosis is a process in which a cell literally eats something which lies outside of itself. The food material may originate within the body as dead cellular material, remnants of a blood clot, or excessive mucous secretions. Or it may come from outside the body by way of the skin or mucosa — through a wound or through the respiratory, digestive, excretory, or reproductive tracts.

Phagocytes are present in or near all tissues of the body. They are of two types: fixed and freely mobile.

SINUSOID. *A large capillary space (10 to 40 micrometers in diameter) through which blood passes in a visceral organ.*

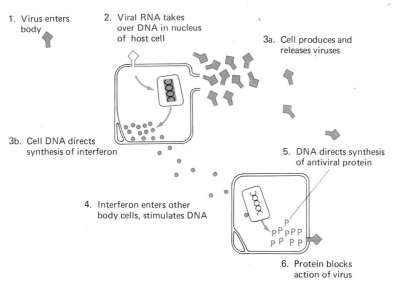

FIG. 31.2. *The interferon defense system.*

The Body's Third Line of Defense

Fixed phagocytes are, for the most part, the reticular cells (macrophages) that form a framework within loose connective tissues throughout the body. They are especially abundant in lymphoid tissues (thymus, spleen, lymph nodes, tonsils, Peyer's patches in the intestine) and myeloid tissue (red and yellow bone marrow).

MYELOID. Myelin-like (G. myelos, *fatty or marrow;* eidos, *resembles*).

The *freely mobile phagocytes* are divided into two classes: all of the *white blood cells* in blood and lymph, and the *histiocytes* (mobile macrophages) of connective tissues. Some of the histiocytes are seen in connective tissues as a result of having moved in from the blood, where they formerly existed as monocytes (blood macrophages).

During every moment of one's life, trillions of macrophages are actively disposing of tissue debris and foreign invaders. Ordinarily, we are totally unaware of their activities.

Occasionally, however, when infectious agents enter the body, a massive army of phagocytic cells is called into action, and certain special activities of mast cells, basophils (one of the three types of granulated while blood cells), and blood platelets come into play. Altogether, the activities of these cells usher in an inflammatory response, a response we usually become quite aware of.

The Inflammatory Response — Helpful and Adverse Consequences. The chief characteristics of an inflammation are: (1) redness, heat, and swelling; (2) pain; and (3) sometimes a reduction or complete loss of function. All of these signs may be observed in some instances but no one of them is always necessarily present.

The redness, heat, and swelling result from an increased amount of blood in the affected tissue due to the release of various kinds of secretory materials from several types of cells (Figure 31.3). The three principal materials are histamine, serotonin, and heparin.

HISTAMINE. A metabolite produced from the amino acid histidine by removing a carboxyl (COO^-) group.

Histamine is released from mast cells in the connective tissues and from basophil cells and platelets in the bloodstream. *Serotonin* is also released from the mast cells and from the basophil cells (Figure 31.3). Both of these

FIG. 31.3. Secretions involved in the inflammation process.

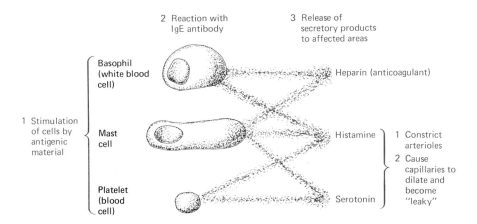

substances cause the capillaries to "loosen at their seams," allowing materials to exchange more readily between the bloodstream and connective tissues, at the same time creating an edema. They also dilate the capillaries, allowing a greater flow of blood at the site of inflammation. In addition, they stimulate the contraction of smooth muscle, allowing small arteries to send larger quantities of blood into the adjacent capillary networks. This increase in blood flow creates the redness of an inflammation.

Although the inflammatory response is a natural one—and helpful when carried out early enough and terminated rapidly enough—it can also be unduly long, discomforting, and deleterious in the long run. Consequently, a person may seek relief by taking antihistamine drugs.

Heparin, like histamine, is released from mast cells in connective tissues and basophils in the bloodstream. Unlike histamine, however, it is an anticoagulant. It is ordinarily released to offset any tendency toward internal clotting, as might happen when we bump forcefully against some object and thus injure ourselves internally. So too, during an inflammatory response, heparin is released to prevent the clotting of blood which leaks out of capillaries into connective tissue spaces.

The *pain* which is felt in an inflamed tissue may be due to the release of prostaglandins from platelets and other cells. Pain may also be caused by pressure on nerve endings, or changes in pH that often result due to the accumulation of carbon dioxide, lactic acid, and other acidic metabolites that are not readily flushed away by the circulatory system.

The *reduction in function* that often accompanies an inflammation may be due to discomfort in movement—as happens in various forms of arthritis. Alternatively, it may be due to the actual destruction of tissues, as happens to some extent in most forms of inflammation, such as pancreatitis, tonsillitis, or hepatitis.

Inflammatory responses to infectious agents are associated with a rapid movement of neutrophils to the site of infection. As the infection continues, large numbers of monocytes come into the picture, and sometimes eosinophils. The total number of white blood cells may reach 100,000 per microliter (normally there are about one tenth as many cells per microliter). This increase occurs in certain infections, such as osteomyelitis, empyema, and septicemia.

An inflammation may arise rapidly and reach a peak quickly, whereupon the individual may recover or die. Some viral and bacterial infections bring about such *acute inflammations*. In the event of recovery, the phagocytic cells rid the area of the infection and in doing so, may form a *pus*—a liquid containing white blood cells and the debris of dead cells and tissue elements that were liquefied during the inflammation. If the inflammation had been great enough, the person may have developed a *fever*. In the final stages of inflammation, new capillaries and connective tissues are formed.

In contrast to the acute case, an inflammation may arise rapidly or slowly, but in either case persist for several months or years. This condition, a *chronic inflammation*, results when the cause of the inflammation, or something produced in the tissue as a result, cannot be overcome.

SEROTONIN. *A neurotransmitter in the brain and a vasoconstrictor.*

HEPARIN. *A glycoprotein which prevents blood from coagulating.*

ARTHRITIS. *An inflammation in a joint (G. arthros, joint; itis, disease).*

SEPTICEMIA. *A systemic disease produced by microbes and their toxins in the blood (G. sepsis, decay; haima, blood).*

PUS. *A yellow product of inflammation composed mainly of serum and white blood cells (G. pyon, pus).*

Summary

The Role of Prostaglandins and Aspirin in Inflammation and Anti-Inflammation. During an inflammatory response, several kinds of chemical substances are liberated from cells into the site of inflammation. Among these substances are: (1) histamine and serotonin; (2) antibodies and complement, which will be discussed soon; and (3) prostaglandins.

The *prostaglandins* are a family of lipids which were originally discovered in 1935 in semen and reproductive tissues. They have since been found in most tissues of the body.

The role of each of the dozen or so prostaglandins has not been firmly established. There is some evidence that certain members of the family may inhibit inflammation. In this regard some of the prostaglandins have been shown to be inhibitors of gastric secretions from the stomach glands. This finding has raised the promise of making some of the prostaglandins useful as drugs for certain people with intestinal or stomach ulcers.

Another example of an anti-inflammatory action of prostaglandins is their effect on the bronchi—relaxation and dilation. For this reason, some prostaglandins may possibly be useful for treating asthma.

In general, however, most prostaglandins appear to be associated with the promotion, or mediation, of inflammations. Two points may be brought up to support this concept. First, through experiments with sheep and rodents it is fairly clear that the disintegration of the corpus luteum—following the completion of its role in the reproductive cycle—is due to prostaglandins.

The second observation has to do with aspirin, which is notable for three actions: it is a pain reliever (analgesic); it combats fever (antipyretic); and it reduces inflammation. It turns out that aspirin also is able to prevent the synthesis of prostaglandins. Moreover, the blood platelets of individuals who are taking large doses of aspirin are relatively depleted of prostaglandins. Altogether it is believed that prostaglandins mediate the pain and fever of inflammation—and that aspirin prevents such mediation by inhibiting the synthesis of prostaglandins.

PROSTAGLANDINS. *A family of fatty acid derivatives which stimulate smooth muscle, elicit pain, and very likely provoke an inflammation.*

ANTIPYRETIC. *An agent which reduces fever (G. anti, against; pyretos, burning heat).*

SUMMARY

1. The keratinized epidermis provides the skin with a certain amount of protection against damage by such items as solvents, acids, and sharp objects.
2. A mucosal membrane lines body passageways that lead to the outside of the body. By virtue of its mucous secretions, the mucosa provides the body with a certain amount of protection against invasion by foreign materials.
3. The blood clotting system is another body defense system. The principal steps in blood clotting consist of: (a) constricting arterioles that supply blood to the injured area, and (b) forming a fibrin clot through the actions of thrombin and various other clotting factors.
4. Abnormal clotting may be due to a deficiency of vitamin K, a liver

disease, a platelet deficiency, or a hereditary disorder. Vitamin K is essential for the synthesis of four clotting factors; the liver synthesizes fibrinogen and prothrombin, which are used in forming a clot; platelets contain additional clotting factors; and hemophilia is a hereditary disease.
5. Heparin normally prevents internal clotting. It is released from mast cells and basophils. Coumarin drugs which inhibit vitamin K activity are useful anticoagulants in certain medical procedures.
6. Xenometabolism refers to the conversion of certain foreign compounds (usually within the liver) to salts that are readily excreted (usually by the kidney).
7. Interferon is a substance that is released by a virus-infected cell; after its release it serves to interfere with the ability of a virus to infect other healthy cells.
8. The reticuloendothelial system, or macrophage system, is a system of phagocytes which are found: (a) as part of sinusoidal capillaries in viscera; (b) as histiocytes in connective tissues; and (c) as monocytes and other phagocytes in the blood.
9. The chief characteristics of an inflammation are: (a) redness, heat, and swelling, due primarily to histamine and heparin; (b) pain, due to physical and chemical interactions with nerves; and (c) reduction in function, due to discomfort or tissue destruction.
10. Some of the benefits of aspirin may be due to "antiprostaglandin" properties.

REVIEW QUESTIONS

1. What are the functions of the epidermis and the mucosal membranes in body defense?
2. Define: intrinsic system, extrinsic system, fibrinogen, prothrombin, xenometabolism, monocyte, histiocyte, mast cells, basophils.
3. Describe the process of blood clotting.
4. What are the roles of vitamin K, heparin, and coumarin drugs in blood clotting?
5. Describe the interferon system.
6. What is the role of the reticuloendothelial system?
7. What are the chief telltale signs of an inflammation?
8. Can you actually feel an increase in temperature in an inflamed area of tissue?
9. What properties does aspirin have that tend to counteract those of the prostaglandins?

32 Body Defense Systems: Immunological

THE IMMUNE PROCESS
    Immunological tissues
    Major kinds of cells in the immune system
    Gamma globulin
    Principal organs of the immunological system
    Abnormalities of immune tissues
IMMUNITY
    Two kinds of immune systems
    Destruction of antigens
    Active and passive circulatory immunity
    Active and passive cell-mediated immunity
    Suppressing or abolishing an immune response
    Tolerance
        Natural tolerance
        Tolerance due to swamping
ANTIGENS
    Classification of antigens
ANTIBODIES
    Classification of antibodies
TYPE I HYPERSENSITIVITY
TYPE II HYPERSENSITIVITY
TYPE III HYPERSENSITIVITY
TYPE IV HYPERSENSITIVITY
    Tissue transplantation
DRUGS AND HYPERSENSITIVITY
CANCER AND IMMUNITY

In the preceding chapter we discussed various methods used by the body in defending itself against foreign invasions. Among these was the use of the reticuloendothelial system, in which phagocyte cells (macrophages) consume foreign materials. Some foreign substances, however, pass into and out of macrophage cells without being destroyed. These substances, known as antigens, initiate an immune response from the immunological system. Other kinds of substances, also known as antigens, may perhaps stimulate an immune response without first being processed within a macrophage cell.

In the past, whenever a beneficial result was obtained through an immunological response, we said the individual received immunity. If the result was harmful, we referred to it as an allergy or hypersensitivity. We didn't appreciate then that the principles of the ABO blood system, the Rh factor, and the rejection of certain foreign tissues are in fact identical to the principles of immunity and allergy.

Today, immunity refers to any process which conforms to the three points described in the following section. Before discussing these points, however—in view of the complexity of the subject of immunology—it will be helpful to provide the reader with an introduction to the subject.

First we describe the immune process in terms of how a substance called an *antigen* induces the body to destroy it through processes which involve *antibodies*. Next, the key immunological tissues, cells, and organs are discussed, followed by brief descriptions of the more common abnormalities which occasionally afflict them.

ANTIGEN. *A substance that provokes the synthesis of antibodies, which subsequently destroy the antigen.*

With the concept of the immune process as background, along with a clear picture of the anatomy involved, it is appropriate to elaborate on the body's two kinds of immune systems—the circulatory immune system and the cell-mediated system. We describe the methods used by these systems in destroying an antigen, and also discuss two forms of immunity—active and passive.

It is possible to suppress or abolish the action of the immune system with respect to a would-be antigen. This topic is discussed next, after which the reader may find it meaningful to learn more about the nature of antigens and antibodies and their classification.

In concluding this chapter, four main classes of antigen-antibody reactions are described. Here you will be exposed to the topic of hypersensitivity, which includes allergic reactions and reactions that occur when mismatched blood is given in a blood transfusion.

THE IMMUNE PROCESS

This process, which is shown schematically in Figure 32.1, can be reduced to three points. First, a foreign material stimulates the immune system. In doing so, certain so-called "uncommitted" lymphocytes "commit" themselves "to get rid of the foreign material." The lymphocytes may be B cells or T cells:

LYMPHOCYTES. *Cells which produce and secrete antibodies or contain antibodies in their cell membrane.*

The *B cells* (bone marrow-derived lymphocytes) produce *antibodies*—specialized proteins which subsequently react chemically with the *antigen*. The antibodies are sent into the blood where they circulate as members of the gamma globulin family of proteins. The antibodies produced by the B

The Immune Process

THYMUS. An organ in the anterior superior mediastinum, behind the breastbone.

cells react with a variety of infectious organisms, including the polio virus, influenza viruses, and various bacteria.

The *T cells* (thymus-derived lymphocytes) contain a variety of weapons (proteins), including antibody-like molecules in their plasma membrane. (Like antibodies, they react with certain kinds of antigens.) The T cells are designed to react directly—as individual cells—against certain bacteria and viruses, against cancer cells and foreign tissue transplants, and against a host of other agents.

In the second step of the immune process, the antibodies, or the T cells—

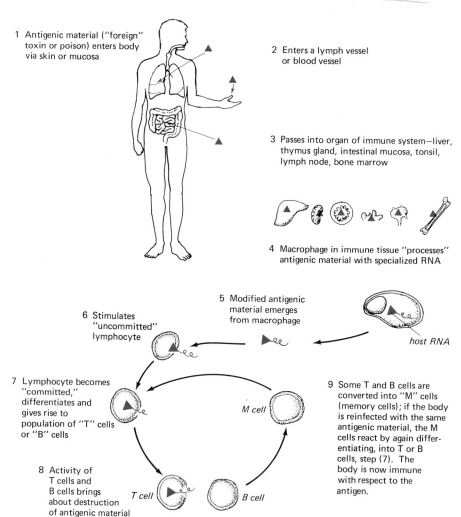

FIG. 32.1. The mechanism of the immune process.

1 Antigenic material ("foreign" toxin or poison) enters body via skin or mucosa

2 Enters a lymph vessel or blood vessel

3 Passes into organ of immune system—liver, thymus gland, intestinal mucosa, tonsil, lymph node, bone marrow

4 Macrophage in immune tissue "processes" antigenic material with specialized RNA

5 Modified antigenic material emerges from macrophage

6 Stimulates "uncommitted" lymphocyte

7 Lymphocyte becomes "committed," differentiates and gives rise to population of "T" cells or "B" cells

8 Activity of T cells and B cells brings about destruction of antigenic material (see Figures 32.2 and 32.3)

9 Some T and B cells are converted into "M" cells (memory cells); if the body is reinfected with the same antigenic material, the M cells react by again differentiating, into T or B cells, step (7). The body is now immune with respect to the antigen.

host RNA

M cell

T cell

B cell

Body Defense Systems: Immunological

depending on the nature of the antigen — react with the antigen to initiate one of several kinds of processes through which the antigen may be disposed of. These processes are shown in Figures 32.2 and 32.3.

In the third step, in response to the presence of antigens, the immune system also converts some of its B cells and T cells into "memory cells." As a result, when the same antigen invades the body at some later time, it is "recognized" by these "memory cells." In turn, additional antibodies or T cells are produced. And thus, once again, a process is initiated through which the body might be able to rid itself of the newly-arrived, but now "familiar," antigen.

Immunological Tissues

B cells, T cells, and memory cells are produced in the immunological tissues of the body. These tissues are the myeloid tissue in the bone marrow, and lymphoid tissue in the lymph nodes, spleen, thymus gland, tonsils, appendix, and in the wall of the digestive tract, particularly as Peyer's patches.

PEYER'S PATCHES. *Lymphoid nodules in the mucosa of the intestinal tract.*

1a Antigens (toxic)

1b Committed T cell (lymphocyte)
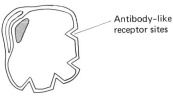
Antibody-like receptor sites

2 Interaction binds antigens to receptor sites

3 Phagocytosis destroys antigen
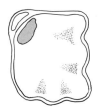

FIG. 32.2. *The cell-mediated immune system is initiated by T cells, as shown here. Other phagocytes — monocytes, histiocytes, and neutrophils — actually dispose of the bulk of the antigenic material.*

634
The Immune Process

Major Kinds of Cells in the Immune System

The important cells of the immune system are: reticular cells; macrophages; mast cells, basophils, and blood platelets; neutrophils and eosinophils; and lymphocytes. Several of these cells also play important roles in connection with phagocytosis and inflammation, as described earlier.

Reticular cells. These cells form a connective tissue "scaffolding" in the bone marrow, lymphoid organs, and widely scattered lymphoid tissues. They are, for the most part, fixed phagocytes. They produce and secrete collagenous, reticular, and elastic fibers (Figure 3.4.) Collectively, the reticular cells constitute a webbing in which antigens may be trapped and reacted with other components of the immune system.

Macrophages. These cells perform at least three immune functions. First, in the course of feeding upon certain kinds of foreign materials (including certain bacteria and viruses), they engulf the material, digest it partially, and somehow add RNA to another part of the material. The new complex of RNA and the foreign material is then sent out of the cell where it serves—as an antigen—to stimulate a lymphocyte.

A second immune function of macrophages is to dispose of antigen-antibody complexes, as shown in Figure 32.3.

The third immune function of macrophages requires some preliminary information. Ordinarily, when an infectious agent—such as a streptococcal bacterium—enters the body for the first time, there are, of course, no antibodies present to cope with it. Soon after, however, antibodies are formed, but often far more slowly than the rate at which the bacteria are multiplying. In the absence of macrophages, there would soon be a great excess of

RETICULAR. *Resembling a net, as a network of cells (L. reticulum, net).*

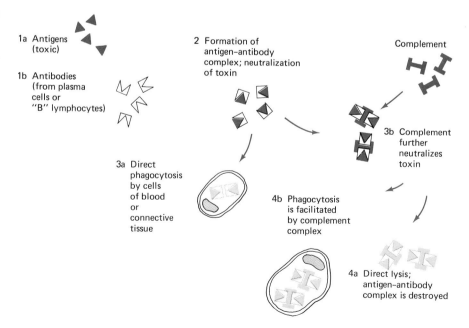

FIG. 32.3. *The circulatory immune system. Neutralization of toxin may occur at either (2) or (3b).*

Body Defense Systems: Immunological

the antigen (bacteria) over the antibody. A phenomenon known as "tolerance" would arise, wherein the immune system would fail to cope with the bacteria and one of two things would usually follow. Either the bacteria would take over and the afflicted individual would succumb, or the body would somehow overcome the bacteria—but immunity would not be granted.

The third immunological function of macrophages is believed to be the trapping of excessive antigen, whereby tolerance is prevented.

Mast cells, basophils, and blood platelets. These cells—through the release of their granular contents (histamine, serotonin, heparin)—promote the movement of blood, lymph, and tissue macrophages into regions of tissue where antigens lie (see page 626 and Figure 31.3).

BASOPHIL. *A white blood cell whose cytoplasmic granules have a strong affinity for alkaline dyes.*

Neutrophils and eosinophils. These cells are involved in feeding upon antigen-antibody complexes and thus disposing of them. They share this role with basophils, other macrophages, and to a lesser extent, perhaps, lymphocytes.

Lymphocytes. These cells circulate in the blood and lymph and wander around freely in connective tissues throughout the body. They are particularly abundant in lymphoid and myeloid tissues, and on occasion are seen wandering between epithelial cells (tissue surface cells and cells that make up the body of glands throughout the body). They differentiate into B cells, T cells, and memory cells.

LYMPHOCYTE. *An agranular white blood cell with scarcely any cytoplasm.*

Gamma Globulin

Gamma globulin is another important component of the immune system. It is a special class of proteins present in blood—in blood serum. *Serum* is the liquid which is exuded when plasma is allowed to clot. Serum contains many kinds of chemicals, including two important kinds of proteins: albumin and globulin. These two classes of proteins can be separated from one another through a process called *electrophoresis*. Using this same process, three fractions of globulins can be separated from one another. They are termed *alpha, beta,* and *gamma* globulin, in the order of their appearance upon being separated from one another (Figure 32.4).

Antibodies are found to a very minor extent in the alpha and beta globulins. Most antibodies are present in gamma globulin. In fact, this globulin fraction increases considerably, in total bulk, sometime after an antigen has entered the blood.

ELECTROPHORESIS. *Migration of charged particles through a salt solution between the negative and positive poles of a battery (G. phoresis, to carry).*

Principal Organs of the Immunological System

The principal organs in the immunological system are the bone marrow, spleen, lymph nodes, thymus gland, tonsils, intestine, and liver.

Bone marrow. There are two major immunological functions of this organ. It supplies the thymus gland with lymphocytes which can be modi-

The Immune Process

fied into T cells, and it supplies the body with B cells, which differentiate into plasma cells and produce antibodies.

The bone marrow is extremely important to the immunological system from another point of view, a viewpoint based upon experiments with laboratory animals. In these experiments, when an animal is radiated with a lethal dose of X rays, the entire hematopoietic system—including the lymphocyte-forming tissue of the lymph nodes and spleen—is destroyed. However, this system may be restored, and the animal may be saved. For this to happen, healthy bone marrow must be transplanted into the animal. No other tissue will do. The effectiveness of bone marrow rests upon the

HEMATOPOIESIS. *Formation of blood cells (G. haima, blood; poiesis, formation).*

FIG. 32.4. *Separation of serum proteins by electrophoresis (1) and profiles (2) of serum proteins from healthy and diseased individuals. Intensity of color in (2) indicates the proportions of the different proteins.*

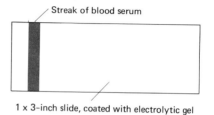

1 Electric current is passed along the material that contains the serum sample; the current causes certain classes of protein to separate into distinct streaks

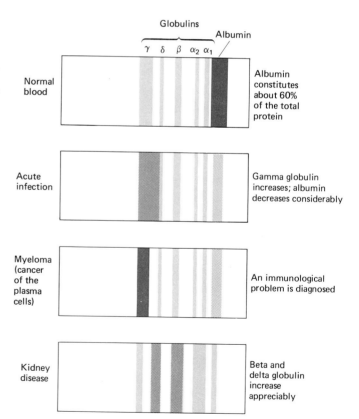

2 A protein stain is then applied to the slide, permitting diagnosis of certain disorders by measurement of albumin and the globulins

fact that it alone has the sufficient amount and kind of *stem cells* that are needed to repopulate the other lymphoid tissues. Any kind of blood cell may be derived from a stem cell. See Figure 3.5.

Spleen. This organ is located just beneath the diaphragm, between the stomach and left kidney. A cross-sectional diagram and histological section of the spleen is given in Figure 32.5. From this figure you may note two things. First, as blood passes from the splenic artery to the splenic vein, it first passes through regions known as white pulp, then through regions known as red pulp, and then through venous sinuses (passageways). The *white pulp* is basically a lymph nodule which forms a globular collar around an arteriole. The *red pulp* consists primarily of cords of tissue which restrain or sequester (imprison) red blood cells—wherein old and damaged blood cells may be destroyed. The red pulp also acts as an emergency reserve of red blood cells, in the event of hemorrhage, shock, or sudden transfer to a high altitude where the availability of oxygen is considerably reduced.

The second thing to note is that the white pulp, which is an aggregate mainly of lymphocytes, contains a *germinal center* and a *marginal zone*. See Figure 32.5. There is now fairly strong evidence that memory cells arise in the germinal center. There is also evidence that the maintenance of the marginal zone—and the proliferation of lymphocytes within that zone—is highly dependent upon either a hormone which originates in the

STEM CELL. *A nonspecialized mother cell or blast cell capable of differentiating into a specialized cell.*

SPLEEN. *The largest lymphoid organ in the body, situated on the left side, beneath the diaphragm (G.).*

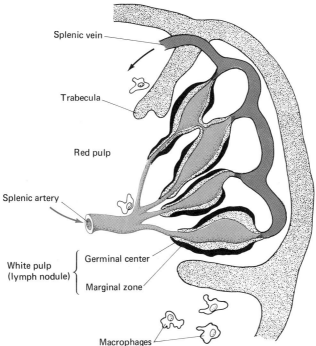

FIG. 32.5. *Microanatomy of a section of the spleen.*

thymus gland, or upon interactions with lymphocytes which enter the marginal zone from the thymus gland.

The spleen has three principal immunological functions: to proliferate B cells within the germinal centers; to store T cells—which it acquires from the thymus gland—probably keeping them in the marginal zones; and to produce memory cells within the germinal centers.

In infants, the spleen comprises a major fraction of the body's total lymphoid tissue. Without it, the infant stands little chance of survival. Even children have a dire need for it. The removal of this organ from children of less than about three years old has been associated with severe, sometimes fatal, bacterial infections. But the removal of the spleen from fully mature adults, particularly the middle-aged and elderly, appears usually to have no harmful effects.

Lymph nodes. These glands are widespread in their distribution, as shown in Figure 25.9. They occur along the lymphatic circulatory route, where they function mainly to filter lymph. They receive a supply of blood, as shown in Figure 32.6. This blood, like the lymph itself, is monitored here, and cleansed of debris and worn-out cells.

A cross-sectional picture of a lymph node is given in Figure 32.6. From this figure you may note that, like the spleen, the lymph node contains lymph nodules. Correspondingly, the germinal centers of these nodules give rise to B cells and memory cells.

You may also note the lymphoid tissue at the base of the cortex—near the medulla. This tissue, like the marginal zones in the spleen, is maintained either by a hormone or by interactions from cells derived from the thymus gland. Also like the marginal zones, it is rich in small lymphocytes—the T cells which it acquires from the thymus gland.

Lymph nodes may enlarge up to fivefold when they are active in produc-

LYMPH NODES. *Masses of lymphatic tissue, 1–25 millimeters long, often bean-shaped, situated along the route of lymph vessels* (L. lympha, *water*).

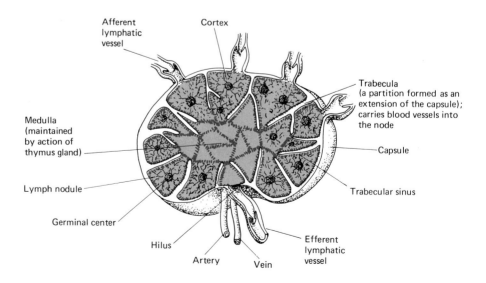

FIG. 32.6. *Anatomy of a lymph node.*

ing antibodies. On the other hand, they may decrease to about a third of their size when the thymus gland is removed.

Thymus gland. This organ is located behind the breastbone, in front of the heart. It is almost completely developed before birth, in contrast to the bone marrow, spleen, lymph nodes, and other lymphoid tissues. In fact, at the time of birth it is as large as the heart.

THYMUS. *An organ, situated behind the sternum, which produces and secretes hormones that stimulate the differentiation of lymphocytes into immunological cells.*

In children, the thymus gland produces lymphocytes that are released into the bloodstream. In the adult, the thymus is a much smaller organ. It produces very few lymphocytes on its own. Most of the lymphocytes in it are derived from bone marrow. And even so, more than 90% of them never leave the thymus gland again.

The thymus gland possesses very little lymphoid character. It is more "epithelial-like." Its "epitheloid" cells—and not reticular cells as would be the case in the other lymphoid tissues—serve as a supporting framework for the lymphocytes which it harbors.

What then, is the immunological function of the thymus gland? There are at least three functions. One is to stimulate the lymph nodes and spleen to produce lymphocytes. This is due to thymic hormone (thymosin) that comes from the thymus gland. When the thymus gland is removed, fewer lymphocytes are produced in the inner region of the lymph nodes and the marginal zones of the spleen.

A second function is to convert lymphocytes—which are derived certainly from the bone marrow, and perhaps also from the spleen, lymph nodes, and other lymphoid tissues—into T cells. The T cells, in turn, are sent out to populate the spleen, lymph nodes, and other lymphoid tissues. It is noteworthy, in this regard, that the thymus gland diminishes considerably in size during certain severe illnesses. The decrease in size occurs as a result of sending lymphocytes out of the organ to cope with the illness.

LYMPHOID. *Like lymphatic vessels or lymph nodes (L. lympha, water; G. eidos, resembles).*

The third function is to possibly allow the proper development of the spleen. When the thymus gland is removed from newborn mice, the spleen fails to reach its normal size.

Tonsils, appendix, and intestine. No one knows the specific immune functions of lymphoid tissue that occurs as lymph nodules in the intestinal wall (Peyer's patches), in the appendix, and in the tonsils.

It is interesting to note that the tonsils form a ring of tissue around the pharynx. You may recall that the nose, mouth, ears, and trachea open into the pharynx. It is also of interest that the tonsils are not covered on their surface by a thick capsule, as are the lymph nodes, thymus, and spleen.

PHARYNX. *Throat (G.).*

It can be reasoned that the tonsils serve to monitor any material which might enter the body—with food from the mouth, or with mucus from the nose, trachea, bronchi, and middle ear. If the material is antigenic, the tonsils could initiate an immune response. If the antigenic material should get past the tonsils and enter the gut, the lymphoid tissues of the intestinal mucosa could serve as the site at which the immune response is elicited.

It is of further interest that the tonsils achieve their maximum development early in childhood, when the immunological machinery is busiest in establishing immunity. Thereafter, the size of these tissues declines—assuming the tissue is not removed in a tonsillectomy operation.

The Immune Process

FAUCES. *The passage from the back of the mouth to the throat (L. throat, gullet).*

TONSILLECTOMY. *Removal of part of the tonsils (G. tome, cut).*

Exactly where are the tonsils? If you look into a mirror with your mouth wide open you'll see the uvula that was discussed in the chapter on digestion. Also present are the pillars of fauces that were discussed then—one of them arches forward and downward, to the sides of the mouth; the other backward and downward, also to the sides of the mouth. Between these arches, on both sides of the mouth, lie the *palatine tonsils*—the tonsils which are most commonly removed during a tonsillectomy. Further back, along the rearmost wall of the oral opening is the pharyngeal wall. Along this wall are the *pharyngeal* or *adenoid tonsils*. These tonsils may be removed when they become severely and repeatedly inflamed (adenoiditis), causing a chronic blockage of the eustachian tube. A third kind of tonsils—the *lingual tonsils*—lie at the root of the tongue. These tonsils extend along the pharyngeal passageway—the passageway taken by food moving into the esophagus and air moving into the trachea.

Liver. Until recently, it was believed that most antigens usually initiated an immune process in a lymph node or some other lymphoid tissue or in the myeloid tissue of bone. It is now believed that the process may perhaps be initiated in a macrophage anywhere within the body, including the liver. Exactly what initial steps are involved is not fully known, but there is considerable evidence that when antigens enter the body, the bulk of the material is sequestered (imprisoned) within the macrophages of the liver (Kupffer cells) while the antibody machinery is being set up. In this way, it is believed that a large quantity of antigenic material is prevented from being circulated, and thus the body does not enter a state of immunological tolerance.

KUPFFER CELLS. *Macrophage cells which line the sinusoidal passages of the liver and monitor the blood which flows by.*

Abnormalities of Immune Tissues

The phagocytic cells and tissues of the reticuloendothelial system and the circulating blood are subject to a variety of abnormalities. Some of the more common problems are tonsillitis, adenoiditis, splenomegaly, leukemia, Hodgkin's disease, multiple myeloma, and infectious mononucleosis.

Tonsillitis is an inflammation of the lymph nodules within the tonsil tissues. *Adenoiditis* is an inflammation of the pharyngeal tonsils. *Splenomegaly* is an enlargement of the spleen due to infectious agents, hypertension in the liver portal system, and other diseases. This problem is usually accompanied by anemia, a low white blood cell count, a low platelet count, or combinations of these conditions.

Leukemia is a disease characterized by an excessive proliferation of white blood cells in the bone marrow. There may also be a proliferation of cells in the spleen, lymph nodes, and liver—among the blood-managing tissues; and in the meninges, intestinal tract, kidney, and skin—tissues which, neither in their embryology nor maturity, are connected with forming or destroying blood cells.

LEUKEMIA. *An uncontrolled proliferation of white blood cells; a disease of the reticuloendothelial system (G. leukos, white; haima, blood).*

There are many forms of leukemia. Each form can be classified as acute or chronic.

Acute leukemia appears suddenly, often with symptoms not unlike those of a cold. It progresses rapidly, bringing about enlargement of the lymph nodes, spleen, and liver, due to infiltration with white blood cells. It is often accompanied by bone pain, pale skin color, a tendency to bleed easily, and a high susceptibility to infection. If not treated, death occurs in about three months, most commonly from hemorrhaging and infection.

Chronic leukemias appear more slowly than the acute varieties. Often, they are discovered during a routine blood examination. One or more years may pass before the symptoms appear. The symptoms resemble those of acute leukemias, though life expectancy in untreated individuals is generally about three years from the onset of the disease.

An individual with untreated acute leukemia may have a normal, low, or high count of white blood cells. When white blood cells are difficult to find in a smear, the disease is called *aleukemic leukemia.*

Acute lymphocytic leukemia is the most common leukemia of childhood, achieving a peak among children three and four years old. Acute lymphocytic and *acute monocytic leukemias* occur more commonly in adults.

An individual with chronic leukemia in almost all cases has a high count of white blood cells, well beyond the normal of 10,000 per microliter, sometimes as high as 500,000. *Chronic granulocytic leukemia* occurs most commonly in young adults. *Chronic lymphocytic leukemia* occurs mostly in people over 60, rarely in people under 40.

The exact causes of leukemia are not usually known. In numerous cases of lymphatic leukemia there is compelling evidence that the cause of the disease is a mutation of chromosome number 21.

Other causes of leukemia may be viruses, though this has never been proved; damage to bone marrow, as for example, through radiation; and an immunological deficiency, as is characteristic of lymphocytic leukemia.

Three abnormalities are common in leukemia: (1) bone marrow failure, which leads to anemia, thrombocytopenia, and—except in chronic granulocytic leukemia—a decrease in granulocytes; (2) enlargement of the lymph nodes, spleen, and bone marrow (causing bone pain); and (3) an immunological deficiency.

The third abnormality occurs less frequently than the first two, but is common in lymphocytic leukemias. The individual's lymphocytes lack the ability to differentiate into plasma cells capable of producing sufficient amounts of antibody. In addition, the lymphocytes are inadequate in mustering up cellular immunological responses, which are described later. Third, the immunological system produces a relatively high amount of autoantibodies, which serve to destroy tissues that are now foreign—no longer native—to the body.

There is no known cure for leukemia. What is hoped for is a *remission,* a temporary condition in which the symptoms of the disease disappear. Remission may be brought about through natural biological causes, often after treating an individual with radiation (for example, the spleen or the whole body), with drugs (such as antibiotics, vincristine, or methotrexate) or with adrenal corticoid hormones.

LYMPHOCYTIC LEUKEMIA. *A cancer of the lymphocytes.*

GRANULOCYTES. *The white blood cells which contain visible granules in their cytoplasm—the neutrophils, basophils, and eosinophils.*

THROMBOCYTOPENIA. *A deficiency in blood platelets (G. thrombos,* clot*; kytos,* cell*; penia,* deficiency*).*

Immunity

HODGKIN *(1778–1866). An English physician who first (1832) described a lymphocyte tumor (lymphoma) now called Hodgkin's disease.*

Hodgkin's disease is a cancerous disease which usually begins in a single lymph node. If it is not discovered soon enough, it generally spreads to other lymph nodes, to the spleen, liver, bone marrow, and throughout the body.

Hodgkin's disease may begin at any age, but is found particularly in two age populations, one that peaks around age 15 and another that peaks around age 50. Its chief undeniable telltale sign is the presence of Reed-Sternberg cells, in a biopsy specimen. These are large cells which may appear in several forms. Reed-Sternberg cells typically have two overlapping nuclei with conspicuous nucleoli.

As the disease progresses, the individual becomes less and less capable of mustering up a cellular-immunological defense in which the lymphocytes dispose of antigenic substances. This is so despite the fact that there is usually a greater than normal number of lymphocytes.

In contrast to leukemia, attempts are made to cure Hodgkin's disease. When radiation is applied early enough, the affected lymph node can be destroyed. Provided the cancer has not spread to other tissues, there is less than a 5% chance of recurrence.

MYELOMA. *A tumor of bone marrow; often also called a plasma cell tumor (G. myelos, marrow; oma, tumor).*

Multiple myeloma, which rarely occurs in people under 40, is a fatal cancer of the plasma cells. Plasma cells are believed by many investigators to be derived from B cells upon being stimulated antigenically. They differ from B cells in having a more highly developed endoplasmic reticulum and Golgi apparatus—organelles involved in the production of antibodies.

The cancerous plasma cells proliferate throughout the bone marrow, giving rise to bone pain, which is usually the first complaint of the afflicted individual. The neoplastic (cancerous) cells also proliferate in the spleen, liver, and lymph nodes. When they pass in larger numbers out into the blood, the individual is said to have *plasma cell leukemia*.

MONONUCLEOSIS. *A disease characterized by the appearance of an excess number of monocytes (G. monos, alone; L. nucleus).*

Infectious mononucleosis, which is believed to be due to the EB virus, is characterized by swollen lymph nodes, spleen, and liver; inflammation of the pharynx (pharyngitis); and a large number of atypical lymphocytes (possibly monocytes) in the bloodstream. The disease can be spread through kissing and other intimate contacts. After a "silent" incubation period of about a week, the individual experiences a sore throat, general weakness, and fever. These characteristics come and go over a variable time period. Usually the disease terminates spontaneously after about three weeks to three months. Once the person has recovered, he is generally immune to future attacks.

At present, nothing specific can be done to cure the disease. There may eventually be an antiserum available against the EB virus.

IMMUNITY

Immunity is granted to an individual when two criteria are fulfilled: (1) An antigen is encountered, specific B cells or T cells appear, and the antigen is disposed of. This is called the *primary immune response*. A few minutes to a few days are usually required before a measurable quantity of antibody appears in the blood.

(2) At a later date—anytime during the person's lifetime—when the same kind of antigen appears, specific B cells or T cells appear in response. This response is called the secondary immune response.

In a *secondary, or anamnestic, immune response*, antibodies or T cells are produced within a shorter latency; that is, less time passes between antigenic stimulation and the appearance of specific antibodies or active T cells. In addition, a larger number of antibodies and T cells are produced, and the antigen is disposed of in a shorter period.

Because of the shorter latency during the secondary immune response, the immune system is said to have *"memory."* Memory is due to the retention, persistence, or subsistence of lymphocytes with specific antibody in or on their plasma membrane. When an antigen provokes a secondary immune response, the antigen reacts with these particular antibodies. The antigen-antibody complex then moves into the interior of the cell—as can be seen directly through an electron microscope using special radioactive isotope techniques. It is believed that shortly after this the lymphocyte begins to divide. In doing so it gives rise to numerous specific B or T cells, depending on whether the dividing cell is a B or T cell. It also gives rise to memory cells.

PLASMA MEMBRANE. *The outer cell membrane.*

Two Kinds of Immune Systems

The immune system reacts to foreign invasion through one of two kinds of systems: a circulatory immune system and a cell-mediated system.

The first system produces B cells, which in turn, produce antibodies. The antibodies circulate throughout the blood—mainly as components of the gamma globulin family of proteins. They constitute the *circulatory immune system,* which attacks antigens "immediately."

The second system produces T cells—lymphocytes which contain antibodylike material in their plasma membrane. These cells constitute the *cell-mediated system,* which attacks antigens "after a delay." These T cells need time to multiply and muster up an attack against the antigen. Usually the symptom of attack—an inflammation—does not show up until 18 to 24 hours or more. Accordingly, the cell-mediated immunity is also referred to as *delayed-hypersensitivity*.

HYPERSENSITIVITY. *In common usage, an allergy.*

Destruction of Antigens

In the *circulatory immune system,* the initial step for ridding the body of an antigen is a reaction between the antigen and an antibody. Thereafter, one of two things usually happens: either the antigen-antibody complex is eaten by a phagocytic cell, or complement comes into the picture.

COMPLEMENT. *A class of serum proteins that participate in immune reactions.*

Complement is a system consisting of eleven different proteins—see Figure 32.7. These proteins are carried in the blood within the beta globulin fraction. You can see from the figure that the eleven different proteins can be combined to form a complex. However, before this complex is formed, various combinations of the components are formed first. You should also note that only nine of the eleven complement components are shown in the figure; this is because component C1 is actually a complex in itself, formed from three of the eleven proteins.

The entire complex of eleven proteins and the various combinations of smaller complexes play distinct roles in immune reactions. Some of these roles are:

(1) The entire complex of eleven proteins attaches itself to an antigen-antibody complex. If the antigen is part of a cell—as is the case for certain bacteria or for the ABO blood system, where antigens A, B, or AB are present in the plasma membrane of red blood cells—the entire cell becomes lysed. Holes are punched into the membrane and the cellular constituents become dissolved and digested.

(2) A complex of six of the components of complement—on being attached to an antigen-antibody complex—renders the entire structure more susceptible to phagocytosis. Here, the antigen need not be part of a cell.

(3) Additional proteins—see Figure 32.7—when added to the structure described in (2) above, renders the entire structure *chemotaxic*. In the case considered here, lymphocytes, macrophages, platelets, and basophils are attracted to the chemotactic complement.

(4) A complex of three components of complement (5, 6, and 7)—not shown in the figure—also serves to bring immunologically-active cells into regions of tissue where antigen-antibody complexes occur.

CHEMOTAXIS. *Attraction or repulsion of a microbe, blood cell, or some biological organism, by a chemical.*

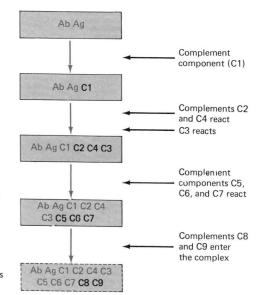

FIG. 32.7. *The complement system. After step (5) organic materials in the complex become available as nutrients for the body cells.*

(5) Also not shown in Figure 32.7 is a combination of components 4 and 2, which neutralizes the toxicity of certain bacterial and viral antigens.

Other possibilities exist wherein single components of the complement system are involved in antigen-antibody reactions.

In short, complement facilitates the disposal of antigen-antibody complexes by dissolving the complex in cases where the antigen is part of a foreign cell; attracting defensive cells into the vicinity of the complex; adding some property to the complex which makes it more likely to be phagocytized; and neutralizing the toxins of certain microbes.

In the *cell-mediated immune system*, lymphocytes (T cells) react directly with the antigen. The reaction occurs in connective tissues, where there is always an available supply of mast cells, basophils, and platelets. An inflammatory response is also part of the picture. Finally, complement may be involved in the process. Altogether, the region of attack—as, for example, the skin in poison ivy, or the bronchi and bronchioles in asthma—becomes warm, swollen, and painful. Tissue destruction occurs, but, assuming the antigenic material is disposed of, the inflammation subsides and the tissues may recover their health.

TOXICITY. *The quality of being poisonous (G. toxikon, poison).*

Active and Passive Circulatory Immunity

Circulatory immunity may be provided actively or passively. *Active immunity* is initiated when an antigen is inoculated into the connective tissues. The antigen may be any one of the following: (1) a vaccine, such as the virus of cowpox, which is used to confer immunity against smallpox; (2) attenuated (weakened) viruses other than vaccine; (3) attenuated or dead bacteria.

Attenuation is accomplished in one of several ways—by treating the toxic microorganism with formaldehyde or heat, or by transferring the organism from one animal (usually a monkey) to another animal of the same species, and then successively to a third, fourth—sometimes up to a fortieth or sixtieth animal.

The antigen is usually inoculated into or below the skin. Here it enters a blood or lymph vessel (see Figure 32.1) and is conveyed into a lymphoid tissue. It becomes trapped within the reticular scaffolding of macrophages in the lymphoid tissue. A macrophage devours the antigen and very likely digests a portion of it. The macrophage then adds RNA to the remainder of the material. The antigen-RNA complex is then sent out into the extracellular fluid.

Soon the antigen-RNA complex comes into contact with a lymphocyte—a lymphocyte which was conferred the status of a B cell following its production in bone marrow and further maturation in some other lymphoid tissue. The B cell commits itself to the task of getting rid of the antigen-RNA complex. It divides and differentiates into plasma cells (so it is believed) and memory cells. Immunity is thus now actively conferred upon the individual with respect to the specific antigen that was introduced.

INOCULATE. *To inject, implant, or insert (L. inoculatus, engraft).*

ATTENUATE. *To weaken (L.).*

RNA. *Ribonucleic acid, several forms of which are essential in the synthesis of protein.*

Immunity

ANTISERUM. Serum which contains an antibody to a particular antigen.

Passive immunity, in contrast to active immunity, is rendered by injecting a person with gamma globulin or an antiserum. An *antiserum* is a serum which is removed from an animal (usually cattle) which was treated "actively" with an antigen. Of course, the particular gamma globulin or antiserum must contain the desired antibody. After one of these substances is injected, the antibody reacts with the antigen which is in the individual or is suspected of being in the individual.

Passive immunity is resorted to only when time will not permit an individual to develop active immunity to an antigen—when the individual might otherwise succumb or suffer some harmful consequence. For example, if a woman is in her first three months of pregnancy, and she has not had rubella (German measles) or is uncertain about it, she may be given a whopping dose of gamma globulin if she has come into contact with someone who has the disease. Time will not permit her to develop a beneficial immune status through which she—and especially the developing fetus—may be safeguarded. Passive immunity may be obligatory in this case.

RUBELLA. German measles, an eruptive skin disease accompanied by fever and swollen lymph glands (L. rubellus, reddish).

However, even with this treatment—though the mother may not suffer the disease—the fetus may be severely afflicted. Should this happen, the mother may have a miscarriage or stillbirth, or the baby may be born with a congenital defect—such as harelip, cleft palate, eye cataracts, deafness, or heart problems. For this reason, the possibility of a therapeutic abortion may be considered once a pregnant woman is known to have contracted, or to have passive immunity to, rubella.

ALLERGY. Hypersensitivity to certain substances that in similar amounts are harmless to other people.

Passive immunity is of limited usefulness because the serum or gamma globulin is difficult to obtain and store and may not even be effective. In addition, the donor's serum may contain viruses that cause hepatitis and/or antigen to which the recipient is allergic. On the other hand, some antiserums (serums with desired antibodies) have proven to be lifesavers. These include antigangrene, antibotulism, and antitetanus.

Active and Passive Cell-Mediated Immunity

Cell-mediated immunity, like circulatory immunity, can be rendered actively or passively. *Active cell-mediated immunity* occurs, for example, in the case of tissue transplantation. If the transplantation involves tissue taken from one person and implanted in another, the tissue may be rejected. Should another transplant be attempted later—between the same recipient and donor—a secondary response will follow. As for circulatory antibodies, this secondary response is characterized by a shorter latency and a more rapid reaction.

Passive cell-mediated immunity is rendered by transferring lymphocytes from the donor to the recipient. Thereafter, when a tissue is transplanted, a reaction occurs which is characteristic of the active secondary cell-mediated response.

Suppressing or Abolishing an Immune Response

The immunological system may be prevented from reacting with an antigen. If this is done artificially, the system is said to be suppressed.

Immunosuppression can be accomplished through drugs or X rays. Cortisone and related steroid drugs, for example, have a powerful inhibitory effect on the production of antibodies. Such steroids are known to promote the formation of glucose from amino acids. In view of this, they tend to reduce the availability of amino acids for the synthesis of proteins, and thus the production of lymphocytes.

X rays also exert an effective immunosuppressive action. When used at an appropriate dose level, the irradiation kills the immunological tissues of the macrophage system as well as the circulating white blood cells.

Tolerance

Immunosuppressive drugs and/or X rays are used in many transplant operations. The aim is to wipe out the immunological system, so that the body will be *tolerant* to foreign invasion.

Tolerance has a technical connotation in immunology. It refers to the state of the immune system when the system fails to recognize a would-be antigen as being foreign. It is often descriptively referred to as immunological paralysis or unresponsiveness.

TOLERANCE. *Refers to a condition in which the immune system does not react with a would-be antigen.*

Tolerance may arise through one of three known ways: (1) through the use of certain drugs and X rays as in immunosuppression; (2) through natural processes—before the macrophage system becomes fully developed; and (3) through swamping the system with excessive quantities of antigen.

Natural Tolerance. The macrophage system is probably fully mature by the sixth month of fetal life. The thymus appears around the twelfth week, at which time the lymphoid and myeloid tissues are capable of producing antibodies. At about this time the placental wall becomes increasingly more effective as a barrier to the movement of antibodies from the mother to the fetus. Thus, it becomes increasingly important for the fetus to form its own antibodies.

If a would-be antigen enters the immunological system before this system becomes "competent," the would-be antigen may be looked upon not as being foreign, but as being native. A subsequent invasion by this would-be antigen—even long after the immunological system matures—is not regarded by the system as being a foreign invasion. Instead, the invader is tolerated.

Tolerance Due to Swamping. *Swamping* is a phenomenon that has been observed in mice. When a small dose of, say, a virulent bacterium is injected, the mouse responds immunogenically. However, when a dose ten times greater is administered, and provided the mouse does not die, toler-

VIRULENT. *Marked by a rapid, severe, and malignant course (L. virulentus, poisonous).*

ance is rendered. That is, the bacterium is not recognized as being foreign thereafter, and antibodies are not formed in response to its invasion.

ANTIGENS

By definition, any substance that is capable of triggering the immunological system into the production of antibody or T cells is said to be an antigen. In general, antigens display three characteristics:

Foreign nature. This property can be satisfied in one of two ways. In one way, the substance is strictly foreign in that it is brought into the body from the outside. In the second way, the substance is a component of the body, but has been modified by radiation, drugs, age, or some unknown factor. Thus, a newly-arising cancer cell may be antigenic.

Large size. Most antigens are large substances, usually exceeding a molecular weight of several thousand. Small substances are usually not antigenic, but upon attaching themselves chemically to a large native substance, they make the larger substance become antigenic. Thus, aspirin—whose molecular weight is only 180—is not an antigen. However, aspirin can become attached to a protein in the bloodstream and thus become an antigenic substance.

Hydrolyzability. Antigens are usually substances that can be decomposed into smaller units through lysis with water. Most antigens are proteins, nucleic acids, polysaccharides, or combinations of these, with or without additional chemically reactive groups.

MOLECULAR WEIGHT. *The weight of a molecule based on a system in which the weight of every kind of atom is related to the weight of carbon.*

HYDROLYSIS. *Cleavage of a chemical bond by water (G. hydros, water; lyse, tear apart).*

Classification of Antigens

Most antigenic materials may be classified into the following four groups:

(1) *Antigen*—any substance which, without modification, is able to elicit an immune response.

(2) *Haptenic-antigen*—a natural body constituent which binds to a small foreign molecule, such as aspirin or penicillin, and thus becomes an antigen.

The agent which chemically modifies a substance in this way is called a *hapten.* Aspirin and penicillin are haptenic agents for some individuals. Phenothiazines (tranquilizers) are also haptenic. By itself, the hapten is not able to induce the production of an antibody or to stimulate a T cell to "commit" itself.

(3) *Isoantigen*—an antigen that is present in some, but not all, members of a species. A classical example is the ABO blood system. Isoantigen A is present on the surface of the red blood cells of individuals with types A and AB blood; isoantigen B is on the red blood cells of people with blood types B and AB, and isoantigen H is present on the red blood cells of persons with blood type O.

HAPTEN. *A substance which, when combined with a high molecular weight compound, behaves like an antigen (G. haptein, to fasten).*

ISOANTIGEN. *A substance in a species that is antigenic to certain members of the species but not to other members.*

(4) *Allergen*—this type of antigen is one which: (a) fails to precipitate with its antibody in vitro, and (b) stimulates the production of blocking antibodies.

Blocking antibodies are antibodies which compete in the body with reagins for the allergen. *Reagins* are another kind of antibody. They are attached to epithelial or connective tissues here and there in the body, particularly in the skin, intestine, and respiratory system. If there is an excess of blocking antibodies, the allergens fail to react with the reagins and the individual is thus spared an uncomfortable, if not deleterious, allergic reaction.

BLOCKING ANTIBODY. *An antibody that can react with an allergin and thus block out an adverse reaction between the allergen and a reagin.*

ANTIBODIES

An *antibody* is a protein that is produced by lymphocytes or plasma cells and is capable of reacting chemically, as described earlier.

Classification of Antibodies

Five main classes of antibodies and numerous subclasses have been isolated and distinguished chemically from one another. Here we shall describe three of the classes—I_gG, I_gM and I_gE. The symbol I_g stands for immunoglobulin (antibody). These are the only classes whose functions are known.

IMMUNOGLOBULIN. *An antibody; a globular serum protein.*

I_gG *antibodies.* These antibodies are the ones most frequently and abundantly involved in immune reactions. They also are the longest lived, with a half-life of about three weeks.

During the first three months of fetal life—before the placenta has become critically thick, and at which time the fetus has not developed its own immune system—I_gG antibodies are able to pass through the placenta from the mother to the fetus. Passive immunity is thus rendered against a wide spectrum of infectious agents.

HALF LIFE. *The time required for half of a population of substances to become altered, to die, to disappear, and the like.*

I_gG antibodies are able to "fix complement." This means they have the chemical capability of forming a complex with complement—after first combining with their antigen. A description of how antigen antibody-complement complexes are disposed of was given earlier.

I_gM *antibodies.* These are the largest antibodies, about five times larger than the I_gG antibodies. They are found only in the bloodstream, where they react mainly with bacteria and mismatched red blood cells (obtained through a transfusion). Following their reaction, the antigen-antibody complex may be disposed of by complement and phagocytosis.

I_gE *antibodies.* These antibodies are by far the least abundant type of antibody in normal individuals. But in allergic people they are present in relatively high concentration, about five to ten times the normal amount.

I_gE antibodies, when present in high concentration, are found chiefly in the dermis of skin and in the mucosa of the respiratory and intestinal tracts. Here they are attached mainly to mast cells within the connective tissues. Known also as *reagins*, they interact with antigens (called *allergens*) to which an individual is allergic. This adverse reaction, which brings on an inflammation, may often be prevented, or reduced in severity, by desensitizing the individual. Desensitization is described below.

TYPE I HYPERSENSITIVITY

REAGIN. *An antibody which reacts with an allergin to produce an uncomfortable allergic reaction.*

ALLERGEN. *An antigen which provokes an allergic reaction.*

RHINITIS. *An inflammation of the mucosal lining of the nasal passageway (G. rhis, nose; itis, inflammation).*

DESENSITIZE. *To make a person less susceptible to an allergic attack.*

This type of hypersensitivity is the one which is most frequently referred to as *allergy*. Its main feature is the presence of I_gE antibodies (reagins) on cell surfaces, particularly on the surface of mast cells. The antibodies may be found in response to one or more of a variety of antigens (allergens), including pollen, dust, fur, certain foods. The allergens and reagins react in an allergic reaction to produce a deleterious effect.

When an individual suffers an allergic reaction, certain tissues are particularly susceptible to damage. The skin may react by forming hives (wheals or blisters), a condition called urticaria. Or the nasal mucosa may be inflamed, giving rise to a form of rhinitis with sneezing and a copious flow of a watery mucus. Or the intestine may be vulnerable, resulting in diarrhea and/or intestinal cramps. Combinations of these disturbances, as well as other problems, may occur.

The specific kinds of allergens to which an individual may be allergic can be determined through a *skin test*. Suspect allergens are scratched into the skin. An inflammation (swelling and reddening) will appear around the scratches that carry substances to which the individual is allergic.

An allergic person may be *desensitized*. This consists of administering small levels of allergen over a period of time ranging from weeks to months, or even indefinitely. Increasing amounts of so-called blocking antibodies (I_gG antibodies) appear as a result.

Later, when the allergic individual is challenged again, the blocking antibodies compete with the reagins for the allergens. The blocking antibodies are favored, if sufficiently high in number. As a result, the allergen is reacted (with the blocking antibodies) and disposed of before the allergen can react with a reagin to produce an inflammation.

TYPE II HYPERSENSITIVITY

Type II hypersensitivity is characterized by the presence of an antigen on the surface of a cell and an antibody in the body fluid. Whereas the I_gG and I_gE antibodies are of major concern in Type I hypersensitivity, the I_gG and I_gM antibodies are of major concern in Types II and III hypersensitivity.

There are five notable examples of Type II hypersensitivity: microbial, the ABO blood system, the Rh factor, drugs, and autoimmunity.

Microbial hypersensitivity. Microorganisms, whether bacterial, viral, or protozoan—usually display antigenic properties. After being introduced into the body—and provided they elude the phagocytic action of neutrophils—an immune response occurs, as a result, and the organism is destroyed and phagocytized.

ABO blood typing. Every person has two of the following genes, in any combination: A, B, O. Six genotypes are therefore possible: AA, AO, BB, BO, AB, and OO. That is, every person has one of these six sets of genes.

Genes A and B are each dominant over gene O. Four blood types are therefore possible: A, B, AB, and O. Every person has one of these four possible types of blood.

Each gene directs the synthesis of a specific isoantigen: A, B, or H. Therefore, if you have blood type A, B, or AB, you also have the corresponding isoantigens in the membranes of your red blood cells; that is (in the same order) A, B, or A *and* B isoantigens. There is no antibody for isoantigen H.

If you have blood type A, your serum contains antibodies B; if you have blood type B, your serum contains antibodies A; blood type AB has neither antibody; and blood type O has both kinds of antibodies—A and B. Therefore, blood type O, with neither isoantigen A nor B, can be donated to anyone, *provided it matches in all other respects.* This type of person is often called a "universal donor."

Persons with blood type A can donate blood to individuals with blood types A or AB. Blood type B can be donated to individuals with blood types B or AB. Blood type AB can be donated only to persons with blood type AB. Because an individual with blood type AB has neither antibody A nor antibody B, that person can accept blood from anyone. Such a person is called a "universal acceptor."

The Rh factor. This factor is an antigen on the red blood cells of about 85% of the human population. It is important from the viewpoint of blood transfusions and pregnancy.

An Rh-negative individual, who might accidentally receive Rh-positive blood, would react to the transfusion by producing antibodies. These antibodies, in turn, would cause agglutination (clumping) of the red blood cells—with dire consequences.

In the case of an Rh-negative woman who is pregnant, the following situation may arise. In the event of damage to the uterus, the Rh antigen may pass from the fetal component of the placenta into the mother's blood. A primary immune response may follow, during which memory cells and antibodies are produced. (Ordinarily, the red blood cells of a fetus do not pass across the placenta into the mother's blood. On occasion, however— due perhaps to an accident or drugs, but more often due to the breakage of placental and uterine tissues during childbirth—the Rh factor can enter the mother's tissues.)

Rarely during the first pregnancy, and only in a low percentage of the second and subsequent pregnancies, difficulties arise. In these cases, the Rh antibody passes through the placenta from the mother to the fetus. If

GENOTYPE. *The hereditary constitution of an individual (G. gignesthai, to be produced; typos, type).*

AGGLUTINATE. *To aggregate cells into clumps or masses (L. ad, to; glutin, glue).*

the father has two genes (is homozygous) for the Rh factor the fetus will surely be Rh-positive. If the father has only one gene (is heterozygous) the fetus has a 50% chance of being Rh-positive. In either case, if the fetus is Rh positive, the mother's Rh antibodies will react with the red blood cells of the fetus, causing deleterious results, sometimes death.

Today, however, due to available information on blood type, precautions are taken to prevent such a mishap. The Rh-negative mother is injected with Rh antibodies on the delivery table at the time of each delivery. She is thus given *passive immunity*, in which any Rh antigens from the fetal placenta are destroyed, and a primary immune response is not established.

Drug hypersensitivity. Drugs of various sorts such as the phenothiazines, which are used as tranquilizers), may induce a type II hypersensitivity (as well as types I, III and IV). Although by itself the drug may be a hapten, its combination with proteinaceous molecules on cell surfaces may result in an antigen. If so, antibodies will be produced. The antibody then usually directs the destruction of the cell. It does so either by activating the complement system or the macrophages.

Two of the more common results of drug-induced hypersensitivities are agranulocytosis (destruction of lymphocytes and monocytes) and hemolytic anemia (anemia due to destruction of red blood cells).

AGRANULOCYTOSIS. *Pronounced reduction in white blood cells, especially the neutrophils (G. a, without; L. granulum, granules; G. kytos, cell; G. osis, activity).*

Autoimmunity. In autoimmunity, a form of type II hypersensitivity, the immune system reacts to an individual's own tissues as antigenic material. Throughout this text we have referred to various diseases that can be attributed to autoimmunity. Among them were Addison's disease, pernicious anemia, myxedema, myasthenia gravis, and rheumatoid arthritis. As a rule, almost any disease of unknown cause, in which there is degeneration of tissues, could have a basis in autoimmunity. Here we will describe one very common autoimmune disease—rheumatic heart disease—which we have not discussed earlier.

Rheumatic heart disease hits about 1% or 2% of school-age children and is the prime cause of heart problems in childhood and early adult life. It is usually triggered by a streptococcal infection.

The result of the infection is often seen as a sore throat, although it may show up as scarlet fever or a skin infection. This is phase I of the disease. A fever may or may not occur, and the heart may or may not be damaged.

SCARLET FEVER. *An acute contagious disease caused by certain streptococci and characterized by a scarlet rash, sore throat, and fever.*

If subsequent heart damage is to be avoided, the streptococci must be eradicated at this point. If they are not, an illusory remission occurs. This is phase II of the disease. During this phase, the individual *appears* to be free of the infection for a period that may last as little as a few weeks or as long as several decades. However, an examination of the blood may reveal that the red blood cells sediment more rapidly than normal in a test tube. Also, antibodies to various streptococcal poisons (tissue-destroying enzymes) are present.

In phase III, the individual may learn that he or she has a heart murmur or some other abnormality. The joints may become inflamed and painful, the pulse rate may rise, and an electrocardiogram may suggest that some of the heart tissue has become scarred and that the valves have become narrowed.

In short, the tissues of the heart and joints have become antigenic as a result of the bacteria's action in modifying the tissues. Over a period of time—which varies from one person to another—the specific antibodies interact with these tissues, causing some destruction.

TYPE III HYPERSENSITIVITY

Type III hypersensitivity differs from types I and II in that neither an antibody (as in type I) nor an antigen (as in type II) is attached to tissue. Both components are free in the body fluids. Together, as a complex, they exert a toxic effect.

The outstanding example of type III hypersensitivity is *serum sickness* — a sickness which may follow the administration of passive immunity. It does not occur very often nowadays because large amounts of serum (usually from cattle) are rarely used. In the early days of immunology it was customary to administer serum for passive immunity where warranted, such as in cases of tetanus. Today, gamma globulin or purified preparations of antitoxin are used instead.

Serum sickness arises when a person forms antibodies to the foreign proteins in the administered serum. Often, however, excessive amounts of antigens are in the serum, relative to the amount of antibody. About two weeks later, the symptoms appear. Such symptoms may include a crop of hives, a large giant hive, or fluid in the joints.

GAMMA GLOBULIN. *A family of serum proteins that include most antibodies.*

HIVES. *An allergic skin condition characterized by blisters, itching, burning, and stinging.*

TYPE IV HYPERSENSITIVITY

Type IV hypersensitivity differs from the others in that it is cell-mediated; that is, free antibodies are not formed. Known also as *delayed hypersensitivity*, its symptoms do not show up for a prolonged period, usually not before 12 hours and very often after about a day or so. The symptoms always include damage to blood vessels and tissue necrosis (death). These symptoms appear at the site of invasion. The invader may be a bacterium, fungus, parasitic worm, venom from an insect, toxin from poison ivy (contact dermatitis) or a foreign tissue transplant.

NECROSIS. *Death of a cell or group of cells in contact with living cells (G. mortification).*

Tissue Transplantation

Tissue transplants may be classified as autografts, isografts, homografts, and heterografts.

An *autograft* is a tissue grafted from one place on an individual to another place on the same individual. An *isograft* is a tissue grafted from one identical twin to another. A *homograft (allograft)* is tissue grafted from one individual to another individual other than an identical twin. A *heterograft* is a tissue grafted from an individual of one species to an individual of another species, for example, from a monkey to a human.

Autografts and isografts are usually transplanted successfully. Homografts are often rejected, and heterografts are usually rejected.

The process of *rejection* is believed to be as follows. First, T cells in the recipient's circulation come into contact with antigens on the foreign tissue. The T cells interact with the antigens and then migrate to a local lymph node. There they multiply, giving rise to memory cells as well as many more T cells. Upon being brought through the circulation back to the transplant, the T cells begin to attack the foreign tissue. In doing so, they release a number of substances into the vicinity. Each of these substances, as indicated below, facilitates the process of rejecting the transplant.

Migration inhibitory factor is a protein which inhibits the migration of macrophages and thus prevents them from leaving the area of the transplant. *Chemotactic factors* are small proteins that attract various sorts of phagocytes into the area. *Recruitment factor* is a protein believed to interact with "uncommitted" lymphocytes, the result of which is a transformation of the lymphocyte into a "committed" T cell. *Lymphotoxin* is a protein which inhibits mitosis. Its main function is to prevent the growth of tissues that are destined to be disposed of.

MITOSIS. Reproduction of body cells.

The attack and disposal of a foreign graft is facilitated by complement, neutrophils, monocytes, and blood platelets.

DRUGS AND HYPERSENSITIVITY

Each of the four types of hypersensitivity can be brought about through drugs. In type I hypersensitivity, the drugs stimulate the production of reagins. These antibodies become attached to tissues. Urticaria (hives, rashes, wheal, and flare responses), bronchial asthma (inflammation of the bronchi), or severe anaphylactic shock may result.

ANAPHYLAXIS. A state of hypersensitivity in which, eventually upon exposure to antigen, the individual enters a state of shock (G. ana, without; phlaktos, defense).

In type II hypersensitivity, the drugs combine loosely with cellular elements, particularly of the blood. The cellular elements become antigenic, and antibodies are formed in response. The antigenic cells thus become subject to destruction, as in the cases of thrombocytopenia (destruction of platelets) and hemolytic anemia.

HEMOLYTIC ANEMIA. A disease due to excessive destruction of red blood cells.

In type III hypersensitivity, the drugs combine with plasma proteins. These proteins thus acquire the properties of an antigen. As in serum sickness, the macrophage system responds by producing antibodies. When the proper balance between antigen and antibody is not achieved, such as when the antigen far exceeds the antibody, a harmful result occurs. This result has one or more of the usual characteristics of allergy, such as a disturbance of the intestinal or respiratory mucosa or skin surface.

In type IV hypersensitivity, the drugs can combine with lymphocytes,

Body Defense Systems: Immunological

render sensitivity to them, and evoke a cell-mediated immune response. One example of this response is a form of contact dermatitis. This dermatitis can be induced by uroshiol or toxidendrol (the active agents of poison ivy), picric acid, and various other chemicals.

CONTACT DERMATITIS. *An allergic or nonallergic skin reaction elicited upon contact with some adverse physical, chemical, or biological agent.*

CANCER AND IMMUNITY

The exact cause of cancer in humans is not known, although it appears to be precipitated by a variety of factors, including excessive sun exposure; radioactive radiations; X rays; chemical carcinogens (e.g., benzpyrene, methylcholanthrene, coal tar dyes); and spontaneous cell mutations.

CARCINOGEN. *An agent which induces cancer.*

No matter what the cause may be, the immune system may possibly recognize the foreign character of the cancerous cells and dispose of them accordingly. There is no direct proof of this, but it is plausible. Moreover, considering the fact that billions of cells are dividing and reproducing in our body every hour of our lives, there is a strong possibility that one or more cancerous cells arise in each of us at one time or another. Very likely it is the immune system that disposes of these cells.

On occasion, however, the cancerous cells evade the immune system, or the immune system is not able to cope with the cancer. These occasions arise more frequently with age. So too, the incidence of autoimmune diseases increases with age. There is much we do not know of the immune system, but it appears that this system becomes less effective as we approach the biological limit of 70 to 100 years.

SUMMARY

1. Immunity is a process in which a foreign material (antigen) stimulates lymphocytes to differentiate into cells which produce antibodies, and memory cells. The antibodies react with the antigenic material and help get rid of it. The memory cells linger on after the reaction, in lymphoid and myeloid tissues, where they stay and proliferate more antibody-producing lymphocytes and memory cells, should the antigen invade the body later on.
2. Lymphocytes which produce antibodies that are secreted into the blood are called B cells. Lymphocytes that attack antigens by way of antibodies or antibody-like material contained in their plasma membrane are called T cells.
3. B cells, T cells, and memory cells are produced in myeloid tissue, which is found in bone marrow, and lymphoid tissue, which is found in such organs as the lymph nodes, spleen, and intestine.

4. Each kind of blood cell, in addition to macrophages and mast cells, plays a special role in the immune system.
5. Antibodies secreted by B cells travel in the blood mainly as members of the gamma globulin family of proteins.
6. The bone marrow's chief immune functions are to supply the body with B cells and to supply the thymus gland with lymphocytes that are modified into T cells.
7. The principal immune functions of the spleen and lymph nodes, and perhaps the tonsils and Peyer's patches in the intestines, are to produce B cells and memory cells and to store T cells.
8. The major immune functions of the thymus gland are: (a) to produce hormones which stimulate the lymph nodes to produce lymphocytes, and (b) to convert certain lymphocytes into T cells.
9. Antigens are destroyed by phagocytic cells with or without the prior intervention of complement, which is a family of serum proteins.
10. In active immunity, antibodies and memory cells are produced in response to antigenic stimulation. Passive immunity, in contrast, is a form of immunity which is rendered to an individual by administering gamma globulin or antibodies.
11. Immune responses may be abolished or suppressed with X rays and certain drugs, such as cortisone.
12. Antigens can be classified as antigens *per se*, haptenic-antigens, or isoantigens.
13. The two most important classes of antibodies are the IgG antibodies, which are involved most often in antigen-antibody reactions, and the IgE antibodies, which bring about harmful reactions and are involved in allergies.
14. Type I hypersensitivity is characterized by the presence of IgE antibodies on cell surfaces. When these antibodies react with their antigens (allergens), they cause pain, discomfort, and tissue destruction. In type II hypersensitivity, the antigenic material is on the surface of a cell and it reacts with antibodies which circulate in the blood. In the ABO blood system for example, in a case of a mismatched blood transfusion, the antibodies in a recipient's serum react with antigens on the surface of the donor's red blood cells. In type III hypersensitivity, which does not occur often, the antigen and antibody circulate freely in the serum during a reaction. Type IV hypersensitivity is a cell-mediated reaction in which lymphocyptes with antibodies or antibody-like materials in their cell membrane react with foreign transplants, or with tissues that take on a foreign character, such as cancer.

REVIEW QUESTIONS

1. Define: B cells, T cells, lymphocytes, memory cells, lymphoid, myeloid, gamma globulin, hypersensitivity, autoimmunity.

2. Describe, in its entirety, the process of immunity.
3. Describe the principal specific functions of the following organs in immunity: bone marrow, spleen, lymph nodes, tonsils, Peyer's patches.
4. Explain how the following diseases differ from each other: leukemia, Hodgkin's disease, multiple myeloma.
5. Describe some of the functions of complement.
6. How is active immunity distinguished from passive immunity? Give two useful applications of passive immunity.
7. How can we abolish or suppress the immune system?
8. Compare I_gG antibodies with I_gE antibodies in terms of their relevance to health and disease.
9. List the distinguishing characteristics of the four types of hypersensitivity. Describe at least one example of each of the four types.
10. Why is it possible for a person with type AB blood to receive type O blood in spite of the presence of antibodies A and B in type O blood?

Special Topics in Physiology

33 Aging
34 Pain, Anesthesia, and Acupuncture
35 Stress and its Relief

PART EIGHT

Aging 33

CHARACTERISTICS OF AGING
THEORIES OF AGING
    Theories based on a fixed lifetime
        The "Genetic Program" theory
        The "DNA-RNA Error" theory
    Other theories of aging
        The cross linking theory
        The autoimmune theory
        The free radical theory
CONTROL OF AGING FACTORS

The average life expectancy in the United States and other developed nations is about 70 to 75 years. Another two years could be added if cancer could be cured, and an additional 17 years could be added by eradicating cardiovascular disease. Thus, by eliminating the major death-causing diseases, we could expect an average lifetime of 100 years.

In this textbook we will not discuss the social, political, or medical aspects of aging. What we will do, however, is look at the biological aspects of the aging process.

CHARACTERISTICS OF AGING

Aging may be defined as a process in which protoplasm gradually and continuously deteriorates, becoming less and less capable of carrying out its functions. In this sense, aging varies greatly among different organs and within different people. For example, the thymus gland may begin to degenerate shortly after one reaches puberty, the skeletal muscle shortly after reaching adulthood, the prostate gland in the 40's, the kidneys in the early 50's, and the brain in the third, fourth, or even seventh decade.

Even within organs, certain cells or tissues may start aging before other kinds of cells and tissues. In general, the labile tissues of the body, such as the blood cells and mucous membranes, do not show degenerative changes

THYMUS. *One of the organs of the immune system, situated behind the breastbone* (G. thymos, mind).

LABILE. *Liable to change; unstable* (L. labilis, to slip).

even late in life. To the cytologist and histologist samples of these cells and tissues appear the same whether they are taken from healthy teen-agers or octagenarians.

Aging of the permanent tissues of the body—nerves, skeletal muscle, and sensory organs, for example—is much more apparent. Fewer nerve cells are present in older people, pointing to the failure of these cells to maintain themselves. The muscle cells of older people contain a much smaller quantity of their contractile proteins, actin and myosin. And the sensory organs, such as the eyes and ears, show degenerative changes which interfere with their functioning.

A third category of tissues includes the endocrine glands, exocrine glands, the liver, and the kidneys. Assuming the absence of disease, these moderately stable tissues do not appear degenerate, though they become lighter in weight and do not function as well.

Although we cannot see many tissue changes, certain characteristics of aging are very obvious. The skin thins out and becomes wrinkled, joints swell or become hardened, cataracts may appear in the lens of the eye, the vertebral column becomes less flexible, there is a decrease in stature as well as graying of the hair or balding, and there is an accumulation of lipofuscin in the skin and various organs throughout the body. (Known as "old age pigment," lipofuscin is a colored complex of fat, carbohydrate, and protein.)

LIPOFUSCIN. *A skin pigment; it forms "liver spots"* (G. lipos, *fat;* L. fuscus, *dark*).

MELANIN. *The natural skin pigment.*

What we see as obvious characteristics of aging are nonetheless only outward expressions of subtle and invisible processes occurring within cells. For example, reduced rates of cell division make wound healing a longer process. Similarly, the failure to produce melanin gives rise to colorless (gray) hair.

Perhaps the most dramatic aging characteristics are related to losses of body protein and potassium, and a retention, or relative accumulation, of fat. Based on body weight, a 70-year-old person has a greater percentage of fat than a 20-year-old. At the same time, the older individual contains about 20% less potassium. Since potassium is the dominant mineral within cells, a reduction of potassium signifies a reduction of other solid materials in cells, especially proteins.

The greatest loss of protein occurs in skeletal muscle cells. This has been shown through comparisons of the weights of various organs in young and old people. At age 70, the heart, lungs, and brain are 5% to 10% heavier than they were at age 20—relative to the total body weight. The spleen, liver, and kidney are 10% to 30% lighter. Skeletal muscle, however, is about 400% lighter.

THEORIES OF AGING

Anyone who wishes to theorize about aging must first take note that there is a genetic limitation to longevity. The strongest evidence to support this statement is the observation that all carefully studied animal species have a definite life span. Hamsters, for example, live an average of two years, rats live 3.5 years, and chimpanzees live about 18 years.

Another kind of evidence on the limitation of life span has resulted from studies on identical and fraternal twins. Statistics show that the difference in longevity between identical twins dying of natural causes is much less than the difference between fraternal twins who die of natural causes.

IDENTICAL TWINS. *Twins that developed from a single egg; they have an exactly identical genetic makeup.*

Theories Based on a Fixed Lifetime

Based on the fact that there is a species-specific life span, scientists have proposed various theories to account for aging. Two of these theories are especially noteworthy because they indicate the great difficulty we face in trying to understand the underlying genetic mechanism.

The "Genetic Program" Theory. The "genetic program" theory of aging states that the biological organism simply plays out its "genetic script." This theory proposes the existence of specific "aging genes." These genes are thought to direct the cell to produce specific types of proteins which either passively inhibit one or more operations of the cell or actively bring about cellular standstill, followed by cell death.

To test this theory, it is necessary to find the "aging genes." To do so, scientists must look for some specific measurable trait which would appear in successive generations of people as the aging process begins. For a variety of reasons this theory is impossible to test in humans. Perhaps the best reason is that there are many traits of aging, and that no single trait—such as protein depletion, graying of the hair, or the development of lipofuscin—signals the initiation of the aging process. Aging, after all, is a gradual process, and it is very probable that it shifts into and out of gear many times in various organs throughout the body before death occurs.

The "DNA-RNA Error" Theory. The second theory deals with DNA and various kinds of RNA. These macromolecules are absolutely necessary if a cell is to produce protein or undergo cell division. During cell division, DNA replicates itself; and during protein synthesis, segments of DNA (genes) produce a complementary copy of themselves, called messenger RNA.

DNA. *The genetic material of a cell.*

According to this theory of aging, errors are made when DNA replicates itself, when DNA transcribes a message in the form of messenger RNA, and when messenger RNA translates its message in producing a protein. These errors cause enzymes and structural materials of cells to be formed improperly. As time goes by, these faulty materials accumulate and the cells, tissues, and organs suffer functional losses.

RNA. *Nucleic acid macromolecules of various forms needed for synthesizing proteins.*

The "DNA-RNA error" theory of aging is perhaps more plausible than the "genetic program" theory. All of the body's living cells are continuously renewing their DNA and RNA, and errors are bound to occur. (Even mature nerve cells, which never undergo cell division, renew half of their DNA content about every two years.)

Like the "genetic program" theory, however, the "DNA-RNA error" theory is difficult, if not impossible, to test. Moreover, although it is a theory for which some supporting experimental data have been obtained, it is doubtful that the data will be of any practical value.

Other Theories of Aging

Many other theories on aging have been proposed. Some of them may be classified as philosophical, since they cannot be tested experimentally. Included in this group are the suggestions that there is a depletion of matter or vitality, that there is an accumulation of stresses, and that there is an adaptive need to increase the survival value of a species by eliminating the older individuals. Another group of theories may be classified as descriptive, since experimental data show that the observed phenomena appear increasingly or cumulatively with age. The theories in this group deal with such topics as hormone imbalance, enzyme deterioration, and somatic (body cell, not sex cell) mutations.

A third group of theories may be classified as basic or causal, since they are designed to probe into the mechanisms that lead progressively to the death of cells, tissues, and organs. Included in this group are the cross linking theory, the autoimmune theory, and the free radical theory.

AUTOIMMUNITY. An "allergy" to one's own tissues.

The Cross Linking Theory. The body has many kinds of molecules that are made up of pairs, trios, or quartets of macromolecules. The individual macromolecules are often attached to each other through hydrogen bonding. Genes, for example, are double helices of nucleic acids; collagen is a triple helix of protein; and hemoglobin is made up of four protein units.

METABOLITE. A substance undergoing a chemical change in a living organism. (G. metabole, change).

The cross linking theory of aging proposes that there are many opportunities for interposing small metabolites between the strands of macromolecules that make up the dynamic or structural components of the body. Cross links are thus formed where spaces should exist. As a result, the cellular component containing the macromolecule is unable to function properly. According to the theory, an increasing number of cross links accumulate with age and produce certain aging characteristics, such as wrinkled skin and degeneration of muscle tissue. Collagen, the principal fibrous protein of connective tissue, is especially vulnerable to cross linking because it has a triple helix structure. Collagen accounts for about 25% of the body's protein.

ANTIBODY. A protein produced by a lymphocyte for destroying an antigen.

The Autoimmune Theory. When certain foreign substances get into the body, they behave as antigens. Antigens stimulate the production of antibodies, and in doing so they bring about their own destruction.

Autoimmunity is a type of immunity in which one or more of a person's body constituents are seen by the body as foreign substances. As a result, the body's immune system is marshalled into action and the "foreign" tissue becomes subject to attack and disposal.

The autoimmune theory of aging points out that the body is subject to a series of infections throughout life. In addition, the body tissues are exposed to attack by environmental radiation, errors are made by DNA and RNA in producing proteins, and alterations in the structure of protoplasm, including the formation of cross links, are caused by many kinds of chemical events. The immune system is called upon to dispose of the foreign material. It does so repeatedly, and thereby brings about tissue destruction. In the course of aging this process occurs here and there throughout the body. It produces arthritic joints, it damages glands, it destroys nerve cells, and so on.

The Free Radical Theory. All matter is made up of atoms, which combine with one another to form molecules or compounds. In order to combine with one another, atoms can form either a covalent bond or an ionic bond. In a covalent bond, one atom must share one or more electrons with another atom. To form an ionic bond, it must acquire a charge opposite in sign to that of the other atom.

For reasons which are not known, an atom can acquire an extra electron or lose an electron without immediately forming a bond with another atom. When this happens, the atom is said to be a *free radical*.

Free radicals are believed to induce aging because they alter the chemical structure of protoplasmic components. Free radicals are "free" to combine with any of a wide variety of substances in the body. They commonly combine with components of cell membranes and prevent the membranes from functioning properly. The membranes acquire an altered permeability to various metabolites, causing abnormalities in bioelectrical events, cell secretion, cell absorption, and the production of cell products, such as hormones. Mitochondrial membranes may not be able to provide the cell with necessary amounts of ATP. The Golgi apparatus may not be able to produce cell secretory products, and lysosomes may fail to keep in or release hydrolytic enzymes in accord with proper cell function.

Free radicals may arise from various kinds of substances, such as hydrogen atoms or iodine molecules. However, it is primarily with respect to oxygen that free radicals are dangerous as an aging factor. Oxygen gives rise to superoxide radicals (O_2^-) and several other kinds of radicals, including the hydroperoxyl radical (OOH$\cdot$) and the hydroxyl radical (OH$\cdot$). These radicals are especially notorious for their reactions with lipids, resulting in lipid peroxides. Lipid peroxides may become components of cell membranes and prevent proper cell functioning.

The body has three protective enzymes that offset the destructive actions of oxygen: catalase, peroxidase, and superoxide dismutase. Catalase and peroxidase have been known to scientists for nearly a century. Their functions are not entirely understood, but it is certain that they consume hydrogen peroxide molecules (HOOH) that arise through various metabolic reactions. Through their consumption of hydrogen peroxide, catalase and peroxidase reduce the amount of hydroxyl radicals that otherwise form when hydrogen peroxide reacts with superoxide radicals.

Superoxide dismutase, which was discovered in 1967, catalyzes a reaction in which two superoxide radicals are converted to oxygen and hydrogen peroxide. Thus, the value of superoxide dismutase in aging is the action it

FREE RADICAL. *An atom or atomic group carrying an unpaired electron (L. radix, root).*

MITOCHONDRIA. *Cell organelles in which most of the body's ATP is formed; named for their appearance (G. mitos, thread; chondrion, granule).*

SENESCENT. Growing old
(L. senex, old).

exerts against superoxide radicals. The hydrogen peroxide that appears as a by-product is disposed of by catalase or peroxidase.

Research is being carried out to find ways in which superoxide radicals may be involved in senescence and the ways in which superoxide dismutase counteracts or at least slows this process. Superoxide radicals also appear to be involved in inflammatory processes such as that associated with certain types of arthritis; a number of investigators are exploring the anti-inflammatory properties of superoxide dismutase in such disorders.

CONTROL OF AGING FACTORS

Any agent or process which shortens life expectancy—other than disease or injury—may be called an aging factor. Theoretically, all of the aging processes we have discussed so far may be called aging factors. Only a few of these factors, however, are potentially controllable.

A person may somehow reduce the frequency and severity of stresses in life and thus may conceivably increase his longevity. So too, eating a proper diet, breathing clean, unpolluted air, and following an adequate exercise program may help. Nevertheless, in spite of observing all of the precautions necessary for good health, aging is bound to occur. Accordingly, the question we should raise is what can be done beyond these measures to achieve a greater realization of life expectancy—an age of 100 years or so?

Little can be done about the immune system since the function of this system—to scavenge foreign materials—is essential for health. What must be done instead, is to minimize the conversion of native tissue to foreign tissue. With respect to other aging factors, it is doubtful that scientists will ever be able to undo errors at the DNA level. It is perhaps only in the area of free radical chemistry, and in the prevention of damage to DNA, that science can hope to make practical advances.

In order to reduce free radical formation in the body it is necessary to consider the chief source of free radicals, namely, oxygen. Considering that fats, carbohydrates, and proteins are combusted (oxidized) by oxygen, and that protoplasm is rich in these combustible nutrients, it is not surprising that oxygen is a toxic substance. This is so in spite of its importance as a nutrient. However, we cannot control the composition of the atmosphere, so it is pointless to attack the free radical problem by reducing the amount of oxygen we consume. Moreover, it is absolutely essential that we have adequate quantities of oxygen to breathe.

There are perhaps two things we can do to cope with the free radical problem. First, we may monitor the levels of superoxide dismutase in tissues that appear to be degenerating, for example, in arthritic tissue. We may also monitor this enzyme in tissues that appear to be failing in their function in spite of the apparent absence of disease. Conceivably, superoxide dismutase or some other by-product of research on this enzyme will provide a means of boosting inadequate levels of activity, and thus of enhancing the level of protection which this enzyme provides.

Second, we may consider vitamins C and E, and certain other metabolites, such as the amino acid cysteine. These substances are antioxidants. It has been proposed by several scientists that one of the chief functions of vitamin E is to work against free radicals—to work as an antiaging factor. Ongoing research may reveal additional or related substances—natural or synthetic—that are somewhat effective against aging.

SUMMARY

1. Aging is a subtle process whose chief characteristics include a relative retention or accumulation of fat, and a loss of potassium and protein. The greatest loss of protein occurs in skeletal muscle cells.
2. The free radical theory of aging—tied in with the enzymes catalase, peroxidase, and superoxide dismutase, and with antioxidants such as vitamins C and E—offers the best hope of increasing man's life expectancy in the immediate future.

REVIEW QUESTIONS

1. What are perhaps the most prominent characteristics of aging?
2. Describe two theories of aging in which recognition is made of the genetic limitations to longevity.
3. Describe the cross linking theory of aging and argue whether it is a valid theory.
4. What is autoimmunity and how is it implicated in aging?
5. Explain the free radical theory of aging.
6. Tell what each of the following enzymes does: catalase, peroxidase, superoxide dismutase.

34 Pain, Anesthesia, and Acupuncture

THEORY OF PAIN
RELIEF OF PAIN BY DRUGS
    Analgesics
    Anesthetics
RELIEF OF PAIN BY ACUPUNCTURE
History of acupuncture
Philosophical basis for acupuncture
Practice of acupuncture
Advantages and disadvantages of acupuncture
Scientific experiments with acupuncture

BARBITURATE. *Any of several drugs derived from barbituric acid; they are used as hypnotic and sedative drugs.*

Everyday, millions of people are afflicted with some form of pain, from a deep throbbing foot pain or gripping headache to the pain which follows surgery. Certain kinds of drugs, such as aspirin and morphine, are often helpful in relieving the pain. Other kinds of drugs, such as ether or barbiturates, are used during surgery to make a person unconscious of the pain which accompanies the operation.

In spite of what can be done to lessen pain, or prevent it from registering in our mind, very little is known about its nature. We cannot even define pain with words or phrases that would be valid for all individuals in all situations. Usually, however, each of us knows what it is. Like emotional feelings and thinking, pain is perceived and interpreted in the brain.

In this chapter we will first discuss a modern theory on pain. We will then deal briefly with the use of drugs to relieve pain. We will conclude with a description of acupuncture and its role in the relief of pain.

THEORY OF PAIN

Pain stimuli arise in the skin, joints, muscles, and viscera. From here they pass through tracts of nerve fibers into the brain's thalamus and cortex, where pain is registered and brought to consciousness. Since there are no pain receptors in the brain itself—the cutting of brain tissue itself is painless—we may conclude that: (1) specific sensory receptors bring pain stimuli to the brain, or (2) a particular combination of sensory fibers must be activated for pain to be perceived. Actually, both of these ideas are involved.

The current, widely accepted theory on pain is called the gate control theory. It suggests that nerve fibers of very thin diameter from gamma neurons carry pain stimuli from the periphery of the body to the substantia gelatinosa (Table 5.1) of the spinal cord, whereas fibers of thicker diameter from alpha neurons carry other kinds of sensory information to the substantia gelatinosa. When the discharge of bioelectricity (action potentials) from the gamma neurons sufficiently exceeds the discharge of alpha neurons, a "gate" is opened in the substantia gelatinosa. Most of the pain information is then transmitted into nerve tracts that are separate from the nerve tracts which carry other kinds of sensory information (Table 5.2).

SUBSTANTIA GELATINOSA. *A gelatinous substance composed chiefly of nerve satellite cells and ganglion cells, forming the apex of the gray matter in the dorsal region of the spinal cord.*

Second-order sensory neurons in the tracts convey the pain information to the thalamus. From there, third-order neurons carry the information further, to the cerebral cortex. The pain is perceived in the thalamus and cerebral cortex. In spite of being perceived, however, the information may not be defined as pain at this point. To be defined as pain, the person must be moved emotionally and mentally to recognize the sensation as pain. This can only be done in context with the person's own psychological makeup.

There is much experimental evidence to support the gate control theory. Simple evidence to support the theory is provided from such experiences as applying pressure to some region of the body to relieve pain arising in another region. Consider, for example, clenching your fists while the dentist drills your teeth. In this case, an excessive amount of alpha activity is generated in the pressure receptors of the wrist relative to the gamma activity arising in the pain receptors of the teeth. Massage, incidentally, which also involves the activation of many alpha neurons, helps relieve pain that arises from aching, cramped muscles.

RELIEF OF PAIN BY DRUGS

It can be reasoned that pain is an adaptive feature, beneficial to the survival of a species. It is true that a painful stimulus will make you withdraw from the source of the pain, as, for example, when you touch a hot stove. Often, however, a residual or lasting pain remains, as happens when tissue is destroyed. In these situations, which commonly occur after a burn or cut, a pain relieving remedy may be needed. The application of pressure, ice, or heat may be helpful in some cases, but for many forms of pain, drugs are needed.

Any drug which relieves pain without affecting consciousness is called an *analgesic*. If the drug also abolishes sensory perception, while still allowing consciousness, it may be classified as a *local anesthetic*. On the

ANALGESIC. *A drug that relieves pain without causing loss of consciousness* (G. a, *without;* algos, *pain*).

Relief of Pain by Drugs

other hand, if the drug produces a loss of consciousness, it is classified as a *general anasthetic*.

ANESTHETIC. *A drug that produces a loss of sensation* (G. anaisthesia, *without feeling*).

Although numerous kinds of analgesics and anesthetics are known and can be classified on the basis of their action, much less is known about the site of action or their mechanism of action. All that can be said for certain is that these drugs block impulse propagation along a nerve fiber, or they reduce or prohibit the transmission of neural messages at synapses, or they bring about both kinds of inhibitory actions.

Analgesics

An analgesic drug may be classified as narcotic or non-narcotic, depending on whether it induces drowsiness. Morphine and codeine are narcotic, while aspirin and colchicine are non-narcotic. Simply by inducing sleep, narcotics raise the threshold of pain.

NARCOTIC. *A drug that causes drowsiness* (G. narkoun, *to numb*).

Individual analgesics may be classified also on the basis of their strength as pain relievers. Among the narcotics, for example, morphine is more powerful than meperidine, which in turn, is stronger than codeine. In general, the non-narcotic analgesics are effective against headache, muscle ache, and pain which originates in the skin or a mucous membrane. Narcotic analgesics are usually restricted for intense pain or deep pain originating in the viscera.

VISCERA. *The soft internal organs of the body* (L.).

Various analgesics are also useful for relieving certain problems other

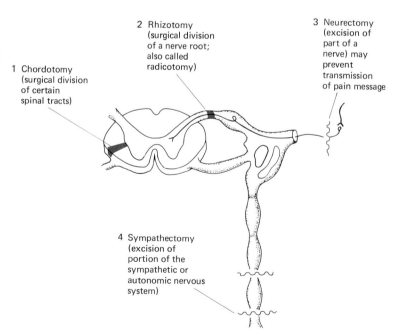

FIG. 34.1. *A variety of surgical techniques may be used to give relief from severe pain.*

than pain. Aspirin, for example, is an anti-inflammatory drug that also reduces fever. Colchicine is used to treat gout itself (to reduce the level of uric acid in the blood) as well as to relieve the pain of gout. Morphine depresses respiration, while codeine is useful as a cough suppressant.

COLCHICINE. *An alkaloid drug derived from the plant* Colchicum autumnale.

Anesthetics

Local, or topical, anesthetics are used to produce anesthesia in a limited region of the skin or a mucous membrane. Cocaine, for example, is used to anesthetize the eyes, nose, or throat. Benzocaine, the active ingredient in certain sprays for the relief of sunburn pain, is also useful against the pain of insect bites. Ethyl chloride, which produces local anesthesia when sprayed on the skin, is also useful as a general anesthetic, because it is gaseous at a rather low temperature (13°C).

In general anesthesia, a variety of drugs is usually used to bring about a series of actions, proceeding from mild sedation to deep unconsciousness. In a balanced general anesthesia, for example, a person may receive premedication to induce relaxation, additional drugs to induce anesthesia, and still other drugs to maintain anesthesia. Premedication may include a barbiturate, a narcotic analgesic, and a drug such as atropine to inhibit the excitatory effect of the tenth cranial nerve (vagus nerve) on the digestive system. An additional dose of barbiturates may be used to induce a deep state of unconsciousness, and ether or nitrous oxide may be used to maintain unconsciousness. If surgery is to be performed on skeletal muscles, gallamine may be administered to block the synapses at the neuromuscular junctions, thus preventing muscular contractions.

PREMEDICATION. *Drugs administered before an operation.*

NITROUS OXIDE. *A colorless anesthetic gas; also called laughing gas.*

RELIEF OF PAIN BY ACUPUNCTURE

In recent years, much attention has been focused on acupuncture as a method of relieving pain. This ancient Chinese practice, which dates back to about 3000 B.C., involves the insertion of needles into certain spots on the body. In addition to using needles, acupuncturists may also apply heat and/or pressure to certain sites on the body.

Most practitioners of acupuncture, especially Chinese doctors who have not been trained in basic science, use acupuncture not only to relieve pain, but also to treat a variety of diseases, including diabetes, encephalitis, and gout. However, it should be stated that acupuncture theory and practice is based on the principle that basic organic disorders can only be given temporary relief, and that degenerated tissues cannot be treated. Thus, the "cure" of diabetes by acupuncture is possible only if the insulin-producing cells of the pancreas have not degenerated and are able to respond to treatment.

GOUT. *An arthritic disease caused by faulty metabolism of uric acid* (L. gutta, *drop*).

Doctors who have been trained in modern Western medicine are perhaps justifiably skeptical about the curative claims of acupuncture. However, many physicians and scientists believe that it may be of value as a way of relieving pain.

History of Acupuncture

ACUPUNCTURE. *An originally Chinese medical technique in which thin needles are inserted into various parts of the body (L. acus, needle + puncture).*

The earliest known document on acupuncture is the "Yellow Emperor's Classic of Internal Medicine," written around 300 B.C. According to legend, acupuncture treatments were established in the course of observing that an injury or the application of pressure to one region of the body would relieve an ongoing pain previously found elsewhere in the body. An injury to the tip of the right index finger, for example, could relieve pain from a toothache; or pressure applied to the crease between an index finger and thumb would provide some relief from a headache. Over a period of many years, numerous observations were made, and maps were made showing points of pain and their corresponding points of relief. Crude objects, such as pointed sticks and stones, were used by the earliest acupuncturists to stimulate the sensitive relief points. Later on, with the development of metals, needles were used instead.

Until about the end of the nineteenth century there was little communication between Western and Chinese medicine. In 1911 the Chinese Emperor Sonbun, who had received training in Western culture, increased the communication gap by forbidding the practice of acupuncture. He thus implied that acupuncture was inferior to Western medicine. Because of the value of acupuncture to the Chinese revolutionaries in 1934–35, however, Mao TseTung reinstated the practice shortly after he came to power in China in 1949.

Acupuncture was finally introduced into the United States in 1972, when American doctors and journalists visiting China witnessed surgical operations, including chest surgery and appendectomies, carried out without anesthesia, but under the pain-relieving effects of acupuncture. By the mid-1970's acupuncture was legalized in more than a dozen states.

Philosophical Basis for Acupuncture

YIN *and* **YANG,** *in Chinese cosmology, are the passive and active principles, respectively, which combine to produce all living and nonliving things.*

According to ancient Chinese theory, there are two major forces: yin and yang. These two forces are complete opposites, but nothing is purely either yin or yang. All females, for example, have some maleness and all males have a certain amount of femaleness.

With regard to health, the Chinese see pain or illness as caused by an excess of either yin or yang. Acupuncture is aimed at restoring the proper proportions of these vital forces in the body. Whether Western doctors acknowledge the yin and yang philosophy is unimportant. What is important is that the practice of acupuncture, may be more beneficial to certain individuals than some forms of Western medicine. Acupuncture is not

Pain, Anesthesia, and Acupuncture

a form of quackery. Nor is it a form of hypnosis, because it works with infants and animals.

Practice of Acupuncture

Acupuncture therapy begins with an examination of the patient's ailment. Unlike some of the sophisticated techniques used in Western medicine, however, the diagnostic techniques are essentially the same from case to case, and no special instruments are used. Facial expressions and voice qualities are examined, as are the nails, skin, tongue, breath, and feces. Questions are asked about the location, date, and origin of the illness, as well as the patient's eating, sleeping, and bowel habits. One of the most important parts of the examination is a simple but subtle procedure in which six different pulse "readings" are taken from each wrist. You should note that doctors trained in modern medicine view the pulse-reading information of acupuncturists with deep skepticism.

When the examination is over, a diagnosis is made, and the acupuncturist begins treatment. Treatment consists of inserting needles into certain select spots on the skin. One or two needles may be used for some illnesses and 30 or more needles for other ailments. The needles range in thickness from about 0.3 mm to 0.5 mm and are 2 to 10 cm long. Short, thin needles are used for certain regions, such as the face and hands, while longer, thicker needles are used for the thighs, shoulders, and buttocks. If treatment is to be effective the needles must "take." That is, the patient must feel a tingling sensation mixed with feelings of numbness or heaviness. The needles are left in place for 10 to 30 minutes, often with periodic twirling. For certain ailments, only one or two treatments may be necessary. For other ailments, 20 or 30 treatments are called for.

The effectiveness of acupuncture treatment depends not only on "take" but on the location of the needles. Depending upon the training and point of view of the acupuncturist, one or more of as few as 70 points or as many as 800 points may be selected for the treatment of any kind of pain or disease. These points run along lines that are called *meridians*. Numbering 14 in all, these meridians are believed to carry "vital energy" from one region of the body to another and from one internal organ to another (Figure 34.2).

There are six pairs of yin meridians, six pairs of yang meridians, a single conception meridian, and a single governor meridian. The yang meridians run along the dorsal surface of the body and head, and along the outer surfaces of the limbs. Three of them carry "vital energy" from the fingertips to the face. The large intestine meridian, for example, begins at the root of the index fingernail and terminates at the side of the nostril. According to basic acupuncture theory, there are 20 points along this meridian. Pressure upon the first point, for example (or the insertion of a needle), may relieve a facial pain.

The other three yang meridians run from the face to the toes. With regard to the yin meridians, three run from the feet to the chest and three run from

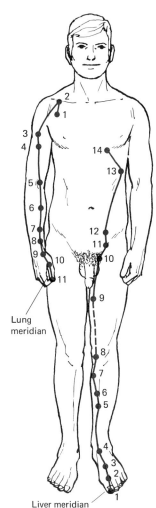

FIG. 34.2. *Two of the six "yin meridians" used in acupuncture and acupressure.*

MERIDIAN. *A line passing from one pole to another pole* (L. meridianus).

the chest to the fingertips. The governor and conception meridians run along the back and front medians of the body, respectively.

Advantages and Disadvantages of Acupuncture

Although modern physicians are generally skeptical about treating diseases with acupuncture, there is a growing acceptance of this practice for the relief of pain and as a substitute for local or general anesthesia. Acupuncture has certain advantages over the use of drugs. Allergic reactions are not provoked, and there are no undesirable side effects, as may occur with drugs. Also, there are no major physiological changes in the body, as in general (inhalation) anesthesia, and the patient is fully conscious during surgery and is sometimes able to assist the surgeon. In addition, the surgeons do not have to wait for the effects of anesthesia to wear off in order to evaluate the consequences of an operation, and no complicated gas-delivering and monitoring devices are needed.

Acupuncture also has some disadvantages and complications, however. As a substitute for inhalation anesthesia, it may be only partially effective as a pain remover. Also, there is no muscle relaxation, and this may be troublesome in certain operations. In addition, inflammatory reactions may arise where the needles are inserted, and a nerve may be punctured accidentally. Also, the patient may faint or enter a state of shock.

Scientific Experiments With Acupuncture

Ever since 1972, when acupuncture was displayed to Western medicine, there has been an intense effort to explain its analgesic effects in scientific terms. Here we will describe some of the correlations that have been made, though as with pain itself, no one can explain exactly how or why acupuncture works.

In one type of experiment, needles were inserted into pairs of acupuncture points on both sides of the body. Novocaine, a local anesthetic, had been previously applied to one side and not to the other. As a result, the sensation of "take" was felt only on the side without the Novocaine, suggesting that neural processes are involved. This suggestion is supported by the observation that Novocaine blocks the ability to induce acupuncture analgesia at a distant site.

HEMIPLEGIA. *Paralysis of one side of the body* (G. hemi, *half;* plege, *stroke*).

PARAPLEGIA. *Paralysis of the lower limbs* (G. paraplēgiē).

In experiments with hemiplegics, the sensation of "take" occurred only on the unparalyzed side of the body. So too, in paraplegics the sensation of "take" was experienced if the patient was able to sweat below the waistline, in the paralyzed region of the body. In nonsweaters there was no "take," indicating that neural elements are involved if acupuncture is to work effectively.

A third kind of experiment involved the electrical stimulation of patients in pain. The electrical stimuli relieved the pain more effectively when they were applied to acupuncture points than to nearby points, even points only about 5 mm away. Through anatomical studies it has been learned that effective acupuncture points coincide with areas that are relatively rich in nerve fibers. Through electrical measurements, it has been shown that these same points offer less resistance to the flow of current than other points.

Still other experiments show that electrical stimulation of acupuncture points on the skin sends more impulses into the central nervous system than can be sent by stimulating other points. It can be theorized that the stimulation of acupuncture points generates a predominance of alpha messages relative to gamma messages, thus reducing or eliminating pain. So too, the manipulation of needles in the absence of an applied electrical current may result in the generation of an excessive amount of alpha activity. It may yet prove to be an irony that certain forms of Western treatments of pain—such as the use of massage, vibrations, the application of cooling or heating agents, and the electrical stimulation of the dorsal spinal column (in modern electroanalgesia)—are essentially the same used by acupuncturists. It may also prove to be ironical that electro acupuncture—a modern form of acupuncture in which electricity is passed into the skin-inserted needles—was practiced by several enterprising physicians as long ago as the mid-1800's. W. G. Oliver, a dentist in Buffalo, New York, used a technique similar to electro acupuncture in 1858. In spite of all we have said, however, you should bear in mind that the claims made for acupuncture have not been proven scientifically, and that the practice is not generally accepted in Western nations.

SUMMARY

1. Theoretically, pain is perceived when gamma neurons discharge an excessive number of impulses—relative to alpha neurons—into the thalamus and higher regions of the brain.
2. Pain may be relieved through analgesic or anesthetic drugs, or through acupuncture, massage, electroanalgesia, heat, or cold.

REVIEW QUESTIONS

1. Describe the gate control theory of pain and give an example, from your own experiences, which might support the theory.
2. Define: analgesic, anesthetic, narcotic.
3. How do pain relieving drugs carry out their action?
4. How can you relate your answer to question 3 to the gate control theory of pain?
5. Upon what two factors is the effectiveness of acupuncture most dependent?

Review Questions

6. List several advantages of acupuncture as a pain reliever substitute for anesthesia. What problems may be associated with use of acupuncture?
7. Do you believe modern science has provided sufficient evidence to indicate that acupuncture may be a valid medical procedure for relieving pain?

Stress and Its Relief 35

MENTAL ATTITUDES, EMOTIONS, AND STRESS
UNCONSCIOUS RELIEF OF STRESS
    Sleep and dreams
        Stages of sleep
CONSCIOUS RELIEF OF STRESS
    Biofeedback
    Transcendental meditation

During every moment of our waking lives we are conscious of experiences of one kind or another. These experiences may be stressful or nonstressful, depending upon many factors, including genetic disposition, former psychological conditioning, physical health, and mental attitude.

A nonstressful experience may provide pleasure or joy; it may put us in a mood of relaxation; or it may allow us to be helpful and productive. In any case, the nonstressful experience is one in which there is freedom from tension or strain.

In stressful situations, the tension or strain that is present leads a person toward, or into, a state called anxiety. In this state, lactic acid accumulates in the blood, the heartbeat is "felt" in or by the chest wall, emotional sweating occurs, and there may be a wide array of other physiological changes.

In severe cases of stress a person may suffer a nervous breakdown. The nervous system collapses somewhat, resulting in an impairment of skeletal muscle action, blood circulation, hormonal balance, and so on. Examinations of individuals suffering from a nervous breakdown usually reveal (1) a history of good health and satisfactory work performance, and (2) an exposure over several weeks or months to severe stress, leading to what is commonly called the individual's "breaking point." No one can predict whether, how, or when this point will be reached in ourselves or others, nor can anyone specify in advance which stressful situations (stressors) can lead to a nervous breakdown. Common stressors include poor social relations with one's peer group, excessive intellectual demands, and too many work responsibilities.

Assuming that anything which creates anxiety is a stressor, let us concentrate in this chapter on measures which provide relief from stress. First, however, we will say something about the dependency of stress on mental and emotional attitudes.

ANXIETY. *A state of being apprehensive, fearful, or worried; anxiety symptoms may be both emotional and physical* (L. anxietas, anxious).

MENTAL ATTITUDES, EMOTIONS, AND STRESS

Some people believe that severe physical work, in the absence of emotional or mental involvement, can precipitate stress. However, this is seldom true. Long before gametime, for example, athletes are far less stressed than when gametime approaches. Moreover, the greatest amount of stress usually occurs on the day of the event, often during the few moments before the start of the game, rather than after the contest has begun. So too, soldiers in intense basic training during peacetime suffer far less stress than combat-ready soldiers who are not involved in training but are located in a war zone. Construction and factory workers engaged in heavy labor are actually less prone to stress than are business executives, writers, artists, and teachers, and a patient awaiting major surgery may be undergoing enormous stress even though he or she is quite inactive. Were it not for our emotions and our mindfulness of social and economic matters, stress would be no more than a minor health problem.

It should not be thought, however, that in the absence of mental attitudes we would be completely spared of stress. Symptoms of stress, as we know them for humans, can be found in the lowest vertebrate animals. Overpopulation of an area by fishes imposes a stressful situation which results in curtailment of reproduction. Among populations of certain northern birds, severe winter conditions induce an enlargement of the adrenal gland and a secretion of excessive amounts of the hormones adrenaline and cortisol. This leads to an imbalance among other hormones, bringing about less vigor and making the bird less able to adapt to environmental demands.

ADRENALINE. *A hormone which sets up a person to encounter energy-demanding tasks.*

It is mainly in mammals, however—and particularly in humans—that we see symptoms of stress in response to mental and physical strain. It is also in mammals that we see the highest development of the limbic cortex and hypothalamus. The limbic cortex is the region of the cerebral cortex which borders portions of each of the sensory areas of the brain. Together with the hypothalamus, it constitutes the focal region of the brain from which decisive commands are issued in setting up emotional responses to environmental stimuli.

LIMBIC. *Refers to the medial cerebral cortex, bordering several sensory areas, and concerned largely with emotionality (L.* limbus, *border).*

VISCERA. *The soft internal organs in the abdominal and thoracic cavities (L.).*

Since the effects of stress are directed primarily against the viscera, which are regulated by the autonomic nervous system, it is logical to conclude that one's mentality and emotionality are the main avenues to the accumulation of stress. By the same token, through a suitable use of mental faculties and an appropriate control over emotions, one may be able to recognize stress situations—even the very subtle ones—and take measures to prevent or relieve them.

UNCONSCIOUS RELIEF OF STRESS

Stressful situations are inevitable in everyone's life, though the intensity or longevity of individual episodes are bound to vary from person to person.

In some situations, it is possible to reduce one's exposure to stress, as for example, by leaving a job, spouse, or environmental climate. In other cases, stress cannot be avoided, but it may be relieved by "talking out" the causes. This may be done through a spouse, psychiatrist, clergyman, neighbor, or a group of persons, as in Alcoholics Anonymous. Even simple occasional conversations with acquaintances at one's job or in a club may help iron out the strains of stress.

Sleep and Dreams

Sleeping, and particularly dreaming, may be looked upon as an unconscious or subconscious way of relieving stress. Very little was known about the physiology of sleep until about 1940, when scientists began to explore this phenomenon through the use of electroencephalography. This procedure, like electrocardiography, measures electrical potentials from the surface of the body, but the potentials that are measured come from the brain, instead of the heart. To obtain an electroencephalogram (EEG) six to 12 or more electrodes are usually fastened to the surface of the scalp. These electrodes pick up electrical waves that are continuously being generated by the nerve cells in the brain. These waves, only about 50 to 150 microvolts in magnitude, are sent into special devices where they are amplified a millionfold and recorded.

SUBCONSCIOUS refers to ideas or feelings that are in the mind but that do not ordinarily reach the threshold of awareness (L. sub-, beneath).

Through EEG studies, along with other physiological measurements, it is now possible to characterize sleep. Moreover, it is possible to show that the normal characteristics are altered when an individual is treated with certain drugs, such as LSD or amphetamines, or when he has certain kinds of mental or emotional problems.

LSD. Lysergic acid diethylamide, a hallucinogenic drug.

Stages of Sleep. Sleep consists of a series of recurring cyclic events in which a rapid eye movement (REM) phase alternates with a nonrapid eye movement (nonREM) phase throughout the night. The nonREM phase is encountered first. It consists of four stages. In the first stage, which lasts seconds to minutes, somewhat unsteady low voltage brain waves of mixed frequency are picked up by the EEG. The individual feels drowsy during this period. During the next 15 to 45 minutes or so, the individual passes into the second stage, a deeper state of subconsciousness. In this stage, brain waves begin to come through more steadily and at a lower frequency. In the third stage of nonREM sleep, moderate amounts of high amplitude waves appear. The waves are lower in frequency than in stage 2, and the individual begins to slip into a deep state of unconsciousness. In the fourth stage, beginning within an hour after stage 1, very large amounts of high amplitude waves are generated as the person passes into his or her lowest level of consciousness. During this stage it is most difficult to arouse the sleeping individual. Moreover, autonomic nerve functions are at their lowest level of activity during this period.

AMPHETAMINES. Stimulatory drugs closely related to adrenaline in chemistry and action.

EEG. Electroencephalograph; an instrument for detecting and recording electrical impulses from within the brain (G. enkephalos, brain).

In certain people, such as those who encounter considerable anxiety in their daytime activities, the EEG of stages 3 and 4 are often abnormal and the autonomic nerve activities are not as severely reduced. Often in people

BARBITURATES. *A class of hypnotic and sedative drugs.*

who are seriously depressed or are taking sleeping pills (barbiturates) stage 4 is extremely short or lacking altogether, and is replaced somewhat by an extended stage 2.

Another notable fact about stage 4 of nonREM sleep is that considerable amounts of growth hormone are released from the pituitary gland during this period. It is believed that this is related to metabolic repair processes that are going on at this time.

After stage 4 of sleep a person successively re-enters stages 3 and 2. Instead of passing into stage 1 again, however, the individual enters the REM phase of sleep.

The REM phase of sleep is characterized by rapid movements of the eyes, by an EEG pattern which is not the same as that of stage 1, and by occasional twitches in the muscles of the limbs and trunk. In spite of the twitches, all of the skeletal muscle fibers are usually completely relaxed, as are the motor neurons which activate them.

The REM phase of sleep is encountered roughly every 90 minutes in a night of sleep, though occasionally a person may pass from stage 2 back to stage 3 without entering the REM phase. Very often (but not always), when an individual is awakened during the REM phase he or she is able to describe a dream that was in progress. It is believed that dreaming rarely occurs in any of the nonREM stages of sleep.

Through observations on many individuals it appears that the intensity of the eye movements during the REM phase of sleep reflects the emotional intensity of an ongoing dream. Most sleep researchers believe that dreams are necessary as outlets for the stresses and anxieties that are built up during the day. When volunteer subjects are awakened each time they enter the REM phase of sleep, they become irritable and aggressive, whereas repeated arousal of the same people during one of the nonREM stages is not nearly as distressing and disturbing. Moreover, when normal experimental subjects are repeatedly aroused during REM or stage 4 nonREM, they enter REM or stage 4 nonREM sleep more frequently when they are allowed to do so thereafter. The research data in this regard suggest that a certain minimum of stage 4 nonREM and REM sleep is needed each night to relieve a person's body and mind from the tensions and anxieties that were built up during the day. In general, normal teen-agers and adults spend 25% of their sleep in the REM phase and a definite amount of time in stage 4 nonREM sleep, depending on age. The need for stage 4 sleep decreases steadily with age. Patients suffering from certain forms of mental illness generally spend much less time in the REM phase, as do people who take amphetamines (pep pills). Often such individuals also make rough transitions from the REM phase to stage 2 of the nonREM phase as a new cycle of sleep begins.

RETICULAR SYSTEM. *A complex network of nerve fibers and cells in the brain stem.*

Contrary to former belief, sleep is not a passive process in which all of the body functions become less active. Rather, a considerable amount of activity is generated in certain regions of the brain, particularly in the reticular system of the brain stem. It is thought that these processes serve to reduce nervous activities elsewhere in the brain, and to inhibit the outflow of nervous impulses which otherwise activate skeletal muscles, smooth muscles, and certain glands.

So too, some regions of the brain become more active during dreaming than at other times. A group of nerve cells in the pons of the brain stem, called the locus coeruleus, has been singled out by some scientists as one of these regions.

How these regions elicit dreams is no more understandable than what a dream itself actually is. All that can be suggested at present is that because of dreaming the brain is able to return to a less strained, rested condition. It has been shown by scientists who have studied the nightmare that this frightening experience occurs during stage 4 of sleep, whereas "anxiety dreams"—dreams that disturb but do not frighten—occur during the REM phase of sleep. Some of these scientists conclude that the nightmare represents a failure of the ego to control anxiety, whereas "anxiety dreams" represent a successful way of dealing with anxiety. These conclusions go far beyond the raw data that scientists gather through EEG measurements and studies of case histories, but at the same time they provide plausible explanations about experiences we have all had at one time or another.

NIGHTMARE. *A terrifying dream characterized by feelings of great distress and helplessness.*

EGO. *The thinking and feeling part of the personality, which is directly conscious of the world outside of one's self (L. ego, "I").*

CONSCIOUS RELIEF OF STRESS

At present, various kinds of drugs are available for treating stress. Some of these drugs, such as sedatives and tranquilizers, are helpful in treating emotionally high strung, or hyperactive persons. Other drugs, such as analgesics or antihypertensive agents, are useful when pain or high blood pressure, respectively, accompany the stress. Dreams, too, may be helpful, as indicated above, and "talking it out" often proves to be beneficial.

There are two additional stress relief mechanisms, each of which is related somehow to "talking it out." One of these mechanisms, known as biofeedback, is available only through certain experimental laboratories. The other mechanism, known as transcendental meditation, is practiced by people throughout the world.

SEDATIVE. *Any agent which calms the nerves and is conducive to sleep.*

TRANQUILIZER. *Any agent which calms the mind without depressing alertness or mental activity.*

Biofeedback

Biofeedback is a process in which a person tries to alter a specific mental or autonomic function. While he is doing this, special instruments are monitoring the particular function, and he is constantly receiving feedback from the instrument regarding his success. All that is required from the subject is a relaxed frame of mind and an attentiveness and desire to achieve the intended change. The person must concentrate on the particular mental or autonomic process and the instrument must immediately and continuously provide information about whether or not the objective is being achieved.

For example, a person may attempt to change her normal EEG pattern from one which is predominated by beta waves (frequency of 13 to 35 cycles per second) to one which is predominated by alpha waves (8 to 12 cycles per second). Or the subject may attempt to reduce or increase her heart rate or blood pressure. In none of these cases is it possible to achieve instant success in one experimental period. Training is required, and the volunteer must be willing to undergo repeated sessions in which his mental concentration is focused on the desired change.

AUTONOMIC *refers to the parts of the nervous system (parasympathetic and sympathetic) that are not normally under conscious control.*

The possibility of being able to control one's autonomic activities is reported to be a relatively old concept among Zen Buddhists and certain other Oriental groups. In Western nations, however, this possibility was largely ignored for many years. Only in recent years has there been a growing awareness that it is possible for some individuals to effectively apply a certain degree of conscious control over certain autonomic functions. It is possible, for example, to consciously reduce blood pressure, increase or decrease the frequency of heartbeat, or shift blood from the head to the hand. What is needed to achieve such control is to put into operation the necessary reflexes. It is important to realize that a reflex is simply an anatomical circuit of nerve fibers through which a motor action can be affected. In the course of our lives we develop many of our reflexes, such as those which are involved in walking or eating, but we fail to develop many others, especially those concerned with conscious regulation of autonomic functions.

CURARE. *A muscle-relaxing drug that is extracted from any of several tropical plants (Carib kurari).*

Various kinds of experiments have shown that there are reflex pathways linking conscious control systems with autonomic nerve functions. In the following experiment, research scientists were able to induce rats to impose a conscious regulation over their heart rate. First, an electrode was implanted into a certain region of the hypothalamus which, when stimulated, provided the rats with the sensation of pleasure. Next, the rats were treated with curare, a drug which acts on neuromuscular junctions, bringing about paralysis of the skeletal muscles. The rats were thus unable to operate their skeletal muscles and therefore could not create a highly variable heartbeat to satisfy varying oxygen requirements in accord with varying skeletal muscle activities. With these preliminaries taken care of, the rats were trained to increase the rate of their heartbeat. Whenever the rate increased in an interval of time, the pleasure center was stimulated. When the frequency was not increased, the pleasure center was not stimulated. In the course of time the rats learned to increase their heartbeat frequency more often than not. In additional experiments with other rats it was possible, through the same procedure, to induce conscious regulation for decreasing the heartbeat.

Various kinds of biofeedback experiments on human subjects confirm what was learned in these experiments. For example, a person is asked to concentrate on reducing the frequency of his heartbeat. Success is rewarded with a beep or a flash which, like the stimulation of the rat's pleasure center, provides the individual with a feeling of satisfaction. Through this procedure of positive reinforcement, behavior experimentalists claim that experimentally cooperative and willing humans are fully capable of achieving in a period of several hours to days what Yogis, Zen Buddhists, and certain other Oriental groups have learned through rigorous discipline over periods of years.

Varying degrees of success have been claimed by research scientists in biofeedback for such things as training a person to regularize an irregular heartbeat, to shift blood from the brain to the hands in an effort to reduce or eliminate migraine headache, or to become somewhat oblivious to intractable pain. In the area of stress physiology, the aim of biofeedback research is to train people to control their autonomic actions in situations that are known to cause stress.

MIGRAINE. *A severe recurring headache, often accompanied by nausea (L.* hemicrania, *pain in the side of the head).*

Transcendental Meditation

Like biofeedback, transcendental meditation (or TM) is a conscious effort to alter mental activities and autonomic functions. Unlike biofeedback, however, no instrumental devices are used, and the person's concentration is directed toward a *mantra*, or incantation. Most often, the mantra is a nonsense word or sound, such as "mo" or "ohmm," which the person says repeatedly throughout the meditation period. In other forms of meditation, the meditator may simply form a mental image of an abstract mantra or of some real object, such as a flickering candle.

MANTRA. *Literally, a mystical word or formula used in incantations* (*Sanskrit* manyate, *a sacred counsel*).

Whatever the form of practice, the object of transcendental meditation is always the same—to direct one's energy and attention inward rather than to send it out. Except for periods spent in sleep, wakeful relaxation, or meditation, most of our activities involve sending energy and attention outward, as happens when we are studying, competing in sports, working, and the like. (As used in TM, the word "transcendental" suggests, literally, a "going outside" of both the self and the usual patterns of thought.)

The following instructions will give you some idea of how to practice transcendental meditation: Sit down with your spine and head erect. Breathe in and out slowly and deeply three times. Focus your attention on the body muscles, from head to toes, making a conscious effort to relax any tenseness that may be present. Close your eyes, but remain awake at all times, and repeat your mantra over and over. For your mantra, you can use the number "one." Whenever a distracting thought crosses your mind, such as thinking about tomorrow's activities, direct your attention toward the mantra in an effort to eliminate the distraction.

When properly done over a period of 15 to 20 minutes twice a day, transcendental meditation will bring about a beneficial effect on the autonomic nervous system. Usually within two weeks after starting, meditators find that they have less anxiety and a more relaxed but constructive attitude toward their daily activities.

Various kinds of controlled experiments have been conducted to determine the physiological effects of transcendental meditation. In general, when an individual successfully achieves a deep state of meditation, in which there is complete relaxation, there are certain physiological benefits. These include a lower basal metabolic rate; a reduction in breathing rate, heart rate, cardiac output, and lactic acid in the blood; an increase in electrical skin resistance (indicating a reduction in sweating); and alterations

LACTIC ACID. *A metabolite that accumulates as a result of inefficient oxidation of certain nutrients.*

in EEG patterns which are interpreted by investigators as a state of alertness along with restfulness.

All of the above changes occur also in sleep, and although claims have been made that the changes are more profound during transcendental meditation, studies have not yet determined whether this is true of all the stages of sleep, or only certain stages. (Very recent evidence, collected in 1975, shows that a considerable amount of transcendental meditation time is spent actually in stages 2, 3, and 4 of nonREM sleep.)

It is possible, though it has not been proved by experiment, that transcendental meditation may be scarcely beneficial, if at all, to individuals who spend several hours a day in wakeful relaxation. Even very productive and hard working individuals, whether engaged in intellectual or physical activity, may encounter very little stress. For these people, brief periods of wakeful relaxation are all that are needed for relief. For goal-oriented individuals who feel they do not have several hours a day for relaxation, 30 minutes of transcendental meditation may be an efficient way of warding off impending ulcers, hypertension, and other maladies that are associated with stress.

SUMMARY

1. Stressful experiences may produce anxiety or lead to a nervous breakdown.
2. Mental attitudes and emotions are more important factors in stress than physical work.
3. Stress may be relieved by sleep, especially in stage 4 of the nonREM phase of sleep, and by the dreams that commonly occur in the REM phase of sleep.
4. Stress may be released consciously by "talking it out" with another person, by group therapy, perhaps through special training in biofeedback, by transcendental meditation, and by suitable periods of wakeful relaxation.

REVIEW QUESTIONS

1. Define: stress, tension, strain, anxiety, nervous breakdown, nightmare, anxiety dream.
2. Are certain mental attitudes more important than excessive physical work in precipitating stress or a nervous breakdown? Explain your answer to this question.
3. What methods, mechanisms, or procedures are available for relieving stress?

4. Describe the two phases of sleep and the four stages of nonREM sleep.
5. Formulate a theory to explain how dreams might provide a mechanism for relieving stress.
6. Describe biofeedback. Suggest a procedure through which you may be able to work with a companion on determining whether you can consciously override some autonomic activity in your body.
7. What is transcendental meditation? Why might transcendental meditation be more effective than simple wakeful relaxation as a stress relieving procedure?

Suggested Readings

In this text we have presented a very broad overview of anatomy, physiology, and pathology, and it is not possible, therefore, to cite all of the authorities from whom facts and concepts have been borrowed. Nor is it possible to present an encyclopedic bibliography on all of the topics covered. Instead, the author has made a relatively small selection from the tremendous amount of literature available to the student of physiology and anatomy. Each of the references given in the following pages has been carefully evaluated as to its level of difficulty; most of the books and articles listed should be within your comprehension as you master the contents of this text.

Relatively recent publications are cited wherever possible. These publications will lead the interested student to older material, which in some cases may be equally valuable if not more so. The list of suggested readings has been organized according to the different Sections of this book. Some of the references cited for a particular chapter are also appropriate for other chapters in the same section or elsewhere throughout the text.

SECTION ONE: INTRODUCTION TO THE HUMAN BODY (CHAPTERS 1-4)

Chapter 2. Cells

Brown, W. V., and E. M. Bertke. *Textbook of Cytology.* 2d ed. C. V. Mosby Co., St. Louis, 1973 (528 pp). An elementary but comprehensive textbook on cell structure, function, and biology.

Bulger, R. E., and J. M. Strum. *The Functioning Cytoplasm.* Plenum Press, New York, 1974 (127 pp). For the serious beginner with an investigative aptitude, this textbook provides an introduction to laboratory approaches to cell biology.

DeRobertis, E. D. P., F. A. Saez, and E. M. F. DeRobertis, Jr. *Cell Biology.* 6th ed. W. B. Saunders Co., Philadelphia, 1975 (615 pp). A comprehensive introduction to cell biology, proceeding from elementary to upper level.

Dingle, J. T. (ed.). *Lysosomes: A Laboratory Handbook.* North Holland Pub., Amsterdam, 1972 (247 pp). Although this is a research-oriented text, it fully explains the importance of lysosomes to overall health.

Lentz, T. L. *Cell Fine Structure.* W. B. Saunders Co., Philadelphia, 1971 (437 pp). This is perhaps the best text on the structure and functions of individual types of cells.

Montgomery, R., R. L. Dryer, T. W. Conway, and A. A. Spector. *Biochemistry: A Case-Oriented Approach.* C. V. Mosby, St. Louis, 1974 (637 pp). Biochemistry, by its very nature, can never be an easy subject, but the authors of this text have made their presentation as practical as can be, aiming the subject matter at health-related problems.

Chapter 3. Tissues

Arey, L. B. *Human Histology.* 4th ed. W. B. Saunders Co., Philadelphia, 1974 (338 pp). An excellent textbook in outline form, which permits the reader to look at general concepts and at some very fine details of these concepts. Includes a small but sufficient number of figures to guide the reader.

Beresford, W. A. *Lecture Notes on Histology.* Blackwell Scientific Pub., Oxford, 1969 (296 pp). An excellent outline on the histology of human tissues. There are no figures, but the excellent descriptions perhaps compensate.

Bloom, W., and D. W. Fawcett. *A Textbook of Histology.* 10th ed. W. B. Saunders Co., Philadelphia, 1975 (1033 pp). This is a thick medical textbook on histology which could be a very helpful reference book, even to the beginner.

Constantinides, P. *Functional Electronic Histology.* Elsevier Pub., Amsterdam, 1974 (244 pp). The author has attempted to correlate electron microscopic structures of cells to their functions in all tissues of the mammalian body, with emphasis on the human.

Greep, Roy O., and L. Weiss. *Histology.* 3d ed. McGraw Hill, New York, 1973. One of the most readable books on the subject.

Hughes, H. E., and T. C. Dodds. *Handbook of Diagnostic Cytology.* E & S Livingstone Ltd., London, 1968 (255 pp). A practical text with a considerable amount of useful information on exfoliative cytology.

Koss, L. G. *Diagnostic Cytology and Its Histopathologic Basis.* 2d ed. J. B. Lippincott Co., Philadelphia, 1968 (653 pp). A comprehensive reference text, rich in photomicrographs. Contains 92 pages on general cell biology; the remainder of the book is concerned with diagnostic problems.

Kulonen, E., and J. Pikharainen (eds.). *Biology of Fibroblasts.* Academic Press, New York, 1972 (689 pp). Sixty-two articles are presented in this book on all aspects of fibroblasts, including their differentiation into various cell types, their role in synthesis, inflammation, repair, etc.

Naib, Z. M. *Exfoliative Cytopathology.* Little, Brown and Co., Boston, 1970 (398 pp). Line drawings and illustrations, along with short definitions and descriptions, make this book a very helpful reference, even to the nonpractitioner who is interested only in basic information of applied nature.

Reith, E. J., and M. H. Ross. *Atlas of Descriptive Histology.* 2d ed. Harper & Row, New York, 1970 (243 pp). This is perhaps the best published collection of photomicrographs on mammalian tissues.

Werner, H. J. *Synopsis of Histology.* 2d ed. McGraw-Hill, New York, 1967 (193 pp). Line drawings and brief descriptions make this a useful text for a beginner.

Chapter 4. Organs and Body Regions

Basmajian, J. V. *Primary Anatomy.* 6th ed. Williams & Wilkins Co., Baltimore, 1970 (404 pp). A basic human anatomy textbook for undergraduate students.

Christensen, J. B., and I. R. Telford. *Synopsis of Gross Anatomy.* 2d ed. Harper & Row, New York, 1972 (304 pp). A concise, straightforward text which should be useful to a student of anatomy engaged in laboratory work.

Ellis, H. *Clinical Anatomy.* 5th ed. Blackwell Scientific Pub., London, 1971 (389 pp). The author wrote this text because he believed that, in spite of the intensive training of medical students, there is a big gap between what they learned and what they should know for practical purposes. This is an excellent text in outline form with clear informative line drawings.

Francis, C. C. *Introduction to Human Anatomy.* 6th ed. C. V. Mosby, St. Louis, 1973 (463 pp). An introductory text for college students, richly illustrated with 325 figures, 25 color plates, and 7 color transparencies.

Gardner, W. D., and W. A. Osborn. *Structure of the Human Body.* 2d ed. W. B. Saunders Co., Philadelphia, 1973 (516 pp). A textbook for the beginning college student.

SECTION TWO: THE NERVOUS SYSTEM (CHAPTERS 5–11)

Barr, M. L. *The Human Nervous System.* Harper & Row, New York, 1972 (405 pp). An introduction to the functional anatomy of the nervous system.

Chalmers, N., R. Crawley, and S. P. R. Rose (eds.). *The Biological Bases of Behaviour.* Harper & Row, New York, 1971 (318 pp). This text, which was written by 45 people, contains a series of articles ranging from the nerve cell to the whole brain. It is aimed at gaining an insight into the neurological basis of behavior.

Essman, W. B. *Neurochemistry of Cerebral Electroshock.* Spectrum Pub., Flushing, New York, 1973 (181 pp). A cursory reading of this book, along with readings from Rose's, "The Conscious Brain," should suggest that mankind sometimes adopts certain neurological practices which are not founded on a sound body of knowledge.

Everett, N. B. *Functional Neuroanatomy.* 6th ed. Lea & Febiger, Philadelphia, 1971 (357 pp). A clearly written, nicely illustrated text on the functional anatomy of the human nervous system.

Eyzaguirre, C., and S. J. Fidone. *Physiology of the*

Nervous System. 2d ed. Year Book Medical Pub., Chicago, 1975 (418 pp). A second-year level text suitable for advanced college students. The emphasis is on principles, and an assumption is made that the student has perhaps as much background as provided by this text.

Ferguson, M. *The Brain Revolution.* Taplinger Pub., New York, 1973 (380 pp). Every student who is being introduced to studies of the nervous system ought to read at least one book of this sort, which deals with brain functions and brain research.

Gazzaniga, M. S. *The Bisected Brain.* Appleton-Century-Crofts, New York, 1970 (172 pp). Although much more research has been done on split brains since this book first appeared, the text will nevertheless expose the value of research on split brains toward an understanding of normal brain function.

Guyton, A. C. *Structure and Function of the Nervous System.* W. B. Saunders Co., Philadelphia, 1972 (254 pp). An introductory text on the entire nervous system, with emphasis on function rather than anatomy. This book was adopted almost unchanged from A. C. Guyton's fourth edition of "Textbook of Medical Physiology."

Mathews, W. B. *Practical Neurology.* 2d ed. Blackwell Scientific Pub., Oxford, 1970 (267 pp). A highly readable account of various neurological problems including headache, blackouts, epilepsy, giddiness, strokes, dementia, and pain.

Noback, C. R. *The Human Nervous System.* McGraw-Hill, New York, 1967 (281 pp). A lucid introductory text on neuroanatomy. It has perhaps the finest published illustrations on this subject.

Rose, S. *The Conscious Brain.* Weidenfeld and Nicolson, London, 1973 (351 pp). Rose presents what one might get from a formal text on the human nervous system, but he does so in an interesting informal way.

Schacter, S. *Emotion, Obesity and Crime.* Academic Press, New York, 1971 (195 pp). This is an excellent example of psychological research which has shed understanding on neural functions which could not be learned about strictly through physiological approaches.

SECTION THREE: THE ENDOCRINE SYSTEM (CHAPTERS 12 AND 13)

Danowski, T. S. *Outline of Endocrine Gland Syndromes.* 2d ed. Williams & Wilkins Co., Baltimore, 1968 (437 pp). A logically and clearly organized treatment of the physiology and diseases of the endocrine system. The level of presentation is suitable for the reader of this text, though the book is meant for medical students and professionals.

Kasirsky, G. *Vasectomy, Manhood, and Sex.* Springer Pub. Co., New York, 1972 (128 pp). This book is meant for lay readers and professionals. It provides a description of the vasectomy operation, relevant psychological commentary, and means of reversing the vasectomy.

Levine, S. (ed.). *Hormones and Behavior.* Academic Press, New York, 1972 (363 pp). Eighteen authors wrote this book, which deals with such topics as the effects of hormones on the central nervous system, hormones and reproductive behavior, hormones and maternal behavior, and the effects of hormones on aggression.

Martin, L. *Clinical Endocrinology.* 5th ed. J & A Churchill Ltd., London, 1969 (338 pp). An introductory but comprehensive and concise coverage of human endocrinology.

Sciarra, J. J., C. Markland, and J. J. Speidel (eds.). *Control of Male Fertility.* Harper & Row, New York, 1974 (316 pp). This is a research-level book in which about 50 authors present various points of view on male reproductive physiology, vasectomy, and nonsurgical approaches to male sterilization.

SECTION FOUR: THE STRUCTURAL AND SUPPORT SYSTEMS (CHAPTERS 14–18)

Chapter 14. The Skin

Champion, R. H., T. Hillman, A. J. Rook, and R. T. Sims (eds.). *An Introduction to the Biology of Skin.* Blackwell Scientific Pub., Oxford, 1970 (450 pp). A comprehensive second-level text by 17 authors on the epidermis, dermis, and accessory structures of the skin.

Hall-Smith, P., R. J. Cairns, and R. L. B. Beare. *Dermatology.* 2d ed. Grune & Stratton, New York, 1973 (374 pp). Following some introductory information on the basic anatomy and biology of skin, the authors present very well-organized discussions on all the major skin problems.

Kuno, Y. *Human Perspiration.* C. C. Thomas Pub., Springfield, Illinois, 1956 (416 pp). A detailed but excellent monograph with plenty of information on experimental approaches to the study of sweat glands.

Milne, J. A. *An Introduction to the Diagnostic Histopathology of the Skin.* Williams & Wilkins Co., Baltimore, 1972 (363 pp). An excellent collection of photomicrographs on various skin diseases.

Nicola, P. D., M. Morsiani, and G. Zavaglia. *Nail Diseases in Internal Medicine.* C. C. Thomas Pub., Springfield, Illinois, 1974 (113 pp). This monograph brings together a collection of descriptions on nail diseases that arise in conjunction with deeply seated internal medical problems. The book includes a good 28-page introduction on nail anatomy and physiology.

Samman, P. D. *The Nails in Disease.* 2d ed. C. C. Thomas Pub., Springfield, Illinois, 1972 (177 pp). After some introductory material on the anatomy and physiology of nails, the author proceeds with descriptions of all of the common nail diseases.

Solomons, B. *Lecture Notes on Dermatology.* 3d ed. Blackwell Scientific Pub., London, 1973 (256 pp). An excellent text in outline form concerning the anatomy, physiology, and diseases of skin.

Chapter 15. The Skeletal System

Brookes, M. *The Blood Supply of Bone.* Appleton-Century-Crofts, London, 1971 (338 pp). A specialized textbook on the circulation of blood through bone, with introductory material on bone structure and biology.

Hancox, N. M. *Biology of Bone.* Cambridge University Press, London, 1972 (199 pp). A comprehensive text on the fundamentals of bone structure, chemistry, and growth.

Little, K. *Bone Behaviour.* Academic Press, New York, 1973 (464 pp). An advanced, comprehensive text on the biology, physiology, and diseases of bone. Also see listing for Chapter 4.

Chapter 16. The Joints

Adams, J. C. *Arthritis and Back Pain.* Medical & Technical Pub., Oxford, 1972 (200 pp). A practical approach to the causes and management of arthritic diseases.

Huskisson, E. C., and F. D. Hart. *Joint Disease: All the Arthropathies.* Williams & Wilkins Co., Baltimore, 1973 (139 pp). A very useful reference text in which the joint diseases are arranged alphabetically, in outline form. The information includes a description of the disease, its incidence, signs and symptoms, course, laboratory findings, and treatment.

Lichtenstein, L. *Diseases of Bone and Joints.* C. V. Mosby Co., St. Louis, 1975 (314 pp). A concise but comprehensive introduction to diseases of bones and joints. Numerous X rays and photographs illustrate the text.

Chapters 17 and 18. Muscle Physiology and Anatomy

Basmajian, J. V. *Muscles Alive.* 3d ed. Williams & Wilkins Co., Baltimore, 1974 (525 pp). A comprehensive text on the involvement of muscles in various kinds of skeletal movements.

Brunnstrom, S. *Clinical Kinesiology.* 3d ed. F. A. Davis Co., Philadelphia, 1972 (349 pp). A textbook on the technique of feeling out and identifying muscles and tendons.

Cooper, K. H. *Aerobics.* Rev. ed. Bantam Books, New York, 1969 (182 pp). Based on sound scientific principles, this text contains specific training programs for physical fitness for people of all ages.

Daniels, L., and C. Worthingham. *Muscle Testing.* 3d ed. W. B. Saunders Co., Philadelphia, 1972 (165 pp). A pictorial approach to the examination of muscles and their ability to execute certain movements.

Falls, H. B. (ed.). *Exercise Physiology.* Academic Press, New York, 1968 (471 pp). Contains several articles on various aspects of physiology with respect to physical training.

Karpovich, P. V., and W. E. Sinming. *Physiology of Muscular Activity.* 7th ed. W. B. Saunders Co., Philadelphia, 1971 (374 pp). A basic, elementary text which relates the energetics of muscular activities to fundamental circulatory, respiratory, and other physiological phenomena.

Kendall, H. O., F. P. Kendall, and G. E. Wadsworth. *Muscles Testing and Function.* 2d ed. Williams & Wilkins Co., Baltimore, 1971 (284 pp). A pictorial approach to the examination of muscles and their capability of executing certain movements.

Morehouse, L. E., and A. T. Miller. *Physiology of Exercise.* 6th ed. C. V. Mosby Co., St. Louis, 1971 (328 pp). A concise, elementary approach to the subject, suitable for laymen as well as students in the health sciences.

Ryan, A. J., and F. L. Allman, Jr. (eds.). *Sports Medicine.* Academic Press, New York, 1974 (735 pp). Nineteen authors collaborated in the writing of this book, which includes chapters on the concept of physical fitness, conditioning for sports, nutrition, the limits of human performance, and special problems of the female athlete.

Schneider, F. R. *Handbook for the Orthopaedic Assistant.* C. V. Mosby Co., St. Louis, 1972 (198 pp). A textbook on the basic principles and techniques in the management of injuries to bones and joints.

Wells, K. F. *Kinesiology.* 5th ed. W. B. Saunders Co., Philadelphia, 1971 (564 pp). A comprehensive text on the anatomy and physiology of muscles, joints, ligaments, joint movements, and the involvement of muscles in athletic activities.

SECTION FIVE: DIGESTION, METABOLISM, AND NUTRITION (CHAPTERS 19–24)

Chapter 19. Digestion and Elimination

Davenport, H. W. *Physiology of the Digestive Tract.* 3d ed. Yearbook Medical Pub., Chicago, 1971 (229 pp). A comprehensive introductory coverage of digestion with scattered discussions on digestion problems.

Toner, P. G., K. E. Carr, and G. M. Wyburn. *The Digestive System—An Ultrastructural Atlas and Review.* Butterworth and Co., London, 1971 (303 pp). A richly illustrated coverage of the microanatomy and functions of the digestive system.

Chapter 20. Absorption of Nutrients

Bargstrom, B., A. Dahlqvist, and L. Hambraeus, L. (eds.). *Intestinal Enzyme Deficiencies and Their Nutritional Implications.* Swedish Nutrition Foundation, Uppsala, 1972 (149 pp). An elementary and concise but thorough discussion of intestinal enzyme deficiencies, with emphasis on lactase. Also a good coverage of our present understanding of stomach lipase and lipid absorption.

Burland, W. L., and P. D. Samuels (eds.). *Transport Across the Intestine.* Williams & Wilkins Co., Baltimore, 1972 (292 pp). A relatively recent picture on the intestinal absorption of water, electrolytes, sugars, amino acids, small peptides, and fats.

Chapters 21–24. Metabolism and Nutrition

Anderson, M. P. H., M. V. Dibble, H. S. Mitchell, and H. J. Rynbergen. *Nutrition in Nursing.* J. B. Lippincott Co., Philadelphia, 1972 (406 pp). After 120 pages of basic information on nutrition, 60 pages are devoted to nutritional needs for growth, pregnancy, lactation, infancy, childhood, and old age. The remainder of the text concerns nutrition in disease.

Berland, T. *Rating the Diets.* Publications International Ltd., Skokie, Illinois, 1974 (385 pp). Descriptions, critiques, and references for all the well-known diets as well as numerous diets that are scarcely known.

Carlson, L. D., and A. C. L. Hsieh. *Control of Energy Exchange.* Macmillan Co., New York, 1970 (151 pp). A brief, nicely illustrated coverage of energy balance and temperature regulation.

Chaney, M. S., and M. L. Ross. *Nutrition.* 8th ed. Houghton Mifflin Co., Boston, 1971 (486 pp). An excellent basic text on the principles of nutrition with useful information on application.

Garrow, J. S. *Energy Balance and Obesity in Man.* North Holland Pub., Amsterdam, 1974 (335 pp). The author of this book attempted to integrate the relevant physiology and psychological data which relate to energy balance and obesity in humans. In doing so, he has refrained as much as possible from references to laboratory studies on other animals.

Marks, J. *The Vitamins in Health and Disease.* Little, Brown and Co., Boston, 1968 (183 pp). A concise elementary exposure to the role of vitamins in nutrition.

Mayer, J. *Human Nutrition.* C. C. Thomas Pub., Springfield, Illinois, 1972 (721 pp). A series of 82 essays by a person who might rightfully be called the worldwide dean of nutrition.

Olsen, I. D. *Metabolism.* Bobbs-Merrill Co., New York, 1973 (214 pp). An introductory text to metabolism, but with chemical formulas throughout. The emphasis is on the fundamentals of metabolism as they pertain to all forms of life in general.

Porter, J. W. G., and B. A. Rolls (eds.). *Proteins in Human Nutrition.* Academic Press, New York, 1972 (560 pp). Forty-five authors provide a comprehensive discussion on protein supplies, requirements, evaluation, science, technology, and use in health and disease.

Shackelton, A. D. *Practical Nurse Nutrition Education.* 3d ed. W. B. Saunders Co., Philadelphia, 1972 (307 pp). Concise, easy to follow principles of nutrition and their application to healthy and sick persons. Specific diets are formulated for various diseases.

Sinclair, H. M. *Hutchinson's Food and the Principles of Nutrition.* 12th ed. Edward Arnold Pub., London, 1969 (644 pp). One of the very best textbooks on this subject.

Upjohn Company. *Vitamin Manual.* Upjohn Co., Kalamazoo, Michigan, 1965 (88 pp). A summary description of the vitamins and their functions.

Watt, R. K., and A. L. Merrill. *Composition of Foods.* Agriculture Handbook #8, USDA, Washington, D.C., 1963. The definitive reference text on the composition and caloric contents of more than 2500 food items.

SECTION SIX: TRANSPORT SYSTEMS: CIRCULATION, RESPIRATION, AND EXCRETION (CHAPTERS 25–30)

Chapters 25–27. The Circulatory System

Astrand, P., and K. Rodahl. *Textbook of Work Physiology.* McGraw-Hill, New York, 1970 (669 pp). This is perhaps the best single reference text or study text which relates work and athletic performance to human physiology.

Fleming, J. S., and M. V. Brainbridge. *Lecture Notes on Cardiology.* 2d ed. Blackwell Scientific Pub., London, 1974 (326 pp). A comprehensive but concise examination of heart problems.

Frank, M. J., and S. V. Alvarez-Mena. *Cardiovascular Physical Diagnosis.* Yearbook Medical Pub., Chicago, 1973 (186 pp). A study-guide approach to the practice of cardiology, with emphasis on individual signs and symptoms of heart problems.

Friedman, M., and R. H. Rosenman. *Type A Behavior and Your Heart.* Alfred A. Knopf, New York, 1974 (276 pp). A point of view worthy of reflection.

Hurst, J. W., and R. J. Myerburg. *Introduction to Electrocardiography.* 2d ed. McGraw-Hill, New York, 1973 (319 pp). A relatively simple text on the subject.

Larson, L. A. (ed.). *Fitness, Health and Work Capacity.* Macmillan Co., New York, 1974 (593 pp). In addition to subject matter dealing with the scientific basis for work capacity and physical fitness, the authors of this book provide techniques and procedures for evaluating an individual's capacity for work and athletic competition.

Marriott, H. J. L. *Practical Electrocardiography.* 5th ed. Williams & Wilkins Co., Baltimore, 1972 (325 pp). An elementary text, highly illustrated, and truly intended for the beginner in this area.

Phibbs, B. *The Human Heart.* 2d ed. C. V. Mosby Co., St. Louis, 1971 (247 pp). A lucid, concise, elementary, well-illustrated book on heart problems.

Chapter 28. The Respiratory System

Cumming, G., and S. J. Semple. *Disorders of the Respiratory System.* Blackwell Scientific Pub., London, 1973 (564 pp). A comprehensive text on the physiology and pathology of the respiratory system.

Pace, W. R. *Pulmonary Physiology in Clinical Practice.* 2d ed. F. A. Davis Co., Philadelphia, 1970 (177 pp). Excellent, clear, elementary descriptions and discussions on lung physiology and diseases, with emphasis on the diagnosis, understanding, and treatment of respiratory deficiencies.

Schonell, M. *Respiratory Medicine.* Churchill Livingston, London, 1974 (271 pp). After a brief introduction to respiration, the author classifies the major respiratory diseases, describes them, and discusses their treatment and management.

Chapter 29. The Red Blood Cell System

Nour-Eldin, F. *Revision Hematology With Examination Exercises.* Butterworth & Co., London, 1973 (170 pp). A concise explanatory review of hematology.

Rapaport, S. I. *Introduction to Hematology.* Harper & Row, New York, 1971 (403 pp). A nicely illustrated, lucid, elementary textbook on blood.

Simmons, A. *Basic Hematology.* C. C. Thomas Pub., Springfield, Illinois, 1973 (278 pp). An introductory text to hematology for health science students.

Woodliff, H. J., and R. P. Herrman. *Concise Haemotology.* Williams & Wilkins Co., Baltimore, 1973 (216 pp). An elementary text on the study of blood.

Chapter 30. The Excretory System

Barnes, R. W., R. T. Bergman, H. L. Hadley, and E. C. Jacobs. *Urology.* 2d ed. Medical Examination Pub., Flushing, New York, 1974 (535 pp). This text, written in outline form, can be useful as a reference book.

Golden, A., and J. F. Maher. *The Kidney.* Williams & Wilkins Co., Baltimore, 1971 (207 pp). An attempt

Suggested Readings

to integrate the structural aspects of the kidneys in disease.

Harvey, R. J. *The Kidneys and the Internal Environment.* Chapman and Hall Ltd., London, 1974 (167 pp). A basic introduction to the functions and mechanisms of the kidneys.

Newsam, J. E., and J. J. B. Petrie. *Urology and Renal Medicine.* E. & S. Livingstone, London, 1971 (259 pp). An elementary text on the disorders of the kidneys, urinary tract, and male genital tract.

Robinson, J. R. *Fundamentals of Acid-Base Regulation.* 4th ed. Blackwell Scientific Pub., London, 1972 (111 pp). An elementary text for the beginner, with emphasis on application.

Wilson, R. F. *Fluids, Electrolytes, and Metabolism.* C. C. Thomas Pub., Springfield, Illinois, 1973 (138 pp). A study-guide to the topic of acid-base-electrolyte balance.

SECTION SEVEN: THE MECHANISMS OF DEFENSE (CHAPTERS 31 AND 32)

Bellanti, J. A. *Immunology.* W. B. Saunders Co., Philadelphia, 1971 (584 pp). A clear, concise, introductory text, with fine illustrations on inflammation and immunology in health and disease.

Bellanti, J. A., and D. H. Dayton. *The Phagocytic Cell in Host Resistance.* Raven Press, New York, 1974 (348 pp). This is a research-level text, but it contains persuasive information that the macrophage and other defense cells evolved several enzyme systems for coping with infectious agents.

Feingold, B. F. *Introduction to Clinical Allergy.* C. C. Thomas Pub., Springfield, Illinois, 1973 (380 pp). A well-illustrated introductory book on the causes and treatment of allergy.

Gottlieb, A. A. (ed.). *Developments in Lymphoid Cell Biology.* CRC Press, Cleveland, 1974 (176 pp). Nine authors attempt to give the reader a glimpse into the various functions of various kinds of lymphoid cells in immunity.

Luckey, T. D. (ed.). *Thymic Hormones.* University Park Press, Baltimore, 1973 (376 pp). This is an advanced text, but a skim through it will reveal some of the problems and uncertainties which researchers face even in modern times.

Medical Communications Inc. *Immunology.* Upjohn Co., Kalamazoo, Michigan, 1972 (60 pp). Seven authors contributed articles on immunoglobulins, hypersensitivity, transplants, tumor immunology, immunosuppression, and injury due to antibodies. Nicely illustrated.

Ramwell, P. W. (ed.). *The Prostaglandins.* Vol. I. Plenum Press, New York, 1973 (400 pp). After several introductory chapters on the biochemistry of prostaglandins, the authors proceed into organ systems, including skin, lungs, kidneys, and the reproductive system.

Rebuck, J. W. (ed.). *The Reticuloendothelial System.* Williams & Wilkins Co., Baltimore, 1975 (328 pp). Twenty-three authors wrote this book, which contains articles on the macrophages in various kinds of cancer, including Hodgkin's disease and lymphomas.

Roitt, I. M. *Essential Immunology.* Blackwell Scientific Pub., London, 1971 (220 pp). An introductory text, but one which emphasizes experimental approaches and mechanisms of immunology.

Rubin, J. M., and N. S. Weiss. *Practical Points in Allergy.* Medical Examination Pub., Flushing, New York, 1974 (208 pp). A concise, clearly written text on the causes and treatment of allergies.

Sell, S. *Immunology, Immunopathology and Immunity.* Harper & Row, New York, 1972 (277 pp). A well-organized, concise introduction to immunity in health and disease, with emphasis on principles for biology and health science students. The topics range from molecular to organismic, from basic to applied.

Solo'ev, V. D., and T. A. Bektemirov. *Interferon Theory and Applications.* Plenum Press, New York, 1973 (304 pp). This is a translation of a Russian text which covers the chemistry, biology, and clinical use of interferon.

Somekh, E. *A Parent's Guide to Children's Allergies.* C. C. Thomas Pub., Springfield, Illinois, 1972 (189 pp). A practical guide to the understanding and management of allergy for laymen and professional people.

Vernon-Roberts, B. *The Macrophage.* Cambridge University Press, London, 1972 (242 pp). A clearly written text covering all of the fundamentals of macrophage biology at an elementary level.

Weiss, L. *The Cells and Tissues of the Immune System.* Prentice-Hall, Englewood Cliffs, New Jersey, 1972 (252 pp). A nicely illustrated introductory text, divided into three sections. The first deals with tissues and organs, the second with cells, and the third with antibodies and cell-mediated immune systems.

Williams, R. T. *Detoxication Mechanisms.* 2d ed. John Wiley & Sons, New York, 1959. A classic book, which shows how earlier investigators systematically screened all of the major classes of organic compounds to determine how they were metabolized in vertebrate bodies. A prerequisite in organic chemistry is required for an understanding.

SECTION EIGHT: SPECIAL TOPICS IN PHYSIOLOGY (CHAPTERS 33–35)

Austin, M. *Acupuncture Therapy*. ASI Pub., New York, 1972 (276 pp). One of the most clearly written and useful texts on acupuncture. Tests, provided throughout, lend some self-teaching character to the book.

Bourne, P. G. (ed.). *The Psychology and Physiology of Stress*. Academic Press, New York, 1969 (242 pp). A casual study of this book, written by eight authors, will show how our attitudes have changed over the years regarding stressed men in combat and the kinds of information that were considered desirable in assessing stress among Vietnam soldiers.

Duke, M. *Acupuncture*. Pyramid House, New York, 1972 (223 pp). A comprehensive book dealing with the historical, philosophical, and descriptive aspects of acupuncture, intended for the nonpracticing but interested layman.

Gutmann, E., and V. Hanzlikova. *Age Changes in the Neuromuscular System*. Scientechnica Pub., Bristol, England, 1972 (195 pp). A comprehensive, clear discussion on changes that occur in motor nerves and skeletal muscles in the course of aging.

Hartmann, E. L. *The Functions of Sleep*. Yale University Press, New Haven, Connecticut, 1973 (198 pp). Written by a psychiatrist who has spent many years as a director of sleep experiments, this book provides the reader with exactly what the title says.

Lowe, W. C. *Introduction to Acupuncture Anesthesia*. Medical Examination Pub. Co., Flushing, New York, 1973 (101 pp). An explanation of the techniques and a partial explanation of the scientific basis on which acupuncture may be effective as an anesthetic.

Manaka, Y., and I. A. Urquhart. *The Layman's Guide to Acupuncture*. Weatherhill Pub., New York, 1972 (143 pp). A clearly written, richly illustrated text.

Matsumoto, T. *Acupuncture for Physicians*. C. C. Thomas Pub., Springfield, Illinois, 1974 (204 pp). The author presents a scientific explanation of acupuncture as well as some historical background. In addition, he presents 70 of the most common and effective acupuncture points.

Melzack, R. *The Puzzle of Pain*. Basic Books, New York, 1973 (232 pp). Written by one of the formulators of the gate control theory of pain, this book provides an interesting accounting of various forms of pain, including referred pain, phantom sensations, neuralgias, and causalgias (severe, burning pain).

Mendels, J. (ed.). *Biological Psychiatry*. John Wiley & Sons, New York, 1973 (527 pp). A background in pharmacology and psychology is required to understand some of the chapters in this book, written by 25 authors, but there is enough sound information on emotion, dreams, sleep, and neurotransmitter physiology to make it worthwhile reading material.

Rockstein, M. (ed.). *Theoretical Aspects of Aging*. Academic Press, New York, 1974 (192 pp). Sixteen specialists in the area of basic biological research on aging present the reader with a summary view on what is now known about aging.

Williams, R. L., I. Karacan, and C. J. Hursch. *EEG of Human Sleep*. John Wiley & Sons, New York, 1974 (169 pp). The authors' purpose in writing this book was to present an atlas of the electroencephalogram of human sleep. Excellent EEGs from infants to people of old age are given along with appropriate descriptions.

Woodruff, D. S., and J. E. Birren (eds.). *Aging*. D. Van Nostrand, New York, 1975 (420 pp). Twenty-two authors discuss some of the sociological, psychological, and biological aspects of aging.

Glossary

abduct to draw a body part away from the median line of the body; opposite of adduct

acetabulum a cup-shaped receptacle in the hipbone for the head of the thighbone

acetate the salt (CH_3COO^-) of acetic acid (CH_3COOH)

acetone an end product of metabolism which may reach high levels in untreated diabetics and persons on a prolonged carbohydrate-free diet

acetylcholine a substance that relays the information of a nerve impulse from a nerve cell to a target cell

acetyl coenzyme A a coenzyme synthesized from acetate and a vitamin B derivative

acid any substance which can give up or release hydrogen ions

acidosis a condition in which there is an excess of carbon dioxide and a deficiency of bicarbonate ions in the blood; opposite of alkalosis

acromegaly a progressive, abnormal enlargement of the body extremities, the face, and certain organs; caused by an excess of growth hormone

ACTH see adrenocorticotropin

actin a muscle protein; together with myosin it forms the bulk of the contractile apparatus

action potential a pulse of bioelectricity which travels from one part of a cell to the extreme opposite end(s) of the cell

active transport the movement of a solute across a cell membrane at the expense of ATP

adduct to draw a body part toward the median line of the body; opposite of abduct

adenoids pharyngeal tonsils

adenosine diphosphate see ADP

adenosine triphosphate see ATP

ADP adenosine diphosphate, an end product of the breakdown of ATP; this breakdown provides the energy needed for physiological processes

adrenaline a hormone, produced by the adrenal medulla, whose effect is to prepare the body for certain energy-demanding activities; also called epinephrine

adrenal medulla the central, or inner, region of the adrenal gland

adrenocorticotropin (*ACTH*) a pituitary hormone which stimulates the adrenal cortex to release cortisol and cortisone

adventitia an outer covering of a visceral organ or tube

aerobic refers to any process that requires free oxygen; opposite of anaerobic

affective related to an emotional feeling

afferent to bring toward

albumin a serum protein which transports fatty materials through the blood

albuminuria the presence of albumin and other proteins in the urine

aldosterone a hormone, secreted by the adrenal cortex, which regulates the metabolism of potassium and sodium

alexia a loss of the ability to understand written words

alkalosis a condition in which there is an excess of bicarbonate ions and a deficiency of hydrogen ions in the blood; opposite of acidosis

allergy a pathological response by the body's immune system to a foreign material

allergen an antigen which provokes an allergic reaction

alveolus a small sac, such as those in the lungs

amino acid a compound which contains an amino (NH_2) group and an acid (H^+); amino acids are the units of which proteins are made

ampulla a saclike dilation, or swelling, in a duct

amygdala the region of the brain which is closely associated with smelling, eating, and emotionality

amyloid a carbohydrate-protein complex found in certain kinds of diseased tissue

ammonia a toxic waste product of nitrogen metabolism, most of which is converted to urea, a less toxic waste product

anabolism the process of synthesizing complex organic compounds from simpler ones; opposite of catabolism

anaphylaxis a state of hypersensitivity in which, upon repeated exposure to the antigen, the individual eventually enters a state of shock

anaplasia a pathological loss of the usual distinctive tissue characteristics

anaerobic refers to a process or condition lacking in oxygen; opposite of aerobic

anatomy the science of the structure of an organism

androgen a male sex hormone, such as testosterone

anemia an abnormal deficiency in red blood cells, resulting in a decrease in the oxygen-carrying capacity of blood *anesthesia* a loss of sensation

aneurism a bulging weakened area in the wall of a blood vessel

analgesia an insensitivity to pain

angina pectoris a spasmodic chest pain accompanied by a feeling of suffocation; caused by a deficiency of oxygen to the heart muscle

angiotensin a small blood protein which increases blood pressure by constricting the arteries and stimulating aldosterone from the adrenal cortex

ankle the joint between the lower leg and the tarsus or instep, consists of the seven tarsal bones

anorexia an abnormal loss of appetite

antibody a protein whose production is elicited by an antigen; the protein subsequently destroys the antigen

antidiuretic hormone or *ADH.* pituitary hormone that inhibits or prevents the excretion of urine

antigen a substance which elicits an immune response

aorta the large artery which distributes blood from the left ventricle to the systemic circulatory arteries

aortic body a neural tissue in the aortic arch which responds to changes in blood levels of oxygen and carbon dioxide

aortic sinus refers to dilations in the aorta that contain pressure-receptor nerve fibers

apocrine gland an exocrine gland whose secretion includes the apical region of the cell itself

aponeurosis an expanded, flattened tendon

appendix any appendage, especially the small worm-like projection arising from the cecum of the large intestine

aphasia any language communication disorder

apraxia a pathological inability to perform a purposeful muscular movement

arachnoid the central of the three membranes covering the brain and spinal cord

arcuate fasciculus nerve fibers which run in an arc-shaped path from the frontal lobe to the temporal lobe

arrector pilorum a bundle of smooth muscle which erects a hair and creates a goose bump

arrhythmia any variation of the normal heart rhythm

arterial pressure usually refers to the systemic blood pressure which can be measured with a sphygmomanometer

arteriosclerosis a hardening of the arteries

arteriovenous anastomosis a direct passageway between an artery and a vein, bypassing the capillary circuit

artery a vessel which conveys blood away from the heart

arthritis an inflammation in a joint

articulate unite by means of a joint

assimilation the process of transforming food into protoplasm or energy

astereognosis a loss of the sense of three-dimensional reality

astigmatism faulty vision due to an irregularity in the curvature of the cornea or lens

asthma a disorder in which episodes of violent, painful breathing occur due to inflammation of the bronchi and other air passageways

astrocyte a star-shaped neuroglial cell

ataxia a loss of power of muscular movement

atherosclerosis a hardening of the arteries caused by depositing fatty and fibrous materials within and beneath the intima

athetosis a disorder characterized by repetitious, slow, continuous changes in position of various parts of the body

atonia a lack of strength or tone in a muscle

ATP adenosine triphosphate, the direct or immediate source of energy for most physiological processes

atrioventricular node a group of specialized bio-electrical fibers in the right atrium of the heart at the border of the right ventricle

atrium either of the two upper chambers of the heart

atrophy a decrease in size of a body part; a wasting away

aura a premonition which precedes a convulsion

autoimmunity an immunological reaction in which one's own tissue appears foreign to the immune system and, accordingly, is attacked and destroyed

autonomic nervous system a division of the nervous system which operates at all times

autopsy an examination of the body after death

axilla an armpit

axis a straight line around which something may revolve, or a straight line passing between two poles

axon the main process of a nerve cell

base any substance capable of combining with hydrogen ions

basement membrane the secretory product of epithelial cells; it is used to anchor the epithelium to underlying connective tissue

basophil a blood cell which produces serotonin, heparin, and histamine

Benemid the trade name for probenecid, a drug which promotes the excretion of uric acid

beta oxidation a metabolic process in which fatty acids are converted into acetate

bicarbonate reserve the amount of bicarbonate present in the blood

bile fluid secreted by the liver and sent into the small intestine to aid in the digestion and absorption of fats

bile salts steroids which are produced from cholesterol and are a major component of bile; they emulsify or solubilize fat

bile wastes mainly the breakdown products of hemoglobin, the oxygen-carrying pigment in red blood cells

bilirubin a red bile pigment formed from biliverdin; in people with jaundice it is found excessively in the blood and tissues

biliverdin a green bile pigment formed from hemoglobin

biopsy a sample of tissue taken from a living person to diagnose a disease

Bowman's capsule one of many cup-shaped structures in the kidney; important in the formation of an ultrafiltrate

bradycardia an abnormal slowing of the heartbeat

bradykinin a small protein which induces dilation of blood vessels

bronchiole any of the small air tubes branching from the bronchi

bronchus either of the two air passageways that extend from the trachea to the bronchioles in the lungs

buffering the process of keeping the hydrogen ion concentration constant

bunion an abnormal enlargement and thickening of the metatarsophalangeal joint of the great toe

bursa a small fluid-filled sac located between certain parts of the body that move upon one another

butyric acid a short-chain fatty acid ($CH_3CH_2CH_2COOH$)

calciferol vitamin D_2, resulting from the irradiation of ergosterol

calcitonin a thyroid hormone which tends to reduce calcium levels in the blood

calorie the amount of heat required to raise the temperature of 1 gram of water 1 degree C

capillary a small tubule (about 8 to 10 micrometers in diameter) which carries blood or lymph

carbohydrate a compound whose carbon atoms carry, on the average, one equivalent of a water molecule; starches and sugars are carbohydrates

carbuncle an inflammation of the skin and subcutaneous tissue; similar to a furuncle, but much larger

carcinogen an agent which induces cancer

carcinoma a malignant tumor that occurs in epithelial tissue

cardia the region of the stomach adjoining the esophagus

cardiac impulse the electrical signal that originates in the heart and is recorded on an electrocardiograph (ECG)

cardiac output the volume of blood ejected in one minute by either of the heart's two ventricles

cardiomegaly a pathological enlargement of the heart

cardiovascular system the circulatory system

carotid body a tissue located at the fork of the carotid artery; it responds to changes in blood levels of oxygen and carbon dioxide

carotid sinus a tissue at the fork of the carotid artery; it picks up information on blood pressure

carpals the eight bones of the wrist

cartilage a tough nonvascular connective tissue made of cartilage cells, fibers, and a ground substance

casein a protein obtained from milk through the action of the enzyme rennin or an acid

catalysis an increase in the speed of a chemical reaction whereby the catalyst itself is unchanged by the reaction

catabolism the process of breaking down complex compounds into simpler ones, releasing energy

cataract a partial or complete opaqueness of the lens of the eye

catheter a tube used for withdrawing the contents of a cavity of the body

cation reserve the ions in the blood which are positively charged, such as sodium (Na^+) or calcium (Ca^{++})

cautery the application of heat or caustic chemicals to destroy body tissue

cecum the blind pouch located where the small intestine joins the large intestine

celiac refers to the abdominal region

cellulose an indigestible polysaccharide made entirely from glucose; the chief component of plant cell walls

cephalic pertaining to the head

cerebellum a coordination center in the brain for voluntary movements, posture, and equilibrium

cerebral cortex the highly convoluted outer tissue of the cerebral hemispheres of the brain; the cerebral cortex is involved in all conscious activities

cerebrum the largest part of the brain; it is divided into two hemispheres

cerumen the soft, brownish ear wax secreted by the ceruminous glands of the outer ear canal

cervical pertaining to the neck

cervix any necklike structure, especially the neck of the uterus

chiasma the crossing of two lines, as happens for example, where the optic nerves cross each other before becoming the optic tracts

cholecystitis an inflammation of the gallbladder

cholelithiasis the presence of gallstones

cholesterol a fatty alcohol found in animal fats and many body tissues; high levels in the blood are related to arteriosclerosis, heart attack, and other disorders

cholinergic refers to synapses in which acetylcholine is liberated as the neurotransmitter

chondroblast an embryonic cell which can mature into a cartilage cell or a bone cell

chondrocyte a mature cartilage cell

chorea a disorder marked by involuntary, irregular muscular actions of the extremities and face

chorion the outermost of two membranes which enclose the fetus

chorionic gonadotropin a placental hormone which stimulates the ovary to produce and release sex hormones *choroid plexuses* the tissues in which cerebrospinal fluid is produced

chyle a milky emulsion of fat formed in the small intestine during digestion

chyle cistern a saclike structure at the beginning of the thoracic duct, at the level of the twelfth thoracic vertebra

chylomicron a microscopic globule of fat and protein that appears in the blood after a meal

chyme the stomach contents before being digested in the small intestine

cilia hairlike extensions from certain epithelial cells; the back-and-forth movements of cilia help move mucus and other materials along body passageways

cingulum a bundle of fibers which run from the base of the brain to the hippocampal fold

circumcision a surgical procedure in which part or all of the foreskin (prepuce) of the penis is removed

cirrhosis a disease in which liver cells are converted into functionless fibers

cistern a closed space that serves as a reservoir for fluids

clearance the maximum rate at which the kidneys can clear the blood of some particular substance

clitoris a small cylindrical erectile structure in the anterior region of the vulva

clonus a series of alternate contractions and relaxations

cochlea the spiral-shaped receptor organ in the inner ear

cognition the process of knowing, becoming, or perceiving

colchicine a drug commonly used in treating gout

collagen a gelatinous fibrous substance found in connective tissues

colliculus a small mound or hill

colloid a substance made of insoluble particles that are small enough to remain permanently suspended in a fluid medium

colon the first part of the large intestine

colostrum the thin milky fluid secreted by the mammary glands after a woman gives birth; secretion of colostrum precedes the secretion of milk

coma a state of deep unconsciousness brought on by injury or disease

commissure a bundle of nerve fibers that crosses over from one side of the body to the other

complement a family of proteins which act as agents in helping rid the body of antigens

conduction deafness a loss of hearing due to some blockage of the normal hearing passageway

cones the short, conelike, light-sensitive cells in the retina responsible for color vision

congenital refers to a condition that exists at birth and usually has occurred earlier in the uterus

congestion an excessive accumulation of blood in a part of the body, usually in the heart or lungs

conjunctiva the mucous membrane which covers the front of the eyeball and the underside of the eyelids

contact dermatitis a skin disorder sometimes caused by an allergy

convulsion an involuntary, forceful contraction of large masses of skeletal muscles

cornea the transparent membrane which forms the anterior sixth of the outer coat of the eyeball

corneum the horny layer of the epidermis

coronary occlusion a blockage of the coronary arteries, leading to a heart attack

corpus albicans a white body formed from a corpus luteum

corpus callosum a large nerve tract which crosses the brain from one side to the other

corpus luteum a yellow cell mass in the ovary which is formed from a follicle that has released its egg; produces and secretes estrogen and progesterone

cortex the outer, or peripheral, portion of an organ

costal pertains to the ribs

cranial nerves the twelve pairs of nerves which originate in the brain

cretinism a disorder in which a child's physical and mental growth are stunted due to a severe thyroid deficiency

crista a projection; the inner ear has three cristae, one on the inner surface of the ampulla of each semicircular canal

crossing-over the process through which genes are shuffled to give different combinations to the offspring

cryptorchism the failure of the testes to descend into the scrotum

Cushing's syndrome a disease due to excess adrenal cortical hormones in the blood

cystic fibrosis a severe congenital disease characterized by fibrosis of epithelial tissue, especially the pancreas

cytology the study of cell biology

cytoplasm the protoplasm which lies outside a cell's nucleus

decidua the mucous lining of the pregnant uterus; it is cast off at childbirth

dendrite any one of the many branching processes of a nerve cell

density the quantity of material which is present in a unit volume of material

deoxyribonucleic acid see DNA

depolarization the process by which some of the negative charges are taken away from the interior surface of a membrane

dermatome an area of the skin which receives sensory nerve fibers from a single spinal nerve

dermis the lower of the two layers that make up skin

diabetes insipidus a metabolic disorder due to a deficiency of antidiuretic hormone, characterized by intense thirst and excessive urination

diabetes mellitus a metabolic disorder characterized

by high levels of sugar in the blood and urine; it is usually caused by a lack of insulin

diaphragm, in anatomy a dome-shaped organ formed mainly of muscle at its circumference and of tendon at the center; separates the abdominal cavity from the chest cavity

diaphragm, in contraception a dome-shaped plastic or rubber device worn over the cervix to prevent sperm from entering the uterus

diaphysis the shaft of a long bone

diarthrosis a freely movable joint

diastase an enzyme that catalyzes the splitting of starch into small units

diastole the relaxation phase of the heart's pumping cycle

diffusion the process by which the random movement of molecules in solution or suspension results in a net movement of substances from regions of higher to regions of lower concentration

digestion the process of breaking down foods into units that can be absorbed and used by the body

digitalis a drug used for correcting heart rhythm problems by stimulating contractility and lengthening the refractory period

dipeptide a molecule made of two molecules of amino acid

diploe the spongy bone between the two layers of compact bone in the flat cranial bones

diplopia a visual disorder in which two images are seen of a single object; double vision

disaccharide a carbohydrate formed from the condensation of two monosaccharides

distal refers to something that is situated further from a point of reference; opposite of proximal

DNA deoxyribonucleic acid, the material from which genes are made

duodenum the first portion of the small intestine

dura mater the hard outermost covering of the brain and spinal cord

dysphagia an abnormal difficulty in swallowing

dyspnea a condition in which breathing is difficult and painful

dysrhythmia a disturbance in rhythm

dystonia a disorder of normal muscle tone

eccrine gland an exocrine gland that secretes sweat which is chemically different from apocrine sweat

ectopic pregnancy the development of a fetus outside the uterus, usually in the oviduct or abdominal cavity

eczema an inflammatory skin disorder

edema the abnormal accumulation of fluid in the intercellular spaces, resulting in swelling

efferent leading away from; efferent neurons carry impulses away from the central nervous system; opposite of afferent

effluent something that flows out; usually a waste material

electrocardiograph an instrument for recording the electrical activity of the heart; the record is called an electrocardiogram or ECG

electrophoresis the movement of charged particles through a salt solution between the negative and positive poles of a battery

emaciation an abnormal extreme loss of flesh

embolus a clot or plug in a blood vessel

embryo see Fetus

emotion a strong feeling; a sentiment

emphysema a pathological accumulation of air in the lungs due to an overdistension of the tiny air sacs (alveoli)

emulsifier an agent used to assist in maintaining an emulsion

emulsion a substance in which one liquid is dispersed throughout the body of a second liquid

endocarditis an inflammation of the inner wall of the heart

endocardosis an abnormal condition in the inner wall of the heart

endocrine gland a gland whose secretion—a hormone—travels by way of the bloodstream to stimulate its target cells; also called a ductless gland

endocytosis the process in which a cell absorbs material into special vesicles which it forms from its plasma membrane

endometrium the mucous membrane lining the nonpregnant uterus

endomysium a connective tissue sheath around individual muscle fibers

endoneurium a connective tissue sheath around individual nerve fibers

endoplasmic reticulum a cell organelle consisting of a network of membranes

endosteum the membrane which lines the marrow cavity in a bone

endothelial cells cells which line the lumen, or bore, of blood vessels, lymphatic vessels, serous cavities, and the heart

enzyme a protein which serves as a catalyst in digestion, metabolism, and certain other physiological processes

eosinophil a white blood cell whose cytoplasmic granules pick up the acid stain eosin

ependyma the cells which line the spinal canal and cavities of the brain

epidermis the epithelial or outer layer of skin; overlies the dermis

epididimys the portion of the seminal duct in which sperm are stored

epidural refers to tissues that lie on or outside the dura mater

epiglottis the cartilaginous structure which covers the opening (glottis) to the larynx during swallowing

epimysium a connective tissue sheath around an entire muscle

epinephrine see adrenaline

epineurium a connective tissue sheath around a group of nerve fibers

episiotomy a surgical incision of the perineum to facilitate childbirth

epithelium the tissue which forms glands and covers all body surfaces

ergosterol a substance which, when exposed to ultraviolet rays, becomes vitamin D_2, or calciferol

erythema a redness of the skin, as occurs in an inflammation

erythrocyte a red blood cell

erythropoiesis the formation of red blood cells

erythropoietin a hormone which stimulates the production of red blood cells

esophagus the tube which carries food from the pharynx to the stomach

estrogen any of a number of female sex hormones; any substance that promotes estrus and stimulates the development of secondary sex characteristics in the female

exocrine gland a gland that delivers its secretions through a duct

exocytosis the process in which a cell discharges material by means of vesicles which it forms from its plasma membrane

exophthalmos an abnormal protrusion of the eyeballs

extension the movement by which the angle in a joint is increased; opposite of flexion

extrapyramidal refers to a descending nerve tract which lies outside of the pyramids of the medulla oblongata

extrasystole a heartbeat which occurs before its normal time in the cardiac cycle

fainting becoming temporarily unconscious because of a decrease in blood supply to the brain

fascia a connective tissue which binds the skin to the body and wraps itself around muscles and other organs

fascicle bundle, as of nerve, muscle, or tendon fibers

fasciculation an involuntary contraction of a bundle of muscle fibers

fauces the passage from the back of the mouth to the throat

fats a class of combustible, ether-soluble, water-insoluble organic molecules

fatty acid a fat with an acid (H^+) component

femur the thigh bone

ferritin a protein which binds specifically with iron and serves to store the iron in body tissues

fetus a developing offspring in the uterus; during the first two months of development it is usually called an embryo

fibrin the principal fibrous protein in a blood clot

fibrinogen a plasma protein needed for blood clotting; after an injury it is converted to fibrin

fibroblast a connective tissue cell which produces fibers and gives rise to other kinds of cells

fibromyositis an inflammation of skeletal muscle

fistula an abnormal passageway or opening between two internal structures, or between an internal structure and the surface of the body

flaccid lacking normal muscle tone; relaxed, flabby

flexion the movement by which the angle in a joint is decreased; opposite of extension

follicle a small solid or hollow spherical mass of cells

fontanel one of the several membrane-covered spaces between the cranial bones in an infant's skull

foramen magnum the large opening in the skull through which the spinal cord joins the brain

force the cause of motion or the stopping or changing a motion

fructose sometimes called levulose, the sugar that forms a major component of honey

furuncle a boil; a localized skin inflammation which originates in a hair follicle

galactose a sugar which combines with glucose to form lactose (milk sugar)

gallop rhythm an abnormal heart rhythm in which three loud sounds occur in successive cardiac cycles

ganglion an aggregation of nerve cell bodies, or a small cystic tumor

gastric glands stomach glands which produce and secrete a mixture of hydrochloric acid, pepsin, and mucus

gastric ulcer a lesion, or sore, in the stomach lining

gene a unit of hereditary material, made of DNA

genotype the hereditary makeup of an individual

glenoid fossa the surface of the scapula (shoulder blade) which articulates with the head of the humerus

glomerular filtration rate the rate at which an ultrafiltrate is formed in the glomeruli of the kidneys

glomerulonephritis a kidney inflammation primarily affecting the glomeruli

glomerulus any berrylike anatomical structure, such as the tuft of capillaries in a renal corpuscle or the clump of gland cells in a sweat gland

glottis the opening between the vocal folds through which air passes into the trachea

glucagon a pancreatic hormone which tends to raise the blood sugar level

glucocorticoids compounds, such as cortisol and cortisone, which increase blood sugar and liver glycogen by increasing gluconeogenesis

glucokinase an enzyme which specifically catalyzes the phosphorylation of glucose

gluconeogenesis the process by which glucose is produced in the body from noncarbohydrate materials, such as amino acids, glycerol, and lactic acid

glucose a simple sugar, or monosaccharide; the main sugar found in human blood

glycerol a three-carbon compound; each carbon bears an alcohol (OH) group

glycogen animal starch

glycoprotein a protein-carbohydrate compound, such as mucin, the chief constituent of mucus *glycosuria* the presence of an abnormally large amount of sugar in the urine

goblet cell a vase-shaped epithelial cell which produces and secretes mucus

goiter an abnormal enlargement of the thyroid gland, often visible as a large swelling at the front of the neck

Golgi apparatus a cell organelle which combines carbohydrates and proteins to form glycoproteins

gout a metabolic disease in which uric acid accumulates in the bloodstream, setting up conditions for pain in the joints

grand mal a severe form of epilepsy

granulocytes white blood cells which contain visible granules in their cytoplasm; there are three kinds: neutrophils, basophils, and eosinophils

gravity a force that tends to draw all matter in the earth's sphere toward the center of earth

groin the inguinal region: the crease between the abdomen and thigh

ground substance a cellular secretory material which, together with minute fibers, forms the matrix of connective tissue

growth hormone a pituitary hormone that regulates growth

gyrus a convoluted ridge on the surface of the cerebral cortex

hair cells epithelial cells which have hairlike projections on their free surface; the hairs respond to sound waves in the ear, odors in the nose, or taste in the tongue

haploid having a single set of chromosomes as normally occurs in sex cells

hapten a substance which, when combined with a compound of high molecular weight, behaves like an antigen

heart attack technically known as a myocardial infarct, which means death to some segment of heart muscle due to a stoppage of blood flow

heart block a condition in which the cardiac impulse is blocked in its normal path from the sinoatrial node to the ventricles

heartburn a burning sensation beneath the breastbone, usually related to a spasm of the esophagus

heart disease any abnormal condition of the heart

heart failure the inability of the heart to provide blood to the tissues of the body

hemangioma an elevated purplish or reddish area of the skin sometimes called a strawberry mark

hematocrit the percentage of blood occupied by red blood cells

hematopoiesis the process by which blood cells are formed

hemiplegia a paralysis of one side of the body

hemoglobin the red oxygen-carrying pigment found in red blood cells

hemoglobinemia the presence of hemoglobin in the blood but outside of red blood cells

hemolysis the process in which red blood cells are destroyed and their hemoglobin is released

hemophilia a sex-linked hereditary clotting disease in which there is prolonged bleeding

hemorrhage an escape of blood from the blood vessels

hemorrhoid a painful vascular disorder, located at or in the anal margin; the hemorrhoid consists of enlarged, twisted veins, which are liable to discharge blood

heparin a glycoprotein which prevents blood from clotting

hepatitis a liver inflammation, usually caused by a virus

hepatomegaly an abnormal enlargement of the liver

hernia the protrusion of a part of an organ through an abnormal opening

heterozygous having different members of a pair of genes; hybrid

hexokinase an enzyme that catalyzes the attachment of a phosphate group on a six-carbon (hexo) sugar, such as glucose, galactose, or fructose

hiatus a gap or opening

hidrosis the excretion of sweat

hilum or hilus an opening or depression in an organ, usually where vessels and nerves enter and leave

hippocampus the region of the brain which is closely associated with smelling, eating, and emotions

histiocyte a macrophage cell connective tissue

histology the study of microscopic anatomy of tissues

hives an allergic skin condition in which there is itching, burning, stinging, and the formation of wheals

holocrine gland a gland, such as a sebaceous gland, whose secretion consists of altered cells of the gland itself

homology a similarity traceable to common origin; a bird's wing is homologous to a human arm

homozygous having like members of a pair of genes

hormone a chemical secretory product of one cell which stimulates another kind of cell

horny layer the few uppermost layers of skin cells

hydrolysis a reaction in which a chemical bond is split by the addition of water

hydrostatic pressure the pressure exerted by a column of water

hymen the thin membrane which partly or totally blocks the vaginal opening in a virgin

hyoid a U-shaped bone in the neck

hypercarbia an abnormally high level of carbon dioxide in blood

hyperglycemia an abnormally high level of sugar in the blood

hyperosmolar refers to a solution which has a greater osmotic pressure than some other solution

hyperplasia an abnormal increase in cell number, as in a tumor

hyperpolarization the process by which negative

charges are added to the interior surface of a membrane
hyperpnea an abnormal increase in depth and/or rate of breathing
hypertension abnormally high blood pressure
hyperthyroidism an abnormal condition marked by an excessive output of thyroxine and triiodothyronine from the thyroid gland
hypertrophy an abnormal increase in the size of an organ due to enlargement of its cells
hyperventilation abnormally deep breathing which may arise involuntarily in response to rising levels of carbon dioxide in the blood
hypoglycemia an abnormally low level of sugar in the blood
hypogonadism an abnormal condition marked by decreased activity of the sex organs
hypophysectomy the surgical removal of the pituitary gland
hypophysis another name for the pituitary gland
hypothalamus the area of the brain concerned primarily with emotional and autonomic nervous activities
hypotonia an abnormal condition in which muscle tone is below normal
hypoxemia an abnormally low level of oxygen in the blood
hysterectomy the surgical removal of all or part of the uterus
iatrogenic refers to any problem that is caused by medical treatment
idiopathic disease a disease for which a cause is not known; or a disease which is not the result of any other disease
ileum the terminal portion of the small intestine
iliac refers to the area of the ileum, around the upper part of the hip
immunoglobulin a globular serum protein; an antibody
infarct a region of dead tissue resulting from a deficiency of blood to that area
inguinal region the region where the abdomen joins the thigh; the groin
inguinal canal the passage which contains the spermatic cord (in the male) or the round ligament of the uterus (in the female)
inguinal hernia an abnormal protrusion of tissue into the scrotum
innervate to supply a tissue or organ with nerves
insulin a pancreatic hormone, which tends to lower blood sugar level
intercalated discs the pleated, intertwined membranous endings of two adjacent heart muscle cells
interferon a class of proteins that inhibit viral multiplication inside the body
interstitial pertains to the spaces and matter between cells
intima the inner wall of a blood or lymph vessel
introitus the entrance into a canal or hollow organ, such as the vagina
in vitro refers to a biological event which occurs in an artificial environment, such as in a culture dish or test tube
in vivo refers to an event occurring inside a living body
iris the colored portion of the eye, the circular disc suspended in the aqueous humor by the ciliary body
ischemia a deficiency of blood to an area due chiefly to a constricted blood vessel
ischium one of the three bones that form the hipbone
isometric pertains to a movement in which opposing muscles contract at the same time with no change in dimension
isoosmotic refers to two solutions, such as an ultrafiltrate and blood plasma, which have an approximately equal osmotic pressure
isotonic pertains to a muscle contraction in which the muscle visibly shortens
jaundice an abnormal yellow coloration of the skin, the sclera of the eyeball, and mucosal membranes; due to the presence of large amounts of the breakdown products of hemoglobin
jejunum the middle portion of the small intestine
jugular pertains to the neck above the collarbone
keratin a hard proteinaceous substance found in hair, nails, and the surface of the epidermis
keratohyalin a deeply-staining substance which forms the granules of the granular layer in the epidermis
ketone a chemical substance with the characterizing group C=O
ketone bodies acetoacetic acid and acetone
ketosis an abnormal condition in which excess ketone bodies are found in the body
kinesiology the study of muscle movements
Kupffer cells macrophage cells in the liver
kwashiorkor a protein deficiency disease
labia majora the outer folds of the vulva
labia minora the inner highly vascular folds of the vulva
lacrimal pertaining to the tears
lacteal a lymphatic channel which picks up chyle from the small intestine
lactic acid a three-carbon sugar which is an intermediate product in the metabolism of carbohydrate; may cause discomfort when it accumulates in working muscles
lactose a disaccharide, also called milk sugar; yields glucose and galactose on hydrolysis
lacuna a small depression, gap, or hollow
lamina propria the connective tissue beneath a mucous or serous membrane
larynx the voice box; the vocal apparatus situated in the Adam's apple
latency a state of seeming inactivity; a concealed property or action which subsequently expresses itself

laxative an agent that relieves constipation; a cathartic; a purgative

lecithin any of a group of phospholipids containing the nitrogen-containing alcohol choline

lemniscus a band of longitudinal fibers running from the crossover point in the medulla and pons, carrying point-to-point information from the periphery of the body to the thalamus

lens the transparent structure that bends light waves as they enter the eye and focuses them on the retina

leukemia a malignant disease marked by an excess of white blood cells in the tissues and often in the blood

leukotomy the surgical cutting of the white matter which connects the gray matter of the frontal lobe with the thalamus

libido conscious or unconscious sexual desire

ligament a band of fibrous connective tissue which connects the articular ends of bones and sometimes envelops them in a capsule

ligate to constrict a vessel with a thread (ligature)

limbic lobe the part of the cerebral cortex which borders several sensory, motor, and emotional regions of the brain

linoleic acid a polyunsaturated fatty acid

lingual pertaining to the tongue

lipase an enzyme which catalyzes the cleavage of fatty acids from triglycerides (neutral fats) and phospholipids

lipoproteins complexes formed from lipids and globulin proteins in the blood

lobotomy an operation in which one or more cerebral nerve tracts are severed

local potential a movement of bioelectricity over a very short distance on a cell membrane

loin the lateral and posterior region of the body between the false ribs and the hipbone

lumbar pertaining to the loins

lumen the bore, or opening, in a tube

lutenization the yellowing of an ovarian follicle as the follicular cells begin to produce increasing amounts of sex hormones

lymphocyte a white blood cell important in immunity

lymphoid pertains to any tissue of the lymphatic system

lymphoma any malignancy of a lymphoid tissue

lysosomes cell organelles which contain hydrolytic enzymes

macrophage a phagocytic cell of the reticuloendothelial system

macula a small spot or area differing from the surrounding tissues

macula densa thickening of the epithelium of the ascending limb of the loop of Henle where this limb comes into contact with the renal corpuscle

malabsorption the faulty absorption of food from the intestine

malformation defective or abnormal development

malignant tending to become progressively worse and result in death

maltase an enzyme that catalyzes the splitting of maltose into two glucose units

mamillary bodies two small masses of gray matter at the base of the brain

marasmus a gradual wasting of the body due to a deficiency of protein and calories

marrow the soft fatty material that fills the cavities of bone

mass a quantitative measure of inertia; commonly used as a measure of the amount of material a body contains

mast cells connective tissue cells which have similar functions to basophil cells in the blood

mastication the chewing of food

matrix the intercellular cement plus connective tissue fibers

media the middle layer of the wall of a blood vessel

median the middle

mediastinum the space in the chest between the two pleurae

medulla the middle or central core of an organ or other structure

medulla oblongata the extension of the spinal cord into the brain stem; partly controls certain visceral functions, such as breathing, heartbeat, and vomiting

meiosis process through which chromosome number is halved during the formation of sex cells (gametes)

melanin a brown-black pigment which colors the skin and protects it against harmful radiations of the sun

melanocytes skin cells which produce melanin

menarche the beginning of a woman's menstrual life

meninges connective tissue sheath around the brain and spinal cord

menopause the ending of a woman's menstrual life

mesenchyme the primordial tissue which has the potential of giving rise to any one of several kinds of mature tissues

mesentery a fold of peritoneum which connects the intestine with the posterior abdominal wall

metabolism all of the chemical changes that occur in the body

metastasis the moving of a diseased tissue from one part of the body to another part, as may occur in mumps, cancer, and bacterial diseases

microglia branching phagocytic cells which support nerve cells

microvilli tiny nipplelike projections which increase the surface area of an epithelial cell

micturition urination

migraine headache on one side of the head

mineralcorticoid a natural or synthetic compound, such as aldosterone, that acts to retain sodium

mitochondria the organelles in which the bulk of a cell's directly usable source of energy (ATP) is formed *mitosis* the method of cell division by which body cells reproduce themselves

mitral valve the two-cusped valve between the left atrium and left ventricle of the heart

mole 6×10^{23} molecules, ions, or atoms

monocyte a white blood cell which, like a lymphocyte, does not have large visible granules outside its nucleus

monomer a molecule that can be built up by repetition to form a polymer

mononucleosis a disease characterized by an excess number of monocytes

monoplegia the paralysis of a single muscle, a single group of muscles, or a single limb

motor nerve cells nerve cells which stimulate muscles and glands

mucopolysaccharide a polysaccharide that contains an amino sugar; found between cells or as a component of mucous secretions

mucosa a mucous membrane

mucous membrane a wet membrane which lines body passageways that lead to the exterior of the body, such as the digestive tract

mucus a viscid liquid consisting of water, inorganic salts, epithelial cells, and other substances held in suspension by glycoproteins

muscle a contractile organ to which a name is usually assigned

muscular dystrophy a hereditary disease characterized by a progressive wasting of the afflicted muscles

myasthenia gravis a muscle disease marked by progressive paralysis

myelin a white fatty sheath surrounding certain nerve fibers

myeloid refers to tissue that is fatty or myelin-like

myeloma a tumor of bone marrow cells; often called a plasma cell tumor

myocardial hypertrophy an abnormal enlargement of the heart due to an increase in size of the individual heart muscle cells

myocardium the heart muscle

myoepithelium contractile cells found among epithelial cells in glands

myogenic cell a muscle cell which generates its own action potentials without being stimulated by a neurotransmitter

myoglobin a respiratory protein in muscle cells which functions like hemoglobin

myometrium the smooth muscle wall of the uterus

myosin a muscle protein; together with actin it forms the bulk of the contractile apparatus

myxedema a condition associated with severe thyroid deficiency (hypothyroidism)

nasolachrymal pertaining to the nose and lacrimal apparatus: the tear glands and ducts

naris a nostril

NADPH a derivative of vitamin B_1; the reduced form of nicotinamide adenine dinucleotide phosphate, as opposed to NADP, the oxidized form

necrosis the death of a cell or group of cells

nephron a renal unit; consists of a glomerulus, Bowman's capsule, and a uriniferous tubule

nerve a cable of the fibrous axonic processes from nerve cells

nerve deafness a loss of hearing due to a defect in the nervous tissue of the inner ear or in the hearing centers in the brain

nerve root the beginning, or proximal portion, of a nerve

neuritis an inflammation of a nerve

neurogenic refers to muscle contractions that are triggered by nervous tissue

neurotransmitter a chemical, such as acetylcholine, which transmits a stimulus from a nerve cell across a synaptic cleft to a target cell

neutral fat a fat which has no acidic (H^+) components; a triglyceride

neutrophil a white blood cell whose cytoplasmic granules do not pick up deep red or blue coloration from stains

nevus a birthmark; a colored skin lesion containing an aggregation of melanocytes (pigment cells)

nitroglycerine a drug which dilates blood vessels and reduces blood pressure

nociceptor a pain receptor

nodal rhythm the abnormal heart rhythm that occurs when the heart is controlled by the atrioventricular node

nucleus the region of a cell which contains the chromosomes; or an aggregate of nerve cell bodies in the central nervous system

nutrition the sum of the metabolic processes involved in the growth, maintenance, and repair of body tissues

nystagmus involuntary movements of the eyeballs

obesity a condition marked by excess weight

occipital region the region at the back of the head

odontoid process a toothlike extension on the axis (the second vertebra of the neck)

olfaction the sense of smell

omentum a fold of peritoneum which connects certain abdominal organs and tissues with the stomach

optic chiasma the region of the brain where the optic nerves cross over each other and enter the brain as the optic tracts

optic disc the circular area in the retina where the ganglion cells converge to form the optic nerve

optic tract either of the two cables carrying nerve fibers from the eye

orbit the bony cavity that contains the eye

organ a structure made up of cells and tissues performing one or more specific functions

organelles the minute organs within a cell

orthopnea a condition in which breathing is difficult except when sitting up *osmoreceptors* nerve cells which are stimulated in proportion to the concentration of molecules that are dissolved in body fluids

osmosis the passage of a solvent through a semipermeable membrane from a dilute solution to a more concentrated one

osteoarthritis an abnormal inflammation of a joint, often due to an autoimmune reaction

osteoarthrosis a degenerative joint disease whose cause is unknown

osteoblast a bone-forming cell

osteoclast a cell which destroys and absorbs bony tissue

oteocyte a mature bone cell

osteomalacia a disease characterized by bone softening because the osteoid tissue has failed to mineralize sufficiently

osteomyelitis an inflammation of bone marrow

osteoporosis an abnormal enlargement of bone marrow at the expense of the hard mineralized regions

osteosclerosis an abnormal hardening of bone tissue

otolith a calcareous particle in the labyrinth of the ear

oxygen debt the deficit of oxygen and accumulation of lactic acid in skeletal muscle during a period of strenuous activity

ovulation the release of a mature egg from a follicle in the ovary

oxidation the chemical process in which electrons are removed from a substance, usually by way of oxygen; opposite of reduction

oxytocin a pituitary hormone that stimulates uterine contractions during childbirth and mammary gland muscle contractions during lactation

pacemaker the sinoatrial node; the area of the right atrium which initiates the heartbeat

palatine refers to the roof of the mouth, the palate

pancreas an abdominal gland which secretes digestive enzymes and the hormones insulin and glucagon

pancreozymin a hormone from the small intestine which stimulates the secretion of enzymes from the pancreas

panniculus adiposus the layer of fat beneath the skin

papilla a small nipplelike or pimplelike structure

papule a small solid elevation on the skin

parageusia an abnormal or false sense of taste

paralysis a loss of skeletal muscle activity due to a fault in the nervous system

paranasal region the area around the nose

paraplegia a paralysis of the lower limbs

parasympathetic nervous system the cranial and sacral nerves of the autonomic nervous system; it tends to stimulate secretion, activate the smooth muscle of the gut, and dilate blood vessels

parathyroid glands the 4 to 6 small glands that are situated on the thyroid gland and are concerned with calcium metabolism

paraventricular nuclei the clusters of nerve cell bodies in the anterior hypothalamus, whose axons merge with those of the supraoptic nuclei

parenchyma the essential or functional part of an organ as opposed to the supporting framework, the stroma

passive process a process which does not require metabolic energy

patella the kneecap

parietal pertains to a wall

pathogen a disease-producing agent

peduncle a stemlike part, such as the stalk of nerve fibers in the brain stem

pelvis any basinlike or cup-shaped cavity, such as the bony pelvis of the skeleton or the renal pelvis of the kidney

penis the male organ of copulation

perfusion the act of passing a liquid through spaces

pericardium the connective tissue sheath which forms the cavity containing the heart

perichondrium a connective tissue which covers cartilage

perimetrium the serous membrane covering the uterus

perimysium a connective tissue sheath around a fascicle of skeletal muscle fibers

perineum the region between the anus and scrotum in the male; between the anus and urethra in the female

perineurium a connective tissue sheath around an entire nerve

periphery the outer part or surface of a structure; the part of a body away from the center

peristalsis the wavelike contractions of the respiratory, excretory, reproductive, and digestive tracts

periosteum the fibrous membrane which covers bone along most of its surface

peripheral away from the center of a structure; outer, external, or distal

peripheral resistance the resistance afforded by blood, arteries, and capillaries to the flow of blood

peritoneum the serous membrane lining the inside of the abdomen and covering the abdominal organs

permeability the property of a membrane which permits the transport of substances from one side to the other

pernicious anemia a deadly form of anemia due to a deficiency of vitamin B_{12} which brings about destruction of the nervous system

peroxidase an enzyme which uses hydrogen peroxide to oxidize a reactant

petit mal a mild form of epilepsy

pH refers to the concentration of hydrogen ions in solution

phagocyte a cell that keeps tissues free of foreign material

phagocytosis the engulfing of matter by special cells, called phagocytes

phalanges the bones of the fingers and toes

pharynx the throat

pheochromocytoma a tumor in the sympathetic nervous system, which leads to the overproduction of noradrenaline and adrenaline

phlebitis an inflammation in a vein

phospholipid a fat which has phosphorus in its chemical structure; for example, lecithin

phrenic pertaining to the diaphragm

physiology the study of the functions and processes of living organisms

pia mater the soft meningeal covering directly over the brain and spinal cord

pituitary a small endocrine gland above the roof of the mouth, beneath the hypothalamus of the brain

placenta the spongy organ that attaches a fetus to the uterine wall

plasma the liquid portion of the blood; blood minus its cells

platelets blood cells which contain clotting factors

pleura the serous membrane which covers the lungs and lines the chest cavity

pleurisy an inflammation of the pleurae, the serous membranes which envelop the lungs and line the chest cavity

plexus a network of blood vessels, lymph vessels, or nerves

plica a fold, as in the mucous membrane of the small intestine

pneumoconiosis a chronic inflammation of the lungs caused by inhaling particulate material, such as carbon dust

pneumonia an inflammation of the lungs, usually caused by infectious bacteria or viruses

poliomyelitis a virus disease of nerves and muscle tissue

polycythemia an abnormal condition marked by an excess of red cells in the blood

polymer a molecule that can be broken into one or more groups of repeating parts

polymorphonuclear cell a white blood cell whose nucleus is divided into complex lobes

polyunsaturated fatty acid a fatty acid that can add on four or more hydrogen atoms before it becomes saturated

popliteal pertaining to the hind part of the knee

portal system a system of blood vessels from one organ which carry partly deoxygenated blood to another organ

portal vein the vein which carries partly deoxygenated blood from the gut, pancreas, and spleen into the liver

precocious unusually early in development

precordium the region over the heart and stomach; part of the chest

prefrontal refers to the anterior part of the frontal lobe of the brain

prepuce the skin over the glans penis or clitoris

pressure force exerted against an opposing body

process an extension, protuberance, elevation, or projection

progesterone a steroid hormone produced by the corpus luteum; its chief function is to prepare and maintain the lining of the uterus during early pregnancy

prolactin a pituitary hormone which promotes lactation, the production of milk

pronate to lie face down, or to turn the palm of the hand downward

prophylactic a precautionary device or procedure used to guard against disease

prostaglandins a family of fatty acid derivatives which stimulate smooth muscle, elicit pain, and very likely provoke an inflammation

prostate gland the small gland, found only in males, which secretes a fluid component of semen

prothrombin a plasma protein needed for blood clotting

protein a polymer made of amino acids

protoplasm the substance of living material

proximal refers to something that is situated nearer to a reference point; opposite of distal

pruritus itching

psoriasis a chronic skin disease marked by silver-colored scaly pimples

psychomotor pertaining to voluntary movements

psychomotor epilepsy an epilepsy in which short-lived mental disturbances replace the typical convulsions of epilepsy

pubis refers to the outer arch of the abdominal pelvis in the anterior region of the body

pulmonary circuit the pathway of the blood through the ventilatory tissues in the lungs

pulmonary edema an abnormal accumulation of liquid in the lungs

pulmonary valve a valve consisting of three semilunar cups situated between the right ventricle and pulmonary artery

pulsus alternans an abnormal pulse marked by a regular alternation between weak and strong beats

pupil the circular opening in the center of the iris through which light waves enter the eye

pus a yellowish material produced by an inflammation; composed mainly of serum and white blood cells

pylorus the circular opening of the stomach into the small intestine

quadriceps the four-headed muscle on the front of the thigh

quadriplegia paralysis of all four limbs

quinidine a drug used to correct faulty heart rhythm

radial muscle the muscle which is arranged in spoke-like fashion in the iris of the eye

reabsorb to get back into the bloodstream a substance which had previously been sent from the blood into the interstitial spaces

reagin an antibody which reacts with an allergin to produce an uncomfortable allergic reaction

rectum the straight terminal portion of the large intestine; opens into the anus *reduction* the process in which one or more electrons are added to a substance

refractory period the time during which a nerve or muscle fiber fails to respond to a stimulus

renal pertaining to the kidneys

renal failure failure of the kidneys to clear the body properly of toxic waste products

renin a kidney hormone or enzyme which activates the hormone angiotensin

rennin or *rennet* an enzyme which coagulates and digests milk

reticular cells macrophage cells which form a network along blood passageways in certain visceral organs

reticuloendothelial system a widespread system of macrophage cells that keep tissues free of foreign material

reticulocyte an immature red blood cell

reticulum a fine network

retina the area of the eyeball which responds to visual stimuli

rheumatism any of several disabling, painful diseases of muscles, tendons, joints, bones, or nerves

rheumatoid arthritis a disease in which the tissues of one or more movable joints become inflamed

rheumatic fever a disease caused by streptococcal bacteria, which produces an inflammation in such tissues as the heart, joints and serous membranes

rhinitis an inflammation of the mucous membrane lining the nasal passageway

ribonucleic acid see RNA

ribose a five-carbon monosaccharide

ribosomes cell organelles in which the terminal steps of protein production take place

rickets childhood osteomalacia

RNA ribonucleic acid; several forms of RNA are essential if a cell is to synthesize protein

rods the rod-shaped light-sensitive cells in the retina

rugae wrinkles, folds, or elevations, as in the mucosa of the stomach, vagina, and palate

sacral region the spinal region between the lumbar and coccygeal vertebrae

saccule the smaller of the two membranous sacs in the vestibule of the labyrinth of the inner ear

salpingectomy the surgical removal of a fallopian tube

saltatory pertaining to a dancing or leaping movement

saphena either of two large superficial veins of the leg

sarcolemma the cell membrane enclosing a muscle fiber

sarcoma a malignant tumor in connective tissue

sarcomere one of the segments into which a muscle fibril is divided by Z discs

sarcoplasm the cytoplasm of a muscle cell

sarcoplasmic reticulum the endoplasmic reticulum of a muscle fiber

sarcosome a mitochondrion of a muscle cell

saturated fatty acid a fatty acid that cannot incorporate any more hydrogen atoms than it has

sclera the opaque white outer covering of the eyeball

scleroderma a disease in which the skin becomes hard, thickened, and rigid

sclerosis an abnormal hardening of a tissue

scrotum the sac which contains the testes

scurvy a disease caused by a deficiency of vitamin C

seborrhea an excessive flow of sebum from the sebaceous glands

sebum the fatty material secreted by sebaceous glands as a lubricant for the hair and skin

secrete to produce, and usually discharge, a substance

seizure a sudden attack; an epileptic convulsion

sella turcica the bony structure protecting the pituitary gland

semen the fluid that contains sperm and secretions from the prostate gland and seminal vesicles

semilunar crescent-shaped

seminal vesicle a gland which produces some of the liquid in which sperm are carried out of the body

seminiferous tubules the coiled sperm-producing tubes inside the testes

septicemia a disease characterized by the presence of bacteria and their toxins in the blood

septum a wall which divides two cavities or masses of softer tissue

serosa a serous membrane

serous membrane the membrane which lines the body cavities and covers their contents

serum blood plasma minus the clotting factors

sinus a hollow, cavity, or channel

sinus rhythm the normal cardiac rhythm proceeding from the sinoatrial node

sinusoid a large capillary space (10 to 40 micrometers in diameter) through which blood passes in a visceral organ

smegma an oily secretion from the sebaceous glands of the prepuce and labia minora

solar plexus the celiac plexus behind the stomach between the adrenal glands; consists of ganglia and sympathetic nerve fibers

somesthetic pertaining to general body feelings, such as pain, pressure, and temperature

spasm an involuntary contraction of skeletal muscle

spastic pertaining to spasmic contractions characterized by exaggerated reflexes and muscle tension

sphincter a ring-shaped muscle which opens or closes a circular passageway

sphygmomanometer an instrument for measuring arterial blood pressure in the systemic circuit

spleen the largest lymphoid organ in the body; situated on the left side, beneath the diaphragm

splenomegaly an abnormal enlargement of the spleen

squamous pertaining to cells that are flat in shape

starch a digestible polysaccharide made entirely from glucose

stellate star-shaped

stem cell a nonspecialized mother cell or blast cell capable of differentiating into a specialized cell

stenosis an abnormal constriction or narrowing of a passageway

steroid a generic name for such compounds as cholesterol, sex hormones, bile salts, vitamin D, and cortisol

stethoscope an instrument for studying sounds coming from the body

stimulus any agent that causes a change in the activity of a cell, tissue, or organism

stroke a sudden loss of a bodily function as a result of a hemorrhage or embolus in the brain

stroke volume the volume of blood ejected by either ventricle during a single contraction of the heart

stroma the supporting framework of an organ; as opposed to the parenchyma

subclavian located below the collarbone (clavicle)

subcutaneous located beneath the skin

sucrose a sugar made from fructose and glucose; table sugar

sulcus a groove or channel

superficial located on or near the surface

supinate to lie on the back, face upward; or to turn the palm of the hand upward

supraoptic nuclei clusters of nerve cell bodies situated laterally to where the optic nerves cross over

suture a thread or fiber used for sewing living tissue

sympathetic nervous system the thoracic and lumbar nerves of the autonomic nervous system; tends to decrease secretion, relax the smooth muscle of the gut, and constrict blood vessels

symphysis a union, such as the pubic symphysis between the pubic bones

synapse transmission of neural activity from a nerve cell to its target cell(s); the actual transmission is effected by a chemical called a neurotransmitter

synarthrosis an immovable or slightly movable joint

syncope a faint; a sudden, temporary loss of consciousness

syncytium a multinucleated protoplasmic mass without distinct cell boundaries

syndrome a group of symptoms that occur together

synovial fluid a lubricating fluid found in joint cavities, bursae, and tendon sheaths

systemic affecting the entire body

systemic circuit the circuit of blood to all regions of the body except the lungs

systemic edema a disease in which there is a widespread accumulation of fluid between the cells

systole the contraction phase of the cardiac cycle

tachycardia an abnormally rapid heartbeat

taste buds minute structures in the tongue which pick up taste stimuli

tectum any roof or covering

telodendria the fine terminal branches of a nerve axon

temporal refers to the sides of the head, the region where graying usually begins

tendon a tough connective tissue which connects muscle to bone, cartilage, ligament, or fascia

testes the paired male reproductive glands

tetanus in physiology, a spastic muscle; in pathology, a disease, often fatal, caused by the toxins of the tetanus bacterium, *Clostridium tetani*

thalamus a sensory relay station from the brainstem to the cerebral cortex

theca a sheath

thoracic region the chest

thrombin an enzyme that induces clotting by catalyzing the conversion of fibrinogen to fibrin

thrombocytes blood platelet cells

thrombocytopenia an abnormal deficiency of blood platelets

thrombophlebitis an inflammation of a vein and subsequent formation of a blood clot

thrombosis the formation of a blood clot (thrombus)

thrombus a blood clot

thymus a gland that produces and secretes hormones that stimulate the differentiation of lymphocytes into immunological cells

thyroid pertains to a shield-shaped gland or cartilage in the neck

thyroxine a thyroid hormone which stimulates metabolism

tidal volume the volume of air which comes in and out of the lungs during a single normal inspiration and expiration

tinnitus an abnormal ringing in one or both ears

tissue an assembly of similar cells

tone the normal tension in a muscle

tonic denotes either normal tension or continuous contraction

tonsils masses of lymph nodules and lymph vessels in the mucosa of the pharynx

toxin a poisonous product of animal or plant cells which, upon entering the body, elicits the formation of antibodies

trachea the cartilaginous tube which conveys air from the larynx to the bronchi

transferrin a serum beta globulin protein for transporting iron through the blood

transitional epithelium an epithelium which lines the ureter, urinary bladder, and urethra

trauma any mechanical, physical, or psychic injury usually of shock-producing nature

tremor an involuntary trembling, shaking, or quivering

tricuspid valve a three-cusped valve, situated between the right atrium and right ventricle *triiodothyronine* a thyroid hormone similar to thyroxine

triglycerides neutral fats, made up of three fatty acids and glycerol

trophoblast the outer cell layer around an embryo

troponin a muscle protein

tuberculosis an inflammatory infectious disease caused by *Mycobacterium tuberculosis*

ultrafiltrate serum from which blood cells and large proteins have been filtered out

uncus the hooklike anterior region of the hippocampal fold of gray matter in the brain

unilateral one-sided

urea the major end product of nitrogen metabolism, formed in the liver and excreted in urine; the most important medium of nitrogen excretion in humans

uremia an abnormal accumulation of urea in the blood

ureter either of the two tubes which convey urine from a kidney to the urinary bladder

urethra the canal through which urine leaves the body

urinalysis an examination of the physical and chemical properties of a sample of urine

uriniferous tubule one of the many tubules which convert an ultrafiltrate into urine; also called renal tubule

urobilinogen a colored material in urine that forms from bilirubin when urine decays

urticaria an allergic skin disease marked by itching red patches on the skin or mucous membranes; also called hives

uterus the organ in which a fetus develops

utricle the larger of the two membranous sacs in the vestibule of the labyrinth of the inner ear

uvula the cone-shaped appendage which hangs from the posterior wall of the soft palate

vagi the paired tenth cranial nerves

vagina a short muscular tube that extends from the uterus to the external orifice of the genital canal in females

varicose vein a dilated, twisted, knotted vein, usually in the leg

vasa recta blood capillaries which run parallel to uriniferous tubules deeply into the medulla of the kidney

vasa vasora the small arteries and veins that supply the walls of larger blood vessels

vascular pertaining to, consisting of, or provided with vessels

vascular occlusion a blockage of a blood vessel by an embolus

vasculature the body's blood vessels, lymph vessels, and heart; also called vascular system

vas deferens the excretory duct of each testis

vasectomy a surgical operation to remove a segment of the vas deferens; carried out, in most cases, to bring about permanent sterility in males

vasoconstriction the narrowing of an arterial vessel, causing a decrease in the flow of blood

vasoconstrictor a chemical substance which causes vasoconstriction

vasopressin another name for antidiuretic hormone

vena cava either of the two large veins which enter the right atrium of the heart

vein a vessel which carries blood toward the heart

ventricle one of the two lower chambers of the heart; or one of the several cavities in the brain

vesicle a small saclike receptacle, usually filled with fluid

villi tiny nipplelike protrusions from a surface, such as the villi of the intestinal mucosa

virilism the development of masculine traits in a girl or woman

virulent extremely poisonous or noxious

viscera the soft inner organs of the major body cavities

viscosity the property of a liquid which makes it resist the tendency to flow

vital capacity the total volume of inspiratory and expiratory air of the lungs

vocal cords muscular sound-producing folds in the passageway of the larynx

volar pertaining to the palm of the hand or sole of the foot

vulva the labia majora and external female sex organs which they enclose

wheal a water blister in the skin

xenobiotic a substance that does not furnish energy to a cell or provide a source of matter for protoplasm

xenometabolism the metabolism of substances that are of no economic value to a cell

zoster another name for herpes zoster, a virus disease commonly called shingles

zygote a fertilized egg, formed by the fusion of a sperm and an egg cell

Index

Page numbers in *italics* indicate illustrations.

Abdomen, 64–65, *69*, 71
   muscles and movements, 352, *359, 360, 361*, 362
Abdominal circulatory system, 521
Abduction, joints, *298*
Abductor muscles, 367
ABO blood types, 651
   in immunity, 644
   isoantigens and, 648
Absorption, epithelial, 143
   intestinal, 402–404
   of nutrients, 401–412
Accommodation, visual, *145*–150
Acetabulum, 694
   pelvic, 278, 301
Acetate, 417, 425, 694, (see also Acetyl coenzyme A)
Acetone, 694
Acetone breath, 213–214
Acetylcholine, 82–84, 135, 149, 161, 694
   in blood vessel dilation, 246
   in digestive tract, 381
   heart effects of, 544
Acetylcholinsterase, 82
Acetyl coenzyme A, *415*, 417–418, 424, 425, 694 (see also Krebs cycle)
Achalasia, 384
Achilles reflex, 124
Acid(s), 17–18, 694
Acid–base balance, 605–606
Acidosis, 215, 437–438, 461, 576, 605, 606, 694
Acne, 255–256
Acoustic aphasia, 173
Acromegaly, 201, 214, 287, 694
ACTH (Adrenocorticotropin), 184, 201, 444, 694
Actin, 312, 316, 495, 694
Action potential, 694 (see also Bioelectrical activity)
   in cell membranes, *15*–*16*, 76, 83
   in heart muscle fibers, 326–327

in muscular contraction, 313–315, 322
myocardial, 495–496, 499, 500
in nerve cell stimulation, 76–77, 79–81
in skeletal muscle, 312–316
saltatory propagation of, 87
Active transport, 408, 694
Acupuncture, 671–675
Addison's disease, 211
Adduction, muscular, *298*, 340–341, *348*
Adductor muscles, 367
Adenine, 605
Adenoid tonsils, 143, 640
Adenosine diphosphate (see ADP)
Adenosine triphosphate (see ATP)
ADH (Antidiuretic hormone), 198, 200, 444, 551, 600, 601, 607, 608, 694
Adipose cells, 472
Adipose tissue, *49*
   metabolic role of, 432–433
ADP (Adenosine diphosphate), 13, 28, 204, 313, 694
Adrenal cortex, 184, 209–216
Adrenal glands, 135, 207–212
Adrenaline, 82, 135, 139, 192, 208, 246, 444, 694
   blood vessel effects of, 543
   cardiac output effects of, 543–544
   metabolic functions of, 442–443
   skin and visceral effects of, 544
Adrenal medulla, 207–209, 211, 216, 543, 545, 694
Adrenal sex hormones, 212
Adventitia, *380*–381, 694
Aerobic exercise, 329
Afferent, 85, 694
Afferent arterioles, 591, *592*
Afferent neurons, 84–85, 123 (see also Sensory neurons)
Ageusia, 154, 171
Aging, 661–667

basal metabolism and, 452
cancer and, 655
control of, 666
lysosome function in, 31
oxygen consumption in, 537
theories of, 662–666
Agnosia, 159, 174
Agranulocytes, 49–50
Agranulocitosis, 652
Agraphia, 174
Air pressure, 7, 20
Albumin, 431, 438, 610, 635, 694
Albuminuria, 611, 694
Alcoholic beverages, metabolism of, 486
Aldosterone, 184, 188, 193, 209, 211, 396, 551, 553, 601, 607, 609, 694
Aleukemic leukemia, 641
Alexia, 173, 694
Alkali, 17
Alkalosis, 575, 576, 605, 694
Allergens, 694, 650, 694
Allergy, 631, 650, 694
Alopecia, 252
Alpha motor neurons, 125, 192
Alveolar ducts, 565, 571
Alveolar process, skeletal, 261, 268, 269
Alveoli, mammary, 233–234
   respiratory, 565, 571–572, 575
Amino acid hormones, 192
Amino acids, 6, 30, 192, 402, 404
   absorption and utilization of, 420–421
   blood levels of, 438–439
   essential, 418, 467
   hormonal regulation in blood, 440
   types of, 418–419
Amino groups, charges on, 14
Ammonia reaction, 605–606
Amnesic aphasia, 174
Ampullas, 157–158
Amygdala, 694

709

Amylase, 390
Amyloid bodies, 223
Anabolic processes, 8
Anaerobic exercizes, 329
Anal canal, 398
Analgesics, 669–671
Anal sphincter, 357, 362, 398
Anamnestic immune response, 643
Anaphase, 32, 35–36
Anaphylaxis, 694
Anaplasia, 56, 694
Anastomosis(es), 517
Anatomical orientations, 61–63, 62
Anatomical regions, 63, 64
Anatomy, 1–2, 694
Anconeus muscle, 367
Androgens, 212, 612, 694 (see also Sex hormones)
Anemia, 580, 640, 694
  characteristics of, 583
  hemolytic, 583
  iron deficiency, 584
  physiological adjustments for, 585–586
  types of, 583–585
Anasthesia, 119, 174, 669–671, 695
Aneurysm, 558, 695
Angina pectoris, 134, 558, 560, 695
Angiotensin, 184, 246, 431, 553, 609, 695
Angular gyrus, 171
Anhydrosis, 250
Ankles, ligaments, 304
  muscles and movements, 359, 365, 366
Anopsia, 174
Anorexia, 695
Anterior, 61
Anterior pituitary gland, 184, 216, 218–220, 224, 234
Anterior thalamic nuclei, 170
Antibodies, 50, 63, 631, 695
  intestinal absorption of, 405
  tolerance for, 647
  types of, 649–650
Anticoagulants, 621–622
Antidiuretic hormone (see ADH)
Antigen–antibody complex, 643–645
Antigen–antibody reaction, 52
Antigen–RNA complex, 645
Antigens, 48, 50, 655 (see also Foreign Bodies; Immunity)
  destruction of, 31, 643–645
  immune reactions to, 631–632
  production of, 48
  sensitivity to, 650–651
  types of, 648
Antihistamine, 161, 627
Antioxydant, 667
Antiserum, 646
Anxiety, and dreaming, 681
Anxiety syndrome, 545
Aortic sinus, 517, 546, 695
  valve, 494

Aphasia, 171, 173, 695
Aplastic anemia, 584
Apocrine glands, 247
Aponeurosis, 293, 333, 695
Appendix, vermiform, 379, 396, 639
Appetite, and obesity, 457
Appetite-satiety center, 434, 456, 457
Apraxia, 174, 695
Aqueous humor, 147
Arachnoid, 108, 110, 695
  villi, 110
Arcuate artery, 591, 592
  fascicle, 171, 173, 695
Arm, 275, 277
  muscle, 344
Arrector pilorum, 250, 695
Arterial circuits (see also specific arteries)
  abdominal, 521
  coronary, 520–521
  digestive system, 521–522
  head and neck, 517, 519
  lower extremities, 522, 523
  thoracic, 519–520
  upper extremities, 518–519
Arterial pressure, 528, 562 (see also Blood pressure)
Arteries (see also under specific names)
  distributive, 511
  elastic, 510
  peripheral resistance in, 531
  structure of, 511
  systemic circuits, 512
Arterioles, 512, 620
Arteriosclerosis, 531, 534–535, 558, 695
Arteriovenous anastomosis, 515, 531, 695
Arthritis, 305
Articulation, of joints, 294–296
Ascorbic acid, 427, 474, 477
Aspirin, 32, 628, 648, 671
Association areas, 164, 166
Astereognosis, 159, 695
Asthma, 574
Astigmatism, 150
Astrocytes, 87, 110, 243
Astroglia, 112
Ataxia, 159, 161
Athetosis, 129
Atkins, Robert, 458
Atlas, vertebral, 270–271, 301
Atmospheric force, 20
Atomic structure, 7, 10
ATP (Adenosine triphosphate), 13, 19, 25
  in active transport, 408
  in glucose synthesis, 414
  in Krebs cycle, 417, 425
  in muscle contraction, 313
  production of, 204, 316, 421, 425, 694
Atrial cells, 325

Atrial fibrillation and flutter, 505, 506
Atrial systole, 498
Atrial vacuum pump, 534
Atrioventricular node, 494–499, 695
Atrium, heart, 493
Auditory cortex, 142
Auditory meatus, 263
Auditory nerve, 142
Auditory ossicles, 140
Auditory receptor cells, 141
Auditory system, 140–144
Auerbach's plexus, 97, 381, 384, 395
Autoimmunity, 584, 652, 695
  aging theory of, 664–665
Autonomic nervous system, 75, 76, 88, 132–138, 695
  afferent information to, 132
  cardiac regulation, 53, 242
  conscious control, 681–682
  embryonic development, 107
  functional characteristics, 134–135
  in heart action, 242, 327
  hypothalamic regulation, 105
  in lower respiratory tract, 570
  motor neurons, 86
  myogenic system and, 497
  nerve centers, 138
  pupil adjustment by, 149
  sensory operations, 132–134
  shock and, 550
  in stress, 678
AV (Atrioventricular) nodal rhythm, 506, 507
Axillary veins, 513–514
Axis, body types defined, 297
  cervical, 271, 272, 301
Axons, 77–84, 78, 79, 695
  motor, 121
  myelination of, 87
  optic, 147
  tracts of, 89

Babinsky reflex, 128
Bainbridge reflex, 545
Balanced diet, 480
Baldness, 252
Barany caloric test, 160
Barometer, 20
Bartholin's glands, 225
Basal cells, 33, 34, 37
Basal ganglia, 101–103, 102, 106, 112
Basal layer, epidermis, 243
Basal metabolism, 452–454
Base, chemical, 17–18
Basement membrane, 42, 380, 381, 695
Basophils, 48–50, 635, 695
B cells, 50, 631, 638, 643
Bence–Jones proteins, 611
Benemid, 695

Index

Beta oxidation, 424–426, 695
Betz cells, 99, 126–127
Bicarbonate (*see also* Carbon dioxide)
  buffering, 576, 605–606
  daily production, 576
  reserve, 695
Bicuspid valve, 494
Bile, 390, 695
Bile duct, common, 390, *391*, 393
Bile salts, 392, 409, 432, 695
Bile wastes, 695
Bilirubin, 393, 581–582, 695
Biliverdin, 582, 696
Bioelectrical activity, 13–16, 78–84 (*see also* Action potential)
  cardiac and circulatory, 495–503, *496*
  in nerve cells, 76–80, *79, 80*, 112
  propagation of, *80*
  in skeletal muscle contraction, 312–316
  in smooth muscle contraction, 322, 323
Biofeedback, 681–682
Biopsy, 56, 696
Biotin, 427
Bipolar cells, 77–78, 151
Birth control, 234–236, 423
Birthmark, 254
Bladder, urinary, 591, 595, *596*
Blast cells, 33 (*see also* Fibroblasts, Mesenchymal cells)
Blastocysts, 232
Blocking antigens, 694
Blood, 49, 492, 579
  buffering (*see* Blood pH)
  calcium regulation, 195, 206
  carbon dioxide in, 571, 572–573
  cell types, 49 (*see also* Red blood cells *and other specific types*)
  clotting of (*see* Blood clotting)
  composition of, 49
  count, 580
  fat level in, 438
  flow of (*see* Blood flow; Circulatory system)
  glucose levels in, 213, 434–438, 441
  oxygen saturation, *571*–572
  insulin level in, 441
  peripheral, 49
  plasma (*see* Blood plasma)
  platelets, 49, 50, 620–622, 635
  pressure of (*see* Blood pressure)
  serum, 635, *636*
  smear, *581*
  as tissue, 49
  types, 389, 651
  vessels (*see* Blood vessels)
Blood–brain cell barrier, 88
Blood clot, 554, 558
Blood clotting, 50, 432, *620*–623
Blood flow (*see also* Cardiac output; Circulatory system)
  arterial pressure, 534
  capillary resistance, 531
  disorders of, 531–532
  elastic resistance, 531
  hydrostatic pressure in, 527
  metabolic resistance, 532
  oxygen supply and, 534–538
  Starling's law, 542
  thoracic pump, 534
  vascular pressure, 527
  venous return, 532–534
Blood pH buffering, 17, 546, 575–576
  chemical factors in, 603–606
  kidney's role in, 606
  lung's role in, 606
Blood plasma, 492
  glomerular filtration rate, 602–603, 699
  renal filtration of, 590–603
Blood pressure, 21, 527–529
  arterial, 528, 562
  disorders of, 552–553
  heart failure and, 562
  hypertensive, 552
  measuring, 528
  normal values, 528
  sudden drop in, 550–551
  system variations, *533*
  U-tube principle, 553–554
Blood sugar (*see* Glucose)
Blood vessels (*see also* Arteries; Capillaries; Veins)
  adrenaline effects on, 543
  anatomy of, 509–514
  constriction and dilation of, 245
  wall structures of, 509–510
Body cells, 32–34, *33* (*see also under* Cells)
Body defenses
  immunological, 630–655
  reticuloendothelial, 625–628, 706
Body fluids, 492, 603–604 (*see also* Blood, Blood plasma; Lymph; Metabolites *and other specific types*)
Body regions, 63, 64–71
Body surface area, 453
Body temperature:
  adrenaline effects on, 246, 254
  hypothalamic regulation of, 104
  perspiration and, 244–245
  sweat glands and, 247–248
  thyroxine in regulation of, 204, 443–444
  vasoconstrictor regulation of, 246
Body wall, 66, 67
Body weight, oxygen consumption and, 537
Boils, 254
Bone(s):
  blood circulation, 283
  calcium deficiency in, 607
  classification of, 259–260
  compact, 282
  development of, 283–287, 607
  disorders of, 206, 287–289
  long, *281*
  marrow of (*see* Bone marrow)
  numbers of, 259
  processes, 260–261
  structure, 280–282
  tissue, *51*
Bone marrow, 51, 282
  cell differentiation in, 47
  damage to, 584
  disorders of, 641
  immunological functions, 635–636
  lymphocytes from, 631
  red blood cell production in, 579–580
Bowman's capsule, *593*, 609, 696
Bradycardia, 503
Brachialis muscle, 367
Brachioradialis muscle, 367
Brain, 98–106 (*see also* Cerebral cortex *and other specific parts*)
  astrocytes role in, 87
  components of, 98
  motor area, 127, 167
  pain perception, 669
  sensory areas, 165
  visual pathway, 147
Brainstem, 106
Breathing, 566–569
  diaphragm action in, 64, 568
  disorders of, 574
  inspiration in, 564, 567
  lung capacity, 573–574
  muscles used in, 568
  pressure and volume in, 567
  regulation of, 566
  voluntary controls, 569
Broca's area, *171*–174
Bronchi, 570, 696
Bronchioles, 565, 570–571, 696
Bruise, 621
Brunner's glands, 393
Buccinator muscle, 367
Buffering, 17, 575–576, 605–606, 695 (*see also* Blood pH buffering)
Bulbocavernosus muscle, 357, 367
Bulbourethral glands, 219
Bulla, 253
Bundle of His, 496, 499
Bursa, *294, 295*, 696

Cachexia, 57
Calciferol, 207, 696
Calcification, 288
Calcitonin, 192, 196, 205, 607, 696
Calcium:
  body levels, 205, 206
  dietary need, 478
  food sources, 485
  ion balance, 608
  ion deficiency effects, 607

muscular biochemistry of, 315–316, 495
in muscular weakness, 320
Callus, 287
Caloric intake, balanced diet, 480–481
Caloric requirements, determination of, 455
Calories, 447–448, 696
basal metabolism rates, 452
expenditure in activities, 451, 454–455
food value measurement of, 448–449
human consumption, 449–452, 480–481
Calorimeter, *448*
Calorimetry, 448–450
Canaliculi, bone, 282
Canaliculus, bile, 392
Canal of Schlemm, 147, 150
Cancer, 54–58, 221
in bone, 289
immunity toward, 655
lymphatic movements and, *510*
types of, 54
Capacity vessels, 513–514
Capillaries, 512–513, 696
in intestinal absorption, 402
in nutrient waste exchange, 532
peripheral resistance in, 531
red blood cell movement in, 531
renal, 591–595
structure of, *512*
Carbohydrates, 9, 465–466, 696
absorption of, 404
caloric content, 448
digestion of, 394
food sources of, 485
metabolism of, 414–418
nutritional role of, 465, 466
respiratory quotient of, 450
storage of, 465
Carbon dioxide:
in blood, 571, 572–573
blood level controls, 546
in buffering action, 573 (*see also* Bicarbonate)
daily production, 575
lung's removal of, 606
in metabolism, 417
in oxygen exchange, 571–573
Carboxyl(ic) groups, 14, 18
Carbuncle, 254, 696
Carcinogens, 54, 655, 696
Carcinomas, 55, 696
Cardia, 386, 696
Cardiac abnormalities, 501–509
Cardiac arrhythmias, 559–560
Cardiac cycle, 497–500, *498*
Cardiac enlargement, 55
Cardiac impulses, 497–500, *503*, 506, 696
Cardiac ischemia, 509
Cardiac muscle, 53, 324–329, 326
and skeletal muscle distinguished, 326
Cardiac output, 528, 534–539, 696
(*see also* Blood flow)
adrenaline effect on, 543–544
bodily activity and, 538–539
oxygen requirements and, 534–538
Cardiac sphincter, 384–385
Cardiac waves, 503–509
Cardio-acceleratory center, 545
Cardio-inhibitory center, 545, 546
Cardiomegaly, 555, 696
Cardiovascular system, 492, 696
(*see also* Circulatory system)
Carotid arteries, 518
Carotid body, 518, 696
Carotid sinus, 517, 518, 546, 696
pressure receptor, 518
syncope, 550
Carpals, 276, 277, 301, 696
Carrier molecules, 407
Cartilage, 283, 293, 696
bone development from, 283–284
nonrepair of, 293–294
Cartilaginous cells, in bone, 284
Casein, 696
Catabolic processes, 8, 696
Catalysis, 26, 696
Cataracts, 150, 696
Catechol methyl transferase, 23
Catheter, 696
Caudal (direction), 61
Caudal anesthesia, 108, *109*
Cecum, *379*, 396, 696
Celiac, 696
Celiac disease, 411
Cell(s), 25–37, *26*, 54 (*see also* specific types)
abnormalities of, 54 (*see also* Cancer)
basal, 33, 34, 37
Betz, 99, 126–127
cardiac muscle types, 325
in connective tissue, 44, *45*
endocrine gland, 192
epithelial (*see* Epithelial cells)
excitation of (*see* Action potential; Bioelectrical activity)
fat, 34, 457
fibroblast, 46–48
immune system types, *633*, 634–635
immunity mediation by, 643
membranes of (*see* Cell membranes)
mesenchymal, 33, 37, 46
mucous, 619
nerve (*see* Nerve cells)
neuroendocrine, 192
nucleus, 28, 310
pacemaker, 323
replacement, 34
reproduction of (*see* Cell reproduction)
sex cells, 34–35
skeletal muscle, 310–313, *311*
smooth muscle, 321
soluble fraction, 32
thyroid, 202, 204
Cell membrane(s)
biolelectrical charges on, 13–15, *16*, 78–79
depolarization and hyperpolarization of, 14, *15*, 16, 79
inorganic ion effects on, 607–609
protein composition of, 14
Cell reproduction, 32–36
body cell mitosis, 32–33
sex cell meiosis, 34–37, *35*
specialized cell inabilities, 34
Cells of Leydig, 219–223
Cellular transport, 402, 405–409
Cellulose, 414, 696
Centimeter, 6
Central input, circulatory reflex, 545
Central nervous system (CNS), 76, 97–111 (*see also* Brain; Sensory system; Spinal cord *and other specific parts*)
development of, 97–98
disorders of, 128–130
hormonal controls, 197–198
in metabolism, 434
myogenic system affected by, 497
reflex activities, 117
Cephalad, direction, 61
Cephalic phase, gastric secretion, 388
Cerebellum, 99, 106, 156–159, 161, 696
Cerebral aqueduct, 110
Cerebral cortex 99–*101*, 115, 163–174, *165*, 696
associative functions, 164, *165*
disorders of, 164, 174
emotional functions, 166–170
motor functions, 164
in pain perception, 669
sensory functions, 165–166
verbal functions, 171, *172*, *173*
visual reflexes, 149
Cerebral hemispheres, *98*, *99*, *100*
Cerebral thrombosis, 554–555
Cerebrospinal fluid, *110*, 114
Cerebrum, 99–105, 696 (*see also* Cerebral cortex *and other specific regions*)
Ceruloplasmin, 431
Cerumen, 247, 696
Cerumen glands, 143
Cervical diaphragm, 234
Cervical vertebrae, 269–272, *271*
Cervix, 226, 696
Cheese, nutritional value, 487
Chemical bond, 10–20
Chemistry, basic concepts, 7–20
Chemotaxis, 644, 654

Index

Chest, disorders of, 558–560
Chiasma, 89, 696
Childbirth, 217, 218, 224, 262
Cholelithiasis, 393, 696
Cholecystitis, 393, 696
Cholesterol, 192, 219, 393, 418, 696
    in atherosclerosis, 554, 558
    intake and blood levels, 472
Cholycystokinin, 394
Chondroblasts, 47, 51, 294, 696
Chondrocytes, 47, 51, 284, 696
Choreiform motion, 129
Chorion, 232, 696
Chorionic gonadotropin, 192, 224, 231, 232, 696
Choroid plexuses, 110, 697
Chromatids, 32
Chromosomes, 28, 29, 32, 36, 37
Chyle cistern, 515, 697
Chylomicrons, 424, 433, 439, 697
Chyme, 387, 388, 390, 395, 697
Chymotrypsin, 390
Cilia, 42, 697
Ciliary body, 146
Ciliated cells, 619
Circuit of Willis, 517, 519
Circulation:
    in bone, 283–285
    fetal, 232
    routes of, 493
    skin, 246
Circulatory immunity, 643–646
Circulatory system, 491–562 (*see also* Blood; Blood flow; Blood pressure; Blood vessels *and other specific parts*)
    blood pressure variations, 533
    chemical receptor controls, 546
    disorders, 549–563
    functions, 525–526
    nervous control, 542–544
    osmotic flow, 530
    parasympathetic effects, 544
    pressures in, 526–531
    pulmonary circuit, 516
    reflex regulation, 545–546
    regulation of, 541–546
    renal branch, 591–593
    routes of, 515–523
Circumcision, 697
Circumduction, 299
Cirrhosis, 393, 697
Citric acid, 418
Clavicle, 275, 276, 344, 348
Clearance (Cm) test, 60
Cleft palate, 269
Clitoris, 225, 226, 697
Clotting (*see* Blood clotting)
Coenzymes, 416, 426
Coccygeus muscle, 367
Coccyx, 259, 269, 272, 273, 277
Cochlea, 140, 157, 158, 697
Cochlear duct, 141
Cognitive factors, 167
Colchicine, 671, 697

Collagen, 42, 44, 697
Collagenase, 52
Collagen fibers, 44, 45, 48, 51, 284
Collecting tubules, 595
Colle's fracture, 289
Colloid, 202, 697
Colon, 396, 398, 697
    spastic, 399
Colorimetry, 581
Colostrum, 234, 697
Columnar epithelial cells, 41–44
    intestinal, 394
Coma, 214, 697
    diabetic, 214, 437–438
Commissures, 89, 100, 697
Common bile duct, 390, 391, 393
Compact bone, 282
Complement system, 644, 697
Compounds, chemical, 7–9
Conchae, 264, 266, 267
Condyle(s), of bone, 260
Cones, 144, 147, 696
Conjunctiva, 147
Conjunctivitis, 150
Connective tissue(s), 44–52, 58
    cells, 45
    dense, 48
    dermis as, 67
    fascia as, 68
    general types of, 47
    malignancy in, 54
    mensenchymal cells in, 33
    occurrences of, 44
    in organs, 60
    structure of, 44
    special types of, 49
    subcutaneous, 68
Consciousness, loss of, 550–551
Conscious perception, 119
    of equilibrium, 159
    of smell, 153
    of sound, 142
    of taste, 154
    of vision, 147
Contractile proteins, 312
Contractility, cardiac, 325
    muscular, 309
Contractions, skeletal muscles, 313–318, 316
    smooth muscles, 322
Contralateral dominance, 116
Coordination, 159
Cornea, 145, 697
Corneum, epidermal, 243, 244, 697
Coronal section, 62, 297
Coronary heart disease, 558–559
Coronary occlusion, 558, 697
Coronary system, 520–521
Coronary thrombosis, 554–555
Corpora cavernosa, 221
Corpora quadrigemina, 105
Corpus albicans, 228, 229, 697
Corpus callosum, 89, 100, 106, 169, 171, 173, 697
Corpus luteum, 228–230, 628, 697
Cortical areas, 101, 148

Corticoids, deficiency of, 211
Corticospinal tracts, 89, 92, 125, 127
Corticosteroids, 210
Cortisol, 184, 193, 209, 443
Cortisone, 32, 193, 209, 443, 647
Coumarin drugs, 621–622
Countercurrent multiplier, 598–601
Covalent bonds, 11, 13, 23
Cowper's glands, 218, 219
Cranial (direction), 61
Cranial cavity, 70, 263
Cranial nerves, 94–95, 106, 121, 147
Cranial sutures, 261–262
Cranium, 70
    bones of, 262, 263, 264–265
    cavities, 263
Creatine, 313, 605
Cretinism, 204, 697
Cribriform plate, 264
Crista, 158, 697
Cross bridges, 315, 316
Crossing-over, in meiosis, 35, 697
Cross linking, age theory, 664
Cryptorchism, 222
Cuboidal epithelial cells, 41–44, 202
Cushing's syndrome, 202, 206, 210, 212, 214, 611, 697
Cyclic AMP, 444–445
Cystic fibrosis, 411, 697
Cystinuria, 411
Cytology, 25, 697
Cytoplasm, 26, 32, 697

Deafness, 143
    conduction, 143, 144, 697
Decidua, 232, 697
Deep fascia, 68
Deep touch, perception of, 114, 115
Defecation, muscles of, 356, 362
Defense mechanisms, nonimmunological, 617–628
Dehydration, 551
Demobilizer hormone, 197, 198
Dendrites, 77, 79, 81, 85, 112, 114, 227, 697
Density, 7, 697
Deoxycorticosterone, 209
Deoxyribonucleic acid (*see* DNA)
Depolarization, 13–15, 16, 78, 79, 80, 84, 697 (*see also* Action potential; Bioelectrical activity; Hyperpolarization)
Depression, 167
Dermal papillae, 251, 252
Dermatitis, 254
Dermatology, 253
Dermatomes, 117, 697
Dermis, 67, 242, 245–246, 697
Desensitization, 650
Detoxification, 430

Diabetes insipidus, 200, 551, 600, 697
Diabetes mellitus, 201, 211–214, 437, 611, 697
Diabetics, vision defects of, 147
Diaphragm, cervical, 234, 698
Diaphragm, visceral, 64, 354, *361*, 398
　in breathing, *566, 567, 568*
　herniation of, 385
Diaphysis, 259, *281–283*, 698
Diarrhea, 398, 406, 410
Diarthrodial joints, 293, 295, 299, 303–305
Diastase, 214, 698
Diastole, 497, 698
Diastolic pressure, 528, 552–553, 562
Diet, balanced, 480
Diffusion, 22, 405–408, 698
Digestion, 378–398, 698 (*see also* Digestive system; Foods; Metabolism)
　bile in, 392
　of carbohydrates, 394
　defects of, 410
　of fats, 390–392
　enzymes of, 30, 382, 386–388, 389–390, 392, 394, 397–398
　of fats, 390–392
　gallbladder in, 391, 392
　gastric juices, 386–387
　hormonal regulation, 387, 397
　hydrolytic process in, 19, 378
　liver in, 390–392
　pancreas in, 389–390
　of proteins, 390
　salivary juices in, 382
　in small intestine, 394–395
　in stomach, 386–388
Digestive system, 378–398
　blood supply to, 515, 521–522
　component organs, 378–379, *380, 385*
　hormonal regulation, *397*
　nervous regulation, *395*
Digitalis, 506, 698
Dilantin, 175
Dipeptidase, 394
Diploe, 260, 284, 698
Diploid number, chromosomes, 32, 34
Discs, intervertebral, 273–274, 293
Distal, direction, 63, 698
Distal tubule, 594, 609
Diverticula, 396, 411
DMPEA, 169
DNA (Deoxyribonucleic acid), 28, 29, 698
DNA–RNA error, 663, 665
DOPA, 169
Dopamine, 82, 167, 169
Dorsal, defined, 61
Dorsal horns, 107
Dramamine, 161
Dreaming, 680–681

Drugs, filtration of, 598
　hypersensitivity to, 652, 654–655
　in pain relief, 669–671
　in stress, 681
Drunkenness, 160
Ducts of Bellini, 595
Duodenum, *389,* 390, 393, 698
Dura mater, *108,* 698
Dust cells, 52, 571
Dwarfism, 201, 222
Dyads, chromosomal, 35
Dysphagia, 384, 698
Dyspnea, 575, 698
Dysrhythmias, 505, 506, 698
Dystonia, 129, 698

Ear, 140, *141,* 142
　inner, 263
EB (Epstein–Barr) virus, 642
Eccrine glands, 247, 698
Ectoderm, embryonic, 97
Ectopic pregnancy, 227, 698
Eczema, 254, 698
Edema, 23, 492, 530, 531, 551, 553, 698
Efferent, defined, 86, 698
　arteriole, 592, 595
　neurons, 85 (*see also* Motor neurons)
Egg cells, 32, 224, 227, *228,* 231
Ejaculatory ducts, 219
Elastic arteries, 510, 528
Elastic fibers, 44, 45, 48
Elasticity, muscular, 309
Elbow, muscles and movements, 347, *353*
Electrical activity (*see* Action potential; Bioelectrical activity)
Electrocardiogram (ECG), 497, *499,* 500–503
Electrocardiograph, 500–503, *502,* 698
Electroencephalogram (EEG), 175, 679
Electrolytes, 12
Electromyographic findings, 292
Electrons, 11–12
Electrophoresis, 431, 635, *636,* 698
Elements, 7
　in living tissue, 8
Elimination, 398–399
Emaciation, 456
Embolus, 554, 698
Embryo (*see* Fetal development)
Emeiocytosis, 27, 406
Emotional activities, 166–167, 170
　cognitive factors, 167
　hypothalamic regulation, 105
Emphysema, 574, 575, 698
Emulsification, 18, 409, 698
Endocardial diseases, 555–556, 698
Endocardium, 495
Endocrine cells, 182, 192

Endocrine glands, *182,* 184–190
Endocrine organs, *183* (*see also* specific kinds)
Endocrine substances (*see* Hormones)
Endocrine systems, in hypertension, 553
　one-level type, 188–189
　two-level type, 186–187
　three-level type, 184–185 (*see also* specific glands and functions)
Endocytosis, 27, 31, 406, 698
Endometrium, 226, 227, 232, 698
Endomysium, 333, 698
Endoneurium, 89, 698
Endoplasmic reticulum, 28, 37, 312, 698
Endosteum, 282, 285, 286, 698
Endothelium, 495, 509, 698
Energy: (*see also* Metabolism)
　biological production of, 13, 18
　carbohydrate sources, 465–466
　central nervous system requirements, 434
　expenditure by activities, 451, 454–455
　mitochondrial generation of, 28
　muscle cell production, 313
　nervous tissue consumption, 434, 436
　protein as source, 467
　release of, 28
Enterogastrone, 388
Enterokinase, 390, 397
Enzymes, 13, 698
　coenzymes, 416, 425, 426
　digestive, 30, 394, 395
　functions of, 26, 31
　intracellular, 28
　lysosomal, 31
　pancreatic, 390
　in starch digestion, 414
Eosinophils, 48–50, 635, 698
Ependymal cells, *88,* 110, 112, 698
Ephase, 322
Epicardium, 495
Epidermal cells, 34
Epidermis, 67, 242–245, 698
　anatomy of, 242, 243
　defensive mechanisms of, 619
　growth rate, 244
　insensible perspiration and, 244–245
　layers of, 242–243
　protective functions, 244
Epidymes, *218–221*
Epidural space, 108, 698
Epiglottis, 383, 698
Epilepsy, 175
Epimysium, 333, 698
Epinephrine (*see* Adrenaline)
Epineurium, 89, 699
Epiphyseal plate, 286
Epiphyses, 259, 283
Episiotomy, 255, 699

Epithelial cells, 41–42, 48, 88
Epithelial tissues, 40, 41, 42–44, 58, 699
  basal cells in, 33
  cancer in, 54
  digestive mucosa as, 380
  epidermis as, 67
  functions of, 43
Equation division, in meiosis, 33, 35
Equilibrium, 157–159
Erection, muscles of, 357
Ergosterol, 207, 699
Erythrocytes (see Red Blood cells)
Erythropoiesis, 579, 699
Erythropoietin, 584, 609, 699
Esophageal varices, 523
Esophagus, 384, 699
Estrogen, 184, 193, 218, 224, 227, 229, 699 (see also Sex hormones)
  female, 224, 234
  functions of, 229
  male, 222
  production of, 209
Estrone, 224
Ethmoid bone, 263–264
Eustachian tube, 60
Eversion, 299
Excitory stimulus, neural, 79–82
Excretion, epithelial, 43
  process of, 595–596, 613
  skin's role in, 249
Excretory ducts, 247
Excretory system, 589–612, 590
Exercise, maximum, 538–539
Exfoliative cytology, 55
Exhaustion, 537–538
Exocrine function:
  of pancreas, 389
  of stomach, 387
Exocrine glands, 182, 247, 699
Exocytosis, 27, 31, 406, 699
Exophthalamos, 205, 503, 699
Expiration, 567
Extensibility, muscular, 309
Extension, 298, 699
Extensor reflexes, 123–124, 131
Exteroceptor neurons, 85
Extracellular fluid, 492
Extrapyramidal system, 127–130, 699
Eye, external muscles, 343
Eyeball, 144, 145–148
  movements of, 147–148
  muscles of, 146, 147
  protection of, 147
Eyelids, muscles of, 147

Face, bone structure, 265, 266–269
  muscles of, 339
Facet, bone, 261
Facilitated diffusion, 406–407
Fainting, 550–551, 699

Fallopian tubes, 223–224, 227
Farsightedness, 150
Fascia, 48, 68, 72, 333, 699
Fascicle, 333, 699
  nerve, 89
Fasciculation, 318, 699
Fat, in aging, 662
  metabolism of, 423, 424, 425
Fat cells, 34, 457
Fats (lipids), 422–425
  absorption of, 404
  absorption defects, 410
  blood levels of, 439
  caloric content of, 448–449
  digestion of, 390–392
  emulsification of, 409
  food sources of, 485
  in glucose metabolism, 418
  metabolism of, 423, 424, 425, 428
  nutritional role, 404, 470
  respiratory quotient, 450
  values of, in specified foods, 482–484
Fatty acids, 422, 699
  essential, 470
  hormonal regulation, 440
  metabolism of, 439–440, 443
Feces, elimination of, 398
Feet, muscles operating, 360
Female hormones, 224
Female sex characteristics, 185–186, 229
Femur, 279, 699
Fertilization, 217, 224, 227
Ferritin, 431, 479, 699
Fetal development, immunity in, 647, 649
  long bones, 284
  nervous system, 97–98
  red blood cells, 579
  spinal cord, 106
  skull fontanels, 262
Fetus, 232, 699
Fever, 627
Fibers, connective tissue, 44, 45–48
Fibrillation, 503–506
Fibrin, 621, 699
Fibrinogen, 432, 621, 699
Fibrinolysin, 621
Fibroblasts, 45–48
Fibula, 279–280
Fick's law, 22, 535, 562
Fine muscular control, 125, 127
Fingers, muscles and movements, 336, 348, 358
First cranial nerve, 151–153
Fissures, epidermal, 254
Fistula, 411, 699
Flexion, 297–298, 699
Flexor reflexes, 124, 131
Flutter, heart, 503–506
Folic acid, 427, 435
Follicles, 699
  ovarian, 227, 228

thyroid, 202–203
Food intake, dietary allowances recommended, 468–469
  nutritional standards for, 480–487
  obesity and, 456–457
  psychological factors in, 457–458
Foods, caloric values of, 448–449
  classes of, 481–482
  digestion of, 9 (see also Digestion)
  nutritive values of, 482–484
  protein values of, 420
Foot, 280
  ligaments, 304
  muscles and movements, 366
Foramen, of bone, 261
  vertebral, 269, 270
Foramen magnum, 246–265, 699
Forearm, 275, 277
  movements and muscles, 334, 347, 354
Forebrain, 99
Foreign bodies (see Antigens; Foreign substances)
Foreign substances, xenobiotic, 623–624
  body defenses against, 623, 624
  elimination of, 624
Fossae, of bones, 261
Fourth ventricle, 110
Fractures, bone, 275, 276, 288
  repair of, 287, 288
Freckle, 253
Free radical theory of aging, 665
Frontal, bone, 263
  lobe, 164
  section, 62
  sinus, 263
FSH (Follicle stimulating hormone), 201, 219–224, 229, 231
Fructose, 414, 699
Fundus, 386
Fungal diseases, 255
Furuncle, 254, 699

Galactose, 414, 699
  absorption disorder, 411
Gallbladder, 379, 389, 390–393
  digestive gland regulation, 397
  disorders of, 393
Gallstones, 393
Gametes, 217
Gamma globulin, 431–432, 631, 635, 646
Gamma reflex loop, 125, 131
Ganglia, 88, 699
Gas exchange, alveolar, 575
  respiratory system, 571, 573
Gastric glands, 387–388, 699
Gastric juices, 387–388
Gastric lipase, 387
Gastrin, 385, 388

716
Index

Gate control theory of pain, 669
Genes, 28–30, 37, 699
Genetic programming, life span, 663
Genital organs, external female, 224–225
male, 221
Germ cells, 32, 217
Germinal epithelium, 44
Gigantism, 201
Gestation, 231–232
Glands, apocrine, 247
eccrine, 247
exocrine, 182, 247, 699
intestinal, 394
mammary, 224, 233–234, 247
sweat, 247–250
Glaucoma, 147, 150
Glenoid fossa, 275, 301
Glial cells, 86, 112
Globulins, 431
Glomular filtration rate (GFR), 602, 699
Glomeruli, 247–248, 515, 591–593, 699
Glomerulonephritis, 612, 699
Glottis, 569
Glucagon, 192, 212, 390, 440–444, 699
Glucocorticoids, 201, 209–210, 443–444, 699
Glucokinase, 431, 699
Gluconeogenesis, 415, 436, 439, 442–443, 699
Glucose, 9, 414, 433, 699
blood levels, 213–214, 438–440
brain tissue supply, 434
defective absorption, 411
hormonal regulation, 196, 440
liver conversion, 431
in metabolism, 414–415
nutritional disorders, 436–437
release and oxidation, 208
synthesis, 421
tolerance curves, 438
tolerance test, 436
in urine, 611
Glucose-6-phosphate synthesis, 414, 431
Glucosuria, 611
Glycerol phosphate, 417, 432–433
Glycogen, 9, 415, 421, 699
liver synthesis of, 431
storage of, 465
utilization of, 436
Glycolysis, 415
Glycoproteins, 30, 31, 37, 431, 699
Glycosuria, 214, 700
Goblet cells, 394, 699 (see also Mucous cells)
Goiter, 203, 204, 700
Golgi, Camillo, 30
Golgi apparatus, 30–31, 37, 619, 700
Golgi tendon receptor, 125, 126, 319

Golgi tendon reflex loop, 125, 128–131
Gonads, 217, 219
Gonadotropin mobilizer, 184, 186, 219–224
Gonadotropins, 219, 222
Graafian follicle, 228, 231
Grafts, tissue, 654–655
Gram weight, 5, 6
Gram-atomic weight, 10
Gram-molecular weight, 10
Grand mal, 175, 700
Granular layer, of epidermis, 243
Granulocytes, 49, 700
Granulocytic leukemia, 641
Gray matter, 169
Great cistern, 110
Gross anatomy, 25
Ground substance, 44, 45, 46–48, 51
Growth hormone, 197, 198, 201, 443, 700
Gyrus, 99, 700

Hair, 251–252
cells, 700
follicle, 251
gray, 662
growth cycle, 252
Hallucinogens, 168, 169
Hands, bones of, 276, 277
joints, 301
muscles and movements, 348–349, 355, 356, 357
Haploid number, 32, 34, 700
Haptenic antigen, 648, 700
Haptoglobin, 582
Hartnup's disease, 411
Haversian systems, 282–284, 287
Head, 70–70 (see also Cranium)
circulatory system, 517, 518
movements, 344, 346
muscles, 339–343
Hearing, 140–144
cerebral center for, 165–166
disorders of, 143, 166
Heart, 493–506 (see also under Cardiac; Coronary)
anatomy, 493–495
beat rate and rhythms, 497–506 (see Heartbeat)
bioelectric system, 495, 496–497
blocks, 507–508
cardiac cycle, 497, 498–500
chambers, 493–494
disorders of (see Heart disorders)
electro-conduction system, 496
failure of (see Heart failure)
force of contraction, 541, 542
muscle fibers, 325–326
myogenic contraction, 541
nervous system effects, 542, 544
reflex regulation, 545–546

self-regulation, 541–542
Starling's law, 54
valves, 494
vasoconstrictor effects on, 543
walls, 495
Heartbeat, 494–500 (see also Electrocardiogram; Pulse)
action potentials for, 495–496
arrhythmias, 503–504
atrial systole, 497–498
beat rates, 503
bradycardia, 503
cardiac cycle, 497, 498, 499
distole, 497–498
dysrhythmias, 505, 506, 695
disorders of, 503–509
electric impulses in, 495–496
fibrillation, 503–507
flutter, 503–507
nervous system factors in, 497
oxygen supply and, 535–537
pacemaker in, 495–496
PQRST waves, 503–509
refractory period, 327, 500
self-regulation, 541–542
systoles, 498
tachycardia, 503
Heart disorders and diseases, 501–507, 552–562, 700
arrest, emergency treatment, 561
arteriosclerotic, 531, 534–535, 558, 695
blood clots, 554, 558
congenital defects, 556–557
congestive failure, 560, 700
coronary disease, 558–559
coronary thrombosis, 554–555
endocardial diseases, 555–556, 695
hypertension, 558
left-sided failure, 560
mural, 555
myocardial infarction, 501, 558
pulmonary embolism, 502
Heart failure, 560, 561
Heat stroke, 247
Helical molecule, 28
Hematocrit, 522, 580, 700
Heme, 432, 582
Hemoglobin, 246, 432, 700
blood smear examination, 581
in carbon dioxide–oxygen exchange, 572
oxygen carrying by, 571–572
recycling of, 582
waste discharge, 392, 430
Hemoglobinuria, 583
Hemolytic anemias, 584, 652
Hemophilia, 623, 700
Hemorrhage, 57, 700
internal, 621
Hemorrhoids, 523, 700
Heparin, 48, 620, 627, 700
Hepatic artery, 392
Hepatic ducts, 389, 391

717
Index

Hepatitis, 393, 700
Hermatemesis, 523
Hernia, 354, 700
    diaphragm, 385
    inguinal, 354, 356
Hexokinase, 431, 700
Hiatus hernia, 385
Hiccupping, 569
Hilum, 592, 594, 700
Hindbrain, 105, 112
Hip, bone, 277–279
    ligaments, 303
    musculature, 357, 362, 363, 364
Hippocampus, 150, 700
Hirschsprung's disease, 399
Histamine, 48, 246, 626
Histiocytes, 48–52, 626, 700
Histology, 25, 700
Hives, 253, 254, 700
Hodgkin's disease, 642
Holocrine glands, 253, 700
Hormones, 27, 182–215, 700 (see also Endocrines; Glands and specific types)
    adrenal cortex production, 184, 186, 188, 209
    anterior pituitary system, 184, 186, 198–200
    in body temperature control, 105
    in childbirth, 224
    classes of, 192–193
    in digestion, 397
    follicle stimulating, 186, 219–222
    gonadotropic, 184, 186, 219, 222 (see also Sex hormones)
    growth, 197, 198, 201, 443, 700
    luteinizing (see Luteinizing hormone)
    mobilizing, 197
    in metabolism, 434–444
    negative feedback, 195
    neuron secretion of, 86
    in obesity, 457
    parathyroid, 205
    pathways and targets, 184–191, 193, 194
    primary messengers, 445
    protein utilizing, 443
    regulatory mechanisms, 194–198
    release inhibiting, 197
    in stress reactions, 678
Horn, 107
Horny layer, epidermis, 243, 244, 700
Humerus, 275, 277, 301
Huntington's chorea, 129
Hydrochloric acid, 387
Hydrogen bonding, 12, 17, 18
Hydrolysis, 9, 19, 23, 192, 700
Hydrostatic pressure, 527, 700
Hydroxyl ions, 17
Hymen, 225, 700
Hyoid, bone, 265

muscles and movements, 343, 345
Hypocarbia, 575, 700
Hyperacidity, 410
Hyperesthesia, 119
Hyperglycemia, 214, 436–438, 611, 700
Hyperhidrosis, 249
Hyperopia, 150
Hyperparathyroidism, 206
Hyperpituitarism, 202
Hyperplasia, 56, 175, 204, 700
Hyperpnea, 575, 701
Hyperpolarization, 14–16, 79–84
Hypersensitivity, 650–655
    delayed, 653
    drugs and, 654–655
Hypertension, 211, 552–553, 558, 701
Hyperthyroidism, 203, 204, 214
Hypertrophy, cell, 204
Hyperventilation, 575, 701
Hypoadrenocorticism, 211
Hypoesthesia, 119
Hypogeusia, 154
Hypoglycemia, 436–438, 701
Hypogonadism, 222, 701
Hypoparathyroidism, 206
Hypophysectomy, 201, 701
Hypopituitarism, 202
Hypoplasia, 584
Hypothalamus, 102, 104, 112, 215–224, 232, 234
    autonomic activity regulation, 105
    autonomic nerve centers in, 138
    body temperature regulation, 104
    emotional activity regulation, 105
    hormone production, 184, 198
    hormone regulation, 198–199
    nuclei, functions of, 104
    in sweat control, 248
    and stress, 678
Hypothyroidism, 204
Hypotonia, 161
Hypoxemia, 575, 701
Hysterectomy, 226, 701

Ileum, length, 393, 394
Iliac veins, 513, 514
Ilium, 278
Immunity, 618, 642–648
    active system, 645, 646
    attenuated, 645
    cancer and, 655
    cell-mediated system, 633, 643, 646
    cell types in, 634–635
    circulatory, 634, 645–646
    complement system, 644
    conditions for, 642–643
    delayed hypersensitivity, 643
    lysosomal enzyme and, 31

organs of, 635–640
passive, 645, 646
process of, 630, 655
suppression of, 647
tissues in, 663
tissue abnormalities, 640–642
tolerance in, 647
Immunoglobin (see Antibodies)
Immunosuppression, 647
Implantation, 232
Inferior, direction, 61
Inferior colliculi, 105
Inflammation, 57, 58, 210, 625, 627
Informational molecules, 30, 31
Infundibulum, 227
Inguinal canal, 355, 356
Inguinal hernia, 354, 356, 701
Inguinal ligament, 354
Inhaling, muscles and movements, 344, 347
Inhibitory stimulus, 79, 83
Innominate bone, 278–279
Inorganic compounds, 8, 9
Insensible perspiration, 244–248
Inspiration, 564, 567
Insulin, 192, 213, 390, 701
    blood sugar levels, 435–438, 440–441
    effects of, 213
    in glucose synthesis, 414
    insufficiency of, 214
    in metabolism 433, 434, 440–441
Insulin shock, 215
Insulin treatment, 214
Intercalated discs, 53, 324–325, 701
Intercostal muscles, 254, 369, 567–568
Interneurons, 86, 125, 132, 147, 149, 152
Interphase, mitosis, 32
Interstitial fluid, 492
Interstitial spaces, 30, 529–530, 701
Interferon, 618, 624–625, 701
Interlobular arteries, 591–592
Internal capsule, 100
Interceptors, 85
Intestinal absorption, 402–404
    defects in, 410–412
Intestinal glands, regulation by, 397
Intestinal lumen, 398, 402
Intestines, 589 (see also Anal canal; Colon; Small intestine and other specific parts)
Intima, 509, 554
Intracartilaginous bone development, 284–285, 287
Intracellular fluid, 492
Intramembranous bone development, 284
Intrauterine device (IUD), 234, 235, 423

Intrinsic factor, 388, 405
Introitus, 225
Intussusceptions, 411
Inulin, 602
Involuntary muscle, 53
Iodine, 8, 480
Iodide, 203
Ionic bond, 11–13, 23
Ionization, 16–17
Ipsilateral representation, 116
Iris, 144, 146
Iron, deficiency, 479–480
  metabolic role, 479
  sources in food, 486
Ischiocavernosus muscle, 357, 369
Ischium, 278, 701
Islets of Langerhans, 212, 390
Isoantigen, 648
Isometric contraction, 316–317
Isometric exercises, 328–329
Isotonic contraction, 316–317, 701
Isthmus, limbic, 150
Itching, 254
IUD (see Intrauterine device)

Jaundice, liver, 393, 701
Jejunum, 393, 394, 701
Joints, 291–307
  abnormalities of, 306
  articulation of, 294–296, 301
  diathrodial, 299–305
  disorders and diseases, 305–307
  movements of, 297–298, 301
  synarthrodial, 297
  types of 296, 297–305
Jugular vein, 513–514

Keratin, 243, 701
Ketone bodies, 213, 460, 611, 701
Ketonemia, 214
Ketonuria, 214
Ketosis, 214, 701
Kidneys, 590, 591–595
  blood flow through, 591–593
  body fluid regulation, 603–609
  buffering blood pH, 586, 596, 606
  disorders of, 206, 211, 611–612
  endocrine functions, 609
  excretory functions, 589–613
  filtration process, 598
  functions evaluation, 603–604
  glomerulus circulation, 515
  metabolic functions, 433
  nutrient–waste exchange, 591–592
  parathyroid effects, 206
Kilocalories, 448
Kinins, 246
Klinefelter's syndrome, 36–37
Knee, 279
  ligaments, 303
  muscles and movements, 358, 365

Knee jerk, 123, 124
Krebs cycle, 415–416, 417, 421, 424
Kupffer cells, 52, 392, 395, 403, 430, 640, 701
Kwashiorkor, 438, 701
Kymograph, 317
Kyphosis, 272

Labia majora, 224, 701
Labia minora, 224, 701
Labor, 226, 232
Lacrimal bones, 266
Lacrimal glands, 147
Lacrimal sulcus, 266
Lactase deficiency, 410
Lactation, 233–234
Lacteals, 402, 701
Lactic acid, 327, 329, 417, 677
Lactose, 414, 418, 701
Lamina propria, 66, 381, 701
Large intestine, 396–397
Larynx, 570
Lateral, body direction, 62
Lateral horns, 107
Lateral ventricals, 110
Lecithin, 478, 702
Leg, 279, 280
Lemniscus, 104, 106
Lens, 146, 702
Lesions, skin, 253–254
Leucocytes, 49 (see also White blood cells)
Leukemia, 54, 585, 640–641, 702
Leukotomy, 150
Life span, 661–666
Ligaments, 48, 292, 302, 303
  ankle, 304
  plantar, 304
  repair of, 293–294
  spinal column, 304, 305
Light, perception of, 145
Light touch, perception of, 114
Limbic cortex, 169, 170
  stress and, 678
Lipase, 390, 392, 702
Lipids (see Fats)
Lipid storage disease, 423
Lipofuscin, 662
Lipoproteins, 27, 433, 438, 554
Littoral cells, 52
Liver, 379, 390, 391–393
  blood supply to, 392
  clotting hormones, 622
  cord, 391–392
  digestive gland regulation, 397
  disorders, 393, 431
  glucose supply by, 435–436
  immunological functions, 640
  metabolic functions, 430–432
  nutrient screening by, 403
Lobotomy, 150, 702
Local potential, 80, 702
long bone, 281, 284
Longevity, 662–663

Longitudinal arch, 280
Loop of Henle, 594, 599
Lordosis, 272
Low carbohydrate diet, 458–461
Lower appendages, 279–280, 304, 366
  disorders of, 523
Lower motor neurons, 122–131
Lucid layer, epidermis, 243
Lumbago, 274
Lumbar cistern, 109, 110
Lumbar vertebrae, 269–272
Lungs, 565–566, 571–572 (see also Breathing; Respiratory system)
  buffering role of, 606
  capacity of, 573
  congestion in heart failure, 560
  diseases of, 574
  excretory role, 589
  vital capacity, 574
Luteinizing hormone, 184, 202, 219–224, 231, 444, 702
Lymph, capillaries, 514, 515
  composition of, 514–515
  in fat absorption, 424
Lymphatic ducts, 510, 515
Lymphatic system, 510, 514–515
  fat absorption, 402–403
  nutrient circulation in, 404
Lymph nodes, 51, 515, 638
  disorders of, 641
Lymphocytes, 49–52, 515 (see also White blood cells)
  disorders of, 641
  in immunity, 631–635, 655
Lymphocytic leukemia, 641
Lymphoid tissue, 47, 51, 702
Lymphoma, 54, 702
Lymphotoxin, 654
Lysosomes, 31
Lysozyme, 49

Macrophage cells, 47–49, 406, 626, 634, 702 (see also Phagocytes; Reticuloendothelial system)
Macula densa, 609, 702
Maculae, 158, 702
Macule, 253, 254
Magnesium, 8, 480, 486
  in muscular biochemistry, 316, 607
Male reproductive system, 217–223, 218, 220, 221
Male sex characteristics (see also Masculine qualities)
  endocrine stimulus, 184–185
  secondary, 222
Malignancy, 54
Malleus, 265
Malpighian layer, 243
Maltase, 394, 414
Mammary glands, 224, 233–234, 247

Index

Mammillary bodies, 104, 702
Mandible, 268, 269, 301
Mandibular fossa, 301
Marasmus, 438, 702
Marrow (*see* Bone marrow)
Masculine qualities, 220–223
Mass, 5, 526, 702
Mast cells, 48, 635, 702
Mastication, muscles for, 343, *344*
Mastoid, 262–263
Matrix, connective tissue, 44, 702
Maxillary bones, 266, 267, 268
Measurement, scientific, 4–5
Meatus, bone, 261
Mechanical receptor neurons, 85
Medial section, 62
Medial sclerosis, 553, 555
Medial sagittal plane, 297
Mediastinum, 65, 702
Medulla, hair, 250–251
    kidney, 591, 594
    oblongata, 106, 112, 115, 127, 142, 154, 545–546, 702
    ovarian, 227
Megaloblastic anemias, 585
Meiosis, 32–36, 702
Meissner's corpuscles, 19
Meissner's plexus, 97, 381, 384, 385
Melanin, 201, 242, 243, 702
Melanocytes, 48, 68, 243
Melanocyte-stimulating hormone, 201
Melanoma, 256
Melatonin, 192
Membrane(s) (*see also* Cell membrane(s))
    longitudinally oriented, 312
    moist, 66
    mucous, 65–66
    osmotic, 22–23
    polarity, 78–84
    potentials, bioelectric, 16
    semipermeable, 22–23
    serous, 65–66
    synovial, 294–295
    transversely oriented, 312
Memory, 165
    disorders and loss of, 74
Memory cells, 633, 643
Menarche, 233, 702
Meningeal sheath, 108
Meninges, 108, 113, 702
Menopause, 233, 702
Menstrual cycle, 229, 230
    "the pill" and, 235
Menstrual flow, 225
Menstruation, 230
Mental foramen, 268, 269
Mental illness, 168
Meridians, acupuncture, 673
Mesangial cells, 52
Mesenchymal cells, 33, 37, 46
Mesentery, 69, 72, 382, 702
Messenger-RNA molecules, 28, 30
Metabolic energy, cellular, 78

Metabolic processes, 8
Metabolic resistance, in blood flow, 532
Metabolism, 413, 702
    absorptive period, 435
    of acetyl coenzyme A, 425 (*see also* Krebs cycle)
    acidity reactions in, 575
    adipose tissue in, 432–433
    adrenaline effects on, 208
    of carbohydrates, 414
    central nervous system in, 434–435
    cyclic AMP in, *444–445*
    disorders of, 213, 469–470, 472
    endocrine stimulus of, 185
    of fats, 423–425, *424*
    glucocorticoids in, 210
    hormonal regulation, 434–440
    in intestinal cells, 402
    kidneys in, 433
    liver in, 430–432
    in muscles, 432
    oxidation–reduction in, 20
    of proteins, *421*
    specific activity requirements, 454–455
    specific dynamic effect, 454
    thyroxine stimulation of, 204
    vitamins in, 426, 427
Metabolites, 8, 23, 28
Metacarpals, 276, 301
Metaphase, 32, 35
Metastasis, 56, 702
Metatarsal arch, *280*
Metatarsals, *280*
Methemoglobin, 582
Metric system, 4–7
Microbial sensitivity, 651
Microglia, 52, 86, 98, 112, 702
Microvilli, *42*, 702
Micturation syncope, 551
Midbrain, 105, 106, 112, 149
Midsagittal section, 62
Migration inhibition factor, 654
Milk, 486–487
    defective absorption of, 410
    production of, 224, 234
    secreting cells, 223
Mineral layer, long bone, 281
Mineralcorticoids, 209, 211, *601*, 702
Minerals, dietary needs, 478–480
    recommended daily allowances, 469
    regulatory functions, 426–428
    values of, for specific foods, 482–484
Mineral salts, absorption of, 402–405
Mitochondria, 28, 37, 417, 594, 702
Mitosis, 21–32, 37, 703
Mobilizers, hormone, 197
Molarity, 10
Molecular weight, 10

Monoamine oxidase, 82
Monocytes, 49–52, 703
Monocytic leukemia, 641
Monomers, 9, 703
Mononucleosis, infectious, 642
Morphine, 670, 671
Motion sickness, 160
Motor area, cerebral cortex, 164–165
    damage to, 129
Motor end plates, *122*
Motor information, 123, 125, 127–128, 131
Motor neurons, 85–86, 121, 135 (*see also* Neurons)
Motor units, 122–123
Mouth, 70, 263
    digestive functions, 382
Mucin, 619
Mucosa, *380*–381, 386, 394, 619
Mucous cells, 31, *379*, 387, 394, 619
Mucous membranes, 65–66, 406, 703
Multiple sclerosis, 175
Murphey's point, 134
Muscle cells (*see* Skeletal muscle cells; Smooth muscle cells)
Muscles, 308–330, 367–373 (*see also* Skelatal muscles; Smooth muscles)
    ciliary, 146
    exercise of, 328–329
    eyeball, *146*, 147
    eyelids, *146*
    iris, *146*
    naming of, 336–337
Muscle spindle, 123
Muscle tissues, 52–54 (*see also specific* types)
Muscle tone, 124, 128, 318
Muscular arteries, 511
Muscular contraction, 82, 312
Muscular controls, 125, 127
Muscular dystrophy, 320, 703
Muscular exercise, 328–329
Muscularis externa, *380*, 381, 395
Muscularis mucosa, 66, 381, 395
Myasthenia gravis, 384, 319, 320
Mycrocytic anemia, 585
Myelin, 87, 703
Myelin sheath, 106
Myeloid tissue, 47, 51
Myeloma, 289, 642, 703
Myocardial diseases, 558–561
Myocardial hypertrophy, 552, 703
Myocardial infarct, 558
Myocardium, 495
    diseases of, 559–561
Myoepithelial cells, 41, 53, 703
Myofibrils, 312
Myofilaments, 312
Myogenic action, 53, 703
Myogenic contraction, 541–542
Myogenic system, 495–497
Myoglobin, 559, 572, 703

720
Index

Myometrium, 226, 703
Myopia, 150
Myosin, 312, 316, 495, 703
Myxedema, 204

NADP, 416, 426
NADPH, 416, 433, 703
Nails, 250
Narcotics, 670
Nasal bones, 267
Nasal cavity, 70, 263
Nasal passage, 569
Nasal septum, 70
Nearsightedness, 151
Neck, 70, 71
  blood circulation, 517, 518
  movements, 346
  muscles, 344
Negative feedback, hormonal, 195
Neoplasia, 54
Nephron, 593, 703
Nephrotic syndrome, 612
Nerve(s), 89, 93, 706 (see also Central nervous system; Neurons; Sensory system; Skeletal motor system and specific parts and types)
  cells (see Neurons)
  cranial, 94–96, 106, 121, 147
  deafness, 143
  endings, 85
  energies, doctrine of specific, 114
  impulses, 76, 80 (see also Action Potential)
  motor (see under Motor)
  nuclei, 89, 90, 103, 104, 106
  spinal, 96–97, 107
  tracts, 89, 91, 92, 116, 127–128
Nervous breakdown, 67
Nervous system(s), 76–111
  anatomical components, 88
  autonomic (see Autonomic nervous system)
  basic elements, 76
  commissures, 89
  damage to, 128
  development, 97–98
  main divisions, 111
  parasympathetic division, 135
  pyramidal tract, 126
  in smooth muscles, 320–323
  sympathetic division, 135
  visceral sensory, 132
  visceral motor (see Autonomic nervous system)
Neural crests, 243
Neuralgia, 119
Neural plate, embryonic, 97, 98
Neural tube, embryonic, 97, 98
Neuritis, 120
Neuroendocrine cells, 192
Neuroendocrine hormones, 197
Neuroepithelia, 44
Neuromuscular junctions, 82, 122

Neuron(s), 76–88 (see also specific parts and types by name)
  adaptation in, 85
  bipolar, 77
  cell body, 77–78
  development of, 107
  endocrine, 86
  electrical activity in, 76, 78–84 (see also Action potential; Bioelectrical activity)
  functions of, 84–86
  ganglia, 88
  myelin production, 87
  nonreproductability of, 34, 86
  polar types, 77–78
  polarization in, 78–82
  response to stimuli, 76
  satellite types, 86
  self-repair, 88
  sensory, 77 (see also Proprioceptors; Sensory neurons)
  synapses of, 81–84
Neurotransmitters, 76, 81–83, 112, 135, 167, 703 (see also specific kinds)
Neutral fats (see Triglycerides)
Neutrophils, 49, 635, 703
Nevus, 254, 703
Niacin, 427, 474
Nightmares, 681
Ninth cranial nerve, 154
Nitrogen, 8
  balance in proteins, 419
  in caloric computations, 449, 450
  removal of excess, 430
Nitrogenous waste, in sweat, 249 (see also Urine)
Nociceptor, 85, 703
Nodal rhythm, 505, 703
Nodes of Ranvier, 87
Nodule, skin, 253
Nomogram, 453
Nonstriated muscle (see Smooth muscle)
Noradrenaline, 82, 135, 167
  in digestive tract, 381
  heart and blood vessel effects, 543
  in metabolism, 444
Normoblast, 580
Nuclease, 390
Nucleic acid, 9
Nucleolus, 28
Nucleus, cell, 28, 310, 703
Nucleus pulposus, 274
Nurse cells, 52
Nutrients, 8–9, 401–404, 413–426
  absorption of, 402–412
  interconversion in the liver, 430
Nutrient–waste exchange, 591–592
Nutrition, 413–486, 703
Nystagmus, 159, 162, 703

Obesity, 433, 456–457, 703

Occipital condyles, 265, 301
Oddi, sphincter of, 390
Olfaction, 151–153, 165, 703
  aparatus for, 152
  reflex pathways, 153
Oligodendrocytes, 87–88, 106, 112
Oliver, W. G., 675
Omentum, 69, 382, 703
Oogenesis, 227, 231
Operator genes, 31
Ophthalmoscope, 147
Optic chiasma, 89, 104, 147, 703
Optic disc, 147, 703
Optic nerve, 147
Optic tract, 147, 149
Oral cavity, 70, 263, 382
Oral contraceptives, 234–235
Orbital cavity, 263
Orbitals, 11
Orbits, cranial, 70
Organic compounds, 9
Organelles, 26–32, 703
Organs, 59–61, 71, 703
Orgasm, 219, 226
Orthostatic hypotension, 528
Os coxae, 278
Osmolality, 597
Osmole, 597
Osmotic gradient, 599
Osmotic pressure, 23, 529–531, 596–598
Osmosis, 22–23, 406, 704
Osmotic disorders, 530
Ossification, 283–287
  sutures and fontanels, 261
Osteoid, 52
Osteitis deformans, 288
Osteitis fibrosa, 289
Osteoarthrosis, 306, 704
Osteoblasts, 47, 51, 52, 281, 284
Osteoclasts, 51, 52, 206, 286
Osteocytes, 47, 51, 52, 60, 282, 284
Osteodystrophy, 288
Osteomalacia, 206, 288, 704
Osteomyelitis, 288, 704
Osteoporosis, 203, 289, 704
Osteosarcoma, 289
Osteosclerosis, 289
Otoliths, 159, 704
Ovary, 184, 224, 227–228
Oviducts, 227
Ovulation, 231, 704
Oxidation, 19, 704
  beta, fatty acids, 424–425
  metabolic, 204, 417
  in mitochondria, 28
Oxidation–phosphorylation, 28
Oxidation–reduction, 23
Oxygen, 8
  in blood, 571
  blood level controls, 546
  body requirements, 534–538
  carbon dioxide exchange, 571–573
  partial pressure, 572

Oxygen consumption, 534–538
  age and, 537
  heart failure demands, 562
  work and, 537, 538
Oxygen debt, in muscle action, 327, 329, 417, 704
Oxytocin, 198, 200, 224, 226, 234, 704

Pacemaker, 704
  artificial, 542
Pacemaker cells, 323
Pacemaker system, 495–496
Pacinian corpuscles, 114
Paget's disease, 288
Pain, 688–675
  acupuncture in relief of, 671–675
  drug relief of, 669–671
  lysosomal enzymes and, 32
  perception of, 114, 669
  phantom sensations of, 134
  referred, 132
  surgical relief techniques, 670
  theory of, 668–669
Pancreas, 212–213, 379, 389–390, 704
  in digestion, 397
  glucagon stimulation, 213, 442
  insulin stimulation, 213
Pancreatic duct, 389, 390, 393
Pancreozymin, 394, 397, 704
Panniculus adiposus, 68
Pantothenic acid, 427, 474
Papillary layer, dermis, 242, 245–246
Pap smear, 57
Papule, 253, 704
Parageusia, 154, 704
Paralysis, muscular, 129
Parasympathetic nerves, 497
Parasympathetic nervous system, 135, 137, 139, 544
Parasympathetic stimulation, 221, 570
Parathyroid glands, 205–207, 216
  disorders of, 203, 206
Parathyroid hormone, 192, 195–196, 205–206, 444, 608
Paraventricular nuclei, 89, 198, 704
Parenchyma, 40, 43, 60, 704
Parietal bones, 262, 263
Parietal layer, 68, 70
Parietal membrane, 66
Parkinson's disease, 102, 129, 149
Pars recta, 594, 599, 600
Parturition, 233
Passenger molecules, 407–408
Passive immunity, 646, 652
Patella, 259, 279
Pauling, Linus, 477
Pedicles, bone, 261
Peduncles, brain, 89
Pelvic cavity, 65, 356, 362, 364

Pelvic girdle, 277, 278
Pelvis, 704
  joints of, 301
  urinary, 595, 596
Penicillin, 648
Penis, 218–221, 357, 544, 704
Pepsin, 387
Peptic ulcer, 389
Peptide bonds, 30
Perception, 163 (see also Proprioception; Sensory information)
Pericardial cavity, 65, 70, 493
Peristalsis, 381, 704
Pericardium, 559, 704
Pericardosis, 559
Perichondrium, 284, 704
Periosteum, 281–284
Perimetrium, 226, 704
Perimysium, 333, 704
Perineum, 224, 356–357, 362, 704
Perineurium, 89, 704
Peripheral blood, 49
Peripheral circulatory input reflex, 545
Peripheral nervous system, 76
Peripheral resistance, heart failure, 561
Peristalsis, 395, 396, 704
Peritoneum, 68, 69, 72, 226, 704
Pernicious anemia, 388, 585, 704
Peroxidase, 49, 382, 704
Perpendicular plate, 264
Perspiration, insensible, 244–245 (see also Sweating)
Petechia, 253
Pertibular plexus, 592, 595
Petit mal, 175, 704
Peyer's patches, 51, 626, 633, 639
pH, 17 (see also Blood pH buffering)
Phagocytes, 47–49, 52, 86, 98, 112, 625, 626, 631, 634, 704 (see also Macrophage cells; Reticuloendothelial system; White blood cells)
Phagocytosis, 27, 49–50, 243, 405, 406, 513, 625, 644, 704
  cell types in, 634–635
  dermis in, 67
  in inflammation, 57
Phalanges, foot, 280
  hand, 276
Phantom sensations, 134
Pharyngeal passage, 569
Pharynx, 382
Phasic contractions, 318
Pheochromocytoma, 214, 611, 705
Phlebitis, 523, 705
Phosphatase, 52
Phosphate buffering, 605
Phosphocreatine, 313
Phospholipids, 409, 422, 440, 705
Phosphorus, 8
  body level, 205
  food sources, 485
  metabolic need, 479

Phosphorylation, 204
Physical endurance, 329
Physical exercise, 328–329
Physiological systems, 61
Physiology, 3, 705
Pia mater, 108, 110, 705
Pillars of fauces, 383
Pills, birth control, 234–235
Pinna, 140
Pinocytosis, 27, 406, 407
Pisiform bones, 259
Pituicytes, 198
Pituitary gland, 198, 200–202, 218, 705
Placenta, 224, 705
  fetal, 232
  hormone production, 231
  maternal, 232
Plantar ligaments, 304
Planes of body, defined, 297
Plasma, 49
Plasma cells, 50
Plasma membrane, 26–27, 37, 705
Pleura, thoracic, 69, 72, 705
Pleural cavities, 65
Plexus, 705
  brachial, 97
  choroid, 110
  spinal nerve, 96
Plica, 66, 705
Plicae circulares, 394, 402
Pneumoconiosis, 575
Pneumonia, 574, 705
Polar body, egg cell, 227
Polar cushion, 609
Polio virus, 128, 135
Polycythemia, 532, 580, 583, 705
Polyhedral cells, 41
Polymers, 9, 705
Polypeptide hormones, 192
Pons, 105, 106, 116, 154
Portal area, 391, 392
Portal system, 199, 403, 515, 705
Portal vein, 391, 392, 402, 513–514
Posterior (direction), 61
Posterior pituitary gland, 224
Postganglionic fibers, 135, 136
Postural syncope, 550
Potassium, in aging, 662
  disorders through loss, 607
  elimination of, 396
  inbalance effects, 605
  ion balance, 607
  in muscle weakness, 320
  reabsorption, 598
Precordial leads, 502
Prefrontal area, 164
Preganglionic fibers, 135, 136
Pregnancy, 217, 226, 229–234
  ectopic, 227
  passive immunity and, 646
  Rh factor in, 651–652
Premotor area, cerebral cortex, 164
Prepuce, 221, 705
Pressure receptors, 546

721
Index

Index

Prickle cell layer, epidermis, 243
P-R interval, heartbeat, 507
Progestation, 231
Progesterone, 184, 193, 209, 224, 227–234
Prolactin, 202, 224, 234, 705
Proliferation phase, 230
Pronation, 299, 300, 705
Pronormblast, 580
Prophase, mitosis, 32
Prophylactic, 234, 705
Proprioception, 156–161
Proprioceptive information, 156–157
Proprioceptors, 263
  evaluation of system, 159–161
Prostaglandins, 235, 423, 628, 705
Prostate gland, 218–223, 705
Protease, 390
Protein, 9, 14, 26, 418–422, 705
  aging loss, 662
  ammonia conversion, 603
  beta globulin complement, 644
  blood levels, 438–439
  blood serum, 636
  buffering, 605
  caloric value, 449
  chemical score, 419–420
  contractile, 312, 495
  digestion of, 390, 394
  essential, 467
  food sources, 482–485
  genetic factor in, 28
  intercontravertability, 467
  osmotic concentrations, 530
  nutritional value, 420, 466–467
  respiratory quotient, 450
  shortage disturbances, 438, 467
  synthesis, 29–30
Prothrombin, 432, 620, 621, 705
Protons, 11
Protoplasm, 8, 23
Proximal, direction, 62, 705
Proximal tubule, 594, 598
Pruritis, 254, 705
Pseudostratified epithelium, 40, 43
Psoriasis, 255, 705
Psychomotor epilepsy, 175, 551, 705
Puberty, 221, 222, 224
Pubic symphysis, 288
Pubis, 278, 705
Pulmonary circuit, 527, 705
Pulmonary circulation, 494, 516
Pulmonary edema, 552, 574, 705
Pulmonary embolism, 502
Pulmonary valve, 494, 705
Pulmonary veins, 513–514
Pulse, 503, 540–541
  palpitation points, 540
  pressure, 528
  rate and rhythm, 541
Pulsus alternans, 560, 705
Pupil, 145, 146, 149, 705
Pure speech aphasia, 174
Pure word deafness, 173

Purkinje fibers, 325, 496, 499
Purpura, 253, 621
Pus, 253, 627, 705
Pustule, 253
P wave, 507
Pyloric sphincter, 384, 385, 388
Pylorus, 385
Pyramidal system, 126, 129
Pyradoxine, 427, 474
Pyruvic acid, 415–417

Quinidine, 506, 705
QRS complex, 508

Radius, 275, 277, 301
Ranvier, nodes of, 87
Reabsorption, kidney, 597
Reagins, 649, 650, 705
Receptor neurons, 84, 114, 115
Receptor potential, 79
Recognition triplet, protein synthesis, 30
Rectum, 356, 362, 396, 398, 705
Recruitment factor, tissue transplant, 654
Red blood cells, 579–587, 580
  breakdown of, 582, 584
  deficiencies in, 584
  excess concentration, 586
  physiology of, 580–583
  production of, 579–580
Reduction division, 32, 34
Reduction reaction, 19, 706
Reed–Sternberg cells, 642
Referred pain, 133
Reflex actions, 117, 123
  in circulatory regulation, 545–546
  disorders of, 128
  for equilibrium, 159
  proprioceptive, 156
  to sound, 142
  in tasting, 154
Refractory period, 500, 706
  in cardiac fibers, 327
  muscular, 318
Rejection of tissue graft, 654
Relaxation, 124, 317–318, 684
Release-inhibiting hormones, 197
REM (Rapid eye movements), 679–680
Remission, pathological, 641
Renal artery, 591
Renal corpuscle, 593
Renal excretory system, 590
Renal vein, 591
Renal failure, 612, 706
Renin, 387, 609, 706
  endocrine effects, 184–185
  in hypertension, 553
  kidney production, 184
Repetition aphasia, 173
Repressor genes, 29
Reproductive process, human, 217

Reproductive system, 217–237
  female, 223–234
  male, 218–223
  male problems, 221–223
Reserpine, 168
Respiratory quotient, 450
Respiration (see also Breathing)
  endocrine stimulation, 184–185
  external, 564, 565–569
  internal, 564
  muscles of, 354, 347, 361, 568
  operations and functions, 574–575
Respiratory system, 564–577
  anatomy of, 565–566
  disorders of, 575
  functions and operations, 574–575
  gas exchange, 571–573
  lower tract, 566, 571–572
  upper tract, 565
Retching, 385
Reticular cells, 47, 51, 634, 706
Reticular fibers, 44–45
Reticular formation, 106
Reticular layer, dermis, 242
Reticulocytes, 580
  count of, 581
Reticuloendothelial system, 47, 49, 50, 51, 406, 618, 625–626, 634–635, 706 (see also Immunity; Lymphocytes; Macrophage cells; Phagocytes)
  body defense functions, 47, 618, 625
  in inflammation, 57, 626
Retina, 144, 145, 147, 706
Rheumatic heart disease, 652
Rh factor, 651
Riboflavin, 427, 474, 477
Ribose, 9, 417, 706
Ribosome cells, 28–31, 706
Ribs, 274–275
  muscles and movements, 344, 347
Rickets, 206, 208, 706
Rinne test, 143
RNA (Ribonucleic acid), 28, 30, 634, 645, 706
Rotation, 298–299
  scapular, 348
Rubella, immunity to, 646
Rugae, 66, 706

Saccharides, 414
Saccule, 158, 160
Sacroiliac, 278
Sacrum, 259, 269, 272–273, 277
Sagittal plane, 61, 297
Saliva, 383–384
Salivary enzymes, 382
Salivary glands, 544
Salt, chemical type, 17
Salpingectomy, 227, 706

Salpingitis, 226
Saltatory propogation, 87
Saphenous varicosities, 523
Sarcolemma, 310, 312, 706
Sarcoma, 55, 706
Sarcoplasmic reticulum, 312, 706
Sarcosomes, 312–313, 706
Satellite nerve cells, 77, 86
Saturation kinetics, 408
Scalene muscles, 344, 347, 567, 568
Scapula, 275, 276
  development of, 287
  movements, 348, 349
  muscles, 344
Scar, 254
Schizophrenia, 168
Schwann cells, 87–88, 106, 112
Sciatica, 274
Sclera, 145, 706
Sclerosis, 553, 706
Scoliosis, 272
Scopolamine, 161
Scrotum, 222, 706
Sebaceous glands, 250, 252–253
Sebum, 253, 256, 619, 706
Secretin, 367
Secretion, 597, 706
  apocrine mode, 247
  eccrine mode, 247
  holocrine mode, 253
  in inflammation, 626
Sella turcica, 198, 265, 706
Semen, 221, 706
Semicircular canals, 157
Semilunar valves, 494
Seminal fluid, 221
Seminal vesicles, 218, 219
Seminiferous tubules, 219, 220, 222
Sensory association areas, 104, 105, 106, 165
Sensory information, 114–119, 156–161, 165–174 (see also Cerebral cortex; Proprioceptive information; Sensory system and specific organs and systems)
  cerebrum and, 163
  contralateral transfer, 116
  consciousness (see Conscious perception)
  dissociation of, 116
  emotional, 166–171, 168, 169
  on equilibrium, 157–159
  in extensor reflexes, 124
  in hearing, 142, 165–166
  ipsilateral representation of, 116
  nerve tract specificity, 116
  olofactory, 152, 165
  reception and integration, 104
  somatotopic registration, 116–117, 118
  somesthetic area identification, 165
  transmission of, 112

  visceral, 132
  visual, 144–147, 166
Sensory neurons, 77, 84–85
  embryonic development, 107
  in extensor reflex, 123
  message transmission in, 112
  pathways of, 115–116
Sensory system, 114–120
  abnormalities, 119
  pain receptors, 669
Septal nuclei, 170
Septum, 264, 706
Serosa, 380, 381, 706
Serotonin, 48, 82, 167, 169, 626
Serous fluid, peritoneal, 69–70
Serous membranes, 65–66, 406
Sertoli cells, 219–220, 223
Serum (see Blood serum)
Serum sickness, 653
Sesamoid bones, 259, 260
Seventh cranial nerve, 154
Sex cells, 32, 34, 36, 217
Sex chromosomes, 36
Sex hormones, 218
  androgens, 612, 694
  disorders of, 212
  female, 218
  gonadotropins, 219, 222
  male, 217
  and obesity, 457
  production of, 209
Sexual deficiencies, male, 222–223
Sexual intercourse, 217, 221
Shingles, 254–255
Shock, 550
Shoulder, bones of, 275–276
  girdle, 275–276
  muscles and movements, 344, 349, 350, 351, 352
Sight (see Vision)
Sinoatrial cells, 325
Sinoatrial (SA) node, 496
Sinus(es), cranial, 262, 264
  maxillary, 269
  paranasal, 269
Sinus arrhythmia, 503–504
Sinus rhythm, 503, 706
Sinusoids, 392, 513, 707
Skeletal motor system, 111, 121–131
  damage and disorders, 128, 129
Skeletal movements, abnormal, 126, 128
  fine control of, 125
Skeletal muscle(s), 52–53, 68, 310–320, 311, 323–374 (see also Smooth muscle)
  abdominal–pelvic, 354, 359, 360, 363
  attachments of, 333
  basic movements listed, 340–341
  and cardiac muscle compared, 324–328
  circular, 334
  contractions of, 122, 313–316, 317–318, 334

  descriptive listing of, 367–373
  disorders of, 318–320
  energy production in, 313, 466
  facial, 339–340
  motor unit, 313
  numbers of, 335
  origins and insertions, 333
  structure of, 333
Skeletal muscle cells, 312–316
Skeletal system, 258–290, 261
  appendicular, 275–280
  axial, 261–275
Skin, 67, 241–257 (see also Epidermis and other specific parts)
  blood coloration of, 246
  blood supply to, 246
  defense role of, 218–219, 618–619
  disorders of, 253–256
  excretory role, 249, 589
  functions of, 241
  pigmentation of, 48, 68, 243
  receptor areas, 117, 118
  receptor nerve endings, 114
Skin cancer, 256
Skin test, for allergies, 650
Skull, 262–265
Sleep, 679–680
Small intestine, 393–396
  nutrient absorption by, 402
Smegma, 221, 706
Smell (see Olfaction)
Smooth muscle, 53, 320–324 (see also Muscles; Skeletal muscles)
  activity regulation by, 323–324
  contraction of, 322–323
  cross-section of, 321
  nerve supply to, 323
  in organs and systems, 321–322
  types of, 322–323
Smooth muscle cells, 41, 53
  in atherosclerosis, 554
Sneezing, 568
Sodium ion (see also Bioelectrical activity; Membrane polarity; Potassium)
  balance of, 605, 607
  body's retention of, 211
  depletion of, 511
  disorders through loss of, 607
  hormonal regulation, 185–186
  in osmotic pressure, 607
  in hypertension, 553
  potassium ion ratio, 396
  retention of, 553
  water retention and, 552
Solar plexus, 551, 706
Solvation, 18
Somatic cells, 32
Somatic chromosomes, 36
Somatotopical motor organization, 126, 127
Somatotopical sensory registration, 116, 117

Somesthetic area, 116, *117*, 157, 165
Sound, perception of (*see* Hearing
Spastic constipation, 410
Specific nerve energies, 114
Speech disorders, 171, 173
Sperm cells, 32, 217, 218, 220, 222
Spermatids, 220
Spermatocytes, *220*
Sphenoid bone, *265*
Sphincter, 384, 706
Sphincter muscles, 334
Sphincter of Oddi, 390
Sphingolipids, 422
Sphygmomanometer, 528–*529*, 706
Spinal anesthesia, 110
Spinal column, 269–274, *270* (*see also* Vertebra, Vertebral column *and specific sections*)
Spinal cord, 106, *107*, *108*
  damage determination, 117
  development of, 106–107
  nerve root connections, 107–*108*
  skin receptor areas, 117–118
Spinal nerves, 77, *93*, *96*, *97*, 107
Spinal reflexes, 123, 128
Spinothalamic tract, 89, 116
Spiny layer, epidermis, 243
Spleen, *379*, 389, 706
  blood platelet destruction, 622–623
  immunity functions, 637–638
Splenomegaly, 640, 706
Spongy bone, 282
Sprain, 306
Squamous epithelium, 30, *41*, 43, 66
Stapes, 140, 265
Starch, 6, 414, 707
Starling's law, 542
Steatorrhea, 412
Stem cells, 33, *46*, 637, 707
Stenosis, of heart, 556
Sterility, male, 222
Sterilization, 236
Sternum, 65, 344, *348*
Steroid hormones, *193*, 209, 418, 423 (*see also* Hormones; Sex hormones *and specific kinds by name*)
Stillman, Irwin, 458
Stimulus, bioelectrical, 112
  in flexor reflex, 125
  general perception of, 114
  muscular response to, 309
  neural, 77
  of target cells, 27
Stomach, *385*–389
  digestive activities, 386–388
  disorders, 389
Strain, 306
Strap (hyoid) muscles and movements, 343, 345
Stratified epithelium, 40–43
Stress, 677–684
  corticosteroid production in, 210

mental attitudes in, 678
relief of, 678–684
work and, 678
Stretch receptors, 545
Striated muscle, 310 (*see also* Skeletal muscle(s))
Stroke, 87, 164, 555
Stroma, of organs, 60, 707
Structural gene, 30
ST segment, 508–509
Subclavian veins, 519
Subcutaneous tissue, 68
Subluxation, of joints, 306
Submucosa, digestive tract, *380*–381
Substantia gelatinosa, 669
Sucrase, 394
Sucrose, 414, 707
Sugars, types of, 414
Superficial, as direction, 63, 707
Superficial fascia, 68
Superior, as direction, 61
Superior colliculi, 105
Supernatant, 32
Supination, 299, *300*, 707
Supraoptical nuclei, 89, 198, 707
Surgical neck, of humerus, 275
Sutures, *261*–262
Swallowing, *383*, 568–569
  disorders of, 384
  muscles and movements, 343, 345
Swamping, immunological, 647–648
Sweat glands, 247–250
  physiological classification, 247–248
  thermal, 248
Sweating, 241–251
Sydenham's chorea, 129
Sympathetic nervous system, 135, 139, 707
  blood vessel effects, 543–544
  cardiac output effects, 543
  heart action effects, 327, 497
  lower respiratory stimulation, 570
Synapses, 72, 80–84, 112, 115, 322
Synarthrodial joints, 297
Syncope, 550–551, 707
Syncytial trophoblasts, 44
Synovial joints, 297
Synovial membranes, 294–295
Synovitis, 306
Syphilis, 119, 128
Systolic pressure, 528, 552

Tabes dorsalis, 119
Tachycardia, 205, 503, 506, 707
Talus, 301
Target cells, 76, 78, 111
  of autonomic nervous system, 135
  multiple innervation of, 84
  stimulation of, 27, 80, 81
Tarsals, *280*, 301

Taste, 153, 166
Taste area, cerebral cortex, 165
Taste buds, 153, 154, 707
T cells, 50, 631, 633, 638, 639, 643, 654
Tears, 147
Tectorial membrane, 141
Telodendria, 77–81, 86
Telophase, 32, 35
Temperature, perception of, 114
Template, genetic, 30
Temporal bones, *263*
Temporal lobe, 166
Tendons, *48*, 293–294, 333
Tenth cranial nerve, 154
Testes, 184, 218–223, 707
Testosterone, 184, 188, 193, 209, 218–223
Tetanus, 317–318
Tetany, 319
Tetrads, chromosonal, 34–35
Thalamic nuclei, 170
Thalamus, 103, 106, 112, 116, 142, 148, 153, 707
  in pain reception, 669
  in proprioceptive executions, 157
Theca interna, 228, 229, 707
Thermoregulation, 241–242
Thiamine, 427, 474
Thigh, bone, *279*
  muscle operations, 357
Third ventrical, 110
Thirst, 607
Thoracic circuit, 519–520
Thoracic duct, 515, 519
Thoracic cage, 274–275
Thoracic cavity, 69, 71, 567
  arterial circuits, 519–520
  viscera of, 65
Thoracic pump, 534
Thoracic vertebrae, 269–272, *271*
Thorax, 64–65
  in breathing, *566*–567
Throat, 382
Thrombin, *620*–621
Thrombocytes, 49–50, 620–622, 635, 707
Thrombocytopenia, 641, 707
Thrombocytopenic purpura, 622–623
Thrombophlebitis, 523, 707
Thrombosis (*see* Blood clotting)
Thrombosthenin, 621
Thumb, muscles and movements, *336*, 348
Thymus gland, 635–636, 639, 707
Thyroglobin, 203
Thyroid gland, *182*, 184, 202, 203–205, 216, 707
  disorders, 399, 503
Thyrotoxicosis, 399
Thyrotropin, 184, 201, 203, 444
Thyrotropin mobilizer, 184
Thyroxine, 184, 192, 201, 203, 204, *440*, 443, 707

725
Index

Tibia, 279–280, 301
Tibialis anterior, 373
Tissues, 39–57 (see also Connective, Epithelial, Muscle and other specific types)
  adipose, 49, 432–433
  abnormalities of, 54–57
  age differences and, 661–662
  grafts, 654–655
  immunity producing, 633
  malignancies in, 54
  of organs, 59–60
  transplantation of, 653–654
TM (Transcendental meditation), 683–684
Tolerance, immunological, 635, 647
Tongue, 153, 343
Tonic contractions, 318
Tonsilitis, 640
Tonsils, palatine, 383, 639–640
Toxin, neutralization of, 634
Touch, perception of, 114
Trachea, 383, 570, 707
Trancellular fluids, 492
Transferrin, 479, 707
Transglutaminase, 621
Transfer-RNA, 30
Transitional epithelium, 40, 44, 707
Transverse section, 62, 297
Tremors, muscular, 129, 707
Triceps reflex, 124
Tricuspid valve, 494
Trigeminal neuralgia, 119
Triglycerides, 417–422, 432–433, 708
Triiodothyronine, 192, 708
Tripsinogen, 390
Trochanter, 261
Trochlea, 261
Trophoblast, 232, 708
Troponin, 315, 316, 708
Trunk, muscles and movements, 351–357, 359, 360, 361
Trypsin, 390
Tubal ligation, 236
Tubercle, of bone, 261
Tuberculosis, 575, 708
Tuberosity, of bone, 261
Turner's syndrome, 36, 37
Twitches, 317
T wave, 509
Tympanic membrane, 140
Tympanic duct, 141
Tyrosine, 203

Ulcers, 254
  peptic, 389
Ulna, 275, 277, 301
Ultrafiltrate, 591–599, 708
Umbilical cord, 232
Uncus, 169, 708
Universal donor, 651
Unipolar nerve cell, 77

Upper appendages, 275–277
  muscles of, 344–346
Upper extremities, circulation, 518–519
Upper motor neurons, 121, 126, 128, 131
Urea, 421, 430, 433, 603
Ureters, 591, 595
Urethra, 218–221, 357
Urinalysis, 610
Urinary bladder, 591, 596
Urinary pelvis, 595–596
Urinary nitrogen, in caloric computations, 449–450
Urination, muscles of, 357
  nervous regulation, 595–596, 613
Urine, 591, 596, 597, 598
  analysis of, 610, 708
  casts in, 611
  concentration of, 599–601
  production of, 593–601
Uriniferous tubules, 591, 593, 599
Urobilinogin, 582
Urticaria, 254
Uterus, 217–218, 224–227
  in pregnancy, 231–233
  prostglandin effects, 423
Utricle, 158, 160, 708
Uvula, 269, 283

Vaccine, 645
Vagina, 221, 224–226, 357
Vagus nerve, 497, 544
Valves, heart, 494
Valvular heart diseases, 555
Valvular incompetence, 556
Varicocele, 523
Varicose veins, 523
Vasa recta, 595
Vasa vasora, 514
Vasum rectum, 592
Vasculature, 492, 509–532, 708
Vas deferens, 218, 219, 708
Vasectomy, 236, 708
Vasoconstrictors, 246, 708
Vasodilators, 246
Vasopressin (see ADH)
Vasovagal syncope, 551
Veins, 513–514
  structure of, 511
  varicosities, 523, 708
Vena cava, 513–514
Venous pump, 534
Ventral, direction, 61
Ventral horns, 107
Ventilation, 575
Ventricles, heart, 493
Ventricular cells, 325
Ventricular fibrillation, 506–507
Ventricular flutter, 506
Ventricular hypertrophy, 501
Ventricular systole, 498–499, 527
Venular constrictions, 534
Venules, 513

Verbal communication centers, 170–174
Vertebra(e), 269–270
  body of, 269
  structural features, 271
Vertebral column, 65, 269–274, 270
  curvatures, 270, 272
  disorders of, 272–274
  ligaments, 304, 305
Vertebral discs, 269, 271
Vertebral pedicles, 269, 271
Vertebral processes, 270
Vesicles, 27, 253
Vestibular ganglion, 159
Vestibular membrane, 141
Vestibular system, 142, 157–158
Vestibule, 224–225
Villi, 66, 232, 394, 402
Viruses, 625
Viscera, 59
  muscles supporting, 354, 359, 360, 362
  thoracic cavity, 65
Visceral layer, pleural, 70
Visceral motor system (see Autonomic nervous system)
Visceral muscle, 53 (see Smooth muscle)
Visceral organs, 133, 544
Visceral sensory system, 111, 132, 138
Vision, 144–150, 166
Visual area, cerebral cortex, 165–166
Visual information, 147
Visual reflector neurons, 84
Visual reflexes, 149
Vital capacity, 574
Vitamin A, 427, 475, 476
Vitamin B_1, 416, 427
Vitamin B_{12}, 388, 427, 475, 581, 583, 585
Vitamin C (ascorbic acid), 427, 474, 477
Vitamin D, 196, 205, 207, 241, 427, 475, 476, 607, 609
Vitamin E, 472, 475
Vitamin K, 472, 475, 621
Vitamins, 426–427 (see also specific types)
  absorption of, 405
  allowances, recommended daily, 468–469, 473–474
  deficiencies and symptoms, 474–475
  essential, 473
  food sources, 486
  health roles, 477
  in large intestine, 397
  metabolic functions, 416, 427, 477
  storage of, 476
  values of, in specified foods, 482–484
Vitreous humor, 146

Vocal folds (cords), 570, 708
Volkmann's canals, 283
Voluntary activities, 75
Voluntary muscles, 310 (*see also* Skeletal muscles)
Vomer, nasal, 266–267
Vomiting, 386, 410
Vulva, 224, 225, 708

Water, 6, *11*
　body balance, 607
　body loss, 551
　body retention, 607 (*see also* Edema)
　excess accumulation, 551–552
　metabolic end product, 417
　osmotic movement, 530–531
　renal extraction, 586–601
Water intoxication, 552
Warts, 254–255

Weber test, 144
Weight, 5, 6, 526
Weight control, 204
Weight reducing, 458–462
Wernicke's area, 171–174
Wetting, *18*
Wheal, 253, 708
White blood cells, 49–50, 626, 634–635 (*see also* Basophils, Lymphocytes, Phagocytes *and other specific types*)
　damage to, 7, 584
　immunity functions, 631–633
　proliferation of, 640
　scavenger functions, 30
Windaus, Adolf, 206
Withdrawal reflex, 125
Work, cardiac output and, *539*
　oxygen consumption and, 537, *538*
　stress and, 678
Wright's stain, 49

Wrist, 276–277
　ligaments, 303
　muscles and movements, *354, 355,* 347–348

X-chromosomes, 36, 201
Xenobiotics, 623–624
Xenometabolism, 618, 633, 708
X-rays, 636, 647

Yan and ying, in acupuncture, 672–673
Y-chromosomes, 36
Yellow marrow, 51
Yudkins, John, 458

Z-line, 313, 316
Zygomatic bones, 267
Zygote, 36, 224, 231, 708

METRIC CONVERSION FACTORS

Metric units	U.S. equivalents	U.S. units	Metric equivalents
1 meter (m)	3.2808 feet	1 foot	0.3048 meter
1 centimeter (cm)	0.3937 inch	1 inch	2.5400 centimeters
1 millimeter (mm)	0.0394 inch	1 inch	25.4000 millimeters
1 square meter (m^2)	10.7639 square feet	1 square foot	0.0929 square meter
1 square centimeter (cm^2)	0.1550 square inch	1 square inch	6.4516 square centimeters
1 square millimeter (mm^2)	0.0016 square inch	1 square inch	645.1600 square millimeters
1 cubic centimeter (cm^3 or cc)	0.0610 cubic inch	1 cubic inch	16.3871 cubic centimeters
1 liter (l)	1.0567 quarts	1 quart	0.9464 liter
1 liter (l)	2.1134 pints	1 pint	0.4732 liter
1 centiliter (cl)	0.3382 fluid ounce	1 fluid ounce	2.9573 centiliters
1 milliliter (ml)	0.0338 fluid ounce	1 fluid ounce	29.5729 milliliters
1 kilogram (kg)	2.2046 pounds	1 pound	0.4536 kilogram
1 gram (g)	0.0353 ounce	1 ounce	28.3495 grams
1 centigram (cg)	0.1543 grain	1 grain	6.4799 centigrams
1 milligram (mg)	0.0154 grain	1 grain	64.7989 milligrams

Sizes below one millimeter (mm) are usually expressed only in metric units:

1 micrometer (μ; also called *micron*) = 0.001 millimeter = 0.000 039 inch

1 nanometer (nm; also called *millimicron* and abbreviated mμ) = 0.001 micrometer = 0.000 000 039 inch

1 angstrom (Å) = 0.1 nm = 0.000 000 000 100 meter = 0.000 000 003 937 inch

Structures such as cells and cell nuclei are measured in micrometers; structures such as cell membranes, small organelles, and large organic molecules are measured in nanometers. In some older publications light waves and single atoms were measured in angstroms; these are also now measured in nanometers.